Studies in Logic

Mathematical Logic and Foundations

Volume 89

Essays on Set Theory

Volume 79
Games Iteration Numbers. A Philosophical Introduction to Computability Theory
Luca M. Possati

Volume 80
Logics of Proofs and Justifications
Roman Kuznets and Thomas Studer

Volume 81
Factual and Plausible Reasoning
David Billington

Volume 82
Formal Logic: Classical Problems and Proofs
Luis M. Augusto

Volume 83
Reasoning: Games, Cognition, Logic
Mariusz Urbański, Tomasz Skura and Paweł Łupkowski, eds.

Volume 84
Witness Theory. Notes on λ-calculus and Logic
Adrian Rezuş

Volume 85
Reason to Dissent. Proceedings of the 3rd European Conference on Argumentation. Volume I
Catarina Dutilh Novaes, Henrike Jansen, Jan Albert van Laar and Bart Verheij, eds.

Volume 86
Reason to Dissent. Proceedings of the 3rd European Conference on Argumentation. Volume II
Catarina Dutilh Novaes, Henrike Jansen, Jan Albert van Laar and Bart Verheij, eds.

Volume 87
Reason to Dissent. Proceedings of the 3rd European Conference on Argumentation. Volume III
Catarina Dutilh Novaes, Henrike Jansen, Jan Albert van Laar and Bart Verheij, eds.

Volume 88
Belief Attitudes, Fine-Grained Hyperintensionality and Type-Theoretic Logic
Jiří Raclavský

Volume 89
Essays on Set Theory
Akihiro Kanamori

Studies in Logic Series Editor
Dov Gabbay dov.gabbay@kcl.ac.uk

Essays on Set Theory

Akihiro Kanamori

ISBN 978-1-84890-357-9

College Publications
Scientific Director: Dov Gabbay
Managing Director: Jane Spurr

http://www.collegepublications.co.uk

For Margaux, Ari, and Danny, and

dedicated to Juliet Floyd and our time together

Contents

0 Introduction

Over the years, besides solely technical papers in set theory I have published wide-ranging essays having to do with its history and development or with philosophical understandings that it has animated. This volume gathers together most of those essays, and through aggregation serves to set out ways of thinking about set theory and mathematics in terms of historical process, mathematical practice, and research activity. The essays are organized into thematically distinct parts within which they are presented in order of date of publication. Here in the introduction I briefly cast a thematic net across the parts and incidentally refer to further essays not included in this volume to round out the contexts.

Part I, History, has the essays on the development and involvements of set theory in the early 20th Century. My *The mathematical development of set theory from Cantor to Cohen* [4] had set out a systematic overall account. With that, the essays delve deeper into early coordinations of set theory over a broad swath. *The emergence of descriptive set theory* sets out the first applications of set-theoretic techniques in the topological, definability investigation of the continuum. *Hilbert and set theory* pursues the dialectical engagement with set theory of the preeminent mathematician of the early 20th Century. *The mathematical import of Zermelo's Well-Ordering Theorem* weaves together seminal results of set theory in terms of a unifying mathematical motif, one that eventually finds expression in fixed-point theorems. *The empty set, the singleton, and the ordered pair* lays out how these basic concepts were at first sources of intensional confusion before they become distinctively set-theoretic. And at the other end of axiomatization, *In praise of Replacement* historically sets out how this most sophisticated of the axioms is crucial for mediating indifference to identification, rigorizing transfinite recursion, and affirming the heuristic of reflection. Bracketing according to time frame are two further essays not included in this volume: *Cantor and continuity* [10] takes up Cantor's topologically directed work in the late 19th Century, and *Large cardinals with forcing* [7] provides a detailed technical survey of broadly advancing work on strong hypotheses and consistency results in the latter 20th Century.

Part II, Philosophy, has the essays which, with set-theoretic approaches in the background, take up issues and considerations in the philosophy of mathematics and its practice. *The mathematical infinite as a matter of method* discusses the infinite in mathematics, drawing on the experience of set theory, not so much as a matter of ontological commitment but of epistemological articulation. *Mathematical knowledge: motley and complexity of proof* discusses proofs across mathematics as importantly the carriers of mathematical knowledge. *Aspect-perception and the history of mathematics* takes up Wittgenstein's meta-concept and casts its net across the history of mathematics for hermeneutic interpretation and epistemic access. *Putnam's constructivization argument* roundly discusses and analyzes this best known deployment of set theory constructs in the service of philosophical advocacy. *Kreisel and Wittgenstein* sets out the historical interactions between these two well-known philosophers of logic and thereby illuminates the first's lifelong engagement with constructivity and proofs. For reasons of space and overlap. *Gödel vis-à-vis Russell: From logic to philosophy* [2] (with Juliet Floyd) was not included in this volume.

Part III, Mathematics, has three of my prominent mathematical papers, the most expository and conceptually substantial so as to be plausibly termed "essays". *Strong axioms of infinity and elementary embeddings* (with Robert Solovay and William Reinhardt) was the seminal paper that motivated and set out the range of the strong large cardinals, like the paradigmatic supercompact and extendible cardinals, that have become standard fare in the broad investigation of strong hypotheses. *On Gödel Incompleteness and finite combinatorics* (with Kenneth McAloon) investigated a combinatorial proposition of Paris-Harrington type that allows a remarkably simple proof of independence from Peano Arithmetic. That proposition featured regressive partitions, and *Regressive partition relations, n-subtle cardinals and Borel Diagonalization* projected the theme into the transfinite to incorporate the strong n-subtle cardinals into the independence work by Harvey Friedman. My book *The Higher Infinite* [6] subsequently charted out the whole canopy of large cardinal hypotheses and played a significant role in further stimulating the investigation of strong hypotheses in set theory. That these pervade modern set theory is evident from the three volume *Handbook of Set Theory* [3] (edited with Matthew Foreman).

Part IV, Lives in Set Theory, has most of the essays in a series which could be viewed as a *Le Vite* of prominent set theorists. The essays are all follow-throughs to either invited addresses at conferences or solicitations for commemorative journal articles. Be that as it may, they substantiate the development of set theory through the lens of an individual's work and thereby affirm the importance of research activity. For logistic reasons the following essays in the series were not included in this volume: *Zermelo and set theory* [5], *Bernays and set theory* [8], and *Erdős and set theory* [9]. Much of [5] was superseded in any case by the introductory notes to *Ernst Zermelo. Collected Works* [1].

References

[1] Heinz-Dieter Ebbinghaus, Craig G. Fraser, and Akihiro Kanamori, editors. *Ernst Zermelo. Collected Works*, volume I. Springer, Berlin, 2010.

[2] Juliet Floyd and Akihiro Kanamori. Gödel vis-à-vis Russell: From logic and set theory to philosophy. In Gabriella Crocco and Eva-Maria Engelen, editors, *Kurt Gödel: Philosopher-Scientist*, pages 243–326. Publications de l'Université de Provence, Aix-en-Provence, 2012.

[3] Matthew Foreman and Akihiro Kanamori, editors. *Handbook of Set Theory*. Springer, Berlin, 2010.

[4] Akihiro Kanamori. The mathematical development of set theory from Cantor to Cohen. *The Bulletin of Symbolic Logic*, 2:1–71, 1996.

[5] Akihiro Kanamori. Zermelo and set theory. *The Bulletin of Symbolic Logic*, 10:487–553, 2004.

[6] Akihiro Kanamori. *The Higher Infinite*. Springer, Berlin, 2009. Second edition, subsequently reprinted in 2012 by World Scientific, Beijing. First edition, 1994.

[7] Akihiro Kanamori. Large cardinals with forcing. In Dov M. Gabbay, Akihiro Kanamori, and John H. Woods, editors, *Sets and Extensions in the Twentieth Century*, pages 359–413. Cambridge University Press, Cambridge, 2012.

[8] Akihiro Kanamori. Bernays and set theory. *The Bulletin of Symbolic Logic*, 15:43–60, 2014.

[9] Akihiro Kanamori. Erdős and set theory. *The Bulletin of Symbolic Logic*, 20:449–496, 2014.

[10] Akihiro Kanamori. Cantor and continuity. In Stewart Shapiro and Geoffrey Hellman, editors, *The History of Continua: Philosophical and Mathematical Perspectives*, pages 219–254. Oxford University Press, Oxford, 2020.

Part I

History

1 The Emergence of Descriptive Set Theory

Descriptive set theory is the definability theory of the continuum, the study of the structural properties of definable sets of reals. Motivated initially by constructivist concerns, a major incentive for the subject was to investigate the extent of the regularity properties, those properties indicative of well-behaved sets of reals. With origins in the work of the French analysts Borel, Baire, and Lebesgue at the turn of the century, the subject developed progressively from Suslin's work on the analytic sets in 1916, until Gödel around 1937 established a delimitative result by showing that if $V = L$, there are simply defined sets of reals that do not possess the regularity properties. In the ensuing years Kleene developed what turned out to be an effective version of the theory as a generalization of his foundational work in recursion theory, and considerably refined the earlier results.

The general impression of the development of set theory during this period is one of preoccupation with foundational issues: analysis of the Axiom of Choice, emerging axiomatics, hypotheses about the transfinite, and eventual formalization in first-order logic. Descriptive set theory on the other hand was a natural outgrowth of Cantor's own work and provided the first systematic study of sets of reals building on his methods, and as such, how it developed deserves to be better known. This article provides a somewhat selective historical account, one that pursues three larger theses: The first is that the transfinite ordinals became incorporated into mathematics, Cantor's metaphysical bent and the ongoing debate about the actual infinite notwithstanding, because they became necessary to provide the requisite length for the analysis of mathematical concepts, particularly those having to do with sets of reals. The second is that later work in recursion theory and set theory emanating from Gödel's results had definite precursors in pre-formal but clearly delineated settings such as descriptive set theory. The third, related to the second, is that as metamathematical methods became incorporated into mathematics, they not only led to extra-theoretic closure results about earlier problems but to intra-theoretic advances to higher levels. The text Moschovakis [1980] serves as the reference for the mathematical development of descriptive set theory; the historical bearings established there are elaborated in certain directions here.[1]

As Cantor was summing up his work in what were to be his last publications, the *Beiträge*, it was the French analysts Emile Borel, René Baire, and Henri Lebesgue who were to carry the study of sets of reals to the next level of complexity. As is well-known, they, perhaps influenced by Poincaré, had considerable reservations about the extent of permissible objects and methods in mathematics. And as with later constructivists, their work led to careful anal-

Republished from Jaakko Hintikka (editor), *From Dedekind to Gödel: Essays on the Development of the Foundations of Mathematics*, Synthèse Library 251, Kluwer, Dordrecht, 1995, pp.241-262, with permission from Springer Nature.

[1]Moore [1982] also provided some historical guidance.

yses of mathematical concepts and a body of distinctive mathematical results. But significantly, the denumerable ordinals, Cantor's second number class, became necessary in their work, as well as the Countable Axiom of Choice, that every countable set of nonempty sets has a choice function.

Soon after completing his thesis Borel in his book [1898:46-47] considered for his theory of measure those sets of reals obtained by starting with the intervals and closing off under complementation and countable union. The formulation was *axiomatic* and in effect *impredicative*, and seen in this light, bold and imaginative; the sets are now known as the *Borel* sets and quite well-understood.

Baire in his thesis [1899] took on a dictum of Dirichlet's that a real function is any arbitrary assignment of reals, and diverging from the 19th Century preoccupation with pathological examples, sought a constructive approach via pointwise limits. He formulated the following classification of real functions: *Baire class 0* consists of the continuous real functions, and for countable ordinals $\alpha > 0$, *Baire class* α consists of those functions f not in any previous class, yet for some sequence $f_0, f_1, f_2, \ldots$ of functions in previous classes f is their pointwise limit, i.e. $f(x) = \lim_{n\to\infty} f_n(x)$ for every real x. The functions in these classes are now known as the *Baire* functions, and this was the first analysis in terms of a *transfinite hierarchy* after Cantor. Baire mainly studied the finite levels, particularly classes 1 and 2; he pointed out in a note [1898:70-71] toward his thesis that Dirichlet's function that assigns 1 to rationals and 0 to irrationals is in class 2. He did observe [1899:70-71] that the Baire functions are closed under pointwise limits (with an implicit use of the Countable Axiom of Choice), and that an appeal to Cantor's cardinality arguments would imply that there are real functions that are not Baire.

Lebesgue's thesis [1902] is, of course, fundamental for modern integration theory as the source of his concept of measurability. Inspired in part by Borel's ideas, Lebesgue's concept of measurable set subsumed the Borel sets, and his analytic definition of measurable function had the simple consequence of closure under pointwise limits, thereby subsuming the Baire functions (and so Dirichlet's old example). See Hawkins [1975] for more on the development of Lebesgue measurability; Lebesgue's first major work in a distinctive direction was to be the seminal paper in descriptive set theory:

In the memoir [1905] Lebesgue investigated the Baire functions, stressing that they are exactly the functions definable via analytic expressions (in a sense made precise). He first established a correlation with the Borel sets by showing that they are exactly the preimages $\{x \mid f(x) \in O\}$ of open intervals O by Baire functions f. With this he introduced the first hierarchy for the Borel sets (differing in minor details from the now standard one from Hausdorff [1914]) with his open sets of class α being those pre-images of some open interval via some function in Baire class α that are not the pre-image of any open interval via any function in a previous class. After verifying various closure properties and providing characterizations for these classes, Lebesgue established two main results. The first demonstrated the necessity of exhausting the countable ordinals:

(1) The Baire hierarchy is proper, i.e. for every countable ordinal α there is a Baire function of class α, and consequently the corresponding hierarchy for the Borel sets is analogously proper.

The second established transcendence beyond countable closure for his concept of measurability:

(2) There is a Lebesgue measurable function which is not in any Baire class, and consequently a Lebesgue measurable set which is not a Borel set.

The hierarchy result (1) was the first of all such results, and a definite precursor of fundamental work in mathematical logic in that it applied Cantor's *universal enumeration* and *diagonal argument* to achieve a transcendence to a next level. What was missing of course was the formalization in first-order logic of Gödel's Incompleteness Theorem, but what was there was the prior extent of the ordinals, as in Gödel's later construction of L. For the first time, Cantor's second number class provided the necessary length for an individuated analysis of a class of simply defined sets of reals.

Baire [1899] had provided a characterization of Baire class 1, one elaborated by Lebesgue [1905], and had found examples of "effective" functions in class 2 [1898] and class 3 [1906] with a systematic presentation in [1909]. In a formal sense, it is necessary to use higher methods to establish the existence of functions in every class.[2] Lebesgue regarded the countable ordinals as an indexing system, "symbols" for classes, but nonetheless he exposed their basic properties, giving probably the first formulation [1905:149] of the concept of proof by transfinite induction. To Borel's credit, it was he (cf. his [1905:note III]) who had broached the idea of applying Cantor's diagonal method; Lebesgue incorporated definability considerations to establish (1).

The transcendence result (2) was also remarkable in that Lebesgue actually provided an explicitly defined set, one that was later seen to be the basic example of an analytic, non-Borel set. For this purpose, the reals were for the first time construed as codes for something else, namely countable well-orderings, and this not only further incorporated the transfinite into the investigation of sets of reals, but foreshadowed the later coding results of mathematical logic.

[2]It turns out that every real function is of Baire class at most 3 if $\aleph_1$ is a countable union of countable sets, and this proposition is consistent with ZF by forcing as first observed by Feferman-Levy [1963].

Lebesgue's results, along with the later work in descriptive set theory, can be viewed as pushing the mathematical frontier of the actual infinite past $\aleph_0$, which arguably had achieved a mathematical domesticity through increasing use in the late 19th Century, to $\aleph_1$. The results stand in elegant mathematical contrast to the metaphysical to and fro in the wake of the antimonies and Zermelo's 1904 proof of the Well-Ordering Theorem. Baire in his thesis [1899:36] had viewed the denumerable ordinals and hence his function hierarchy as merely *une façon de parler*, and continued to view infinite concepts only in potentiality. Borel [1898] took a pragmatic approach and seemed to accept the denumerable ordinals. Lebesgue was more equivocal but still accepting, perhaps out of mathematical necessity, although he was to raise objections against arbitrary denumerable choices. (For his (2) above, the example is explicitly defined, but to establish the transcendence the Countable Axiom of Choice was later seen to be necessary.) Poincaré [1906], Shoenflies [1905] and Brouwer [1907] (his dissertation) all objected to the existence of $\aleph_1$, although at least the latter two did accept the denumerable ordinals individually. In any case, mathematics advanced in Hausdorff's work [1908] on transfinite ordertypes: Objecting to all the fuss being made over foundations and pursuing the higher transfinite with vigor, he formulated for the first time the Generalized Continuum Hypothesis, introduced the η_α sets—prototypes for saturated model theory—and broached the possibility of a uncountable regular limit cardinal—the beginning of large cardinal theory. The mathematical advances of the period in set theory were soon codified in the classic text Hausdorff [1914].

During these years, Lebesgue measure became widely accepted as a *regularity property*, a property indicative of well-behaved sets of reals. Two others were discussed: the Baire property and the perfect set property. All three properties were to be some of central concern in descriptive set theory for, unlike the Borel sets, there did not seem to be any hierarchical analysis, and indeed the extent of the sets of reals possessing these properties was quite unclear.

The Baire property evolved from the other important concept in Baire's thesis [1899], that of category: A set of reals is *nowhere dense iff* its closure under limits contains no open set; a set of reals is *meager* (or *of first category*) *iff* it is a countable union of nowhere dense sets; and a set of reals has the *Baire property* if it has a meager symmetric difference with some open set. Straightforward arguments show that every Borel set has the Baire property.

The second regularity property has its roots in the very beginnings of set theory: A set of reals is *perfect iff* it is nonempty, closed and contains no isolated points; and a set of reals has the *perfect set property* if it is either countable or else has a perfect subset. Using his notion of derived set emerging out of his work on the convergence of trigonometric series, Cantor [1883,1884] and Bendixson [1883] established that every closed set has the perfect set property. Since Cantor [1884] established that every perfect set has the cardinality of the continuum, this provided a more concrete approach to his Continuum Problem: at least no closed set of reals can have an intermediate cardinality between $\aleph_0$ and the cardinality of the continuum. William Young [1903] extended the Cantor-Bendixson result by showing that every G_δ set[3] of reals has the perfect set property. However, unlike for the other regularity properties it was by no means clear that every Borel set has the perfect set property, and the verification of this was only to come a decade later with the shifting of the scene from Paris to Moscow.

The subject of descriptive set theory emerged as a distinct discipline through the initiatives of the Russian mathematician Nikolai Luzin. Through a focal seminar that he began in 1914 at the University of Moscow, he was to establish a prominent school in the theory of functions of a real variable.[4] Luzin had become acquainted with the work and views of the French analysts while he was in Paris as a student, and from the beginning a major topic of his seminar was the "descriptive theory of functions". Significantly, the young Polish mathematician Waclaw Sierpiński was an early participant; he had been interned in Moscow in 1915, and Luzin and his teacher Egorov interceded on his behalf to let him live freely until his repatriation to Poland a year later. Not only did this lead to a decade long collaboration between Luzin and Sierpiński, but undoubtedly it encouraged the latter in his efforts toward the founding of the Polish school of mathematics[5] and laid the basis for its interest in descriptive set theory.

In the spring of 1915 Luzin described the cardinality problem for Borel sets (operatively whether they have the perfect set property) to Pavel Aleksandrov, an early member of Luzin's seminar and later a pioneer of modern topology. By that summer Aleksandrov [1916] had established his first important result:

(3) Every Borel set has the perfect set property.

Hausdorff [1916] also established this, after getting a partial result [1914:465]. The proof of (3) required a new way of comprehending the Borel sets, as underscored by the passage of a decade after Lebesgue's work. It turns out that collection of sets having the perfect set property is not closed under complementation, so that an inductive proof of (3) through a hierarchy is not possible. The new, more direct analysis of Borel sets broke the ground for a dramatic development:

---

[3] G_δ sets are countable intersections of open sets in the cumulative hierarchy for Borel sets from Hausdorff [1914].

[4] See Phillips [1978]. Uspenskii [1985] and Kanovei [1985] are recent, detailed surveys of the work of Luzin and his school in descriptive set theory. Uspenskii [1985:98] wrote: "... in his days the descriptive theory, distinguished by his work and that of Suslin, Aleksandrov, Kantorovich, Keldysh, Kolmogorov, Lavrent'ev, Lyapunov, and Novikov, was the fame of mathematics in our country ..."

[5] See Kuzawa [1968] and Kuratowski [1980].

Soon afterwards another student of Luzin's, Mikhail Suslin (often rendered Souslin in the French transliteration), began reading Lebesgue [1905]. Memoirs of Sierpiński [1950:28ff] recalled how Suslin then made a crucial discovery in the summer of 1916. For $Y \subseteq R^{k+1}$, the *projection* of Y is

$$pY = \{\langle x_1, \dots x_k\rangle \mid \exists y(\langle x_1, \dots, x_k, y\rangle \in Y)\}.$$

Suslin noticed that at one point Lebesgue asserted ([1905:191-192]) that the projection of a Borel subset of the plane is also a Borel set. This was based on the mistaken claim that given a countable collection of subsets of the plane the projection of their intersection equals the intersection of their projections. Suslin found a counterexample to Lebesgue's assertion, and this led to his inspired investigation of what are now known as the analytic sets. (Lebesgue later ruefully remarked that his assertion was "simple, short, but false" (Luzin [1930:vii]); however, it did not affect the main results of his memoir.)

Suslin [1917] formulated the analytic sets as the A-sets (les ensembles (A)), sets resulting from an explicit operation, the Operation (A): A *defining system* is a family $\langle X_s\rangle_s$ of sets indexed by finite sequences s of integers. $A(\langle X_s\rangle_s)$, the result of the Operation (A) on such a system, is that set defined by:

$$x \in A(\langle X_s\rangle_s) \quad \textit{iff} \quad \exists f\colon \omega \to \omega \forall n \in \omega (x \in X_{f \restriction n}).$$

For X a set of reals,

$$\text{X is } \textit{analytic} \quad \textit{iff} \quad X = A(\langle X_s\rangle_s) \text{ for some defining system } \langle X_s\rangle_s \text{ consisting of closed sets of reals.}$$

As Suslin essentially noted, this implies that *a set of reals is analytic iff it is the projection of a G_δ subset of the plane.*[6] He announced three main results:

(4) Every Borel set is analytic.

In fact:

(5) A set of reals is Borel *iff* both it and its complement are analytic;

and:

(6) There is an analytic set that is not Borel.

These results are analogous to later, better known results with "recursive" replacing "Borel" and "recursively enumerable" replacing "analytic". [1917] was

[6]See footnote 3. The Borel subsets of the plane, R^2, and generally those of R^k, are defined analogously to those of R.

to be Suslin's sole publication, for he succumbed to typhus in a Moscow epidemic in 1919 at the age of 25. (The whole episode recalls a well-known equivocation by Cauchy and the clarification due to the young Abel that led to the concept of uniform convergence, even to Abel's untimely death.)

In an accompanying note, Luzin [1917] announced the regularity properties for the analytic sets:

(7) Every analytic set is Lebesgue measurable, has the Baire property, and has the perfect set property.

He attributed the last to Suslin.[7]

Whether via the geometric operation of projection of G_δ sets or via the perfect set property for Borel sets and the explicit Operation (A) on systems of closed sets, the Russians had hit upon a simple procedure for transcending the Borel sets, one that preserves the regularity properties. Paradigmatic for later hierarchy results, Suslin's (5) provided a dramatically simple characterization from above of a class previously analyzed from below in a hierarchy of length $\aleph_1$, and held the promise of a new method for generating simply defined sets of reals possessing the regularity properties.[8]

The notes Suslin [1917] and Luzin [1917] were to undergo considerable elaboration in the ensuing years. Proofs of the announced results (4)-(6) appeared in Luzin-Sierpiński [1918,1923]; as for (7), the Lebesgue measurability result

[7] This attribution is actually a faint echo of a question of priority. According to the memoirs of Aleksandrov [1979:284-286] it was he who had defined the A-sets, and Suslin proposed the name, as well as "Operation (A)" for the corresponding operation, in Aleksandrov's honor. This eponymy is not mentioned in Suslin [1917], but is supported by recollections of Lavrent'ev [1974:175] and Keldysh [1974:180] as well as Kuratowski [1980:69]. Aleksandrov recalled that it was he who had shown that every Borel set is an A-set and that every A-set has the perfect set property, although this is not explicit in his [1916]. He then tried hard in 1916 to show that every A-set is Borel, only ceasing his efforts when it became known that in the summer Suslin had found a non-Borel A-set. According to Aleksandrov: "Many years later Luzin started to call A-sets analytic sets and began, contrary to the facts, which he knew well, to assert that the term 'A-set' is only an abbreviation for 'analytic set'. But by this time my personal relations with Luzin, at one time close and sincere, were estranged." Luzin [1925,1927] did go to some pains to trace the term "analytic" back to Lebesgue [1905] and pointed out that the original example there of a non-Borel Lebesgue measurable set is in fact the first example of an non-Borel analytic set. See also the text Luzin [1930:186-187], in which the Operation (A) is conspicuous by its absence. Aleksandrov also wrote: "This question of my priority in this case never made much difference to me; it was just my first result and (maybe just because of that) the one dearest to me."

[8] Although there may have been growing acceptance of the denumerable ordinals, it was still considered hygienic during this transitional period to eliminate the use of the transfinite where possible. The emphasis of Suslin [1917] was on how (5) does this for the definition of the Borel sets. Sierpiński [1924] featured a "new" proof without transfinite ordinals of the perfect set property for Borel sets. And earlier his younger compatriot Kuratowski [1922] had offered what is now known as Zorn's Lemma primarily to avoid the use of transfinite ordinals.

was established in the former, the Baire property result in the latter, and the perfect set property had to await Luzin [1926]. Luzin-Sierpiński [1923] was a pivotal paper, in that it shifted the emphasis toward *co-analytic* sets, complements of analytic sets, and provided a basic representation for them from which the main results of the period owed. With it, they established:

(8) Every analytic set is both a union of $\aleph_1$ Borel sets and an intersection of $\aleph_1$ Borel sets.

The representation of co-analytic sets had an evident precedent in Lebesgue's proof of (2); the idea can be conveyed in terms of the Operation (A): Suppose that $Y \subseteq R$ is co-analytic, i.e. $Y = R - X$ for some $X = A(\langle X_s \rangle_s \rangle)$, so that

$$x \in Y \quad \textit{iff} \quad x \notin X \quad \textit{iff} \quad \forall f \colon \omega \to \omega \, \exists n (x \notin X_{f \restriction n}) \, .$$

For finite sequences s_1 and s_2 , define: $s_1 \prec s_2$ *iff* s_2 is a proper initial segment of s_1. For a real x define: $T_x = \{s \mid x \in X_t$ for every initial segment t of $s\}$. Then

(9) $x \in Y$ *iff* $\prec$ on T_x is a well-founded relation.

i.e. there is no infinite descending sequence $\ldots \prec s_2 \prec s_1 \prec s_0$.

Thus did well-founded relations enter mathematical praxis. The well-known analysis by von Neumann [1925] and Zermelo [1930] prefigured by Mirimanoff [1917] was particular to the membership relation; this of course led to the Axiom of Foundation and the cumulative hierarchy view of the universe of sets, and crucial as this development was, the main thrust was in the direction of axiomatization of an underlying structural principle.

Luzin and Sierpiński [1918,1923] linearized their well-founded relations, submerging well-foundedness under the better known framework of well-ordering and getting an ordinal analysis of co-analytic sets. This was natural to do for their results, as various technical aspects became simplified by appeal to the linear comparability of well-ordered sets. The linearization was through none other than what is now known as the "Kleene-Brouwer" ordering. But already in Luzin [1927:50] well-founded relations on the integers were defined explicitly because of their necessary use in his proof of the Borel separability of analytic sets,[9] one route to Suslin's (5). It was only later through the fundamental collapsing isomorphism result of Andrzej Mostowski [1949] that well-founded relations were seen to have a canonical representation via the membership relation that well-orderings have in the correlation with (von Neumann) ordinals. Dana Scott's celebrated result [1961] that if a measurable cardinal exists, then

[9] If X and Y are disjoint analytic sets, there is a Borel set B such that $X \subseteq B$ and $Y \cap B = \emptyset$.

$V \neq L$ can be viewed as a well-founded version of a previous result about well-orderings due to Keisler [1960]. Through such reorientations, well-foundedness has achieved a place of prominence in current set theory shoulder-to-shoulder next to well-ordering.

The next conceptual move, a significant advance for the theory, was to extend the domain of study by taking the operation of projection as basic. Luzin [1925a] and Sierpńiski [1925] defined the projective sets as those sets obtainable from the Borel sets by the iterated applications of projection and complementation. We have the corresponding projective hierarchy in modern notation: For $A \subseteq R^k$,

$$A \text{ is } \mathbf{\Sigma}^1_1 \quad \textit{iff} \quad A \text{ is analytic}\,.$$

(defined as for $k = 1$ in terms of a defining system consisting of closed subsets of R^k) and inductively for integers n,

$$A \text{ is } \mathbf{\Pi}^1_n \quad \textit{iff} \quad R^k - A \text{ is } \mathbf{\Sigma}^1_n\,,$$

and

$$A \text{ is } \mathbf{\Sigma}^1_{n+1} \quad \textit{iff} \quad A = pY \text{ for some } \mathbf{\Pi}^1_n \text{ set } Y \subseteq R^{k+1}\,.$$

We also define

$$A \text{ is } \mathbf{\Delta}^1_n \quad \textit{iff} \quad A \text{ is both } \mathbf{\Sigma}^1_n \text{ and } \mathbf{\Pi}^1_n\,.$$

Luzin [1925] and Sierpiński [1925] recast Lebesgue's use of the Cantor diagonal argument to show that the projective hierarchy is proper, and soon its basic properties were established by various people, e.g. each of the classes $\mathbf{\Sigma}^1_n$ and $\mathbf{\Pi}^1_n$ is closed under countable union and intersection.

On the other hand, the investigation of projective sets encountered basic obstacles from the beginning. For one thing, unlike for the analytic sets the perfect set property for the $\mathbf{\Pi}^1_1$, or co-analytic, sets could not be established. Luzin [1917:94] had already noted this difficulty, and it was emphasized as a major problem in Luzin [1925a]. In a confident and remarkably prophetic passage, he declared that his efforts towards its resolution led him to a conclusion "totally unexpected", that "one does not know and *one will never know*" of the family of projective sets, although it has the cardinality of the continuum and consists of "effective" sets, whether every member has the cardinality of the continuum if uncountable, has the Baire property, or is even Lebesgue measurable. This speculation from mathematical analysis stands in contrast to the better known anticipation by Skolem [1923:229] of the independence of the Continuum Hypothesis based on metamathematical considerations. Luzin

[1925b] pointed out another problem, that of establishing the Lebesgue measurability of $\mathbf{\Sigma}^1_2$ sets. Both these difficulties at $\mathbf{\Pi}^1_1$ and $\mathbf{\Sigma}^1_2$ were also observed by Sierpiński [1925:242], although he was able to show:

(10) Every $\mathbf{\Sigma}^1_2$ set is the union of $\aleph_1$ Borel sets.

The first wave of progress from Suslin's results having worked itself out, Luzin provided systematic accounts in two expository papers [1926,1927] and a text [1930]. In [1927] and more generally in [1930] he introduced the concepts of *sieves* and *constituents*, implicit in earlier papers. Loosely speaking, a sieve is a version of a defining system for Operation (A), and a constituent is, in terms of (9), a set of form

$$C_\alpha = \{x \in Y \mid \prec \text{ on } T_x \text{ has rank } \alpha\}\,.$$

for some ordinal. (Every well-founded relation has a rank, its "height", defined by transfinite recursion. These constituents turn out to be Borel sets if the defining system consists of Borel sets, and so the first half of (8) is already evident in $Y = \bigcup_{\alpha<\aleph_1} C_\alpha$.) Sieves and constituents not only became the standard tools for the classical investigation of the first level of the projective hierarchy, but also became the subjects of considerable study in themselves.

Most extensively in his classic text [1930], Luzin aired the constructivist views of his French predecessors. Not only did he contrive self-effacingly to establish definite precedents for his own work in theirs,[10] but he also espoused their distrust of the unbridled Axiom of Choice and advocated their views on definability, especially analyzing Lebesgue's informal concept of nameability (qu'on peut nommer). He regarded his investigations as motivated by these considerations, as well as by specific new intuitions. For instance, he considered the complementation operation used in the formulation of the projective hierarchy to be "negative" in a sense that he elaborated, its use equivalent to that of all the denumerable ordinals, and that this led to the difficulties. He wrote ([1930:196]): "Thus, the transfinite can be profoundly hidden in the form of a definition of a negative notion." Related to this, Lebesgue [1905] had ended with question, "Can one name a non-measurable set?", and taking this as a starting point for his own work, Luzin [1930:323] wrote sagaciously: "...the author considers the question of whether all projective sets are measurable or not to be unsolvable [insoluble], since in his view the methods of defining the projective sets and Lebesgue measure are not comparable, and consequently, not logically related."

If the projective sets proved intractable with respect to the regularity properties, significant progress was nonetheless made in other directions. In Luzin [1930a] the general problem of *uniformization* was proposed. For $A, B \subseteq R^2$,

[10] But see footnote 7.

$$A \text{ is } \textit{uniformized} \text{ by } B \quad \textit{iff} \quad B \subseteq A \text{ and } \forall x(\exists y(\langle x,y\rangle \in A) \leftrightarrow \exists! y(\langle x,y\rangle \in B)).$$

(As usual, $\exists!$ abbreviates the formalizable "there exists exactly one".) Since B is in effect a choice function for an indexed family of sets, asserting the uniformizability of arbitrary $A \subseteq R^2$ is a version of the Axiom of Choice. Taking this approach to the problem of definable choices, Luzin announced several results about the uniformizability of analytic sets by like sets. One was affirmed in a sharp form by Novikov [1931], who showed that there is a closed set that cannot be uniformized by any analytic set. It was eventually shown by Yankov [1941] that every analytic set can be uniformized by a set that is a countable intersection of countable unions of differences of analytic sets. Interestingly enough, von Neumann [1949:448] also established a less structured uniformization result for analytic sets as part of an extensive study of rings of operators. Presumably because of his difficulties with $\mathbf{\Pi}^1_1$ sets, Luzin [1930a:351] claimed that there were $\mathbf{\Pi}^1_1$ sets that could not be uniformized by any "distinguishable" set, and gave a purported example. Notwithstanding, Sierpiński [1930] asked whether every $\mathbf{\Pi}^1_1$ set can be uniformized by a projective set, and a result of Petr Novikov in Luzin-Novikov [1935] implied that they can, by sets that are at least $\mathbf{\Sigma}^1_2$. Building on this, the Japanese mathematician Motokiti Kondô [1937,1939] established the $\mathbf{\Pi}^1_1$ *Uniformization Theorem*:

(11) Every $\mathbf{\Pi}^1_1$ subset of R^2 can be uniformized by a $\mathbf{\Pi}^1_1$ set.

This was the culminating result of the ordinal analysis of $\mathbf{\Pi}^1_1$ sets. As Kondô noted, his result implies through projections that every $\mathbf{\Sigma}^1_2$ set can be uniformized by a $\mathbf{\Sigma}^1_2$ set, but the question of whether every $\mathbf{\Pi}^1_2$ set can be uniformized by a projective set was left open.

There was also systematic elaboration. Of Luzin's school,[11] Lavrent'ev and Keldysh carried out a deeper investigation of the Borel hierarchy in terms of topological invariance, canonical sets, and constituents of sieves. And Selivanovskii, Novikov, Kolmogorov, and Lyapunov pushed the regularity properties to a stage intermediate between the first and second levels of the projective hierarchy with their study of the C-sets and especially the R-sets. The Poles, who had redeveloped the basic theory through various shorter papers in *Fundamenta Mathematicae* through the 1920's, emphasized topological generalization, lifting the theory to what are now known as *Polish spaces*.[12] Baire [1909] had already stressed the economy of presentation in switching from the reals to what is now known as Baire space, $\{f \mid f\colon \omega \to \omega\}$, essentially the "fundamental domain" of Luzin [1930]. Soon the development of the theory in axiomatically presented topological spaces became popular. As for Luzin himself, he returned to the problem of the perfect set property for $\mathbf{\Pi}^1_1$ sets, first

[11] For details and references, see Kanovei [1985].

[12] Complete separable metric spaces.

broached in his [1917]; in a lecture in 1935, anticipating Gödel's delimitative result Luzin stated several *constituent problems*, each of which would establish the existence of a $\mathbf{\Pi}^1_1$ set without the perfect set property.[13]

The next advances were to be made through the infusion of metamathematical techniques. Kuratowski-Tarski [1931] and Kuratowski [1931] observed that in the study of the projective sets, the set-theoretic operations correspond to the logical connectives, and projection to the existential quantifier, and consequently, the basic manipulations with projective sets can be recast in terms of logical operations. This move may seem like a simple one, but one must recall that it was just during this period that first-order logic was being established as the canonical language for foundational studies by the great papers of Skolem [1923] and Gödel [1930,1931]. This sacred tradition established a precise notion of "definable", and so in retrospect, prudent was the profane choice of the term "descriptive".

The total impasse in descriptive set theory with respect to the regularity properties was to be explained by Gödel's work on the consistency of the Axiom of Choice (AC) and the Continuum Hypothesis (CH). This work can be viewed as a steady intellectual development from his celebrated Incompleteness Theorem, and with respect to our theme of the mathematical necessity of the transfinite, the prescient footnote 48a of Gödel [1931] is worth quoting:

> The true reason for the incompleteness inherent in all formal systems of mathematics is that the formation of ever higher types can be continued into the transfinite ... while in any formal system at most denumerably many of them are available. For it can be shown that the undecidable propositions constructed here become decidable whenever appropriate higher types are added .... An analogous situation prevails for the axiom system of set theory.

Gödel of course established his consistency results by formulating the inner model L, still one of the most beautiful constructions in set theory, and showing that if $V = L$, then AC and CH holds. His main breakthrough can be loosely described as taking the extent of the ordinals as *a priori* and carrying on a kind of Gödel numbering of definable sets through the transfinite. Here we see in an ultimate form how having enough length turns negative or paradoxical assertions to positive ones. Russell's paradox became the proposition about the existence of the next aleph, and Gödel's Incompleteness Theorem became a rectification of Russell's ill-fated Axiom of Reducibility with the proof, making an ironic use of Skolem's paradox argument, of the consistency of the Continuum Hypothesis.

In his initial article [1938] on L, Gödel announced:

[13]See Uspenskii [1985:126], Kanovei [1985:162], and especially Uspenskii-Kanovei [1983].

(12) If $V = L$, then there is a $\mathbf{\Delta}^1_2$ set of reals which is not Lebesgue measurable and a $\mathbf{\Pi}^1_1$ set of reals which does not have the perfect set property.

Thus, the classical descriptive set theorists were up against an essential obstacle of ZFC. The importance that Gödel attached to these results can be evinced from his listing of each of them on equal footing with his AC and CH results. Gödel did not publish proofs, and more than a decade was to pass before proofs first appeared in Novikov [1951]. In the meantime, Gödel in the second edition [1951:67] of his monograph on L had sketched a more basic result:

(13) If $V = L$, then there is a $\mathbf{\Sigma}^1_2$ well-ordering of the reals.

(According to Kreisel [1980:197], "... according to Gödel's notes, not he, but S. Ulam, steeped in the Polish tradition of descriptive set theory, noticed that the definition of the well-ordering ... of subsets of ω was so simple that it supplied a non-measurable [$\mathbf{\Sigma}^1_2$] set of real numbers ...") Mostowski had also established the result in a manuscript destroyed during the war, but it is not apparent in Novikov [1951]. Details were eventually provided by John Addison [1959] who showed that in L every projective set can be uniformized by a projective set:

(14) If $V = L$, then for $n > 1$ every $\mathbf{\Sigma}^1_n$ subset of R^2 can be uniformized by a $\mathbf{\Sigma}^1_n$ set.

Gödel's incisive metamathematical analysis not only provided an explanation for the descriptive set theorist in terms of the limits of formal systems, but also provided explicit counterexamples at the next level, once logical operations were correlated with the classical concepts. Perhaps Wittgenstein would have found congenial the theme of the mathematical necessity of the transfinite ordinals through their increasing use, but no friend of set theory, in his railings against metamathematics he would have frowned at its inversion into mathematics *par excellence*, owing ultimately to the coding possibilities afforded by infinite sets.

Looking ahead, just a year after Paul Cohen's invention of forcing, Robert Solovay [1965,1970] established the following relative consistency result, showing what level of argument is possible with the method.

(15) Suppose that in ZFC there is an inaccessible cardinal. Then there is a ZFC forcing extension in which every projective set of reals is Lebesgue measurable, has the Baire property, and has the perfect set property.

The existence of an inaccessible cardinal is the least in the hierarchy of "large cardinal" axioms adding consistency strength to ZFC that have been extensively studied. Solovay himself noted that that consistency strength is necessary for the perfect set property, and rather unexpectedly, it was eventually shown by Saharon Shelah [1984] that it was also necessary for Lebesgue measurability, but not for the Baire property. These beautiful results in terms of relative consistency provide a mathematically satisfying resolution of the universal possibilities for the projective sets. Solovay was actually able to get a model of ZF in which every set of reals whatsoever has the regularity properties; this is well- known to contradict AC, but on the other hand Solovay's model satisfied the Axiom of Dependent Choices, a weak form of AC adequate for carrying out all of the arguments of descriptive set theory.

As for forcing and uniformization, Levy [1965] observed that in the original Cohen model ("adding a Cohen real over L") there is a $\mathbf{\Pi}^1_2$ set that cannot be uniformized by any projective set, in contradistinction to (14) and establishing that (11) is the best possible (cf. the paragraph after).

Scott's result that if there is a measurable cardinal, then $V \neq L$ was already mentioned. The existence of a measurable cardinal is the paradigmatic large cardinal hypothesis, much stronger in consistency strength than the existence of an inaccessible cardinal. In 1965, building on (15) Solovay reactivated the classical program of investigating the extent of the regularity properties by providing characterizations at the level of Gödel's delimitative results (see Solovay [1969] for the perfect set property), and establishing the following direct implication:

(16) If there is a measurable cardinal, then every $\mathbf{\Sigma}^1_2$ set is Lebesgue measurable, has the Baire property, and has the perfect set property.

Natural inductive arguments were later to establish that, under hypotheses about the determinateness of certain infinitary games, every projective set possesses the regularity properties. These results focused attention on the Axiom of Determinacy and its weak versions, and led in the latter 1980's to remarkable advances in the investigation of strong hypothesis and relative consistency.[14] But an adequate discussion of these matters would go far beyond the scope of this paper.

Returning to much earlier developments based on Gödel's work, after his fundamental work on recursive function theory in the 1930's Stephen Kleene expanded his investigations of effectiveness and developed a general theory of definability for relations on the integers. In [1943] he studied the *arithmetical relations*, those relations obtainable from the recursive relations by application of number quantifiers. Developing canonical representations, he classified these relations into a hierarchy according to quantifier complexity and showed that the hierarchy is proper. In [1955,1955a,1955b] he studied the *analytical relations*, those relations obtainable from the arithmetical relations by applications

[14] See Martin-Steel [1988,1989] and Woodin [1988].

of function quantifiers. Again, he worked out representation and hierarchy results, and moreover, established an elegant theorem that turned out to be an effective analogue of Suslin's characterization (5) of the Borel sets.

Kleene was developing what amounted to the effective content of the classical theory, unaware that his techniques had direct antecedents in the papers of Lebesgue, Luzin, and Sierpiński. Certainly, he had very different motivations: with the arithmetical relations he wanted to extend the Incompleteness Theorem, and analytical relations grew out of his investigations of notations for recursive ordinals. On the other hand, already in [1943:50] he did make elliptic remarks about possible analogies with the classical theory. Once the conceptual move was made to the consideration of relations on functions of integers and with the classical switch to Baire space already in place, it was Kleene's student Addison who established the exact analogies: the analytical relations are analogous to the projective sets, and the arithmetical relations are analogous to the sets in the first ω levels of the Borel hierarchy.

Another mathematical eternal return: Toward the end of his life, Gödel regarded the question of whether there is a linear hierarchy for the recursive sets as one of the big open problems of mathematical logic. Intuitively, given two decision procedures, one can often be seen to be simpler than the other. Now a set of integers is recursive *iff* both it and its complement are recursively enumerable. The pivotal result of classical descriptive set theory is Suslin's, that a set is Borel *iff* both it and its complement are analytic. But before that, a hierarchy for the Borel sets was in place. In an ultimate inversion, as we look back into the recursive sets, there is no known hierarchy.

References

Addison, John W. [1959], Some consequences of the axiom of constructibility, *Fundamenta Mathematicae* 46, 337-357.

Aleksandrov, Pavel S. [1916], Sur la puissance des ensembles mesurables B, *Comptes Rendus de l'Académie des Sciences, Paris* 162, 323-325.

Aleksandrov, Pavel S. [1972], Pages from an autobiography, *Russian Mathematical Surveys* 34(6), 267-302.

Baire, René [1898], Sur les fonctions discontinues qui se rattachment aux fonctions continues, *Comptes Rendus de l'Académie des Sciences, Paris* 129, 1621-1623.

Baire, René [1899], Sur les fonctions de variables réelles, *Annali di Matematica Pura ed Applicata* (3)3, 1-122.

Baire, René [1906], Sur la représentation des fonctions discontinues. Première partie, *Acta Mathematica* 30, 1-48.

Baire, René [1909], Sur la représentation des fonctions discontinues. Deuxième partie, *Acta Mathematica* 32, 97-176.

Bendixson, Ivar [1883], Quelques théorèmes de la théorie des ensembles de points, *Acta Mathematica* 2, 415-429.

Borel, Emile [1898], *Leçons sur la théorie des fonctions*, Paris, Gauthier-Villars.

Borel, Emile [1905], *Leçons sur les fonctions de variables réelles et les développements en series de polynomes*, Paris, Gauthier-Villars.

Brouwer, Luitzen E.J. [1906], Over de Grondslagen der Wiskunde. Amsterdam, Maas & van Suchtelen. Translated in [1975] below, 11-101.

Brouwer, Luitzen E.J. [1975], Heyting, Arend (ed.), *Collected Works*, vol. 1, Amsterdam, North-Holland.

Cantor, Georg [1883], Über unendliche, lineare Punktmannichfaltigkeiten. V, *Mathematicshe Annalen* 21, 545-591. Reprinted in [1966] below, 165-209.

Cantor, Georg [1884], Über unendliche, linear Punktmannichfaltigkeiten. VI, *Mathematicshe Annalen* 23, 453-488. Reprinted in [1966] below, 210-246.

Cantor, Georg [1966], Zermelo, Ernst (ed.) *Gesammelte Abehandlungen*, Hildesheim, Georg Olms Verlag. Reprint of the original 1932 edition, Berlin, Springer.

Feferman, Solomon, and Azriel Levy [1963], Independence results in set theory by Cohen's method. II (abstract), *Notices of the American Mathematical Society* 10, 593.

Gödel, Kurt F. [1930], Die Vollstandigkeit der Axiome des logischen Funktionenkalküls, *Monatshefte für Mathematik und Physik* 37, 349-360. Reprinted and translated in [1986] below, 102-123.

Gödel, Kurt F. [1931], Über formal unentscheidbare Sätze der Principia Mathematica und verwandter Systeme I, *Monatshefte für Mathematik und Physik* 38, 173-198. Reprinted and translated in [1986] below, 145-195.

Gödel, Kurt F. [1938], The consistency of the Axiom of Choice and of the Generalized Continuum-Hypothesis, *Proceedings of the National Academy of Sciences U.S.A.* 24, 556-557. Reprinted in [1990] below, 26-27.

Gödel, Kurt F. [1951], *The Consistency of the Axiom of Choice and of the Generalized Continuum Hypothesis with the Axioms of Set Theory*, Annals of Mathematics Studies #3, Princeton, Princeton University Press. Second printing. Reprinted in [1990] below, 33-101.

Gödel, Kurt F. [1986], *Collected Works*, vol. 1, edited by Feferman, Solomon, *et al.*, New York, Oxford University Press.

Gödel, Kurt F., 1990], *Collected Works*, vol. 2, edited by Feferman, Solomon, *et al.*, New York, Oxford University Press.

Hausdorff, Felix [1908], Grundzüge einer Theorie der geordneten Mengen, *Mathematische Annalen* 65, 435-505.

Hausdorff, Felix [1914], *Grundzüge der Mengenlehre*, Leipzig, de Gruyter. Reprinted in New York, Chelsea 1965.

Hausdorff, Felix [1916], Die Mächtigkeit der Borelschen Mengen, *Mathematische Annalen* 77, 430-437.

Hawkins, Thomas W. [1975], *Lebesgue's Theory of Integration. Its Origins and Development.* Second edition, New York, Chelsea.

Kanovei, V.G. [1985], The development of the descriptive theory of sets under the influence of the work of Luzin, *Russian Mathematical Surveys* 40(3), 135-180.

Keisler, H. Jerome [1962], Some applications of the theory of models to set theory, in Nagel, Ernest, Patrick Suppes, and Alfred Tarski (eds.) *Logic, Methodology and Philosophy of Science*, Proceedings of the International Congress, Stanford. Stanford, Stanford University Press.

Keldysh, Ljudmila V. [1974], The ideas of N.N. Luzin in descriptive set theory, *Russian Mathematical Surveys* 29(5), 179-193.

Kleene, Stephen C. [1943], Recursive predicates and quantifiers, *Transactions of the American Mathematical Society* 53, 41-73.

Kleene, Stephen C. [1955], On the forms of predicates in the theory of constructive ordinals (second paper), *American Journal of Mathematics* 77, 405-428.

Kleene, Stephen C. [1955a], Arithmetical predicates and function quantifiers, *Transactions of the American Mathematical Society* 79, 312-340.

Kleene, Stephen C. [1955b], Hierarchies of number-theoretic predicates, *Bulletin of the American Mathematical Society* 61, 193-213.

Kondô, Motokiti [1937], L'uniformisation des complémentaires analytiques, *Proceedings of the Imperial Academy of Japan* 13, 287-291.

Kondô, Motokiti [1939], Sur l'uniformisation des complémentaires analytiques et les ensembles projectifs de la seconde classe, *Japanese Journal of Mathematics* 15, 197-230.

Kreisel, Georg [1980], Kurt Gödel, 28 April 1906–14 January 1978. *Biographical Memoirs of the Fellows of the Royal Society* 26, 149-224. Corrections 27 (1981), 697, and 28 (1982), 718.

Kuratowski, Kazimierz [1922], Une méthode d'elimination des nombres transfinis des raisonnements mathématiques, *Fundamenta Mathematicae* 3 (1922), 76-108.

Kuratowski, Kazimierz [1931], Evaluation de la classe Borélienne ou projective d'un ensemble de points à l'aide des symboles logiques, *Fundamenta Mathematicae* 17, 249-272.

Kuratowski, Kazimierz [1966], *Topology*, vol. 1, New York, Academic Press.

Kuratowski, Kazimierz [1980], *A Half Century of Polish Mathematics. Remembrances and Reflections* Oxford, Pergamon Press.

Kuratowski, Kazimierz, and Alfred Tarski [1931], Les opérations logiques et les ensembles projectifs, *Fundamenta Mathematicae* 17, 240-248. Reprinted in Givant, Steven R., and Ralph N. McKenzie (eds.) *Alfred Tarski. Collected Papers,* Basel, Birkhäuser 1986, vol. 1, 551-559.

Kuzawa, Mary G. [1968], *Modern Mathematics. The Genesis of a School in Poland*, New Haven, College & University Press.

Lavrent'ev, Mikhail A. [1974], Nikolai Nikolaevich Luzin, *Russian Mathematical Surveys* 29(5), 173-178.

Lebesgue, Henri [1902], Intégrale, longueur, aire, *Annali di Matematica Pura ed Applicata* (3)7, 231-359.

Lebesgue, Henri [1905], Sur les fonctions représentables analytiquement, *Journal de Mathématiques Pures et Appliquées* (6)1, 139-216. Reprinted in (1972) below, vol. 3, 103-180.

Lebesgue, Henri [1972], *Oeuvres Scientifiques*, Geneva, Kundig.

Levy, Azriel [1965], Definability in axiomatic set theory I, in Bar-Hillel, Yehoshua (ed.) *Logic, Methodology and Philosophy of Science*, Proceedings of the 1964 International Congress, Jerusalem. Amsterdam, North-Holland, 127-151.

Luzin, Nikolai N. [1917], Sur la classification de M. Baire, *Comptes Rendus de l'Académie des Sciences, Paris* 164, 91-94.

Luzin, Nikolai N. [1925], Sur un problème de M. Emile Borel et les ensembles projectifs de M. Henri Lebesgue; les ensembles analytiques, *Comptes Rendus de l'Académie des Sciences, Paris* 180, 1318-1320.

Luzin, Nikolai N. [1925a], Sur les ensembles projectifs de M. Henri Lebesgue, *Comptes Rendus de l'Académie des Sciences, Paris* 180, 1572-1574.

Luzin, Nikolai N. [1925b], Les propriétés des ensembles projectifs, *Comptes Rendus de l'Académie des Sciences, Paris* 180, 1817-1819.

Luzin, Nikolai N. [1925c], Sur les ensembles non mesurables B et l'emploi de diagonale Cantor, *Comptes Rendus de l'Académie des Sciences, Paris* 181, 95-96.

Luzin, Nikolai N. [1926], Mémoires sur les ensembles analytiques et projectifs, *Matematicheskii Sbornik* 33, 237-290.

Luzin, Nikolai N. [1927], Sur les ensembles analytiques, *Fundamenta Mathematicae* 10, 1-95.

Luzin, Nikolai N. [1930], *Leçons sur Les Ensembles Analytiques et Leurs Applications.* Paris, Gauthier-Villars. Reprinted with corrections in New York, Chelsea 1972.

Luzin, Nikolai N. [1930a], Sur le problème de M.J. Hadamard d'uniformisation des ensembles, *Comptes Rendus de l'Académie des Sciences* 190, 349-351.

Luzin, Nikolai N., and Petr S. Novikov [1935], Choix éffectif d'un point dans un complémentaires analytique arbitraire, donné par un crible, *Fundamenta Mathematicae* 25, 559- 560.

Luzin, Nikolai N., and Waclaw Sierpiński [1918], Sur quelques propriétés des ensembles (A). *Bulletin de l'Académie des Sciences Cracovie, Classe des Sciences Mathématiques, Série A*, 35-48.

Luzin, Nikolai N., and Waclaw Sierpiński [1923], Sur un ensemble non mesurable B, *Journal de Mathématiques Pures et Appliquées* (9)2, 53-72.

Martin, Donald A., and John R. Steel [1988], Projective determinacy, *Proceedings of the National Academy of Sciences U.S.A.* 85, 6582-6586.

Martin, Donald A., and John R. Steel [1989], A proof of Projective Determinacy, *Journal of the American Mathematical Society* 2, 71-125.

Mirimanoff, Dmitry [1917], Les antinomies de Russell et de Burali-Forti et le problème fondamental de la théorie des ensembles, *L'Enseignment Mathématique* 19, 37-52.

Moore, Gregory H. [1982], *Zermelo's Axiom of Choice. Its Origins, Development and Infiuence*, New York, Springer-Verlag.

Moschovakis, Yiannis N. [1980], *Descriptive Set Theory*, Amsterdam, North-Holland.

Mostowski, Andrzej M. [1949], An undecidable arithmetical statement, *Fundamenta Mathematicae* 36, 143-164. Reprinted in Kuratowski, Kazimierz *et al.* (eds.) *Foundational Studies. Selected Works,* Amsterdam, North-Holland 1979, vol. 1, 531-552.

Novikov, Petr S. [1931], Sur les fonctions implicites mesurables B, *Fundamenta Mathematicae* 17, 8-25.

Novikov, Petr S. [1951], On the consistency of some propositions of the descriptive theory of sets (in Russian), *Trudy Matematiěskogo Instituat imeni V.A. Steklova* 38, 279-316. Translated in *American Mathematical Society Translations* 29, 51-89.

Phillips, Esther R. [1978], Nicolai Nicolaevich Luzin and the Moscow school of the theory of functions, *Historia Mathematica* 5, 275-305.

Poincaré, Henri [1906], Les mathématiques et la logique, *Revue de Métaphysique et de Morale* 14, 17-34.

Schoenfies, Arthur [1905], Über wohlgeordnete Mengen, *Mathematische Annalen* 60, 181- 186.

Scott, Dana S. [1961], Measurable cardinals and constructible sets, *Bulletin de l'Académie Polonaise des Sciences, Série des Sciences Mathématiques, Astronomiques et Physiques* 9, 521-524.

Sierpiński, Waclaw [1925], Sur une classe d'ensembles, *Fundamenta Mathematicae* 7, 237- 243.

Sierpiński, Waclaw [1930], Sur l'uniformisation des ensembles mesurables (B), *Fundamenta Mathematicae* 16, 136-139.

Sierpiński, Waclaw [1950], *Les ensembles projectifs et analytiques*, Mémorial des Sciences Mathématiques #112, Paris, Gauthier-Villars.

Skolem, Thoralf [1923], Einige Bemerkung zur axiomatischen Begründung der Mengenlehre, in *Matematikerkongressen i Helsingfors 4-7 Juli 1922, Dem femte skandinaviska matematikerkongressen, Redogřelse*, Helsinki, Akademiska Bokhandeln, 217- 232. Reprinted in [1970] below, 137-152. Translated in van Heijenoort [1967], 290-301.

Skolem, Thoralf [1970], Fenstad, Jens E. (ed.) *Selected Works in Logic*, Oslo, Universitetsforlaget.

Shelah, Saharon [1984], Can you take Solovay's inacessible away? *Israel Journal of Mathematics* 48, 1-47.

Solovay, Robert M. [1965], The measure problem (abstract), *Notices of the American Mathematical Society* 13, 217.

Solovay, Robert M. [1969], The cardinality of $\mathbf{\Sigma}^1_2$ sets of reals. In Bullof, Jack J., Thomas C. Holyoke, and S.W. Hahn (eds.) *Foundations of Mathematics*, Symposium papers commemorating the sixtieth birthday of Kurt Gödel, Berlin, Springer-Verlag, 58-73.

Solovay, Robert M. [1970], A model of set theory in which every set of reals is Lebesgue measurable, *Annals of Mathematics* 92, 1-56.

Suslin, Mikhail Ya. [1917], Sur une définition des ensembles mesurables B sans nombres transfinis, *Comptes Rendus de l'Académie des Sciences, Paris* 164, 88-91.

Uspenskii, Vladimir A. [1985], Luzin's contribution to the descriptive theory of sets and functions: concepts, problems, predictions, *Russian Mathematical Surveys* 40(3), 97-134.

Uspenskii, Vladimir A., and Kanovei, V.G. [1983], Luzin's problem on constituents and their fate, *Moscow University Mathematics Bulletin* 38(6), 86-102.

van Heijenoort, Jean [1967], (ed.) *From Frege to Gödel: A Source Book in Mathematical Logic, 1879-1931*, Cambridge, Harvard University Press.

von Neumann, John [1925], Eine Axiomatisierung der Mengenlehre, *Journal für die reine und angewandte Mathematik* 154, 219-240. Reprinted in [1961] below, vol. 1, 34-56. Translated in van Heijenoort [1967], 393-413.

von Neumann, John [1949], On rings of operators. Reduction theory, *Annals of Mathematics* 50, 401-485. Reprinted in [1961] below, vol. 3, 400-484.

von Neumann, John [1961], Taub, Abraham H. (ed.) *John von Neumann. Collected Works*, New York, Pergamon Press.

Woodin, W. Hugh [1988], Supercompact cardinals, sets of reals, and weakly homogeneous trees, *Proceedings of the National Academy of Sciences U.S.A.* 85, 6587-6591.

Yankov, V. [1941], Sur l'uniformisation des ensembles A, *Doklady Akademiia Nauk SSSR* 30, 597-598.

Young, William H. [1903], Zur Lehre der nicht abgeschlossenen Punktmengen, *Berichte über die Verhandlungen der Königlich Sächsischen Gesellschaft der Wissenschaften zu Leipzig, Mathematisch-Physische Klasse* 55, 287-293.

Zermelo, Ernst [1930], Über Grenzzahlen und Mengenbereiche: Neue undersuchungen über die Grundlagen der Mengenlehre, *Fundamenta Mathematicae* 16, 29-47.

2 Hilbert and Set Theory (with Burton Dreben)

David Hilbert (1862-1943) was the preeminent mathematician of the early decades of the 20th Century,[1] a mathematician whose pivotal and penetrating results, emphasis on central problems and conjectures, and advocacy of programmatic approaches greatly expanded mathematics with new procedures, initiatives, and contexts. With the emerging extensional construal of mathematical objects and the development of abstract structures, set-theoretic formulations and operations became more and more embedded into the basic framework of mathematics. And Hilbert specifically championed Cantorian set theory, declaring [1926: 170]: "From the paradise that Cantor has created for us no one will cast us out."

On the other hand, Hilbert did not make direct mathematical contributions toward the development of set theory. Although he liberally used non-constructive arguments, his were still the concerns of mainstream mathematics, and he stressed concrete approaches and the eventual solvability of every mathematical problem. After its beginnings as the study of the transfinite numbers and definable collections of reals, set theory was becoming an open-ended, axiomatic investigation of *arbitrary* collections and functions. For Hilbert this was never to be a major concern, but he nonetheless exerted a strong influence on this development both through his broader mathematical approaches and through his specific attempt to establish the Continuum Hypothesis.

What follows is a historical and episodic account of Hilbert's results and initiatives and their ramifications and extensions, in so far as they bear on set theory and its development.[2] The emphasis on set theory presents a tangential view of Hilbert's main mathematical endeavors, but one that illuminates their larger themes and motivations. Because of its basic interplay with set theory, we deal at length with Hilbert's program for establishing the consistency of mathematics by "finitary reasoning". Section 1 discusses Hilbert's use of non-constructive existence proofs, with the focus on his first major result; section 2 discusses his axiomatization of Euclidean geometry, with the focus on his Completeness Axiom; and then section 3 discusses questions about the real numbers and their arithmetic that Hilbert would later approach through his proof theory. With this as a backdrop, section 4 considers Hilbert's involvement in the early development of set theory, and section 5 considers both his mathematical logic as a reaction to Russell's and the two crucial new questions

Republished from *Synthèse* 110 (1997), pp.77-125, with permission from Springer Nature BV.

This article grew out of an invited talk given by Kanamori on 9 November 1993 at a symposium on Hilbert's Philosophy of Mathematics held as part of the Boston Colloquium for Philosophy of Science, for which he would like to thank the organizers, Jaakko Hintikka and Alfred Tauber. The authors are very grateful to Volker Peckhaus, Jose Ruiz, and Christian Thiel for numerous helpful comments and corrections.

[1]Henri Poincaré, Hilbert's only rival for preeminence, died in 1912.

[2]See Kanamori [1996] for the development of set theory from Cantor to Cohen.

that Hilbert raised. Section 6 describes Hilbert's approach to establishing the consistency of mathematics, and section 7 its application to the Continuum Hypothesis. Then section 8 discusses Gödel's work, particularly on the consistency of the Continuum Hypothesis, in relation to Hilbert's. In the appendix, Hilbert's consistency program is reconsidered in light of recent developments in "reverse" mathematics.

1 Basis Theorem

Hilbert's work from the beginning greatly accelerated the move away from the traditional constructive moorings, being driven by strong impulses: the solution of focal problems by intuitively clear though not necessarily constructive means, and the drive for systematization with its emerging concern with consistency. When Hilbert was in his late twenties, he [1890] established his first major result, *Hilbert's basis theorem*, which cast in current terms is the assertion:

> Suppose that F is a field and $F[x_1, \ldots, x_n]$ the ring of polynomials over F in $x_1, \ldots, x_n$. Then every ideal in $F[x_1, \ldots, x_n]$ is finitely generated.

Invariant theory, the subject of Hilbert's doctoral dissertation and Habilitationsschrift, was the bridge between geometry and algebra in 19th Century mathematics, and the basis theorem was the key ingredient in his solution [1890] of invariant theory's then central problem.[3] Moreover, Hilbert's [1890] with its new structural approach can be considered the first paper of modern algebra: In straightforward generalizations in terms of algebraic varieties the basis theorem serves as a foundation for algebraic geometry.

[3]That central problem emanated from the work of Arthur Cayley. It had been known that for a polynomial $ax^2 + 2bxy + cy^2$ in x and y, if $x = \alpha x' + \beta y'$ and $y = \gamma x' + \delta y'$ are substituted to get $a'x'^2 + 2b'x'y' + c'y'^2$, then

$$b'^2 - a'c' = (b^2 - ac)(\alpha\delta - \beta\gamma)^2 ,$$

i.e. the new *discriminant* $b'^2 - a'c'$ equals the old discriminant $b^2 - ac$ times a constant factor (in fact the square of the *determinant* of the transformation).

Generalizing, a *form* (Cayley's *quantic*), is a polynomial in $x_1, \ldots, x_n$ which is homogeneous (i.e. there is a fixed k such that the sum of the powers of the variables in each summand is k). A *linear transformation* of $x_1, \ldots, x_n$ to $x'_1, \ldots, x'_n$ is given by a system of equations, each x_i being equated to a linear form in $x'_1, \ldots, x'_n$. For a form P, a polynomial Q in the coefficients and variables of P is an *invariant* if and only if for every linear transformation, if the corresponding substitutions are made to get a corresponding form P' in $x'_1, \ldots, x'_n$ and a Q' corresponding to Q, then Q and Q' differ only by a constant factor. A *complete system of invariants* for P is a collection C of such invariants such that every invariant is a linear combination of members of C. Finally, the central problem of invariant theory solved by Hilbert [1890] was for any form P to find a *finite*, complete system of invariants.

Of course, it is straightforward to generalize the foregoing in modern terms to polynomials over a field and groups of linear transformations, and then to vector spaces on which groups act linearly, and this is how invariant theory was eventually reactivated.

The basis theorem caused a sensation since it argued for a finite number of generators, yet provided no explicit construction. Moreover, in the form that it was actually established by Hilbert, that for any appropriate sequence of polynomials every polynomial in the sequence is a linear combination of the first few, it was a widely applicable result. Paul Gordan [1868] had solved the central invariant theory problem for the special case of two variables by an ingenious but tedious construction which was a culmination of what came to be called the "symbolic method". After seeing Hilbert's basis theorem Gordan quipped (Max Noether [1914: 18], Felix Klein [1926: 330]): "This is not mathematics; this is theology!" Hilbert had carried out a streamlining double induction (or rather, finite descent), first putting the case of n variables into a simple form, and then effecting a reduction to $n-1$. He had established a startling result by a convincing argument, one that was soon accepted by the mathematical community. Not only was the proof reasonably surveyable, but it made a large array of algebraic constructions manageable and introduced simplicity where there had been none.

Nonetheless, Hilbert [1893] soon provided an even more informative proof of his invariant theory result. It was for this purpose that he established his well-known Nullstellensatz,[4] which like the basis theorem had a non-constructive proof and has become fundamental in modern algebra. Applying the non-constructive Nullstellensatz Hilbert provided an otherwise constructive algorithm for computing complete systems of invariants, building on a technique due to Arthur Cayley. This was a striking instance of what was becoming a major trend in mathematics: the development of contextually appropriate proofs for results established by apparently less informative means, leading to a further enrichment of mathematics. Indeed, Gordan (Klein [1926: 331]) conceded that "even theology has its merits", and soon provided his own proofs [1893, 1899] of Hilbert's basis theorem.[5] Both Hilbert's basis theorem as well as his Nullstel-

[4]In Hilbert's original [1893] form the Nullstellensatz states that if $f, f_1, \ldots, f_r$ are in $C[x_1, \ldots, x_n]$, the ring of polynomials in $x_1, \ldots, x_n$ over the complex field, and f vanishes at all the common roots of $f_1, \ldots, f_r$, then some power f^k is a linear combination $f^k = h_1 f_1 + \ldots + h_r f_r$. The assertion is equivalent to the special case when $f = 1$ and the f_i's have no common roots. In modern terms, this special case amounts to the assertion that if F is a field, I is the ideal of $F[x_1, \ldots, x_n]$ generated by $\{f_1, \ldots, f_r\}$ (and all ideals of that polynomial ring are generated by some such finite collection by Hilbert's basis theorem), and the f_i's have no common roots in the algebraic closure of F, then the ideal is the unit ideal, i.e. the whole ring.

[5]Hilbert's basis theorem would stimulate the search for algebraic generalizations, with an optimistic one suggested by the 14th of Hilbert's [1900] problems, and much progress would be made. §3 discusses the first two of Hilbert's [1900] problems; see Mumford [1976] for the 14th problem.

As for invariant theory itself, Hilbert's comprehensive result there was to leave the field fallow for most of Hilbert's lifetime, only revived by his brilliant student Hermann Weyl [1939] for the classical Lie groups as part of their representation theory. The subject was then fully reactivated by David Mumford [1965] with his incisive investigation of groups of

lensatz would be precisely analyzed in terms of formal systems. In particular, the double induction in Hilbert's proof of the basis theorem would turn out to be a remarkable foreshadowing of how a close variant of the theorem would be shown equivalent to a proposition (the provable totality of Ackermann's function) that just transcends one common characterization of Hilbert's later finitistic viewpoint. (See in the appendix Theorem 3 and remarks following.)

In addition to non-constructive existence proofs Hilbert championed the use of "ideal elements". Well-established were the imaginary i and the points at infinity for projective geometry, and emerging into prominence were the ideals of algebraic number fields, to the theory of which Hilbert made fundamental contributions. The imaginary i had stimulated the inaugural use of non-constructive existence proofs in algebra: The fundamental theorem of algebra, that every polynomial in complex coefficients has a root, was first established by Gauss in his doctoral dissertation [1799] by a proof that provided no means of algebraically calculating a root. Weierstrass and Dedekind carried out involved constructive extensions of Gauss's work in the 1880's; but significantly, Hilbert [1896] considerably streamlined this work by applying his Nullstellensatz, later claiming [1928] (see van Heijenoort [1967: 474]) that its (non-constructive) proof "uncovers the inner reason for the validity of the assertions adumbrated by Gauss and formulated by Weierstrass and Dedekind."

A remarkable example of the use of non-constructive existence proofs is Hilbert's ingenious solution to Waring's Problem. Broached by Edward Waring in 1770, it asks of natural numbers whether for every positive k there is a fixed r such that for every n,

$$n = n_1^k + \cdots + n_r^k \text{ for some } n_1, \ldots, n_r\,.$$

In that same year Lagrange had established the result for $k = 2$ with $r = 4$, but for no other $k > 2$ was the result known until Hilbert [1909] completely solved the problem by establishing the existence for every k of a corresponding r. However, taking $g(k)$ to be the least possible such r, Hilbert's proof provided no way of calculating $g(k)$. Hilbert's result spurred extensive activity in analytic number theory, in part to determine the values $g(k)$, and they are "almost" completely known today.[6]

However non-constructive Hilbert's approach, he himself never seemed to have entertained sets of arbitrary choices as formalized by the Axiom of Choice,

---

automorphisms on algebraic varieties (see also Mumford-Fogarty-Kirwan [1994]). Notably it was the approach of [1893] rather than the initial [1890] that was to inspire Mumford [1965], which can be considered as perpetuating in geometric terms the 19th Century view of invariant theory as a constructive theory.

[6]See Ellison [1971] for a history of Waring's Problem. The conjecture is that $g(k) = [(\frac{3}{2})^k] + 2^k - 2$, and according to recent research literature this has been verified for $k \leq 471{,}600{,}000$ and for sufficiently large k.

an axiom first made explicit by Ernst Zermelo [1904]. The expansion of mathematics to this level of abstraction was initiated by Felix Hausdorff in his classic *Grundzüge der Mengenlehre* [1914] which broke the ground for a generation of mathematicians in both set theory and topology. Of particular interest was Hausdorff's use of the Axiom of Choice (in [1914: 469ff.] and also in [1914a]) to get what is now known as Hausdorff's Paradox, an implausible decomposition of the sphere; this was a dramatic synthesis of classical mathematics and the new set-theoretic view.

Of those directly influenced by Hilbert, Georg Hamel, whose doctoral work was supervised by Hilbert, made [1905] an early and explicit use of the Axiom of Choice to provide what is now known as a Hamel basis, a basis for the real numbers as a vector space over the rational numbers. The full exercise of the Axiom of Choice in ongoing mathematics first occurred in the pioneering work of Ernst Steinitz [1910], who made systematic use of well-orderings to establish the abstract theory of fields, their algebraic and transcendental extensions, and algebraic closures. Zermelo [1914] modified Hamel's basis to get one for the complex numbers and with a further use of the Axiom of Choice answered a question about the existence of a collection of complex numbers with special closure and basis conditions. Presaging her later work Emmy Noether [1916] axiomatically characterized those integral domains satisfying Zermelo's conditions.

Noether's mathematical roots were in invariant theory and in [1915] she brought together Hilbert's basis theorem arguments with those of Steinitz's field theory. Going to Hilbert's Göttingen, Noether became the leading figure in algebra there through her work on the theory of ideals in commutative rings. In her incisive [1921] she lifted the finiteness properties emanating from Hilbert's basis theorem to a general axiomatic setting by introducing the *ascending chain condition,* and rings satisfying this condition are now known as *Noetherian rings.* Similarly abstracting another finiteness property, Noether [1927] extended Dedekind's unique factorization theory for ideals of rings of algebraic numbers to the general setting. She [1927: 45ff.] applied the Axiom of Choice without much ado, but only a weak version, the so-called Axiom of Dependent Choices, is needed for the general formulations of her basic results. The full exercise of the Axiom of Choice entered Noether's axiomatic ring theory when Wolfgang Krull [1929] investigated rings not necessarily satisfying the ascending chain condition, specifically in the general assertion that every ideal in a ring can be extended to a maximal ideal. Ring theory today is often presented at this level of generality, but Hilbert's basis theorem remains a palliative in the crucial cases for algebraic geometry, where the theorem's applicability renders any appeal to the Axiom of Choice unnecessary.

In terms of his later consistency program Hilbert's advocacy of nonconstructive existence proofs and the use of ideal elements necessarily raised the stakes involved. Not only did the issue of consistency become more critical when explicit constructions were not available or ideal elements seamlessly introduced, but the weight was shifted from algebraic calculations to logical deductions, which, however, increasingly took on the spirit of calculations not unlike those in the "symbolic method" used by Gordan. The existential quantifier assumed a pivotal role, both in its interplay with the Law of Excluded Middle and the extent to which it could be construed as instrumental in the generation of terms through instantiation. Such issues became central for Hilbert in his mathematical investigation of formalized proofs (see §6), and his early work, which assumed an increasingly abstract and logical form from invariant theory to algebraic number theory, undoubtedly predisposed him to this later development.

2 Geometry

Hilbert's new conception of the role of axiomatization as not reflecting an antecedently given subject matter and his resulting concern for consistency first appeared in print in his *Grundlagen der Geometrie* [1899], based on lectures given in the 1890's and especially on those in the winter of 1898-9. In the introduction to the *Grundlagen* Hilbert wrote of his investigation as "a new attempt to establish for geometry a *simple* and *complete* [*vollständiges*] system of axioms *independent* of one another." What *vollständiges* was to mean would become a central concern of mathematical logic in later decades. He proceeded to provide a rigorous axiomatization of Euclidean geometry with five groups I-V of axioms, for *incidence, order, congruence, parallelism,* and *continuity* respectively. Previous and venerable work had already established the consistency of non-Euclidean geometries via models in Euclidean geometry. Hilbert in a groundbreaking move raised the question of the consistency of Euclidean geometry itself as given by his axioms, and proceeded to establish it via a *countable* arithmetical model. Then, as with the work on the Parallel Axiom, Hilbert went on to use various models of subcollections of his axioms to establish the independence of axioms and theorems.

Hilbert's model for the consistency of his full list of axioms took as its "points" the countable collection of ordered pairs of real numbers generated from 1 by the arithmetical operations and the taking of square roots of positive numbers. While fitting into the development of algebraic number fields, this model is notable as arguably the first instance of the Löwenheim-Skolem phenomenon, a "Skolem's Paradox" for the continuum. Hilbert had accentuated the reliance on arithmetic by reducing geometry to a countable domain of ordered pairs of algebraic real numbers; Skolem's [1923] argument for generating a countable model using Skolem terms would give for any (countable first-order) theory a countable model. To distinguish a countable substructure of the continuum as Hilbert had done was the most informative type of "application" that the Löwenheim-Skolem Theorem could have had before Skolem's own application in [1923] to get his "paradox" in set theory. However, despite his professed indifference to whether his axioms were about points or tables,[7] Hilbert did not dwell on this model and soon moved to secure Euclidean space.

Hilbert's axiom group V for continuity initially consisted of a single axiom, the Archimedean Axiom,[8] but he soon added another, the Completeness [Vollständigkeit] Axiom V,2:

> It is impossible to adjoin further elements to the system of points, lines, and planes in such a way that the system thus extended forms a new geometry satisfying all the axioms in groups I-V; in other words, the elements of the geometry form a system which is not susceptible to extension, if all of the stated axioms are to be maintained.

An arithmetical version of this axiom first appeared in *Über den Zahlbegriff* [1900b], of which more in §3. The axiom itself is mentioned first in the French translation [1900a: 25] of the *Grundlagen* and then in the English translation [1902: 25], prior to its incorporation into the second edition [1903: 16]. In the original *Grundlagen* [1899: 39] (see also [1971: 58ff.]) Hilbert had shown that every "geometry" satisfying I-IV and the Archimedean Axiom is faithfully embeddable into the "ordinary analytic geometry", i.e. Euclidean space.[9] The Completeness Axiom amounted to making this maximal geometry the unique geometry.

A set of axioms is *categorical* if it has a unique model up to isomorphism. Having investigated his axioms for geometry with models, Hilbert with his Completeness Axiom simply posited categoricity with the maximal geometry. Hilbert's professed aim in the introduction to the *Grundlagen* had been to get "a *simple* and *complete* system of axioms", yet today his axiom would be considered neither simple nor immediately related to notions of completeness later studied by Hilbert. With the Completeness Axiom Hilbert had come to an axiom about models of axioms and thereby raised the sort of issues that would

[7]According to Blumenthal [1935: 403] Hilbert already in 1891 uttered his aphorism portending his axiomatic and formalist leanings: "One must always be able to say for *points, line, plane*: *table, chair, beer-mug*."

[8]The Archimedean Axiom asserts that for any two line segments s and t a finite number of contiguous copies of s along the ray of t will subsume t.

[9]The embeddability of an axiomatically presented geometry into Euclidean space was Hilbert's first "meta" result in mathematics. In his [1895], appearing as Appendix I from the second edition [1903] on of the *Grundlagen*, what amounts in modern terms to a homeomorphism of a "general geometry" with a finite convex part of Euclidean space played a crucial role. [1895] dealt with the problem of "the straight line as the shortest distance between two points", and a general version of this became the fourth of Hilbert's [1900] problems. See Busemann [1976] for the fourth problem.

become amenable to mathematical investigation only decades later. (See §8, especially footnote 51.)

The Completeness Axiom had specific antecedents in the tradition leading to the development of set theory. In the well-known formulations of the real numbers by Georg Cantor [1872] as fundamental sequences and by Richard Dedekind [1872] as cuts, the correlation with "the straight line" was not regarded as automatic. Cantor [1872: 128] wrote:

> In order to complete the connection ... with the geometry of the straight line, one must only add an *axiom* which simply says that conversely every numerical quantity also has a determined point on the straight line, whose coordinate is equal to that quantity .... I call this proposition an *axiom* because by its nature it cannot be universally proved. A certain objectivity is then subsequently gained thereby for the quantities although they are quite independent of this.

Dedekind [1872: III] wrote:

> If all points on the straight line fall into two classes such that every point of the first class lies to the left of every point of the second class, then there exists one and only one point that produces this division ... The assumption of this property of the line is nothing else than an axiom by which we attribute to the line its continuity, by which we find continuity in the line.

At such an interface one finds what one seeks: Henri Poincaré [1902: 40] commended Dedekind's cuts as reflecting the "intuitive truth that if a straight line is cut into two rays their common border is a point." On the other hand, Bertrand Russell [1919: 71] decried Dedekind's approach of postulating what one wants as having the same advantages as "theft over honest toil". Russell's genetic approach of building up from the natural numbers to the rational numbers and then *defining* the real numbers as the cuts is congenial to his logicist reductionism,[10] but obscures the antecedent sense of the continuum that both Cantor and Dedekind were trying to accommodate. They both had recognized the need for a sort of Church's Thesis, a thesis of adequacy for their new construals of the continuum.

Dedekind [1872: II] wrote of the "connection [Zusammenhang]" between the rational numbers and points on the straight line when an origin and a unit

[10] After *defining* the real numbers as the cuts, Russell [1919: 73] continued: "The above definition of real numbers is an example of 'construction' as against 'postulation', of which we had another example in the definition of cardinal numbers. The great advantage of this method is that it requires no new assumptions, but enables us to proceed deductively from the original apparatus of logic."

of length have been selected. This "connection" is accomplished in Hilbert's axiomatization through the Archimedean Axiom. Hilbert's Completeness Axiom then ensures through maximality that Dedekind's cuts actually correspond to points. Conversely, Dedekind's postulation of points corresponding to cuts entails the Completeness Axiom by an argument given in §9 of later editions of the *Grundlagen*: If to the contrary a new point could be added, it would induce a Dedekind cut of old points which would then have an old dividing point; but then, a simple argument using the Archimedean Axiom implies that there would be another old dividing point, which is a contradiction.

Although the Completeness Axiom would stir interest as an axiom about (models of) axioms, it could thus have been replaced by a continuity axiom along the lines of its antecedents. In remarks accompanying the first appearance of the Completeness Axiom, Hilbert [1902: 25ff.] opined that "the value of [the Completeness Axiom] is that it leads indirectly to the introduction of limiting points." Today the view would be opposite: securing limit points directly through some axiom like Cantor's or Dedekind's would be considered more simple than to introduce an axiom about axioms. Not only does formalizing continuity axioms require second-order quantification over the real numbers, the Completeness Axiom has the added complication of having to formalize the second-order satisfaction relation. But with the central role that he accorded axiomatization, Hilbert thought that he had readily positioned continuity into the heart of his axioms with his Completeness Axiom, and upon its incorporation into the second edition of the *Grundlagen* he [1903: 17] wrote that it "*forms the cornerstone of the entire system of axioms.*" Nevertheless, years later in a popular book on geometry, Hilbert and Stephan Cohn-Vossen [1932: §34] noted that the ways in which the axioms of continuity are formulated varies a great deal, and the Completeness Axiom is simply replaced by "Cantor's axiom", that every infinite sequence of nested segments has a common point.

3 Arithmetic

Before the Completeness Axiom appeared in any version of his *Grundlagen*, Hilbert in his *Über den Zahlbegriff* [1900b], dated 12 October 1899, provided an axiomatization of the real numbers as an ordered field satisfying arithmetical versions of the Archimedean Axiom and the Completeness Axiom.[11] Just as for geometry, Hilbert had in effect posited categoricity through maximality, for it must have been immediately seen that any system satisfying the [1900b] axioms, except possibly that arithmetical version of Completeness, is faithfully

[11]The arithmetical version of the Archimedean Axiom for ordered fields states that for any $a > 0$ and $b > 0$, a can be added to itself a (finite) number of times so that: $a+a+\ldots+a > b$. Ordered fields having this property are now called *Archimedean*. The arithmetical version of the Completeness Axiom in [1900b] states that the reals cannot be properly extended if the Archimedean ordered field properties are to be maintained.

embeddable into the real numbers.[12]

Although Hilbert [1900b] acknowledged the pedagogical value of the "genetic method" by which one builds up from the natural numbers through the rational numbers to the real numbers, he contended that only an axiomatic presentation of the real numbers all at once can be logically secure. Just as for geometry, Hilbert in [1900b] reduced arithmetic to the workings of a few axioms. Today "arithmetic" most often refers to number theory, i.e. the structure of addition and multiplication for the natural numbers, but for Hilbert "arithmetic" would remain what he would also refer to as analysis, i.e. the structure of addition, multiplication, and continuity for the real numbers. He initially expressed confidence that he could easily establish the consistency of his axioms.[13] However, this was to become a major and lifelong concern for him, and he was soon to promulgate it as the second of his famous problems.

Hilbert's main program for mathematics was launched by his famous declaration [1900] of 23 central problems for the 20th Century at the 1900 International Congress of Mathematicians at Paris.[14] Not only did he advance the basic picture of mathematical practice as driven by the force of problems and conjectures, but he inspired progress with his firm belief that every problem can ultimately be solved, that "in mathematics there is no *ignorabimus*". Fermat's last theorem, although unsolved, had already stimulated great developments in mathematics, and now the gauntlet was thrown to the coming generations, one that would gradually result in the development of new fields of mathematics.

Hilbert made the first of his [1900] problems the problem of establishing Cantor's Continuum Hypothesis, and the second, the problem of establishing "the consistency of the arithmetical axioms", referring to the axioms of his *Über den Zahlbegriff* [1900b]. Both of these problems dealt with basic questions about numerical construals of the traditional continuum: the first about

---

[12]Hilbert in [1900b] actually asserted that his axioms characterize the real numbers since its version of the Completeness Axiom implies the existence of limit points; this was the first statement along these lines. In connection with the discussion above at the end of §2 but shifting from geometry to the real numbers, the later polemic of Hilbert [1905: 185] (as translated in van Heijenoort [1967: 138]) is notable: "[The Completeness Axiom] expresses the fact that the totality [Inbegriff] of real numbers contains, in the sense of one-to-one correspondence between elements, any other set whose elements satisfy also the axioms that precede; thus considered, the completeness axiom, too, becomes a stipulation expressible by formulas constructed like those constructed above, and the axioms for the totality of real numbers do not differ qualitatively in any respect from, say, the axioms necessary for the definition of the integers. In the recognition of this fact lies, I believe, the real refutation of the conception of the foundations of arithmetic associated with L. Kronecker and characterized at the beginning of my lecture as dogmatic."

[13]Hilbert [1900b: 184] wrote: "In order to prove the consistency of the given axioms all that is needed is a suitable modification of known methods of inference." When [1900b] appeared as Appendix VI in later editions of the *Grundlagen*, this sentence is missing.

[14]Browder [1976] is a compendium on the mathematical developments arising from Hilbert's problems.

the possibility of enumerating the real numbers using the countable ordinal numbers, and the second about the consistency of an arithmetical axiomatization. It is quite remarkable that over two decades later Hilbert himself would use a specific strategy in his proof theory to attack *both* problems (see §6 and §7).

In his [1900] discussion of his second problem, Hilbert remarked that the consistency of the geometrical axioms had been reduced to that of the arithmetical axioms, but that "a direct method is needed for the proof of the consistency of the arithmetical axioms." In the *Grundlagen* his axiomatically presented geometry can be shown consistent by taking as the "points" ordered pairs of real numbers and relying on their arithmetic. However, no such model-theoretic recourse is available for arithmetic itself, and what is left is a direct investigation of its axioms and their consequences. Hilbert argued (as translated in [1902a: 446]):

> The totality of real numbers, i.e. the continuum ... is not the totality of all possible series of decimal fractions, or of all possible laws according to which elements of a fundamental sequence may proceed. It is rather a system of things whose mutual relations are governed by the axioms set up and for which all propositions, and only those, are true which can be derived from the axioms by a finite number of logical processes.

This view of the continuum as axiomatically given would later be reflected in Hilbert's own attempt to solve his first problem, the Continuum Hypothesis, through the use of definable functions, and the emphasis on deductive consequences of axioms would later animate his metamathematics. With an arithmetical axiomatization of the continuum whose consequences are exactly the true propositions of arithmetic consistency may be established through the finiteness of proofs without any reference to an antecedent geometric continuum, increasingly the bugbear of 19th Century mathematics.

Upon incorporating his Completeness Axiom into his *Grundlagen* Hilbert [1903: 17] observed that it presupposes the Archimedean Axiom, in the sense that "it can be shown" that there are geometries satisfying I-IV, and not that axiom, that can be properly extended.[15] In the tradition of Hilbert [1900b], Hans Hahn [1907] introduced into the theory of ordered fields a completeness condition analogous to the Completeness Axiom, which however did not presuppose the Archimedean condition, and provided an incisive analysis of the

[15]This contention did not become clear until the development of the theory of real-closed fields by Emil Artin and Otto Schreier in their [1926, 1927]. This development was resonating: Real-closed fields have a maximal property analogous to Hilbert's Completeness Axiom, and the theory was crucial for Artin's [1927] non-constructive solution of Hilbert's 17th Problem (constructive solutions were later given). See Pfister [1976] for more on Hilbert's 17th Problem.

resulting structures.[16] The connection to be made here is with Kurt Gödel who as a student and friend of Hahn's much admired his work.[17] It would only be through Gödel's epochal results, themselves responses to questions later raised by Hilbert, that the concepts of categoricity and completeness would become clarified (see §8).

4 Set Theory

Although Hilbert did not himself pursue axiomatic set theory, he fostered its development through his encouragement of Ernst Zermelo.[18] Zermelo began his investigations of Cantorian set theory at Göttingen under Hilbert's influence. Zermelo soon found Russell's Paradox independently of Russell and communicated it to Hilbert. Zermelo then established the Well-Ordering Theorem in a letter to Hilbert, the relevant part of which soon appeared as Zermelo [1904]. This seminal paper introduced the Axiom of Choice and stirred considerable controversy. In the tradition of Hilbert's axiomatization of geometry, Zermelo [1908] subsequently provided the first substantial axiomatization of set theory, partly to establish set theory as a discipline free of paradoxes, and particularly to put his Well-Ordering Theorem on a firm footing. Zermelo's axiomatization shifted the emphasis from Cantor's transfinite numbers to an abstract view of sets as structured solely by $\in$ and simple operations. In addition to generative axioms corresponding to these operations and the Axiom of Choice, Zermelo with his Separation [Aussonderung] Axiom incorporated a means of generating sets corresponding to properties that seemed to avoid paradoxes. The Separation Axiom asserted that given a set M, for each *definite* property [definite Eigenschaft] a set can be formed of those elements of M having that property. The vagueness of definite property would invite Skolem's [1923] proposal to base it on first-order logic, and this would tie in with Hilbert's later development of mathematical logic (see §5).

For Hilbert himself much of what today would be regarded as the subject matter of set theory would remain largely embedded in mainstream mathematics or be intermixed with the emerging mathematical logic. In what was to be his only publication on logic when he was still in his mathematical prime, Hilbert [1905] addressed the recent paradoxes of logic and set theory with remarks that prefigured his later work in metamathematics and his finitistic viewpoint. Hilbert [1905] advocated an axiomatic approach, observing that (as translated in van Heijenoort [1967: 131])

> In the traditional exposition of the laws of logic certain fundamental arithmetical notions are already used, for example, the notion of

[16]See Ehrlich [1995].
[17]Wang [1987] describes Gödel's admiration for Hahn.
[18]See Moore [1982: 89ff.] for more about Zermelo.

> set, and to some extent, also that of number. Thus we find ourselves turning in a circle, and that is why a partly simultaneous development of the laws of logic and of arithmetic is required if paradoxes are to be avoided.

Significantly, "the notion of set" for Hilbert here is an "arithmetical notion", and this is connected with his second [1900] problem, to establish the consistency of the "arithmetical axioms". As mentioned earlier, these axioms were to be those of *Über den Zahlbegriff* [1900b] including its version of the Completeness Axiom.

Hilbert [1905] provided only a tentative sketch of how he would carry out such a "simultaneous development", but intriguingly it has some anticipation of Zermelo's [1908] generative view of sets. Schematizing a process proceeding by stages, Hilbert [1905] stated five principles, the first three of which are (see van Heijenoort [1967: 135ff.]): (I) "a further proposition is true as soon as we recognize that no contradiction results if it is added as an axiom to the propositions previously found true"; (II) at any stage the "all" in the axioms is to range over only those "thought-objects" then taken to be primitive; and (III) a set is a "thought-object" and "the notion of element of a set appears only as a subsequent product of the notion of set itself."

Hilbert would become associated with the "consistency implies truth and existence" view behind principle I. First set out by him in correspondence with Frege about the axiomatization of geometry, the view is similar to that of Cantor but opposite to Frege's "truth implies consistency" view.[19] Principle II foreshadowed Hilbert's later advocacy of Russell's theory of types. As for the somewhat cryptic principle III, Hilbert went on to develop its sense by deducing what amounts to a version of Zermelo's [1908] Separation Axiom: From the thought-objects taken to be primitive at a given stage, propositions determine subcollections that are then further thought-objects.

[19] Hilbert wrote to Frege (see Frege [1980: 39ff.]):

> You write: "I call axioms propositions that are true but are not proved because our knowledge of them flows from a source very different from the logical source, a source which might be called spatial intuition. From the truth of the axioms it follows that they do not contradict each other." I found it very interesting to read this sentence in your letter, for as long as I have been thinking, writing and lecturing on these things, I have been saying the exact opposite: if the arbitrarily given axioms do not contradict each other with all their consequences, then they are true and the things defined by the axioms exist. For me this is the criterion of truth and existence.

Cantor [1883: §8] had written:

> Mathematics is completely free in its development and only bound by the self-evident consideration that its concepts must on the one hand be consistent in themselves and on the other stand in orderly relation, fixed through definitions, to the previous formed concepts already present and tested.

Despite Zermelo's association with Hilbert, it is notable that Hilbert's later lectures [1917] on set theory were imbued with the Cantorian initiatives on number and relatively unaffected by the Zermelian emphasis on abstract set-theoretic operations and axiomatization. Hilbert first discussed the real numbers, giving a detailed account of the transcendental numbers and his [1900b] axiomatization for an ordered field. He then developed Cantor's cardinal numbers, and after discussing well-orderings, Cantor's ordinal numbers. Without much ado Zermelo's Axiom of Choice is stated and his Well-Ordering Theorem proved. The approach is reminiscent of Hausdorff's *Grundzüge der Mengenlehre* [1914], with set theory presented as a new initiative *within* mathematical practice, one providing a new number context and new approaches to mathematical problems. Hilbert's lectures concluded with a discussion of the paradoxes, both set-theoretic and so-called semantic, and the Dedekind-Peano axioms for the natural numbers.

Given his own axiomatization of geometry and with Zermelo in his circle, one might have thought that Hilbert would have jumped at the issue of specific axiomatizations of set theory. Zermelo's axiomatization had for example been the setting for the incisive work of Friedrich Hartogs [1915] on Cardinal Comparability, cited by Hilbert in the [1917] lectures. However, not Hilbert but Abraham Fraenkel [1922] would investigate the independence of Zermelo's axioms, particularly the Axiom of Choice, in the style of Hilbert's *Grundlagen* with the liberal use of various models. Hilbert [1918: 411] did point out how the paradoxes were avoided by Zermelo's axiomatization. But significantly Hilbert [1918: 412] continued:

> The question of the consistency of the axiom system for the *real numbers* is reduced, through the use of set-theoretic concepts, to the same question for the natural numbers: This is the merit of the theories of irrational numbers of Weierstrass and Dedekind.
>
> Only in two cases, namely when it is a question of the axioms for the *natural numbers* themselves, and when it is a question of the foundations of *set theory*, is the method of reduction to another specific field of knowledge obviously unavailable, since beyond logic there is no further discipline to which an appeal is possible.
>
> Since however the proof of consistency is a task that cannot be dismissed, it seems necessary to axiomatize logic itself and then to demonstrate that number theory as well as set theory are only parts of logic.

This attitude would presumably have precluded any model-theoretic analysis of axioms for set theory, or indeed any detailed investigation of axiomatizations of set theory separate from axiomatizations of logic. The passage is consistent with the previously displayed passage from [1905]. However, it does suggest a softening of both the *Über den Zahlbegriff* [1900b] attitude that a direct axiomatic presentation of the real numbers is more logically secure than the genetic method of set-theoretic building up from the natural numbers, and the attitude from his discussion of his second [1900] problem that a "direct method is needed" to establish the consistency of the axioms for the real numbers, in that Hilbert now acknowledges a reduction to number theory and set theory.[20]

Subsequently, Hilbert [1929: 136] did come to appreciate the importance of firmly establishing the underlying assumptions of Zermelo's axioms. But as with Gödel later, Hilbert would be more influenced by Russell than by Zermelo, and whatever the affinity of Hilbert's [1905] picture to Zermelo's [1908], Hilbert's investigation of purely set-theoretic notions would largely remain part of his investigations of the underlying logic. The Axiom of Choice would be positioned in logic (see §6), and the Continuum Hypothesis would be approached through a hierarchy of definable functions (see §7).

5 Logic

Hilbert only began to carry out a systematic investigation of mathematical logic over a decade after his precursory [1905] and after the appearance of the three tomes of Whitehead and Russell's *Principia Mathematica*.[21] This work was, in the words of Gödel [1944: 126], the "first comprehensive and thorough going presentation of a mathematical logic and derivation of Mathematics". Much of the further development of logic would turn on reactions to and simplifications for this system, but its two basic interlocking hierarchical features, *types* and *orders*, would be crucial to the development of set theory.

To the modern eye there are two main sources for the great complexity – and even greater obscurity – of the *Principia*. First, it is, as Gödel [1944: 126] went on to write, "greatly lacking in formal precision in [its] foundations ... What is missing, above all, is a precise statement of the syntax of the formalism." This lack of formal precision is exacerbated by Russell's elucidatory accounts of his key logical notions, especially of "propositional function", which when taken literally are peculiarly opaque.[22]

[20] Hallett [1995: §3] corroborates, through notes to Hilbert's lectures during this period, his more favorable attitude toward the genetic method of building up mathematical objects.

[21] See Goldfarb [1979] and the note of Dreben and van Heijenoort in Gödel [1986: 44-59] for a discussion of logic in the 1910's and 1920's. And Hylton [1990] for a discussion of the metaphysics underlying Russell's logic.

[22] Russell wrote (Whitehead-Russell [1910: 41]):

> By a 'propositional function' we mean something which contains a variable x, and expresses a *proposition* as soon as a value is assigned to x. That is to say, it differs from a proposition solely by the fact that it is ambiguous: it contains a variable of which the value is unassigned. It agrees with the ordinary functions of mathematics in the fact of containing an unassigned variable; where it differs

The second source of difficulty, not unrelated to the first, is the complexity of Russell's "theory of logical types", his way of avoiding (Whitehead-Russell [1910: vii]) "the contradictions and paradoxes which have infected logic and the theory of aggregates [sets]." Russell first diagnosed the paradoxes as resulting from the "vicious circle" of "supposing that a collection of objects may contain members which can only be defined by means of the collection as a whole", and then adopted as a remedy the *vicious-circle principle,* "Whatever involves *all* of a collection must not be one of the collection" (Whitehead-Russell [1910: 39-40]). Moreover, he recognized that his own concept of propositional function represents "perhaps the most fundamental case" of the principle.[23] Adhering to the vicious-circle principle Russell insisted that the universe of *Principia* be viewed as ramified into *orders*. Speaking anachronistically, we may say that this universe consists of *objects*, where those of the lowest order are the *individuals*, and both the objects and the (formalized) language of *Principia* are to satisfy at least the following three conditions:

> (i) each object S "consists" of objects of some one fixed order, an order lower than the order of S;
>
> (ii) all values of each variable are of some one fixed order, called the *order* of the variable; and
>
> (iii) the *order* of any notational specification N of an object S is the least order (number) greater than the orders of all the bound variables in N and not exceeded by the orders of any free variables in N.

> is in the fact that the values of the function are propositions. ... The question as to the nature of a [propositional] function is by no means an easy one. It would seem, however, that the essential characteristic of a [propositional] function is *ambiguity*.

A few pages on Russell declared (Whitehead-Russell [1910: 50]): "A [propositional] function, in fact, is not a definite object ...; it is a mere ambiguity awaiting determination"

In a later book Russell [1919: 157] wrote: "We do not need to ask, or attempt to answer, the question: 'What *is* a propositional function?' A propositional function standing all alone may be taken to be a mere schema, a mere shell, an empty receptacle for meaning, not something already significant."

[23]Russell wrote (continuing in Whitehead-Russell [1910: 41-42] after the quotation from there in the previous note):

> A [propositional] function is not ... well-defined unless all its values are already well-defined. It follows from this that no [propositional] function can have among its values anything which presupposes the function .... This is a particular case, perhaps the most fundamental case, of the vicious-circle principle. A [propositional] function is what ambiguously denotes some one of a certain totality, namely the values of the [propositional] function; hence this totality cannot contain any members which involve the [propositional] function, since, if it did, it would contain members involving the totality, which, by the vicious-circle principle, no totality can do.

These are the essential features of what came to be called the *ramified theory of types*, and guided by them a full formalization up to modern standards can be carried out.[24] In the ramified theory, objects of different orders can have constituents of the same order. The collection of such constituents (objects) Russell also called a *type*. In particular, by conditions (i) and (iii), there could be objects consisting of individuals but of orders differing according to definitional complexity. But then, by condition (ii), it is impossible to quantify over all objects having individuals as constituents. Analogous situations will occur for objects whose constituents are of higher types, and this makes the formulation of numerous mathematical propositions at best cumbersome and at worst impossible. Consequently, Russell was led to introduce the *Axiom of Reducibility*:

> For each object there is a predicative object
> consisting of exactly the same objects,

where Russell called an object *predicative* if its order is the least greater than that of its constituents. Clearly, Russell did not think that objects having exactly the same constituents need be identical; in his jargon, they were *intensional* and not *extensional*.

The order hierarchy becomes greatly simplified if it were restricted to just predicative objects. There would only be individuals, predicative objects consisting of individuals, predicative objects consisting of predicative objects consisting of individuals, and so on. In this simplified hierarchy, *the simple theory of types*, the orders are just the types.[25] For Russell, it was obvious that there could only be finite orders and types, that is, only natural numbers could index orders and types.

The subsequent simplifications introduced into the system of *Principia* have mostly amounted to adopting a purely extensional version of the simple theory of types in which polyadic relations are reduced to sets through the Wiener-Kuratowski definition of ordered pair.[26] The Axiom of Reducibility, only germane for the ramified theory, would become moot. However, for Gödel the axiom would be considered both the basis of comprehension axioms in set theory as well as the antecedent to his argument for the relative consistency of the Continuum Hypothesis (see §8).

Hilbert enthusiastically espoused the *Principia*, saying (Hilbert [1918: 412]) "should Russell's impressive undertaking to *axiomatize logic* be carried to fruition

[24] See for example Church [1976].

[25] It is from the simple theory that the terms "first-order logic", "second-order logic", and so forth evolved, with "order" retained instead of "type". For example, with the zeroth order comprised of individuals and the first order consisting of the (predicative) objects consisting of individuals, first-order logic treats quantification over individuals of a domain. Similarly, second-order logic treats in addition quantification over objects consisting of individuals.

[26] The first such presentation of *Principia* in print was Gödel's system P in his Incompleteness paper [1931].

it would be the crowning achievement of axiomatization". But by "fruition [Vollendung]" Hilbert meant something utterly unlike what Russell would have meant. At a minimum, Hilbert meant showing "the consistency of the arithmetical axioms", i.e. solving his second [1900] problem.

The book [1928] by Hilbert and Wilhelm Ackermann, originating in Hilbert's [1917a] lectures, reads remarkably like a recent text. In marked contrast to the formidable works of Frege and Russell with their forbidding notation and all-inclusive approach, it proceeded pragmatically and upward to probe the extent of structure, making those moves emphasizing syntactic forms and axiomatics typical of modern mathematics. After a thorough analysis of sentential logic, it distinguished and focused on first-order logic as already the source of significant problems. While Frege and Russell never separated out first-order logic, Hilbert would establish it as a subject in its own right. Nevertheless, for the formalization required to investigate the foundations of mathematical theories, Hilbert thought that an "extended calculus is essential" (Hilbert-Ackermann [1928: 86]). In the [1917a] lectures on logic, this extended calculus is evidently Russell's ramified theory of types, and in it Hilbert constructed the real numbers as the Dedekind cuts using an extensional version of Russell's Axiom of Reducibility. The book Hilbert-Ackermann [1928] continued to use Russell's ramified theory of types and the Axiom of Reducibility. However, in the course of his development of mathematical logic Hilbert, like Ramsey,[27] would come to regard Russell's ramifying orders and the Axiom of Reducibility as unnecessary, as is stated on the last two pages of Hilbert-Ackermann [1928].

While Hilbert was lecturing on set theory and logic his former student Weyl brought out a notable monograph, *Das Kontinuum* [1918]. Waxing philosophical, Weyl railed against the "vicious circle" involved in even such basic concepts as the least upper bound for a bounded set of real numbers. That its definition presupposes its existence among the possible upper bounds would become the standard example of an *impredicative definition*, definitions that Weyl would banish (as did of course Russell through his ramified theory). Reasoning that he could not avoid presupposing the natural numbers, Weyl took these as the individuals and considered what is essentially a version of that part of the ramified theory of types in which quantification is restricted to variables ranging over the individuals.[28] The consequences of Weyl's system for the real numbers

[27]Ramsey is not mentioned in the text of Hilbert-Ackermann [1928], but his paper [1926] in which he suggested that the ramified theory be replaced by the simple theory and Axiom of Reducibility be dropped is cited in their bibliography.

[28]Significantly, Weyl [1910: 112ff.] had begun his foundational investigations by trying to provide a satisfactory formulation for Zermelo's definite property for the Separation Axiom and had suggested building up the concept from $\in$ and $=$ by a finite number of generating principles. It was in the course of developing these principles that Weyl [1918: 36] found that he could not avoid presupposing the natural numbers – a primordial vicious circle. Weyl [1918: 35] acknowledged that his hierarchy "corresponds" to Russell's, but rejected the Axiom of Reducibility.

is the same as the system ACA_0, formulated in the appendix below. Weyl went on to show that the basic theory of continuous functions could be adequately developed in his system. This was a remarkable accomplishment at such an early stage, both in the formulation of a parsimonious formal system to mirror mathematical practice and in the use of coding procedures to adequately develop a surprisingly large part of analysis. The key ingredient was to revert from continuity in terms of *sets* as given by Dedekind cuts to continuity in terms of *sequences* in the spirit of Cantor's fundamental sequences, where however real functions and sequences of real numbers are simulated by just sets of natural numbers.

Hilbert to be sure was to inspire the development of subsystems of number theory and of analysis. However, he reacted vigorously against what he regarded as Weyl's emasculation of mathematics. The difference between the two is that Weyl was advocating his system as what mathematical analysis ought to be, whereas Hilbert was investigating formal systems for specific purposes, primarily to carry out proofs of consistency.

In spirited response to Weyl's constructivism and also to Brouwer's intuitionism, which would banish the Law of Excluded Middle and non-constructive existence proofs, Hilbert [1922, 1923] developed *metamathematics* and proposed, most fully in [1926], his program of establishing the consistency of ongoing mathematics by *finitary reasoning* [das finite Schliessen]. Metamathematics would grow to be a broad, ultimately mathematical, investigation of the content and procedures of ongoing mathematics through its formalization; for Hilbert, metamathematics was primarily his *proof theory*, the investigation of formalized proofs as objects of study. Elaborating on two motifs, the primacy of logical deduction and the finiteness of formal proof, Hilbert argued that the mathematical investigation of proofs would secure the reduction of the consistency of mathematics to a bedrock of finitary and incontrovertible means.

Hilbert-Ackermann [1928: 65ff.,72ff.] raised two crucial questions with respect to first-order logic: the semantic completeness of its axioms, that is, whether a formula holding in every model of the axioms is provable from the axioms; and its decision problem [Entscheidungsproblem], that is, whether there is an algorithm for deciding whether any formula has a model or not. The first figured in the last of the five problems raised in Hilbert's lecture [1929] at the 1928 International Congress of Mathematicians at Bologna, the main theme of which however was still his program for establishing the consistency of mathematics. Hilbert thus generated all the major problems of mathematical logic that would be decisively informed by Gödel's work (see §8). As with his [1900] problems, Hilbert was again to stimulate major developments through the formulation of pivotal questions, questions that are contextually specific yet set a new frontier. Such questions, especially weighted as conjectures, became increasingly significant for the progress of modern mathematics, and it is Hilbert whom one acknowledges as pioneer and exemplar for this new development.

6 Metamathematics

Much has been written about Hilbert's metamathematics. Here we restrict ourselves to describing his specific strategy for settling his own second problem from his [1900], namely the problem of showing the "consistency of the arithmetical axioms". In §7 we show how this strategy was a starting point for his attempt to solve his first problem from [1900], that of establishing the Continuum Hypothesis.

Pursuing the analogy with the introduction of ideal elements in mainstream mathematics Hilbert [1926] distinguished between numerical formulas communicating *contentual* [inhaltlich] propositions and those communicating *ideal* propositions. Quantifiers are contentual as long as they range over specified finite domains, in which case they can be replaced by finite disjunctions or conjunctions. Hilbert [1923: 154] had noted that the first time "something beyond the concretely intuitive and finitary" enters logic is in (unrestricted) quantification and this he [1926] took to be characteristic of ideal propositions, undertaking his metamathematics as an investigation toward establishing the consistency of their use. That investigation itself would be conducted in contentual mathematics with formalized proofs as objects of study, and indeed Hilbert [1926] wrote of metamathematics as "the contentual theory of formalized proofs."

Hilbert [1926] (see van Heijenoort [1967: 382]) stated several axioms for quantifiers, and then asserted that they can be derived from a single axiom, one that "contains the core" of the Axiom of Choice:

$$A(a) \to A(\varepsilon(A)),$$

"where ε is the transfinite logical choice function." The symbol ε serves as a logical operator, taking formulas A as arguments and producing terms $\varepsilon(A)$; the more specific $\varepsilon_x A(x)$ was soon deployed to handle A's with several free variables. The ε-terms had an engaging indeterminism: they serve as syntactic witnesses to A if $\exists x A(x)$, but are *bona fide* terms even if $\neg\exists x A(x)$.[29] Like the ideal points at infinity of projective geometry, Hilbert had in effect introduced new ideal elements into first-order logic.

Hilbert [1928] (see van Heijenoort [1967: 466]) spelled out how the quantifiers can be defined in terms of ε-terms:

$$\forall a A(a) \textit{ iff } A(\varepsilon(\neg A))\,, \quad \text{and} \quad \exists a A(a) \textit{ iff } A(\varepsilon(A))\,.$$

[29] As the Athenians were wont to say of Aristides, if there is an honest man, then it is he.

The usual quantifier rules follow immediately, e.g.

$$\neg\forall a A(a) \longrightarrow \exists a \neg A(a) .$$

From Frege on, this rule had been regarded as an immediate consequence of the definitions of $\exists$ and $\forall$. For Hilbert, it is only immediate for specified finite domains as an instance of *tertium non datur*, the Law of Excluded Middle, and is otherwise a substantial manipulation on ideal propositions as an infinitary form of the Law.

Hilbert [1922: 157] had already expressed the need to formulate the Axiom of Choice so that it is as evident as $2 + 2 = 4$. However, to say that $A(a) \rightarrow A(\varepsilon(A))$ "contains the core" of the Axiom of Choice is misleading from the modern perspective, for it is after all just a variant of existential generalization.[30] However, it is indeed as a "choice function" that Hilbert had a particular *use* for his innovation in mind as part of a specific strategy for establishing consistency that he advanced along with his development of proof theory itself.

That strategy was first broached by Hilbert in his [1923], where before the ε-operator he had introduced his τ-operator through what he called the Transfinite Axiom:

$$A(\tau(A)) \longrightarrow A(a) .$$

The logical τ-operator encapsulated the universal quantifier as his later ε-operator would the existential quantifier. From this one axiom he derived all the quantifier rules, which he considered the source of non-finitary or "transfinite" reasoning.[31] Focusing on number-theoretic functions f, i.e. functions from the natural numbers into the natural numbers, he then extended the logical τ-operator with $\tau(f) = \tau_a(f(a) = 0)$, specifying the free variable a in the

[30]The Second ε-Theorem of Hilbert-Bernays [1939] would establish that in first-order logic with ε-terms, if neither the premises nor the conclusion of a deduction contains such terms, then there is a deduction not using such terms. In order to derive the Axiom of Choice using ε-terms, the crucial set-theoretic feature of the Axiom, the existence of a *set* of choices, or concomitantly a choice function, must be incorporated. One approach is to allow ε-terms in the Replacement Axiom, an essential feature of modern set theory. Hilbert [1923: 164] himself used an informal variant of this approach to argue for the Axiom of Choice for sets of reals. (Wang [1955] discusses the interplay of ε-terms and the Axiom of Choice in axiomatic set theory.) Interestingly, Zermelo [1930: 31] in his final axiomatizations of set theory also regarded the Axiom of Choice as a logical principle and did not list it explicitly among his axioms. In later years a fully Tarskian semantics was developed by Günter Asser [1957] and Hans Hermes [1965] for the ε-operator with its interpretations being global choice functions for the structure at hand. More in the spirit of Hilbert's intention was Rudolf Carnap's [1961] indeterminate use of the ε-operator as an interpretation of his T-, or theoretical, terms.

[31]Hilbert [1923: 161] specifically asserted that transfinite reasoning was necessary for his solution of the central invariant theory problem discussed in §1, and that although Gordan thought that he had removed this "theological" aspect of the argument with his own version of the proof, it remained embedded in his "symbolic" approach. Hilbert's view of the complexity of his proof was substantiated; see Theorem 3 in the appendix.

formula $f(a) = 0$, so that from the Transfinite Axiom we have

$$f(\tau(f)) = 0 \longrightarrow f(a) = 0 .$$

He interpreted $\tau(f)$ as a mathematical "function-of-functions", a *functional* we would now say, that had already appeared at the end of his [1922]: This mathematical functional κ took number-theoretic functions f as arguments, with $\kappa(f) = 0$ if $f(a) = 0$ for every natural number a, and otherwise $\kappa(f)$ is the least a such that $f(a) \neq 0$. Evidently, the admissibility of κ rests on an infinitary form of *tertium non datur* and embodies Hilbert's use of non-constructive existence proofs. From the very beginning of his work on metamathematics Hilbert emphasized number-theoretic functions and substantial functionals operating on them, and this emphasis would soon extend to his attempt to establish the Continuum Hypothesis.

In terms of τ, Hilbert [1923: 159ff.] gave for a very weak subsystem of analysis an example of his strategy for establishing consistency: Starting with a putative proof of $0 \neq 0$, successive substitutions of numerals were made for the τ-terms appearing in the proof so that only a deductive sequence of true numerical formulas was left, and hence $0 \neq 0$ could not have appeared at the end after all. Hilbert had thus shown how to exploit the finiteness of proofs in a specific way, eliminating the "transfinite" τ-terms in favor of finitely many numerical instances. Ackermann [1925] undertook to carry out Hilbert's plan to apply this substitution strategy to the full system with quantification over number-theoretic functions; this would establish the consistency of analysis, with the number-theoretic functions construed as the real numbers. Hilbert had by then switched from τ-terms to ε-terms, which in the new rendition of his strategy were indeed interpreted as finite "choice functions". At the beginning of his career Hilbert had established a fundamental finiteness property with his basis theorem; he would now effect a new reduction to a "finite basis" to establish the consistency of mathematics.

Hilbert's strategy of eliminating ε-terms encountered a basic difficulty in the general setting: the possible nestings of ε-terms corresponding to quantifier dependence. In carrying out the substitution procedure, a numerical choice made for an ε-term t might typically conflict with a later choice made for an ε-term within which t occurs, necessitating a new substitution for t. This process can cycle in complicated ways, with the possibility that successive substitutions may not terminate. Ackermann's [1925] argument fell far short, failing to handle number-theoretic functions and even full induction for the natural numbers. John von Neumann [1927] then carried out a complex argument, based on Hilbert's [1905] approach to consistency as developed by Julius König [1914],[32] to establish the consistency of quantifier-free induction for the natural

[32]von Neumann [1927: 22] acknowledges König [1914].

numbers. Thereupon Ackermann established the same result with his original approach. In [1928] Hilbert sketched this new argument of Ackermann's, and in succeeding comments Bernays [1928] elaborated on it.

Hilbert and his school (mainly Ackermann, Bernays, and von Neumann) believed at this time that Ackermann's new argument in fact established the consistency of full number theory (first-order Peano Arithmetic).[33] At the end of [1928] Hilbert wrote (as translated in van Heijenoort [1967: 479]):

> For the foundations of ordinary analysis [Ackermann's] approach has been developed so far that only the task of carrying out a purely mathematical proof of finiteness [of the number of necessary substitutions of numerals for ε-terms] remains.

Thus Hilbert was also confident that his second [1900] problem, "the consistency of the arithmetical axioms" for the real numbers, would be solved. In his lecture at the 1928 International Congress of Mathematicians at Bologna, Hilbert [1929] assumed that the finiteness condition for the elimination of ε-terms had been established for number theory and made his first problem that of establishing the analogous finiteness condition for analysis. In a lecture given in December 1930, Hilbert [1931: 490] still thought that the consistency of number theory had been established.

However, in a lecture given in September 1930, Gödel [1930a] had announced his First Incompleteness Theorem, the existence of formally undecidable propositions of number theory. Von Neumann who was in the audience saw not only its broad significance but its particular relevance to the work of the Hilbert school. Some weeks after his lecture Gödel established his Second Incompleteness Theorem, the unprovability of consistency, and soon afterwards in November heard from von Neumann that he too had established this result.[34] The Second Incompleteness Theorem of Gödel [1931] implies in particular that for any theory subsuming the addition and multiplication of the natural numbers and for any putative proof of $0 \neq 0$ in that theory, no "proof of finiteness" (as in the quotation above) is formalizable in that theory. Thus, there had to be something wrong with the assumption of the Hilbert school that Ackermann's new argument established the consistency of full number theory, and von Neumann soon produced an example for which the argument failed.[35] Beyond the common impression that Gödel's Second Incompleteness Theorem largely precluded Hilbert's consistency program, this close interplay between Gödel and von Neumann brings out the specific *mathematical* impact that Gödel's result had on a concerted effort then being made by the Hilbert school.

---

[33]This was corroborated in oral communication from Bernays to Dreben in 1965, and in a letter from Bernays to the editor of a projected Spanish translation of van Heijenoort [1967], dated 15 June 1974.

[34]See Gödel [1986: 137].

[35]The example is given in Hilbert-Bernays [1939: 123ff.].

Gerhard Gentzen [1936, 1938, 1943] would show that there *is* a "purely mathematical proof" of the consistency of number theory. However, his method necessarily relied on a mathematical principle presumably non-finitary by Hilbert's standards, the principle of transfinite induction up to the ordinal ϵ_0.[36] Later Ackermann [1940] showed that for number theory Hilbert's original substitution method also provides a "purely mathematical proof of finiteness" and thereby establishes the consistency of number theory, but again by invoking transfinite induction up to ϵ_0. For number theory, Hilbert's goal of establishing consistency has been accomplished and through his substitution method – only the mathematical means were not finitary.[37]

7 Continuum Hypothesis

In [1923: 151] Hilbert had indicated that not only could his proof theory establish the consistency of analysis and set theory, but that it could also provide the means to solve "the great classical problems of set theory such as the Continuum Problem", the first of his [1900] problems. In [1926] Hilbert claimed to have established the Continuum Hypothesis with his "continuum theorem" and proceeded to sketch a proof. It is a failure,[38] but a notable one both for exhibiting the extent to which Hilbert thought he could extract mathematical content from formal proofs and for stimulating Gödel's work with L.

The Continuum Hypothesis would be established if the number-theoretic functions, functions from the natural numbers into the natural numbers, can be put into one-to-one correspondence with the countable ordinals. Hilbert apparently thought[39] that if he could show that from any given formalized putative disproof of the Continuum Hypothesis, he could prove the Continuum

---

[36] ϵ_0 is the supremum of the ordinals $\omega, \omega^\omega, \omega^{\omega^\omega}, \ldots$. There is a primitive recursive ordering $\prec$ of the natural numbers which is isomorphic to ϵ_0. The *principle of transfinite induction up to* ϵ_0 asserts that for any formula $\varphi(v)$,

$$\forall n(\forall m(m \prec n \to \varphi(m)) \to \varphi(n)) \to \forall n \varphi(n) .$$

This assertion is formalizable as a schema in any first-order number theory that subsumes primitive recursion, of which the minimal is Primitive Recursive Arithmetic described in the appendix. Gentzen showed that a single instance of the schema for a certain quantifier-free φ implies the consistency of number theory.

[37] A more intuitive, constructive model-theoretic version of Hilbert's substitution method was provided by Jacques Herbrand [1930] with his Fundamental Theorem; in particular, Herbrand gave a much simpler proof of the result of von Neumann [1927]. Expanding on Dreben and John Denton's analysis in their [1966, 1970] of Herbrand's Theorem, Thomas Scanlon [1973] provided a Herbrand-style proof for the full number theory result of Ackermann [1940].

[38] In the reprintings of [1926] and the related [1928] in the seventh edition [1930] of the *Grundlagen*, Hilbert excised all reference to his purported proof of the Continuum Hypothesis.

[39] Paul Lévy [1964: 89] remarked, as pointed out by van Heijenoort [1967: 368]: "Zermelo told me in 1928 that even in Germany nobody understood what Hilbert meant".

Hypothesis, then the Continuum Hypothesis would have been established. (At best, Hilbert's argument could only establish the *consistency* of the Continuum Hypothesis, but for him consistency is (mathematical) truth.[40])

According to Hilbert, the only way that the Continuum Hypothesis could be false is if there are non-constructively defined number-theoretic functions, i.e. functions defined using *tertium non datur* over existential quantifiers. A favorite example of Hilbert of such a function is $\varphi(a) = 0$ or 1 according to whether $a^{\sqrt{a}}$ is rational or not.[41] Hence, any proof of a proposition contradicting the Continuum Hypothesis would have to make use of such definitions of functions. Hilbert then asserted that the solvability of every well-posed mathematical problem is a "general lemma" of his metamathematics,[42] and that a "part of the lemma" is the following (as translated in van Heijenoort [1967: 385]):

> Lemma I. If a proof of a proposition contradicting the continuum theorem is given in a formalized version with the aid of functions defined by means of the transfinite symbol ε (axiom group III), then in this proof these functions can always be replaced by functions defined, without the use of the symbol ε, by means merely of ordinary and transfinite recursion, so that the transfinite appears only in the guise of the universal quantifier.

For establishing the consistency of arithmetic, Hilbert had started with a putative proof of $0 \neq 0$ and outlined a substitution procedure for replacing in effect its ε-terms by finite choice functions and showing that $0 \neq 0$ could not have appeared at the end after all. With Lemma I he would now start with a "proof of a proposition contradicting the continuum theorem", and presumably carry out a similar but more complex substitution procedure, this time replacing number-theoretic functions defined using ε symbols by a collection

[40]See note 19 for Hilbert's attitude about consistency and truth. With the metamathematical viewpoint slow to filter into mathematical practice only Nikolai Luzin [1933] among the early commentators saw that Hilbert's argument was really aimed at the consistency of the Continuum Hypothesis. To Gödel [1939b: 129] this was clear: "the first to outline a *program* for a consistency proof of the continuum hypothesis was *Hilbert*".

[41]This example occurred in Hilbert's lectures and in his [1923]. For natural numbers a with $\sqrt{a}$ irrational, it was unknown then whether $a^{\sqrt{a}}$ is rational or not. The seventh of Hilbert's [1900] problems was to establish that if α is an algebraic number and β an algebraic irrational, then α^{β} is transcendental, or at least irrational. This problem was to stimulate the development of transcendental number theory. Aleksander Gel'fond [1934] and Theodor Schneider [1934] independently solved the problem by showing that under the hypotheses (and excluding the trivial cases $\alpha \neq 0, 1$) α^{β} is in fact transcendental. See Tijdeman [1976] for more on Hilbert's seventh problem.

[42]However, Hilbert never claimed that there is an algorithm, a general method, for solving every mathematical problem. Indeed, he asserted in [1926] (as translated in van Heijenoort [1967: 384]) that there is no "general method for solving every mathematical problem; that does not exist." Presumably neither Hilbert nor any of his school thought that a positive solution to the decision problem for first-order logic would yield such an algorithm.

of functions defined by various forms of recursion. (Hilbert, we assume, was not making the stronger claim that for each given non-constructively definable function one can find an equivalent recursively definable function.) Hence, for Hilbert it remained to examine and to handle the functions so defined because (as translated in van Heijenoort [1967: 387]):

> In order to prove the continuum theorem, it is essential to correlate those definitions of number-theoretic functions that are free from the symbol ε one-to-one with Cantor's numbers of the second number class [the denumerable ordinals].

Hilbert was the first to consider number-theoretic functions defined through recursions more general than primitive recursion. He not only allowed definitions incorporating transfinite recursions through countable ordinals, but also higher type functionals. These are themselves defined recursively, a functional being a function whose arguments and values are previously defined functionals, and were classified by Hilbert into a hierarchy. Hilbert's logical beginnings in Russell's ramified theory of types is arguably discernible both in the preoccupation with definability, here reduced to recursions by Lemma I, and the introduction of a type hierarchy, though one extended into the transfinite.

In his hierarchy Hilbert classified functionals according to their *variable-type* by recursively considering their complexity of definition. He then recursively defined the *height* of a variable-type as the supremum of the heights plus 1 of the variable-types of the arguments and values. He argued that all definitions of functionals can be reduced to *substitution*, i.e. composition of functionals, and to *recursion*, i.e. primitive recursion allowing functionals. Hilbert next described how heights of certain variable-types, the *Z-types*, can be correlated with countable ordinals. The Z-types are those variable-types generated by the two processes of substitution and enumeration of a countable sequence of Z-types. Hilbert pointed out that in his correlation of heights with countable ordinals he had "presupposed" the theory of the latter. But he argued that only a formalization of the process of generating countable ordinals is necessary for his overall argument, and for that only those countable ordinals corresponding to Z-types matter. Hilbert then went on to describe how new variable-types, and therefore new ordinals, are generated by recursive enumeration of the variable-types up to a certain height and an application of "Cantor's diagonal procedure".

Hilbert next pointed out how his correlation of heights with countable ordinals was based on two apparent restrictions. First, he had only considered "ordinary recursion", not transfinite recursion directly through infinite ordinals, and second, he had only considered Z-types, those variable-types generated by enumeration of *countably* many variable-types. But he then claimed in his remarkable Lemma II that *all* number-theoretic functions defined by recursion can "also be defined by means of ordinary recursions and the exclusive use of Z-types".[43] But then, Hilbert has done what he said had to be done "in order to prove the continuum theorem". To recapitulate, from a formalized disproof of the Continuum Hypothesis Hilbert has "given" a proof of the Continuum Hypothesis!

The basic underlying difficulty with Hilbert's argument lies in his use of his Lemma I. Hilbert apparently thought that he can restrict his attention to only those number-theoretic functions that appear in purported disproofs of the Continuum Hypothesis. Whether such functions can be put in one-to-one correspondence with the countable ordinals gets us no closer to establishing even the consistency of the Continuum Hypothesis. However, Hilbert seems to have believed that there can be no number-theoretic functions unless definable in some formal proof. This is borne out by his later remark in [1928] (see van Heijenoort [1967: 476]) that Lemma I is "useful in fixing the train of thought, but it is dispensable for the proof itself." He noted that the introduction of ε-terms does not affect the denumerability of the possible recursions in higher type functionals up to any particular height. Moreover, the ε-terms can be systematically "normalized", e.g. for those acting on number-theoretic functions, the functional κ (defined in §6) from his earliest paper [1922] in metamathematics can be used. The difficulty with Hilbert's attempted proof of the Continuum Hypothesis can arguably be reduced to his attempt to capture the force of functionals like κ in some constructive way by a collection of recursively defined functionals, whereas ironically κ, as mentioned earlier, embodies Hilbert's use of non-constructive existence proofs.

There is a sense in which Hilbert's Lemma II is correct. Let us suppose, as his discussion would indicate, that the possible transfinite recursions that he speaks about are those given by recursive well-orderings. Then the number-theoretic functions that he was considering coincide with what today are called the general recursive functions. This is so because the class of general recursive functions is closed under recursions along any recursive well-ordering and is also closed under recursions in higher type functionals generated by primitive recursion using previously defined higher type functionals. But then, the conclusion of Lemma II was established independently by Myhill [1953] and

[43]The lemma states in full (as translated in van Heijenoort [1967: 391]):

Lemma II. In the formation of functions of a number-theoretic variable transfinite recursions are dispensable; in particular, not only does ordinary recursion (that is, the one that proceeds on a number-theoretic variable) suffice for the actual formation process of the functions, but also the substitutions call merely for those variable-types whose definition requires only ordinary recursion. Or, to express ourselves with greater precision and more in the spirit of our finitist attitude, if by adducing a higher recursion or a corresponding variable-type we have formed a function that has only an ordinary number-theoretic variable as argument, then this function can always be defined also by means of ordinary recursions and the exclusive use of Z-types.

Routledge [1953], who proved that every general recursive function is generated by recursion along primitive recursive well-orderings of ordertype ω.[44]

Hilbert broke fertile ground for the later, broad investigation of recursions. Ackermann [1928] showed that a scheme given in Hilbert [1926] does indeed define a non-primitive recursive function, now well-known as the Ackermann function. The association of ordinals with recursive definitions has become common place, with Gentzen's [1936, 1938, 1943] analysis of the consistency of number theory paradigmatic. And recursion in higher type functionals up to height ω in Hilbert's scheme was used by Gödel in his *Dialectica* interpretation [1958], already worked out in his [1941], to give a consistency proof of intuitionistic number theory and hence because of his [1933a] a consistency proof of (classical) number theory.[45]

8 Gödel

Kurt Gödel virtually completed the mathematization of logic by submerging metamathematical methods into mathematics.[46] The main vehicle was of course the direct coding, "the arithmetization of syntax", in his celebrated Incompleteness Theorem [1931], which transformed Hilbert's consistency program and led to the undecidability of the Decision Problem from Hilbert-Ackermann [1928] and the development of recursion theory. But starting an undercurrent, the earlier Completeness Theorem [1930] from his thesis answered affirmatively the Hilbert-Ackermann [1928] question about semantic completeness, clarified the distinction between the formal syntax and model theory (semantics) of first-order logic, and secured its key instrumental property with the Compactness Theorem. This work would establish first-order logic as the canonical language for formalization because of its mathematical tractability, and higher order logics would become downgraded, now viewed as the workings of the power set operation in disguise. Skolem's earlier suggestion in [1923] that Zermelo's axiomatic set theory be based on first-order logic would be generally adopted, thus vindicating Hilbert's emphasis on first-order logic.

---

[44] Both Myhill [1953] and Routledge [1953] pointed out that the natural hierarchy generating the recursive functions already terminates in ω stages. Kleene [1958] formulated a hierarchy of recursive functions which may be closer to Hilbert's intentions. Hilbert had argued that his scheme leads to new functions by applying "Cantor's diagonal procedure" on a recursive enumeration of the functions previously constructed. Kleene's hierarchy is based on enumeration and diagonalization, the former according to a fixed system of primitive recursive codes for well-orderings ("Kleene's $\mathcal{O}$"). Feferman [1962] showed that Kleene's hierarchy encompasses all the recursive functions. He showed moreover that such hierarchies terminate rather quickly so that they do not provide an informative hierarchical analysis of the general recursive functions. In his later years Gödel considered providing such an analysis to be a major problem of mathematical logic.

[45] Clifford Spector [1962] extended the *Dialectica* interpretation to full analysis, bringing in certain basic ideas of Brouwer.

[46] Alfred Tarski shares the honor.

To pursue our earlier discussion of categoricity in connection with Hilbert's Completeness Axiom in geometry, say that a theory is *deductively complete* if each sentence of its language or its negation is provable from the axioms. In the Königsberg lecture [1930a] where Gödel discussed his Completeness Theorem and announced his First Incompleteness Theorem, he observed that the former implies that *for first-order theories categoricity implies deductive completeness.* The argument is simple: if there were a sentence such that neither it nor its negation can be proved from the axioms, then there would be two (non-isomorphic) models of the theory.[47] Now Hilbert's axioms for geometry inclusive of the Completeness Axiom and the Dedekind-Peano axioms for the natural numbers *are* categorical, but as second-order theories. However, Gödel's First Incompleteness Theorem established that *no* (decidable) set of axioms for first-order or higher order theories, which subsumes the arithmetic of the natural numbers and only proves true sentences of that arithmetic, can be deductively complete. Thus, the Incompleteness Theorem makes a distinction between first-order and higher order theories in terms of categoricity and deductive completeness. Although Gödel in his Incompleteness paper [1931] did not mention this distinction, he had made it the *motivation* for the Incompleteness Theorem in his Königsberg lecture [1930a: 29].

Footnote 48a of Gödel's [1931] was as follows:

> As will be shown in Part II of this paper, the true reason for the incompleteness inherent in all formal systems of mathematics is that the formation of ever higher types can be continued into the transfinite (cf. D. Hilbert, "Über das Unendliche", Math. Ann. 95, p. 184), while in any formal system at most denumerably many of them are available. For it can be shown that the undecidable propositions constructed here become decidable whenever appropriate higher types are added (for example, the type ω to the system P [the simple theory of types superposed on the natural numbers as individuals satisfying the Peano axioms]). An analogous situation prevails for the axiom system of set theory.

This prescient note would be an early indication of a steady intellectual progress on Gödel's part that would take him from the Incompleteness Theorem through pivotal relative consistency results for set theory to speculations about its fur-

[47] Actually, the assertion that for first-order theories categoricity implies deductive completeness is largely vacuous, since a now well-known consequence of the Compactness Theorem is that *any first-order theory with infinite models is not categorical.* However, call a first-order theory $\aleph_0$-*categorical iff* it has a unique countably infinite model up to isomorphism. Then by the argument given in the text as sharpened by the Löwenheim-Skolem Theorem, *for first-order theories* $\aleph_0$-*categoricity implies deductive completeness.* This assertion is not vacuous, and also applicable to the distinction to be made in the text between first-order and higher order logics.

ther possibilities. The reference to Hilbert [1926] and Russell's theory of types foreshadows the strong influence that they would have on this progress.

In a subsequent lecture [1933], Gödel expanded on the theme of footnote 48a. He regarded the axiomatic set theory of Zermelo, Fraenkel, and von Neumann as "a natural generalization of the [simple] theory of types, or rather, what becomes of the theory of types if certain superfluous restrictions are removed."[48] First, instead of having separate types with sets of type $n+1$ consisting purely of sets of type n, sets can be *cumulative* in the sense that sets of type n can consist of sets of *all* lower types. If S_n is the collection of sets of type n, then: S_0 is the type of the individuals, and inductively, $S_{n+1} = S_n \cup \{X \mid X \subseteq S_n\}$. Second, the process can be continued into the transfinite, starting with the cumulation $S_\omega = \bigcup_n S_n$, proceeding through successor stages as before, and taking unions at limit stages. Gödel [1933: 46] credited Hilbert for pointing out the possibility of continuing the formation of types beyond the finite types. As for how far this cumulative hierarchy of sets is to continue, the "first two or three types already suffice to define very large ordinals" ([1933: 47]) which can then serve to index the process, and so on. Gödel observed that although this process has no end, this "turns out to be a strong argument in favor of the theory of types" ([1933: 48]). Implicitly referring to his incompleteness result Gödel noted that for a formal system S based on the theory of types a number-theoretic proposition can be constructed which is unprovable in S but becomes provable if to S is adjoined "the next higher type and the axioms concerning it" ([1933: 48]).

In 1938 modern set theory was launched by Gödel's formulation of the model L of "constructible" sets, a model of set theory that established the consistency of the Axiom of Choice and the (Generalized) Continuum Hypothesis. In his first announcement Gödel [1938: 556] described L as a hierarchy "which can be obtained by Russell's ramified hierarchy of types, if extended to include transfinite orders." Indeed, with L Gödel had refined the cumulative hierarchy of sets described in his [1933] to a cumulative hierarchy of definable sets which is analogous to the orders of Russell's *ramified* theory. This hierarchy of definable sets was in the spirit of Hilbert [1926] as was the extension of the hierarchy into the transfinite. However, Gödel's further innovation was to continue the indexing of the hierarchy through *all* the ordinals to get a model of set theory.[49] The extent of the ordinals was highlighted in his monograph

[48]For this view Gödel [1933: 46] mainly acknowledged von Neumann [1929], although Zermelo [1930] would have been a better source.

[49]Years later in 1968 Gödel wrote to Hao Wang [1974: 8ff.]: "there was a special obstacle which *really* made it *practically impossible* for constructivists to discover my consistency proof. It is the fact that the ramified hierarchy, which had been invented *expressly for constructive purposes,* had to be used in an *entirely nonconstructive way.*" Gödel [1947: 518] mentioned in a footnote that the transfinite iteration of the procedure for constructing sets in Weyl [1918] results exactly in the real numbers of L.

[1940], based on lectures in 1938, in which he formally generated L set by set using a sort of Gödel numbering in terms of ordinals. As with his proof of the Incompleteness Theorem, Gödel's careful coding of metamathematical features may have precluded any misinterpretations; however, it also served to purge the intuitive underpinnings and historical motivations. In his [1939a], Gödel presented the hierarchy whose cumulation is L essentially as it is today:

$$M_0 = \{\emptyset\};\ M_\beta = \bigcup_{\alpha<\beta} M_\alpha \text{ for limit ordinals } \beta;\text{ and } M_{\alpha+1} = M'_\alpha\,,$$

where M' is "the set of subsets of M defined by propositional functions $\phi(x)$ over M," these propositional functions having been precisely defined. Significantly, footnote 12 of [1939a] revealed that Gödel viewed his axiom A, that every set is constructible (now written $V = L$ following Gödel [1940]), as deriving its sense from the cumulative hierarchy of sets regarded as an extension of the simple theory of types: "In order to give A an intuitive meaning, one has to understand by 'sets' all objects obtained by building up the simplified hierarchy of types on an empty set of individuals (including types of arbitrary transfinite orders)."

The recent publication of hitherto unpublished lectures of Gödel on the Continuum Hypothesis has dramatically substantiated the strong influence of both Russell and Hilbert on him. Both figures loom large in Gödel's lecture [1939b] given at Hilbert's Göttingen. Gödel recalled at length Hilbert's work on the Continuum Hypothesis and cast his own as an analogical development, one leading however to the constructible sets as a model for set theory. Gödel [1939b: 131] pointed out that "*the model ... is by no means finitary*; in other words, the transfinite and impredicative procedures of set theory enter into its definition in an essential way, and that is the reason why one obtains only a relative consistency proof [of the Continuum Hypothesis]".

To motivate the model Gödel referred to Russell's ramified theory of types. Gödel first described what amounts to the orders of that theory for the simple situation when the members of a countable collection of real numbers are taken as the "individuals" and new real numbers are successively defined via quantification over previously defined real numbers, and emphasized that the process can be continued into the transfinite. He then observed that this procedure can be applied to sets of real numbers, and the like, as "individuals", and moreover, that one can "intermix" the procedure for the real numbers with the procedure for sets of real numbers "by using in the definition of a real number quantifiers that refer to sets of real numbers, and similarly in still more complicated ways" ([1939b: 135]). Gödel called a *constructible* set "the most general [object] that can at all be obtained in this way, where the quantifiers may refer not only to sets of real numbers, but also to sets of sets of real numbers and so on, *ad transfinitum*, and where the indices of iteration ... can also be arbitrary transfinite ordinal numbers". Gödel considered that although this definition of constructible set might seem at first to be "unbearably complicated", "the *greatest generality yields,* as it so often does, at the same time the *greatest simplicity*" ([1939b: 137]). Gödel was picturing Russell's ramified theory of types by first disassociating the types from the orders, with the orders here given through definability and the types represented by real numbers, sets of real numbers, and so forth. Gödel's intermixing then amounted to a recapturing of the complexity of Russell's ramification, the extension of the hierarchy into the transfinite allowing for a new simplicity.

Gödel went on to describe the universe of set theory, "the objects of which set theory speaks", as falling into "a transfinite sequence of Russellian [simple] types" ([1939b: 137]), the cumulative hierarchy of sets that he had described in [1933]. He then formulated the constructible sets as an analogous hierarchy, the hierarchy of [1939a], in effect introducing Russellian orders through definability. In a comment bringing out the intermixing of types and orders, Gödel pointed out that "there are sets *of lower type* that *can* only *be defined* with the help of *quantifiers for sets of higher type*" ([1939b: 141]). This lecture of Gödel's is a remarkably clear presentation of both the mathematical and historical development of L.

Gödel's argument for the Continuum Hypothesis in the model L rests on [1939] "a generalization of Skolem's method for constructing enumerable models". It is arguably the next significant application of the Löwenheim-Skolem Theorem after Hilbert's anticipatory one with his countable interpretation for Euclidean geometry (*sans* the Completeness Axiom) and Skolem's own [1923] to get his "paradox" for set theory. Gödel showed that every subset of M_ω in L belongs to M_α for some $\alpha < \omega_1$. (Thus, every real number in L belongs to M_α for some $\alpha < \omega_1$.) In [1939b: 143] he asserted that "this fundamental theorem constitutes the corrected core of the so-called Russellian axiom of reducibility." Thus, Gödel established another connection between L and Russell's ramified theory of types. But while Russell had to *postulate* his Axiom of Reducibility for his finite orders, Gödel was able to *derive* an analogous form for his transfinite hierarchy. In his first announcement Gödel [1938: 556] had written: "The extension to transfinite orders has the consequence that the model satisfies the impredicative axioms of set theory, because an axiom of reducibility can be proved for sufficiently high orders." The beginnings of this was already hinted at in Gödel's Incompleteness paper [1931: 178], where he wrote of its Axiom IV: "This axiom plays the role of the axiom of reducibility (the comprehension axiom of set theory)." For Gödel, Russell's Axiom of Reducibility with its capability of replacing notationally specified objects of any order by equivalent objects of the lowest order of the same type was the direct antecedent to "the comprehension axiom of set theory". As he said [1939b: 145]:

> This character of the fundamental theorem as an axiom of reducibility is also the reason why *the axioms of classical* mathematics hold for the model of the constructible sets. For after all, as Russell

> showed, the axioms of reducibility, infinity and choice are the only axioms of classical mathematics that do not have a tautological character. To be sure, one must observe that the axiom of reducibility appears in different mathematical systems under different names and in different forms, for example, in Zermelo's system of set theory as the axiom of separation, in Hilbert's systems in the form of recursion axioms, and so on.

Hilbert and Russell also figure prominently in a later lecture [1940a] at Brown University on the Continuum Hypothesis. Gödel began by announcing that he had "succeeded in giving the [consistency] proof a new shape which makes it somewhat similar" to Hilbert's [1926] attempt, and proceeded to sketch the new proof, considering it "perhaps the most perspicuous". First, Gödel reviewed his construction of the model L. Once again he emphasized that his argument showing that the Continuum Hypothesis holds in L proves an axiom of reducibility.[50] Then Gödel turned to his new approach to the consistency proof, and introduced the concept of a relation being "recursive of order α" for ordinals α. This concept is a generalization of the notion of definability, a generalization obtained by interweaving the operation M', given five paragraphs above, with a recursion scheme akin to Hilbert's for his [1926] hierarchy of functionals. As Gödel [1940a: 180] said: "The difference between this notion of recursiveness and the one that Hilbert seems to have had in mind is chiefly that I allow quantifiers to occur in the definiens. This makes one [Lemma I] of Hilbert's lemmas superfluous and the other [Lemma II] demonstrable in a certain modified sense". Using this new concept of recursiveness – better, new concept of definability – Gödel gave a model of Russell's *Principia*, construed as his system P of his incompleteness paper [1931], in which the Continuum Hypothesis holds. (The types of this model were essentially coded versions of $M_{\omega_{n+1}} - M_{\omega_n}$.)

In his monograph [1940] Gödel had provided a formal presentation of L using an axiomatization of set theory with an antecedent in von Neumann [1925]. Gödel's formalization not only recalled von Neumann's [1925: II] analysis of "subsystems", but also shed light on von Neumann's main concern: the categoricity of his axiomatization. Fraenkel [1922] had expressed the desirability of closing off the Zermelian generative axioms through an "axiom of restriction"; this required that there should be no further sets than those generated by the axioms, a notable move antithetical to the role played by Hilbert's Completeness Axiom in geometry. It was to pursue this that von Neumann had investigated subsystems for his axiomatization, but he concluded that there

[50] He further said [1940a: 178]: "So since an axiom of reducibility holds for constructible sets it is not surprising that the axioms of set theory hold for the constructible sets, because the axiom of reducibility or its equivalents, e.g., Zermelo's Aussonderungsaxiom, is really the only essential axiom of set theory."

was probably no way to *formally* achieve Fraenkel's idea of a minimizing, and hence categorical, axiomatization. Gödel's axiom A, that every set is constructible, can be viewed as formally achieving this sense of categoricity, since, as he essentially showed in [1940], in axiomatic set theory L is a definable class that together with the membership relation restricted to it is a model of set theory, and L is a submodel of every other such class.[51] In his first description of L Gödel wrote ([1938: 557]): "The proposition A added as a new axiom seems to give a natural completion of the axioms of set theory, in so far as it determines the vague notion of an arbitrary infinite set in a definite way."

However, Gödel came to regard L as primarily a contrivance for establishing relative consistency results. In his [1947] he suggested that the Continuum Hypothesis is false and in footnote 22 that a new axiom "in some sense directly opposite" to A might entail this. In a revision [1964: 266] of [1947], he expanded the footnote: "I am thinking of an axiom which (similar to Hilbert's completeness axiom in geometry) would state some maximum property of the system of all sets, whereas axiom A states a minimum property. Note that only a maximum property would seem to harmonize with the concept of [arbitrary set]." This is related to Gödel's speculations with large cardinal hypotheses;[52] whereas his axiom A had enforced a kind of categoricity through minimization, large cardinals as maximum properties might establish the negation of the Continuum Hypothesis. Although the historical connection is now admittedly faint, just as the addition of the Completeness Axiom in geometry precludes Hilbert's countable interpretation, so maximum properties in set theory may preclude versions of Gödel's Skolem function argument for the consistency of the Continuum Hypothesis.

In an earlier letter to Ulam (see Ulam [1958: 13]) Gödel had written of von Neumann's axiom [1925] that a class is proper exactly when it can be put into one-to-one correspondence with the entire universe:

> The great interest which this axiom has lies in the fact that it is a maximum principle, somewhat similar to Hilbert's axiom of completeness in geometry. For, roughly speaking, it says that any set which does not, in a certain well-defined way, imply an inconsistency exists. Its being a maximum principle also explains the fact that this axiom implies the axiom of choice. I believe that the basic problems of abstract set theory, such as Cantor's continuum

[51] In [1940a: 176], Gödel wrote: "One may at first doubt that this assertion [A] has a meaning at all, because A is apparently a metamathematical statement since it involves the manifestly metamathematical term 'definable' or 'constructible'. But now it has been shown in the last few years how metamathematical statements can be translated into mathematics, and this applies also to the notion of constructibility and the proposition A, so that its consistency with the axioms of mathematics is a meaningful assertion."

[52] See Kanamori [1994] for the recent work in set theory on large cardinal hypotheses.

> problem, will be solved satisfactorily only with the help of stronger axioms of *this* kind, which in a sense are opposite or complementary to the constructivistic interpretation of mathematics.

Hilbert's Completeness Axiom thus fueled speculations about maximization for set theory, speculations resonating with his "consistency implies existence" view, speculations still being investigated to this day.

Appendix[53]

Recent developments have not only led to a precise logical analysis of Hilbert's basis theorem but to results that can be regarded as affirmatory for Hilbert's consistency program. In this appendix some of these developments are briefly described to recast Hilbert's results and initiatives in a new light.

Harvey Friedman [1975] observed that when a theorem of "ordinary" mathematics is proved from a very economical comprehension (or "set existence") axiom, then it should be possible to "reverse" the process by proving the axiom from the theorem over a weak ambient theory. Together with initial and continuing results by Friedman, Stephen Simpson and his collaborators since the late 1970's proceeded to carry out a program analyzing theorems in this spirit, the program of *reverse mathematics.* We first set the stage:

Primitive Recursive Arithmetic is the system in the language with the logical connectives (but no quantifiers), the constant 0, a unary function symbol for the successor function, and a function symbol for each (definition of a) primitive recursive function, where the axioms are the recursive defining equations for the functions symbols. First presented in Skolem [1923a] and extensively investigated in Hilbert-Bernays [1934], Primitive Recursive Arithmetic has been widely regarded as a characterization of Hilbert's "finitary" methods.

The *language of second-order arithmetic*[54] is a two-sorted language with *number variables* $i, j, m, n, \ldots$ and *set variables* $X, Y, Z, \ldots$. The number variables are intended to range over the natural numbers, and the set variables to range over sets of natural numbers. Numerical terms are generated as usual from the number variables, the constants 0 and 1, and the binary operations $+$ and $\times$. The atomic formulas are $t = u$, $t < u$, and $t \in X$, where t, u are numerical terms. Finally, formulas are generated from the atomic formulas via logical connectives, number quantifiers $\forall n$ and $\exists n$, and the set quantifiers $\forall X$ and $\exists X$.

[53] This appendix is mostly drawn from Simpson [1985], to which we refer for more details and references. See also Simpson [1988].

[54] "arithmetic" here refers to number theory, the structure of addition and multiplication of the natural numbers. As mentioned in §3, Hilbert used "arithmetic" to refer to analysis, which in the present setting corresponds to "second-order arithmetic" if sets of natural numbers are construed as real numbers.

All the formal systems to be considered include the familiar axioms about $+, \times, 0, 1, <$ as well as the *induction axiom:*[55]

$$(0 \in X \wedge \forall n(n \in X \rightarrow n+1 \in X)) \rightarrow \forall n(n \in X).$$

Full *second-order arithmetic,* or *analysis,* consists of these axioms together with the *full comprehension scheme*: For all formulas φ,

$$\exists X \forall n(n \in X \leftrightarrow \varphi(n)).$$

As shown in Hilbert-Bernays [1939], a great deal of classical mathematics can be faithfully recast in second-order arithmetic with codes for the real numbers. In what follows, certain subsystems are considered that exactly capture the strength of several basic mathematical results. We begin with an analysis of the complexity of formulas:

A formula is Δ^0_0 if it has no set quantifiers and all of its number quantifiers are *bounded*, i.e. can be rendered in form $\forall m(m < t \rightarrow \ldots)$ or $\exists m(m < t \wedge \ldots)$. A formula is Σ^0_1 if it is of form $\exists m\varphi$ where φ is Δ^0_0, and Π^0_1 if it is of form $\forall m\varphi$ where φ is Δ^0_0. For each natural number n, a formula is Σ^0_{n+1} if it is of the form $\exists m\varphi$ where φ is Π^0_n, and a formula is Π^0_{n+1} if it is of the form $\forall m\varphi$ where φ is Σ^0_n. A formula is *arithmetical* if it contains no set quantifiers, i.e. its prenex form is for some n a Σ^0_n or Π^0_n formula. Finally, a formula is Π^1_1 if it is of the form $\forall X\varphi$ where φ is arithmetical.

RCA_0 (Recursive Comprehension Axiom)[56] is the subsystem of second-order arithmetic consisting of the axioms of the Σ^0_1-induction scheme, i.e. for each Σ^0_1 formula φ,

$$(\varphi(0) \wedge \forall n(\varphi(n) \rightarrow \varphi(n+1))) \rightarrow \forall n\varphi(n),$$

and axioms of the Δ^0_1-comprehension scheme, i.e. for Σ^0_1 formulas φ and Π^0_1 formulas ψ,

$$\forall n(\varphi(n) \leftrightarrow \psi(n)) \rightarrow \exists X \forall n(n \in X \leftrightarrow \varphi(n)).$$

RCA_0 just suffices to establish the existence of the (general) recursive sets and also to develop some basic theory of real-valued continuous functions and of countable algebraic structures. However, with its parsimonious form of induction it can only establish the totality of number-theoretic functions in a restricted class. It is essentially a result of Charles Parsons [1970] that the provably total general recursive functions of RCA_0 are exactly the primitive

[55]The full induction scheme, which is *not* assumed, is: For all formulas φ,

$$(\varphi(0) \wedge \forall n(\varphi(n) \rightarrow \varphi(n+1))) \rightarrow \forall n\varphi(n).$$

The subscript 0 in the acronyms for the subsystems distinguished below is an evolutionary artifact, indicating that only the induction axiom is being assumed and not the full scheme.

[56]See the previous note for the use of the subscript 0.

recursive functions.[57] RCA_0 proves that the ordinal ω^n is well-ordered for each particular natural number n, but not that ω^ω is.[58] For RCA_0 proves that ω^ω is well-ordered implies the totality of Ackermann's function, the paradigmatic non-primitive recursive function.

WKL_0 (Weak König's Lemma) is the subsystem consisting of the axioms of RCA_0 together with: Every infinite tree of finite sequences of 0's and 1's ordered by extension has an infinite path. WKL_0 provides a better theory of continuous functions and suffices for the development of ideal theory for countable commutative rings.

ACA_0 (Arithmetical Comprehension Axiom) is the subsystem consisting of the axioms of the arithmetical comprehension scheme, i.e. for each arithmetical formula φ,

$$\exists X \forall n (n \in X \leftrightarrow \varphi(n)).$$

(In what follows, other comprehension schemes based on formula complexity have analogous formulations.) ACA_0 subsumes WKL_0. Since RCA_0 can encode functions as sets of ordered pairs, it follows that over this base theory ACA_0 is equivalent to the Σ^0_1-comprehension scheme. In terms of well-orderings, ACA_0 proves that every ordinal less than ϵ_0 is well-ordered, but not ϵ_0 itself.[59] ACA_0 has the same consequences for analysis as the system explored by Weyl [1918].

Theorem 1 (Friedman, Simpson). The following are equivalent over RCA_0:

(a) WKL_0.

(b) The Heine-Borel Theorem: Every covering of the unit interval of reals by a countable sequence of open sets has a finite subcover.

(c) Every continuous real function on the unit interval has a supremum.

(d) Every countable commutative ring has a prime ideal.

(e) The Gödel Completeness Theorem.

(f) The Hahn-Banach Theorem for separable Banach spaces.

Theorem 2 (Friedman, Simpson). The following are equivalent over RCA_0:

(a) ACA_0.

(b) The Bolzano-Weierstrass Theorem: Every bounded sequence of real numbers has a convergent subsequence.

(c) Every bounded sequence of real numbers has a least upper bound.

---

[57] A recursive function $f\colon \omega \to \omega$ has the Kleene normal form $f(i) = U(\mu m(T(i,m) = 0))$ where U and T are primitive recursive functions and μ is the least number operator, specifying the least m such that $T(i,m) = 0$. That f is *total* is the assertion $\forall i \exists m T(i,m) = 0$, and f is *provably total* in a system of arithmetic if that system proves this assertion.

[58] Let $\prec$ be a primitive recursive ordering of the natural numbers which is isomorphic to the ordinal ϵ_0 (cf. note 36). For an ordinal $\alpha \leq \epsilon_0$, "α is well-ordered" is the Π^1_1 assertion that every set consisting of natural numbers corresponding via $\prec$ to ordinals less than α has a $\prec$-least element.

[59] See notes 36 and 58 for the terminology.

(d) Every countable commutative ring has a maximal ideal.

(e) König's Lemma: Every infinite, finitely branching tree consisting of finite sequences of natural numbers ordered by extension has an infinite path.

Theorem 2(e) highlights the new strength beyond WKL_0, which draws the same conclusion for finite sequences of 0's and 1's.

Simpson [1988a] provided the following analysis of Hilbert's basis theorem:

Theorem 3 (Simpson). The following are equivalent over RCA_0:

(a) Hilbert's basis theorem in the following sense: For countable fields K and $x_1, \ldots, x_n$, the (commutative) ring of polynomials $K[x_1, \ldots, x_n]$ is finitely generated.

(b) The ordinal ω^ω is well-ordered.

The proof incidentally is similar to Gordan's [1899] proof of the basis theorem. By our previous remarks about RCA_0, (a) thus just transcends RCA_0 and implies the totality of Ackermann's function.

Friedman (unpublished) has in fact established an equivalence between a variant of Hilbert's basis theorem and the totality of Ackermann's function. Friedman showed: *For any natural number k there is a natural number n such that for every sequence of n polynomials in k variables over any field, where the ith term of the sequence has degree at most i, some polynomial is in the ideal generated by the previous polynomials. With $h(k)$ denoting the least such n, the function h is essentially Ackermann's function.* Note that h does not depend on the field. The assertion cast for polynomials over the two-element field is formalizable as a Π^0_2 sentence, as is the assertion of the totality of Ackermann's function. This is a remarkable historical confluence of Hilbert's mathematics and metamathematics, in that a variant of his first major result is seen to be equivalent to the totality of the first recursive function that he [1926] had considered for transcending primitive recursion, and hence just transcends Primitive Recursive Arithmetic, the common characterization of Hilbert's "finitary" methods.

Hilbert's Nullstellensatz has also been analyzed, though in a different setting. Following a major reduction of the theorem to an effective form by W. Dale Brownawell [1987], Michael Shub and Stephen Smale in their [1995] observed that that effective form is equivalent to an algebraic version for the real numbers of the well-known NP $\neq$ P assertion in theoretical computer science.

Perhaps the main triumphs of reverse mathematics are the following two conservation results:

Theorem 4

(a) (Friedman; Kirby and Paris [1976]) WKL_0 is a conservative extension of Primitive Recursive Arithmetic with respect to Π^0_2 sentences, i.e. every Π^0_2

sentence provable in WKL_0 is already provable in Primitive Recursive Arithmetic.

(b) (Harrington) For every model of RCA_0 there is a model of WKL_0 with the same "natural numbers". In particular, (Friedman [1975: 238]) WKL_0 is a conservative extension of RCA_0 with respect to Π_1^1 sentences, i.e. every Π_1^1 sentence provable in WKL_0 is already provable in RCA_0.

As emphasized by Simpson [1985: 469], (a) provides a significant advance towards the realization of Hilbert's consistency program in the sense that strong ideal propositions can be eliminated from the proofs of substantial assertions of Primitive Recursive Arithmetic. One can apply the powerful methods of Riemann integration, the ideal theory for countable commutative rings, and Gödel's Completeness Theorem available in WKL_0 to establish results of a rich logical complexity as in (a) and (b). In the simplest case, one cannot derive $0 \neq 0$ in WKL_0 if one cannot already derive it in Primitive Recursive Arithmetic.

Theorem 4 was established by model-theoretic means; Sieg [1991] provided systematic proof-theoretic proofs based on Herbrand and Gentzen. Feferman [1988] gives a detailed account of constructive consistency proofs for various powerful subsystems of analysis.

References

Ackermann, Wilhelm [1925], Begründung des "tertium non datur" mittels der Hilbertschen Theorie der Widerspruchsfreiheit, Mathematische Annalen **93**, 1–36.

Ackermann, Wilhelm [1928], Zum Hilbertschen Aufbau der reellen Zahlen, Mathematische Annalen **99**, 118–133; translated in van Heijenoort [1967], 493–507.

Ackermann, Wilhelm [1940], Zur Widerspruchsfreiheit der Zahlentheorie, Mathematische Annalen **117**, 162–194.

Artin, Emil [1927], Über die Zerlegung definiter Functionen in Quadrate, Abhandlungen aus dem Mathematischen Seminar der Hamburgischen Universität **5**, 100–115; reprinted in [1965], 273–288.

Artin, Emil [1965], Lang, Serge, and John T. Tate (eds.), *Collected Papers,* Reading, Addison Wesley.

Artin, Emil, and Otto Schreier [1926], Algebraische Konstruktion reeller Körper, Abhandlungen aus dem Mathematischen Seminar der Hamburgischen Universität **5**, 85–99; reprinted in Artin [1965], 258–272.

Artin, Emil, and Otto Schreier [1927], Eine Kennzeichnung der reell abgeschlossenen Körper, Abhandlungen aus dem Mathematischen Seminar der Hamburgischen Universität **5**, 225–231; reprinted in Artin [1965], 289–295.

Asser, Günter [1957], Theorie der logischen Auswahlfunktionen, Zeitschrift für mathematische Logik und Grundlagen der Mathematik **3**, 30–68.

Bernays, Paul [1928], Zusatz zu Hilberts Vortrag über "Die Grundlagen der Mathematik", Abhandlungen aus dem Mathematischen Seminar der Hamburgischen Universität **6**, 89–92; translated in van Heijenoort [1967], 486–489.

Blumenthal, Otto [1935], Lebensgeschichte, in Hilbert [1932–5], vol. 3, 388–429.

Browder, Felix E. (ed.) [1976], *Mathematical Developments Arising from Hilbert's Problems,* Proceedings of Symposia in Pure Mathematics vol. 28, Providence, American Mathematical Society.

Brownawell, W. Dale [1987], Bounds for the degrees in the Nullstellensatz, Annals of Mathematics **126**, 577–591.

Busemann, Herbert [1976], Problem IV: Desarguesian Spaces, in Browder [1976], 131–141.

Cantor, Georg [1872], Über die Ausdehnung eines Satzes aus der Theorie der trignometrischen Reihen, Mathematische Annalen **5**, 123–132; reprinted in [1932] below, 92–102.

Cantor, Georg [1883], Über unendliche, lineare Punktmannigfaltigkeiten. V. Mathematische Annalen **21**, 545–591; published separately as *Grundlagen einer allgemeinen Mannigfaltigkeitslehre: Ein mathematisch-philosophischer Versuch in der Lehre des Unendlichen*; Leipzig, Teubner 1883; reprinted in [1932] below, 165–209.

Cantor, Georg [1932], Zermelo, Ernst (ed.) *Gesammelte Abhandlungen mathematischen und philosophischen Inhalts,* Berlin, Julius Springer; reprinted in Berlin, Springer-Verlag 1980.

Carnap, Rudolf [1961], On the use of Hilbert's ε-operator in scientific theories, in: Bar-Hillel, Yehoshua, E.I.J. Poznanski, Michael O. Rabin, and Abraham Robinson (eds.) *Essays on the Foundations of Mathematics*, Jerusalem, Magnes Press, 156–164.

Church, Alonzo [1976], Comparison of Russell's resolution of the semantical antimonies with that of Tarski, The Journal of Symbolic Logic **41**, 747–760.

Dedekind, Richard [1872], *Stetigkeit und irrationale Zahlen,* Braunschweig, F. Vieweg; fifth, 1927 edition reprinted in [1932] below, vol. 3, 315–334; translated in [1963] below, 1–27.

Dedekind, Richard [1932], Fricke, Robert, Emmy Noether, and Öystein Ore (eds.) *Gesammelte mathematische Werke*, Braunschweig, F. Vieweg; reprinted in New York, Chelsea 1969.

Dedekind, Richard [1963], *Essays on the Theory of Numbers*, translations by Wooster W. Beman, New York, Dover (reprint of original edition, Chicago, Open Court 1901).

Dreben, Burton, and John Denton [1966], A supplement to Herbrand, The Journal of Symbolic Logic **31**, 393–398.

Dreben, Burton, and John Denton [1970], Herbrand-style consistency proofs, in Myhill, John R., Akiko Kino, and Richard E. Vesley (eds.) *Intuitionism and Proof Theory*, Amsterdam, North-Holland, 419–433.

Ellison, W.J. [1971], Waring's Problem, American Mathematical Monthly **78**, 10–36.

Ehrlich, Philip [1995], Hahn's *Über die Nichtarchimedischen Grössensysteme* and the Development of the Modern Theory of Magnitudes and Numbers to Measure Them, in: Hintikka, Jaakko (ed.) *From Dedekind to Gödel. Essays on the Development of the Foundations of Mathematics,* Synthese Library volume 251, Dordrecht, Kluwer, 165–213.

Feferman, Solomon [1962], Classifications of recursive functions by means of hierarchies, Transactions of the American Mathematical Society **104**, 101–122.

Feferman, Solomon [1988], Hilbert's Program relativized: Proof-theoretical and foundational reductions, The Journal of Symbolic Logic **53**, 364–384.

Fraenkel, Abraham A. [1922], Zu den Grundlagen der Cantor-Zermeloschen Mengenlehre, Mathematische Annalen **86**, 230–237.

Frege, Gottlob [1980], Gabriel, G., *et al.* (eds.) *Philosophical and Mathematical Correspondence,* Chicago, University of Chicago Press.

Friedman, Harvey M. [1975], Some systems of second order arithmetic and their uses, in *Proceedings of the International Congress of Mathematicians,* Vancouver 1974, vol. 1, Canadian Mathematical Congress, 235–242.

Gauss, Carl F. [1799], Demonstratio nova theorematis omnem functionem algebraicam rationalem integram unius variabilis in factores reales primi vel secundi gradus resolvi posse, doctoral dissertation, University of Helmstedt.

Gel'fond, Aleksander O. [1934], On Hilbert's Seventh Problem, Doklady Akademii Nauk SSSR **2**, 1–6.

Gentzen, Gerhard [1936], Die Widerspruchsfreiheit der reinen Zahlentheorie. Mathematische Annalen **112**, 493–565; translated in [1969] below, 132–213.

Gentzen, Gerhard [1938], Neue Fassung des Widerspruchsfreiheitsbeweises für die reine Zahlentheorie, Forschungen zur Logik und zur Grundlegung der exakten Wissenschaften, Leipzig, S. Hirzel, 19–44; translated in [1969] below, 252–286.

Gentzen, Gerhard [1943], Beweisbarkeit und Unbeweisbarkeit von Anfangsfällen der transfiniten Induktion in der reinen Zahlentheorie. Mathematische Annalen **119**, 140–161; translated in [1969] below, 287–308.

Gentzen, Gerhard [1969], Szabo, M.E. (ed.) *The Collected Papers of Gerhard Gentzen.* Amsterdam, North-Holland.

Gödel, Kurt F. [1930], Die Vollständigkeit der Axiome des logischen Funktionenkalküls. Monatshefte für Mathematik und Physik **37**, 349–360; reprinted and translated in [1986] below, 102–123.

Gödel, Kurt F. [1930a], Vortrag über Vollständigkeit des Funktionenkalküls; text and translation in [1995] below, 16-29, and the page references are to these.

Gödel, Kurt F. [1931], Über formal unentscheidbare Sätze der *Principia Mathematica* und verwandter Systeme I, Monatshefte für Mathematik und Physik **38**, 173–198; reprinted and translated with minor emendations by the author in [1986] below, 144–195.

Gödel, Kurt F. [1933], The present situation in the foundations of mathematics, in [1995] below, 45–53, and the page references are to these.

Gödel, Kurt F. [1933a], Zur intuitionistischen Arithmetik und Zahlentheorie, Ergebnisse eines mathematischen Kolloquiums **4**, 34-38; reprinted and translated in [1986] below, 286–295.

Gödel, Kurt F. [1938], The consistency of the Axiom of Choice and of the Generalized Continuum-Hypothesis, Proceedings of the National Academy of Sciences U.S.A. **24**, 556–557; reprinted in [1990] below, 26–27.

Gödel, Kurt F. [1939], The consistency of the generalized continuum hypothesis, Bulletin of the American Mathematical Society **45**, 93; reprinted in [1990] below, 27.

Gödel, Kurt F. [1939a], Consistency-proof for the Generalized Continuum-Hypothesis, Proceedings of the National Academy of Sciences U.S.A. **25**, 220–224; reprinted in [1990] below, 28–32.

Gödel, Kurt F. [1939b], Vortrag Göttingen; text and translation in [1995] below, 126–155, and the page references are to these.

Gödel, Kurt F. [1940], *The Consistency of the Axiom of Choice and of the Generalized Continuum Hypothesis with the Axioms of Set Theory,* Annals of Mathematics Studies #3, Princeton, Princeton University Press; reprinted in [1990] below, 33–101.

Gödel, Kurt F. [1940a], Lecture [on the] consistency [of the] continuum hypothesis, Brown University, in [1995] below, 175–185, and the page references are to these.

Gödel, Kurt F. [1941], In what sense is intuitionist logic constructive?, in [1995] below, 189–200.

Gödel, Kurt F. [1944], Russell's mathematical logic, in Schilpp, Paul A. (ed.) *The Philosophy of Bertrand Russell,* The Library of Living Philosophers, vol. 5, Evanston, Northwestern University, 123–153; reprinted in Gödel [1990] below, 119–141.

Gödel, Kurt F. [1947], What is Cantor's Continuum Problem? American Mathematical Monthly **54**, 515–525; errata **55** (1948), 151; reprinted in [1990] below, 176–187; see also [1964] below.

Gödel, Kurt F. [1958], Über eine bisher noch nicht benützte Erweiterung des finiten Standpunktes, Dialectica **12**, 280–287; translated by Wilfrid Hodges and Bruce Watson as On a hitherto unexploited extension of the finitary standpoint, Journal of Philosophical Logic **9** (1980), 133–142; reprinted and translated in [1990] below, 240–251.

Gödel, Kurt F. [1964], revised and expanded version of [1947] in Benacerraf, Paul, and Hilary Putnam (eds.) *Philosophy of Mathematics. Selected Readings.* Englewood Cliffs, N.J., Prentice Hall, 258–273; this version reprinted with minor emendations by the author in [1990] below, 254–270.

Gödel, Kurt F. [1986], Feferman, Solomon, *et al.* (eds.) *Collected Works, Volume I: Publications 1929–1936,* New York, Oxford University Press.

Gödel, Kurt F. [1990], Feferman, Solomon, *et al.* (eds.) *Collected Works, Volume II: Publications 1938–1974*, New York, Oxford University Press.

Gödel, Kurt F. [1995], Feferman, Solomon, *et al.* (eds.) *Collected Works, Volume III: Unpublished Essays and Lectures*, New York, Oxford University Press.

Goldfarb, Warren D. [1979], Logic in the Twenties: the nature of the quantifier, The Journal of Symbolic Logic **44**, 351–368.

Gordan, Paul [1868], Beweis, dass jede Covariante und Invariante einer binären Form eine ganze Funktion mit numerischen Coefficienten einer endlichen Anzahl solcher Formen ist, Journal für die reine und angewandte Mathematik (Crelle's Journal) **69**, 323–354.

Gordan, Paul [1893], Über einen Satz von Hilbert, Mathematische Annalen **42**, 132–142.

Gordan, Paul [1899], Neuer Beweise des Hilbertschen Satzes über homogene Funktionen, Nachrichten von der Königlichen Gesellschaft der Wissenschaften zu Göttingen, 240–242.

Hahn, Hans [1907], Über die Nichtarchimedischen Grössensysteme, Sitzungsberichte der Kaiserlichen Akademie der Wissenschaften, Wien, Mathematisch-Naturwissenschaftliche Klasse **116**, 601–655.

Hallett, Michael [1995], Hilbert and logic, in: Marion, Mathieu, and Cohen, Robert S. (eds.) *Québec Studies in the Philosophy of Science I*, Amsterdam, Kluwer, 135–187.

Hamel, Georg [1905], Eine Basis aller Zahlen und die unstetigen Lösungen der Funktionalgleichung: $f(x+y) = f(x) + f(y)$, Mathematische Annalen **60**, 459–462.

Hartogs, Friedrich [1915], Über das Problem der Wohlordnung, Mathematische Annalen **76**, 436–443.

Hausdorff, Felix [1914], *Grundzüge der Mengenlehre*, Leipzig, de Gruyter; reprinted New York, Chelsea 1965.

Hausdorff, Felix [1914a], Bemerkung über den Inhalt von Punktmengen, Mathematische Annalen **75**, 428–433.

Herbrand, Jacques [1930], Recherches sur la théorie de la démonstration, thesis, University of Paris; translated in [1971] below, 44–202; chapter 5 translated in van Heijenoort [1967], 529–581.

Herbrand, Jacques [1931], Sur le problème fondamental de la logique mathématique, Sprawozdania z posidzeń Towarzystwa Naukowego Warszawskiego WydziałIII **24**, 12–56; translated in [1971] below, 215–271.

Herbrand, Jacques [1971], Goldfarb, Warren D. (ed.) *Logical Writings*, Cambridge, Harvard University Press.

Hermes, Hans [1965] *Eine Termlogik mit Auswahloperator*, Berlin, Springer-Verlag; revised and translated as *Term Logic with Choice Operator*, Lecture Notes in Mathematics #6, Berlin, Springer-Verlag 1970.

Hilbert, David [1890], Über die Theorie der algebraischen Formen, Mathematische Annalen **36**, 473-534; reprinted in [1932–5] below, vol. 2, 199–257; translated in [1978] below, 143–224.

Hilbert, David [1893], Über die vollen Invariantensysteme, Mathematische Annalen **42**, 313–373; reprinted in [1932–5] below, vol. 2, 287–344; translated in [1978] below, 225–301.

Hilbert, David [1895], Über die gerade Linie als kürzeste Verbindung zweier Punkte, Mathematische Annalen **46**, 91–96.

Hilbert, David [1896], Zur Theorie der aus n Haupteinheiten gebildeten komplexen Grössen, Nachrichten von der Königlichen Gesellschaft der Wissenschaften zu Göttingen, 179–183; reprinted in [1932–5] below, vol. 2, 371–374.

Hilbert, David [1899], *Grundlagen der Geometrie. Festschrift zur Feier der Enthüllung des Gauss–Weber Denkmals in Göttingen*, Leipzig, Teubner; reprinted as *Grundlagen der Geometrie*, Leipzig, Teubner 1899; translated in [1900a] and [1902] below; see also [1903], [1930], [1968], and [1971].

Hilbert, David [1900], Mathematische Probleme. Vortrag, gehalten auf dem internationalem Mathematiker-Kongress zu Paris. 1900. Nachrichten von der Königlichen Gesellschaft der Wissenschaften zu Göttingen, 253–297; reprinted in Archiv für Mathematik und Physik **(3)1** (1901), 44–63 and 213–237; in the third, 1909 edition of [1899] above, 263–279; in [1930] below, 247–261; and in [1932–5] below, vol. 3, 290–329; see also [1902a].

Hilbert, David [1900a], *Les Principes Fondamentaux de la Géometrie*, French translation by L. Laugel of an enlarged version of [1899] above, Gauthier-Villars, Paris; reprinted in Annales Scientifiques de L'école Normal Supérieure **7**, 103–209.

Hilbert, David [1900b], Über den Zahlbegriff, Jahresbericht der Deutschen Mathematiker-Vereinigung **8**, 180–184.

Hilbert, David [1902], *The Foundations of Geometry*, English translation by E.J. Townsend of an enlarged version of [1899] above, La Salle, Open Court.

Hilbert, David [1902a], Mathematical Problems, translation of [1900], Bulletin of the American Mathematical Society **8**, 437–479; reprinted in Browder [1976], 1–34.

Hilbert, David [1903], second edition of [1899] above.

Hilbert, David [1905], Über die Grundlagen der Logic und der Arithmetik, in Krazer, A. (ed.) *Verhandlungen des Dritten Internationalen Mathematiker-Kongresses in Heidelberg vom 8. bis 13. August 1904*, Leipzig, Teubner, 174–185; translated in van Heijenoort [1967], 129–138.

Hilbert, David [1909], Beweis für die Darstellbarkeit der ganzen Zahlen durch eine feste Anzahl n-ter Potenzen (Waringsches Problem), Mathematische Annalen **67**, 281–300; reprinted in [1932–5] below, vol. 1, 510–527.

Hilbert, David [1917], Mengenlehre, unpublished notes to a course given at Göttingen in the summer semester of 1917; Mathematisches Institut, Göttingen.

Hilbert, David [1917a], Prinzipien der Mathematik und Logik, unpublished notes to a course given at Göttingen in the winter semester of 1917-8; Mathematisches Institut, Göttingen.

Hilbert, David [1918], Axiomatisches Denken, Mathematische Annalen **78**, 405–415; reprinted in [1932–5] below, vol. 3, 146–156.

Hilbert, David [1922], Neubegründung der Mathematik (Erste Mitteilung), Abhandlungen aus dem Mathematischen Seminar der Hamburgischen Universität **1**, 157–177; reprinted in [1932–5] below, vol. 3, 157–177.

Hilbert, David [1923], Die logischen Grundlagen der Mathematik, Mathematische Annalen **88**, 151–165; reprinted in Hilbert [1932-5] below, vol. 3, 178–191.

Hilbert, David [1926], Über das Unendliche, Mathematische Annalen **95**, 161–190; translated into French by André Weil in Acta Mathematica vol. 48 (1926), 91-122; translated in van Heijenoort [1967], 367-392.

Hilbert, David [1928], Die Grundlagen der Mathematik, Abhandlungen aus dem Mathematischen Seminar der Hamburgischen Universität **6**, 65–85; translated in van Heijenoort [1967], 464–479.

Hilbert, David [1929], Probleme der Grundlegung der Mathematik, Atti del Congresso Internazionale dei Matematici, Bologna 3–10 Settembre 1928, Bologna, Zanichelli, I, 135–141.

Hilbert, David [1930], seventh edition of [1899] above with new supplements.

Hilbert, David [1931], Die Grundegung der elementaren Zahlenlehre, Mathematische Annalen **88**, 485–494; partially reprinted in [1932–5] below, 192–195.

Hilbert, David [1932-5], *Gesammelte Abhandlungen,* in three volumes, Berlin, Julius Springer; reprinted New York, Chelsea 1965; reprinted Berlin, Springer 1970.

Hilbert, David [1968], tenth edition of [1899] above with supplements by Paul Bernays but also deleting some previous appendices, Stuttgart; translated in [1971] below.

Hilbert, David [1971], translation of [1968] above by Leo Unger, La Salle, Open Court.

Hilbert, David [1978], Ackerman, Michael (trans.) *Hilbert's Invariant Theory Papers*, Brookline, Math Sci Press.

Hilbert, David, and Wilhelm Ackermann [1928], *Grundzüge der theoretischen Logik,* Berlin, Julius Springer; second edition, 1938; third edition, 1949; second edition was translated by Hammond, Lewis M., George G. Leckie, and F. Steinhardt as *Principles of Mathematical Logic*, New York, Chelsea 1950.

Hilbert, David, and Paul Bernays [1934], *Grundlagen der Mathematik,* volume 1, Berlin, Springer-Verlag; second edition, Die Grundlehren der mathematischen Wissenschaften in Einzeldarstellungen #40, New York, Springer-Verlag 1968.

Hilbert, David, and Paul Bernays [1939], *Grundlagen der Mathematik,* volume 2, Berlin, Springer-Verlag; second edition, Die Grundlehren der mathematischen Wissenschaften in Einzeldarstellungen #50, New York, Springer-Verlag 1970.

Hilbert, David, and Stephan Cohn-Vossen [1932], *Anschauliche Geometrie,* Berlin, Julius Springer; translated as *Geometry and the Imagination,* New York, Chelsea 1952.

Hylton, Peter [1990], *Russell, Idealism, and the Emergence of Analytic Philosophy*, Oxford, Clarendon Press.

Kanamori, Akihiro [1994], *The Higher Infinite,* Berlin, Springer-Verlag.

Kanamori, Akihiro [1996], The mathematical development of set theory from Cantor to Cohen, The Bulletin of Symbolic Logic **2**, 1–71.

Kirby, Lawrence, and Jeffrey Paris [1977], Initial segments of models of Peano Arithmetic, in: *Set Theory and Hierarchy Theory V,* Lecture Notes in Mathematics #619, Berlin, Springer-Verlag, 211-226.

Kleene, Stephen C. [1958], Extension of an effectively generated class of functions by enumeration, Colloquium Mathematicum **6**, 67–78.

Klein, Felix [1926], *Vorlesungen über die Entwicklung der Mathematik im 19. Jahrhundert,* Berlin, Julius Springer, reprinted New York, Chelsea; translated by Michael Ackerman in *Developments of Mathematics in the 19th Century,* Brookline, Math Sci Press 1979.

König, Julius [1914], *Neue Grundlagen der Logik, Arithmetik und Mengenlehre*, Leipzig, Veit.

Krull, Wolfgang [1929], Die Idealtheorie in Ringen ohne Endlichkeitsbedingungen, Mathematische Annalen **101**, 729–744.

Lévy, Paul [1964], Remarques sur un théorème de Paul Cohen, Revue de Métaphysique et de Morale **69**, 88–94.

Luzin, Nikolai N. [1933], Sur les classes des constituantes des complémentaires analytiques, Annali della Scuola Normale Superiore de Pisa, Scienze fisiche e matematiche **2**, 269-282.

Moore, Gregory H, [1982], *Zermelo's Axiom of Choice. Its Origins, Development and Influence,* New York, Springer-Verlag.

Mumford, David [1965], *Geometric Invariant Theory*, Berlin, Springer-Verlag.

Mumford, David [1976], Hilbert's Fourteenth Problem – the finite generation of subrings such as rings of invariants, in Browder [1976], 431–444.

Mumford, David, John Fogarty and Frances C. Kirwan [1994], *Geometric Invariant Theory*, third enlarged edition, Berlin, Springer-Verlag.

Myhill, John R. [1953], A Stumbling block in constructive mathematics (abstract), The Journal of Symbolic Logic **18**, 190–191.

Noether, Emmy [1915], Körper und Systeme rationaler Funktionen, Mathematische Annalen **76**, 161–191.

Noether, Emmy [1916], Die allgemeinsten Bereiche aus ganzen transzendenten Zahlen, Mathematische Annalen **77**, 103–128.

Noether, Emmy [1921], Idealtheorie in Ringbereichen, Mathematische Annalen **83**, 24–66.

Noether, Emmy [1926], Abstrakter Aufbau der Idealtheorie in algebraischen Zahl- und Funktionenkörpen, Mathematische Annalen **96**, 26–61.

Noether, Max [1914], Paul Gordan, Mathematische Annalen **75**, 1-41.

Parsons, Charles [1970], On a number-theoretic choice schema and its relation to induction, in: Myhill, John R., Akiko Kino, and Richard E. Vesley (eds.) *Intuitionism and Proof Theory,* Amsterdam, North-Holland, 459-473.

Pfister, Albrecht [1976], Hilbert's Seventeenth Problem and related problems on definite forms, in Browder [1976], 483–489.

Poincaré, Henri [1902], *La Science et L'hypothèse*, Paris, Flammarion; translated as *Science and Hypothesis*, New York, Dover 1952.

Ramsey, Frank P. [1926], The foundations of mathematics, Proceedings of the London Mathematical Society **25**, 338–384; reprinted in Ramsey [1931], 1–61.

Ramsey, Frank P. [1931], Braithwaite, Richard B. (ed.) *The Foundations of Mathematics and Other Logical Essays*, London, Paul, Trench, Trubner and New York, Harcourt, Brace.

Routledge, N.A. [1953], Ordinal recursion, Proceedings of the Cambridge Philosophical Society **49**, 175–182.

Russell, Bertrand A.W. [1908], Mathematical logic as based on the theory of types, American Journal of Mathematics **30**, 222–262; reprinted in van Heijenoort [1967], 150–182.

Russell, Bertrand A.W. [1919], *Introduction to Mathematical Philosophy,* London, Allen and Unwin.

Scanlon, Thomas M. [1973], The consistency of number theory via Herbrand's Theorem, The Journal of Symbolic Logic **38**, 29–58.

Schneider, Theodor [1934], Transzendenzuntersuchungen periodischer Funktionen, Journal für die reine und angewandte Mathematik (Crelle's Journal) **172**, 65–74.

Shub, Michael, and Stephen Smale [1995], On the intractability of Hilbert's Nullstellensatz and an algebraic version of "$NP \neq P$", Duke Mathematical Journal **81**, 47–54.

Sieg, Wilfried [1991], Herbrand analyses, Archive for Mathematical Logic, **30**, 409–441.

Simpson, Stephen [1985], Reverse mathematics, in: Nerode, Anil, and Richard A. Shore (eds.) *Recursion Theory*, Proceedings of Symposia in Pure Mathematics vol. 42, Providence, American Mathematical Society, 461-471.

Simpson, Stephen [1988], Partial realizations of Hilbert's Program, The Journal of Symbolic Logic **53**, 349–363.

Simpson, Stephen [1988a], Ordinal numbers and the Hilbert Basis Theorem, The Journal of Symbolic Logic **53**, 961-974.

Skolem, Thoralf [1923], Einige Bemerkungen zur axiomatischen Begründung der Mengenlehre, in: *Matematikerkongressen i Helsingfors den 4–7 Juli 1922, Den femte skandinaviska matematikerkongressen, Redogörelse,* Helsinki, Akademiska-Bokhandeln, 217-232; reprinted in [1970] below, 137–152; translated in van Heijenoort [1967], 290–301.

Skolem, Thoralf [1923a], Begründung der elementaren Arithmetik durch die rekurrierende Denkweise ohne Anwendung scheinbarer Veränderlichen mit unendlichem Ausdehnungsbereich, Videnskaps-selskapts Skrifter, I, Mathematisk-Naturvidenskabelig Klasse, no. 6; reprinted in [1970] below, 153–188; translated in van Heijenoort [1967], 302–333.

Skolem, Thoralf [1970], Fenstad, Jens E. (ed.) *Selected Works in Logic,* Universitetsforlaget, Oslo.

Spector, Clifford [1962], Provably recursive functionals of analysis: a consistency proof of analysis by an extension of principles formulated in current intuitionistic mathematics, in Dekker, Jacob C.E. (ed.) *Recursive Function Theory*, Proceedings of Symposia in Pure Mathematics vol. 5, Providence, American Mathematical Society, 1-27.

Steinitz, Ernst [1910], Algebraische Theorie der Körper, Journal für die reine und angewandte Mathematik (Crelle's Journal) **137**, 167–309.

Tijdeman, Robert [1976], Hilbert's Seventh Problem: On the Gel'fond-Baker method and its applications, in Browder [1976], 241–268.

Ulam, Stanisław M. [1958], John von Neumann, 1903–1957, Bulletin of the American Mathematical Society **64**, 1–49.

van Heijenoort, Jean [1967], (ed.) *From Frege to Gödel. A Source Book in Mathematical Logic, 1879–1931,* Cambridge, Harvard University Press 1967.

von Neumann, John [1925], Eine Axiomatisierung der Mengenlehre, Journal für die reine und angewandte Mathematik (Crelle's Journal) **154**, 219–240; Berichtigung **155**, 128; reprinted in [1961] below, 34–56; translated in van Heijenoort [1967], 393–413.

von Neumann, John [1927], Zur Hilbertschen Beweistheorie, Mathematische Zeitschrift **26**, 1–46; reprinted in [1961] below, 256–300.

von Neumann, John [1929], Über eine Widerspruchfreiheitsfrage in der axiomatischen Mengenlehre, Journal für die reine und angewandte Mathematik (Crelle's Journal) **160**, 227–241; reprinted in [1961] below, 494–508.

von Neumann, John [1961], Taub, Abraham H. (ed.) *John von Neumann. Collected Works,* Vol. 1; New York, Pergamon Press.

Wang, Hao [1955], On denumerable bases of formal systems, in Skolem, Thoralf, *et al.* (eds.) *Mathematical Interpretations of Formal Systems,* Amsterdam, North-Holland, 57–84.

Wang, Hao [1974], *From Mathematics to Philosophy,* New York, Humanities Press.

Wang, Hao [1987], *Reflections on Kurt Gödel,* Cambridge, The MIT Press.

Weyl, Hermann [1910], Über die Definitionen der mathematischen Grundbegriffe, Mathematisch-naturwissenschaftliche Blätter **7**, 93–95, 109–113; reprinted in [1968] below, vol. 1, 298–304.

Weyl, Hermann [1918], *Das Kontinuum: Kritische Untersuchungen über die Grundlagen der Analysis,* Leipzig, Veit; reprinted New York, Chelsea Publishing 1960, 1973; translated by Stephen Pollard and Thomas Bole as *The Continuum: A Critical Examination of the Foundation of Analysis*, Kirksville, Thomas Jefferson University Press, 1987; this translation reprinted New York, Dover 1994.

Weyl, Hermann [1939], *The Classical Groups – Their Invariants and Representations,* Princeton, Princeton University Press; second edition 1946.

Weyl, Hermann [1944], David Hilbert and his Mathematical Work, Bulletin of the American Mathematical Society **50**, 612-654; reprinted in [1968] below, vol. 4, 130–172.

Weyl, Hermann [1968], *Gesammelte Abhandlungen,* in four volumes, Berlin, Springer-Verlag.

Whitehead, Alfred N., and Bertrand A.W. Russell [1910], *Principia Mathematica.* vol. 1. Cambridge, Cambridge University Press.

Zermelo, Ernst [1904], Beweis, dass jede Menge wohlgeordnet werden kann (Aus einem an Herrn Hilbert gerichteten Briefe), Mathematische Annalen **59**, 514–516; translated in van Heijenoort [1967], 139–141.

Zermelo, Ernst [1908], Untersuchungen über die Grundlagen der Mengenlehre I, Mathematische Annalen **65**, 261–281; translated in van Heijenoort [1967], 199–215.

Zermelo, Ernst [1914], Über ganze transzendente Zahlen, Mathematische Annalen **75**, 434–442.

Zermelo, Ernst [1930], Über Grenzzahlen und Mengenbereiche: Neue Untersuchungen über die Grundlagen der Mengenlehre, Fundamenta Mathematicae **16**, 29–47.

3 The Mathematical Import of Zermelo's Well-Ordering Theorem

Dedicated to Frau Gertrud Zermelo on the occasion of her 95th birthday

Set theory, it has been contended,[1] developed from its beginnings through a progression of *mathematical* moves, despite being intertwined with pronounced metaphysical attitudes and exaggerated foundational claims that have been held on its behalf. In this paper, the seminal results of set theory are woven together in terms of a unifying mathematical motif, one whose transmutations serve to illuminate the historical development of the subject. The motif is foreshadowed in Cantor's diagonal proof, and emerges in the interstices of the inclusion vs. membership distinction, a distinction only clarified at the turn of this century, remarkable though this may seem. Russell runs with this distinction, but is quickly caught on the horns of his well-known paradox, an early expression of our motif. The motif becomes fully manifest through the study of functions $f\colon \mathcal{P}(X) \to X$ of the power set of a set into the set in the fundamental work of Zermelo on set theory. His first proof in 1904 of his Well-Ordering Theorem is a central articulation containing much of what would become familiar in the subsequent development of set theory. Afterwards, the motif is cast by Kuratowski as a fixed point theorem, one subsequently abstracted to partial orders by Bourbaki in connection with Zorn's Lemma. Migrating beyond set theory, that generalization becomes cited as the strongest of fixed point theorems useful in computer science.

Section 1 describes the emergence of our guiding motif as a line of development from Cantor's diagonal proof to Russell's Paradox, fueled by the clarification of the inclusion vs. membership distinction. Section 2 engages the motif as fully participating in Zermelo's work on the Well-Ordering Theorem and as newly informing on Cantor's basic result that there is no bijection $f\colon \mathcal{P}(X) \to X$. Then Section 3 describes in connection with Zorn's Lemma the transformation of the motif into an abstract fixed point theorem, one accorded significance in computer science.

1 Cantor's Diagonal Proof to Russell's Paradox

Georg Cantor in [1891] gave his now famous diagonal proof, showing in effect that for any set X the collection of functions from X into a two-element set is

---

Republished from *The Bulletin of Symbolic Logic* 3 (1997), pp.281-311, with permission from The Association for Symbolic Logic.

The author gratefully acknowledges the generous support of the Volkswagen-Stiftung during his stay at the Mathematischen Forschungsinstitut Oberwolfach under the Research in Pairs program. This paper was partly inspired by a conversation with the late George Boolos and a reading of Moschovakis [1994]. The author would like to express his gratitude to Burton Dreben for his careful reading and numerous suggestions for improvement.

[1] See Kanamori [1996] for the mathematical development of set theory from Cantor to Cohen.

of a strictly higher cardinality than that of X. Much earlier in [1874], the paper that began set theory, Cantor had established the uncountability of the real numbers by using their completeness under limits. In retrospect the diagonal proof can be drawn out from the [1874] proof, but in any case Cantor could now dispense with its topological trappings. Moreover, he could affirm "the general theorem, that the powers [cardinalities] of well-defined sets have no maximum."

Cantor's diagonal proof is regarded today as showing how the power *set* operation leads to higher cardinalities, and as such it is the root of our guiding motif. However, it would be an exaggeration to assert that Cantor himself used power sets. Rather, he was expanding the 19th Century concept of *function* by ushering in arbitrary functions. His theory of cardinality was based on one-to-one *correspondence [Beziehung]*, and this had led him to the diagonal proof which in [1891] is first rendered in terms of *sequences* "that depend on infinitely many coordinates". By the end of [1891] he did deal explicitly with "all" *functions* with a specified domain L and range $\{0, 1\}$; regarded these as being enumerated by one super-function $\phi(x, z)$ with enumerating variable z; and formulated the diagonalizing function $g(x) = 1 - \phi(x, x)$. In his mature presentation [1895] of his theory of cardinality Cantor defined cardinal exponentiation in terms of the set of all functions from a set N into a set M, but such arbitrary functions were described in a convoluted way, reflecting the novelty of the innovation.[2]

The recasting of Cantor's diagonal proof in terms of sets could not be carried out without drawing the basic distinction between $\subseteq$, inclusion, and $\in$, membership. Surprisingly, neither this distinction nor the related distinction between a class a and the class $\{a\}$ whose sole member is a was generally appreciated in logic at the time of Cantor [1891]. This was symptomatic of a general lack of progress in logic on the traditional problem of the copula (how does "is" function?), a problem with roots going back to Aristotle. The first to draw these distinctions clearly was Gottlob Frege, the greatest philosopher of logic since Aristotle. Indeed, the inclusion vs. membership distinction is fundamental to the development of logic in Frege's *Begriffsschrift* [1879], and the a

[2]Cantor wrote [1895: §4]: "... by a '*covering* [Belegung] of N with M,' we understand a law by which with every element n of N a definite element of M is bound up, where one and the same element of M can come repeatedly into application. The element of M bound up with n is, in a way, a one-valued function of n, and may be denoted by $f(n)$; it is called a 'covering function [Belegungsfunktion] of n.' The corresponding covering of N will be called $f(N)$."

A convoluted description, one emphasizing the generalization from one-to-one correspondence [Beziehung]. Arbitrary functions on arbitrary domains are now of course commonplace in mathematics, but several authors at the time referred specifically to the concept of covering, most notably Zermelo [1904] (see Section 2). Jourdain in the introduction to his English translation [1915: 82] of Cantor's [1895, 1897] wrote: "The introduction of the concept of 'covering' is the most striking advance in the principles of the theory of transfinite numbers from 1885 to 1895 ...."

vs. $\{a\}$ distinction is explicit in his *Grundgesetze* [1893]. These distinctions for sets are also basic for Cantor's theory of cardinality and are evident from the beginning of his [1895], starting with its oft-quoted definition of set [Menge].[3]

Of other pioneers, Ernst Schröder in the first volume [1890] of his major work on the algebra of logic held to a traditional view that a class is merely a collection of objects (without the { }, so to speak), so that inclusion and membership could not be clearly distinguished and e.g. the existence of a null class was disputable. Frege in his review [1895] of Schröder's [1890] soundly took him to task for these shortcomings.[4] Richard Dedekind in his classic essay on arithmetic *Was sind und was sollen die Zahlen?* [1888: §3] used the same symbol for inclusion and membership and subsequently identified an individual a with $\{a\}$.[5] In a revealing note found in his *Nachlass* Dedekind was to draw attention to the attendant danger of such an identification and showed how this leads to a contradiction in the context of his essay.[6]

Giuseppe Peano in his essay [1889] distinguished inclusion and membership with different signs, and it is to him that we owe "$\in$" for membership. In the preface he warned against confusing "$\in$" with the sign for inclusion. However, at the end of part IV he wrote, "Let s be a class and k a class contained in s; then we say that k is an individual of class s if k consists of just one individual. Thus," and proceeded to give his formula 56, which in modern terms is:

$$k \subseteq s \rightarrow (k \in s \leftrightarrow (k \neq \emptyset \;\&\; \forall x \in k \forall y \in k(x = y)))\,.$$

Unfortunately, this way of having membership follow from inclusion undercuts the very distinction that he had so emphasized. For example, suppose that a is

---

[3] While we use the now familiar notation $\{a\}$ to denote the class whose sole member is a, it should be kept in mind that the notation varied through this period. Cantor [1895] wrote $M = \{m\}$ to indicate that M consists of members *typically* denoted by m, i.e. m was a variable ranging over the possibly many members of M. Of those soon to be discussed, Peano [1890: 192] used ιa to denote the class whose sole member is a. Russell [1903: 517] followed suit, but from his [1908] on he used ι'a. It was Zermelo [1908a: 262] who introduced the now familiar use of $\{a\}$, having written just before: "The set that contains only the elements $a, b, c, \ldots, r$ will often be denoted briefly by $\{a, b, c, \ldots r\}$."

[4] Edmund Husserl in his review [1891] of Schröder's [1890] also criticized him for not distinguishing between inclusion and membership (cf. Rang-Thomas [1981: 19] and the beginning of Section 2 below).

[5] For a set [System] S and transformation [Abbildung] ϕ: $S \rightarrow S$, Dedekind [1888: §37] defined K to be a chain [Kette] *iff* $K \subseteq S$ and for every $x \in K$, $\phi(x) \in K$; for any $A \subseteq S$, he [1888: §44] then defined A_o to be the intersection of all chains $K \supseteq A$. In the crucial definition of "simply infinite system", one isomorphic to the natural numbers, Dedekind [1888: §71] wrote $N = 1_o$, where 1 is a distinguished *element* of N. Hence, we would now write $N = \{1\}_o$.

[6] See Sinaceur [1971]. In the note Dedekind proposed various emendations to his essay to clarify the situation; undercutting a comment in the essay ([1888: §2]) he pointed out the necessity of having the empty set [Nullsystem]. He also mentioned raising these issues, quaint as this may now seem, in conversation with Felix Bernstein on 13 June 1897 and with Cantor himself on 4 September 1899. However, the emendations were never incorporated into any later editions of [1888].

any class, and let $s = \{a\}$. Then formula 56 implies that $s \in s$. But then $s = a$, and so $\{a\} = a$. This was not intended by Peano; in [1890: 192] he carefully distinguished between a and $\{a\}$.[7]

The equivocation between inclusion and membership in the closing years of the 19th Century reflected a traditional reluctance to comprehend a collection as a unity and was intertwined with the absence of the liberal, iterative use of the set formation $\{\ \}$ operation. Of course, set theory as a mathematical study of that operation could only develop after a sharp distinction between inclusion and membership had been made. This development in turn would depend increasingly on rules and procedures provided by axiomatization, an offshoot of the motif being traced here (see Section 2).

The turn of the century saw Bertrand Russell make the major advances in the development of his mathematical logic. As he later wrote in [1944]: "The most important year in my intellectual life was the year 1900, and the most important event in this year was my visit to the International Congress of Philosophy in Paris." There in August he met Peano and embraced his symbolic logic, particularly his use of different signs for inclusion and membership. During September Russell extended Peano's symbolic approach to the logic of relations. Armed with the new insights Russell in the rest of the year completed most of the final draft of *The Principles of Mathematics* [1903], a book he had been working on in various forms from 1898. However, the sudden light would also cast an abiding shadow, for by May 1901 Russell had transformed Cantor's diagonal proof into Russell's Paradox.[8] In reaction he would subsequently formulate a complex logical system of orders and types in Russell [1908] which multiplied the inclusion vs. membership distinction many times over and would systematically develop that system in Whitehead and Russell's *Principia Mathematica* [1910-3].

Soon after meeting Peano, Russell prepared an article singing his praises, writing [1901: 354] of the symbolic differentiation between inclusion and membership as "the most important advance which Peano has made in logic."[9] At

[7]Having a such that $a = \{a\}$ can serve certain technical purposes. For W.V. Quine [1940: 136] such a are the individuals for a theory of sets, now known as Mathematical Logic. Paul Bernays [1954] based a proof of the independence of the Axiom of Foundation on such a. For Ernst Specker [1957] such a serve as the atoms of his Fraenkel-Mostowski permutation models for independence results related to the Axiom of Choice. Since Dana Scott [1962] "Quinean atoms" $a = \{a\}$ have figured in the model-theoretic investigation of Quine's best known set theory, New Foundations; see Forster [1995: chapter 3].

[8]See Garciadiego [1992] and Moore [1995] for the evolution of Russell's Paradox.

[9]Russell had first mentioned Peano in a letter dated 9 October 1899 to the philosopher Louis Couturat. There Russell had expressed agreement with Couturat's review of Peano's work, the main thrust of which was Couturat's contention [1899: 628-9] that Peano's introduction of $\in$ was an unnecessary complication beyond Schröder's system of logic. Russell's [1901] was meant to be a counterpart to Couturat [1899], though it was never published. See Moore's comments in Russell [1993: 350-1].

Years later, Russell [1959: 66-7] wrote: "The enlightenment that I derived from Peano

first, it seems anomalous that Russell had not absorbed the basic inclusion vs. membership distinction from either Frege or Cantor. However, Russell only became fully aware of Frege's work in 1902.[10] Also, Russell had rejected Cantor's work on infinite numbers when he had first learned of it in 1896 and came to accept it only after meeting Peano.[11]

Much can be and has been written about Russell's predisposition in 1900 to embrace Peano's "ideography" and Cantor's theory, both in terms of Russell's rejection a few years earlier of a neo-Hegelian idealism in favor of a Platonic atomism,[12] and in terms of Leibniz's *lingua characteristica* for logical reasoning.[13] Newly inspired and working prodigiously, Russell used Peano's symbolic approach to develop the logic of relations, to define cardinal number, and to recast some of Cantor's work. However, in the course of this development a fundamental tension emerged, as we shall soon see, between Cantor's one-to-one correspondences and Peano's inclusion vs. membership distinction, a tension fueled by Russell's metaphysical belief in the existence of the class of all classes.

At first, Russell was convinced that he had actually found an error in Cantor's work. In a letter to the philosopher Louis Couturat dated 8 December 1900 Russell wrote:

> I have discovered an error in Cantor, who maintains that there is no largest cardinal number. But the number of classes is the largest number. The best of Cantor's proofs to the contrary can be found in [Cantor [1891]]. In effect, it amounts to showing that if u is a

came mainly from two purely technical advances of which it is very difficult to appreciate the importance unless one has (as I had) spent years in trying to understand arithmetic ... The first advance consisted in separating propositions of the form 'Socrates is mortal' from propositions of the form 'All Greeks are mortal' [i.e. distinguishing membership from inclusion] ... neither logic nor arithmetic can get far until the two forms are seen to be completely different ... The second important advance that I learnt from Peano was that a class consisting of one member is not identical with that one member."

On this last point however, Russell of *The Principles* [1903] was not so clear; see the discussion of the book toward the end of this section.

[10]Frege does not appear in Russell's reading list through March 1902, *What shall I read?* [1983: 347ff.]. Russell's first and now famous letter to Frege of 16 June 1902, informing him of an inconsistency in his mature system, starts: "For a year and a half I have been acquainted with your *Grundgesetze der Arithmetik*, but it is only now that I have been able to find the time for the thorough study I intended to make of your work."

[11]See Moore [1995: §3]. Russell in a letter to Jourdain of 11 September 1917 (see Grattan-Guinness [1977: 144]) reminisced: "I read all the articles in 'Acta Mathematica' [mainly those in vol. 2, 1883, French translations of various of Cantor's papers] carefully in 1898, and also 'Mannigfaltigkeitslehre' [for example Cantor [1883]]. At that time I did not altogether follow Cantor's arguments, and I thought he had failed to prove some of his points. I did not read [Cantor [1895] and Cantor [1897]] until a good deal later." Cantor [1895] only appears in Russell's reading list *What shall I read?* [1983: 364] for November 1900, when Russell was suffused with the new insights from Peano.

[12]See Hylton [1990: 103ff.], and generally for the metaphysics underlying Russell's logic.

[13]Russell had just completed a book on Leibniz.

> class whose number is α, the number of classes included in u (which is 2^α) is larger than α. The proof presupposes that there are classes included in u which are not individuals [i.e. members] of u; but if $u =$ *Class* [i.e. the class of all classes], that is false: every class of classes is a class.[14]

Also, in a popular article completed in January 1901, Russell [1901a: 87] wrote:

> Cantor had a proof that there is no greatest number, and if this proof were valid, the contradictions of infinity would reappear in a sublimated form. But in this one point, the master has been guilty of a very subtle fallacy, which I hope to explain in some future work.[15]

Russell was shifting the weight of the argumentation away from (Cantor's) result through (Cantor's) 'error' to (Russell's) paradox. By May 1901 Russell had formulated a version of his now famous paradox in terms of self-predication in a draft of his book *The Principles of Mathematics*.[16] In the published book Russell discussed the paradox extensively in various forms, and described [1903: 364ff.] in some detail how he had arrived at his paradox from Cantor's [1891] proof. Having developed the logic of relations Russell made the basic move of correlating subclasses of a class with the relations on the class to 0 and 1. By this means he converted Cantor's functional argument to one about inclusion and membership for classes, concluding that [1903: 366] "the number of classes contained in any class exceeds the number of terms belonging to the class."

---

[14]This passage, originally in French, is quoted in Moore [1995: 231], and by Moore in Russell [1993: xxxii].

In a subsequent letter to Couturat dated 17 January 1901, Russell pointedly praised Peano's introduction of $\in$ and then wrote: "...there is a concept *Class* and there are classes. Hence *Class* is a class. But it can be proved (and this is essential to Cantor's theory) that every class has a cardinal number. Hence there is a number of classes, i.e. a number of the class *Class*. But this does not result in a contradiction, since the proof which Cantor gives that

$$\alpha \,\varepsilon\, \mathrm{Nc}\,.\supset.\, 2^\alpha > \alpha$$

presupposes that there is at least one class contained in the given class u (whose number is α) which is not itself a member of u .... If we put $u = Cls$ [*Class*, the class of all classes], this is false. Thus the proof no longer holds."

This letter, originally in French, is quoted in large part in Russell [1992: 210-2]; see also Moore [1995: 233].

[15]However, as a foretaste of things to come we note that Russell added the following footnote to this passage in 1917: "Cantor was not guilty of a fallacy on this point. His proof that there is no greatest number is valid. The solution of the puzzle is complicated and depends upon the theory of types, which is explained in *Principia Mathematica*, Vol. 1 (Camb. Univ. Press, 1910)."

[16]See Russell [1993: 195].

It is here that our mathematical motif emerges and begins to guide the historical description. Cantor's argument is usually presented nowadays as showing that no function $f\colon X \to \mathcal{P}(X)$ is bijective, since the set $\{x \in X \mid x \notin f(x)\}$ is not in the range of f. Cantor [1891] himself first established (in equivalent terms with characteristic functions as we would now say) the *positive* result that for any $f\colon X \to \mathcal{P}(X)$ there is a subset of X, namely the set just defined, which is not in the range of f.[17] Russell's remark starting "The proof presupposes ..." in the penultimate displayed quotation above may at first be mystifying, until one realizes what concerned Russell about the class of all classes. In modern terms, if U is that class and $\mathcal{P}(U)$ the class of its subclasses, then $\mathcal{P}(U) \subseteq U$. Thus, the *identity* map on $\mathcal{P}(U)$ is an injection of $\mathcal{P}(U)$ into U. However, Cantor's argument also shows that *no* function $F\colon \mathcal{P}(X) \to X$ is injective.[18] This will be our guiding mathematical motif, the study of functions $F\colon \mathcal{P}(X) \to X$.

For the Russell of *The Principles* mathematics was to be articulated in an all-encompassing logic, a complex philosophical system based on universal categories.[19] He had drawn distinctions within his widest category of "term" (but with "object" wider still[20]) among "propositions" about terms and "classes" of various kinds corresponding to propositions. Because of this, Russell's Paradox became a central concern, for it forced him to face the threat of both the conflation of his categories and the loss of their universality.

Discussing his various categories, Russell [1903: 366-7] first described the problem of "the class of all terms": "If we are to assume ... that every constituent of every proposition is a term, then classes will be only some among terms. And conversely, since there is, for every term, a class consisting of that term only, there is a one-one correlation of all terms with some classes. Hence, the number of classes should be the same as the number of terms." For this last, Russell explicitly appealed to the Schröder-Bernstein Theorem.

[17]As emphasized by Gray [1994], Cantor [1874] similarly established first a positive result, that for any countable sequence of real numbers there is a real number not in the sequence, and only then drew the conclusion that the reals are uncountable. Cantor's main purpose in [1874] was actually to show that the algebraic real numbers are countable and then to apply his positive result to get a new proof that there are transcendental numbers.

[18]Generally, from an injection $g\colon A \to B$ a surjection $B \to A$ can be defined by inverting g on its range and extending that inversion to all of B. However, Cantor's result established that there is no surjection $X \to \mathcal{P}(X)$.

[19]Russell [1903: 129] wrote: "The distinction of philosophy and mathematics is broadly one of point of view: mathematics is constructive and deductive, philosophy is critical, and in a certain impersonal sense controversial. Wherever we have deductive reasoning, we have mathematics; but the principles of deduction, the recognition of indefinable entities, and the distinguishing between such entities, are the business of philosophy. Philosophy is, in fact mainly a question of insight and perception."

[20]Russell [1903: 55n] wrote: "I shall use the word *object* in a wider sense than *term*, to cover both singular and plural, and also cases of ambiguity, such as 'a man.' The fact that a word can be framed with a wider meaning than *term* raises grave logical problems."

However, classes consist of terms, so Cantor's argument shows that there are more classes than terms! To (most) contemporary eyes, this is a remarkable mixing of mathematics and metaphysics. Russell then observed with analogous arguments that: there are more classes of objects than objects; there are more classes of propositions than propositions; and there are more propositional functions than objects.

Russell next made the closest connection in *The Principles* [1903: 367] between Cantor's argument and Russell's Paradox: Let V be the class of all terms, and U the class of all classes; for Russell U is a proper subclass of V. Russell defined a function $f\colon V \to U$ by stipulating that if x is not a class then $f(x)$ is the class $\{x\}$, and if x is a class then $f(x) = x$, i.e. f restricted to U is the identity. What is now seen as the Cantorian $\{x \in U \mid x \notin f(x)\}$ then becomes the Russellian $w = \{x \in U \mid x \notin x\}$. However, Cantor's argument implies that w is not in the range of f, yet for Russell $f(w) = w$ *is* in the range.[21]

As emphasized above, Cantor's argument has a positive content in the generation of sets not in the range of functions $f\colon X \to \mathcal{P}(X)$. For Russell however, $\mathcal{P}(U) \subseteq U$ with the identity map being an injection, and so the Russellian $w \subseteq U$ must satisfy $w \in U$, arriving necessarily at a *contradiction*. Having absorbed the inclusion vs. membership distinction, Russell had to confront the dissolution of that very distinction for his universal classes.

Russell soon sought to resolve his paradox with his theory of types, adumbrated in *The Principles*. Although the inclusion vs. membership distinction was central to *The Principles*, the issues of whether the null-class exists and whether a term should be distinct from the class whose sole member is that term became part and parcel of the considerations leading to types. First, Russell [1903:68] distinguished between "class" and "class-concept", and asserted that "there is no such thing as the null-class, though there are null class-concepts ... [and] that a class having only one term is to be identified, contrary to Peano's usage, with that one term." Russell then distinguished between "class as one" and "class as many" (without the { }, so to speak), and asserted [1903: 76] "an ultimate distinction between a class as many and a class as one, to hold that the many are only many, and are not also one."

In an early chapter (X, "The Contradiction") discussing his paradox Russell decided that propositional functions, while defining classes as many, do not always define classes as one, else they could participate *qua* terms for self-predication as in the paradox. There he first proposed a resolution by resorting to a difference in type [1903: 104-5]:

> We took it to be axiomatic that the class as one is to be found wherever there is a class as many; but this axiom need not be

[21]The formal transition from $\{x \mid x \notin f(x)\}$ to $\{x \mid x \notin x\}$ was pointed out by Crossley [1973]. Though not so explicit in *The Principles*, the analogy is clearly drawn in 1905 letters from Russell to G.H. Hardy and to Philip Jourdain (as quoted in Grattan-Guinness [1978]).

> universally admitted, and appears to have been the source of the contradiction. ... A class as one, we shall say, is an object of the same *type* as its terms .... But the class as one does not always exist, and the class as many is of a different type from the terms of the class, even when the class has only one term ...

He consequently decided [1903: 106]: "that it is necessary to distinguish a single term from the class whose only member it is, and that consequently the null-class may be admitted."

In an appendix to the *Principles* devoted to Frege's work, Russell described an argument of Frege's showing that a should not be identified with $\{a\}$ (in the case of a having many members, $\{a\}$ would still have only one member), and wrote [1903: 513]: "...I contended that the argument was met by the distinction between the class as one and the class as many, but this contention now appears to me mistaken." He continued [1903: 514]: "...it must be clearly grasped that it is not only the collection as many, but the collection as one, that is distinct from the collection whose only term it is." Russell went on to conclude [1903: 515] that "the class as many is the only object that can play the part of a class", writing [1903: 516]:

> Thus a class of classes will be many many's; its constituents will each be only many, and cannot therefore in any sense, one might suppose, be single constituents. Now I find myself forced to maintain, in spite of the apparent logical difficulty, that this is precisely what is required for the assertion of number.

Russell was then led to infinitely many types [1903: 517]:

> It will now be necessary to distinguish (1) terms, (2) classes, (3) classes of classes, and so on *ad infinitum*; we shall have to hold that no member of one set is a member of any other set, and that $x \in u$ requires that x should be of a set of a degree lower by one than the set to which u belongs. Thus $x \in x$ will become a meaningless proposition; in this way the contradiction is avoided.

And he wrote further down the page:

> Thus, although we may identify the class with the numerical conjunction of its terms [class as many], wherever there are many terms, yet where there is only one term we shall have to accept Frege's range [Werthverlauf] as an object distinct from its only term.

Today, these shifting metaphysical distinctions concerning classes and worries focusing on the difference between a and $\{a\}$ may seem strange and convoluted. But for us logic is mathematical, and we are heir to the development of set theory based on the iterated application of the { } operation and axioms governing it. Of Russell's concerns and formulations, his theory of types has found technical uses in set theory.[22] But ultimately, the ontological question "What *is* a class?", like the ontological questions "What *is* a set?" and "What *is* a number?", has little bearing on mathematics and has not contributed substantially to its development.

2 Zermelo's Well-Ordering Theorem

The first decade of the new century saw Ernst Zermelo at Göttingen make his major advances in the development of set theory.[23]His first substantial result was his independent discovery of the argument for Russell's Paradox. He then established the Well-Ordering Theorem, provoking an open controversy about this initial use of the Axiom of Choice. After providing a second proof of the Well-Ordering Theorem in response, Zermelo also provided the first full-fledged axiomatization of set theory. In the process, he ushered in a new abstract, generative view of sets, one that would dominate in the years to come.

Zermelo's independent discovery of the argument for Russell's Paradox is substantiated in a note dated 16 April 1902 found in the *Nachlass* of the philosopher Edmund Husserl.[24]According to the note, Zermelo pointed out that any set M containing all of its subsets as members, i.e. with $\mathcal{P}(M) \subseteq M$, is "inconsistent" by considering $\{x \in M \mid x \notin x\}$. Schröder [1890: 245] had argued that Boole's "class 1" regarded as consisting of everything conceivable is inconsistent, and Husserl in a review [1891] had criticized Schröder's argument for not distinguishing between inclusion and membership. Zermelo was pointing out an inherent problem when inclusion *implies* membership as in the case of a universal class, but he did not push the argument in the direction of paradox as Russell had done. Also, Zermelo presumably came to his argument independently of Cantor's diagonal proof with functions. That $\mathcal{P}(M)$ has higher cardinality than M is evidently more central than $\mathcal{P}(M) \nsubseteq M$, but the connection between sub*sets* and characteristic *functions* was hardly appreciated then, and Zermelo was just making the first moves toward his abstract view of sets.[25]

---

[22]See Dreben-Kanamori [1997] for the line of development from Russell's theory of types to Gödel's constructible universe.

[23]Peckhaus [1990] provides a detailed account of Zermelo's years 1897-1910 at Göttingen.

[24]See Rang-Thomas [1981].

[25]In the earliest notes about axiomatization found in his *Nachlass*, written around 1905, Zermelo took the assertion $M \notin M$ as an axiom, as well as the assertion that any "well-defined" set M has a subset not a member of M (see Moore [1982: 155]). In Zermelo's axiomatization paper [1908a], the first result of his axiomatic theory was just the result in the Husserl note, that every set M has a subset $\{x \in M \mid x \notin x\}$ not a member of M, with the consequence that there is no universal set. Modern texts of set theory usually take the opposite tack, showing that there is no universal set by *reductio* to Russell's Paradox. Zermelo [1908a] applied his first result positively to generate specific sets disjoint from given sets for his recasting of Cantor's theory of cardinality.

Reversing Russell's progress from Cantor's correspondences to the identity map inclusion $\mathcal{P}(U) \subseteq U$, Zermelo considered functions $F\colon \mathcal{P}(X) \to X$, specifically in the form of *choice functions*, those F satisfying $F(Y) \in Y$ for $Y \neq \emptyset$. This of course was the basic ingredient in Zermelo's [1904] formulation of what he soon called the Axiom of Choice for the purpose of establishing his Well-Ordering Theorem. Russell the metaphysician had drawn elaborate philosophical distinctions and was forced by Cantor's diagonal argument into a dialectical confrontation with them, as well as with the concomitant issues of whether the null class exists and whether a term should be distinct from the class whose sole member is that term. Zermelo the mathematician never quibbled over these issues for sets and pushing the Cantorian extensional and operational view proceeded to resolve the problem of well-ordering sets mathematically. As noted in Footnote 3, in describing abstract functions Cantor had written [1895: §4]: "... by a '*covering* [Belegung] of N with M,' we understand a law ...", and thus had continued his frequent use of the term "law" to refer to functions. Zermelo [1904: 514] specifically used the term "covering", but with his choice functions any residual sense of "law" was abandoned by him [1904]: "... we take an arbitrary covering γ and derive from it a definite well-ordering of the elements of M." It is here that *abstract* set theory began.

That part of Zermelo's proof which does not depend on the Axiom of Choice can be isolated in the following result, the central articulation of our guiding motif. The result establishes a basic correlation between functions $F\colon \mathcal{P}(X) \to X$ and canonically defined well-orderings. For notational convenience, we take well-orderings to be strict, i.e. irreflexive, relations.

2.1 Theorem. *Suppose that* $F\colon \mathcal{P}(X) \to X$. *Then there is a unique* $\langle W, < \rangle$ *such that* $W \subseteq X$, $<$ *is a well-ordering of* W, *and:*

(a) *For every* $x \in W$, $F(\{y \in W \mid y < x\}) = x$, *and*

(b) $F(W) \in W$.

Remarks. The picture here is that F generates a well-ordering of W which according to (a) starts with

$$\begin{aligned} a_0 &= F(\emptyset)\,, \\ a_1 &= F(\{a_0\}) = F(\{F(\emptyset)\})\,, \\ a_2 &= F(\{a_0, a_1\}) = F(\{F(\emptyset), F(\{F(\emptyset)\})\}) \end{aligned}$$

and so continues as long as F applied to an initial segment of W constructed thus far produces a new element. W is the result when according to (b) an old element is again named. Note that if X is *transitive*, i.e. $X \subseteq \mathcal{P}(X)$, and F is the identity on at least the elements in the above display, then we are generating the first several von Neumann ordinals.[26] But as was much discussed in Section 1, F cannot be the identity on all of $\mathcal{P}(X)$. Whereas Russell's Paradox grew out of the insistence that inclusion implies membership, membership in a transitive set implies inclusion in that set. This later and positive embodiment of the inclusion vs. membership distinction became important in set theory after the work of John von Neumann [1923] on ordinals, and central to the subject since the work of Kurt Gödel [1938] on the constructible universe L.

The claim that 2.1 anticipates later developments is bolstered by its proof being essentially the argument for the Transfinite Recursion Theorem, the theorem that justifies definitions by recursion along well-orderings. This theorem was articulated and established by von Neumann [1923, 1928] in his system of set theory. However, the argument as such first appeared in Zermelo's [1904]:

Proof of 2.1. Call $Y \subseteq X$ an *F-set iff* there is a well-ordering R of Y such that for each $x \in Y$, $F(\{y \in Y \mid yRx\}) = x$. The following are thus F-sets (some of which may be the same):

$$\{F(\emptyset)\}; \quad \{F(\emptyset), F(\{F(\emptyset)\})\}; \quad \{F(\emptyset), F(\{F(\emptyset)\}), F(\{F(\emptyset), F(\{F(\emptyset)\})\})\}.$$

We shall establish:

(∗) If Y is an F-set with a witnessing well-ordering R and Z is an F-set with a witnessing well-ordering S, then $\langle Y, R\rangle$ is an initial segment of $\langle Z, S\rangle$, or conversely.

(Taking $Y = Z$ it will follow that any F-set has a unique witnessing well-ordering.)

For establishing (∗), we continue to follow Zermelo: By the comparability of well-orderings, we can assume without loss of generality that there is an order-preserving injection $e\colon Y \to Z$ with range an S-initial segment of Z. It then suffices to show that e is in fact the identity map: If not, let t be the R-least member of Y such that $e(t) \neq t$. It follows that $\{y \in Y \mid yRt\} = \{z \in Z \mid zSe(t)\}$. But then,

$$e(t) = F(\{z \in Z \mid zSe(t)\}) = F(\{y \in Y \mid yRt\}) = t\,,$$

a contradiction.

To conclude the proof, let W be the union of all the F-sets. Then W is itself an F-set by (∗) and so, with $<$ its witnessing well-ordering, satisfies (a). For (b), note that if $F(W) \notin W$, then $W \cup \{F(W)\}$ would be an F-set, contradicting the definition of W. Finally, that (a) and (b) uniquely specify $\langle W, <\rangle$ also follows from (∗). ⊣

[26] Notably, Zermelo in unpublished 1915 work sketched the rudiments of the von Neumann ordinals. See Hallett [1984: 278ff.].

Zermelo of course focused on choice functions as given by the Axiom of Choice to well-order the entire set:

2.2 Corollary (The Well-Ordering Theorem)(Zermelo [1904]). *If $\mathcal{P}(X)$ has a choice function, then X can be well-ordered.*

Proof. Suppose that $G\colon \mathcal{P}(X) \to X$ is a choice function, and define a function $F\colon \mathcal{P}(X) \to X$ to "choose from complements" by: $F(Y) = G(X - Y) \in X - Y$ for $Y \neq X$, and $F(X)$ some specified member of X. Then the resulting W of the theorem must be X itself. ⊣

It is noteworthy that 2.1 leads to a new proof and a positive form of Cantor's basic result that there is no bijection between $\mathcal{P}(X)$ and X:

2.3 Corollary. *For any $F\colon \mathcal{P}(X) \to X$, there are two distinct sets W and Y both definable from F such that $F(W) = F(Y)$.*

Proof. Let $\langle W, <\rangle$ be as in 2.1, and let $Y = \{x \in W \mid x < F(W)\}$. Then by 2.1(a) $F(Y) = F(W)$, yet $F(W) \in W - Y$. ⊣

This corollary provides a *definable* counterexample $\langle W, Y\rangle$ to injectivity. In the $F\colon \mathcal{P}(X) \to X$ version of Cantor's diagonal argument, one would consider the definable set

$$A = \{x \in X \mid \exists Z(x = F(Z) \ \wedge\ F(Z) \notin Z)\}.$$

By querying whether or not $F(A) \in A$, one deduces that there must be some $Y \neq A$ such that $F(Y) = F(A)$. However, no such Y is provided with a *definition.* This is also the main thrust of Boolos [1997], in which the argument for 2.1 is given *ab initio* and not connected with Zermelo [1904].

Another notable consequence of the argument for 2.1 is that since the F there need only operate on the *well-orderable* subsets of X, the $\mathcal{P}(X)$ in 2.3 can be replaced by the following set:

$$\mathcal{P}_{\mathrm{WO}}(X) = \{Z \subseteq X \mid Z \text{ is well-orderable}\}.$$

That this set, like $\mathcal{P}(X)$, is not bijective with X was first shown by Alfred Tarski [1939] through a less direct proof. Tarski [1939] (Theorem 3) did have a version of 2.1; substantially the same version appeared in the expository work of Nicolas Bourbaki [1956: 43] (chapter 3, §2, lemma 3).[27]

[27] Bourbaki's version is weighted in the direction of the application to the Well-Ordering Theorem (cf. 2.2). It supposes that for some $Z \subseteq \mathcal{P}(X)$, $F\colon Z \to X$ with $F(Y) \notin Y$ for every $Y \in Z$, and concludes that there is a $\langle W, <\rangle$ as in 2.1 except that its (b) is replaced by $W \notin Z$. From this version Bell [1995] developed a version in a many-sorted first-order logic and used it to recast Frege's work on the number concept.

Zermelo's main contribution with his Well-Ordering Theorem was the introduction of choice functions, leading to the postulation of the Axiom of Choice. But besides this, 2.1 brings out Zermelo's delineation of the power set as a sufficient domain of definition for generating well-orderings. Also, 2.1 rests on the argument for establishing the Transfinite Recursion Theorem; here however, it is the well-ordering itself that is being defined. The argument is avowedly impredicative: After specifying the collection of F-sets its union is taken to specify a member of the collection, namely the largest F-set. All these were significant advances, seminal for modern set theory, especially when seen against the backdrop of how well-orderability was being investigated at the time.

Cantor [1883: 550] had propounded the basic principle that every "well-defined" set can be well-ordered. However, he came to believe that this principle had to be established, and in 1899 correspondence with Dedekind gave a remarkable argument.[28] He first defined an "absolutely infinite or inconsistent multiplicity" as one into which the class Ω of all ordinal numbers can be injected and proposed that these collections be exactly the ones that are not sets. He then proceeded to argue that every *set* can be well-ordered through a presumably recursive procedure whereby a well-ordering is defined through successive choices. The set must get well-ordered, otherwise Ω would be injected into it. G.H. Hardy [1903] and Philip Jourdain [1904, 1905] also gave arguments involving the injection of all the ordinal numbers, but such an approach would only get codified at a later stage in the development of set theory in the work of von Neumann [1925].

Consonant with his observation on Schröder's inconsistent classes that no X can satisfy $\mathcal{P}(X) \subseteq X$, Zermelo's advance was to preclude the appeal to inconsistent multiplicities by shifting the weight away from Cantor's well-orderings with their *successive* choices to the use of functions on power sets making *simultaneous* choices. Zermelo, when editing Cantor's collected works, criticized him for his reliance on successive choices and the doubts raised by the possible intrusion of inconsistent multiplicities. Zermelo noted that "it is precisely doubts of this kind that impelled the editor [Zermelo] a few years later to base his own proof of the well-ordering theorem purely upon the axiom of choice without using inconsistent multiplicities."[29]

Cantor's realization that taking the class Ω of all ordinal numbers as a set is problematic was an early emanation of the now well-known Burali-Forti Paradox, generated *qua* paradox by Russell in *The Principles* [1903: 323] after reading Cesare Burali-Forti's [1897].[30] It is notable that Russell [1906: 35ff.]

[28]The 1899 correspondence appeared in Cantor [1932] and Noether-Cavaillès [1937] and, translated into French, in Cavaillès [1962]. The main letter is translated into English in van Heijenoort [1967: 113ff.].

[29]See Cantor [1932: 451] or van Heijenoort [1967: 117].

[30]Moore-Garciadiego [1978] and Garciadiego [1992] describe the evolution of the Burali-Forti paradox.

later provided a unified approach to both Russell's Paradox and the Burali-Forti Paradox that can be seen as reformulating the heart of the argument for 2.1 to yield a contradiction.[31] He considered the following schema for a property ϕ and a function f:

$$\forall u(\forall x(x \in u \to \phi(x)) \longrightarrow (\exists z(z = f(u)) \,\&\, f(u) \notin u \,\&\, \phi(f(u))))\,.$$

It follows that if $w = \{x \mid \phi(x)\}$ and w is in the variable range of $\forall u$, then both $\phi(f(w))$ and $\neg\phi(f(w))$, a contradiction. Russell's Paradox is the case of $\phi(x)$ being "$x \notin x$" and $f(x) = x$. The Burali-Forti Paradox is the case of $\phi(x)$ being "x is an ordinal number" and $f(x) =$ the least ordinal number greater than every ordinal number in x. Russell went on to describe how to define "a series ordinally similar to that of all ordinals" via Cantor's principles for generating the ordinal numbers:

Starting with a function f and an x such that $\exists z(z = f(x))$, the first term of the series is to be $f(x)$. Having recursively defined an initial segment u of the series and assuming $\exists z(z = f(u)) \,\&\, f(u) \notin u$, the next term of the series is to be $f(u)$. Thus, Russell was describing what corresponds to the defining property of the F-sets in the proof of 2.1 except for insisting that $f(u) \notin u$.

The above schema consequently implies that $w = \{x \mid \phi(x)\}$ contains a series similar to all the ordinal numbers. In particular, as Russell observed, $\{x \mid x \notin x\}$ contains a series similar to all the ordinal numbers, and "the series as a whole does not form a class."[32] Hence, Russell had interestingly correlated the structured 2.1 idea anew with the possibility of injecting all the ordinal numbers. Whereas 2.1 with its positing of a power set domain led to the positive conclusion that there is a well-ordered set W satisfying $F(W) \in W$, Russell's positing that $f(u) \notin u$ illuminated the paradoxes as necessarily generating series similar to all the ordinal numbers.

With its new approach via choice functions on power sets, Zermelo's [1904] proof of the Well-Ordering Theorem provoked considerable controversy,[33] and in response to his critics Zermelo published a second proof [1908] of his theorem. The general objections raised against Zermelo's first [1904] proof of the Well-Ordering Theorem had to do mainly with its exacerbation of a growing conflict among mathematicians about the use of arbitrary functions. But there were also specific objections raised about the possible role of ordinal numbers through rankings in the proof, and the possibility that again the class of all ordinal numbers might be lurking. To preclude these objections Zermelo in his second [1908] proof resorted to an approach with roots in Dedekind [1888].

[31] See also the discussion in Hallett [1984: 180-1]. While Russell [1906] discussed Zermelo [1904], it is unlikely that Russell made a conscious adaptation along the lines of 2.1.

[32] Russell attributed this observation to G.G. Berry of the Bodleian Library, well-known for Berry's Paradox, given in Russell [1906a: 645].

[33] See Moore [1982: Chapter 2].

Instead of initial segments of the desired well-ordering, Zermelo switched to final segments and proceeded to define the maximal reverse inclusion chain by taking an *intersection* in a larger setting:

To well-order a set M using a choice function φ on $\mathcal{P}(M)$, Zermelo defined a Θ-*chain* to be a collection Θ of subsets of M such that: (a) $M \in \Theta$; (b) if $A \in \Theta$, then $A - \{\varphi(A)\} \in \Theta$; and (c) if $Z \subseteq \Theta$, then $\bigcap Z \in \Theta$. He then took the intersection I of all Θ-chains, and observed that I is again a Θ-chain. Finally, he showed that I provides a well-ordering of M given by: $a \prec b$ *iff* there is an $A \in I$ such that $a \notin A$ and $b \in A$. Thus, I consists of the final segments of the desired well-ordering, and the construction is "dual" to the one provided by 2.1 and 2.2.

With the intersection approach no question could arise, presumably, about intrusions by classes deemed too large, such as the class of all ordinal numbers. While his first [1904] proof featured a (transfinite) recursive construction of a well-ordering, Zermelo in effect now took that well-ordering to be inclusion, the natural ordering for sets. He thus further emphasized the various set theoretic operations, particularly the power set. As set theory would develop, however, the original [1904] approach would come to be regarded as unproblematic and more direct, leading to incisive proofs of related results (see Section 3).

The main purpose of Zermelo's [1908a] axiomatization of set theory, the first full-scale such axiomatization, was to buttress his [1908] proof by making explicit its underlying set-existence assumptions.[34] The salient axioms were the generative Power Set and Union Axioms, the Axiom of Choice of course, and the Separation Axiom. These incidentally could just as well have been motivated by the first [1904] proof. With his axioms Zermelo advanced his new view of sets as structured solely by $\in$ and generated by simple operations, the Axiom of Infinity and the Power Set Axiom furnishing a sufficient setting for the set-theoretic reduction of all ongoing mathematics. In this respect, it is a testament to Zermelo's approach that the argument of 2.1 would, with the Axiom of Infinity in lieu of the Power Set Axiom furnishing the setting, become the standard one for establishing the Finite Recursion Theorem, the theorem for justifying definitions by recursion on the natural numbers.[35]

In his axiomatization paper [1908a] Zermelo provided a new proof of the

[34]Moore [1982: 155ff.] supports this contention using items from Zermelo's *Nachlass*.

[35]The Finite Recursion Theorem first appeared in the classic Dedekind [1888: §125], and his argument can be carried out rigorously in Zermelo's axiomatization of set existence principles. The theorem does not seem to appear in the work of Peano, who was mainly interested in developing an efficient symbolic system. Nor does the Finite Recursion Theorem always appear in the subsequent genetic accounts of the numbers starting with the natural numbers and proceeding through the rational numbers to the real numbers. A significantly late example where the theorem does not appear is Landau [1930], with its equivocating preface. We would now say that without the Finite Recursion Theorem the arithmetical properties of the natural numbers would remain at best schemas, inadequate for a rigorous definition of the rational and real numbers.

Schröder-Bernstein Theorem, and it is noteworthy that themes of Kuratowski [1922], to be discussed in Section 3, were foreshadowed by the proof. Zermelo focused on the following formulation: *If* $M' \subseteq M_1 \subseteq M$ *and there is a bijection* $g\colon M \to M'$, *then there is a bijection* $h\colon M \to M_1$. The following is his proof in brief, where $\subset$ denotes *proper* inclusion:

For $A \subseteq M$, define $f(A) = (M_1 - M') \cup \{g(x) \mid x \in A\}$. Since g is injective, f is monotonic in the following sense: if $A \subset B \subseteq M$, then $f(A) \subset f(B)$. Set $T = \{A \subseteq M \mid f(A) \subseteq A\}$, and noting that T is not empty since $M \in T$, let $A_0 = \bigcap T$. Then $A_0 \in T$ (and this step makes the proof impredicative, like Zermelo's argument used for 2.1). Moreover, $f(A_0) \subset A_0$ would imply by the monotonicity of f that $f(f(A_0)) \subset f(A_0) \subset A_0$, contradicting the definition of A_0. Consequently, we must have $f(A_0) = A_0$. It is now straightforward to see that $M_1 = A_0 \cup (M' - \{g(x) \mid x \in A_0\})$, a disjoint union, and so if $h\colon M \to M_1$ is defined by $h(x) = x$ if $x \in A_0$ and $h(x) = g(x)$ otherwise, then h is a bijection.

Zermelo himself did not define the function f explicitly, but he did define T and $A_0 = \bigcap T$, and his argument turned on A_0 being a "fixed point of f", i.e. $f(A_0) = A_0$. This anticipated the formulations of Kuratowski [1922], as we shall see. Another connection is to the general aim of that paper, to avoid numbers and recursion, which Zermelo did for a specific mathematical purpose:

The first correct proof of the Schröder-Bernstein Theorem to appear in print was due to Felix Bernstein and appeared in Borel [1898: 104-6]. However, like proofs often given today Bernstein's proof depended on defining a countable sequence of functions by recursion. One of Henri Poincaré's criticisms of the logicists[36] was that "logical" developments of the natural numbers and their arithmetic inevitably presuppose the natural numbers and mathematical induction, and in connection with this Poincaré [1905: 24ff.] pointed out the circularity of developing the theory of cardinality with the Schröder-Bernstein Theorem based on Bernstein's proof, and therefore on the natural numbers. This point had *mathematical* weight, and in 1906 Zermelo sent his new proof of Schröder-Bernstein to Poincaré. Zermelo in a footnote in [1908a: 272-3] emphasized how his proof avoids numbers and induction altogether. He also observed that the proof "rests solely upon Dedekind's chain theory [1888: IV]", and that Peano [1906] published a proof that was "quite similar".[37] Russell on first reading Zermelo [1908a] expressed delight with his proof of Schröder-Bernstein

[36]See Goldfarb [1988] for more about Poincaré against the logicists.

[37]Zermelo's footnote is footnote 11 of van Heijenoort [1967: 209]. The reference to Dedekind [1888] raises two points: First, Dedekind [1888] had arguments that can also be construed as getting fixed points by taking intersections, but the fixed points were always (isomorphic to) the set of natural numbers; Zermelo's proof of Schröder-Bernstein used the fixed point idea to get a mathematical result, and so can be regarded as mathematically midway between Dedekind [1888] and the explicit fixed point formulations of Kuratowski [1922], to be discussed below. Second, it turns out that Dedekind in fact had a proof like Zermelo's of Schröder-Bernstein already in 1887, but the proof only appeared in 1932, both in a manuscript

but went on to criticize his axiomatization of set theory.[38] In the first volume of Whitehead and Russell's *Principia Mathematica* [1910-13] there was no formal use of the class of natural numbers, and indeed the Axiom of Infinity was avoided; while this would not satisfy Poincaré, the theory of cardinals was developed using Zermelo's proof.[39]

The Zermelian abstract generative view of sets as set forth by his [1908a] axiomatization would become generally accepted by the mid-1930's, the process completed by adjunction of the Axioms of Replacement and Foundation and the formalization of the axiomatization in first-order logic. But as with the Well-Ordering Theorem itself, the [1908a] axiomatization from early on served to ground the investigation of well-orderings, with the incisive result of Friedrich Hartogs [1915] on the comparability of cardinals being a prominent example. The early work also led to a new transformation of our motif.

3 Fixed Point Theorems

Kazimierz Kuratowski [1922] provided a fixed point theorem which can be seen as a refocusing of 2.1, and thereby recast our guiding motif. Fixed point theorems assert, for functions $f\colon X \to X$ satisfying various conditions, the existence of a *fixed point of* f, i.e. an $x \in X$ such that $f(x) = x$. With solutions to equations becoming construed as fixed points of iterative procedures, a wide-ranging theory of fixed points has emerged with applications in analysis and topology.[40] Of the pioneering work, fixed points figured crucially in Poincaré's classic analysis [1884] of the three-body problem. The best-known fixed point theorem in mathematics, due to L.E.J. Brouwer [1909], is fundamental to al-

---

of 11 July 1887 appearing in Dedekind's collected works [1932: 447-9] and in a letter of 29 August 1899 from Dedekind to Cantor appearing in Cantor's collected works [1932: 449]. As editor for the latter, Zermelo noted in a footnote that Dedekind's proof is "not essentially different" from that appearing in Zermelo [1908a]; while we might today regard this to be the case, the fixed point idea is much less evident in Dedekind's formulation.

The reference to Peano [1906] is part of a contretemps involving Poincaré. Poincaré [1906: 314-5] published Zermelo's proof, but then proceeded to make it part of his criticism of Zermelo's work based on the use of impredicative notions, the main front of Poincaré's critique of the logicists. Zermelo in a footnote to his [1908: 118] (footnote 8 of van Heijenoort [1967: 191]) expressed annoyance that Peano when referring to Poincaré [1906] only mentioned Peano [1906], not Zermelo, in connection with the new proof of Schröder-Bernstein but went on the argue against Zermelo's use of the Axiom of Choice.

[38] Russell's letter to Jourdain of 15 March 1908 (see Grattan-Guinness [1977: 109]) began: "I have only read Zermelo's article once as yet, and not carefully, except his new proof of Schröder-Bernstein, which delighted me." Russell then criticized the Axiom of Separation as being "so vague as to be useless". For Russell, the paradoxes cannot be avoided in this way but had to be solved through his theory of types.

[39] The Schröder-Bernstein Theorem in *Principia* is ∗73-88. In ∗94 the Bernstein and Zermelo proofs are compared.

[40] See for example the account Dugundji-Granas [1982].

gebraic topology.[41] And a few pages after Kuratowski's [1922] in the same journal, Stefan Banach [1922] published work from his thesis including what has become another basic fixed point theorem, a result Banach used to provide solutions for integral equations.

Kuratowski's overall purpose in [1922] was heralded by its title, "A method for the elimination of the transfinite numbers from mathematical reasoning". At a time when Zermelo's abstract generative view of sets had shifted the focus away from Cantor's transfinite numbers but the von Neumann ordinals had not yet been incorporated into set theory, Kuratowski provided an approach for replacing definitions by transfinite recursion on ordinal numbers by a set-theoretic procedure carried out within Zermelo's [1908a] axiomatization. Kuratowski's fixed point theorem served as a basis for that approach and can be viewed as a corollary of 2.1:

For a function f and set X, $f``X = \{f(x) \mid x \in X\}$ is the image of X under f. Again to affirm, $\subset$ is *proper* inclusion.

3.1 Theorem (Kuratowski's Fixed Point Theorem [1922: 83]). *Suppose that $X \subseteq \mathcal{P}(E)$ for some E, and whenever $C \subseteq X$, $\bigcup C \in X$. Suppose also that $f\colon X \to X$ satisfies $x \subseteq f(x)$ for every $x \in X$. Then there is a fixed point of f. In fact, there is a unique $W \subseteq X$ well-ordered by $\subset$ satisfying:*

(a) For every $x \in W$, $x = \bigcup f``\{y \in W \mid y \subset x\}$, and

(b) W contains exactly one fixed point of f, namely $\bigcup f``W$.

Proof. Adapting the proof of 2.1, call $Y \subseteq X$ an *f-set iff* Y is well-ordered by $\subset$, and for each $x \in Y$, $\bigcup f``\{y \in Y \mid y \subset x\} = x$. This last condition devolves to two cases: either x has an immediate $\subset$-predecessor y in which case $x = f(y)$, or else x has no immediate $\subset$-predecessor and so inductively $x = \bigcup\{y \in Y \mid y \subset x\}$.

As in the proof of 2.1, the union W of all the f-sets is again an f-set and so satisfies 3.1(a). Corresponding to 2.1(b) we have $\bigcup f``W \in W$, and so $\bigcup f``W$ is the $\subset$-maximum element of W. Next, if a $y \in W$ has an $\subset$-successor in W, then by previous remarks about f-sets, $f(y)$ is the immediate $\subset$-successor of y, and so y cannot be a fixed point. Also, the $\subset$-maximum element $\bigcup f``W$ of $W$ must be a fixed point of $f$, else $W \cup \{f(\bigcup f``W)\}$ would have been an f-set, contradicting the definition of W. Hence, we have 3.1(b). Finally, the uniqueness of W follows as in 2.1. ⊣

This theorem could also have been established by applying 2.1 directly to $F\colon \mathcal{P}(X) \to X$ given by: $F(Y) = \bigcup(f``Y)$. For the resulting $\langle W, < \rangle$ it follows

[41]The Brouwer fixed point theorem asserts that any continuous function from a closed simplex into itself must have a fixed point; to illustrate without bothering to define these terms, any continuous function on a closed triangle into itself must have a fixed point.

by induction along $<$ that $<$ and $\subset$ coincide on W. With that, 3.1(b) can be verified for W. However, as our proof emphasizes, 3.1 is based on an underlying ordering, namely $\subseteq$, which can be used directly.

Kuratowski's 3.1 strictly speaking involves two levels: It is about members of a set X, but those members are also subsets of a fixed set E and so are naturally ordered by $\subseteq$. 2.1 had also featured an interplay between levels: subsets of a set (corresponding to E in 3.1) and its members. However, with maps $f: X \to X$ satisfying $x \subseteq f(x)$ and their fixed points Kuratowski refocused the setting to a single level of sets mediated by $\subseteq$.

Kuratowski actually dealt more with a dual form of 3.1 resulting from replacing "$\bigcup$" by "$\bigcap$"; "$\subset$" by "$\supset$"; and "$x \subseteq f(x)$" by "$f(x) \subseteq x$". His argument for this dual form generalized Zermelo's second [1908] proof of the Well-Ordering Theorem, which together with related work of Gerhard Hessenberg [1909] Kuratowski acknowledged.[42] In terms of the Θ-chains of Zermelo [1908] as described toward the end of Section 2, the generalization corresponds to replacing the condition (b) "if $A \in \Theta$, then $A - \{\varphi(A)\} \in \Theta$" where φ is a choice function by "if $x \in \Theta$, then $f(x) \in \Theta$". Indeed, Kuratowski's first application was to derive Zermelo's Well-Ordering Theorem as a special case.

Kuratowski went on to use 3.1 and its dual form to carry out the "elimination of the transfinite numbers" from various arguments, especially those in descriptive set theory that had depended on explicit transfinite recursions. Most notably, Kuratowski established with the Axiom of Choice the following proposition:

> Let $A \subseteq E$ be sets, $\mathcal{R}$ a property, and $Z = \{X \subseteq E \mid A \subseteq X \;\&\; X$ has property $\mathcal{R}\}$. Suppose that for every $C \subseteq Z$ well-ordered by $\subseteq$, $\bigcup C \in Z$. Then there is a $\subseteq$-maximal member of Z.

This proposition can be seen as a version of Zorn's Lemma as originally formulated by Max Zorn [1935]:

> Suppose that Z is a collection of sets such that for every $C \subseteq Z$ linearly ordered by $\subseteq$, $\bigcup C \in Z$. Then there is an $\subseteq$-maximal member of Z.

Although Kuratowski's proposition is ostensibly stronger than Zorn's Lemma in that it only required $\bigcup C \in Z$ for *well-ordered* C, it is equivalent by the proof of 3.3 below.[43] Zorn's Lemma is more generally accessible in that its statement does not mention well-orderings.

[42] The idea of rendering well-orderings in set theory in terms of $\supseteq$ occurred in Hessenberg [1906] and was pursued by Kuratowski [1921]. See Hallett [1984: 256ff.] for an analysis of Zermelo's [1908] proof in this light.

[43] The first to provide a proposition similarly related to Zorn's Lemma was Felix Hausdorff [1909: 300]. See Campbell [1978] and Moore [1982: 220ff.] for more on Zorn's Lemma and related propositions.

Set theory was veritably transformed in the decades following Kuratowski's work in ways not related to 3.1, but our motif emerged again in connection with Zorn's Lemma. In a now familiar generalization in terms of partially ordered sets Zorn's Lemma soon made its way into the expository work of Nicolas Bourbaki, first appearing in his summary of results in set theory [1939: 37] (§6, item 10) for his series *Eléments de Mathématique.* There Bourbaki in fact formulated a partial order generalization of 3.1 which he called the "fundamental lemma". To set the stage, we first develop some terminology:

Suppose that $\langle P, \leq_P \rangle$ is a partially ordered set, i.e. $\leq_P$ is a reflexive, transitive, and anti-symmetric relation on P. $<_P$ is the strict order derived from $\leq_P$. $C \subseteq P$ is a *chain iff* C is linearly ordered by $\leq_P$. For $A \subseteq P$, $\sup_P(A)$ is the least upper bound of A with respect to $\leq_P$, assuming that it exists.

$$\langle P, \leq_P \rangle \text{ is } \textit{inductive iff} \text{ for every chain } C \subseteq P, \sup_P(C) \text{ exists.}$$

Note that every inductive partially ordered set $\langle P, \leq_P \rangle$ has a $\leq_P$-least element, namely $\sup_P(\emptyset)$. Finally, a function $f\colon P \to P$ is *expansive iff* for any $x \in P$, $x \leq_P f(x)$.

3.2 Theorem (Bourbaki's Fixed Point Theorem [1939: 37]). *Suppose* $\langle P, \leq_P \rangle$ *is an inductive partially ordered set, and* $f\colon P \to P$ *is expansive. Then there is a fixed point of* f. *In fact, there is a unique* $W \subseteq P$ *well-ordered by* $<_P$ *satisfying:*

(a) For every $x \in W$, $x = \sup_P(f``\{y \in W \mid y <_P x\})$, *and*

(b) W *contains exactly one fixed point of* f, *namely* $\sup_P(f``W)$.

Proof. The proof is essentially the same as for 3.1, with $<_P$ replacing $\subset$ and $\sup_P$ replacing $\bigcup$. ⊣

This theorem completed the transformation of our guiding motif begun by Kuratowski's 3.1. With an abstract formulation in terms of partial orders, the two levels of 3.1 (E and $X \subseteq \mathcal{P}(E)$) were left behind. Stressing the innovation of bringing in the fixed point idea, Bourbaki wrote a paper [1949/50][44] giving a proof of 3.2 and showing moreover how both Zermelo's Well-Ordering Theorem and Zorn's Lemma (in Bourbaki's partial order version) follow from 3.2. Nevertheless, we might today regard this work as a straightforward generalization of Kuratowski [1922], even to the extent that Bourbaki's proof of 3.2 was along the lines of Zermelo's second [1908] proof of the Well-Ordering Theorem. In later editions of Bourbaki's 1939 summary, 3.2 is intriguingly deleted, surfacing only as an exercise in his full treatment of set theory [1956: 49] (chapter 3, §2, exercise 6). Interestingly, the exercise gives a formulation in the style of Zermelo's second [1908] proof of the Well-Ordering Theorem, but then suggests a

[44] In [1949/50] Bourbaki referred to himself as "à Nancago".

use of a lemma (chapter 3, §2, lemma 3 and a version of 2.1) along the lines of Zermelo's *first* [1904] proof.

Referring to Bourbaki's [1939] "fundamental lemma" and Zermelo's two proofs for the Well-Ordering Theorem, Helmuth Kneser [1950], a student of David Hilbert, provided a proof of 3.2 much as given above, i.e. in the style of Zermelo's first [1904] proof. Kneser also pointed out that the partially ordered set need not be quite inductive; it suffices to have a function g that chooses for each chain $C \subseteq P$ an upper bound $g(C)$.[45] This was part of Kneser's observation that 3.2 provides a straightforward means to prove (a partial order version of) Zorn's Lemma from the Axiom of Choice. However, Szele [1950] and Weston [1957] pointed out that a more direct proof is possible; essentially, the proof of 2.1 can be adapted, and we give such an argument:

3.3 Theorem (AC). *Suppose that $\langle P, \leq_P \rangle$ is a partially ordered set such that every chain $C \subseteq P$ has a $\leq_P$-upper bound. Then P has a $\leq_P$-maximal element.*

Proof. Fix $x_0 \in P$. Using the Axiom of Choice, let $g\colon \mathcal{P}(P) \to P$ satisfy $g(C) = x_0$, unless C is a chain with $\leq_P$-least element x_0 and a $\leq_P$-upper bound in $P - C$, in which case $g(C)$ is such an upper bound. Adapting the proof of 2.1, call $Y \subseteq P$ a *g-set iff* Y is well-ordered by $<_P$, has x_0 as its $\leq_P$-least member, and for each $x \in Y$, $g(\{y \in Y \mid y <_P x\}) = x$. Then as in the proof of 2.1, the union W of all the g-sets is again a g-set, and $g(W) = x_0$. This last implies that W has a $\leq_P$-maximum element which then is also a $\leq_P$-maximal element of P. ⊣

This short proof with its interplay between $\mathcal{P}(P)$ and P harkens back to 2.1 again. Thus, our motif has come full circle after progressing from 2.1 (sets and members) through 3.1 (sets and inclusion) to 3.2 (partial orders).

The following version of Zorn's Lemma follows analogously:

(†) Suppose that $\langle P, \leq_P \rangle$ is an inductive partially ordered set. Then P has a $\leq_P$-maximal element.

It is now well-known that this version also implies 3.3: Given a $\langle P, \leq_P \rangle$ as in 3.3, let Q be the set of chains of P. Then $\langle Q, \subseteq \rangle$ is an *inductive* partially ordered set. Hence, by (†), Q has an $\subseteq$-maximal element M. But then, any $\leq_P$-upper bound of M must be a $\leq_P$-maximal element.

In a new chapter of the seventh edition of Bartel van der Waerden's classic *Algebra* [1966: 206ff.], 3.2 is established and from it both Zorn's Lemma in the

[45] Bourbaki [1939] defined "inductive" partially ordered sets as we have done. However, with his penchant for generalization and perhaps influenced by Kneser [1950], Bourbaki in the later editions of his full treatment of set theory weakened the definition of "inductive" to chains just having upper bounds, not necessarily least upper bounds. This caused an ambiguity, for the hint to their aforementioned exercise (chapter 3, §2, exercise 6) no longer works unless a function g choosing upper bounds for chains is given beforehand.

(†) version and Zermelo's Well-Ordering Theorem are derived. Van der Waerden acknowledged following Kneser [1950], and according to the foreword this was one of the main changes from the previous edition. In Serge Lang's popular *Algebra* [1971], 3.2 is established with the proof of Bourbaki [1949/50], and from 3.2 both the (†) and 3.3 versions of Zorn's Lemma are derived. Notably, with the argument so structured Lang did not even point out the essential use of the Axiom of Choice.[46]

Moving forward to the present, we see that our motif has spread beyond set theory in several guises and in new roles. In general, the existence of a fixed point as in 3.2 underlies all modern theories of inductive definitions.[47] With the rise of computer science such theories have gained a wide currency, and a variant of 3.2 has become particularly pertinent.

Suppose $\langle P, \leq_P \rangle$ is a partially ordered set and $f\colon P \to P$. f is *monotonic iff* for any $x \leq_P y \in P$, $f(x) \leq_P f(y)$. There can be expansive maps which are not monotonic, and monotonic maps which are not expansive. w is a *least fixed point of* f *(with respect to* $\leq_P$*) iff* w is a fixed point of f, and whenever $f(y) \leq_P y$, $w \leq_P y$. Clearly there is at most one least fixed point of f.

3.4 Theorem. *Suppose that* $\langle P, \leq_P \rangle$ *is an inductive partially ordered set and* $f\colon P \to P$ *is monotonic. Then there is a least fixed point of* f. *In fact, there is a* $W \subseteq P$ *well-ordered by* $<_P$ *satisfying:*

(a) For every $x \in W$, $x = \sup_P(f``\{y \in W \mid y <_P x\})$, *and*

(b) $\sup_P(f``W)$ *is the least fixed point of* f.

Proof. The proof is essentially the same as for 3.1 and 3.2.[48] ⊣

3.4 can also be seen as a corollary of 3.2: From the hypotheses of 3.4 it follows by straightforward arguments that

$$Q = \{x \in P \mid x \leq_P f(x) \ \&\ \forall y(f(y) \leq_P y \to x \leq_P y)\}$$

ordered by $\leq_P$ is an inductive partially ordered subset (with the same suprema of chains as P) such that $f``Q \subseteq Q$. But since f is expansive on Q, the conclusions of 3.4 follow from those of 3.2.

[46]Earlier in his text Lang [1971: 507] had written: "We show how one can prove Zorn's Lemma from other properties of sets which everyone would immediately grant as acceptable psychologically." In the argument corresponding to 3.3, Lang [1971: 510] then blithely wrote: "Suppose A does not have a maximal element. Then for each $x \in A$ there exists an element $y_x \in A$ such that $x < y_x$." It is as if the mighty struggles of the past never took place!

[47]See Aczel [1977] for a survey of the theory of inductive definitions in extensions of recursion theory and Moschovakis [1974] for a detailed development (for the "positive" case) in abstract structures.

[48]Kuratowski [1922: 83] in the context of sets and inclusion had in fact considered monotonic functions leading to least fixed points, but with the standing assumption of expansiveness for functions.

In the early 1970's Dana Scott and Christopher Strachey developed the now standard denotational semantics for programming languages in terms of algebraic structures, devised by Scott, called continuous lattices.[49] The theory was then generalized to what has come to be called complete partially ordered sets (CPO's), but what we called inductive partially ordered sets. In the language of CPO's a weak form of 3.4, where the function f in the hypothesis is required to satisfy a continuity condition subsuming monotonicity, has become a cornerstone of denotational semantics. This weak form is simple to establish because of the imposed continuity condition and suffices for theories of computation. As Scott himself pointed out, this weak form is in fact a "semantic" version of the First Recursion Theorem of Kleene [1952: 348] on least fixed point, recursive solutions for recursive functionals.[50]

Although their full strength is not required, stronger fixed point theorems have nonetheless come to be cited in computer science, perhaps for historical contextualization or in anticipation of mathematical generalizations. One such theorem was established by Alfred Tarski in 1939 (see his [1955: 286]) and is the version of 3.4 for complete lattices (i.e. partially ordered sets in which every subset, not only chains, has a least upper bound, and hence, it can be shown, every subset also has a greatest lower bound). The version of Tarski's result corresponding to 3.1, i.e. for sets and inclusion, had been established earlier by Bronisław Knaster [1928]. But in these results the existence of a well-ordered chain is not necessary for the proofs. On the other hand, well-orderings are intrinsic to the proofs of 3.2 and 3.4 as is made explicit in their (a)'s and (b)'s; in the general set-theoretic context transfinite well-orderings are necessarily involved. Nonetheless, the fixed point results of 3.2 and 3.4 themselves have come to be cited in computer science, even though transfinite well-orderings are only relevant in the higher reaches of abstract theories of computation.[51]

---

[49]See Stoy [1977] for an authoritative account of Scott-Strachey denotational semantics. The algebraic theory of continuous lattices, with motivations from a variety of quarters, has since been considerably elaborated; see Gierz *et al.* [1980].

[50]See Scott [1975] for the interplay of denotational semantics and recursion theory. The (Second) Recursion Theorem of Kleene [1938][1952: 352ff.] is a fixed point theorem much deeper than weak versions of 3.4 in that it provides fixed points for (partial) recursive procedures which even depend on their numerical codes and as such is a central result of Recursion Theory. The theorem is a generalization of the core of Gödel's proof of his Incompleteness Theorem.

[51]Surveying recent treatments of 3.2 and 3.4 by those emphasizing the significance of these results in computer science, the text Davey-Priestley [1990: 94ff.] establishes 3.2 in the style of Zermelo's second [1908] proof of the Well-Ordering Theorem, presumably following Bourbaki [1949/50] though without mentioning well-orderings, and derives 3.4 from 3.2 essentially as described in the paragraph above following 3.4. Moschovakis [1994: 108] points out the same route to get to 3.4 from 3.2, but before that establishes 3.2 (and notes that the proof also gives 3.4) after first establishing Hartogs's Theorem and then applying it. More heavyhandedly, Phillips [1992: §3] establishes 3.4 only after developing the elementary theory of ordinals and implicitly using the Axiom of Replacement.

Today, of course, Zorn's Lemma provides the most general setting for algebra, although in various constructive contexts a closer examination shows that special cases of the lemma suffice. Analogously, our guiding motif as embodied in 3.2 and 3.4 is coming to play a background role for computer science.

References

1977 PETER ACZEL, An introduction to inductive definitions, in: *Handbook of Mathematical Logic*, K. Jon Barwise (editor), North-Holland, Amsterdam, pp. 739–782.

1922 STEFAN BANACH, Sur les opérations dans les ensembles abstraits et leur application aux équations intégrales, *Fundamenta Mathematicae*, vol. 3, pp. 133–181.

1995 JOHN L. BELL, Type reducing correspondences and well-orderings: Frege's and Zermelo's constructions re-examined, *The Journal of Symbolic Logic*, vol. 60, pp. 209–221.

1954 PAUL BERNAYS, A system of axiomatic set theory VII, *The Journal of Symbolic Logic*, vol. 19, pp. 81–96.

1997 GEORGE BOOLOS, Constructing Cantorian counterexamples, *Journal of Philosophical Logic*, vol. 26, pp. 237-239.

1898 EMILE BOREL, *Leçons sur la théorie des fonctions*, Paris, Gauthier-Villars.

1939 NICOLAS BOURBAKI, *Eléments de Mathématique. I. Théorie des Ensembles. Fascicule de Résultats*; Actualités Scientifiques et Industrielles #846, Hermann, Paris.

1949/50 NICOLAS BOURBAKI, Sur le théoreme de Zorn, *Archiv der Mathematik*, vol. 2, pp. 434–437.

1956 NICOLAS BOURBAKI, *Eléments de Mathématique. I. Théorie des Ensembles. Chapter III: Ensembles ordonnés, cardinaux, nombres entiers*, Actualités Scientifiques et Industrielles #1243, Hermann, Paris.

1909 LUITZEN E.J. BROUWER, On continuous one–one transformations of surfaces into themselves, *Koninklijke Nederlandse Akademie van Wetenschappen te Amsterdam, Proceedings*, vol. 11, pp. 788–798; reprinted in [1976] below, 195–206.

1976 LUITZEN E.J. BROUWER, *Collected Works*, Vol. 2, Freudenthal, Hans (editor), North-Holland, Amsterdam.

1897 CESARE BURALI-FORTI, Una questione sui numeri transfini. *Rendiconti del Circolo Matematico di Palermo*, vol. 11, pp. 154–164; translated in van Heijenoort [1967], 104–111.

1978 PAUL J. CAMPBELL, The origin of "Zorn's Lemma", *Historia Mathematica*, vol. 5, pp. 77–89.

1874 GEORG CANTOR, Über eine Eigenschaft des Inbegriffes aller reellen algebraischen Zahlen, *Journal für die reine und angewandte Mathematik* (Crelle's Journal), vol. 77, pp. 258–262; reprinted in [1932] below, pp. 115–118.

1883 GEORG CANTOR, Über unendliche, lineare Punktmannigfaltigkeiten. V. *Mathematische Annalen*, vol. 21, pp. 545–591; published separately as *Grundlagen einer allgemeinen Mannigfaltigkeitslehre. Ein mathematisch-philosophischer Versuch in der Lehre des Unendlichen*, B.G. Teubner, Leipzig 1883; Reprinted in [1932] below, 165–209.

1891 GEORG CANTOR, Über eine elementare Frage der Mannigfaltigkeitslehre, *Jahresbericht der Deutschen Mathematiker-Vereinigung*, vol. 1, pp. 75–78. Reprinted in [1932] below, pp. 278–280.

1895 GEORG CANTOR, Beiträge zur Begründung der transfiniten Mengenlehre. I. *Mathematische Annalen*, vol. 46, pp. 481–512; translated in [1915] below; reprinted in [1932] below, pp. 282–311.

1897 GEORG CANTOR, Beiträge zur Begründung der transfiniten Mengenlehre. II. *Mathematische Annalen* vol. 49, pp. 207–246; translated in [1915] below; reprinted in [1932] below, pp. 312–351.

1915 GEORG CANTOR, *Contributions to the Founding of the Theory of Transfinite Numbers*, including translations of [1895] and [1897] above with introduction and notes by Philip E.B. Jourdain, Open Court, Chicago; reprinted Dover, New York 1965.

1932 GEORG CANTOR, *Gesammelte Abhandlungen mathematicschen und philosophischen Inhalts*, Ernst Zermelo (editor), Julius Springer, Berlin; reprinted by Springer-Verlag, Berlin 1980.

1962 JEAN CAVAILLÈS, *Philosophie mathématique*, Hermann, Paris; includes French translation of Noether-Cavaillès [1937].

1899 LOUIS COUTURAT, La logique mathématique de M. Peano, *Revue de Métaphysique et de Morale*, vol. 7, pp. 616–646.

1973 JOHN N. CROSSLEY, A note on Cantor's theorem and Russell's paradox, *Australasian Journal of Philosophy*, vol. 51, pp. 70-71.

1990 BRIAN A. DAVEY and HILARY A. PRIESTLEY, *Introduction to Lattices and Order*, Cambridge University Press, Cambridge.

1888 RICHARD DEDEKIND, *Was sind und was sollen die Zahlen?* F. Vieweg, Braunschweig; sixth, 1930 edition reprinted in [1932] below, pp. 335–390; second, 1893 edition translated in [1963] below, pp. 29–115.

1932 RICHARD DEDEKIND, *Gesammelte mathematische Werke*, vol. 3, Fricke, Robert, Emmy Noether, and Öystein Ore (editors), F. Vieweg, Braunschweig; reprinted Chelsea, New York 1969.

1963 RICHARD DEDEKIND, *Essays on the Theory of Numbers*, translations by Wooster W. Beman, Dover, New York (reprint of original edition, Open Court, Chicago 1901).

1997 BURTON DREBEN and AKIHIRO KANAMORI, Hilbert and set theory, *Synthese*, vol. 110, pp. 77–125.

1982 JAMES DUGUNDJI and ANDRZEJ GRANAS, *Fixed Point Theory*, Państwowe Wydawnictwo Naukowe, Warsaw.

1995 THOMAS E. FORSTER, *Set Theory with a Universal Set*, Logic Oxford Guides #31, Clarendon Press, Oxford; second edition.

1879 GOTTLOB FREGE, *Begriffsschrift, eine der arithmetischen nachgëbildete Formelsprache des reinen Denkens*, Nebert, Halle; reprinted Hildesheim, Olms 1964; translated in van Heijenoort [1967], pp. 1–82.

1893 GOTTLOB FREGE, *Grundgesetze der Arithmetik, Begriffsschriftlich abgeleitet*, Vol. 1, Hermann Pohle, Jena; reprinted Olms, Hildesheim 1962.

1895 GOTTLOB FREGE, Kritische Beleuchtung einiger Punkte in E. Schröders Vorlesungen über die Algebra der Logik, *Archiv für systematische Philosophie*, vol. 1, pp. 433–456; translated in [1952] below, pp. 86–106.

1952 GOTTLOB FREGE, *Translations from the Philosophical Writings of Gottlob Frege*, Geach, Peter, and Max Black (translators and editors), Blackwell, Oxford 1952; second, revised edition 1960; latest edition, Rowland & Littlewood, Totowa 1980.

1992 ALEJANDRO R. GARCIADIEGO, *Bertrand Russell and the Origins of the Set-theoretic "Paradoxes"*, Birkhäuser, Boston.

1980 GERHARD GIERZ, KARL H. HOFMANN, KLAUS KEIMEL, JIMMIE D. LAWSON, MICHAEL MISLOVE, and DANA S. SCOTT, *A Compendium of Continuous Lattices*, Springer-Verlag, Berlin.

1938 KURT F. GÖDEL, The consistency of the Axiom of Choice and of the Generalized Continuum-Hypothesis, *Proceedings of the National Academy of Sciences U.S.A.*, vol. 24, pp. 556–557; reprinted in [1990 below, pp. 26–27.

1990 KURT F. GÖDEL, *Collected Works*, Vol. 2, Feferman, Solomon, *et al.* (editors), Oxford University Press, New York.

1988 WARREN GOLDFARB, Poincaré against the logicists, in: Aspray, William, and Philip Kitcher (editors) *History and Philosophy of Modern Mathematics*, Minnesota Studies in the Philosophy of Science vol. 11, University of Minnesota Press 1988, Minneapolis, pp. 61–81.

1977 IVOR GRATTAN-GUINNESS, *Dear Russell – Dear Jourdain*, Duckworth & Co., London, and Columbia University Press, New York.

1978 IVOR GRATTAN-GUINNESS, How Bertrand Russell discovered his paradox, *Historia Mathematica*, vol. 5, pp. 127–137.

1994 ROBERT GRAY, Georg Cantor and Transcendental Numbers, *American Mathematical Monthly*, vol. 101, pp. 819–832.

1984 MICHAEL HALLETT, *Cantorian Set Theory and Limitation of Size*, Logic Guides #10, Clarendon Press, Oxford.

1903 GODFREY H. HARDY, A theorem concerning the infinite cardinal numbers, *The Quarterly Journal of Pure and Applied Mathematics*, vol. 35, pp. 87–94; reprinted in [1979] below, vol. 7, pp. 427–434.

1979 GODFREY H. HARDY, *Collected Papers of G.H. Hardy*, Busbridge, I.W., and R.A. Rankin (editors), Clarendon, Oxford.

1915 FRIEDRICH HARTOGS, Über das Problem der Wohlordnung, *Mathematische Annalen*, vol. 76, pp. 436–443.

1909 FELIX HAUSDORFF, Die Graduierung nach dem Endverlauf, *Berichte über die Verhandlungen der Königlich Sächsischen Gesellschaft der Wissenschaften zu Leipzig, Mathematische-Physische Klasse*, vol. 61, pp. 297–334.

1906 GERHARD HESSENBERG, *Grundbegriffe der Mengenlehre*, Göttingen, Vandenhoeck & Ruprecht; reprinted from *Abhandlungen der Fries'schen Schule, Neue Folge* vol. 1 (1906), pp. 479–706.

1909 GERHARD HESSENBERG, Kettentheorie und Wohlordnung, *Journal für die reine und angewandte Mathematik* (Crelle's Journal), vol. 135, pp. 81–133.

1891 EDMUND HUSSERL, Besprechung von E. Schröder, Vorlesungen über die Algebra der Logic (Exakte logik), Vol. I, Leipzig 1890, *Göttingische Gelehrte Anzeigen* 1891, pp. 243–278; reprinted in Edmund Husserl, *Aufsätze und Rezensionen (1890–1910). Husserliana Vol. XXII*, Nijhoff, The Hague 1978, pp. 3–43.

1990 PETER HYLTON, *Russell, Idealism, and the Emergence of Analytic Philosophy*, Oxford, Clarendon Press.

1904 PHILIP E.B. JOURDAIN, On the transfinite cardinal numbers of well-ordered aggregates, *Philosophical Magazine*, vol. 7, pp. 61–75.

1905 PHILIP E.B. JOURDAIN, On a proof that every aggregate can be well-ordered, *Mathematische Annalen*, vol. 60, pp. 465–470.

1996 AKIHIRO KANAMORI, The mathematical development of set theory from Cantor to Cohen, *The Bulletin of Symbolic Logic*, vol. 2, pp. 1–71.

1938 STEPHEN C. KLEENE, On notation for ordinal numbers, *The Journal of Symbolic Logic*, vol. 3, pp. 150–155.

1952 STEPHEN C. KLEENE, *Introduction to Metamathematics*, Van Nostrand, Princeton.

1928 BRONISŁAW KNASTER, Un théorème sur les fonctions d'ensembles, *Annales de la Société Polonaise de Mathématique*, vol. 6, pp. 133–134.

1950 HELMUTH KNESER, Eine direkte Ableitung des Zornschen Lemmas aus dem Auswahlaxiom, *Mathematische Zeitschrift*, vol. 53, pp. 110–113.

1921 KAZIMIERZ KURATOWSKI, Sur la notion de l'ordre dans la théorie des ensembles, *Fundamenta Mathematicae*, vol. 2, pp. 161–171.

1922 KAZIMIERZ KURATOWSKI, Une méthode d'élimination des nombres transfinis des raisonnements mathématiques, *Fundamenta Mathematicae*, vol. 3, pp. 76–108.

1930 EDMUND LANDAU, *Grundlagen der Analysis*, Akademische Verlagsgesellschaft, Leipzig; translated as *Foundations of Analysis*, Chelsea, New York 1951.

1971 SERGE LANG, *Algebra*, Addison-Wesley, Reading (revised printing of original 1965 edition).

1982 GREGORY H. MOORE, *Zermelo's Axiom of Choice. Its Origins, Development, and Influence.* Springer-Verlag, New York.

1995 GREGORY H. MOORE, The origins of Russell's Paradox: Russell, Couturat, and the antimony of infinite number, in: *From Dedekind to Gödel*, Jaako Hintikka (editor) Synthese Library volume 251, Kluwer, Dordrecht, pp.215–239.

1891 EDMUND HUSSERL, Besprechung von E. Schröder, Vorlesungen über die Algebra der Logic (Exakte logik), Vol. I, Leipzig 1890, *Göttingische ind to Gödel*, Jaakko Hintikka (editor), Synthese Library volume 251, Kluwer, Dordrecht, pp. 215–239.

1981 GREGORY H. MOORE and ALEJANDRO R. GARCIADIEGO, Burali-Forti's Paradox: a reappraisal of its origins, *Historia Mathematica*, vol. 8, pp. 319–350.

1974 YIANNIS N. MOSCHOVAKIS, *Elementary Induction on Abstract Structures*, North-Holland, Amsterdam.

1994 YIANNIS N. MOSCHOVAKIS, *Notes on Set Theory*, Springer-Verlag, New York.

1937 EMMY NOETHER and JEAN CAVAILLÈS, *Briefwechsel Cantor-Dedekind*, Hermann, Paris.

1889 GIUSEPPE PEANO, *Arithmetices principia nova methodo exposita*, Bocca, Turin; reprinted in [1957-9] below, vol. 2, pp. 20–55; partially translated in van Heijenoort [1967], pp. 85–97; translated in [1973] below, pp. 101–134.

1890 GIUSEPPE PEANO, Démonstration de l'intégrabilité des équations différentielles ordinaires, *Mathematische Annalen*, vol. 37, pp. 182–228; reprinted in [1957] below, vol. 1, pp. 119–170.

1906 GIUSEPPE PEANO, Super theorema de Cantor-Bernstein, *Rendiconti del Circolo Matematico di Palermo*, vol. 21, pp. 360–366; reprinted in [1957-9] below, vol. 1, pp. 337–344.

1957-9 GIUSEPPE PEANO, *Opera Scelte*, three volumes, U. Cassina (editor), Edizioni Cremonese, Rome.

1973 GIUSEPPE PEANO, *Selected Works of Giuseppe Peano*, Hubert C. Kennedy (translator and editor), University of Toronto Press, Toronto.

1990 VOLKER PECKHAUS, "Ich habe mich wohl gehütet, alle Patronen auf einmal zu verschiessen". Ernst Zermelo in Göttingen. *History and Philosophy of Logic*, vol. 11, pp. 19–58.

1992 I.C.C. PHILLIPS, Recursion Theory, in: *Handbook of Logic in Computer Science*, S. Abramsky, Dov M. Gabbay, and T.S.E. Maibaum (editors), Clarendon Press, Oxford, vol. 1, pp. 80–187.

1884 HENRI POINCARÉ, Sur certaines solutions particulières du problème des trois corps, *Bulletin Astronomique*, vol. 1, pp. 65–74; reprinted in [1952] below, pp. 253–261.

1905 HENRI POINCARÉ, Les mathématiques et la logique, *Revue de Métaphysique et de Morale*, vol. 13, pp. 815–835 and vol. 14 (1906), pp. 17–34.

1906 HENRI POINCARÉ, Les mathématiques et la logique, *Revue de Métaphysique et de Morale*, vol. 14, pp. 294–317.

1952 HENRI POINCARÉ, *Oeuvres de Henri Poincaré*, Vol. 7, Gauthier-Villars, Paris.

1940 WILLARD V.O. QUINE, *Mathematical Logic*, Norton, New York.

1981 BERNHARD RANG and WOLFGANG THOMAS, Zermelo's discovery of the "Russell paradox", *Historia Mathematica*, vol. 8, pp. 15–22.

1901 BERTRAND A.W. RUSSELL, Recent Italian work on the foundations of mathematics, in [1993] below, pp. 350–362 (page references are to these).

1901a BERTRAND A.W. RUSSELL, Recent work on the principles of mathematics, *International Monthly*, vol. 4, pp. 83–101; reprinted as "Mathematics and the Metaphysicians" (with six footnotes added in 1917) in [1918] below, Chapter V, pp. 74–94 (page references are to these); this version reprinted in [1993] below, pp. 363–379.

1903 BERTRAND A.W. RUSSELL, *The Principles of Mathematics*, Cambridge University Press, Cambridge; later editions, George Allen & Unwin, London.

1906 BERTRAND A.W. RUSSELL, On some difficulties in the theory of transfinite numbers and order types, *Proceedings of the London Mathematical Society*, vol. (2)4, pp. 29–53, reprinted in [1973] below, pp. 135–164.

1906a BERTRAND A.W. RUSSELL, Les paradoxes de la logique, *Revue de Métaphysique et de Morale*, vol. 14, pp. 627–650; translated in [1973] below, pp. 190–214.

1908 BERTRAND A.W. RUSSELL, Mathematical logic as based on the theory of types, *American Journal of Mathematics*, vol. 30, pp. 222–262; reprinted in van Heijenoort [1967], pp. 150–182.

1918 BERTRAND A.W. RUSSELL, *Mysticism and Logic, and Other Essays*, Longmans, Green & Co., New York.

1944 BERTRAND A.W. RUSSELL, My mental development, in: *The Philosophy of Bertrand Russell*, Paul A. Schilpp (editor), The Library of Living Philosophers vol. 5, Northwestern University, Evanston, pp. 3–20.

1959 BERTRAND A.W. RUSSELL, *My Philosophical Development*, George Allen & Unwin, London.

1973 BERTRAND A.W. RUSSELL, *Essays in Analysis*, Douglas Lackey (editor), George Brailler, New York.

1983 BERTRAND A.W. RUSSELL, *Cambridge Essays, 1888–99*, Kenneth Blackwell *et al.* (editors), vol. 1 of *The Collected Works of Bertrand Russell*, George Allen & Unwin, London.

1992 BERTRAND A.W. RUSSELL, *The Selected Letters of Bertrand Russell, Volume 1: The Private Years, 1884–1914*, Nicholas Griffin (editor), Houghton Mifflin, Boston.

1993 BERTRAND A.W. RUSSELL, *Toward the "Principles of Mathematics", 1900–1902*, Gregory H. Moore (editor), vol. 3 of *The Collected Works of Bertrand Russell*, Routledge, London.

1890 ERNST SCHRÖDER, *Vorlesungen über die Algebra der Logik (exakte Logik)*, vol. 1, B.G. Teubner, Leipzig; reprinted in [1966] below.

1966 ERNST SCHRÖDER, *Vorlesungen über die Algebra der Logik*, three volumes, Chelsea, New York.

1962 DANA S. SCOTT, Quine's individuals, in: *Logic, Methodology and the Philosophy of Science*, Ernst Nagel (editor), Stanford University Press, Stanford, pp. 111-115.

1975 DANA S. SCOTT, Data types as lattices, in: *Logic Conference, Kiel 1974*, Müller, Gert H., Arnold Oberschelp, and Klaus Potthoff (editors), Lecture Notes in Mathematics #499, Berlin, Springer-Verlag, pp. 579–651.

1971 MOHAMMED A. SINACEUR, Appartenance et inclusion: un inédit de Richard Dedekind, *Revue d'Histoire des Sciences et de leurs Applications*, vol. 24, pp. 247–255.

1957 ERNST SPECKER, Zur Axiomatik der Mengenlehre (Fundierungs- und Auswahlaxiom), *Zeitschrift für mathematische Logik und Grundlagen der Mathematik*, vol. 3, 173–210.

1977 JOSEPH E. STOY, *Denotational Semantics. The Scott-Strachey Approach to Programming Language Theory*, MIT Press, Cambridge; third printing, 1985.

1949-50 T. SZELE, On Zorn's Lemma, *Publicationes Mathematicae Debrecen*, vol. 1, pp. 254–256; cf. Errata, p. 257.

1939 ALFRED TARSKI, On well-ordered subsets of any set, *Fundamenta Mathematicae*, vol. 32, pp. 176–183.

1955 ALFRED TARSKI, A lattice-theoretical fixpoint theorem and its applications, *Pacific Journal of Mathematics*, vol. 5, pp. 285–309.

1966 BARTEL L. VAN DER WAERDEN, *Algebra I*, seventh edition, Springer-Verlag, Berlin 1966; translated into English, Ungar, New York 1970; this translation reprinted by Springer-Verlag, Berlin 1991.

1967 JEAN VAN HEIJENOORT (editor), *From Frege to Gödel: A Source Book in Mathematical Logic, 1879–1931*, Harvard University Press, Cambridge.

1923 JOHN VON NEUMANN, Zur Einführung der transfiniten Zahlen, *Acta Litterarum ac Scientiarum Regiae Universitatis Hungaricae Francisco-Josephinae, sectio scientiarum mathematicarum*, vol. 1, pp. 199–208; reprinted in [1961] below, pp. 24–33; translated in van Heijenoort [1967], pp. 346–354.

1925 JOHN VON NEUMANN, Eine Axiomatisierung der Mengenlehre, *Journal für die reine und angewandte Mathematik* (Crelle's Journal), vol. 154, pp. 219–240; Berichtigung, vol. 155, p. 128; reprinted in [1961] below, pp. 34–56; translated in van Heijenoort [1967], pp. 393–413.

1928 JOHN VON NEUMANN, Über die Definition durch transfinite Induktion und verwandte Fragen der allgemeinen Mengenlehre, *Mathematische Annalen*, vol. 99, pp. 373–391; reprinted in [1961] below, pp. 320–338.

1961 JOHN VON NEUMANN, *John von Neumann. Collected Works*, Vol. 1, Taub, Abraham H. (editor), Pergamon Press, New York.

1954 JEFFREY D. WESTON, A short proof of Zorn's Lemma, *Archiv der Mathematik*, vol. 8, p. 279.

1910-3 ALFRED N. WHITEHEAD and BERTRAND A.W. RUSSELL, *Principia Mathematica*, three volumes, Cambridge University Press, Cambridge.

1904 ERNST ZERMELO, Beweis, dass jede Menge wohlgeordnet werden kann (Aus einem an Herrn Hilbert gerichteten Briefe), *Mathematische Annalen*, vol. 59, pp. 514–516; translated in van Heijenoort [1967], pp. 139–141.

1908 ERNST ZERMELO, Neuer Beweis für die Möglichkeit einer Wohlordnung, *Mathematische Annalen*, vol. 65, pp. 107–128; translated in van Heijenoort [1967], pp. 183–198.

1908a ERNST ZERMELO, Untersuchungen über die Grundlagen der Mengenlehre I, *Mathematische Annalen*, vol. 65, pp. 261–281; translated in van Heijenoort [1967], pp. 199–215.

1935 MAX ZORN, A remark on method in transfinite algebra, *Bulletin of the American Mathematical Society*, vol. 41, pp. 667–670.

4 The Empty Set, the Singleton, and the Ordered Pair

Dedicated to the memory of Burton S. Dreben

For the modern set theorist the empty set $\emptyset$, the singleton $\{a\}$, and the ordered pair $\langle x, y\rangle$ are at the beginning of the systematic, axiomatic development of set theory, both as a field of mathematics and as a unifying framework for ongoing mathematics. These notions are the simplest building blocks in the abstract, generative conception of sets advanced by the initial axiomatization of Ernst Zermelo [1908a] and are quickly assimilated long before the complexities of Power Set, Replacement, and Choice are broached in the formal elaboration of the 'set of' $\{\}$ operation. So it is surprising that, while these notions are unproblematic today, they were once sources of considerable concern and confusion among leading pioneers of mathematical logic like Frege, Russell, Dedekind, and Peano. In the development of modern mathematical logic out of the turbulence of 19th century logic, the emergence of the empty set, the singleton, and the ordered pair as clear and elementary set-theoretic concepts serves as a motif that reflects and illuminates larger and more significant developments in mathematical logic: the shift from the intensional to the extensional viewpoint, the development of type distinctions, the logical vs. the iterative conception of set, and the emergence of various concepts and principles as distinctively set-theoretic rather than purely logical. Here there is a loose analogy with Tarski's recursive definition of truth for formal languages: The mathematical interest lies mainly in the procedure of recursion and the attendant formal semantics in model theory, whereas the philosophical interest lies mainly in the basis of the recursion, truth and meaning at the level of basic predication. Circling back to the beginning, we shall see how central the empty set, the singleton, and the ordered pair were, after all.

1 The Empty Set

A first look at the vicissitudes specific to the empty set, or null class, provides an entrée into our main themes, particularly the differing concerns of logical analysis and the emerging set theory. Viewed as part of larger philosophical traditions the null class serves as the extensional focus for age-old issues about Nothing and Negation, and the empty set emerged with the increasing need for objectification and symbolization.

The work of George Boole was a cresting of extensionalism in the 19th century. He introduced "0" without explanation in his *The Mathematical Analysis of Logic* [1847: 21], and used it forthwith as an "elective symbol" complementary

Republished from *The Bulletin of Symbolic Logic* 9 (2003), pp.273-298, with permission from The Association for Symbolic Logic.

I wish to express my particular thanks to the Dibner Institute for the History of Science and Technology for its support and hospitality and to Juliet Floyd and Nimrod Bar-Am for their many suggestions.

to his "1" denoting the "'Universe". However, both "0" and "1" were reconstrued variously, as predicates, classes, states of affairs, and as numerical quantities. Significantly, he began his [1847: 3] with the assertion that in "Symbolical Algebra ... the validity of the processes of analysis does not depend upon the interpretation of the symbols which are employed, but solely upon the laws of their combination." In his better known *An Investigation into the Laws of Thought* [1854] he had "signs" representing "classes", and incorporating the arithmetical property of 0 that $0 \cdot y = 0$ for every y, assigned [1854: 47] to "0" the interpretation "Nothing", the class consisting of no individuals.[1] However shifting his interpretations of "0", it played a natural and crucial role in Boole's "Calculus", which in retrospect exhibited an over-reliance on analogies between logic and arithmetic. It can be justifiably argued that Boole had invented the empty set. Those following in the algebraic tradition, most prominently Charles S. Pierce and Ernst Schröder, adopted and adapted Boole's "0".

Gottlob Frege, the greatest philosopher of logic since Aristotle, aspired to the logical analysis of mathematics rather than the mathematical analysis of logic, and what in effect is the null class played a key role in his analysis of number. Frege in his *Grundlagen* [1884] eschewed the terms "set" ["Menge"] and "class" ["Klasse"], but in any case the extension of the concept "not identical with itself" was key to his definition of zero as a logical object. Schröder, in the first volume [1890] of his major work on the algebra of logic, held a traditional view that a class is merely a collection of objects, without the { } so to speak. In his review [1895] of Schröder's [1890], Frege argued that Schröder cannot both maintain this view of classes and assert that there is a null class, since the null class contains no objects.[2] For Frege, logic enters in giving unity to a class as the extension of a *concept* and thus makes the null class viable.

Giuseppe Peano [1889], the first to axiomatize mathematics in a symbolic language, used "Λ" to denote both the falsity of propositions (Part II of Logical Notations) and the null class (Part IV). He later [1897] provided a definition of the null class as the intersection of all classes, making it more explicit that there is exactly one null class. More importantly, [1897] had the first occurrence of "$\exists$", used there to indicate that a class is not equal to the null class. Frege had taken the existential quantifier to be derivative from the universal quantifier in the not-for-all-not formulation; in Peano's development, the first indication

[1] Earlier Boole had written [1854: 28]: "By a class is usually meant a collection of individuals, to each of which a particular name or description may be applied; but in this work the meaning of the term will be extended so as to include the case in which but a single individual exists, answering to the required name or description, as well as the cases denoted by the terms 'nothing' and 'universe,' which as 'classes' should be understood to comprise respectively 'no beings,' 'all beings.' " Note how Boole had already entertained the singleton.

[2] Edmund Husserl in his review [1891] of Schröder's [1890] also criticized Schröder along similar lines. Frege had already lodged his criticism in the *Grundgesetze* [1893: 2-3] and written that in Husserl's review "the problems are not solved."

of the existential quantifier is intimately tied to the null class. Thus, in the logical tradition the null class played a focal and pregnant role.

It is among the set theorists that the null class, *qua* empty set, emerged to the fore as an elementary concept and a basic building block. Georg Cantor himself did not dwell on the empty set. Early on in his study of "pointsets" ["Punktmengen"] of real numbers Cantor did write [1880: 355] that "the identity of two pointsets P and Q will be expressed by the formula $P \equiv Q$"; defined disjoint sets as "*lacking intersection*"; and then wrote [1880: 356] "for the absence of points ... we choose the letter O; $P \equiv O$ indicates that the set P contains *no single point.*" So, "$\equiv O$" is arguably more like a predication for being empty at this stage.

Richard Dedekind in his groundbreaking essay on arithmetic *Was sind und was sollen die Zahlen?* [1888: (2)] deliberately excluded the empty set [Nullsystem] "for certain reasons", though he saw its possible usefulness in other contexts.[3] Indeed, the empty set, or null class, is not necessary in his analysis: Whereas Frege the logician worked to get at what numbers *are* through definitions based on equi-numerosity, Dedekind the mathematician worked to define the number *sequence* structurally, up to isomorphism or "inscrutability of reference." Whereas zero was crucial to Frege's logical development, the particular "base element" for Dedekind was immaterial and he denoted it by the symbol "1" before proceeding to define the numbers "by abstraction."

Ernst Zermelo [1908a], the first to provide a full-fledged axiomatization of set theory, wrote in his Axiom II: "There exists a (improper [uneigentliche]) set, the *null set* [*Nullmenge*] 0, that contains no element at all." Something of intension remained in the "(improper [uneigentliche])", though he did point out that because of his Axiom I, the Axiom of Extensionality, there is a single empty set.

Finally, Felix Hausdorff, the first developer of the transfinite after Cantor and the first to take the sort of extensional, set-theoretic approach to mathematics that would dominate in the years to come, unequivocally opted for the empty set [Nullmenge] in his classic *Grundzüge der Mengenlehre* dedicated to Cantor. However, a hint of predication remained when he wrote [1914: 3]: "... the equation $A = 0$ means that the set A has no element, vanishes [verschwindet], is empty." In a footnote Hausdorff did reject the formulation that the set A does not exist, insisting that it exists with no elements in it. The use to which Hausdorff put "0" in the *Grundzüge* is much as "$\emptyset$" is used in modern mathematics, particularly to indicate the extension of the conjunction of mutually exclusive properties.

The set theorists, unencumbered by philosophical motivations or traditions, attributed little significance to the empty set beyond its usefulness. Interestingly, there would be latter-day attempts in philosophical circles to nullify the

[3]But see footnote 11.

null set.[4] Be that as it may, just as zero became basic to numerical notation as a place holder, so also did $\emptyset$ to the algebra of sets and classes to indicate the empty extension.[5]

2 Inclusion vs. Membership

In 19th century logic, the main issue concerning the singleton, or unit class, was the distinction between a and $\{a\}$, and this is closely connected to the emergence of the basic distinction between inclusion, $\subseteq$, and membership, $\in$, a distinction without which abstract set theory could not develop.

Set theory as a field of mathematics began, of course, with Cantor's [1874] result that the reals are uncountable. But it was only much later in [1891] that Cantor gave his now famous diagonal proof, showing in effect that for any set X the collection of functions from X into a two-element set is of a strictly higher cardinality than that of X. In retrospect the diagonal proof can be drawn out from the [1874] proof, but in any case the new proof enabled Cantor to dispense with the earlier topological trappings. Moreover, he could affirm [1891: para.8] "the general theorem, that the powers [cardinalities] of well-defined sets have no maximum."

Cantor's diagonal proof is regarded today as showing how the power *set* operation leads to higher cardinalities. However, it would be misleading to assert that Cantor himself used power sets. Rather, he was expanding the 19th century concept of *function* by ushering in arbitrary functions. His theory of cardinality was based on one-to-one *correspondence [Beziehung]*, and this had led him to the diagonal proof which in [1891] was first rendered in terms of *sequences* "that depend on infinitely many coordinates". By the end of [1891] he did deal explicitly with "all" *functions* with a specified domain L and range $\{0, 1\}$.

The recasting of Cantor's diagonal proof in terms of sets cannot be carried out without drawing the basic distinction between $\subseteq$, inclusion, and $\in$, membership. Surprisingly, neither this distinction nor the related distinction between a class a and its unit class $\{a\}$ was generally appreciated in logic at the time of Cantor [1891]. This was symptomatic of a general lack of progress in logic on the traditional problem of the copula (how does "is" function?), a problem with roots going back at least to Aristotle. George Boole, like his British contemporaries, only had inclusion, not even distinguishing between proper and improper inclusion, in his part-whole analysis, and he conflated individuals with their unit classes in his notation. Cantor himself, working initially

---

[4]See for example Carmichael [1943][1943a].

[5]Something of the need of a place holder is illustrated by the use of {} in the computer typesetting program TEX. To render ${}^\omega 2$, {} must be put in as a something to which ω serves as a superscript:

```
${}^\omega 2$.
```

and mostly with real numbers and sets of real numbers, clearly distinguished inclusion and membership from the beginning for his sets, and in his *Beiträge* [1895], his mature presentation of his theory of cardinality, the distinction is clearly stated in the passages immediately following the oft-quoted definition of set [Menge]. The first to draw the inclusion vs. membership distinction generally in logic was Frege. Indeed, in his *Begriffsschrift* [1879] the distinction is manifest for concepts, and in his *Grundlagen* [1884] and elsewhere he emphasized the distinction in terms of "subordination" and "falling under a concept." The a vs. $\{a\}$ distinction is also explicit in his *Grundgesetze* [1893: §11], in a functional formulation. For Peirce [1885: III] the inclusion vs. membership distinction is unambiguous. Although Schröder drew heavily on Peirce's work, Schröder could not clearly draw the distinction, for he maintained (as we saw above) that a class is merely a collection of objects. Also, Schröder [1890: 35] was concerned about "the representation [Vorstellung] of the representation of horse", and remained undecided about a traditional infinite regress problem. With an intensional approach there is an issue here, but if a is the extensional representation of horse, then $\{a\}$ is representation of the representation and hence avowedly distinct from a. In Zermelo's axiomatization paper [1908a] the inclusion vs. membership distinction is of course basic. In his Axiom II he posited the existence of singletons, and he pointed out in succeeding commentary that a singleton has no subset other than itself and the empty set. Thus, Zermelo would have understood that his Axiom I, the Axiom of Extensionality, ruled out having $a = \{a\}$, at least when a has more than one element.[6]

However acknowledged the a vs. $\{a\}$ distinction, there was still uncertainty and confusion in the initial incorporation of the distinction into the symbolization,[7] as we point out in describing in detail the writings of Dedekind and Peano:

Dedekind in *Was sind und was sollen die Zahlen?* [1888: (3)] appears to have identified the unit class $\{a\}$ with, or at least denoted it by, a — writing in that case (in modern notation) $a \subseteq S$ for $a \in S$.[8] This lack of clarity about

[6] However, Zermelo's [1908a] axiomatization does not rule out having *some* sets a satisfying $\{a\} = a$; see our §4. Zermelo in the earliest notes about axiomatization found in his *Nachlass*, written around 1905, took the assertion $M \notin M$ as an axiom (see Moore [1982: 155]), and this of course rules out having $\{a\} = a$ for any set a.

[7] While we use the now familiar notation $\{a\}$ to denote the singleton, or unit class, of a, it should be kept in mind that the notation varied through this period. Cantor [1895] wrote $M = \{m\}$ to indicate that M consists of members *typically* denoted by m, i.e. m was a variable ranging over the possibly many members of M. Of those to be discussed, Peano [1890: 192] used ιa to denote the unit class of a. Russell [1903: 517] followed suit, but from his [1908] on he used ι'a. Frege [1893: §11] wrote a boldface backslash before a to denote its unit class. It was Zermelo [1908a: 262] who introduced the now familiar use of $\{a\}$, having written just before: "The set that contains only the elements $a, b, c, \ldots, r$ will often be denoted briefly by $\{a, b, c, \ldots, r\}$."

[8] At the beginning of the paragraph [1888: (3)] Dedekind wrote: "A system A is said to be *part* of a system S when every element of A is also an element of S. Since this relation

the unit class would get strained: For a set [System] S and transformation [Abbildung] ϕ: $S \to S$, he formulated what we would now call the closure of an $A \subseteq S$ under ϕ and denoted it by A_o. Then in the crucial definition of "simply infinite system" — one isomorphic to the natural numbers — he wrote $N = 1_o$, where 1 is a distinguished *element* of N. Here, we would now write $N = \{1\}_o$. In his *Grundgesetze* [1893: 2-3] Frege roundly and soundly took Dedekind to task for his exclusion of the null class and for his confusion about the unit class, seeing this to be symptomatic of Dedekind's conflation of the inclusion vs. membership distinction. As in his criticism of Schröder, Frege railed against what he regarded as extensional sophistries, firmly convinced about the primacy of his intensional notion of concept.[9] Dedekind in a revealing note entitled "Dangers of the Theory of Systems" written around 1900 and found in his *Nachlass* drew attention to his own confused paragraph [1888: (3)] in order to stress the danger of confounding a with $\{a\}$:

> Suppose that a has distinct elements b and c, yet $\{a\} = a$; since the principle of extensionality had been assumed, it follows that $b = a = c$, which is a contradiction.

He mentioned raising the singleton problem in conversation, briefly with Felix Bernstein on 13 June 1897 and then with Cantor himself on 4 September 1899 pointing out the contradiction.[10] It is hard to imagine such a conversation taking place between the two great pioneers of set theory, years after their foundational work on arithmetic and on the transfinite!

---

between a system A and a system S will occur continually in what follows, we shall express it briefly by the symbol $A \ni S$." At the end, he wrote: "Since further every element s of a system S by (2) can itself be regarded as a system, we can hereafter employ the notation $s \ni S$." In the immediately preceding paragraph (2), Dedekind had written: "For uniformity of expression it is advantageous to include also the special case where a system S consists of a *single* (one and only one) element a, i.e., the thing a is an element of S, but no thing different from a is an element of S."

[9] Against Dedekind's description of a system (set) as different things "put together in the mind" Frege [1893: 2] had written: "I ask, in whose mind? If they are put together in one mind but not in another, do they form a system then? What is supposed to be put together in my mind, no doubt must be in my mind: then do the things outside myself not form systems? Is a system a subjective figure in the individual soul? In that case is the constellation Orion a system? And what are its elements? The stars, or the molecules, or the atoms?"

[10] See Sinaceur [1971]. In the note Dedekind proposed various emendations to his essay to clarify the situation. Undercutting the exclusion of the empty set [Nullsystem] in the essay ([1888: (2)]) he also pointed out the importance of having the empty set [Nullsystem]. However, the emendations were never incorporated into later editions of [1888].

Interestingly, the second, 1887 draft of Dedekind [1888] had contained the following passage: "A system can consist in *one* element (i.e. in a *single* one, in one and *only* one), it can also (contradiction) be *void* (contain no element)." Ferreirós [1999: 227ff] points this out and also that Dedekind in a manuscript written in the 1890's did introduce the empty set, denoting it by "0", and used "[a]" for the singleton of a (see Dedekind [1932: 450–60]).

The following is the preface to the third, 1911 edition of Dedekind [1888], with some words italicized for emphasis:[11]

> When I was asked roughly eight years ago to replace the second edition of this work (which was already out of print) by a third, I had misgivings about doing so, because in the mean time doubts had arisen about the reliability [Sicherheit] of important foundations of my conception. Even today I do not underestimate the importance, and to some extent the correctness of these doubts. But my trust in the inner harmony of our logic is not thereby shattered; *I believe that a rigorous investigation of the power [Schöpferkraft] of the mind to create from determinate elements a new determinate, their system, that is necessarily different from each of these elements, will certainly lead to an unobjectionable formulation of the foundations of my work.* But I am prevented by other tasks from completing such an investigation; so I beg for leniency if the paper now appears for the third time without changes—which can be justified by the fact that interest in it, as the persistent inquiries about it show, has not yet disappeared.

Thus, Dedekind did draw attention to the distinction between a system and its members, implicitly alluding to his confounding of a and $\{a\}$ in the text. This preface is remarkable for the misgivings it expresses, especially in a work that purports to be foundational for arithmetic.

It was Peano [1889] who first distinguished inclusion and membership with different signs, and it is to him that we owe "$\in$", taken as the initial for the Greek singular copula '$\varepsilon\sigma\tau\iota$.[12] In Part IV of his introduction of logical notations, Peano wrote: "The sign $\in$ signifies *is*. Thus $a \in b$ is read *a is a* [*est quoddam*] b; ...." In the preface he warned against confusing "$\in$" with the sign for inclusion. However, at the end of part IV he wrote, "Let s be a class and k a class contained in s; then we say that k is an individual of class s if k consists of just one individual." He then proceeded to give his formula 56, which in modern notation is:

$$k \subseteq s \rightarrow (k \in s \leftrightarrow (k \neq \emptyset \ \& \ \forall x \in k \forall y \in k(x = y)))\,.$$

Unfortunately, this way of having membership follow from inclusion undercuts the very distinction that he had so emphasized: Suppose that a is any class,

[11] cf. Ewald [1996: 796]. This preface does not appear in Dedekind [1963], which is an English translation of the *second* edition of Dedekind [1888].

[12] Actually, the sign appearing in Peano [1889] is only typographically similar to our "$\in$", and only in Peano [1889a] was the Greek noted. The epsilon "ε" began to be used from then on, and in Peano [1891: fn.8] the '$\varepsilon\sigma\tau\iota$ connection was made explicit.

and $s = \{a\}$. Then formula 56 implies that $s \in s$. But then $s = a$, and so $\{a\} = a$.[13] This was not intended by Peano:

In the paper that presented his now well-known existence theorems for ordinary differential equations, Peano introduced the sign "ι" as follows [1890: 192]:

> Let us decompose in effect the sign = into its two parts *is* and *equal to*; the word *is* is already represented by ε; let us represent also the expression *equal to* by a sign, and let ι (initial of '$\iota\sigma o\varsigma$) be that sign; thus instead of $a = b$ one can write $a \,\varepsilon\, \iota\, b$.

It is quite striking how Peano intended his mathematical notation to mirror the grammar of ordinary language. If indeed $a = b$ can be rendered as $a \,\varepsilon\, \iota\, b$, then the grammar dictates that ιb must be a class, the unit class of b. ι is the initial of '$\iota\sigma o\varsigma$, "same", and any member of $\iota\, b$ is the same as b. That this is how "ιb" is autonomously used is evident from a succeeding passage, the first passage in print that drew the a vs. $\{a\}$ distinction (and also described the unordered pair) [1890: 193]: "To indicate the class constituted of individuals a and b one writes sometimes $a \cup b$ (or $a + b$, following the more usual notation). But it is more correct to write $\iota a \cup \iota b$; ...." Peano had let "=" range widely from the equality of numerical quantities and of classes to the equivalence of propositions. He introduced "ι" to analyze his overworked "=", and this objectification of the unit class emerged from a remarkable analysis of one form of the copula, equality, in terms of another, membership.[14] While today we take equality and membership as primitive notions and proceed to define the singleton, in Peano's formulation the unit class played a fundamental role in articulating these notions, just as the null class did for "$\exists$" (as we saw above).

The lack of clarity about the distinction between inclusion and membership in the closing years of the 19th century reflected a traditional reluctance to comprehend a collection as a unity and was intertwined with the absence of the liberal, iterative use of the "set of" { } operation. Of course, set theory as a mathematical study of that operation could only develop after a sharp distinction between inclusion and membership had been made. This development in turn would depend increasingly on rules and procedures provided by axiomatization. On the logical side, the grasp of the inclusion vs. membership distinction was what spurred Bertrand Russell to his main achievements.

---

[13] Van Heijenoort [1967: 84] alluded specifically to formula 56, but did not point out how it leads to $\{a\} = a$ for every a.

[14] In [1894:§31] Peano again described the "ι" analysis of "=" and moreover emphasized the difference between "0" and "$\iota\, 0$" for arithmetic. Interestingly, Zermelo [1908a] reflected this Peano analysis of equality in his discussion of definite [definit] properties, those admissible for use in his Axiom (Schema) of Separation. Zermelo initially took $a \,\varepsilon\, b$ to be definite and *derived* that $a = b$ is definite: After positing singletons in Axiom II he wrote: "The question whether $a = b$ or not is definite (No. 4), since it is equivalent to the question of whether or not $a \,\varepsilon\, \{b\}$."

3 Russell

The turn of the century saw Russell make the major advances in the development of his mathematical logic. As he later wrote in [1944]: "The most important year in my intellectual life was the year 1900, and the most important event in this year was my visit to the International Congress of Philosophy in Paris." There in August he met Peano and embraced his symbolic logic, particularly his use of different signs for inclusion and membership. During September Russell (see [1901]) extended Peano's symbolic approach to the logic of relations. Armed with the new insights, in the rest of the year Russell completed most of the final draft of *The Principles of Mathematics* [1903], a book he had been working on in various forms from 1898. However, the sudden light would also cast an abiding shadow, for by May 1901 Russell had transformed Cantor's diagonal proof into Russell's Paradox.[15] In reaction he would subsequently formulate a complex logical system of orders and types in Russell [1908] which multiplied the inclusion vs. membership distinction many times over and would systematically develop that system in Whitehead and Russell's *Principia Mathematica* [1910-3].

For Russell of *The Principles* mathematics was to be articulated in an all-encompassing logic, a complex philosophical system based on universal categories.[16] He had drawn distinctions within his widest category of "term" (but with "object" wider still[17]) among "propositions" about terms and "classes" of various kinds corresponding to propositions. Because of this, Russell's Paradox became a central concern, for it forced him to face the threat of both the conflation of his categories and the loss of their universality.

With the inclusion vs. membership distinction in hand, Russell had informatively recast Cantor's diagonal argument in terms of classes,[18] a formulation synoptic to the usual set-theoretic one about the power set. Cantor's argument has a positive content in the generation of sets not in the range of functions $f\colon X \to \mathcal{P}(X)$ from a set into its power set. For Russell however, with U the class of all classes, $\mathcal{P}(U) \subseteq U$ so that the identity map is an injection, and so the Russellian $\{x \in U \mid x \notin x\} \subseteq U$ must satisfy $\{x \in U \mid x \notin x\} \in U$, arriving

[15]See Grattan-Guinness [1978], Garciadiego [1992], Moore [1995] and Kanamori [1997] for more on the evolution of Russell's Paradox.

[16]Russell [1903: 129] wrote: "The distinction of philosophy and mathematics is broadly one of point of view: mathematics is constructive and deductive, philosophy is critical, and in a certain impersonal sense controversial. Wherever we have deductive reasoning, we have mathematics; but the principles of deduction, the recognition of indefinable entities, and the distinguishing between such entities, are the business of philosophy. Philosophy is, in fact mainly a question of insight and perception."

[17]Russell [1903: 55n] wrote: "I shall use the word *object* in a wider sense than *term*, to cover both singular and plural, and also cases of ambiguity, such as 'a man.' The fact that a word can be framed with a wider meaning than *term* raises grave logical problems."

[18]See Russell [1903: 365ff.].

necessarily at a contradiction. Having absorbed the inclusion vs. membership distinction, Russell had to confront the dissolution of that very distinction for his universal classes.

Russell soon sought to resolve his paradox with his theory of types, adumbrated in *The Principles*. Although the inclusion vs. membership distinction was central to *The Principles*, the issues of whether the null class exists and whether a term should be distinct from its unit class became part and parcel of the considerations leading to types. Early on Russell had written [1903: 23]:

> If x is any term, it is necessary to distinguish from x the class whose only member is x: this may be defined as the class of terms which are identical with x. The necessity for this distinction, which results primarily from purely formal considerations, was discovered by Peano; I shall return to it at a later stage.

But soon after, Russell [1903: 68] distinguished between "class" and "class-concept" and asserted that "there is no such thing as the null-class, though there are null class-concepts ... [and] that a class having only one term is to be identified, contrary to Peano's usage, with that one term." Russell then distinguished between "class as one" and "class as many" and asserted [1903: 76] "an ultimate distinction between a class as many and a class as one, to hold that the many are only many, and are not also one."

Next, in an early chapter (X, "The Contradiction") discussing his paradox Russell decided that propositional functions, while defining classes as many, do not always define classes as one, else they could participate *qua* terms for self-predication as in the paradox. There he first proposed a resolution by resorting to a difference in type [1903: 104-5]:

> We took it to be axiomatic that the class as one is to be found wherever there is a class as many; but this axiom need not be universally admitted, and appears to have been the source of the contradiction. ... A class as one, we shall say, is an object of the same *type* as its terms .... But the class as one does not always exist, and the class as many is of a different type from the terms of the class, even when the class has only one term ...

Reversing himself again Russell concluded [1903: 106] "that it is necessary to distinguish a single term from the class whose only member it is, and that consequently the null-class may be admitted."

Russell only discussed Frege's work in an appendix, having absorbed it relatively late. Of this Russell had actually forewarned the reader in the Preface [1903: xvi] revealing moreover that he had latterly discovered errors in the text; "these errors, of which the chief are the denial of the null-class, and the identification of a term with the class whose only member it is, are rectified in the Appendices." In the appendix on Frege, Russell alluded to an argument, from Frege's critical review [1895] of Schröder's work, for why a should not be identified with $\{a\}$: In the case of a having many members, $\{a\}$ would still have only one member – the same and obvious concern Dedekind had, as we saw above. Russell observed [1903: 514]: "... I contended that the argument was met by the distinction between the class as one and the class as many, but this contention now appears to me mistaken." Russell continued [1903: 514]: "... it must be clearly grasped that it is not only the collection as many, but the collection as one, that is distinct from the collection whose only term it is." Russell went on to conclude [1903: 515] that "the class as many is the only object that can play the part of a class", writing [1903: 516]:

> Thus a class of classes will be many many's; its constituents will each be only many, and cannot therefore in any sense, one might suppose, be single constituents. Now I find myself forced to maintain, in spite of the apparent logical difficulty, that this is precisely what is required for the assertion of number.

Russell was then led to infinitely many types [1903: 517]:

> It will now be necessary to distinguish (1) terms, (2) classes, (3) classes of classes, and so on *ad infinitum*; we shall have to hold that no member of one set is a member of any other set, and that $x \in u$ requires that x should be of a set of a degree lower by one than the set to which u belongs. Thus $x \in x$ will become a meaningless proposition; in this way the contradiction is avoided.

And he wrote further down the page:

> Thus, although we may identify the class with the numerical conjunction of its terms [class as many], wherever there are many terms, yet where there is only one term we shall have to accept Frege's range [Werthverlauf] as an object distinct from its only term.

Reading such passages from the *Principles* it does not seem too outlandish to draw connections to and analogies with the high scholastics, relating the "class as many" vs. "class as one" distinction to the traditional pluralism vs. monism debate. The a vs. $\{a\}$ distinction is inextricably connected for Russell, but it is now time to invoke a simple point. The following result is provable just in first-order logic with equality and the membership relation, with the usual definitions of $\subseteq$ and $\{a\}$.

Theorem: $\forall a(a = \{a\})$ is equivalent to: $(*)$ $\forall a \forall b(a \subseteq b \longleftrightarrow a \in b)$.
Proof: Suppose first that $a = \{a\}$. Then for any b, $a \subseteq b$ *iff* $\{a\} \subseteq b$. But $\{a\} \subseteq b$ *iff* $a \in b$, and hence we can conclude that $a \subseteq b$ *iff* $a \in b$.

For the converse, first note that since $a \in \{a\}$, $(*)$ implies that $a \subseteq \{a\}$. But also, since $a \subseteq a$, $(*)$ implies that $a \in a$, so that $\{a\} \subseteq a$. Hence we can conclude that $a = \{a\}$. $\dashv$

In other words, the identification of classes with their unit classes is equivalent to the very dissolution of the inclusion vs. membership distinction that Russell so assiduously cultivated and exploited! And this is arguably near the surface of that by-now notorious paragraph from Dedekind [1888: (3)].

Years later, Russell was firm about the danger of confusing a class with its unit class, but still on the basis of a conflation of type. In his *A History of Western Philosophy* he wrote [1945: 198]:[19]

> Another error into which Aristotle falls ... is to think that a predicate of a predicate can be a predicate of the original subject ... The distinction between names and predicates, or, in metaphysical language, between particulars and universals, is thus blurred, with disastrous consequences to philosophy. One of the resulting confusions was to suppose that a class with only one member is identical with that one member.

One direction of the above theorem is implicit here. However, Russell subsequently wrote in his retrospective *My Philosophical Development* [1959: 66-7]:

> The enlightenment that I derived from Peano came mainly from two purely technical advances of which it is very difficult to appreciate the importance unless one has (as I had) spent years in trying to understand arithmetic ... The first advance consisted in separating propositions of the form 'Socrates is mortal' from propositions of the form 'All Greeks are mortal' [i.e. distinguishing membership from inclusion] ... neither logic nor arithmetic can get far until the two forms are seen to be completely different ... The second important advance that I learnt from Peano was that a class consisting of one member is not identical with that one member.

Russell may still have been unaware, as one might presume from this passage, that the a vs. $\{a\}$ distinction *follows from* the inclusion vs. membership distinction, let alone that they are equivalent, according to the above theorem. In view of the considerable mathematical prowess exhibited by Russell in the *Principia Mathematica* this is surely an extreme case of metaphysical preoccupations beclouding developments based on "purely technical advances."

Whether here, or Peano's formula 56, or the analysis of the ordered pair to be discussed below, there is a simple mathematical argument that crucially

[19]My thanks to Nimrod Bar-Am for bringing this passage to my attention.

informs the situation. It is simple for us today to find such arguments, since for us logic is mathematical, and we are heir to the development of set theory based on the iterated application of the "set of" {} operation and axioms governing it. This development was first fostered by Zermelo; it is noteworthy that he discovered the argument for Russell's paradox independently and that it served as the beginning of a progress that in effect reverses Russell's:

4 Classes and Singletons

The first decade of the new century saw Zermelo at Göttingen make his major advances in the development of set theory.[20]His first substantial result in set theory was his independent discovery of Russell's Paradox. He then established [1904] the Well-Ordering Theorem, provoking an open controversy about this initial use of the Axiom of Choice. After providing a second proof [1908] of the Well-Ordering Theorem in response, Zermelo also provided the first full-fledged axiomatization [1908a] of set theory. In the process, he ushered in a new abstract, generative view of sets, one that would dominate in the years to come.

Zermelo's independent discovery of the argument for Russell's Paradox is substantiated in a note dated 16 April 1902 found in Edmund Husserl's *Nachlass*.[21] According to the note, Zermelo pointed out that any set M containing all of its subsets as members, i.e. $\mathcal{P}(M) \subseteq M$, is "inconsistent" by considering $\{x \in M \mid x \notin x\}$. Schröder [1890: 245] had argued that Boole's "class 1" regarded as consisting of everything conceivable is inconsistent, and Husserl in a review [1891] had criticized Schröder's argument for not distinguishing between inclusion and membership. That inclusion may imply membership is of course the same concern that Russell had to confront, but Zermelo did not push the argument in the direction of paradox as Russell had done. Also, Zermelo presumably came to his argument independently of Cantor's diagonal proof with functions. That $\mathcal{P}(M)$ has higher cardinality than M is evidently more central than $\mathcal{P}(M) \not\subseteq M$, but the connection between sub*sets* and characteristic *functions* was hardly appreciated then, and Zermelo was just making the first moves toward his abstract view of sets.[22]

Reversing Russell's progress from Cantor's correspondences to the identity map inclusion $\mathcal{P}(U) \subseteq U$, Zermelo considered functions $F\colon\ \mathcal{P}(X) \to X$, specifically in the form of *choice functions*, those F satisfying $F(Y) \in Y$ for $Y \neq \emptyset$.

[20]Peckhaus [1990] provides a detailed account of Zermelo's years 1897-1910 at Göttingen.

[21]See Rang-Thomas [1981].

[22]In Zermelo's axiomatization paper [1908a], the first result of his axiomatic theory was just the result in the Husserl note, that every set M has a subset $\{x \in M \mid x \notin x\}$ not a member of M, with the consequence that there is no universal set. Modern texts of set theory usually take the opposite tack, showing that there is no universal set by *reductio* to Russell's Paradox. Zermelo [1908a] applied his first result *positively* to generate specific sets disjoint from given sets for his recasting of Cantor's theory of cardinality.

This of course was the basic ingredient in Zermelo's [1904] formulation of what he soon called the Axiom of Choice, used to establish his Well-Ordering Theorem. Russell the metaphysician had drawn elaborate philosophical distinctions and was forced by Cantor's diagonal argument into a dialectical confrontation with them, as well as with the concomitant issues of whether the null class exists and whether a term should be distinct from its unit class. Zermelo the mathematician never quibbled over these issues for sets and proceeded to resolve the problem of well-ordering sets mathematically. In describing abstract functions Cantor had written in the *Beiträge* [1895: §4]: "...by a '*covering* [Belegung] of N with M,' we understand a law ...", and thus had continued his frequent use of the term "law" to refer to functions. Zermelo [1904: 514] specifically used the term "covering", but with his choice functions any residual sense of "law" was abandoned by him [1904]: "...we take an arbitrary covering γ and derive from it a definite well-ordering of the elements of M." It is here that *abstract* set theory began.

The generative view of sets based on the iteration of the "set of" $\{\}$ operation would be further codified by the assumption of the Axiom of Foundation in Zermelo's final axiomatization [1930], the source of the now-standard theory ZFC. Foundation disallows $a \in a$, and hence precludes $\{a\} = a$, for any set a. In any case, having any sets a such that $\{a\} = a$ is immediately antithetical to the emerging cumulative hierarchy view or "iterative conception" of set, and this scheme would become dominant in subsequent set-theoretic research. Nonetheless, having sets a satisfying $\{a\} = a$ has a striking simplicity in bearing the weight of various incentives, and the idea of having *some* such sets has been pivotal in the formal semantics of set theory.

Paul Bernays [1941: 10] announced the independence of the Axiom of Foundation from the other axioms of his system and provided [1954: 83] a proof based on having a's such that $\{a\} = a$. Ernst Specker in his Habilitationsschrift (cf. [1957]) also provided such a proof. Moreover, he coordinated such a's playing the role of individuals in his refinement of the Fraenkel-Mostowski method for deriving independence results related to the Axiom of Choice. Individuals, also known as urelemente or atoms, are objects distinct from the null set yet having no members and capable of belonging to sets. Zermelo [1908a][1930] had conceived of set theory as a theory of collections built on a base of individuals. However, Abraham Fraenkel [1921][1922: 234ff] from the beginning of his articles on set theory emphasized that there is no need for individuals and generally advocated a minimalist approach as articulated by his Axiom of Restriction [Beschränktheit]. Individuals have since been dispensed with in the mainstream development of axiomatic set theory owing to considerations of elegance, parsimony, and canonicity: Retaining individuals would require a two-sorted formal system or at least predication for sets, with e.g. the Axiom of Extensionality stated only for sets; set theory surrogates were worked out for mathematical objects; and the cumulative hierarchy view based on the empty set provided a uniform, set-theoretic universe. Nevertheless, individuals have continued to be used for various adaptations of the set-theoretic view. Fraenkel [1922] himself, in the Hilbertian axiomatic tradition and notwithstanding his own Axiom of Restriction, concocted a domain of individuals to argue for the independence of the Axiom of Choice, in the inaugural construction of the Fraenkel-Mostowski method.

Stipulating that all individuals are formally to satisfy $\{a\} = a$ has been a persistent theme in the set-theoretic work of the philosopher and logician Willard V. Quine, and is arguably a significant reflection of his philosophy. Already in [1936: 50] Quine had provided a reinterpretation of a type theory that associated individuals with their unit classes. In his [1937] Quine formulated a set theory now known as New Foundations (NF), and in his book *Mathematical Logic* [1940] he extended the theory to include classes.[23] In §22 of [1940] Quine extended the "$x \in y$" notation formally to individuals ("non-classes") y by stipulating for such y that "$x \in y$" should devolve to "$x = y$". In §25 Quine gave a rationale in the reduction of primitive notions: He defined "$x = y$" in terms of $\in$ via extensionality, $\forall z(z \in x \leftrightarrow z \in y)$, and this now encompassed individuals x and y since via $\forall z(z = x \leftrightarrow z = y)$ one gets $x = y$. For individuals y, through the peculiarity of assimilating one form of the copula, "$x \in y$", into another, "$x = y$", there is a formal identification of y with $\{y\}$. While acknowledging that he himself had no occasion for entertaining "non-classes", Quine regarded this as a way of assimilating them into classes: Now every entity is a class, some are "individuals" satisfying $\{a\} = a$, and the Axiom of Extensionality serves all. Thus, Quine's pragmatism in simply stipulating that individuals satisfy $\{a\} = a$ served his extensionalism, with identity defined in terms of membership becoming universal, encompassing even individuals! In his later text *Set Theory and its Logic* Quine [1964: §4] carried out his stipulatory approach and emphasized how it leads to the seamless incorporation of individuals. Stimulated by Quine's discussion of his "individuals" Dana Scott [1962] showed how to transform models of NF into models having a satisfying $\{a\} = a$. Moreover, his technique has led to a spate of recent results.[24]

Many years later in *Pursuit of Truth* [1992: 33] Quine wrote, diffusing ontology: "We could reinterpret 'Tabitha' as designating no longer the cat, but the whole cosmos minus the cat; or, again, as designating the cat's singleton, or unit class." This comes in the wake of his discussion of "proxy functions" as exemplifying the "inscrutability of reference." Such functions are one-to-one

[23] As in von Neumann's set theory [1925] as recast by Bernays [1937], in Quine's system only some classes are membership-eligible, i.e. capable of appearing to the left of the membership relation, and these "elements" correspond to the sets of Quine's NF.

[24] See Forster [1995: 92ff]. He wrote that the technique introduced by Scott is of great importance in set theories with a universal set, "for many of which it is the only independence technique available."

correspondences for the objects of the purported universe, correspondences that preserve predication and so provide different ontologies "empirically on a par." But a transposition of a with $\{a\}$ that keeps all other sets fixed is exactly what Scott [1962] had used to generate a's satisfying $\{a\} = a$ in models of NF. To paraphrase a paraphrase of Quine's, philosophy recapitulates mathematics! There is no longer a confounding of a and $\{a\}$, yet the *a posteriori* identification has served mathematical, even philosophical, purposes.

Finally, despite all the experience with the iterative conception of set, ruminations as to the mysteriousness of the singleton and even the non-existence of the null class have resurfaced in recent mereology, the reincarnated part-whole theory, like an opaque reflection of the debates of a century ago.[25]

5 The Ordered Pair

We now make the final ascent, up to the ordered pair. In modern mathematics, the ordered pair is basic, of course, and is introduced early in the curriculum in the study of analytic geometry. However, to focus the historical background it should be noted that ordered pairs are not explicit in Descartes and in the early work on analytic geometry. Hamilton [1837] may have been the first to objectify ordered pairs in his reconstrual of the complex numbers as ordered "couples" of real numbers. In logic, the ordered pair is fundamental to the logic of relations and epitomized, at least for Russell, metaphysical preoccupations with time and direction. Given the formulation of relations and functions in modern set theory, one might presume that the analysis of the ordered pair emerged from the development of the logic of relations, and that relations and functions were analyzed in terms of the ordered pair. Indeed, this was how the historical development proceeded on the mathematical side, in the work of Peano and Hausdorff. However, the development on the logical side, in the work of Frege and Russell, was just the reverse, with functions playing a key initial role as a fundamental notion. The progression was *to* the ordered pair, and only later was the inverse direction from the ordered pair to function accommodated, after the reduction of the ordered pair itself to classes. While the approach of the logicians is arguably symptomatic of their more metaphysical preoccupations, the approach on the mathematical side was directed toward an increasingly extensional view of functions regarded as arbitrary correspondences.

Frege [1891] had two fundamental categories, *function* and *object*, with a function being "unsaturated" and supplemented by objects as arguments. A *concept* is a function with two possible values, the True and the False, and a *relation* is a concept that takes two arguments. The extension of a concept is its graph or course-of-values [Werthverlauf], which is an object, and Frege [1893: §36] devised an iterated or double course-of-values [Doppelwerthverlauf]

[25]See Lewis [1991].

for the extension of a relation. In these involved ways Frege assimilated relations to functions.

As for the ordered pair, Frege in his *Grundgesetze* [1893: §144] provided the extravagant definition that the ordered pair of x and y is that class to which all and only the extensions of relations to which x stands to y belong.[26] On the other hand, Peirce [1883], Schröder [1895], and Peano [1897] essentially regarded a relation from the outset as just a collection of ordered pairs.[27] Whereas Frege was attempting an analysis of thought, Peano was mainly concerned about recasting ongoing mathematics in economical and flexible symbolism and made many reductions, e.g. construing a *sequence* in analysis as a *function* on the natural numbers. Peano from his earliest logical writings had used "(x, y)" to indicate the ordered pair in formula and function substitutions and extensions. In [1897] he explicitly formulated the ordered pair using "$(x; y)$" and moreover raised the two main points about the ordered pair: First, equation 18 of his Definitions stated the instrumental property which is all that is required of the ordered pair:

$$(*) \qquad \langle x, y\rangle = \langle a, b\rangle \text{ \textit{iff} } x = a \text{ and } y = b\,.$$

Second, he broached the possibility of reducibility, writing: "The idea of a pair is fundamental, i.e. we do not know how to express it using the preceding symbols."

Peano's symbolism was the inspiration and Frege's work a bolstering for Whitehead and Russell's *Principia Mathematica* [1910-3], in which relations distinguished in intension and in extension were derived from "propositional" functions taken as fundamental and other "descriptive" functions derived from relations. They [1910: ∗55] like Frege defined an ordered pair derivatively, in their case in terms of classes and relations, and also for a specific purpose.[28] Previously Russell [1903: §27] had criticized Peirce and Schröder for regarding a relation "essentially as a class of couples," although he overlooked this shortcoming in Peano.[29] Commenting obliviously on *Principia* Peano [1911, 1913]

[26]This definition, which recalls the Whitehead–Russell definition of the cardinal number 2, depended on Frege's famously inconsistent Basic Law V. See Heck [1995] for more on Frege's definition and use of his ordered pair.

[27]Peirce [1883] used "i" and "j" schematically to denote the components, and Schröder [1895: 24] adopted this and also introduced "$i : j$". For Peano see below.

[28]Whitehead and Russell had first defined a cartesian product by other means, and only then defined their ordered pair $x{\downarrow}y$ as $\{x\} \times \{y\}$, a remarkable inversion from the current point of view. They [1910: ∗56] used their ordered pair initially to define the ordinal number 2.

[29]In a letter accepting Russell's [1901] on the logic of relations for publication in his journal *Rivista*, Peano had pointedly written "The classes of couples correspond to relations" (see Kennedy [1975: 214]) so that relations are extensionally assimilated to classes. Russell [1903: §98] argued that the ordered pair cannot be basic and would itself have to be given sense, which would be a circular or an inadequate exercise, and "It seems therefore more correct to take an intensional view of relations ...".

simply reaffirmed an ordered pair as basic, defined a relation as a class of ordered pairs, and a function extensionally as a kind of relation, referring to the final version of his *Formulario Mathematico* [1905-8: 73ff.] as the source.

Capping this to and fro Norbert Wiener [1914] provided a definition of the ordered pair in terms of unordered pairs of classes only, thereby reducing relations to classes. Working in Russell's theory of types, Wiener defined the ordered pair $\langle x, y\rangle$ as

$$\{\{\{x\},\Lambda\},\{\{y\}\}\}$$

when x and y are of the same type and Λ is the null class (of the next type), and pointed out that this definition satisfies the instrumental property $(*)$ above. Wiener used this to eliminate from the system of *Principia* the Axiom of Reducibility for propositional functions of two variables; he had written a doctoral thesis comparing the logics of Schröder and Russell.[30] Although Russell praised Sheffer's stroke, the logical connective not-both, he was not impressed by Wiener's reduction. Indeed, Russell would not have been able to accept it as a genuine analysis. After his break with neo-Hegelian idealism, Russell insisted on taking relations to have genuine metaphysical reality, external to the mind yet intensional in character. On this view, order had to have a primordial reality, and this was part and parcel of the metaphysical force of intension. Years later Russell [1959: 67] wrote:

> I thought of relations, in those days, almost exclusively as *intensions*. ... It seemed to me — as indeed, it still seems — that, although from the point of view of a formal calculus one can regard a relation as a set of ordered couples, it is the intension alone which gives unity to the set.

On the mathematical side, Hausdorff's classic text, *Grundzüge der Mengenlehre* [1914], broke the ground for a generation of mathematicians in both set theory and topology. A compendium of a wealth of results, it emphasized mathematical approaches and procedures that would eventually take firm root.[31] Making no intensional distinctions Hausdorff [1914: 32ff.,70ff.] defined an ordered pair in terms of unordered pairs, formulated functions in terms of ordered pairs, and ordering relations as collections of ordered pairs. (He did not so define an arbitrary relation, for which there was then no mathematical

[30] See Grattan-Guinness [1975] for more on Wiener's work and his interaction with Russell.

[31] Hausdorff's mathematical attitude is reflected in a remark following his explanation of cardinal number in a revised edition [1937:§5] of [1914]: "This formal explanation says what the cardinal numbers are supposed to do, not what they are. More precise definitions have been attempted, but they are unsatisfactory and unnecessary. Relations between cardinal numbers are merely a more convenient way of expressing relations between sets; we must leave the determination of the 'essence' of the cardinal number to philosophy."

use, but he was first to consider general *partial* orderings, as in his maximality principle.[32]) Hausdorff thus made both the Peano [1911, 1913] and Wiener [1914] moves *in* mathematical practice, completing the reduction of functions to sets.[33] This may have been congenial to Peano, but not to Frege nor Russell, they having emphasized the primacy of functions. Following the pioneering work of Dedekind and Cantor Hausdorff was at the crest of a major shift in mathematics of which the transition from an intensional, rule-governed conception of function to an extensional, arbitrary one was a large part, and of which the eventual acceptance of the Power Set Axiom and the Axiom of Choice was symptomatic.

In his informal setting Hausdorff took the ordered pair of x and y to be

$$\{\{x, 1\}, \{y, 2\}\}$$

where 1 and 2 were intended to be distinct objects alien to the situation. It should be pointed out that the definition works even when x or y is 1 or 2 to maintain the instrumental property $(*)$ of ordered pairs. In any case, the now-standard definition is the more intrinsic

$$\{\{x\}, \{x, y\}\}$$

due to Kazimierz Kuratowski [1921: 171]. Notably, Kuratowski's definition is a by-product of his analysis of Zermelo's [1908] proof of the Well-Ordering Theorem. Not only does the definition satisfy $(*)$, but it only requires pairings of x and y.

The general adoption of the Kuratowski pair proceeded through the major developments of mathematical logic: Von Neumann initially took the ordered pair as primitive but later noted [1928: 338][1929: 227] the reduction via the Kuratowski definition. Gödel in his incompleteness paper [1931: 176] also pointed out the reduction.[34] Tarski in his [1931: n.3], seminal for its precise, set-theoretic formulation of a first-order definable set of reals, pointed out the reduction and acknowledged his compatriot Kuratowski. In his recasting of von Neumann's system, Bernays [1937: 68] also acknowledged Kuratowski [1921]

[32]Before Hausdorff and going beyond Cantor, Dedekind was first to consider non-linear orderings, e.g. in his remarkably early, axiomatic study [1900] of lattices.

[33]As to historical precedence, Wiener's note was communicated to the Cambridge Philosophical Society, presented on 23 February 1914, while the preface to Hausdorff's book is dated 15 March 1914. Given the pace of book publication then, it is arguable that Hausdorff came up with his reduction first.

[34]In footnote 18, Gödel blandly remarked: "Every proposition about relations that is provable in [*Principia Mathematica*] is provable also when treated in this manner, as is readily seen." This stands in stark contrast to Russell's labors in *Principia* and his antipathy to Wiener's reduction of the ordered pair.

and began with its definition for the ordered pair. It is remarkable that Nicolas Bourbaki in his treatise [1954] on set theory still took the ordered pair as primitive, only later providing the Kuratowski reduction in the [1970] edition.[35]

As with the singleton, we end here with a return to philosophy through Quine. In *Principia Mathematica* Whitehead and Russell had developed the theory of monadic (unary) and dyadic (binary) relations separately and only indicated how the theory of triadic (ternary) relations and so forth was to be analogously developed. Quine in his 1932 Ph.D. dissertation showed how to develop the theory of n-ary relations for all n simultaneously, by defining ordered n-tuples in terms of the ordered pair. Again, a simple technical innovation was the focus, but for Quine such moves would be crucial in an economical, extensional approach to logic and philosophy, already much in evidence in his dissertation.[36] In its published version Quine acknowledged in a footnote [1934: 16ff.] having become aware of the reduction of the ordered pair itself, which he had taken to be primitive, to classes. Significantly, Quine wrote of the "Wiener-Kuratowski" ordered pair, though their definitions were quite different; the possibility, rather than the particular implementation, was for Quine the point. And unlike for Russell, for whom a reduction was metaphysically unacceptable, for Quine it was merely a question of how the reduction could be worked in.

Quine in his major philosophical work *Word and Object* [1960: §53] declared that the reduction of the ordered pair is a paradigm for philosophical analysis. Perhaps for Quine the Wiener-Kuratowski reduction loomed particularly large because of his encounter with it after the labors of his dissertation. After describing the Wiener ordered pair, Quine wrote:

> This construction is paradigmatic of what we are most typically up to when in a philosophical spirit we offer an "analysis" or "explication" of some hitherto inadequately formulated "idea" or expression. We do not claim synonymy. We do not claim to make clear and explicit what the users of the unclear expression had unconsciously in mind all along. We do not expose hidden meanings, as the words 'analysis' and 'explication' would suggest; we supply lacks. We fix on the particular functions of the unclear expression that make it worth troubling about, and then devise a substitute, clear and couched in terms to our liking, that fills those functions. Beyond those conditions of partial agreement, dictated by our interest and purposes, any traits of the explicans come under the head of "don't cares" . . .

Quine went on to describe the Kuratowski ordered pair but emphasized that

[35]See Mathias [1992] for the ignorance of Bourbaki.

[36]See Dreben [1989] for more on Quine and his dissertation.

as long as a definition satisfies the instrumental property (*), the particular choice to be made falls under the "don't care" category. Quine then wrote:

> A similar view can be taken of every case of explication: *explication is elimination.* We have, to begin with, an expression or form of expression that is somehow troublesome. It behaves partly like a term but not enough so, or it is vague in ways that bother us, or it puts kinks in a theory or encourages one or another confusion. But also it serves certain purposes that are not to be abandoned. Then we find a way of accomplishing those same purposes through other channels, using other and less troublesome forms of expression. The old perplexities are resolved.
>
> According to an influential doctrine of Wittgenstein's, the task of philosophy is not to solve problems but to dissolve them by showing that there were really none there. This doctrine has its limitations, but it aptly fits explication. For when explication banishes a problem it does so by showing it to be in an important sense unreal; viz., in the sense of proceeding only from needless usages.

The ordered pair may be a simple device in set theory, but it is now carrying the weight of what "analysis" is to mean in philosophy.

References

1937 PAUL BERNAYS, A system of axiomatic set theory – Part I, *The Journal of Symbolic Logic 2*, 65–77.

1941 PAUL BERNAYS, A system of axiomatic set theory – Part II, *The Journal of Symbolic Logic 6*, 1–17.

1954 PAUL BERNAYS, A system of axiomatic set theory – Part VII, *The Journal of Symbolic Logic 19*, 81–96.

1847 GEORGE BOOLE, *The Mathematical Analysis of Logic, Being an Essay Towards a Calculus of Deductive Reasoning*, Macmillan, Barclay, and Macmillan, Cambridge; reprinted in Ewald [1996], vol. 1, 451–509.

1854 GEORGE BOOLE, *An Investigation of the Laws of Thought, On Which are Founded the Mathematical Theories of Logic and Probabilities*, Macmillan, London.

1954 NICOLAS BOURBAKI, *Eléments de Mathématique. I. Théorie des Ensembles*, Chapters I and II, Hermann, Paris.

1970 NICOLAS BOURBAKI, *Eléments de Mathématique. I. Théorie des Ensembles*, combined edition, Hermann, Paris.

1874 GEORG CANTOR, Über eine Eigenschaft des Inbegriffes aller reellen algebraischen Zahlen, *Journal für die reine und angewandte Mathematik 77*, 258–262; reprinted in Cantor [1932] below, 115–118; translated in Ewald [1996], vol. 2, 839–843.

1880 GEORG CANTOR, Über unendliche, lineare Punktmannigfaltigkeiten. II. *Mathematische Annalen 17*, 355–358; reprinted in Cantor [1932] below, 145–148.

1891 GEORG CANTOR, Über eine elementare Frage der Mannigfaltigkeitslehre, *Jahresbericht der Deutschen Mathematiker-Vereinigung 1*, 75–78; reprinted in Cantor [1932] below, 278–280; translated in Ewald [1996], vol. 2, 920–922.

1895 GEORG CANTOR, Beiträge zur Begründung der transfiniten Mengenlehre. I. *Mathematische Annalen 46*, 481–512; translated in Cantor [1915] below; reprinted in Cantor [1932] below, 282–311.

1915 GEORG CANTOR, *Contributions to the Founding of the Theory of Transfinite Numbers*, including translations of Cantor [1895] and Cantor [1897] above with introduction and notes by Philip E.B. Jourdain, Open Court, Chicago; reprinted Dover, New York 1965.

1932 GEORG CANTOR, *Gesammelte Abhandlungen mathematischen und philosophischen Inhalts*, Ernst Zermelo (editor), Julius Springer, Berlin; reprinted in Springer-Verlag, Berlin 1980.

1943 PETER A. CARMICHAEL, The null class nullified, *Philosophical Review 52*, 61–68.

1943a PETER A. CARMICHAEL, Animadversion on the null class, *Philosophy of Science 10*, 90–94.

1888 RICHARD DEDEKIND, *Was sind und was sollen die Zahlen?* F. Vieweg, Braunschweig; sixth, 1930 edition reprinted in Dedekind [1932] below, 335–390; second, 1893 edition translated in Dedekind [1963] below, 29–115 and also translated in Ewald [1996], vol. 2, 795–833.

1900 RICHARD DEDEKIND, Über die von drei Moduln erzeugte Dualgruppe, *Mathematische Annalen 53*, 371–403; reprinted in Dedekind [1932], vol. 2, 236–271.

1932 RICHARD DEDEKIND, *Gesammelte mathematische Werke*, vol. 3, Fricke, Robert, Emmy Noether, and Öystein Ore (editors), F. Vieweg, Braunschweig; reprinted Chelsea, New York 1969.

1963 RICHARD DEDEKIND, *Essays on the Theory of Numbers*, translations by Wooster W. Beman, Dover, New York (reprint of original edition, Open Court, Chicago 1901).

1989 BURTON S. DREBEN, Quine, in *Quine in Perspective*, Robert Barrett and Roger Gibson (editors), Blackwell, Oxford.

1996 WILLIAM EWALD, *From Kant to Hilbert: A Source Book in the Foundations of Mathematics*, Clarendon Press, Oxford.

1999 JOSÉ FERREIRÓS, *Labyrinth of Thought: A History of Set Theory and its Role in Modern Mathematics*, Birkhauser Verlag, Basel.

1995 THOMAS E. FORSTER, *Set Theory with a Universal Set*, Oxford Logic Guides #31, Clarendon Press, Oxford; second edition.

1921 ABRAHAM FRAENKEL, Über die Zermelosche Begründung der Mengenlehre, *Jahresbericht der Deutschen Mathematiker-Vereinigung 30*, 97–98.

1922 ABRAHAM FRAENKEL, Zu den Grundlagen der Cantor-Zermeloschen Mengenlehre, *Mathematische Annalen 86*, 230–237.

1879 GOTTLOB FREGE, *Begriffsschrift, eine der arithmetischen nachgëbildete Formelsprache des reinen Denkens*, Nebert, Halle; reprinted Hildesheim, Olms 1964; translated in van Heijenoort [1967], 1–82.

1884 GOTTLOB FREGE, *Die Grundlagen der Arithmetik, eine logisch-mathematische Untersuchung über den Begriff der Zahl.* Breslau, Wilhelm Köbner 1884; translated with German text by John L. Austin, as *The Foundations of Arithmetic, A logico-mathematical enquiry into the concept of number*; Blackwell, Oxford 1950; later editions without German text, Harper, New York.

1891 GOTTLOB FREGE, Function und Begriff, Hermann Pohle, Jena; translated in Frege [1952] below, 21–41.

1893 GOTTLOB FREGE, *Grundgesetze der Arithmetik, Begriffsschriftlich abgeleitet*, vol. 1, Hermann Pohle, Jena; reprinted Olms, Hildesheim 1962.

1895 GOTTLOB FREGE, Kritische Beleuchtung einiger Punkte in E. Schröders Vorlesungen über die Algebra der Logik, *Archiv für systematische Philosophie 1*, 433–456; translated in Frege [1952] below, pp. 86–106.

1952 GOTTLOB FREGE, *Translations from the Philosophical Writings of Gottlob Frege*, Geach, Peter, and Max Black (translators and editors), Blackwell, Oxford 1952; second, revised edition 1960; latest edition, Rowland & Littlewood, Totowa 1980.

1992 ALEJANDRO R. GARCIADIEGO, *Bertrand Russell and the Origins of the Set-theoretic "Paradoxes"*, Birkhäuser, Boston.

1931 KURT GÖDEL, Über formal unentscheidbare Sätze der *Principia Mathematica* und verwandter Systeme I, *Monatshefte für Mathematik und Physik 38*, 173–198; reprinted, together with a translation incorporating minor emendations by Gödel, in Gödel [1986] below, 144–195.

1986 KURT GÖDEL, *Collected Works*, vol. 1, Feferman, Solomon, *et al.* (editors), New York, Oxford University Press.

1975 IVOR GRATTAN-GUINNESS, Wiener on the logics of Russell and Schröder. An account of his doctoral thesis, and of his subsequent discussion of it with Russell, *Annals of Science 32*, 103–132.

1978 IVOR GRATTAN-GUINNESS, How Bertrand Russell discovered his paradox, *Historia Mathematica 5*, 127–137.

1837 WILLIAM R. HAMILTON, Theory of conjugate functions, or algebraic couples: with a preliminary and elementary essay on algebra as the science of pure time, *Transactions of the Royal Irish Academy 17*, 293–422; reprinted in *The Mathematical Papers of Sir William Rowan Hamilton*, H. Halberstam and R.E. Ingram (editors), Cambridge University Press, Cambridge 1931–1967.

1914 FELIX HAUSDORFF, *Grundzüge der Mengenlehre*, de Gruyter, Leipzig; reprinted Chelsea, New York 1965.

1937 FELIX HAUSDORFF, *Mengenlehre*, third, revised edition of Hausdorff [1914]; translated by John R. Auman as *Set Theory*; Chelsea, New York 1962.

1995 RICHARD G. HECK, JR., Definition by Induction in Frege's *Grundgesetze der Arithmetik*, in *Frege's Philosophy of Mathematics*, William Demopoulos (editor), Harvard University Press, Cambridge, 295–333.

1891 EDMUND HUSSERL, Besprechung von E. Schröder, Vorlesungen über die Algebra der Logik (exakte Logik), vol. I, Leipzig 1890, *Göttingische Gelehrte Anzeigen* 1891, 243–278; reprinted in Edmund Husserl, *Aufsätze und Rezensionen (1890–1910), Husserliana Vol. XXII*, Nijhoff, The Hague 1978, 3–43; translated by Dallas Willard in *Edmund Husserl, Early*

Writings in the Philosophy of Logic and Mathematics. Collected Works V, Kluwer, Dordrecht 1994, 52-91.

1997 AKIHIRO KANAMORI, The mathematical import of Zermelo's Well-Ordering Theorem, *The Bulletin of Symbolic Logic 3*, 281–311.

1975 HUBERT C. KENNEDY, Nine letters from Giuseppe Peano to Bertrand Russell, *Journal of the History of Philosophy 13* 205–220.

1921 KAZIMIERZ KURATOWSKI, Sur la notion de l'ordre dans la théorie des ensembles, *Fundamenta Mathematicae 2*, 161–171; reprinted in Kuratowski [1988] below, 1–11.

1988 KAZIMIERZ KURATOWSKI, *Selected Papers,* Borsuk, Karol, *et al.* (editors), Warsaw, Państwowe Wydawnictwo Naukowe.

1991 DAVID LEWIS, *Parts of Classes*, Basil Blackwell, Cambridge.

1992 ADRIAN R.D. MATHIAS, The ignorance of Bourbaki, *Mathematical Intelligencer 14*, 4–13.

1982 GREGORY H. MOORE, *Zermelo's Axiom of Choice: Its Origins, Development, and Influence.* Springer-Verlag, New York.

1995 GREGORY H. MOORE, The origins of Russell's Paradox: Russell, Couturat, and the antinomy of infinite number, in *From Dedekind to Gödel,* Jaakko Hintikka (editor) Synthese Library volume 251, Kluwer, Dordrecht, 215–239.

1889 GIUSEPPE PEANO, *Arithmetices principia nova methodo exposita,* Bocca, Turin; reprinted in Peano [1957-9] below, vol. 2, 20–55; partially translated in van Heijenoort [1967], 85–97; translated in Peano [1973] below, 101–134.

1889a GIUSEPPE PEANO, I principii di geometrica logicamente espositi, Bocca, Turin; reprinted in Peano [1957-9] below, vol. 2, 56–91.

1890 GIUSEPPE PEANO, Démonstration de l'intégrabilité des équations différentielles ordinaires, *Mathematische Annalen 37*, 182–228; reprinted in Peano [1957-9] below, vol. 1, 119–170.

1891 GIUSEPPE PEANO, Principii di logica matematica, *Rivista di matematica 1*, 1–10; reprinted in Peano [1957-9] below, vol. 2, 92–101; translated in Peano [1973] below, 153–161.

1894 GIUSEPPE PEANO, Notations de logique mathématique (Introduction au Formulaire de mathématiques), Guadaguini, Turin.

1897 GIUSEPPE PEANO, Studii di logica matematica. *Atti della Accademia delle Scienze di Torino, Classe di Scienze Fisiche, Matematiche e Naturali 32*, 565–583; reprinted in Peano [195-9] below, vol. 2, 201–217; translated in Peano [1973] below, 190–205.

1905-8 GIUSEPPE PEANO, *Formulario Mathematico,* Bocca, Torin; reprinted Edizioni Cremonese, Rome 1960.

1911 GIUSEPPE PEANO, Sulla definizione di funzione. *Atti della Accademia Nazionale dei Lincei, Rendiconti, Classe di Scienze Fisiche, Matematiche e Naturali 20-I*, 3–5.

1913 GIUSEPPE PEANO, Review of: A.N. Whitehead and B. Russell, *Principia Mathematica*, vols. I,II. *Bollettino di bibliografia e storia delle scienze matematiche (Loria) 15*, 47–53, 75–81; reprinted in Peano [1957-9] below, vol. 2, 389–401.

1957-9 GIUSEPPE PEANO, *Opera Scelte*, three volumes, U. Cassina (editor), Edizioni Cremonese, Rome.

1973 GIUSEPPE PEANO, *Selected Works of Giuseppe Peano*, Hubert C. Kennedy (translator and editor), University of Toronto Press, Toronto.

1990 VOLKER PECKHAUS, "Ich habe mich wohl gehütet, alle Patronen auf einmal zu verschiessen". Ernst Zermelo in Göttingen, *History and Philosophy of Logic 11*, 19–58.

1883 CHARLES S. PEIRCE, A theory of probable inference. Note B. The logic of relatives, *Studies in Logic by Members of the John Hopkins University,* Charles S. Peirce (editor), Little-Brown, Boston, 187–203; reprinted in Peirce [1931-58] below, vol. 3, 195–209.

1885 CHARLES S. PEIRCE, On the algebra of logic: A contribution to the philosophy of notation, *The American Journal of Mathematics 7*, 180–202; reprinted in Peirce [1931-58] below, vol. 3, 210-238, and in Ewald [1996], vol. 1, 608–32.

1931-58 CHARLES S. PEIRCE, *Collected Papers of Charles Sanders Peirce*, Hartshorne, Charles, and Paul Weiss (editors), Harvard University Press, Cambridge.

1934 WILLARD V. QUINE, *A System of Logistic*, Harvard University Press. Cambridge.

1936 WILLARD V. QUINE, Set-theoretic foundations for logic, *The Journal of Symbolic Logic 1*, 45–57; reprinted in Quine [1966] below, 83–99.

1937 WILLARD V. Quine, New foundations for mathematical logic, *American Mathematical Monthly 44*, 70–80.

1940 WILLARD V. QUINE, *Mathematical Logic*, Norton, New York.

1960 WILLARD V. QUINE, *Word and Object*, MIT Press, Cambridge.

1963 WILLARD V. QUINE, *Set Theory and its Logic*, Harvard University Press, Cambridge; revised edition 1969.

1966 WILLARD V. QUINE, *Selected Logic Papers*, Harvard University Press, Cambridge; enlarged edition, 1995.

1992 WILLARD V. QUINE, *Pursuit of Truth*, revised Edition, Harvard University Press, Cambridge.

1981 BERNHARD RANG and WOLFGANG THOMAS, Zermelo's discovery of the "Russell paradox", *Historia Mathematica 8*, pp. 15–22.

1901 BERTRAND A.W. RUSSELL, Sur la logique des relations avec des applications à la théorie des séries, *Revue de mathématiques (Rivista di matematica) 7*, 115–148; partly reprinted in *The Collected Papers of Bertrand Russell*, volume 3 (Gregory H. Moore, editor), Routledge, London 1993, 613–627; translated in same, 310–349.

1903 BERTRAND A.W. RUSSELL, *The Principles of Mathematics*, Cambridge University Press, Cambridge; later editions, George Allen & Unwin, London.

1908 BERTRAND A.W. RUSSELL, Mathematical logic as based on the theory of types, *American Journal of Mathematics 30*, 222–262; reprinted in van Heijenoort [1967], 150–182.

1918 BERTRAND A.W. RUSSELL, *Mysticism and Logic, and Other Essays*, Longmans, Green & Co., New York.

1944 BERTRAND A.W. RUSSELL, My mental development, in: *The Philosophy of Bertrand Russell*, Paul A. Schilpp (editor), The Library of Living Philosophers vol. 5, Northwestern University, Evanston, 3–20.

1945 BERTRAND A.W. RUSSELL, *A History of Western Philosophy*, Simon Schuster, New York.

1959 BERTRAND A.W. RUSSELL, *My Philosophical Development*, George Allen & Unwin, London.

1890 ERNST SCHRÖDER, *Vorlesungen über die Algebra der Logik (exakte Logik)*, vol. 1, B.G. Teubner, Leipzig; reprinted in Schröder [1966] below.

1895 ERNST SCHRÖDER, *Vorlesungen über die Algebra der Logik (exakte Logik)*. vol. 3: *Algebra und Logik der Relative.* B.G. Teubner, Leipzig; reprinted in Schröder [1966] below.

1966 ERNST SCHRÖDER, *Vorlesungen über die Algebra der Logik*, three volumes, Chelsea, New York.

1962 DANA S. SCOTT, Quine's individuals, in: *Logic, Methodology and the Philosophy of Science*, Ernst Nagel (editor), Stanford University Press, Stanford, 111-115.

1971 MOHAMMED A. SINACEUR, Appartenance et inclusion: un inédit de Richard Dedekind, *Revue d'Histoire des Sciences et de leurs Applications 24*, 247–255.

1957 ERNST SPECKER, Zur Axiomatik der Mengenlehre (Fundierungs- und Auswahlaxiom), *Zeitschrift für mathematische Logik und Grundlagen der Mathematik 3*, 173–210.

1931 ALFRED TARSKI, Sur les ensembles définissables de nombres réels, *Fundamenta Mathematicae 17*, 210–239; translated in *Logic, Semantics, Metamathematics. Papers from 1923 to 1938,* translations by J.H. Woodger, second edition, Hackett, Indianapolis 1983, 110–142.

1967 JEAN VAN HEIJENOORT, *From Frege to Gödel: A Source Book in Mathematical Logic, 1879–1931*, Harvard University Press, Cambridge; reprinted 2002.

1925 JOHN VON NEUMANN, Eine Axiomatisierung der Mengenlehre, *Journal für die reine und angewandte Mathematik 154*, 219–240, Berichtigung in *155*, 128; reprinted in von Neumann [1961] below, 34–56; translated in van Heijenoort [1967], 393–413.

1928 JOHN VON NEUMANN, Über die Definition durch transfinite Induktion und verwandte Fragen der allgemeinen Mengenlehre, *Mathematische Annalen 99*, 373–391; reprinted in von Neumann [1961] below, 320–338.

1929 JOHN VON NEUMANN, Über eine Widerspruchfreiheitsfrage in der axiomaticschen Mengenlehre. *Journal für die reine und angewandte Mathematik 160*, 227–241; reprinted in von Neumann [1961] below, 494–508.

1961 JOHN VON NEUMANN, *John von Neumann. Collected Works.* vol. 1, Taub, Abraham H. (ed.), Pergamon Press, New York.

1914 NORBERT WIENER, A simplification of the logic of relations, *Proceedings of the Cambridge Philosophical Society 17*, 387–390; reprinted in van Heijenoort [1967], 224–227.

1910-3 ALFRED N. WHITEHEAD and BERTRAND A.W. RUSSELL, *Principia Mathematica*, three volumes, Cambridge University Press, Cambridge.

1904 ERNST ZERMELO, Beweis, dass jede Menge wohlgeordnet werden kann (Aus einem an Herrn Hilbert gerichteten Briefe), *Mathematische Annalen 59*, 514–516; translated in van Heijenoort [1967], pp. 139–141.

1908 ERNST ZERMELO, Neuer Beweis für die Möglichkeit einer Wohlordnung, *Mathematische Annalen 65*, 107–128; translated in van Heijenoort [1967], 183–198.

1908a ERNST ZERMELO, Untersuchungen über die Grundlagen der Mengenlehre I, *Mathematische Annalen 65*, 261–281; translated in van Heijenoort [1967], 199–215.

1930 ERNST ZERMELO, Über Grenzzahlen und Mengenbereiche: Neue Untersuchungen über die Grundlagen der Mengenlehre, *Fundamenta Mathematicae 16*, 29–47; translated in Ewald [1996], vol. 2, 1208–1233.

5 In Praise of Replacement

This article serves to present a large mathematical perspective and historical basis for the Axiom of Replacement as well as to affirm its importance as a central axiom of modern set theory. The standard ZFC axioms for set theory provide an operative foundation for mathematics in the sense that mathematical concepts and arguments can be reduced to set-theoretic ones, based on sets doing the work of mathematical objects. As is widely acknowledged, Replacement together with the Axiom of Infinity provides the rigorization for transfinite recursion and together with the Power Set Axiom provides for sets of large cardinality e.g. through infinite iterations of the power set operation. And in a broad sense, in so far as the concept of function (mapping, functor) plays a central role in modern mathematical practice, Replacement plays a central role in set theory.

A significant motivation for the writing of this article is to confirm—to rectify, if need be perceived—the status of Replacement among the axioms of set theory. From time to time hesitation and even skepticism have been voiced about the importance and even the need for Replacement.[1] The issues, animated by how, historically, Replacement emerged as an axiom later than the others, have largely to do with ontological commitment, about what sets (there) are or should be. Replacement has been regarded as necessary only to provide for large cardinality sets and, furthermore, unmotivated by an ultimately ontological "iterative conception" of set.

The answer here, as well as the general affirmation, is that Replacement is a subtle axiom, with the subtlety lying in both what it says and what it does. Like the Axiom of Choice, Replacement has to do with functional correlation, but the latter turns on formalization, on definability. And the definability crucially leads to a full range of possibilities for recursive definition. Together with the Axiom of Foundation, Replacement is thus a crucial pillar of modern set theory, which, as a matter of method, is an investigation through and of recursion and well-foundedness. From a modern vantage point, Replacement

Republished from *The Bulletin of Symbolic Logic* 18 (2012), pp.45-90, with permission from The Association for Symbolic Logic.

Most of this article was written whilst the author was a senior fellow at the Lichtenberg-Kolleg of the Georg-August-Universität Göttingen for 2009-2010. He expresses his gratitude for the productive conditions and support of the Kolleg. The article served as the text for an invited address at the Workshop on Set Theory and the Philosophy of Mathematics held at the University of Pennsylvania, October 2010, and the author would like to express his gratitude to the organizer Scott Weinstein. For information or inspiration on some points, many thanks to Juliet Floyd, John Burgess, Heinz-Dieter Ebbinghaus, Colin McLarty, Adrian Mathias, Michael Rathjen, Erich Reck, Wilfried Sieg and the referee.

[1]This is brought out below with respect to the early contributors to the development of set theory and in §6 in connection with writings of George Boolos. The text [Potter, 2004] pursues an axiomatic development of set theory which avoids Replacement as being contextually unmotivated (cf. its pp. 296ff). The recent webpage discussion of Replacement at http://cs.nyu.edu/pipermail/fom/2007-August/ drew out some skeptical attitudes.

also carries the weight of the heuristic of reflection, that any property ascribable to the set-theoretic universe should already be ascribable to a curtailed initial segment, as an underlying and motivating principle of set theory.

But stepping back even from this, Replacement can be seen as a crucial bulwark of *indifference to identification*, in set theory and in modern mathematics generally. To describe a prominent example, several definitions of the real numbers as generated from the rational numbers have been put forward—in terms of the geometric continuum, Dedekind cuts, and Cauchy sequences—yet in mathematical practice there is indifference to actual identification with any particular objectification as one proceeds to work with *the* real numbers. In set theory, one opts for a particular representation for an ordered pair, for natural numbers, and so forth. What Replacement does is to allow for articulations that these representations are not necessary choices and to mediate generally among possible choices. Replacement is a corrective for the other axioms, which posit specific sets and subsets, by allowing for a fluid extensionalism. The deepest subtlety here is also on the surface, that through functional correlation one can shift between tokens (instances, representatives) and types (extensions, classes), and thereby shift the ground itself for what the types are.[2]

In what follows, these themes, here briefly broached, are widely explored in connection with the historical emergence of set theory and Replacement. This account can in fact be viewed as one of the early development of set theory as seen through the prism of Replacement, drawing out how the Replacement motif gauges the progress from 19th Century mathematics to modern set theory. There is recursion emerging as method; early incentives for trans-species coordination, expressive of indifference to identification; provision for sets of relatively large cardinality; the crucial role in transfinite recursion; and the formalization as principle of the heuristic of reflection. §1 casts Replacement as underlying modern mathematics through the motif $\{t_i \mid i \in I\}$, the roots of this involvement going back particularly to the infinitary initiatives of Richard Dedekind. §2 sets out the early history of Replacement in the work of the pioneers of set theory from an adumbration in correspondence of Cantor himself through to the formulations of Fraenkel and Skolem. §3 sets out the middle history of Replacement, largely the work of von Neumann on transfinite recursion and the emergence of the cumulative hierarchy picture. §4 sets out the later history of Replacement, the main focus being on Gödel's work with the constructible universe. §5 discusses the various forms of Replacement, bringing out its many emanations and applications. Finally, §6 sets out skeptical

[2]Willard Van Orman Quine with his theses of indeterminacy of reference and ontological relativity illuminated indifference to identification, and discussed in particular the ordered pair in his *Word and Object* [1960], §53. For recent articulations of indifference to identification see Richard Pettigrew [2008] and John Burgess [2009], from which the phrase was drawn.

reactions about Replacement, recent involvements of the axiom, and affirms on general grounds the importance of Replacement as a central axiom of modern set theory.

1 Replacement Underlying Mathematics

The Axiom of Replacement is expressible in first-order set theory as the following schema: For any formula φ with free variables among $a, x, y, z_1, \ldots, z_n$ but not including b,

$$\text{(Rep)} \qquad \forall z_1 \forall z_2 \ldots \forall z_n \forall a (\forall x \exists ! y \varphi \longrightarrow \exists b \forall y (\exists x (x \in a \wedge \varphi) \longleftrightarrow y \in b)),$$

where $\exists!$ abbreviates the formalizable "there exists exactly one". Seen through unpracticed eyes this formalization suggests complications and fuss; in a historically resonant sense as we shall see, the first-order account conveys a fundamental principle about functional substitution as an inherent feature of modern mathematics. Informally, and as historically presented, Replacement asserts that for any definable class function F ($F(x) = y \longleftrightarrow \varphi$ in the above) and for any set a, the image $F``a = \{F(x) \mid x \in a\}$ is a set. Thus, Replacement is proximate whenever one posits $\{t_i \mid i \in I\}$ according to some specifiable correlation of t_i's to i's in an index set I. The many facets of $\{t_i \mid i \in I\}$ only emerged in mathematics, however, in its 19th Century transformation from a structural analysis of what there is to a complex edifice of conceptual constructions, and it is worthwhile to recapitulate this development. In what follows, "Replacement motif" is used to refer to $\{t_i \mid i \in I\}$ as a broader conceptualization than "Replacement" for versions of the later axiom.

As with the Axiom of Choice, before Replacement was articulated implicit assumptions had been made that now can be seen as dependent on the axiom. The seminal work of Richard Dedekind is particularly relevant, and in what follows we highlight its involvement with Replacement. In the development of set theory as a subject, the Replacement motif first occurred in Cantor's work in connection with cardinality (cf. §2) with the working assumption that $i \mapsto t_i$ was one-to-one, which is more faithful to "replacement". The injectivity would be importantly relaxed in important situations, however, and this is inherent in the consideration of more and more functions in mathematics. To be noted is that in discussing the Replacement motif in the early stages one is focusing on an emergent theme, and retroactive formalization will only require, naturally enough, a small part of the full modern schema.

Surely the first substantive appearance of $\{t_i \mid i \in I\}$ was in analysis, with the t_i's themselves functions and I the natural numbers. Initially, the functions were typically real functions given symbolically as polynomials, and the assignment $i \mapsto t_i$ would not itself have been considered a function. When infinite sequences of infinite series were considered, e.g. for nowhere differentiable functions as limits of continuous functions, new stress was put on the traditional approach to the infinite regarded only in potentiality. The Replacement motif became substantively involved when the infinite in actuality became incorporated into mathematics. This occurred with the emergence of the now basic construction of a set I structured by an equivalence relation E, the consideration of the corresponding equivalence classes $[i]_E$ for $i \in I$ leading to the totality $\{[i]_E \mid i \in I\}$. One sees here that, as the function $i \mapsto [i]_E$ becomes something to be reckoned with, it is quite general and, importantly, not one-to-one. This early move toward modern algebra went hand in hand with the development of set-theoretic conceptualizations, particularly in the work of Dedekind.

Already in the early [1857], Dedekind worked in modular arithmetic with the actually infinite residue classes themselves as unitary objects. His context was in fact $Z[x]$, the polynomials in x with integer coefficients, and with this he was the first to entertain a totality consisting of infinitely many infinite equivalence classes. The roots of these polynomials, first investigated by Dedekind, are of course what came to be called the algebraic numbers. In a telling passage discussing $Z[x]$ modulo a prime number p, Dedekind wrote (cf. [Dedekind, 1930 2, vol. 1, pp. 46-47]):

> The preceding theorems correspond exactly to those of number divisibility, in that the whole system of infinitely many functions of a variable congruent to each other modulo p behaves here as a single concrete number in number theory, as each function of that system substitutes completely for any other; such a function is a representative of the whole class; each class possesses its definite degree, its divisors, etc., and all those traits correspond in the same manner to each particular member of the class. The system of infinitely many incongruent classes—infinitely many, since the degree may grow indefinitely—corresponds to the series of whole numbers in number theory.

One can arguably date the entry of the actual infinite into mathematics here, in the sense of infinite totalities serving as unitary objects within an infinite mathematical system,[3] as well as the beginnings of mathematical indifference to identification for equivalence classes and particular representatives.

[3]Carl Friedrich Gauss, in his *Disquisitiones Arithmeticae* and later, worked in modular arithmetic directly with residues and never entertained equivalence classes as unitary objects.

Bernard Bolzano in his *Paradoxien des Unendlichen* [1851] and elsewhere (see his mathematical works in [Russ, 2004]) had earlier advocated the actual infinite, but his advocacy went only so far as to discuss comparative mappings between infinite collections and moreover was not embedded in mathematical practice.

In the prefaces to both his celebrated essays [1872] and [1888] Dedekind recorded the autumn of 1858 as when he first came to the Dedekind cuts. Thus he had devised his best-known construction of an infinite system consisting of infinite totalities serving as unitary objects just a year after his analysis of $Z[x]$ modulo a prime.

In a fragment "Aus den Gruppen-Studien (1855-1858)" Dedekind [1930 2, vol. 3, pp. 439-445], working notationally with finite groups, gave in effect the Homomorphism Theorem, that given a homomorphism of a group G onto a group H with a corresponding kernel K, there is an isomorphism between the quotient group G/K and H. Allowing for the historical happenstance of working in the finite, Dedekind here was first to *start* with a function and develop corresponding equivalence classes. The Replacement motif $\{t_i \mid i \in I\}$ comes into focus, soon to be put in a new light.

In Dedekind's *Nachlass* can be found sketches, conjectured to be from 1872,[4] of the now-familiar genetic generation of the integers as equivalence classes of pairs of natural numbers, a pair representing their difference, and of the rational numbers as equivalence classes of pairs of integers, a pair representing their ratio. With this approach to achieve the status of definitions in mathematics, one sees here the further beginnings of mathematical indifference to identification. Not only are there to be correlations with antecedent notions of integers and rationals, but also correlations between construals of integers and their reconstruals as rationals. Today one works indifferently with integers and rationals as algebraic systems.

Notably, equivalence classes would *not* be initially involved in the genetic generation of the reals from the rationals. Dedekind's [1872] cuts themselves represented the reals, and Cantor [1872] considered that he had defined the reals as fundamental, i.e. Cauchy, sequences of rationals but did not actually work with equivalence classes of such sequences.

Also in 1872 Dedekind worked his way down to the bedrock of the natural numbers, and the eventual result was his celebrated essay *Was sind und was sollen die Zahlen?* [1888]. Dedekind here took sets [Systeme] and mappings [Abbildungen] as basic notions; worked with unions and intersections of sets and compositions and inversions of mappings; and developed a theory of chains [Ketten], sets with one-to-one mappings into themselves.

Dedekind famously defined (paragraph 64) an infinite set to be a set having a one-to-one mapping into a proper subset. With this he had come to a positive formulation of the actual infinite, one which is evidently the logical negation of Dirichlet's Pigeonhole Principle.[5] Dedekind then (in)famously "proved" (66)

[4] See [Sieg and Schlimm, 2005, pp. 134-138].

[5] One cannot presume that Dedekind consciously negated the Pigeonhole Principle to get his definition of infinite set, but there are interactions and confluences: The Pigeonhole Principle seems to have been first applied in mathematics by Dirichlet in papers of 1842 (cf. [Dirichlet, 1889 1897, 579,636]), one on Pell's equation and another in which the principle is applied to prove a crucial approximation lemma for his well-known Unit Theorem describing the group of units of an algebraic number field. The Pigeonhole Principle occurred in Dirichlet's *Vorlesungen über Zahlentheorie* [1863], edited and published by Dedekind. The occurrence is in the second, 1871 edition, in a short Supplement VIII by Dedekind on Pell's equation, and it was in the famous Supplement X that Dedekind laid out his theory of ideals in algebraic number theory, working directly with infinite totalities.

the existence of a Dedekind-infinite set by invoking "my own realm of thoughts". Working toward the natural numbers, Dedekind defined (71) a "simply infinite system" to be a set N for which there is a one-to-one mapping ϕ into itself and a $1 \in N$ not in the range such that N is the closure of $\{1\}$ under ϕ. From a Dedekind-infinite set he got to a simply infinite system, and abstracting from particulars but proceeding with the same notation, took (73) *the* natural numbers to be the members of N, with base element 1 and successor operation $n' = \phi(n)$. This was the paradigmatic move between token and type; Dedekind soon gave expression to this indifference to identification by affirming recursive definition.

Recursive definition is crucial for modern mathematics, the basic case being the formulation of $\{t_i \mid i \in I\}$ with I the natural numbers and t_{i+1} uniformly given in terms of t_i—with the shifting of the focus to the function $i \mapsto t_i$ itself. One has here the Replacement motif cast with iteration of procedure as a generalization of counting itself. Dedekind proceeded to establish the Recursion Theorem (126): Given a set Ω, a distinguished element $\omega \in \Omega$, and a mapping $\theta\colon \Omega \to \Omega$, there is one and only one mapping $\psi\colon N \to \Omega$ satisfying $\psi(1) = \omega$ and $\psi(n') = \theta(\psi(n))$ for every n.

Dedekind was the first to point out the need to posit the existence of a completed mapping in this way,[6] and with the theorem, he soon provided the now-familiar recursive definition of addition as a mapping, the recursive definition of multiplication in terms of addition, and the recursive definition of exponentiation in terms of multiplication. However, it is instructive to point out that Dedekind's argument at this foundational level, as seen through modern eyes, can be considered to be subtly circular:

To establish (126), Dedekind appealed to the preparatory (125) according to which one can recursively define finite mappings ψ_n on $\{1, \ldots, n\}$ unique in satisfying: $\psi_n(1) = \omega$ and $\psi_n(m') = \theta(\psi_n(m))$ for $m < n$. What is the status of $\{\psi_n \mid n \in N\}$ at this point? Dedekind did consider sets [Systeme] of mappings, e.g. in (131), but to have construed the sequence $\langle \psi_n \mid n \in N \rangle$ as an objectified mapping at (125) would have been circular since it is (126) itself which posits recursively presented mappings. Dedekind then considered the desired mapping for (126) to be given by $\psi(n) = \psi_n(n)$ as a direct definition. He had deftly devolved first to the approximations ψ_n's, but without the ψ_n's collectively comprehended, how is one running through the indexing of the ψ_n? Seen through modern eyes sensitized to the ways of paradox, this "diagonal" stipulation would not be taken to be a contextually internal definition. The thrust of the Recursion Theorem (126) is to be able to pass from a given mapping to another mapping stipulated by recursion; the proof of (126)

[6] Gottlob Frege in his 1893 *Grundgesetze* also established the Recursion Theorem; see Heck [1995] for an account.

itself has this form, but with the givenness of $\langle \psi_n \mid n \in N \rangle$ not quite in hand.[7] However basic Dedekind was taking the notion of mapping, with (126) he had made an existence assertion about mappings, and logically speaking, an existence assertion cannot be arrived at without generative existence principles at work. And if one were to make such principles explicit for (126) as Dedekind had done elsewhere for sets [Systeme], one would be drawing in some version of the Replacement motif $\{\psi_n \mid n \in N\}$.

Looking briefly ahead to compare and contrast, how one articulates recursion would become a crucial theme in the formalization of both arithmetic and set theory. Thoralf Skolem [1923a], developed elementary arithmetic by taking the "recursive mode of thought" as basic and proceeding with a system of equation rules. *Primitive Recursive Arithmetic* is a subsequent formalization of Skolem's approach, in which function symbols are simply introduced for each (primitive) recursion and their generating rules in the manner of Dedekind's (126) are given as axioms. Like Dedekind, Skolem took function as a basic concept, but he made explicit the recursions. Paul Bernays [1941, pp. 11-12], in an exposition of an axiomatic set theory with classes, articulated Dedekind's (126) argument prior to introducing an axiom of infinity by formalizing functions as classes of ordered pairs and applying the "class theorem" or the predicative comprehension schema, which asserts that extensions of formulas without class quantifiers are classes. With this, $\{\psi_n \mid n \in N\}$ is a class and so also the resulting ψ. In ZFC set theory, functions are formalized as sets of ordered pairs; N can be taken to be the natural numbers as given by the Axiom of Infinity; one can appeal to Replacement to get $\{\psi_n \mid n \in N\}$ as a set; and Union establishes $\psi = \bigcup\{\psi_n \mid n \in N\}$ to be a set.[8]

Returning to Dedekind [1888], the first use of the Recursion Theorem (126) was actually to establish (132) that *all simply infinite systems are isomorphic.* In (126) one takes Ω to be an arbitrary simply infinite system and, appealing to "complete induction" in both N and Ω, shows that the resulting θ is an isomorphism. The result is nowadays touted as establishing the second-order categoricity of the Dedekind-Peano axioms, although strictly speaking Dedekind was not proceeding in an axiomatic context. Replacement can again be seen as involved in providing for indifference to identification, to establish both the efficacy of a chosen representative token of a type and a crucial plia-

---

[7]Something of this circularity surfaces in Landau's text *Grundlagen der Analysis* [1930], cf. its preface.

[8]This importantly is the argument when the Recursion Theorem is situated in set theory with Ω not necessarily a set. When Ω is a set, one can appeal to Power Set and Separation, getting $\{\psi_n \mid n \in N\}$ as a subset of $P(N \times \Omega)$. This is often done in expository accounts, but the direct appeal to Power Set is not cardinally parsimonious and has a strange look. Alternately, one can invoke the Dedekind chain theory by defining a function $\phi \subseteq (N \times \Omega) \times (N \times \Omega)$ by $\phi(\langle n, x \rangle) = \langle n', \psi(x) \rangle$ and taking the closure of $\langle 1, \omega \rangle$ under ϕ; Power Set and Separation are used here to get the closure.

bility whereby one can entertain various tokens as efficacious.

Many years later, Ernst Zermelo in his final axiomatization paper [1930b] also established a second-order categoricity result, one for his natural models of set theory. With transfinite recursion he pieced together their cumulative hierarchies, and Replacement—incorporated into his later axiomatization—is crucial for the process. Zermelo in his progress toward his first axiomatization [1908b] had studied *Was sind und was sollen die Zahlen?* carefully and had subsequently pointed out what turned out to be Dedekind's only other logical gap in his essay. Dedekind had established (160) that, as we would now say, Dedekind-finite sets are finite, but [Zermelo, 1909, p. 190, n. 5] pointed out that this requires a use of the Axiom of Choice, to choose from a set of sets of mappings. Also, Dedekind's proof relied on the Recursion Theorem (126). If formalization through reduction to sets is to be worked out, then Dedekind's work needs the Axiom of Choice, and in a methodologically similar way, Replacement, namely to procure certain sequences of mappings. These two principles are thus implicated, for the first time, in the formalization of the infinite in actuality, and they would become part and parcel of its accommodation in set theory.

Although Replacement would seem to have a crucial role underpinning mathematics through $\{t_i \mid i \in I\}$, textbook expositions of ZFC set theory most often introduce the axiom schema rather late in the development. This presumably has to do with several factors: As formally stated in first-order logic, the axiom schema is syntactically complicated, and its paraphrase in terms of classes adds a layer of conceptualization over sets; Replacement, together with the Axiom of Foundation, was adjoined to Zermelo's initial 1908 axioms significantly later; and the set-theoretic reduction of the ordered pair, and thereby function, induced ready appeals to the Power Set Axiom in preference to Replacement. Instead, one can proceed as in the text Bernays [1958] and introduce Replacement early, emphasizing the motif $\{t_i \mid i \in I\}$ as basic. The text sets out a formal apparatus with class terms $\{x \mid \varphi(x)\}$ and the conversion $\varphi(a) \longleftrightarrow a \in \{x \mid \varphi(x)\}$.

But even in an elementary text, one can develop axiomatic set theory to exhibit Replacement as reflective of the use of $\{t_i \mid i \in I\}$ in mathematical practice: After an informal look at first-order logic and class terms, introduce the initial axioms of Extensionality, Separation, Existence, Pairing, and Union and develop the basics through the Kuratowski ordered pair to relations and functions. With the work on Separation as a first-order schema having prepared the way, now introduce Replacement and emphasize the paraphrases in terms of class functions and $\{t_i \mid i \in I\}$. Put Replacement forthwith to use to establish e.g. that given an equivalence relation the equivalence classes together form a set, and that given two sets their Cartesian product exists.[9] Proceed

[9]The expected argument works with any adequate definition of the ordered pair, i.e. any for

to introduce the Axiom of Infinity and the natural numbers. With Replacement, establish Dedekind's Recursion Theorem and with it the arithmetic of the natural numbers as well as Transitive Containment, i.e. that every set is a subset of a transitive set. With precedents set, proceed to the (von Neumann) ordinals, and with Replacement in full play establish the basic theory through to the Transfinite Recursion Theorem. As Cantor and Dedekind had come to see, ordinality should be prior to cardinality, and so only now introduce cardinality and the Power Set Axiom to give the concept heft. One thus sees that the substantive issues about the existence of large cardinality sets arise with Power Set, not Replacement. Finally, there is the Axiom of Foundation and the picture of the cumulative hierarchy of sets; with all sets now to appear in this recursively defined hierarchy, Replacement and Foundation work together to establish results by recursion for all sets.

2 Early History of Replacement

The history of the emergence of Replacement as an integral axiom of modern ZFC set theory, like that of the Axiom of Choice, has to do most importantly with the emergence of set-theoretic methods and their formalization. With Choice, it was well-ordering and through it maximalization as with Zorn's Lemma, and with Replacement it was transfinite recursion and through it closure of the set-theoretic universe under processes involving large cardinality. With the thrust of Replacement involving functional correlation, its history, like that of Choice, has to do initially with the liberalization of the concept of function and the expansion of the concept of set. The motif $\{t_i \mid i \in I\}$ underlying mathematics, as described in the previous section, became rigorized as a set-existence principle, one that became central to set theory.

Replacement first occurred in an early form in terms of one-to-one correspondence in late work of Georg Cantor himself. The motif $\{t_i \mid i \in I\}$ had in any case been congenial to Cantor both in his investigation of cardinality and because of his generation principles for limit and higher cardinality ordinal numbers.[10] Having developed the transfinite landscape for over two decades, Cantor in correspondence with Hilbert and Dedekind newly formulated a crucial juncture for set existence.[11]

In a letter to Hilbert of 26 September 1897, Cantor stressed that "the totality of all alephs cannot be conceived as a determinate, well-defined *finished set* [*fertige Menge*]" since a cardinal number for it cannot be entertained without

which one can uniquely recover the first and second coordinates. Conversely, [Mathias, 2011, sect. C] has observed that if as a schema Cartesian products exist for every adequate definition of ordered pair, then Replacement follows.

[10] Bernays in his book [1958] introduced Replacement early for "general set theory", which was to correspond to Cantor's context for generating the transfinite numbers.

[11] The correspondence appears in [Meschkowski and Nilson, 1991], and some letters are translated in [Ewald, 1996], including those cited below except that of 10 October 1898.

contradiction. He repeated this assertion in a letter to Hilbert of 2 October 1897 where he wrote further that a set can be thought of as *finished* if "it is possible ...to *think of all its elements as existing together*, ..." A year later, in a letter to Hilbert of 10 October 1898 elaborating his concept of finished set, Cantor concluded as a "theorem" the assertion that *if two multiplicities are cardinally equivalent (i.e. in one-to-one correspondence) and one is a finished set, then so is the other.* This is the first substantive anticipation of Replacement in set theory. In that letter Cantor made the observation that the totality of all subsets of the natural numbers is cardinally equivalent to the totality of all functions from the natural numbers into $\{0, 1\}$, a few years before Russell would make this connection in his theory of relations. With this observation applied to the "linear continuum" and a prior "proposition" that "[t]he multiplicity of all the subsets of a finished set M is a finished set", Cantor then derived that the "linear continuum" is a finished set. Thus, Cantor could be seen as adumbrating indifference to identification for the roughest criterion, one-to-one correspondence, and with it, expressing a now-standard plasticity for the real numbers.

In a well-known letter to Dedekind of 3 August 1899, Cantor, after confronting the Burali-Forti Paradox,[12] emphasized his distinction between consistent multiplicities — sets — and inconsistent, absolute multiplicities — like the totality of all ordinal numbers. Cantor in fact argued that if a definite multiplicity does not have an aleph as its cardinal number, then there would be a "projection" of the totality of all ordinal numbers into the multiplicity. Cantor wrote:ă "Two equivalent multiplicities either are both 'sets' or are both inconsistent." Thus his anticipation of Replacement now served less as a set existence principle and more as an articulation of dichotomy based on the possibilities for well-ordering. In any case, Replacement had been anticipated in a pre-formal setting with one-to-one correlation taken as a basic notion.[13]

The early set-theoretic work of Zermelo, formative of course for the development of set theory, has bearing through thematic points of contact.[14] Zermelo [1904] famously made explicit an initial appeal to the Axiom of Choice and established the Well-Ordering Theorem, that every set can be well-ordered.

[12]Cantor came to the "paradox" of the largest ordinal number before the appearance of Burali-Forti [1897]. Whether paradoxical or not, it had to be confronted and analyzed. See [Moore and Garciadiego, 1981] for the history of the Burali-Forti paradox.

[13]The little known A. E. Harward, in an article [1905] about cardinal arithmetic and well-orderings, also came to these issues and ideas entertained by Cantor. Though Harward took a distinction between his "unlimited classes" and "aggregates" as provisional, he emphasized how the class of all ordinal numbers is unlimited and like Cantor, anticipated Replacement as following from the meaning of the terms (p. 440): "Any class of which the individuals can be correlated one to one with the elements of an aggregate is itself an aggregate." Although Harward would remain obscure, he probably has the distinction of being the first to have anticipated Replacement in print. See [Moore, 1976] for more about Harward.

[14]See [Kanamori, 2004] for more on Zermelo.

The proof of the theorem presented a new mode of argument, one that can be viewed as an anticipation of the proof of the later Transfinite Recursion Theorem, the theorem that validates definitions by recursion indexed by well-orderings. This theorem was first properly articulated and established by von Neumann (see below) using Replacement. The difference is only that Zermelo's proof does not require functional correlation, as enabled by Replacement, as one is defining the well-ordering itself.

Zermelo's proof generated controversy about the acceptability of the Axiom of Choice, and largely in response, Zermelo subsequently published both a second proof [1908a] of the Well-Ordering Theorem and the first full-fledged axiomatization [1908b] of set theory. Zermelo's axioms were few and had a remarkable simplicity, except for the one "logical" axiom having to do with properties, the Axiom of Separation. As it provided for the crucial separating out of members of a given set according to a "definite" property, what these properties were came to be seen as in need of clarification as mathematical logic was further developed. As is well-known, the stronger property-based Axiom of Replacement played no role in Zermelo's axiomatization. Zermelo was proceeding pragmatically and parsimoniously to establish set theory as a discipline axiomatically given in the Hilbertian style and to put his Well-Ordering Theorem on a sound footing with a modicum of set-existence principles. Zermelo's second proof [1908a] of the Well-Ordering Theorem indeed coordinated with the axioms, particularly as concerns well-ordering. Taking an approach first used by Gerhard Hessenberg [1906, pp. 674ff], Zermelo cast a well-ordering as the set of its final segments under the reverse-inclusion ordering, thereby situating the concept in his axiomatic framework.

The work of Friedrich Hartogs [1915] on well-orderings and Cardinal Comparability is a conspicuous juncture having interaction with Replacement as motif but still historically and mathematically prior to the use of Replacement as axiom. The thrust of [1915] is that for any set x there is a well-ordered set y not injectible into x. Cardinality Comparability—that for any two sets, one is injectible into the other or vice versa—then implies that x is injectible into y and hence is well-orderable. Thus, Cardinal Comparability implies the Axiom of Choice, a first "reverse mathematics" result establishing the equivalence among the Axiom of Choice, the Well-Ordering Theorem, and Cardinal Comparability.

In what follows, we give the spine of Hartogs' proof to better discuss how he proceeded with the details: Given a set x, let W consist of the well-orderings of subsets of x. Let E be the equivalence relation defined on W by $r\,E\,s$ exactly when r is order-isomorphic to s. Then $<$ on E-equivalence classes is well-defined by: $[r]_E < [s]_E$ exactly when s has an initial segment in $[r]_E$. Moreover, $<$ is a well-ordering of $\{[r]_E \mid r \in W\}$. Finally, this set cannot be injectible into x, else the range of the injection would have a well-ordering belonging to W

and $<$ would be order-isomorphic to an initial segment of itself, which is a contradiction.

This argument, amounting to a positive subsumption of the Burali-Forti paradox, would have to be properly implementable without the Axiom of Choice in order to effect the implication from Cardinal Comparability to Choice. To affirm this, Hartogs emphasized implementation in the [Zermelo, 1908b] system without Choice. This early use of Zermelo's axiomatization is resonant with Zermelo's own, initial use to buttress his proof of the Well-Ordering Theorem.

Pursuing the above sketch, the first implementation issue is how to render a well-ordering and thus to have the set W. Norbert Wiener [1914] and Felix Hausdorff [1914, pp. 32ff,70ff] rendered the ordered pair, and with this a theory of relations and functions can be developed in set theory. Hartogs was quite unlikely to be aware of this, but he had become aware of a known approach for rendering well-ordering with sets. In his appendix Hartogs acknowledged being informed by Hessenberg that well-orderings can be rendered in Zermelo's axiomatization through systems of final segments—which is what in fact Zermelo had done in his second proof [1908a] of the Well-Ordering Theorem. Thus, with Power Set and Separation, one can take W as a set, being a subset of $P(P(x))$ given by a definite property. Proceeding, once the equivalence relation E is formulated, $\{[r]_E \mid r \in W\}$, though an instance of the Replacement motif, can be shown to be a set by Separation and Power Set, separating from $P(W)$, and $<$ can be defined to complete the rendition in Zermelo's system without Choice.

But how is the equivalence relation E to be formulated? Here's the rub. Hartogs noted (p. 438): "The task of checking all notions and theorems ... is made somewhat difficult since the axiomatic presentation [Durcharbeitung] of set theory given by Zermelo does not yet extend to the theory of ordered and well-ordered sets." In particular, the move beyond Cantor's ordinal numbers as type to the concept of order-isomorphism between well-ordered sets, necessary to formulate E, had yet not been made in the axiomatic presentation. It is here, the situation elaborative of indifference to identification, that Replacement can serve. Indeed, in some modern accounts of Hartogs' theorem one simply uses Replacement to associate with each member of W the corresponding (von Neumann) ordinal and then takes the supremum to complete the argument. However, this would be how one would proceed on the other side of this historical cusp for Replacement, which is not actually necessary for the proof. Without an ordered pair and a theory of relations, Zermelo [1908b] did develop a "theory of equivalence" sufficient for rendering the Cantorian theory of cardinality; this theory was based on getting disjoint copies of two sets and in the union of the copies using the unordered pair to render one-to-one correspondence. Such a theory can be developed for rendering the Cantorian theory of ordinality with order-isomorphisms between Hessenberg-style well-ordered sets, thereby effecting a formalization of Hartogs' theorem and thus establishing that Cardinal Comparability implies Choice. Hartogs did not see his argument through to the end, but the techniques were available to do so at the time, techniques that did not require Replacement.

The set-theoretic work of Dimitry Mirimanoff [1917a, 1917b] sits interestingly intermediate between that of Cantor and of von Neumann. As with Hartogs, Mirimanoff was publishing at a time when Zermelo's axiomatization was not widely called upon and the reduction of ordered pair and function to sets was only beginning. His work goes in and out of the modern theory, but in any case led to a *published* appearance of a form of Replacement, one like Cantor's based on equivalence.

In connection with Russell's Paradox, Mirimanoff [1917a] formulated the *ordinary sets* as those sets x for which every descending $\in$-chain $\ldots x_2 \in x_1 \in x$ is finite. These mediated by the Axiom of Choice are, of course, what we now call the well-founded sets. In connection with the Burali-Forti Paradox, Mirimanoff formulated what can be seen to be the (von Neumann) ordinals.[15] He specifically started with an urelement[16] e, getting at what we here call the e-ordinals, and pictured the first three as:

$$(e); (e, (e)); (e, (e), (e, (e))).$$

He (pp. 45-46) motivated these by considering a well-ordering, replacing its members by initial segments, then replacing these by the set of their initial segments, and so forth. Although he thus took every well-ordering to be order-isomorphic to an e-ordinal, the envisioned, repeated replacement process can only get to an e-ordinal for *finite* well-orderings. In [1917b], Mirimanoff more assuredly described how to correlate a well-ordering with an e-ordinal by progressively replacing the ordering relation by membership from the front. But even then, the pivotal von Neumann result that every well-ordering is order-isomorphic to a (von Neumann) ordinal can only be rigorously proved by a recursion requiring Replacement. On the other hand, [Mirimanoff, 1917a, p. 47] offered a characterization of e-ordinal as those ordinary sets with e that are transitive and linearly ordered by membership. Such a characterization for ordinals does not appear in von Neumann's work, and has been attributed much

[15]Zermelo was most probably the first chronologically to have formulated the concept of (von Neumann) ordinal, and this by 1915; the rudiments of the theory appear in his *Nachlass* (cf. [Hallett, 1984, pp. 277ff]) and indications are there of collaboration with Paul Bernays (cf. [Ebbinghaus, 2007, 3.4.3.]).

[16]Urelements, also called atoms or individuals, are objects distinct from the empty set yet having no member. Zermelo allowed for urelements in his "domain" for his [1908b] axiomatization. Mirimanoff used the term node [noyau].

later to Raphael Robinson [1937] as a simplification.[17] In [1917b], Mirimanoff elegantly developed the theory of ordinals much as it is done today. In Mirimanoff's terms, a set [ensemble] can exist or not; the set of all "ordinal numbers of Cantor" does not exist because of the Burali-Forti Paradox, but no decision is made on whether non-ordinary [extraordinary] sets exist or not.

Mirimanoff [1917a] proceeded to his solution of the "fundamental problem" of when sets exist for the case of ordinary sets. To this end he stated three "postulates" corresponding to Union, Power Set, and Replacement. While the first two stated that the ordinary sets are closed under the taking of unions and power sets respectively, the latter stated that *if any set exists, then so does any (cardinally) equivalent ordinary set.* The first 'set' is not qualified, and how he initially applied this principle brings out the conceptual distance from how we now work with Replacement. With his formulation Mirimanoff concluded forthwith that for each ordinal number of Cantor the corresponding e-ordinal exists, and with this he established as a preliminary result that a set of e-ordinals exists exactly when their corresponding ordinal numbers are bounded. Mirimanoff's Replacement thus served as a *trans-species* bridge between "the ordinal numbers of Cantor" and their set representations as e-ordinals, numbers and sets being regarded as of different species at this time. This is interestingly alien to the present sense of Replacement as an axiom within set theory proper but is also consonant with Replacement's role in shifting between tokens and types and Cantor's original use.

For the solution of the "fundamental problem" for ordinary sets in general, Mirimanoff briefly described what can be seen as the cumulative hierarchy indexed by the ordinal numbers, and showed that an ordinary set exists exactly when it has a rank in this hierarchy. In his argument his several uses of Replacement are resonant with how it now serves to affirm the recursive definition of the cumulative hierarchy; however, there is steady coordination, trans-species, with the "ordinal numbers of Cantor". The extent and details of Mirimanoff's work were presumably not fully appreciated in the next, formative years for set theory, and one can only speculate as to why.[18]

Replacement as an axiom henceforth to be reckoned with for the axiomatization of set theory first emerged in 1921 correspondence between Zermelo and Abraham Fraenkel.[19] Fraenkel was the first to investigate Zermelo's 1908

---

[17] Bernays in a letter of 3 May 1931 to Gödel actually provided the first direct definition of ordinal, as a transitive set each of whose members is transitive. This letter will be discussed in some detail in §3.

[18] There seems to have been only two citations of his work, in [Fraenkel, 1922, p. 233] and [von Neumann, 1925, p. 230], and these only in connection with the extraordinary sets. Mirimanoff had published in the French Swiss journal, *L'Enseignement Mathématique*, which may not have been readily seen in Germany in the difficult period after the war.

[19] See [Ebbinghaus, 2007, 3.5] for this correspondence and about Replacement; what follows draws substantially from this source.

axioms with respect to possible independences, and their first exchange had to do with this. In a letter of early May 1921, Fraenkel astutely raised what would be a pivotal issue: With $Z_0 = \{\emptyset, \{\emptyset\}, \{\{\emptyset\}\}, \ldots\}$ being the infinite set given by Zermelo's axiom of infinity and $Z_1 = P(Z_0)$ the power set, $Z_2 = P(Z_1)$, and so forth, how can Zermelo's axioms establish that $\{Z_0, Z_1, Z_2, \ldots\}$ is a set? Without the union of this set, the existence of sets of cardinality $\aleph_\omega$ would not be provable.

Zermelo wrote back forthwith on 9 May (cf. [Ebbinghaus, 2007, p.137]):

> Your remark concerning $Z^* = Z_0 + Z_1 + Z_2 + \ldots$ seems to be justified, and I missed this point when writing my [axiomatization paper [1908b]]. Indeed, a new axiom is necessary here, but which axiom? One could try to formulate it as follows: If the objects $A, B, C, \ldots$ are assigned to the objects $a, b, c, \ldots$ by a one-to-one relation, and the latter objects form a set, then $A, B, C, \ldots$ are also elements of a set M. Then one only needed to assign $Z_0, Z_1, Z_2 \ldots$ to the elements of the set Z_0, thus getting the set $\Theta = \{Z_0, Z_1, Z_2, \cdots\}$ and $Z^* = \bigcup \Theta$. However, I do not like this solution. The abstract notion of assignment it employs seems to be not "definite" enough. Precisely this was the reason for trying to replace it by my "theory of equivalence" [from [1908b]]. As you see, this difficulty is still unsolved. Anyway, I appreciate your having brought it to my attention.

Thus, almost immediately after being confronted with Fraenkel's example Zermelo had formulated Replacement as Cantor had earlier in correspondence, in terms of cardinal equivalence. Zermelo's expressed skepticism about the "abstract notion of assignment" not being " 'definite' enough" has to do with his definite property for the Separation Axiom, perhaps because of the recursive aspect of $\{Z_0, Z_1, Z_2, \cdots\}$. The "abstract notion of assignment" has a resonance in Fraenkel's letter of 19 May, in which he ostensibly argued that the new axiom implies the Axiom of Choice, by replacing each set of a system of non-empty, pairwise disjoint sets by a member. This first unpracticed reaction brings out the conceptual difficulties at the time about what should go for functional correlation.

Soon afterward, in a paper [1922] completed 10 July 1921, Fraenkel described the shortcoming he had found in Zermelo's axioms and introduced: "Axiom of Replacement. If M is a set and each element of M is replaced by 'a thing in the domain', then M turns into a set again." By a "thing in the domain" Fraenkel was referring to either a set or urelement in Zermelo's domain for his axioms. This is the first occasion when injectivity for "replacing" is no longer assumed, but on the other hand what the process of "replacing" is is left undeveloped. On 22 September, at the 1921 meeting of the Deutsche

Mathematiker-Vereinigung at Jena, Fraenkel [1921] announced his results, and it was reported that Zermelo at the meeting accepted the axiom but voiced reservations about its scope.

On 6 July 1922, at the Fifth Congress of Scandinavian Mathematicians at Helsinki, Skolem [1923b] delivered an address on the axiomatization of set theory. Skolem actually sought to devalue axiomatic set theory, considering his most important result, as he remarked at the end, to be that set-theoretic notions are relative. However, his analysis not only featured the now well-known "Skolem's paradox" but was remarkably prescient and far-ranging about how to proceed in set theory. Skolem argued that Zermelo's system should be formalized in first-order logic, so that in particular his Axiom of Separation becomes a schema with first-order formulas rendering Zermelo's "definite" property. Skolem (in 4.) also pointed out the same deficiency that Fraenkel had, that $\{Z_0, P(Z_0), P(P(Z_0)), \ldots\}$ cannot be proved to be a set, but moreover provided a semantic argument, which in modern terms amounts to the argument that the rank $V_{\omega+\omega}$ is a model of Zermelo's axioms without the set in question. Skolem then wrote (cf. [van Heijenoort, 1967, p. 297]): "In order to remove this deficiency of the axiom system, we could introduce the following axiom: *Let U be a definite proposition that holds for certain pairs (a, b) in the domain B; assume, further, that for every a there exists at most one b such that U is true. Then, as a ranges over the elements of a set M_a, b ranges over all elements of a set M_b.*" Taken in Skolem's context as describing a first-order schema of propositions about the domain of all sets, this is the first substantively accurate statement of the modern axiom schema of Replacement.[20] In a succeeding footnote, Skolem sketched how the new axiom actually establishes that $\{Z_0, P(Z_0), P(P(Z_0)), \ldots\}$ is a set, going through a proof for this instance of (Dedekind's) Recursion Theorem. One sees here the first clear account of Replacement and its role in a recursion.

Reviewing Skolem's paper, Fraenkel[21] simply stated that Skolem's considerations about Replacement correspond to his own, though evidently, Fraenkel's initial formulation conveyed only a loose idea of replacing. Later Fraenkel [1925, p. 271] did attempt a more formal account specifying the applicable replacing procedures but did not pursue this, opining that Replacement was too strong an axiom for "general set theory". Fraenkel's reservation now seems quite remarkable, and it is also to be noted that Skolem only wrote that "we could introduce" Replacement (in the above quotation). Whatever is the case, Fraenkel's name is now, of course, firmly entwined with Replacement, and this

[20]Actually, there is a slight variance. Modern Replacement would have "exactly one b" instead of Skolem's "at most one b". Skolem's stronger formulation can differ from Replacement for restricted versions of the schemas. For this, see [Mathias, 2001b, sect. 9, peroration]; its 9.32 also translates the Fraenkel-Skolem observation into an algebraic one nearer to the interests of many mathematicians.

[21]*Jahrbuch über die Fortschritte der Mathematik* 49 (1922), 138-139.

presumably has to do with the thrust of his publications on the axiomatics of set theory as well as the acknowledgments of those engaged with Replacement, von Neumann and Zermelo. The first reference to "Fraenkel's Axiom" appears in [von Neumann, 1923, p. 347] and the first to "Zermelo-Fraenkel" as the axioms of Zermelo augmented with Replacement appears in von [von Neumann, 1928a, p. 374], soon to be followed by [Zermelo, 1930b, p. 30]. It would be the work of von Neumann which would firmly establish Replacement as an important and needed axiom and orient set theory toward the first-order formulation first advocated by Skolem.

3 Middle History of Replacement

Von Neumann effected a counter-reformation of sorts for ordinal numbers. The ordinal numbers had been focal to Cantor as separate entities from sets but peripheral to Zermelo with his emphasis on set-theoretic reductionism; von Neumann reconstrued then as *bona fide* sets, the (von Neumann) ordinals. In connection, von Neumann formalized transfinite recursion, this as he worked out a new axiomatization of set theory. Von Neumann's axiomatization was the first, via his I-objects and II-objects, to allow proper classes, as we would now say, together with sets; the paradoxes were systemically avoided by having only sets be members of classes. In all this work Replacement was a crucial feature from the first.

To briefly set the stage for transfinite recursion, emergent from Cantor's work with transfinite numbers is transfinite induction and transfinite recursion, the first a mode of proof and the second a mode of definition. Transfinite induction on a well-ordering is essentially just a contraposition of the main, least element property of well-ordering. Transfinite recursion, on the other hand, depends on having sufficient resources. In Whitehead and Russell's *Principia Mathematica*, volume 3 [1913], transfinite induction and transfinite recursion were articulated to their purposes, the second workable in the theory of types context. In Hausdorff's *Grundzuge der Mengenlehre* [1914], V§5, transfinite induction and transfinite recursion were just presented as working principles. It would be the axiomatic formalization of transfinite recursion in set theory that draws in Replacement.

Now to progressively describe von Neumann's work, he in [1923] proceeded informally, in part to bring out the lack of dependence on any particular axiomatic system, to set out the concept of ordinal, working the basic idea of taking precedence in a well-ordering just to be membership.[22] A Cantorian ordinal number had often been construed as the order type of the set of its predecessors, and von Neumann just forestalled the abstraction to order type. With this in hand, he set out the fundamental result that every well-ordering

[22] Von Neumann became aware of Zermelo's anticipation of the theory of ordinals; see [Hallett, 1984, p. 280] and [Ebbinghaus, 2007, p. 134].

is order-isomorphic to an ordinal. In [1928b], von Neumann duly formalized the proof, making use of Replacement, and one sees here the first and basic use of Replacement to articulate canonical token representing type. With Replacement he moreover established the fundamental Transfinite Recursion Theorem, that given any class function F in two variables, there is a unique class function G on the ordinals such that $G(\alpha) = F(\alpha, G\restriction\alpha)$. As with Dedekind's Recursion Theorem (cf. §1) one pieces together initial segments; one proves by transfinite induction that for each ordinal β, there is exactly one function g_β with domain β and satisfying for all $\alpha < \beta$ that $g_\beta(\alpha) = F(\alpha, g_\beta\restriction\alpha)$. The shift from Dedekind to von Neumann is the shift from the finite to the transfinite, and Replacement is just what is needed to establish the inductive existence of g_β for limit ordinals $\beta > 0$, in the same way that for Dedekind's Recursion Theorem the existence of the desired function on the natural numbers is secured. The bootstrapping would have been subtle then but is straightforward now, and again it is the infinite in actuality that draws in the Replacement motif. Von Neumann moreover applied transfinite recursion to define (proper) classes, proceeding forthwith as Dedekind had done for the natural numbers to the definition of the ordinal arithmetical operations. The key method of modern set theory is transfinite recursion, and von Neumann thus established the intrinsic necessity of Replacement.[23]

Many years later Fraenkel retrospectively acknowledged the importance of Replacement as brought out by von Neumann. Writing about his [1922], he wrote ([1967, pp. 149ff]): "I did not immediately recognize or take advantage of the full significance of my new axiom. Rather, this was done by von Neumann ..." Fraenkel subsequently wrote (p. 169): "The importance of this axiom was now shown in a wholly unexpected way: von Neumann based the theory of transfinite numbers on my axiom in a way which showed that it is indispensable for this purpose."

[23]There is a local version of transfinite recursion provable without Replacement: If β is an ordinal, A is a set, and $f\colon\ P(A) \to A$, then there is a unique function $g\colon\ \beta \to A$ satisfying $g(\alpha) = f(g``\alpha)$. With a formalization of function as a set of ordered pairs one can apply Separation to the power set $P(\alpha \times A)$ to get the set of approximating functions and then take the union. This approach is indeed analogous to Zermelo's first, [1904] proof of the Well-Ordering Theorem. With this local transfinite recursion, the attempt to define ordinal addition $\gamma + \alpha$ to a fixed γ and for α up to a specified β encounters the difficulty that $\gamma + \alpha$ may surpass β.

Notably, [Potter, 2004, p.183] presents this local version of transfinite recursion in his setting without Replacement. But his proof slides into an implicit appeal to Replacement even though the Power Set, Separation, Union approach was available to him. Moreover, with his theory of ordinal numbers based on equivalence classes of well-orderings he subsequently defines ordinal addition, not recursively because of the above mentioned difficulty, but directly by putting well-orderings in series. But then, he gives (p. 194) the recursion equations as if they were immediate consequences. The very formulation of the limit case requires justification, ordinarily given with Replacement, but here possible with Power Set, Separation, and Union. It is hard to sublimate Replacement.

Von Neumann's axiomatization of set theory, completed by 1923 for his eventual Budapest thesis and aired in [1925] and in full detail in [1928a], has several interactions with Replacement. He actually axiomatized the notion of function rather than the notion of set, noting [1925] (§2) that "every axiomatization uses the notion of function" and citing Replacement. Considering a rigorization of Zermelo's concept of "definite" property to be one of the accomplishments of his axiomatization, von Neumann proceeded to present his axioms for generating classes sufficient so that all definite properties can be correlated with classes according to his Reduction Theorem (§3), essentially the predicative comprehension schema, asserting that extensions of formulas without class quantifiers are classes. This has as a consequence a rigorization of Replacement as put forth by Fraenkel, which von Neumann took (§2, last footnote) in the class form: If F is a (class) function and a is a set, then the image $F``a$ is a set.

Replacement however is not actually an axiom of von Neumann's axiomatization. Rather, the taking of function as primitive was in part to the purpose of formulating his focal axiom IV 2, stated here in terms of sets and classes with V the class of all sets: *A class A is not (represented) by a set exactly when there is a surjection of A onto V.* Von Neumann thus transformed the negative concept of proper class, which had appeared in various guises, e.g. Cantor's inconsistent multiplicities, into the positive concept of having a surjection onto V. In fact, von Neumann had formalized a dichotomy based on possibilities for well-ordering that had been broached by Cantor in his letter of 3 August 1899 to Dedekind (cf. §2). IV 2 is an existence principle that plays the role of regularizing proper classes, much as the Axiom of Choice does for sets, by extending the Cantorian canopy of functional correspondence, and IV 2 appropriately implies both Replacement and Choice in class forms, the latter asserting the existence of a global choice function on V.[24] Both in his development of the ordinals and with his axiomatization, von Neumann rigorized and provided for the full extent of Cantor's later vision of set theory as set out in correspondence with Hilbert and Dedekind at the turn of the century, and Replacement was a basic underpinning for this.

In his last paper on axiomatization von Neumann [1929] provided an incisive analysis of his system by providing as a model the class of well-founded sets, an analysis that showed his axiom IV 2 to have a concrete plausibility. In modern terms, he in effect defined by transfinite recursion the *cumulative hierarchy* of

[24]For Replacement, if F is a class function whose range is not a set, then there is a surjection G of that range onto V; but then F composed with G is a surjection of the domain of F onto V, and consequently the domain is not a set. For Choice, since the class of all ordinals cannot be a set by the Burali-Forti argument, there is a surjection of the class onto V, and an inversion according to least preimages induces a well-ordering of V itself, and so there is a global choice function.

well-founded sets through their stratification into cumulative "ranks" V_α, where

$$V_0 = \emptyset;\ V_{\alpha+1} = P(V_\alpha);\ \text{and}\ V_\delta = \bigcup_{\alpha<\delta} V_\alpha \text{ for limit ordinals } \delta\,.$$

[Mirimanoff, 1917a, pp. 51ff] had been first to study the well-founded sets, and the cumulative hierarchy is distinctly anticipated in his work (cf. §2). In the axiomatic tradition [Fraenkel, 1922], [Skolem, 1923b], and [von Neumann, 1925] had considered the salutary effects of restricting the universe of sets to the well-founded sets. Von Neumann [1929] formulated in his functional terms the Axiom of Foundation, asserting that every non-empty class has a $\in$-minimal element, and observed that it is equivalent to the assertion that the cumulative hierarchy *is* the universe, $V = \bigcup_\alpha V_\alpha$. Moreover, he observed that in the presence of Foundation, Replacement and Choice (in class forms) are together actually *equivalent* to IV 2.[25] [26] This result has the notable thematic effect of localizing the thrust of IV 2 to asserting the existence of just *one* class, a choice function on the universe. To conclude, von Neumann established that his axioms hold in the class of well-founded sets, thereby establishing the relative consistency of Foundation. The result is cited now as the first relative consistency result via "inner models" and about Foundation, but for von Neumann a major incentive was to affirm the plausibility of his axiom IV 2.

As for the underlying logic, von Neumann's work is evidently formalizable in a two-sorted first-order logic with variables for sets and variables for classes. However, as with his Reduction Theorem ([1925] §3), i.e. the predicative comprehension schema, quantification over classes is delimited and the full potency of second-order logic for sets is never invoked. As subsequently emended by Bernays and Gödel, this first-order aspect promoted the move toward Skolem's [1923b] suggestion of basing set theory on first-order logic, and in particular, on regarding Replacement as a first-order schema of axioms.

Zermelo in his remarkable [1930b] offered his final axiomatization of set theory as well as a striking, synthetic view of a procession of natural models. This axiomatization incorporated for the first time both Replacement and Foundation (Zermelo's term), and thus the now-standard ZFC axiomatization is recognizable. However, Zermelo worked in a full second-order context, proceeding in effect with proper classes but not distinguishing a concept of class. Zermelo was actually presenting a dramatically new view of set theory as applicable through processions of models, each having a "basis" of urelements for

[25] To get IV 2, for any class A, A is the union of the layers $A \cap (V_{\alpha+1} - V_\alpha)$ by Foundation. If A is not represented by a set, then these layers are nonempty for arbitrarily large α by Replacement. But each such layer has a well-ordering by Choice, and these well-orderings can be put together to well-order all of A, again by Choice. Hence, there is a one-to-one correspondence between A and the class of all ordinals.

[26] [Levy, 1968] latterly showed by a clever argument that the Union Axiom for sets also follows from von Neumann's IV 2, so that IV 2 is equivalent to Replacement and Choice (in class forms) and Union (for sets) in the presence of the other axioms.

some specific application and a cumulative hierarchy built on the basis, up to a height given by the "characteristic". He established these "*Grenzzahlen*" to be either the (von Neumann) ω or a strongly inaccessible cardinal and moreover established a second-order categoricity of sorts for his models as determined up to isomorphism by the cardinal numbers of the basis and of the characteristic. Significantly, Zermelo pointed out (p. 38) that those models starting with one urelement satisfy von Neumann's axiom IV 2. However, in Zermelo's approach IV 2 would not always hold, this depending on the cardinality of the basis of urelements.[27]

Whatever the relationship to von Neumann's IV 2 and work on Foundation, Zermelo's adoption of both Replacement and Foundation promoted the modern mathematical approach to set theory. In modern, first-order set theory, Replacement and Foundation focus the notion of set, with the first making possible the means of transfinite recursion, and the second making possible the application of those means to get results about *all* sets, they now appearing in the cumulative hierarchy.

Zermelo himself, however, was haphazard on the methodological role of Replacement. First, like Mirimanoff [1917a] Zermelo worked with ordinals starting from an urelement; did not make the von Neumann identification of the Cantorian ordinal numbers with ordinals; and just stated (p. 33) the von Neumann result that every well-ordering is isomorphic to an ordinal without mention of the role of Replacement. Then in his characterization (p. 34) of the characteristics of his models as the inaccessible cardinals, Zermelo did not explicitly associate his ordinals with the subsets of a set, and when finally he appealed explicitly to Replacement it is to a limit case made redundant by a previous assertion. For Zermelo, the importance of Replacement resided in its role in cofinality, as attested to by his mention of Hausdorff's work (pp. 33,34). Getting to his categoricity results, Zermelo established (p. 41) his first isomorphism result, that two models with the same characteristic and bases of the same cardinality are isomorphic, by extending a one-to-one correspondence between the bases through the two cumulative hierarchies. Unbridled second-order Replacement is crucial here, this time to establish that the cumulative ranks of the two models are level-by-level extensionally correlative, and this from one perspective requires a *trans-species* correlation across models. However, Zermelo did not mention Replacement at all here, nor indeed does he ever point out how Replacement is implicated in definitions by transfinite recursion. As mentioned above (cf. §1) there is a historical resonance with [Dedekind, 1888] who had established the categoricity of *his* second-order axioms for arithmetic. A final involvement of Replacement is in Zermelo's postulation of an "unlimited sequence of Grenzzahlen" and how one model "can also be conceived of as a 'set' " in a further model (p. 46). This entails a "reflection principle" which can

[27]See [Kanamori, 2004, p. 525] and generally for Zermelo's work in set theory.

be seen in terms of later developments (cf. §4) as amounting to a strong form of Replacement.[28]

In a note "On the set-theoretic model" [1930a] found in his *Nachlass*, Zermelo provided an analysis of set theory and a motivation for its axioms based on an "iterative conception".[29] He first set out his "set-theoretic model", and it is the cumulative hierarchy based on a totality of urelements. With this as a schematic picture he proceeded to motivate and justify his [1930b] axioms. This note reveals Zermelo to be actually the first who would retroactively motivate the axioms of set theory in terms of an iterative conception of sets as built up through stages of construction. For Replacement, Zermelo simply argued that, in modern terms, if a set's elements are replaced by members of a V_β, then the result would appear in $V_{\beta+1}$. Of this evidently circular argument, one of course bypassing any modern concern about definability, Zermelo wrote: "Of course, the assumption that the replacing elements belong to a *segment* of the development [i.e. a V_β], while being essential here, constitutes no real restriction." With this he was presumably importing his [1930b] picture of always having Grenzzahlen beyond, but circularity is still there at this further remove.

As set theory would develop, Replacement, Foundation, and the cumulative hierarchy picture would provide the setting for a developing high tradition that had its first milestone in Kurt Gödel's development of the constructible universe L. [Zermelo, 1930b] would be peripheral to this development, presumably because of its second-order lens and lack of rigorous detail. Gödel's work newly confirmed Replacement as a central axiom of set theory, and it was featured in a formal presentation of this work for which Gödel adapted an axiomatization of set theory due to Bernays, which itself was a transmutation of von Neumann's axiomatization.

4 Later History of Replacement

Gödel's advances in set theory can be seen as part of a steady intellectual development from his fundamental work on completeness and incompleteness. In the well-known, prescient footnote 48a to his celebrated incompleteness paper [1931], Gödel had pointed out that the formation of ever higher types over the Russellian theory of types can be continued into the transfinite and that his undecidable propositions become decided if the type ω is added. Matters in a footnote, perhaps an afterthought then, took up fully one-third of a summary (cf. [Gödel, 1932, pp. 234ff]) dated 22 January 1931. Gödel pointed out that the enlargement of Z, first-order Peano arithmetic, with higher-order variables and corresponding comprehension axioms leads not only to the decidability of his undecidable propositions but also to new, undecidable propositions, all

[28]The reflection aspect of [Zermelo, 1930b] was emphasized by William Tait [1998].

[29]For more on this note and its significance, see the author's introductory note appearing [Zermelo, 2010].

expressible in Z. He continued:

> In case we adopt a type-free construction of mathematics, as is done in the axiom system of set theory, axioms of cardinality (that is, axioms postulating the existence of sets of ever higher cardinality) take the place of type extensions, and it follows that certain arithmetic propositions that are undecidable in Z become decidable by axioms of cardinality, for example, by the axiom that there exist sets whose cardinality is greater than every α_n, where $\alpha_0 = \aleph_0$, $\alpha_{n+1} = 2^{\alpha_n}$.

This is Gödel's first remark on set theory of substance, and significantly, his example of an "axiom of cardinality" is evidently closely connected to the existence of the set that both [Fraenkel, 1922] and [Skolem, 1923b] had pointed to as the one to be secured by adding Replacement to Zermelo's 1908 axiomatization.

In an incisive lecture [1933o] Gödel expanded on his theme of higher types. He propounded the view that the axiomatic set theory "as presented by Zermelo, Fraenkel and von Neumann ... is nothing else but a natural generalization of the [simple] theory of types, or rather, it is what becomes of the theory of types if certain superfluous restrictions are removed." First, instead of having separate types with sets of type $n+1$ consisting purely of sets of type n, sets can be cumulative in the sense that sets of type n can consist of sets of all lower types. That is, with S_n to consist of the sets of type n newly construed, S_0 consists of the "individuals", and recursively, $S_{n+1} = S_n \cup \{X \mid X \subseteq S_n\}$. Second, the process can be continued into the transfinite, starting with the cumulation $S_\omega = \bigcup_n S_n$, proceeding through successor stages as before, and taking unions at limit stages. Gödel is seen here as promoting the cumulative hierarchy picture, which had been advocated for set theory by Zermelo [1930b, 1930a], as an extension of the simple theory of types.

As for how far this cumulative hierarchy of sets is to continue, Gödel [1933o, p. 47] wrote:

> The first two or three [transfinite] types already suffice to define very large ordinals. So you can begin by setting up axioms for these first types, for which purpose no ordinal whatsoever is needed, then define a transfinite ordinal α in terms of these first few types and by means of it state the axioms for the system, including all classes of type less than α. (Call it S_α.) To the system S_α you can apply the same process again, i.e., take an ordinal β greater than α which can be defined in terms of the system S_α and by means of it state the axioms for S_β including all types less than β, and so on.

Gödel thus envisioned an "autonomous progression", in later terminology, with large ordinals definable in low types leading to higher types.

Gödel, according to Hao Wang [1981, p. 128] reporting on conversations with Gödel in 1976, had already been working on the continuum problem for some time and had devised what he considered to be a transfinite extension of Russell's *ramified* theory of types. He in effect had started working up the constructible hierarchy, where in modern terms, for any set x, $\text{def}(x)$ is the collection of subsets of x definable over $\langle x, \in \rangle$ via a first-order formula allowing parameters from x, and:

$$L_0 = \emptyset;\ L_{\alpha+1} = \text{def}(L_\alpha);\ \text{and}\ L_\delta = \bigcup_{\alpha<\delta} L_\alpha \text{ for limit ordinals } \delta\,.$$

There is evidence however that Gödel could not get far pursuing his envisioned autonomous progression. Wang [1981, p. 129] reported how in Gödel's efforts to go up the constructible hierarchy he "spoke of experimenting with more and more complex constructions [of ordinals for indexing] for some extended period somewhere between 1930 and 1935." More pointedly, Georg Kreisel in his memoir [1980] of Gödel wrote (p. 193): "As early as 1931, Gödel alluded to some reservations [about Replacement]." Kreisel continued (p. 196, with $C_{\omega+\omega}$ his notation for $V_{\omega+\omega}$): "In keeping with his reservations, mentioned on p. 193, Gödel first tried to do without the replacement property, and to describe the constructible hierarchy L_α only for $\alpha <$ card $C_{\omega+\omega}$; in particular, without using von Neumann's canonical well-ordering [i.e. ordinals]. Instead, well-orderings had to be defined (painfully) in $C_{\omega+\omega} \ldots$"

The full embrace of Replacement led to a dramatic development. Set theory reached a new plateau with Gödel's formulation of the class $L = \bigcup_\alpha L_\alpha$ of *constructible* sets with which he established the relative consistency of the Axiom of Choice in mid-1935 and of the Continuum Hypothesis (CH) in mid-1937.[30] Gödel had continued the indexing of his hierarchy through *all* the ordinals as given beforehand to get a class model L of set theory and thereby to achieve *relative* consistency results, by showing that L satisfies Choice and CH. His early idea of using large ordinals defined in low types in a bootstrapping process would not suffice. Von Neumann's ordinals would be the spine for a thin hierarchy of sets, and this would be the key to both the Choice and CH results.

In his monograph [1940], based on 1938 lectures, Gödel provided a specific, formal presentation of L which made evident the methodological importance of Replacement. Gödel relied on an axiomatization of a class-set theory which, save for an economy of presentation, was the one detailed to him in a letter from Bernays of 3 May 1931.[31] Bernays' axiomatization itself was a transmu-

[30]See [Dawson, 1997, pp. 108,122]; in one of Gödel's *Arbeitshefte* there is an indication that he established the relative consistency of CH in the night of 14-15 June 1937.

[31]See [Gödel, 2003, pp. 105ff]. Gödel in [1940] routinely acknowledged Bernays by citing his later, published account [1937], but it is clear from their correspondence that Gödel had assimilated Bernays' axiomatization through the letter from 1931.

tation of von Neumann's axiomatization which differed from it in two respects. First, while von Neumann had taken function as a primitive notion with his I-objects and II-objects being functions, Bernays reverted to collections, sets and classes. More importantly, Bernays like Zermelo [1930b] adopted Foundation, and with von Neumann's [1929] analysis of his characteristic axiom IV 2 as being equivalent under Foundation to the conjunction of Replacement and Choice in class forms, Bernays adopted the latter two instead. Bernays's approach thus had the effect of bringing von Neumann's axiomatization more in line with [Zermelo, 1930b]. But unlike in Zermelo's approach, there were no urelements, and as Bernays wrote in that May 1931 letter, "A *complete formalization*, and in fact in a *first-order* [*ersten Stufe*] framework, can be carried out without difficulty." To summarize, the features of Bernays' axiomatization that would commend its further use and influence were that it recast von Neumann's work to present a viable theory starting with sets and classes as primitive notions, and it incorporated Replacement and Foundation, as did Zermelo's later axiomatization, but in a first-order context and without the relativism of having urelements. Gödel had thus become aware of Replacement early on; his full espousal of the axiom in a first-order context affirmed its importance.

Gödel in his monograph carried out a careful development of "abstract" set theory through the ordinals and cardinals with features that have now become common fare. Gödel then used eight binary operations, producing new classes from old, to generate L set by set via transfinite recursion. This veritable "Gödel numbering" with ordinals bypassed the formalization of the def(x) operation and made evident certain aspects of L. Since there is a direct, definable well-ordering of L, choice functions abound in L, and Choice holds there.

Gödel's proof that L satisfies CH consisted of two separate parts, both depending on Replacement and both to become paradigmatic for inner model theory, that large part of modern set theory with beginnings in Gödel's work. Gödel established the implication $V = L$ implies CH and, in order to apply this implication within L, the absoluteness $L^L = L$—that the construction of L in L again gives L—to establish CH within L.

The absoluteness of L depends inherently, albeit subtly, on Replacement. Replacement, affirming the interplay of token and type, bolsters the (von Neumann) ordinals as tokens for all well-orderings. Being an ordinal is absolute for transitive sets, i.e. being an ordinal in the sense of a set closed under membership is really to be an ordinal. With this, one can construct L, e.g. through Gödel's eight binary operations, in the sense of L and again get L. A basic reason why Gödel's early efforts with general well-orderings could not have succeeded is that one needs this absoluteness; one cannot just work with well-orderings or even equivalence classes of these. The absoluteness of ordinals as canonical for well-orderings is a pre-condition for the substantive arguments in

L and now, in all inner model theory.

As for Gödel's argument for $V = L$ implies CH, it rests on what he termed an "axiom of reducibility"—which properly can be thought of as the fact that for any α the constructible subsets of L_α all belong to some L_β—and what is now known as a Skolem hull argument—by which that β can be taken to have the same cardinality as α when α is infinite. Replacement is crucial for the "axiom of reducibility". In referring to Russell's ill-fated axiom in his ramified theory of types, Gödel took his version to be a rectification. In his first announcement [1938] he wrote:

> [The] 'constructible' sets are defined to be those sets which can be obtained by Russell's ramified hierarchy of types, if extended to include transfinite orders. The extension to transfinite orders has the consequence that the model satisfies the impredicative axioms of set theory, because an axiom of reducibility can be proved for sufficiently high orders.

In his analysis of Russell's mathematical logic Gödel [1944, p. 147] again wrote about how with L he had proved an axiom of reducibility, emphasizing: "... all impredicativities are reduced to one special kind, namely the existence of certain large ordinal numbers (or well-ordered sets) and the validity of recursive reasoning for them." Decades later Gödel wrote in a letter of 7 March 1968 to Wang (cf. [Wang, 1974b, pp. 8-9]):

> ... there was a special obstacle which *really* made it *practically impossible* for constructivists to discover my consistency [of the Continuum Hypothesis] proof. It is the fact that the ramified hierarchy, which had been invented *expressly for constructive purposes,* had to be used in an *entirely nonconstructive way.*

This nonconstructive way was to prolong the ramified hierarchy using arbitrary ordinals for the indexing, and for this their extent had to be sustained by Replacement.

In late conversations, Gödel justified Replacement as follows ([Wang, 1996, p. 259]; see also [Wang, 1974a, p. 186]):

> 8.2.15. From the very idea of the iterative concept of set it follows that, if an ordinal number a has been obtained, the operation P of power set iterated a times from any set y leads to a set $P^a(y)$. But, for the same reason, it would seem to follow that, if instead of P, one takes some larger jump in the hierarchy of types, for example, the transition Q from x to the set obtained from x by iterating as many times as the smallest ordinal [not] of the well-orderings of x, $Q^a(y)$ likewise is a set. Now, to assume this for any conceivable

> jump operation—even for those that are defined by reference to the universe of all sets or by use of the choice operation—is equivalent to the axiom of replacement.

Gödel recalled here the lesson learned about the need for impredicative reference to the universe of all sets. Technically speaking, all that is needed to get Replacement is Gödel's $P^a(y)$'s together with the assertion that the class of all ordinals is regular with respect to (definable) class functions.[32]

With the "axiom of reducibility" Gödel had broached a new aspect of Replacement, that of reflection. Recalling Cantor's Absolute but cast in terms of the cumulative hierarchy, the heuristic of reflection surrounds the idea that the universe $V = \bigcup_\alpha V_\alpha$ cannot be characterized uniquely by a formula so that any particular property ascribable to it must already be ascribable to some rank V_α. In oral, necessarily brief remarks [1946] at a conference, Gödel voiced what would become a prominent way to motivate and formulate "strong axioms of infinity", now called large cardinal axioms, by reflection: "Any proof for a set-theoretic theorem in the next higher system above set theory", i.e. if the satisfaction relation for V itself were available, "is replaceable by a proof from such an axiom of infinity."

Gödel's L stood as a high watermark for set theory for quite a span of years, and during this period the Bernays-Gödel (BG) class-set theory maintained an expository sway. There was then a shift toward the more parsimonious ZFC set theory, especially after BG and ZFC were shown around 1950 to have the same provable consequences for sets. Concomitantly, Replacement became widely seen, as first envisioned by Skolem [1923b], as the first-order schema that it is taken to be today. Forthwith, new model-theoretic initiatives led to the formalization of reflection properties that put Replacement in a new light.

With the basic concepts and methods of model theory being developed by Tarski and his students at Berkeley, Richard Montague [1961] in his 1957 Berkeley dissertation had studied reflection properties in set theory and had shown that the axiom schema of Replacement is not finitely axiomatizable over the other axioms in a strong sense. Azriel Levy [1960a, 1960b] then exploited the model-theoretic methods to establish a broader significance for reflection principles. Sufficient for reflection is ZF, ZFC minus the Axiom of Choice. The ZF *Reflection Principle*, drawn from [Montague, 1961, p. 99] and [Levy, 1960a, p. 234], asserts that for any (first-order) formula $\varphi(v_1, \ldots, v_n)$ in the free variables as displayed and any ordinal β, there is a limit ordinal $\alpha > \beta$ such that for any $x_1, \ldots, x_n \in V_\alpha$,

$$\varphi[x_1, \ldots, x_n] \text{ iff } \varphi^{V_\alpha}[x_1, \ldots, x_n] ,$$

[32] That is, there is no such function with domain an ordinal and cofinal in the class of all ordinals.

where as usual φ^M denotes the relativization of the formula φ to M. The idea is to carry out a Skolem closure argument with the collection of subformulas of φ. Montague showed that this schema holds in ZF, and Levy showed that it is actually equivalent to the axiom schema of Replacement together with the Axiom of Infinity in the presence of the other axioms of ZF. Through this work the ZF Reflection Principle has become well-known as making explicit how reflection is intrinsic to the ZF system and as a new face for Replacement. Levy [1960a] cast the ZF Reflection Principle as motivation for stronger reflection principles, with the first in his hierarchy being the ZF principle but with "limit ordinal α" replaced by "inaccessible cardinal α". This motivated the strongly Mahlo cardinals, and, notably, was implicit in [Zermelo, 1930b] (cf. §3).

Reflection, and thus Replacement, was seen in a further new light through an axiomatic set theory proposed by Wilhelm Ackerman [1956]. His theory A is a first-order theory that can be cast as follows: There is one binary relation $\in$ for membership and one constant V; the objects of the theory are to be referred to as classes, and members of V as sets. The axioms of A are the universal closures of:

(1) Extensionality: $\forall z(z \in x \longleftrightarrow z \in y) \longrightarrow x = y$.

(2) Comprehension: For each formula ψ not involving t,

$$\exists t \forall z(z \in t \longleftrightarrow z \in V \wedge \psi).$$

(3) Heredity: $x \in V \wedge (t \in x \vee t \subseteq x) \longrightarrow t \in V$.

(4) Ackermann's Schema: For each formula ψ in free variables $x_1, \ldots, x_n, z$ and having no occurrence of V,

$$x_1, \ldots, x_n \in V \wedge \forall z(\psi \longrightarrow z \in V) \longrightarrow \exists t \in V \forall z(z \in t \longleftrightarrow \psi).$$

This last, a comprehension schema for sets, is characteristic of Ackermann's system. It forestalls Russell's Paradox, and its motivation was to allow set formation through properties independent of the whole extension of the set concept and thus to be considered sufficiently definite and delimited.

Ackermann [1956] himself argued that every axiom of ZF, when relativized to V, can be proved in A. However, Levy [1959] found a mistake in Ackermann's proof of Replacement, and whether the schema can be derived from Ackermann's Schema remained an issue. Toward a closer correlation with ZF, Levy came to the idea of working with A*: A together with the Axiom of Foundation relativized to V. As for ZF, Foundation focuses the sets with a stratification into a cumulative hierarchy. Levy [1959] showed that, leaving aside the question of Replacement, A* establishes substantial reflection principles. On the other hand, he also showed through a sustained axiomatic analysis that for a sentence σ of set theory (so without V) : *If* σ *relativized to* V *is provable in* A*, *then* σ *is provable in* ZF. The thrust of this work was to show that Ackermann's Schema can be assimilated into ZF—a somewhat surprising result—and that ZF and A* have about the same theorems for sets.

Levy and Robert Vaught in their [1961] later observed by an inner model argument that, as for ZF and Foundation, if A is consistent, then so is A*. They then went on to confirm that the addition of Foundation to A was substantive; they showed that Ackermann's Schema is equivalent to a reflection principle in the presence of the other A^* axioms, and that A* establishes the existence of $\{V\}$ and the power classes $P(V)$, $P(P(V))$, and so forth.

Years later, returning to the original issue about Replacement, William Reinhardt [1970] in his 1967 Berkeley dissertation built on Levy-Vaught [1961] to establish for A* what Ackermann could not establish for A: *Every axiom of* ZF*, when relativized to* V*, can be proved in* A*. Thus, A* and ZF do have exactly the same theorems for sets. Replacement is the central axiom schema to establish in order to get ZF, and once done, Ackermann's set theory exhibited it in a new light.

5 Forms of Replacement

Replacement is central to ZFC, but it also has a complicated form, and as befits the situation it has been analyzed in a variety of ways. In this section we discuss various forms of Replacement, starting from simple modulations and proceeding to more substantive ones. To repeat, we take Replacement to be the following schema: For any formula φ with free variables among $a, x, y, z_1, \ldots, z_n$ but not including b,

$$\text{(Rep)} \qquad \forall z_1 \forall z_2 \ldots \forall z_n \forall a(\forall x \exists ! y \varphi \longrightarrow \exists b \forall y(\exists x(x \in a \wedge \varphi) \longleftrightarrow y \in b)),$$

where $\exists!$ abbreviates the formalizable "there exists exactly one". The Replacement schema easily implies the Separation schema, but in modern presentations of ZFC both are kept by convention, especially as in comparative investigations of subsystems of ZFC one wants to retain versions of Separation.

An initial observation with historical relevance is that Replacement is equivalent to the form requiring the functional correlation to be one-to-one. If F is a class function, then G defined via the ordered pair by $G(x) = \langle x, F(x)\rangle$ is one-to-one. So, for any set a, $G``a = \{\langle x, F(x)\rangle \mid x \in a\}$ is a set by "one-to-one Replacement". But then, one can as usual appeal to Union and Separation to get to the range of this set, i.e. $\{F(x) \mid x \in a\}$.[33]

A syntactic complication of Replacement is that in its formal statement one has to be attentive to passive parameters. Levy [1974] considered the

[33] I do not know whether, without Separation, Replacement is derivable from "one-to-one Replacement".

elimination of parameters. Let R_0 (his notation) be the parameter-free version of Replacement, in the sense that Rep is altered so that the only free variables occurring in φ are x, y and the universal quantification $\forall z_1 \forall z_2 \ldots \forall z_n$ is deleted. Let P_2 be the restriction of the Power Set Axiom asserting that for any set x the totality of two-element subsets if x is a set. Levy established that in the presence of Extensionality, Pairing, Union, and P_2, R_0 implies Replacement. Hence, for a quick presentation of ZFC one can rightly give only the parameter-free R_0. Earlier, Harvey Friedman in his 1967 Ph.D. thesis (cf. [1971b]) had also shown that a restricted, "more explicit" group of axioms including a parameter-free version of Replacement entails ZFC, but indirectly in terms of consistency by going through the construction of the constructible universe L.

Levy called $R^{\leftarrow}$ another variant of Rep where the "$\longleftrightarrow$" is relaxed to "$\longrightarrow$". That is, for any set a one concludes only that there is a set b that subsumes the image $F``a$ under the class function. Of course, with Separation, $R^{\leftarrow}$ is equivalent to full Replacement. ZFC is sometimes presented with $R^{\leftarrow}$ instead of Replacement, giving Separation an independent status. Levy showed that Extensionality, Empty Set, Pairing, Union, Power Set, a parameter-free version of Separation, and $R^{\leftarrow}$ together do *not* imply Replacement. On the other hand, recently Adrian Mathias [2007] showed that Extensionality, Empty Set, Pairing, Union, Power Set, Δ_0-Separation, Foundation, Transitive Containment (that every set is a member of a transitive set), and $R^{\leftarrow}$ *do* imply Replacement. Here, Δ_0-Separation is the Separation schema restricted to Δ_0, or bounded, formulas in the Levy hierarchy, i.e. those formulas of set theory that can be rendered with quantifiers only of form $\forall v \in w$ and $\exists v \in w$.

A more substantive variant of Rep is Collection, where both the "$\longleftrightarrow$" is relaxed to $\longrightarrow$" and the "$\exists! y$" is relaxed to "$\exists y$". That is, one finally gives up the historically given functionality of φ so that for any x there is some witness y and concludes for any set a that there is a set b containing witnesses for every $x \in a$. Collection emerged in the late 1960s with Replacement having become less explicitly a class function principle and more explicitly a schema of formulas, suggestive of the relaxation of the $\forall x \exists! y$ to $\forall x \exists y$. In any case, Collection is equivalent over the other axioms to Replacement: To establish Collection from Replacement, suppose that $\forall x \exists y \varphi$ and a is a set. For each $x \in a$, let α_x be the least α such that φ holds for a $y \in V_{\alpha+1}$. With Replacement we can consider the ordinal $\beta = \sup\{\alpha + 1 \mid x \in a\}$, and V_β can serve as the set b to confirm Collection. Power Set is necessary here, in the sense that Andrzej Zarach [1996] showed with a forcing argument that over a base theory without Power Set, Replacement does not imply Collection.

Replacement and Collection are distinct in intuitionistic set theory IZF and in constructive set theory CZF, theories based on intuitionistic logic in which the appeal to least ranks V_α ("Scott's trick") as above is not available. Nicolas Goodman [1985] showed using Kripke models that if one adjoins a class parameter (that is, a new predicate symbol) then intuitionistically one cannot show that Replacement implies Collection. Inspired by this, Harvey Friedman and Andrej Scedrov [1985] showed that intuitionistic set theory as formulated with Replacement, IZF_R, does not prove Collection, which is used in IZF. It would remain open whether IZF_R and IZF have the same proof-theoretic strength. Michael Rathjen [2005] showed that for the weaker CZF, Replacement and a strong form of Collection do have the same proof-theoretic strength.

Refinements of the various schemas have to do with restricting them according to complexity of formula, as with Δ_0-Separation. For the formulations, we briefly recall the standard Levy hierarchy of set-theoretic formulas: $\Sigma_0 = \Pi_0 = \Delta_0$ are the bounded formulas, where again the quantifiers are only of form $\forall v \in w$ and $\exists v \in w$. Then a formula is Σ_{n+1} if it is of the form $\exists v \varphi$ where φ is Π_n, and Π_{n+1} if it is of the form $\forall v \varphi$ where φ is Σ_n. It is to be pointed out that the very efficacy of concept classification according to this hierarchy depends on Replacement, or more directly, Collection. Collection amounts to having

$$\forall v \in w \exists v_0 \varphi \longleftrightarrow \exists v_1 \forall v \in w \exists v_0 \in v_1 \varphi\,,$$

and this shows how the bounded quantifier $\forall v \in w$ can be inductively absorbed in the complexity analysis. As expected, Σ_n-Replacement refers to the Replacement schema restricted to the Σ_n formulas, and so forth. Π_n-Collection implies Σ_{n+1}-Collection, for if $\exists w \varphi$ is Σ_{n+1} where φ is Π_n, then in $\forall x \exists y \exists w \varphi$ for Collection one can pair y and w. Also, the simple argument getting from Replacement to Separation shows that Σ_{n+1}-Replacement implies Σ_n-Separation.

The prominent set theory with restricted schemas is Kripke-Platek (KP) set theory: Extensionality, Empty Set, Pairing, Union, Δ_0-Separation, and Δ_0-Collection.[34] KP can carry the weight of substantive recursive procedures, and in particular the construction of Gödel's L. On the other hand, Σ_1-Separation is needed over KP to establish the fundamental von Neumann result that every well-ordering of a set is order-isomorphic to an ordinal. Latterly, Mathias [2001b] showed that over KP + Power Set + Choice, Σ_1-Separation is equivalent to every well-ordering of a set being order-isomorphic to an ordinal.[35] Finally, over KP + Power Set, Σ_2-Replacement suffices to define recursively the cumulative hierarchy, as being a V_α is Π_1.

The work on restricted set theories and schemas have certainly brought out the logical dependencies of the classical development surrounding Replacement. In particular, they have brought out the crucial aspects of Replacement as

[34] [Mathias, 2001b] includes Π_1-Foundation in KP.

[35] See [Mathias, 2001b, 3.18]; this paper has much more on restricted schemas of Replacement and variants. See [Mathias, 2006] for such schemas in weaker set theories.

functioning separately from Power Set and so from large cardinality sets. At the same time, much of the work has focused attention on Replacement as the principle of set theory to be reckoned with, especially in its providing the basis for well-founded recursion, the importance of method coming to the fore.

6 Replacement Vindicated

In this last, more freewheeling section, we address issues that have invited skepticism, or at least hesitation, about Replacement based largely on grounds of ontological commitment and we provide an affirmation on general grounds for Replacement's central role in set theory. The large mathematical and historical perspective and basis for Replacement provided in the previous sections should already serve to address issues having to do with thematic importance and historical emergence, and so it is that we here turn to the further issues.

As mentioned in §3, in the *Nachlass* note [1930a] Zermelo first motivated the axioms of set theory in terms of the cumulative hierarchy picture. From the late 1960s, what has come to be regarded as *the iterative conception*, conceiving sets as built up through stages of construction, has become well-known as a heuristic for motivating the axioms of set theory generally.[36] This has opened the door to a metaphysical appropriation in the following sense: It is as if there is some notion of set that is "there", in terms of which the axioms must find justification as being true or false. But set theory has no particular obligations to mirror some "prior" notion of set, especially one like the ultimately ontological iterative conception, arrived at after the fact.

When Replacement has been justified according to the iterative conception, the reasoning has in fact been circular as it was in [Zermelo, 1930a], with some feature of the cumulative hierarchy picture newly adduced solely for this purpose. George Boolos [1971] argued that neither Replacement nor Choice, in their providing for a store of sets, is evident from the iterative conception. Subsequently, he [1989] observed that Replacement and Choice do follow from a certain "limitation of size" conception of set that he called FN; this is evident since FN amounts to an espousal of von Neumann's axiom IV 2 (cf. §3). With the two, the iterative and limitation of size concepts, Boolos ends: "Perhaps one may conclude that there are at least two thoughts 'behind' set theory."

Another thought is that set theory is what it is, a historically given, autonomous field of mathematics proceeding with its own self-fueling methods and procedures. We continue the dialectic with Boolos as the specificities of his writings invite discussion as he intended and provide an opportunity to raise important, general considerations.

In his late "Must we believe in set theory?" [2000], Boolos quite remarkably argued against the existence of the least fixed point $\kappa = \aleph_\kappa$ of the aleph

[36] Joseph Shoenfield [1967, pp. 238ff] and [1977], Hao Wang [1974a], Dana Scott [1974], and George Boolos [1971, 1989] motivate the axioms of set theory in terms of iterative conceptions.

function. According to Boolos,

> The burden of proof should be, I think, on one who would adopt a theory so removed from experience and the requirements of the rest of science (including the rest of mathematics) as to claim that there are κ objects.

Boolos was of course aware that the existence of κ is established with Replacement, and is inclined to jettison Replacement as not following from the "natural" iterative conception.

First, "belief" is a vague notion, and there is little one can profitably discuss about it as a concept, especially in mathematics and as concerns ontology.

Second, there are varieties of "experience". Though Boolos anticipates the charge that asking whether κ exists from an external vantage point is to fall into metaphysical error, this counterpoint is nonetheless quite relevant. From an external vantage point κ could seem a magic mountain to the foot soldier encumbered with dated provisions and determined to climb at some previously ordained pace. For Henri Poincaré $\aleph_1$ was inaccessible, and to one who takes successive counting to be the overwhelming primal act, 2^{100} is analogously inaccessible. But for Cantor, the new transfinite landscape was set out with generating principles for which the climb to κ would have been congenial. And Hausdorff subsumed κ in his schematic investigation of the uncountable transfinite.

κ is approachable in set theory via $\kappa_0 = \aleph_0$, $\kappa_{n+1} = \aleph_{\kappa_n}$, and $\kappa = \sup \kappa_n$. Like 2^{100}, κ is to be understood in terms of the operations and procedures that went into its formulation. What there is to grasp of κ is its recursive definition, legitimized by Replacement, together with the Cantorian notion of the alephs and what further can be proved from these about κ. One works *schematically*, in so far as one can work, the mathematical experience focused by guidelines for proceeding, these amenable to axiomatic presentation.

Once the Cantorian theory of the transfinite is taken in, one sees that $\kappa = \aleph_\kappa$ could be less than $2^{\aleph_0}$, the cardinality of the continuum, the subject of investigation of classical analysis. As mathematics was transmuting to a complex edifice of conceptual constructions, Hilbert [1900] wrote in connection with his axiomatization of the reals:

> Under the conception described above [the axiomatic method], the doubts which have been raised against the existence of the totality of all real numbers (and against the existence of infinite sets generally) lose all justification, for by the set of real numbers we do not have to imagine, say, the totality of all possible laws according to which the elements of a fundamental sequence can proceed, but rather— as just described—a system of things, whose mutual relations are given by the *finite and closed* system of axioms [for complete ordered

> fields], and about which new statements are valid only if one can derive them from the axioms by means of a finite number of logical inferences.

Third, and the most sweeping, of dialectical points vs. Boolos, which we expand to be about Replacement generally, concerns the use and "requirements" for mathematics. For set theory itself, the previous sections have provided ample evidence for the importance of Replacement in various directions. To press the point, in ZF without Replacement neither the (von Neumann) ordinal $\omega + \omega$ nor the rank V_ω nor Zermelo's particular infinite set $\{\emptyset, \{\emptyset\}, \{\{\emptyset\}\}, \ldots\}$ can be shown to exist. Keeping in mind the now-standard formulation of the Axiom of Infinity as the existence of the (von Neumann) ordinal ω, one simple model of ZF minus Replacement having none of these sets is $\langle W, \in \rangle$, where $W_0 = \omega$, $W_{n+1} = P(W_n)$, and $W = \bigcup\{W_n \mid n \in \omega\}$. This is a drastic failure vis-à-vis all of functional substitution, recursion, and indifference to identification.[37] One can of course just posit the existence of any particular V_α and allow for recursions up to some fixed length β, but then analogous issues arise at the level of $V_{\alpha+\beta}$.

Proceeding in the cumulative hierarchy, the ranks $V_{\omega+1}$ and $V_{\omega+2}$ correspond via cardinality to reals and real functions, and this has advanced the presumption *contra* Replacement that classical mathematics, at least, can be accommodated at these low levels of the cumulative hierarchy. However, the roughest indifference to identification, that according to cardinal equivalence, has to be acknowledged and then affirmed by considerable coding and dexterity. With ordered pairs and functions, one starts to move from just ontological considerations to analysis and method, and these require in the set-theoretic context several iterations of the 'set of' operation, which in terms of the cumulative hierarchy amounts to a climb of several ranks.

Moving away from the iterative conception for such analyses of set-theoretic representation, it is cardinally more parsimonious and theoretically more coherent to consider for a regular uncountable cardinal κ the sets hereditarily of cardinality less than κ, i.e.

$$H_\kappa = \{x \mid |\mathrm{tc}(x)| < \kappa\},$$

where $\mathrm{tc}(x)$ is the transitive closure of x, informally $\mathrm{tc}(x) = x \cup \bigcup(x) \cup \bigcup\bigcup(x) \ldots$. The existence of the H_κ cannot be established in ZF without

[37]See [Mathias, 2001a] for constructions of models of ZF minus Replacement exhibiting such weaknesses, but containing all the ordinals. Model 13 of [Mathias, 2006, section 7] shows easily that rank cannot be defined in ZF minus Replacement. For an extreme failure of Replacement, see [Mathias, 2010] for a model of Bourbaki's 1949 version of set theory with ordered pair treated as a primitive notion, in which some unordered pair fails to exist.

Replacement.[38]

The H_κ models ZFC$^-$, the ZFC axioms minus Power Set, whereas V_α for limit $\alpha > \omega$ model the ZFC axioms except Replacement. Set theorists from the 1970s on have regularly and widely appealed to the H_κ as arbitrarily large models of ZFC$^-$ approximating V, especially as Replacement has become so intrinsic to all of set theory.[39] A substantial point is that one can carry out forcing over H_κ as a model of ZFC$^-$, getting forcing extensions of ZFC$^-$; without Replacement it becomes problematic to control the proliferation of forcing terms.

It is the rendering of mathematical proof rather than of mathematical objects which is substantive, and metamathematical investigations have established the necessity of substantial resources. In an example that has become widely cited for calibration with Replacement, H. Friedman [1971a] showed in 1968 that to establish Borel Determinacy would require the use of the cumulative hierarchy up to V_{ω_1}, with a level-by-level analysis revealing that determinacy at each new level of the ω_1-level Borel hierarchy would require one more iteration of the power set operation. Martin [1975] in 1974 duly established Borel Determinacy, later [1985] providing a purely inductive proof that made more evident the correlation of the Borel and the cumulative hierarchies. Borel Determinacy is thus an incisive example of the methodological involvement of Replacement.

Years later H. Friedman [1981] considerably expanded his [1971a] work to establish new independences for propositions about Borel sets, some requiring the strength of large cardinals. In that part of the work closest to [1971a], Friedman established Borel "diagonalization" and "selection" theorems that follow from Borel Determinacy and have a notable simplicity, yet still require the use of the cumulative hierarchy up to V_{ω_1}. With I the unit interval of reals, one such proposition is: *Every symmetric binary Borel relation on I contains or is disjoint from a Borel function on I.*

Martin's Borel Determinacy result itself has through the years found wide-ranging applications. In the latest, Jan Reimann and Theodore Slaman [2010] established the Martin-Löf randomness of non-computable reals with respect to continuous probability measures, with a necessary use of infinitely many iterations of the power set operation. The original Fraenkel-Skolem motivation for Replacement in the first place was thus called upon for getting randomness for reals.

[38]Replacement is needed to proceed generally from regular uncountable cardinals κ to their H_κ, but there is the issue of having regular uncountable cardinals at all without Replacement. One can readily observe though that e.g. in the model $\langle W, \in \rangle$ above of ZFC minus Replacement, there are certainly uncountable well-orderings but not the set H_{ω_1} consisting of the hereditarily countable sets.

[39]See [Foreman and Kanamori, 2010]; All three of its volumes exhibit extensive use of the H_κ.

We next consider set theory not so much in its foundational role but as of a piece with modern mathematics. Since set theory emerged as a sophisticated field of mathematics in the early 1960s with the creation of forcing, both the conceptual space it provides and the methods that it engages have become more and more integrated into the broad fabric of modern mathematics. In this regard the Replacement motif $\{t_i \mid i \in I\}$, when unrestricted in colonizing new domains, implicates Replacement.

The conceptual space provided by the set-theoretic universe as buttressed by Replacement has been increasingly brought into play in modern mathematics to generate examples and counterexamples from analysis to topology. General topology particularly has become imbued with set-theoretic constructions involving large cardinalities, to the extent that set theory and topology have here a large intersection of methods and procedures. Two conspicuous examples are Mary Ellen Rudin's construction [1971] of a Dowker space of cardinality $\aleph_\omega^{\aleph_0}$ and Peter Nyikos's [1980] derivation of the normal Moore space conjecture from a measure extension "axiom", one proved consistent relative to a strongly compact cardinal by Kenneth Kunen. The first result refuted a conjecture posed in 1951, and the second confirmed a conjecture, relative to large cardinals, posed in 1962. Natural questions about general topological spaces had thus been posed early on, and substantive elucidations only emerged as set theory itself became common coin.

The growth of category theory from the middle of the 20th Century has brought to the fore the Replacement motif $\{t_i \mid i \in I\}$, and in so far as this development is to be formalized in axiomatic set theory, it evidently implicates Replacement as immanent.[40] Of course, if one aspires to a categorical foundation for mathematics then a reduction to set theory is not a primary concern. However, it is notable that the converse reduction of set theory to category theory seems to run afoul of the problem of how to handle the axiom schema of Replacement. Replacement is what mainly needs to be accommodated, and category theory seems unable to meet the challenge.

Categorical imperatives have become particularly topical of late because of the involvement of Grothendieck universes in Andrew Wiles's proof of the Shimura-Taniyama modularity for elliptic curves, establishing Fermat's Last Theorem. In terms of set-theoretic resources, the straightforward recasting of the proof in set theory initially set the bar for establishing the famous statement of classical number theory at ZFC plus having many inaccessible cardinals (cf. [McLarty, 2010]). Recently, Mark Kisin [2009] provided a conceptually simpler proof of the modularity, one avoiding the Grothendieck apparatus and straightforwardly renderable in ZFC. In a continuing reduction, McLarty [2011] has avowed (January 2011) that both the Wiles and Kisin proofs of Fermat's

[40]The (Dedekind) Recursion Theorem is directly implicated in the common construction for a functor $F\colon Set \to Set$ of the colimit $0 \to F0 \to FF0 \to \ldots$.

Last Theorem can be formalized in ZFC without Replacement and with only Δ_0-Separation.

The issue of "purity of method" emerges here in significant fashion. The classic example is the Prime Number Theorem having been first proved using sophisticated methods of complex analysis and then decades later having been given an elementary proof formalizable in Peano Arithmetic. One looks similarly for an elementary proof of Fermat's Last Theorem. It would be very interesting and quite significant to find an elementary proof, and one would think this to be a definite new conjecture of ongoing mathematics.

Whatever purity of method would outwardly dictate however, it is also very interesting and quite significant that there are proofs at all of Fermat's Last Theorem and that these can be rendered in ZFC or restricted systems. Replacement in particular is seen to play a conspicuous role both in allowing for the straightforward rendition of proofs and for providing the conceptual space for the comparative and complexity analysis of proofs. The refining or transformation of a proof to one depending on weaker resources is itself a proof process. If for the Fermat a proof is found evidently formalizable in Peano Arithmetic, it would be part of a historical process in which Replacement has played a substantive role. In any case, Replacement was seen to underpin a proof, and transcending specific statements, like the Fermat or the larger Kimura-Taniyana modularity for arithmetic algebraic geometry, proofs as arguments evolving to methods are free-standing and autonomous in modern mathematics.

Modern mathematics is, to my mind, a historically given, complex edifice based on *conceptual constructions*. With its richness, variety, and complexity any discussion of the nature of mathematics cannot but accede to the primacy of its history and practice. Mathematics is in a broad sense self-generating and self-authenticating, and alone competent to address issues of its correctness and authority.

What brings us mathematical knowledge? The carriers of mathematical knowledge are *proofs*, more generally arguments and constructions, as embedded in larger contexts. Mathematical knowledge does not consist of theorem statements and does not consist of more and more "epistemic access", somehow, to "abstract objects" and their workings. Moreover, mathematical knowledge extends not so much into the statements, but back into the means, methods and definitions of mathematics, sometimes even to its axioms. Statements gain or absorb their senses from the proofs made on their behalf. A statement may have several different proofs each investing the statement with a different sense, the sense reinforced by different refined versions of the statement and different corollaries from the proofs. The movement from one proof to another is itself a proof, and generally the analysis of proofs is an activity of proof and argument. In modern mathematics, proofs and arguments have often achieved an independent status beyond their initial contexts, with new spaces of mathematical objects formulated and investigated because certain proofs can be carried out. Proofs are not merely stratagems or strategies; they and thus their evolution are what carries forth modern mathematics.

What about existence in mathematics? Questions like "What is a number?" and "What is a set?"' are not mathematical questions, and any answers would have no operational significance in mathematics. *Within* mathematics, an existence assertion may be an axiom, a goal of a proof, or one of the junctures of a proof. Existence is then submerged into context, and it is at most a matter of whether one regards the manner in which an existence assertion is deduced or incorporated as coherent with the context. An existence axiom, like the Axiom of Infinity, simply sets the stage with a means for proof. An existence proof, especially one that depends on the law of the excluded middle and is not constructive, may call for a different proof, but it is still a proof of some kind.

As for the Axiom of Replacement, it is a central axiom of set theory, indeed the one to reckon with in verifications of the axioms. Replacement is the generative existence principle that legitimizes functional substitution, recursion, and through it, the means for proving results about *all* sets. Replacement is intrinsic to inner model theory, starting with the work of Gödel, as the theory depends on the absolute and canonical nature of the ordinals and the fine details of the many transfinite recursions that define inner models. Replacement is essential for forcing, in both the range of transfinite recursions necessary to build forcing extensions and the ability to control the proliferation of forcing terms. Through functional correlation Replacement gives expression to an expansive indifference to identification, allowing for various canonical tokens to stand for types and articulates the interaction and identification of type and token. In this bolstering of the concept of set through functional substitution, recursion, and indifference to identification, Replacement has become part of the sense of set in modern set theory.

References

[Ackermann, 1956] Wilhelm Ackermann. Zur Axiomatik der Mengenlehre. *Mathematische Annalen*, 131:336–345, 1956.

[Benacerraf and Putnam, 1983] Paul Benacerraf and Hilary Putnam, editors. *Philosophy of Mathematics. Selected Readings.* Cambridge University Press, Cambridge, 1983. Second edition.

[Bernays, 1937] Paul Bernays. A system of axiomatic set theory—Part I. *The Journal of Symbolic Logic*, 2:65–77, 1937. Reprinted in [Müller, 1976, pp. 1-13].

[Bernays, 1941] Paul Bernays. A system of axiomatic set theory—Part II. *The Journal of Symbolic Logic*, 6:1–17, 1941. Reprinted in [Müller, 1976, pp. 14-30].

[Bernays, 1958] Paul Bernays. *Axiomatic Set Theory*. Studies in Logic and the Foundations of Mathematics. North-Holland, Amsterdam, 1958. With a historical introduction by Abraham A. Fraenkel.

[Bolzano, 1851] Bernard Bolzano. *Paradoxien des Unendlichen*. C. H. Reclam, Leipig, 1851.

[Boolos, 1971] George S. Boolos. The iterative concept of set. *Journal of Philosophy*, 68:215–231, 1971. Reprinted in [Boolos, 1998], pp. 13-29.

[Boolos, 1989] George S. Boolos. Iteration again. *Philosophical Topics*, 17:5–21, 1989. Reprinted in [Boolos, 1998], pp. 88-105.

[Boolos, 1998] George S. Boolos. *Logic, Logic, and Logic*. Harvard University Press, Cambridge MA, 1998. Edited by Richard Jeffery.

[Boolos, 2000] George S. Boolos. Must we believe in set theory? In Gila Sher and Richard Tieszen, editors, *Between Logic and Intuition: Essays in Honor of Charles Parsons*, pages 257–268. Cambridge University Press, Cambridge, 2000. Preprinted in [Boolos, 1998, pp. 120-132].

[Burali-Forti, 1897] Cesare Burali-Forti. Una questione sui numeri transfiniti. *Rendiconti del circolo matematico de Palermo*, 11:154–164, 1897.

[Burgess, 2009] John P. Burgess. Putting structuralism in its place, 2009. On website: www.princeton.edu/~jburgess.

[Cantor, 1872] Georg Cantor. Über die Ausdehnung eines Satzes aus der Theorie der trigonometrischen Reihen. *Mathematische Annalen*, 5:123–132, 1872.

[Dawson, 1997] John W. Dawson. *Logical Dilemmas: The Life and Work of Kurt Gödel*. A. K. Peters, Wellesley MA, 1997.

[Dedekind, 1857] Richard Dedekind. Abriss einer Theorie der höheren Kongruenzen in bezug auf einen reellen Primzahl-Modulus. *Journal für reine und Augwandte Mathematik*, 54:1–26, 1857. Reprinted in [Dedekind, 1930 2, vol. 1, pp. 40-67].

[Dedekind, 1872] Richard Dedekind. *Stetigkeit und irrationale Zahlen*. F. Vieweg, Braunschweig, 1872. Translated with commentary in [Ewald, 1996, pp. 765-779].

[Dedekind, 1888] Richard Dedekind. *Was sind und was sollen die Zahlen?* F. Vieweg, Braunschweig, 1888. Third, 1911 edition translated with commentary in [Ewald, 1996, pp. 787-833].

[Dedekind, 1930 2] Richard Dedekind. *Gesammelte mathematische Werke.* F. Vieweg, Baunschweig, 1930-2. Edited by Robert Fricke, Emmy Noether and Öystein Ore; reprinted by Chelsea Publishing Company, New York, 1969.

[Dirichlet, 1863] Gustav Lejeune Dirichlet. *Vorlesungen über Zahlentheorie.* F. Vieweg, Braunschweig, 1863. Edited by Richard Dedekind; second edition 1871, third edition 1879.

[Dirichlet, 1889 1897] Gustav Lejeune Dirichlet. *G. Lejeune Dirichlet's Werke.* Reimer, Berlin, 1889 1897.

[Ebbinghaus, 2007] Heinz-Dieter Ebbinghaus. *Ernst Zermelo. An Approach to His Life and Work.* Springer, Berlin, 2007.

[Ewald, 1996] William Ewald, editor. *From Kant to Hilbert: A Source Book in the Foundations of Mathematics.* Clarendon Press, Oxford, 1996.

[Foreman and Kanamori, 2010] Matthew Foreman and Akihiro Kanamori. *Handbook of Set Theory.* Springer, 2010.

[Fraenkel, 1921] Abraham A. Fraenkel. Über die Zermelosche Begründung der Mengenlehre. *Jahresbericht der Deutschen Mathematiker-Vereinigung*, 30:97–98, 1921.

[Fraenkel, 1922] Abraham A. Fraenkel. Zu den Grundlagen der Cantor-Zermeloschen Mengenlehre. *Mathematische Annalen*, 86:230–237, 1922.

[Fraenkel, 1925] Abraham A. Fraenkel. Untersuchungen über die Grundlagen der Mengenlehre. *Mathematische Zeitschrift*, 22:250–273, 1925.

[Fraenkel, 1967] Abraham A. Fraenkel. *Lebenskreise. Aus den Erinnerungen eines jüdischen Mathematikers.* Deutsche Verlags-Anstalt, Stuttgart, 1967.

[Friedman and Scedrov, 1985] Harvey M. Friedman and Andrej Scedrov. The lack of definable witnesses and provably recursive functions in intuitionistic set theories. *Advances in Mathematics*, pages 1–13, 1985.

[Friedman, 1971a] Harvey M. Friedman. Higher set theory and mathematical practice. *Annals of Mathematical Logic*, 2:325–357, 1971.

[Friedman, 1971b] Harvey M. Friedman. A more explicit set theory. In Dana S. Scott, editor, *Axiomatic Set Theory*, volume 13(1) of *Proceedings of Symposia in Pure Mathematics*, pages 49–65. American Mathematical Society, Providence RI, 1971.

[Friedman, 1981] Harvey M. Friedman. On the necessary use of abstract set theory. *Advances in Mathematics*, 41:209–280, 1981.

[Gödel, 1931] Kurt Gödel. Über formal unentscheidbare Sätze der Principia Mathematica und verwandter System I. *Monatshefte für Mathematik und Physik*, 38:173–198, 1931. Reprinted and translated in [Gödel, 1986, pp. 144-195].

[Gödel, 1932] Kurt Gödel. Über Vollständigkeit und Widerspruchsfreiheit. *Ergebnisse eines mathematischen Kolloquiums*, 3:12–13, 1932. Reprinted and translated in [Gödel, 1986, pp. 24-27].

[Gödel, 1933o] Kurt Gödel. The present situation in the foundation of mathematics, 1933o. Printed in [Gödel, 1995, pp. 45-53].

[Gödel, 1938] Kurt Gödel. The consistency of the Axiom of Choice and of the Generalized Continuum-Hypothesis. *Proceedings of the National Academy of Sciences of the United States of America*, 24:556–557, 1938. Reprinted in [Gödel, 1990, pp. 26-27].

[Gödel, 1940] Kurt Gödel. *The Consistency of the Axiom of Choice and of the Generalized Continuum Hypothesis with the Axioms of Set Theory*, volume 3 of *Annals of Mathematics Studies*. Princeton University Press, Princeton, 1940. Reprinted in [Gödel, 1990, pp. 33-101].

[Gödel, 1944] Kurt Gödel. Russell's mathematical logic. In Paul A. Schilpp, editor, *The Philosophy of Bertrand Russell*, volume 5 of *Library of Living Philosophers*. Northwestern University, Evanston IL, 1944. Reprinted in [Gödel, 1990, pp. 119-141].

[Gödel, 1946] Kurt Gödel. Remarks before the Princeton bicentennial conference on problems in mathematics, 1946. Reprinted in [Gödel, 1990, pp. 150-153].

[Gödel, 1986] Kurt Gödel. *Collected Works, volume I: Publications 1929-1936*. Clarendon Press, Oxford, 1986. Edited by Solomon Feferman *et al.*

[Gödel, 1990] Kurt Gödel. *Collected Works, volume II: Publications 1938-1974*. Oxford University Press, New York, 1990. Edited by Solomon Feferman *et al.*

[Gödel, 1995] Kurt Gödel. *Collected Works, volume III: Unpublished Essays and Lectures.* Oxford University Press, New York, 1995. Edited by Solomon Feferman *et al.*

[Gödel, 2003] Kurt Gödel. *Collected Works, volume IV: Correspondence A-G.* Clarendon Press, Oxford, 2003. Edited by Solomon Feferman *et al.*

[Goodman, 1985] Nicolas D. Goodman. Replacement and collection in intuitionistic set theory. *The Journal of Symbolic Logic*, 50:344–348, 1985.

[Hallett, 1984] Michael Hallett. *Cantorian Set Theory and the Limitation of Size*, volume 10 of *Oxford Logic Guides.* Clarendon Press, Oxford, 1984.

[Hartogs, 1915] Friedrich Hartogs. Über das Problem der Wohlordnung. *Mathematische Annalen*, 76:438–443, 1915.

[Harward, 1905] A. E. Harward. On the transfinite numbers. *Philosoophical Magazine*, (6)10:439–460, 1905.

[Hausdorff, 1914] Felix Hausdorff. *Grundzüge der Mengenlehre.* de Gruyter, Leipzig, 1914.

[Heck, 1995] Richard Heck. Definition by induction in Frege's Grungesetze der Arithmetik. In William Demopoulos, editor, *Frege's Philosophy of Mathematics*, pages 295–333. Harvard University Press, Cambridge MA, 1995.

[Hessenberg, 1906] Gerhard Hessenberg. *Grundbegriffe der Mengenlehre.* Vandenhoeck and Ruprecht, 1906. Reprinted from *Abhandlungen der Fries'schen Schule*, Neue Folge 1 (1906), pp. 479-706.

[Hilbert, 1900] David Hilbert. Über den Zahlbegriff. *Jahresbericht der Deutschen Mathematiker-Vereinigung*, 8:180–184, 1900. Translated with commentary in [Ewald, 1996, pp. 1092-1095].

[Kanamori, 2004] Akihiro Kanamori. Zermelo and set theory. *The Bulletin of Symbolic Logic*, 10:487–553, 2004.

[Kisin, 2009] Mark Kisin. Moduli of finite flat group schemes, and modularity. *Annals of Mathematics*, pages 1085–1180, 2009.

[Kreisel, 1980] Georg Kreisel. Kurt Gödel. *Biographical Memoirs of Fellows of the Royal Society*, 26:149–224, 1980. 27:697, 1981; 28:719, 1982.

[Landau, 1930] Edmund Landau. *Grundlagen der Analysis.* Akademische Verlagsgesellschaft, Leipzig, 1930. Translated as *Foundations of Analysis*, Chelsea Publishing Company, 1951.

[Levy and Vaught, 1961] Azriel Levy and Robert L. Vaught. Principles of partial reflection in the set theories of Zermelo and Ackermann. *Pacific Journal of Mathematics*, 11:1045–1062, 1961.

[Levy, 1959] Azriel Levy. On Ackermann's set theory. *The Journal of Symbolic Logic*, 24:154–166, 1959.

[Levy, 1960a] Azriel Levy. Axiom schemata of strong infinity in axiomatic set theory. *Pacific Journal of Mathematics*, 10:223–238, 1960. Reprinted in *Mengenlehre*, Wissensschaftliche Buchgesellschaft Darmstadt, 1979, pp. 238-253.

[Levy, 1960b] Azriel Levy. Principles of reflection in axiomatic set theory. *Fundamenta Mathematicae*, 49:1–10, 1960.

[Levy, 1968] Azriel Levy. On von Neumann's axiom system for set theory. *American Mathematical Monthly*, pages 762–763, 1968.

[Levy, 1974] Azriel Levy. Parameters in the comprehension axiom schemas of set theory. In *Proceedings of the Tarski Symposium*, volume 25 of *Proceedings of Symposia in Pure Mathematics*, pages 309–324. American Mathematical Society, Providence, RI, 1974.

[Martin, 1975] D. Anthony Martin. Borel determinacy. *Annals of Mathematics*, 102:363–371, 1975.

[Martin, 1985] D. Anthony Martin. A purely inductive proof of Borel determinacy. In *Recursion Theory*, volume 42 of *Proceedings of Symposia in Pure Mathematics*, pages 303–308. American Mathematical Society, Providence RI, 1985.

[Mathias, 2001a] Adrian R.D. Mathias. Slim models of Zermelo set theory. *The Journal of Symbolic Logic*, 66:487–496, 2001.

[Mathias, 2001b] Adrian R.D. Mathias. The strength of Mac Lane set theory. *Annals of Pure and Applied Logic*, pages 107–234, 2001.

[Mathias, 2006] Adrian R.D. Mathias. Weak systems of Gandy, Jensen and Devlin. In Joan Bagaria and Stevo Todorcevic, editors, *Set Theory: Centre de Recerca Matemàtica, Barcelona 2003-4*, Trends in Mathematics, pages 149–224. Birkhäuser Verlag, Basel, 2006.

[Mathias, 2007] Adrian R.D. Mathias. A note on the schemes of replacement and collection. *Archive for Mathematical Logic*, 46:43–50, 2007.

[Mathias, 2010] Adrian R.D. Mathias. Unordered pairs in the set theory of Bourbaki 1949. *Archive for Mathematical Logic*, 94:1–10, 2010.

[Mathias, 2011] Adrian R.D. Mathias. Bourbaki and the scorning of logic, 2011. To appear in *Infinity and Truth*, Lecture Notes Series of the Institute for Mathematical Sciences of the National University of Singapore Vol. 25, World Scientific.

[McLarty, 2010] Colin McLarty. What does it take to prove Fermat's Last Theorem? Grothendieck and the logic of number theory. *The Bulletin of Symbolic Logic*, pages 359–377, 2010.

[McLarty, 2011] Colin McLarty. Set theory for Grothendieck's number theory, 2011. www.cwru.edu/artsci/philo/Groth%20found26.pdf.

[Meschkowski and Nilson, 1991] Herbert Meschkowski and Winfried Nilson. *Georg Cantor. Briefe.* Springer, Berlin, 1991.

[Mirimanoff, 1917a] Dimitry Mirimanoff. Les antinomies de Russell et de Burali-Forti et le problème fondamental de la théorie des ensembles. *L'Enseignement Mathématique*, pages 37–52, 1917.

[Mirimanoff, 1917b] Dimitry Mirimanoff. Remarques sur la théorie des ensembles et les antinomies cantoriennes. I. *L'Enseignement Mathématique*, pages 209–217, 1917.

[Montague, 1961] Richard M. Montague. Fraenkel's addition to the axioms of Zermelo. In Yehoshua Bar-Hillel, E.I.J. Poznanski, Michael O. Rabin, and Abraham Robinson, editors, *Essays on the Foundations of Mathematics*, pages 91–114. Magnes Press, Jerusalem, 1961. Dedicated to Professor A.A. Fraenkel on his 70th birthday.

[Moore and Garciadiego, 1981] Gregory H. Moore and Alejandro R. Garciadiego. Burali-forti's paradox: a reappraisal of its origins. *Historia Mathematica*, 8:319–350, 1981.

[Moore, 1976] Gregory H. Moore. Ernst Zermelo, A. E. Harward, and the axiomatization of set theory. *Historia Mathematica*, 3:206–209, 1976.

[Müller, 1976] Gert H. Müller, editor. *Sets and Classes. On the Work of Paul Bernays*, volume 84 of *Studies in Logic and the Foundations of Mathematics.* North-Holland, Amsterdam, 1976.

[Nyikos, 1980] Peter J. Nyikos. A provisional solution to the normal Moore space problem. *Proceedings of the American Mathematical Society*, 78:429–435, 1980.

[Pettigrew, 2008] Richard Pettigrew. Platonism and Aristoteleanism in mathematics. *Philosophia Mathematica*, 16:310–332, 2008.

[Potter, 2004] Michael Potter. *Set theory and its Philosophy*. Oxford University Press, Oxford, 2004.

[Quine, 1960] Willard V. O. Quine. *Work and Object*. MIT Press, Cambridge, 1960.

[Rathjen, 2005] Michael Rathjen. Replacement versus collection and related topics in constructive Zermelo-Fraenkel set theory. *Annals of Pure and Applied Logic*, 136:156–174, 2005.

[Reimann and Slaman, 2010] Jan Reimann and Theodore A. Slaman. Effective randomness for continuous measures, 2010. Preprint.

[Reinhardt, 1970] William N. Reinhardt. Ackermann's set theory equals ZF. *Annals of Mathematical Logic*, 2:189–249, 1970.

[Robinson, 1937] Raphael M. Robinson. The theory of classes, a modification of von Neumann's system. *The Journal of Symbolic Logic*, 2:29–36, 1937.

[Rudin, 1971] Mary Ellen Rudin. A normal space X for which $X \times I$ is not normal. *Fundamenta Mathematicae*, 73:179–186, 1971.

[Russ, 2004] Steve Russ. *The Mathematical Works of Bernard Bolzano*. Oxford University Press, Oxford, 2004.

[Scott, 1974] Dana S. Scott. Axiomatizing set theory. In Thomas Jech, editor, *Axiomatic Set Theory*, volume 13(2) of *Proceedings of Symposia in Pure Mathematics*, pages 207–214. American Mathematical Society, Providence RI, 1974.

[Shoenfield, 1967] Joseph R. Shoenfield. *Mathematical Logic*. Addison-Wesley, Reading MA, 1967.

[Shoenfield, 1977] Joseph R. Shoenfield. Axioms of set theory. In K. Jon Barwise, editor, *Handbook of Mathematical Logic*, pages 321–344. North-Holland, Amsterdam, 1977.

[Sieg and Schlimm, 2005] Wilfried Sieg and Dirk Schlimm. Dedekind's analysis of number: systems and axioms. *Synthèse*, 147:121–170, 2005.

[Skolem, 1923a] Thoralf Skolem. Begründung der elementaren Arithmetik durch die rekurrierende Denkweise ohne Anwendung scheinbarer Veränderlichen mit unendlichem Ausdehnungsberich. *Skrifter utgit av Videnskabsselskapets i Kristiania, I. Matematisk-naturvidenskabelig klasse*, 6:1–38, 1923. Translated in van [van Heijenoort, 1967, pp. 302-333].

[Skolem, 1923b] Thoralf Skolem. Einige bermerkungen zur axiomatischen Begründung der Mengenlehre. In *Wissenschaftliche Vorträge gehalten auf dem Fünften Kongress der Skandinavischen Mathematiker in Helsingfors vom 4. bis 7. Juli 1922*, pages 217–232, Helsinki, 1923. Akademische Buchhandlung. Translated in van [van Heijenoort, 1967, pp. 290-301].

[Tait, 1998] William W. Tait. Zermelo's conception of set theory and reflection principles. In Matthias Schirn, editor, *The Philosophy of Mathematics Today*, pages 469–483. Oxford University Press, Oxford, 1998.

[van Heijenoort, 1967] Jean van Heijenoort, editor. *From Frege to Gödel: A Source Book in Mathematical Logic, 1879-1931.* Harvard University Press, Cambridge MA, 1967. Reprinted in 2002.

[von Neumann, 1923] John von Neumann. Zur Einführung der transfiniten Zahlen. *Acta Litterarum ac Scientiarum Regiae Universitatis Hungaricae Francisco-Josephinae (Szeged), sectio scientiarum mathematicarum*, 1:199–208, 1923. Reprinted in [von Neumann, 1961, pp. 24-33], translated in van [van Heijenoort, 1967, pp. 346-354].

[von Neumann, 1925] John von Neumann. Eine Axiomatisierung der Mengenlehre. *Journal für die reine und angewandte Mathematik 154*, pages 219–240, 1925. Berichtigung ibid. *155*, 128; reprinted in [von Neumann, 1961, pp. 34-56], translated in [van Heijenoort, 1967, pp. 393-413].

[von Neumann, 1928a] John von Neumann. Die Axiomatisierung der Mengenlehre. *Mathematische Zeitschrift*, 27:669–752, 1928. Reprinted in [von Neumann, 1961, pp. 339-422].

[von Neumann, 1928b] John von Neumann. Über die Definition durch transfinite Induktion und verwandte Fragen der allgemeinen Mengenlehre. *Mathematische Annalen*, 99:373–391, 1928. Reprinted in [von Neumann, 1961, pp. 320-338].

[von Neumann, 1929] John von Neumann. Über eine Widerspruchfreiheitsfrage in der axiomatischen Mengenlehre. *Journal für die reine und angewandte Mathematik*, 160:227–241, 1929. Reprinted in [von Neumann, 1961, pp. 494-508].

[von Neumann, 1961] John von Neumann. *Collected Works, Volume 1.* Pergamon Press, New York, 1961. Edited by Abraham H. Taub.

[Wang, 1974a] Hao Wang. The concept of set. In *From Mathematics to Philosophy*, pages 181–223. Humanities Press, New York, 1974. Reprinted in [Benacerraf and Putnam, 1983, 530-570].

[Wang, 1974b] Hao Wang. *From Mathematics to Philosophy.* Humanities Press, New York, 1974.

[Wang, 1981] Hao Wang. *Popular Lectures on Mathematical Logic.* Van Nostrand Reinhold, New York, 1981.

[Wang, 1996] Hao Wang. *A Logical Journey: From Gödel to Philosophy.* The MIT Press, Cambridge MA, 1996.

[Whitehead and Russell, 1913] Alfred North Whitehead and Bertrand Russell. *Principia Mathematica, volume 3.* Cambridge University Press, Cambridge, 1913.

[Wiener, 1914] Norbert Wiener. A simplification of the logic of relations. *Proceedings of the Cambridge Philosophical Society*, pages 387–390, 1914. Reprinted in van [van Heijenoort, 1967, pp. 224-227].

[Zarach, 1996] Andrzej M. Zarach. Replacement $\nrightarrow$ Collection. In Petr Hájek, editor, *Gödel '96. Logical Foundations of Mathematics, Computer Science and Physics – Kurt Gödel's Legacy*, volume 6 of *Lecture Notes in Logic*, pages 307–322. AK Peters, Natick MA, 1996.

[Zermelo, 1904] Ernst Zermelo. Beweis dass jede Menge wohlgeordnet werden kann. *Mathematische Annalen*, 59:514–516, 1904. Reprinted and translated in [Zermelo, 2010, pp. 114-119].

[Zermelo, 1908a] Ernst Zermelo. Neuer Beweis für die moglichkeit einer Wohlordnung. *Mathematische Annalen*, 65:107–128, 1908. Reprinted and translated in [Zermelo, 2010, pp. 120-159].

[Zermelo, 1908b] Ernst Zermelo. Untersuchungen über die Grundlagen der Mengenlehre I. *Mathematische Annalen*, 65:261–281, 1908. Reprinted and translated in [Zermelo, 2010, pp. 188-229].

[Zermelo, 1909] Ernst Zermelo. Sur les ensembles finis et le principe de l'induction complète. *Acta Mathematica*, 32:185–193, 1909. Reprinted and translated in [Zermelo, 2010, pp. 236-253].

[Zermelo, 1930a] Ernst Zermelo. Über das mengentheoretische Model, 1930. Printed and translated in [Zermelo, 2010, pp. 446-453].

[Zermelo, 1930b] Ernst Zermelo. Über Grenzzahlen und Mengenbereiche: Neue Untersuchungen über die Grundlagen der Mengenlehre. *Fundamenta Mathematicae*, 16:29–47, 1930. Reprinted and translated in [Zermelo, 2010, pp. 400-431].

[Zermelo, 2010] Ernst Zermelo. *Collected Works, Volume 1.* Springer, Berlin, 2010. Edited by Heinz-Dieter Ebbinghaus and Akihiro Kanamori.

Part II

Philosophy

6 The Mathematical Infinite as a Matter of Method

Abstract. I address the historical emergence of the mathematical infinite, and how we are to take the infinite in and out of mathematics. The thesis is that *the mathematical infinite in mathematics is a matter of method.*

The infinite, of course, is a large topic. At the outset, one can historically discern two overlapping clusters of concepts: (1) wholeness, completeness, universality, absoluteness. (2) endlessness, boundlessness, indivisibility, continuousness. The first, the *metaphysical infinite*, I shall set aside. It is the second, the *mathematical infinite*, that I will address. Furthermore, I will address the mathematical infinite by considering its historical emergence in set theory and how we are to take the infinite in and out of mathematics. Insofar as physics and, more broadly, science deals with the mathematical infinite through mathematical language and techniques, my remarks should be subsuming and consequent.

The main underlying point is that how the mathematical infinite is approached, assimilated, and applied in mathematics is not a matter of "ontological commitment", of coming to terms with whatever that might mean, but rather of epistemological articulation, of coming to terms through knowledge. *The mathematical infinite in mathematics is a matter of method.* How we deal with the specific individual issues involving the infinite turns on the narrative we present about how it fits into methodological mathematical frameworks established and being established.

The first section discusses the mathematical infinite in historical context, and the second, set theory and the emergence of the mathematical infinite. The third section discusses the infinite in and out of mathematics, and how it is to be taken.

1 The Infinite in Mathematics

What role does the infinite play in modern mathematics? In modern mathematics, infinite sets abound both in the workings of proofs and as subject matter in statements, and so do universal statements, often of $\forall\exists$ "for all there exists" form, which are indicative of direct engagement with the infinite. In many ways the role of the infinite is importantly "second-order" in the sense that Frege regarded number generally, in that the concepts of modern mathematics are understood as having infinite instances over a broad range.

Reprinted from *The Annals of the Japan Association for Philosophy of Science* 20 (2012), pp.3-15, with permission from The Japan Association for Philosophy of Science.

This article was written whilst the author was a 2009-2010 senior fellow at the Lichtenberg-Kolleg of the University of Göttingen; he expresses his gratitude for the productive conditions and support of the kolleg. The themes of this article were presented in a keynote address at the 2010 annual meeting of the Japan Association for the Philosophy of Science; the author expresses his gratitude for the invitation and the productive discussions that ensued.

But all this has been the case for just more than a century. Infinite totalities and operations on them only emerged in mathematics in a recent period of algebraic expansion and rigorization of proof. It becomes germane, even crucial, to see how the infinite emerged through interaction with proof to come to see that the infinite in mathematics is a matter of method. If one puts the history of mathematics through the sieve of proof, one sees the emergence of methods drawing in the mathematical infinite, and the mathematical infinite came in at three levels: *the countably infinite*, the infinite of the natural numbers; *the continuum*, the infinite of analysis; and *the empyrean infinite* of higher set theory.

As a thematic entrée into the matter of proof and the countably infinite, we can consider the Pigeonhole Principle:

If n pigeons fly into fewer than n pigeonholes,
then one hole has more than one pigeon.

Taken primordially, this may be considered immediate as part of the meaning of natural number and requires no proof. On the other hand, after its first explicit uses in algebraic number theory in the mid-19th Century, it has achieved articulated prominence in modern combinatorics, its consequences considered substantive and at times quite surprisingly so given its immediacy. Rendered as a $\forall\exists$ statement about natural numbers, it however does not have the feel of a basic law of arithmetic but of a "non-constructive" existence assertion, and today it is at the heart of combinatorics, and indeed is the beginning of Ramsey Theory, a field full of non-constructive existence assertions.

So how does one *prove* the Pigeonhole Principle? For 1729 pigeons and 137 pigeonholes one can systematically generate all assignments $\{1,\ldots,1729\} \longrightarrow \{1,\ldots,137\}$ and check that there are always at least two pigeons assigned to the same pigeonhole. But we "see" nothing here, nor from any other particular brute force analysis. With the Pigeonhole Principle seen afresh as being at the heart of the articulation of finite cardinality and requiring proof based on prior principles, Richard Dedekind in his celebrated 1888 essay *Was sind und was sollen die Zahlen?* [8, §120] first gave a proof applying the Principle of Induction on n. Today, the Pigeonhole Principle is regarded as a theorem of Peano Arithmetic (PA). In fact, there is a "reverse mathematics" result, that in the presence of the elementary axioms of PA, the Pigeonhole Principle *implies* the Principle of Induction, and is hence *equivalent* to this central principle. Moreover, the proof complexity of weak forms of the Pigeonhole Principle have been investigated in weak systems of arithmetic.[1]

This raises a notable historical point drawing in the infinite. The Pigeonhole Principle seems to have been first applied in mathematics by Gustav Lejeune Dirichlet in papers of 1842, one to the study of the classical Pell's equation and

[1]See for example [18].

another to establish a crucial approximation lemma for his well-known Unit Theorem describing the group of units of an algebraic number field.[2] Upon Carl Friedrich Gauss's death in 1855 his Göttingen chair went to Dirichlet, but he succumbed to a heart attack a few years later. Richard Dedekind was Gauss's last student, and later colleague and friend of Dirichlet. Already in 1857, Dedekind [6] worked in modular arithmetic with the actually infinite residue classes themselves as unitary objects. His context was in fact $Z[x]$, the polynomials of x with integer coefficients, and so he was entertaining a totality of infinitely many equivalence classes.

The Pigeonhole Principle occurred in Dirichlet's *Vorlesungen über Zahlentheorie* [10], edited and published by Dedekind in 1863. The occurrence is in the second, 1871 edition, in a short Supplement VIII by Dedekind on Pell's equation, and it was in the famous Supplement X that Dedekind laid out his theory of ideals in algebraic number theory, working directly with infinite totalities. In 1872 Dedekind was putting together *Was sind und was sollen die Zahlen?*, and he would be the first to *define* infinite set, with the definition being a set for which *there is* a one-to-one correspondence with a proper subset. This is just the negation of the Pigeonhole Principle. Dedekind in effect had inverted a negative aspect of finite cardinality into a *positive* existence definition of the infinite.

The Pigeonhole Principle brings out a crucial point about method. Its proof by induction is an example of what David Hilbert later called *formal* induction. In so far as the natural numbers do have an antecedent sense, a universal statement $\forall n\varphi(n)$ about them should be correlated with all the informal counterparts to $\varphi(0), \varphi(1), \varphi(2), \ldots$ taken together. *Contra* Poincaré, Hilbert [15] distinguished between *contentual* [*inhaltlich*] induction, the intuitive construction of each integer as numeral,[3] and *formal* induction, by which $\forall n\varphi(n)$ follows immediately from the two statements $\varphi(0)$ and $\forall n(\varphi(n) \rightarrow \varphi(n+1))$ and "through which alone the mathematical variable can begin to play its role in the formal system." In the case of the Pigeonhole Principle, we see the proof by formal induction, but it bears little constructive relation to any particular instance. Be that as it may, the schematic sense of the countably infinite, the infinite of the natural numbers, is carried in modern mathematics by formal induction, a principle used everywhere in combinatorics and computer science to secure statements about the countably infinite. The Pigeonhole Principle is often regarded as surprising in its ability to draw strong conclusions, but

[2] See Dirichlet [11, pp.579,636]. The principle in the early days was called the *Schubfachprinzip* ("drawer principle"), though not however by Dirichlet. The second, 1899 edition of Heinrich Weber's *Lehrbuch der Algebra* used the words "in Faecher verteilen" ("to distribute into boxes") and Edmund Landau's 1927 *Vorlesungen über Zahlentheorie* had "Schubfachschluss".

[3] For Hilbert, the numeral for the integer n consists of n short vertical strokes concatenated together.

one way to explain this is to point to the equivalence with the Principle of Induction. Of course, the Principle of Induction is itself often regarded as surprising in its efficacy, but this can be seen as our reaction to it as method in contrast to the countably infinite taken as primordial. There is no larger mathematical sense to the Axiom of Infinity in set theory other than to provide an extensional counterpart to formal induction, a method of proof. The Cantorian move against the traditional conception of the natural numbers as having no end in the "after" sense is neatly rendered by extensionalizing induction itself in modern set theory with the ordinal ω, with "after" recast as "$\in$".

The next level of the mathematical infinite would be the continuum, the infinite of mathematical analysis. Bringing together the two traditional Aristotelean infinities of infinite divisibility and of infinite progression, one can ask: How many points are there on the line? This would seem to be a fundamental, even primordial, question. However, to cast it as a *mathematical* question, underlying concepts would have to be invested with mathematical sense and a way of mathematical thinking provided that makes an answer possible, if not informative. First, the real numbers as representing points on the linear continuum would have to be precisely described. A coherent concept of cardinality and cardinal number would have to be developed for infinite mathematical totalities. Finally, the real numbers would have to be enumerated in such a way so as to accommodate this concept of cardinality. Georg Cantor made all of these moves as part of the seminal advances that have led to modern set theory, eventually drawing in also the third, empyrean infinite of higher set theory. His Continuum Hypothesis would propose a specific, structured resolution about the size of the continuum in terms of his transfinite numbers, a resolution that would become pivotal where set-theoretic approaches to the continuum became prominent in mathematical investigations.

2 Set Theory and the Emergence of the Infinite

Set theory had its beginnings in the great 19th Century transformation of mathematics, a transformation beginning in analysis. With the function concept having been steadily extended beyond analytic expressions to infinite series, sense for the new functions could only be developed through carefully specified deductive procedures. Proof reemerged in mathematics as an extension of algebraic calculation and became the basis of mathematical knowledge, promoting new abstractions and generalizations. The new articulations to be secured by proof and proof in turn to be based on prior principles, the regress lead in the early 1870s to the appearance of several formulations of the real numbers, of which Cantor's and Dedekind's are the best known. It is at first quite striking that the real numbers as a totality came to be developed so late, but this can be viewed against the background of the larger conceptual shift from intensional to extensional mathematics. Infinite series outstripping sense, it became necessary to adopt an arithmetical view of the continuum given extensionally as a totality of points.

Cantor's formulation of the real numbers appeared in his seminal paper [1] on trigonometric series; proceeding in terms of "fundamental" sequences, he laid the basis for his theorems on sequential convergence. Dedekind [7] formulated the real numbers in terms of his "cuts" to express the completeness of the continuum; deriving the least upper bound principle as a simple consequence, he thereby secured the basic properties of continuous functions. In the use of arbitrary sequences and infinite totalities, both Cantor's and Dedekind's objectifications of the continuum helped set the stage for the subsequent development of that extensional mathematics *par excellence*, set theory. Cantor's formulation was no idle conceptualization, but to the service of specific mathematics, the articulation of his results on uniqueness of trigonometric series involving his derived sets, the first instance of topological closure. Dedekind [7] describes how he came to his formulation much earlier, but also acknowledges Cantor's work. Significantly, both Cantor [1, p.128] and Dedekind [7, III] accommodated the antecedent geometric sense of the continuum by asserting as an "axiom" that each point on the geometric line actually corresponds to a real number as they defined it, a sort of Church's thesis of adequacy for their construals of the continuum. In modern terms, Cantor's reals are equivalence classes according to an equivalence relation which importantly is a congruence relation, a relation that respects the arithmetical structure of the reals. It is through Cantor's formulation that completeness would be articulated for general metric spaces, thereby providing the guidelines for proof in new contexts involving infinite sets.

Set theory was born on that day in December 1873 when Cantor established that *the continuum is not countable*: There is no one-to-one correspondence between the natural numbers $\mathbf{N} = \{0, 1, 2, 3, \ldots\}$ and the real numbers $\mathbf{R}$. Like the irrationality of $\sqrt{2}$, the uncountability of the continuum was an impossibility result established via *reductio ad absurdum* that opened up new possibilities. Cantor addressed a specific *problem*, embedded in the mathematics of the time, in his seminal [2] entitled "On a property of the totality of all real algebraic numbers". Dirichlet's algebraic numbers, it will be remembered, are the roots of polynomials with integer coefficients; Cantor established the countability of the algebraic numbers. This was the first substantive correlation of an infinite totality with the natural numbers, and it was the first application of what now goes without saying, that finite words based on a countable alphabet are countable. Cantor then established: *For any (countable) sequence of reals, every interval contains a real not in the sequence.*[4]

[4]The following is Cantor's argument, in brief: Suppose that s is a sequence of reals and I an interval. Let $a < b$ be the first two reals of s, if any, in I. Then let $a' < b'$ be the first two reals of s, if any, in the open interval (a, b); $a'' < b''$ the first two reals of s, if any, in

By this means Cantor provided a new proof of Joseph Liouville's result [16, 17] that there are transcendental (i.e. real, non-algebraic) numbers, and only afterward did Cantor point out the uncountability of the reals altogether. Accounts of Cantor's existence deduction for transcendental numbers have mostly reserved the order, establishing first the uncountability of real numbers and only then drawing the conclusion from the countability of the algebraic numbers, thus promoting the misconception that his argument is "non-constructive".[5] It depends how one takes a proof, and Cantor's arguments have been implemented as algorithms to generate successive digits of transcendental numbers.[6] The Baire Category Theorem, in its many uses in modern mathematics, has similarly been regarded as non-constructive in its production of examples; however, its proof is an extension of Cantor's proof of the uncountability of the reals, and analogously constructive.

Cantor went on, of course, to develop his concept of cardinality based on one-to-one correspondences. Two totalities have the same cardinality exactly when *there is* a one-to-one correspondence between them. Having made the initial breach with a *negative* result about the lack of a one-to-one correspondence, he established infinite cardinality as a methodologically *positive* concept, as Dedekind had done for infinite set, and investigated the possibilities for *having* one-to-one correspondences. Just as the discovery of the irrational numbers had led to one of the great achievements of Greek mathematics, Eudoxus's theory of geometric proportions, Cantor began his move toward a full-blown mathematical theory of the infinite. By his 1878 *Beitrag* [3] Cantor had come to the focal Continuum Hypothesis—that there is no intervening cardinality between that of the countably infinite and of the continuum—and in his 1883 *Grundlagen* [4] Cantor developed the *transfinite numbers* and the key concept of *well-ordering*, in significant part to take a structured approach to infinite cardinality and the Continuum Hypothesis. At the end of the *Grundlagen*, Cantor propounded this basic well-ordering principle: "It is always possible to bring any *well-defined* set into the form of a well-ordered set." Sets are to be well-ordered, and they and their cardinalities are to be gauged via the transfinite numbers of his structured conception of the infinite.

Almost two decades after his [2] Cantor in a short 1891 note [5] gave his now celebrated diagonal argument, establishing *Cantor's Theorem: For any set X the totality of functions from X into a fixed two-element set has a larger cardinality than X*, i.e. there is no one-to-one correspondence between the two. This result generalized his [2] result that the continuum is not countable, since the totality of functions from $\mathbf{N}$ into a fixed two-element set has the same

(a', b'); and so forth. Then however long this process continues, the intersection of the nested intervals must contain a real not in the sequence s.

[5]Indeed, this is where Wittgenstein ([21], part II, 30-41) located what he took to be the problematic aspects of the talk of uncountability.

[6]See Gray [13].

cardinality as **R**. In retrospect the diagonal argument can be drawn out from the [2] proof.[7]

Cantor had been shifting his notion of set to a level of abstraction beyond sets of reals and the like, and the casualness of his [5] may reflect an underlying cohesion with his [2]. Whether the new proof is really "different" from the earlier one, through this abstraction Cantor could now dispense with the recursively defined nested sets and limit construction, and he could apply his argument to any set. He had proved for the first time that there is a cardinality larger than that of **R** and moreover affirmed "the general theorem, that the powers of well-defined sets have no maximum." Thus, Cantor for the first time entertained the third level of the mathematical infinite, the empyrean level beyond the continuum, of higher set theory.

Nowadays it goes without saying that each function from a set X into a two-element set corresponds to a subset of X, so Cantor's Theorem is usually stated as: *For any set X its power set $\mathcal{P}(X) = \{Y \mid Y \subseteq X\}$ has a larger cardinality than X.* However, it would be an exaggeration to assert that Cantor at this point was working on power sets; rather, he was expanding the 19th Century concept of *function* by ushering in arbitrary functions. Significantly, Bertrand Russell was stimulated by the diagonal argument to come up with his well-known paradox, this having the effect of emphasizing power sets. At the end of [5] Cantor dealt explicitly with "all" functions with a specified domain X and range $\{0, 1\}$; regarded these as being enumerated by one super-function $\phi(x, z)$ with enumerating variable z; and formulated the diagonalizing function $g(x) = 1 - \phi(x, x)$. This argument, even to its notation, would become method, flowing into descriptive set theory, the Gödel Incompleteness Theorem, and recursion theory, the paradigmatic means of transcendence over an established context.

Cantor's seminal work built in what would be an essential tension of methods in set theory, one that is still very much with us today. His Continuum Hypothesis was his proposed answer to the continuum problem: Where is the size of the continuum in the hierarchy of transfinite cardinals? His diagonal ar-

[7]Starting with a sequence s of reals and a half-open interval I_0, instead of successively choosing delimiting *pairs* of reals in the sequence, avoid the members of s *one* at a time: Let I_1 be the left or right half-open subinterval of I_0 demarcated by its midpoint, whichever does not contain the first element of s. Then let I_2 be the left or right half-open subinterval of I_1 demarcated by its midpoint, whichever does not contain the second element of s; and so forth. Again, the nested intersection contains a real not in the sequence s. Abstracting the process in terms of reals in binary expansion, one is just generating the binary digits of the diagonalizing real.

Cantor first gave a proof of the uncountability of the reals in a letter to Dedekind of 7 December 1873 (Ewald [12, pp.845ff]), professing that ". . . only today do I believe myself to have finished with the thing . . .". It is remarkable that in this letter already appears a doubly indexed array of real numbers and a procedure for traversing the array downward and to the right, as in a now common picturing of the diagonal proof.

gument, with its mediation for totalities of arbitrary functions (or power sets), would have to be incorporated into the emerging theory of transfinite number. But how is this to be coordinated with respect to his basic 1883 principle that sets should come well-ordered?

The first decade of the new century saw Ernst Zermelo make his major advances, at Göttingen with Hilbert, in the development of set theory. In 1904 Zermelo [22] analyzed Cantor's well-ordering principle by reducing it to the Axiom of Choice (AC), the abstract existence assertion that every set X has a choice function, i.e. a function f such that for every non-empty $Y \in X$, $f(Y) \in Y$. Zermelo thereby shifted the notion of set away from Cantor's principle that every well-defined set is well-orderable and replaced that principle by an explicit axiom. His Well-Ordering Theorem showed specifically that a set is well-orderable exactly when its power set has a choice function. How AC brought to the fore issues about the non-constructive existence of functions is well-known, and how AC became increasingly accepted in mathematics has been well-documented.[8] The expansion of mathematics into infinite abstract contexts was navigated with axioms and proofs, and this led to more and more appeals to AC.

In 1908 Zermelo [23] published the first full-fledged axiomatization of set theory, partly to establish set theory as a discipline free of the emerging paradoxes and particularly to put his Well-Ordering theorem on a firm footing. In addition to codifying generative set-theoretic principles, a substantial motive for Zermelo's axiomatizing set theory was to buttress his Well-Ordering Theorem by making explicit its underlying set existence assumptions.[9] Initiating the first major transmutation of the notion of set after Cantor, Zermelo thereby ushered in a new more abstract, prescriptive view of sets as structured solely by membership and governed and generated by axioms, a view that would soon come to dominate. Thus, the pathways to a theorem played a crucial role by stimulating an axiomatization of a field of study and a corresponding transmutation of its underlying notions.

In the tradition of Hilbert's axiomatization of geometry and of number, Zermelo's axiomatization was in the manner of an implicit definition, with axioms providing rules for procedure and generating sets and thereby laying the basis for proofs. Concerning the notion of definition through axioms, Hilbert [14, p.184] had written already in 1899 as follows in connection with his axiomatization of the reals:

> Under the conception described above [the axiomatic method], the doubts which have been raised against the existence of the totality of

[8]See [19].

[9]Moore [19, pp.155ff] supports this contention using items from Zermelo's *Nachlass*.

> all real numbers (and against the existence of infinite sets generally) lose all justification, for by the set of real numbers we do not have to imagine, say, the totality of all possible laws according to which the elements of a fundamental sequence can proceed, but rather—as just described—a system of things, whose mutual relations are given by the *finite and closed* system of axioms I-IV [for complete ordered fields], and about which new statements are valid only if one can derive them from the axioms by means of a finite number of logical inferences.

Zermelo's revelation of the Axiom of Choice for the derivation of the Well-Ordering Theorem was just this, the uncovering of an axiom establishing in a "finite number of logical inferences" Cantor's well-ordering principle, and Zermelo's axiomatization set out a "system of things" given by a "system of axioms".

As with Hilbert's later distinction between contentual and formal induction, infinite sets draw their mathematical meaning not through any direct or intuitive engagement but through axioms like the Axiom of Choice and finite proofs. Just as Euclid's axioms for geometry had set out the permissible geometric constructions, the axioms of set theory would set out the specific rules for set generation and manipulation. But unlike the emergence of mathematics from marketplace arithmetic and Greek geometry, infinite sets and transfinite numbers were neither laden nor buttressed with substantial antecedents. Like strangers in a strange land stalwarts developed a familiarity with them guided hand in hand by their axiomatic framework. For Dedekind in *Was sind und was sollen die Zahlen?* it had sufficed to work with sets by merely giving a few definitions and properties, those foreshadowing the axioms of Extensionality, Union, and Infinity. Zermelo provided more rules, the axioms of Separation, Power Set, and Choice. Simply put, these rules revealed those aspects to be ascribed to possibly infinite sets for their methodological incorporation into emerging mathematics.[10]

The standard axiomatization ZFC was completed by 1930, with the axiom of Replacement brought in through the work of von Neumann and the axiom of Foundation, through axiomatizations of Zermelo and Bernays. The Cantorian transfinite is contextualized by von Neumann's incorporation of his ordinals and Replacement, which underpins transfinite induction and recursion as methods of proof and definition. And Foundation provides the basis for applying transfinite recursion and induction procedures to get results about *all* sets, they all appearing in the cumulative hierarchy. While Foundation set the stage with a recursively presented picture of the universe of sets, Replacement, like Choice,

[10] As is well-known, the Axiom of Choice came to be widely used in ongoing mathematics in the methodologically congenial form of Zorn's Lemma. Significantly, Max Zorn entitled the paper [25] in which he presented his lemma, "A remark on method in transfinite algebra".

was seem as an essential axiom for infusing the contextualized transfinite with the order already inherent in the finite.

3 The Infinite In and Out of Mathematics

It is evident that mathematics has been much inspired by problems and conjectures and has progressed autonomously through the communication of proofs and the assimilation of methods. Being socially and historically contingent, mathematics has advanced when individuals could collectively make mathematics out of concepts, whether they involve infinite totalities or not. The commitment to the infinite is thus to what is communicable about it, to the procedures and methods in articulated contexts, to language and argument. Infinite sets are what they do, and their sense is carried by the methods we collectively employ on their behalf.

When considering the infinite as a matter of method in modern mathematics and its relation to a primordial infinite, there is a deep irony about mathematical objects and their existence. Through the rigor and precision of modern mathematics, mathematical objects achieve a sharp delineation in mathematical practice as founded on proof. The contextual objectification then promotes, perhaps even urges, some larger sense of reification. Or, there is confrontation with some prior-held belief or sense about existence that then promotes a skeptical attitude about what mathematicians do and prove, especially about the infinite. Whether mathematics inadvertently promotes realist attitudes or not, the applicability of mathematics to science should not extend to philosophy if the issues have to do with existence itself, for again, existence in mathematics is contextual and governed by rules and procedures, and metaphysical existence, especially concerning the infinite, does not inform and is not informed by mathematical work.

Mathematicians themselves are prone to move in and out of mathematics in their existential assessments, stimulated by their work and the urge to put a larger stamp of significance to it. We quoted Hilbert above, and he famously expressed larger metaphysical views about finitism and generated a program to establish the overall consistency of mathematics. In set theory, Cantor staked out the Absolute, which he associated with the transcendence of God, and applied it in the guise of the class of all ordinal numbers for delineative purposes. Gödel attributed to his conceptual realism about sets and his philosophical standpoint generally his relative consistency results with the constructible universe L.

What are we to make of what mathematicians say and their motivations? Despite the directions in which mathematics has been led by individuals' motivations it is crucial to point out, again, that what has been retained and has grown in mathematics is communicable as proofs and results. We should assess the role of the individual, but keep in mind the larger autonomy of mathematics. A delicate but critical point here is that, as with writers and musicians, we should be dispassionate, sometimes even critical, about what mathematicians say about their craft. As with the surface Platonism often espoused by mathematicians, we must distinguish what is *said* from what is *done*. The interplay between philosophical views of individual mathematicians, historically speaking, and the space of philosophical possibility, both in their times and now, is what needs to be explored.

There is one basic standpoint about the infinite which seems to underly others and leads to prolonged debates about the "epistemic". This is the (Kantian) standpoint of human finitude. We are cast into a world which as a whole must be infinite, yet we are evidently finite, even to the number of particles that makes us up. So how can we come to know the infinite in any substantive way? This long-standing attitude is part of a venerable tradition, and to the extent that we move against it, our approach may be viewed as bold and iconoclastic. Even phenomenologically, we see before us mathematicians working coherently and substantially with infinite sets and concepts. The infinite is embedded in mathematics as method; we can assimilate methods; and we use the infinite through method in proofs. Even those mathematicians who would take some sort of metaphysical stance against the actuality of the infinite in mathematics can nevertheless follow and absorb a proof by induction.

There is a final, large point in this direction. As mathematics has expanded with the incorporation of the infinite, several voices have advocated the restriction of proof procedures and methods. Brouwer and Weyl were early figures and Bishop, a recent one. How are we to take all this? We now have a good grasp of intuitionistic and constructivist approaches to mathematics as various explicit, worked-out systems. We also have a good understanding of hierarchies of infinitistic methods through quantifier complexity, proof theory, reverse mathematics and the like. Commitments to the infinite can be viewed as the assimilation of methods along hierarchies. Be that as it may, an *ecumenical* approach to the infinite is what seems to be called for: There is no metamathematics, in that how we are able to argue about resources and methods is itself mathematical. As restrictive approaches were advocated, they themselves have been brought into the fold of mathematics, the process itself having mathematical content. Proofs about various provabilities are themselves significant proofs. It is interesting to carry out a program to see how far a strictly finitist or predicativist approach to mathematics can go, not to emasculate mathematics or to tout the one true way, but to find new, informative proofs and to gain an insight into the resources at play, particularly with regard to commitments to infinity.

Stepping back, the study of the infinite in mathematics urges us to develop a larger ecumenicism about the role of the infinite. Like the modern ecumenical approach to proofs in all their variety and complexity, proofs *about* resources provide new mathematical insights about the workings of method. Even then, in relation to later "elementary" proofs or formalized proofs in an elementary system of a statement, prior proofs may well retain an irreducible semantic content. In this content aspects of the infinite reside robustly, displaying the autonomy of mathematics as an evolving practice.

References

[1] Georg Cantor. Über die Ausdehnung eines Satzes aus der Theorie der trignometrischen Reihen. *Mathematische Annalen*, 5:123–132, 1872.

[2] Georg Cantor. Über eine Eigenschaft des Inbegriffes aller reellen algebraischen Zahlen. *Journal für die reine und angewandte Mathematik*, 77:258–262, 1874. Translated with commentary in Ewald [12, pp.839–843].

[3] Georg Cantor. Ein Beitrag zur Mannigfaltigkeitslehre. *Journal für die reine und angewandte Mathematik*, 84:242–258, 1878.

[4] Georg Cantor. Über unendlich, lineare Punktmannigfaltigkelten. V. *Mathematische Annalen*, 21:545–591, 1883. Published separately as *Grundlagen eine allgemeinen Mannigfaltigkeitslehre. Ein mathematisch-philosophischer Versuch in der Lehre des Unendlichen*; B.G. Teubner, Leipzig 1883. Translated with commentary in Ewald [12, pp.878–920].

[5] Georg Cantor. Über eine elementare frage der mannigfaltigkeitslehre. *Jahresbericht der Deutschen Mathematiker-Vereinigung*, 1:75–78, 1891. Translated with commentary in Ewald [12, pp.920–922].

[6] Richard Dedekind. Abriss einer Theorie der höheren Kongruenzen in bezug auf einen reellen Primzahl-Modulus. *Journal für reine und Augwandte Mathematik*, 54:1–26, 1857. Reprinted in Dedekind [9, pp.40–67].

[7] Richard Dedekind. *Stetigkeit und irrationale Zahlen.* F. Vieweg, Braunschweig, 1872. Fifth, 1927 edition translated with commentary in Ewald [12, pp.765–779].

[8] Richard Dedekind. *Was sind und was sollen die Zahlen?* F. Vieweg, Braunschweig, 1888. Third, 1911 edition translated with commentary in Ewald [12, pp.787–833].

[9] Richard Dedekind. *Gesammelte mathematische Werke.* F. Vieweg, Baunschweig, 1930-2. Edited by Robert Fricke, Emmy Noether and Öystein Ore; reprinted by Chelsea Publishing Company, New York, 1969.

[10] Gustav Lejeune Dirichlet. *Vorlesungen über Zahlentheorie.* F. Vieweg, Braunschweig, 1863. Edited by Richard Dedekind; second edition 1871, third edition 1879.

[11] Gustav Lejeune Dirichlet. *G. Lejeune Dirichlet's Werke.* Reimer, Berlin, 1889/97. Edited by Leopold Kronecker.

[12] William Ewald, editor. *From Kant to Hilbert: A Source Book in the Foundations of Mathematics.* Clarendon Press, Oxford, 1996. In two volumes.

[13] Robert Gray. Georg Cantor and transcendental numbers. *American Mathematical Monthly*, 101:819–832, 1994.

[14] David Hilbert. Über den Zahlbegriff. *Jahresbericht der Deutschen Mathematiker-Vereinigung*, 8:180–184, 1900. Translated with commentary in Ewald [12, pp.1092–1095].

[15] David Hilbert. Die Grundlagen der Mathematik. *Abhandlungen aus dem mathematischen Seminar der Hamburgischen Universität*, 6:65–92, 1928. Translated in van Heijenoort [20, pp.464–479].

[16] Joseph Liouville. Des remarques relatives 1° à des classes très-étendues de quantités dont la valeur n'est ni rationelle ni même réductible à des irrationnelles algébriques; 2° à un passage du livre des *principes* où newton calcule l'action exercée par une sphère sur un point extérieur. *Comptes Rendus Hebdomadaires des Séances de l'Académie des Sciences, Paris*, 18:883–885, 1844.

[17] Joseph Liouville. Sur des classes très-étendues de quantités dont la valeur n'est ni algébrique ni même réductible à des irrationnelles algébriques. *Journal de mathématiques pures et appliquées*, 16:133–142, 1851.

[18] Alexis Maciel, Toniann Pitassi, and Alan R. Woods. A new proof of the weak pigeonhole principle. *Journal of Computer and System Sciences*, 64:843–872, 2002.

[19] Gregory H. Moore. *Zermelo's Axiom of Choice. Its Origins, Development and Influence.* Springer, New York, 1982.

[20] Jean van Heijenoort, editor. *From Frege to Gödel: A Source Book in Mathematical Logic, 1879–1931.* Harvard University Press, Cambridge, 1967. Reprinted 2002.

[21] Ludwig Wittgenstein. *Remarks on the Foundations of Mathematics.* Basil Blackwell, Oxford, 1967. Edited by von Wright, G. H. and Rhees, R. and Anscombe, G. E. M.. Second printing.

[22] Ernst Zermelo. Beweis, dass jede Menge wohlgeordnet werden kann (Aus einem an Herrn Hilbert gerichteten Briefe). *Mathematische Annalen*, 59:514–516, 1904. Reprinted and translated with commentary in Zermelo [24, pp.81–119].

[23] Ernst Zermelo. Untersuchungen über die Grundlagen der Mengenlehre I. *Mathematische Annalen*, 65:261–281, 1908. Reprinted and translated with commentary in Zermelo [24, pp.160–229].

[24] Ernst Zermelo. *Collected Works I.* Springer, 2010. Edited by Heinz-Dieter Ebbinghaus and Akihiro Kanamori.

[25] Max Zorn. A remark on method in transfinite algebra. *Bulletin of the American Mathematical Society*, 41:667–670, 1935.

7 Mathematical Knowledge: Motley and Complexity of Proof

Modern mathematics is, to my mind, a complex edifice based on *conceptual constructions.* The subject has undergone something like a biological evolution, an opportunistic one, to the point that the current subject matter, methods, and procedures would be patently unrecognizable a century, certainly two centuries, ago. What has been called "classical mathematics" has indeed seen its day. With its richness, variety, and complexity any discussion of the nature of modern mathematics cannot but accede to the primacy of its history and practice. As I see it, the applicability of mathematics may be a driving motivation, but in the end mathematics is autonomous. Mathematics is in a broad sense self-generating and self-authenticating, and alone competent to address issues of its correctness and authority.

What brings us mathematical knowledge? The carriers of mathematical knowledge are *proofs*, more generally arguments and constructions, as embedded in larger contexts.[1] Mathematicians and teachers of higher mathematics know this, but it should be said. Issues about competence and intuition can be raised as well as factors of knowledge involving the general dissemination of analogical or inductive reasoning or the specific conveyance of methods, approaches or ways of thinking. But in the end, *what can be directly conveyed as knowledge are proofs.*

Mathematical knowledge does not consist of theorem statements,[2] and does not consist of more and more "epistemic access", somehow, to "abstract objects" and their workings. Moreover, mathematical knowledge extends not so much into the statements, but back into the means, methods and definitions of mathematics, sometimes even to axioms. Of course, statements are significant as encapsulations or markers of arguments. However, the bare statement, no matter how striking or mundane, remains aloof and often mysterious as to what it conveys. Statements gain or absorb their senses from the proofs made on their behalf. One does not really get to *know* a statement, but rather a proof, the complex of argument taken altogether as a conceptual construction. A statement may have several different proofs each investing the statement with a different sense, the sense reinforced by different refined versions of the

Reprinted from *Annals of the Japan Association for Philosophy of Science* 21 (2013), pp.21-35, with permission from The Japan Association for Philosophy of Science.

Versions of this paper have been given several times, mostly recently at the Ideals of Proof Fellows' Seminar, Paris France in May 2009; the Diverse Views of Mathematics Workshop, the University of Göttingen, Göttingen Germany in November 2009; and the Workshop on Aspects of Infinity, Keio University, Tokyo Japan in June 2010. My thanks to all those who made helpful suggestions, especially Juliet Floyd.

[1] In a perceptive paper advocating similar themes Rav [15] used the term "bearer" instead of "carrier". The former is more passive, as in "bearer of good tidings".

[2] Here, "statement" is being used to suggest mere prose expression, in preference to the weightier "proposition", which often translates Frege's "Satz" and was also used by Russell.

statement and different corollaries from the proofs.[3] In modern mathematics, proofs and arguments often achieve an autonomous status beyond their initial contexts, with new, generalizing definitions formulated and new spaces of mathematical objects devised just so that certain proofs and, more generally, methods can be carried out in a larger context. In this sense knowledge begets knowledge, each new context a playing field for the development of new proofs. Proofs are not merely stratagems or strategies; they and thus their evolution are what carries forth mathematical knowledge.

Based on the autonomy of mathematics and anchored in its practice and history, this emphasis on proof is not to be any particular "theory". It is not to advocate a verificationism with a thick sense of meaning, for it is the myriad of verifications themselves which is the message. Nor is it to advocate a deductivism, since there is no presumption that the content of modern mathematics can be coherently axiomatized, and there is no presumption that mathematical proofs need be converted to formal deductions in a formal system. Nor is it to advocate any form of constructivism, although there is a surface affinity in the use of the word "construction" and with the basic constructive tenet that to assert a statement is to provide a proof. Mathematics is to be accounted for both as a historical and an epistemological phenomenon, with the evolution of proofs and methods at the heart.

What about truth in mathematics? Since proofs are the carriers of mathematical knowledge, whether mathematical statements are true or false is not as central to mathematics as the proofs devised on their behalf. Truth does not have an independent, *a priori* meaning in modern mathematics, but it does operate in mathematics in a primordial way. Truth is normative; statements are incipiently taken to be true or false; and bivalence is intrinsic to its basic grammar. Problems and conjectures get clarified and contextualized, and some are eventually transformed into a proof. When a mathematician proves a statement, it is true mainly in that a proof has been provided, one that can be examined. The picture then is that mathematical truth abounds everywhere, and the directions of investigations, the working out of particular truths, is a matter of historical contingency and the social context of mathematicians working with problems and conjectures. Truth serves at the beginning of investigation to intimate an initial parting of ways, but is not the end, which is

[3]Evidently, "sense" here is being used in a common-sensical way, one not tied to any theory of meaning. Waismann [21] p.109 recorded Wittgenstein on 25 September 1930 on proof: "A proof is not a vehicle for getting anywhere, but is the thing itself. . . . two different proofs cannot lead to the same thing. Two proofs can either meet, like two paths leading to the same destination, or they prove different things: a difference between proofs corresponds to a difference between things proved." A note is added: "A transformation of two proofs into one another is the proof of their proving the same thing." In other words, it may be that one proof can be transformed into another of the same statement, but that very transformation is yet another proof.

the proof. Of course, one can analyze an argument as a web of implications and consider whether each implication as a statement is true or false. Truth again functions primordially here, but invariably this pressing into the interstices comes to an end, and we take in the argument as a whole.

Truth, whether a statement has the property of being true or the property of being false, is itself too primal to be definable or otherwise characterizable. Little can be *done* with truth as such in mathematics.[4] This by no means is to dispense with truth or to advocate a deflationist theory of truth; rather, it is to point out how no general theory of truth will have any bearing on mathematics and its practice.[5]

What, finally, about existence in mathematics? Existence is surely not a predication in mathematics. Whether numbers or sets exist or not in some prior sense is devoid of mathematical interest. Questions like "What is a number?" and "What is a set?" are not mathematical questions, and any answers would have no operational significance in mathematics.[6] *Within* mathematics, an existence assertion may be an axiom, a goal of a proof, or one of the junctures of a proof. Existence is then submerged into context, and it is at most a matter of whether one regards the manner in which the existence assertion is deduced or incorporated as coherent with the context.[7] Finally, there are many junctures in proofs, especially since the turn of the 20th Century and in modern algebra, in which negations of universal statements in a variable are considered. In formal terms, one works with the equivalence of $\neg\forall$ with $\exists\neg$, and in informal terms one searches for counterexamples to deny universality — there are proofs and refutations. Existence in mathematics is embedded

[4]Kant, *Critique of Pure Reason* A58/B83: "What is truth? The nominal definition of truth, that it is the agreement of knowledge with its object, is assumed as granted; the question asked is as to what is the general and sure criterion of the truth of any and every knowledge.

To know what questions may reasonably be asked is already a great and necessary proof of sagacity and insight. For if a question is absurd in itself and calls for an answer where none is required, it not only brings shame to the propounder of the question, but may betray an incautious listener into absurd answers, thus presenting, as the ancients said, the ludicrous spectacle of one milking a he-goat and the other hold a sieve underneath."

[5]Wittgenstein, *Philosophical Investigations* §241: "So you are saying that human agreement decides what is true and what is false?" –It is what human beings *say* that is true and false; that is not agreement in opinions but in form of life [Lebensform].

[6]The Fields medalist Timothy Gowers [9] p.18 aptly wrote: "A mathematical object *is* what it *does*."

[7]Waismann [21] pp.172ff recorded Wittgenstein on 21 September 1931 on proof of existence: "If, on one occasion, I prove that an equation of degree n must have n solutions by giving, e.g. one of the Gaussian proofs, and if, on another, I specify a procedure for deriving the solutions and so prove their existence, I have by no means given two different proofs of the same proposition; I have proved entirely different things. What is common to them is simply the prose proposition 'There are n solutions', and that, taken by itself, means nothing, being a mere abbreviation standing for a proof. If the proofs are different, then this proposition simply *means* different things."

in the calculi of mathematics; to be is to be in the range of a variable in some mathematical system. No theory of mathematical existence, let alone any resolution of any realism vs. anti-realism debate, will have any bearing on mathematics and its practice. Varieties of structuralism may try to deal with mathematical objects as submerged in relations and structures, but there is still the "second-order" question of what is a structure, and in any case, the preoccupation is still with mathematical existence.

Mathematicians themselves have often described a feeling of dealing with independent, autonomous objects, some professing an avowedly realist view of mathematics. The simple reply is that this is psychological, perhaps informatively attributable to our cognitive mechanisms, and the reification, an operational attitude. Objectification is part of the practice of mathematics, the sense of existence here to be described as in any other concerted social activity, and it is particularly fostered by the precision of mathematical language. Mathematicians work with equal vigor in incompatible theories, e.g. non-Euclidean geometries, with objects "existing" in one theory but not in another, and more pointedly, there is the *counterexample phenomenon*: In order to establish a universal $\forall$ statement, one works at length on counterexamples, these "existing" in context in just as full-bodied a sense as examples, until one can dismiss them as not "existing". Some of the most complex and prolonged proofs, e.g. in finite group theory,[8] have this overall character of a *reductio ad absurdum* proof based on the equivalence of $\neg\forall$ and $\exists\neg$.

1 At the Beginnings of Set Theory

Let me give a notable example at the heart of set theory that illustrates several of the foregoing points. In 1904 Ernst Zermelo [23] established the Well-Ordering Theorem, that every set can be well-ordered, applying the Axiom of Choice. Indeed, it was for this purpose that he made explicit the axiom, which he considered a logical principle "used everywhere in mathematical deduction". That part of the argument that does not depend on the axiom can be isolated in the following result, not made explicit by Zermelo, which establishes a basic correlation between functions "type-reducing" $\gamma\colon \mathcal{P}(M) \to M$ from the power set $\mathcal{P}(M) = \{X |\, X \subseteq M\}$ of a set M into the set and definable well-orderings.[9]

Theorem 1. *Suppose that $\gamma\colon \mathcal{P}(M) \to M$. Then there is a unique $\langle W, \prec\rangle$ such that $W \subseteq M$ and $\prec$ is a well-ordering of W satisfying:*

(a) For every $x \in W$, $\gamma(\{y \in W \mid y \prec x\}) = x$, and

(b) $\gamma(W) \in W$.

[8] The concluding section has other examples.

[9] The following analysis is drawn from Kanamori [13, §2]. Tarski [19, Theorem 3] was a version of the theorem. Substantially the same version appeared in the expository work of Bourbaki [5, p.43] (Chapter 3, §2, Lemma 3).

The picture here is that γ generates a well-ordering which according to (a) starts with

$$a_0 = \gamma(\emptyset),\ a_1 = \gamma(\{a_0\}),\ a_2 = \gamma(\{a_0, a_1\}),\ \ldots$$

and so continues as long as γ applied to the initial segment constructed thus far produces a new element. W is the result when according to (b) an old element is again named.

Proof of Theorem 1. Call $Y \subseteq M$ a γ-*set* if and only if there is a well-ordering R of Y such that for each $x \in Y$, $\gamma(\{y \in Y \mid y\ R\ x\}) = x$. We shall establish:

(*) If Y is a γ-set with a witnessing well-ordering R and Z is an γ-set with a witnessing well-ordering S, then $\langle Y, R\rangle$ is an initial segment of $\langle Z, S\rangle$, or vice versa.

Taking $Y = Z$ it will follow that any γ-set has a unique witnessing well-ordering.

For establishing (*), we continue to follow Zermelo: By the comparability of well-orderings we can assume without loss of generality that there is an order-preserving injection $e\colon Y \to Z$ with range an S-initial segment of Z. It then suffices to show that e is in fact the identity map on Y: If not, let t be the R-least member of Y such that $e(t) \neq t$. It follows that $\{y \in Y \mid y\ R\ t\} = \{z \in Z \mid z\ S\ e(t)\}$. But then,

$$e(t) = \gamma(\{z \in Z \mid z\ S\ e(t)\}) = \gamma(\{y \in Y \mid y\ R\ t\}) = t\,,$$

a contradiction.

To conclude the proof, let W be the union of all the γ-sets. Then W is itself a γ-set by (*) and so, with $\prec$ its witnessing well-ordering, satisfies (a). For (b), note that if $\gamma(W) \notin W$, then $W \cup \{\gamma(W)\}$ would be a γ-set, contradicting the definition of W. Finally, that (a) and (b) uniquely specify $\langle W, \prec\rangle$ also follows from (*). ⊣

Zermelo of course focused on choice functions as given by AC to well-order the entire set:

Corollary 2 (The Well-Ordering Theorem)(Zermelo [23]). *If $\mathcal{P}(M)$ has a choice function, then M can be well-ordered.*

Proof. Suppose that $\varphi\colon \mathcal{P}(M) \to M$ is a choice function, i.e. $\varphi(X) \in X$ whenever X is non-empty, and define a function $\gamma\colon \mathcal{P}(M) \to M$ to "choose from complements" by: $\gamma(Y) = \varphi(M - Y)$. The resulting W of the theorem must then be M itself. ⊣

Theorem 1 also leads to a new proof of Cantor's Theorem, a proof that eschews diagonalization altogether and moreover provides a definable counterexample to having a one-to-one correspondence between a set and its power set:

Corollary 3. *For any* $\gamma\colon \mathcal{P}(M) \to M$*, there are two distinct sets* W *and* Y *both definable from* γ *such that* $\gamma(W) = \gamma(Y)$.

Proof. Let $\langle W, \prec\rangle$ be as in Theorem 1, and let $Y = \{x \in W \mid x \prec \gamma(W)\}$. Then by (a) of Theorem 1, $\gamma(Y) = \gamma(W)$, yet $\gamma(W) \in W - Y$. ⊣

In the $\gamma\colon \mathcal{P}(M) \to M$ version of Cantor's diagonal argument, first given by Zermelo himself ([24, Theorem 2]), one would consider the definable set

$$A = \{\gamma(Z) \mid \gamma(Z) \notin Z\} \subseteq M.$$

If $\gamma(A) \notin A$, then we have the contradiction $\gamma(A) \in A$. If on the other hand $\gamma(A) \in A$, then $\gamma(A) = \gamma(B)$ for some B such that $\gamma(B) \notin B$. But then, $B \neq A$. However, no such B is provided with a *definition.*[10]

To emphasize an important contention about the primacy of proof for mathematical knowledge: At the heart of set theory is Cantor's Theorem, its crux being: there is no one-to-one correspondence between a set and its power set. Cantor's diagonal proof provides one sense as a *reductio* argument, one that can also be used to generate a new real from a countable sequence of reals. Corollary 3 provides another sense, with a recursive definition of a well-ordering.

Bringing out the new sense, the proof of Theorem 1 informs Cantor's Theorem with another notable consequence expressible only in a setting where sets are not inherently well-ordered: Since the γ there need only operate on the *well-orderable* subsets of M, the $\mathcal{P}(M)$ in Corollary 3 can be replaced by the following set:

$$\{Z \subseteq M \mid Z \text{ is well-orderable}\}.$$

That this subset of $\mathcal{P}(M)$ is also not in one-to-one correspondence with M was first pointed out by Tarski [19] through a less direct proof.

2 Proofs in Mathematics

If mathematical proofs are the carriers of mathematical knowledge, the nature of mathematical proof as embedded in mathematical practice plays a central role in epistemology. But what *is* a proof? In what ways do proofs validate statements? There can only be *descriptive* answers, as one surveys the motley

[10]This is also the main thrust of Boolos [4], in which the argument for Theorem 1 is given *ab initio* and not connected with Zermelo [23].

of modern mathematics in all its variety and complexity and sees a motley of proofs as avenues to knowledge.

At the outset, a proof is a network of implicational connections between a series or web of statements. The connections can be fused or split, and the juncture statements themselves can be reduced, shifted, or proliferated according to how a mathematician might conceive of or teach a proof. In the direction of refinement, the connections can be split more and more with intermediary statements introduced, but this may jeopardize the surveyability of the proof. A *mathematical* proof has an irreducible semantic content that is carried by the implicational connections and juncture statements at an appropriate level of organization, one that is based on the concepts and methods of the context.

There are crucial aspects of proof beyond just the schematic picture which are brought out by the *communication* of proofs among mathematicians. Proofs ought to be communicable, and so ultimately they should be written down. But before then, proofs—generally methods and schemes—are often communicated in conversations and seminars among those in the subfield. Pictures are drawn; ideas are metaphorically described; there are telling side comments. Just a few gestures and remarks may suffice to convey a proof, as one draws on the common store of knowledge and ways of thinking that have long been assimilated as part of the "language game" of the subfield. Mathematicians in a subfield are very reliable in checking each other's work and have a sharp sense of when a proof is correct based merely on a high-level configuration of ideas and methods, the details to be attended to later. Indeed, the sign of a very good mathematician is the ability to separate the steps which are procedural and unproblematic from those that require something distinctly new.

This last brings out a crucial feature of a *new* proof in the historical, evolutionary sense: Although a proof is a network of implicational connections, one or two particular connections becomes pivotal. These are the keys that unlock, the crucial gaps that are filled, or the new innovations that reorganize the implicational connections. From simple proofs to the intimation of new methods, there is often just one telling connection that clinches the proof, and this is related to the common epiphanous experience of suddenly seeing a proof whole and correct.

The writing down of proofs raises major issues for the conveyance of mathematical knowledge. Research mathematicians can be poor expositors, often adhering to presentations with overemphasis on brevity of argument or generality of applicability. One looks in vain for a discussion of themes, motivations, the wider historical context in which to place the new proofs. A strenuous effort is often required to assimilate a proof presented in a research paper, an effort that amounts to making one's own implicational connections at different conceptual levels. Mathematicians know that published accounts of proofs are often backward in that they are written forward, that the road to discovery was some new connection toward the end. So what actually *was* the proof presented? Even subsequent expositions, however well-intended, sometimes err on the side of simplicity. Simpler and simpler versions of proofs may be presented, but this can be at the the cost of emasculation or mystification; a proof as a larger conceptual construction may be shorn of its richness and consequence through streamlining, leaving an air of mystery as to how it could have been conceived. A sure sign is that structural corollaries that could have been drawn formerly are no longer possible. We are left with the realization that formal proofs are not crucial, that the correctness and authority of the modern mathematical enterprise must lie in a wider sense of proof as embedded in mathematical practice.

Mathematicians have steadily re-proved theorems, and this is a conspicuous feature of modern mathematics and significant for its epistemology. Why re-prove a state if it is already true? Because the fact of the matter does not reside in its truth but in its avenues, the exploration of the conceptual constructions that see it out. Mathematical knowledge is not a roster of true statements but the web of connections among them and the techniques and methods that hold up the edifice. There are a variety of reasons why mathematicians re-prove theorems:[11]

(1) To provide a constructive or more effective demonstration. This includes providing constructive proofs for existence statements and providing faster algorithms, like the recent primality testing in polynomial time.

(2) To eliminate hypotheses or apply methods appropriate to the statement context. This includes eliminating the Axiom of Choice, the Riemann Hypothesis, and so forth, and providing proofs for number-theoretic statements in Peano Arithmetic, like the elementary proof of the Prime Number Theorem.

(3) To provide a regressive reconstruction of historical practices. This includes Hilbert's rigorization of Euclidean geometry and the development of generalized distributions to handle Dirac's delta function. New statements are proved, as the underlying concepts themselves having been transmuted.

(4) To simplify and make more direct and surveyable earlier proofs. This involves reducing the computations and hypotheses that depended on an earlier proof having come out of a complex setting, but comes with the danger of increasing the mystery and a sense of surprise. An example is Hilbert's basis theorem, which actually goes counter to (1).

(5) To demonstrate different approaches. This includes algebraic or topological proofs of geometric statements and proofs of non-standard analysis.

(6) To generalize an earlier result into a larger context. This includes extensions of the Riemann integral and Henkin's proof of the Completeness Theorem.

[11] See Dawson [7]; the following list amounts to a reorganization of his.

Proofs at the complex end of the range of proofs raise further issues, particularly those of *surveyability*, *intelligibility*, *acceptance*, and the dependence on *authority*, i.e. the reliance on experts—and these are the significant and interesting issues, further and further away from abstract preoccupations about the relation of proof to truth. What is most striking about the evolution of mathematics in the last half-century is that it has become such a complex edifice. This complexity presumably came into mathematics with the 1962 proof of Walter Feit and John Thompson resolving the Burnside Problem: *All finite groups of odd order are solvable.*[12] The proof was non-constructive, proceeding from a minimal counterexample to draw a contradiction. This is an important, early example of existence in mathematics as informed by the *counterexample phenomenon*, the working at length on counterexamples, these "existing" in context until they can be dismissed. The publication [8] was well over two hundred pages and had to be farmed out to several referees. And this itself was just a beginning: 100 group theorists working in a cottage industry for the two subsequent decades and publishing 10,000 journal pages were said to have completed the classification of all finite simple groups: *Every finite simple group is either a group of prime order, an alternating group, a group of Lie type, or else one of the 26 sporadic groups.* No single mathematical statement hitherto had such a long proof, a proof variously featuring the counterexample phenomenon. In the 1990s a program was launched to simplify large parts of the proof and to write it all down in one place, but gaps emerged that continue to be filled.[13] This may be increasing the surveyability and intelligibility of the classification, but with the best estimates for the size of the proof still at around 4000 pages, acceptance by the wider mathematical community may remain across a spectrum. "True" applied here may have a longstanding vagueness akin to words like "big". The classification has been applied in subsequent mathematics, and as a juncture statement in a proof it will very much have to be based on authority, on the reliance on experts.

Issues about the concept of proof and its possible liberalization came to fore with the 1976 proof by Kenneth Appel and Wolfgang Haken of the Four-Color Theorem: *Every planer graph is four-colorable.*[14] The counterexample phenomenon at play, they were able to reduce the infinitude of possible counterexamples to 1476 configurations, and then in 1200 hours of computing time they showed that these too are four-colorable. Going beyond finite group theory in terms of authority, computer-assisted proofs manifest reliance on experts no longer for presumably intelligible implicational connections, but also for avowedly un-intelligible connections. Understanding has been relegated to an overview of the myriad of connections mapped out, the computer completing

[12]Equivalently, every finite simple non-Abelian group is of even order.

[13]See Aschbacher [3] for the current status.

[14]See Appel-Haken [1], and the monograph Appel-Haken [2].

the tasks.

On the one hand, there was debate about admitting the Appel-Haken argument into the pantheon of proofs, but on the other hand, there was the feeling that at least the door was closed on the problem. However, subsequent events interestingly provided new knowledge, as for "conventaionl" surveyable proofs. First, the 1989 741-page monograph Appel-Haken [2] significantly diminished skepticism, with its detail and overall careful account. Then in 1995, Neil Robertson, Daniel Sanders, Paul Seymour and Robin Thomas provided a new, still computer-assisted, proof featuring new techniques that initially reduced the configurations to 633 and provided a four-coloring algorithm of quadratic, $O(n^2)$ order[15] — the Appel-Haken proof had led to a quartic, $O(n^4)$ algorithm. Latterly in 2004, Georges Gonthier and Benjamin Werner wrote a formal proof script for the computer code part of the 1995 proof in a language that used logical propositions and applied a proof checking system which mechanically verified the correctness.[16]

In 2005 Gonthier in a singular advance carried out a complete, computer-assisted formalization of a proof of the Four-Color Theorem in its entirety. Gonthier began with the Robertson-Sanders-Seymour-Thomas proof, and in the process implemented new approaches, as befitted the computer-assisted framework, with combinatorial hypergraphs. This remarkable development signaled the establishment of a new, notable field of mathematics devoted to the complete computer-assisted formalization of proofs. A traditional, mathematical proof is first written out in greatly expanded form with all the assumptions made explicit and all the cases treated in full. From the expanded text a computer script is prepared which generates all the logical inferences of the proof. This is done with "computer assistants", computer systems which are either declarative, in which proofs are written out in atomic steps, or procedural, in which proofs are generated interactively through dialogue. Currently, Gonthier is carrying out a complete formalization of the Feit-Thompson result.

This new field of mathematics substantively realizes the ideal of formalization in the direction of actual proof implementation. Instead of "formalizable in principle" we have formalization in realization. Hilbert with his metamathematics emphasized and advanced the formalizability of mathematics, but was not interested in formalizations of actual proofs. Before him Gottlob Frege and Bertrand Russell provided formal proofs to establish the reduction of mathematics to deductive logic. The real precursor was Guiseppe Peano, who devised an efficient symbolic language and sought to provide, with his *Formulario Mathematico*, an encyclopedia of all known formulas and theorems of mathematics. Similarly, a stated goal of the new formalization field is to someday provide a library of Alexandria of complete, formalized proofs, the possibility only newly

[15]See [16].

[16]See http://ralyx.inria.fr/2003/Raweb/moscova/uid23.html.

realizable through computer assistants. Another, new incentive that may become more and more prominent is to verify increasingly complex proofs, so that authority is transferred from humans to computers. This is a remarkable development, one prompted by the new complexities of proof, and it may well turn out that the very issues of surveyability, intelligibility, and authority about proof will affirm a new meta-mathematics, new mathematical knowledge generated by the subject of proof itself.

The issues of surveyability and intelligibility vis-à-vis computer-assisted proofs had become accentuated by the 1998 proof by Thomas Hales of the Kepler Conjecture: *Congruent balls in three dimensions can be packed with the highest density in rows with rectilinear or, equivalently seen from another perspective, hexagonal arrangements.*[17] This was a specific part of Hilbert's 18th Problem, and the proof involved global optimization, linear programming, and interval arithmetic, and like the proof of the Four-Color Theorem, the counterexample phenomenon. At the outset there was a reduction to about 5000 large scale nonlinear optimization problems. Then there was a further reduction to about 100,000 linear programming problems each with 100 to 200 variables. Having appealed to 12 referees the *Annals of Mathematics* [12] published the non-computer part of the argument with a footnote distancing itself from the computer part, appearing in Hales [11].[18]

Although Hales's argument for Kepler's Conjecture may become increasingly accepted as a proof by the mathematical community, it is one *without potentiality*, even more than the solutions to the Four-Color Conjecture, and this perhaps is its most telling feature for epistemology: Nothing larger is learned from the argument, and it does not open up new possibilities for mathematics. But in this case, the gap with computer capability is much more pronounced, and only the future will tell how new developments in formalization will color the conjecture. With his Flyspeck Project Hales aspires to provide a complete formalization of a proof of the Kepler Conjecture and estimates it to be a 7000 work-day project.

In contrast, there is Andrew Wiles' 1994 proof of Fermat's Last Theorem: *There is no solution to $x^n + y^n = z^n$ in positive integers for any $n > 2$.* In a substantial sense, this statement isolated and unto itself has no intrinsic interest whatsoever, and it only grew in historical significance as it withstood more and more techniques, techniques that have enriched mathematics considerably like the Kummer theory of ideals. No, what is tremendous about the Fermat is the Wiles *proof.* He actually established the Shimura–Taniyama Conjecture about elliptic curves in algebraic geometry through a beautiful synthetic proof, and this among mathematicians has been seen as the great advance. The perennial

[17]See Hales [10] for a short, and Szpiro [17] for a book-length, account.

[18]See Hales's website www.math.pitt.edu/~thales for more on the interaction with computation.

pushing for the Fermat had led to a novel observation about a connection to elliptic curves, a connection that was affirmed by the late 1980s to establish that the Shimura–Taniyama Conjecture implies Fermat's Last Theorem. Thus, Wiles's proof had a last implicational connection due to others. This is not to say that Wiles did not covet the prize. This truth among a myriad was definitely stalked, but in any case the proof considerably enriched the theory of elliptic curves, a theory that has its origins centuries ago in the study of planetary motion. There is still the issue of authority, but the understanding brought about by the proof is acknowledged in large part through its potentiality, its opening up of new possibilities.

Wiles's proof [22] has raised interesting issues about proof.[19] Wiles wrote that "the turning point in this and indeed in the whole proof came" when he was led to two cohomology invariants from Grothendieck's duality theory. The Wiles proof in fact depends on the so-called "Grothendieck Universes". In arithmetic algebraic geometry of the last fifty years the algebra and topology of *schemes*, certain spaces, have been investigated through cohomology with respect to how they are situated in large categories. A straightforward formalization of Wiles' proof thus requires several levels "above" the usual cumulative hierarchy of ZFC. However, at the expense of elegance and a surveyable level of organization, the Wiles's proof can evidently be established in ZFC alone, and here, set-theoretic reflection would be the evident approach. Recently, Angus MacIntyre has claimed that the central Modularity Thesis in Wiles's proof is provable in Peano Arithmetic. If so, then there could a new proof of the Fermat Last Theorem in Peano Arithmetic, possibly by bypassing through further analysis the Modularity Thesis.

The issues aired above about proof figured fresh in the recent excitement about Grigori Perelman's proof of the Poincaré's Conjecture: *Every simply connected closed three-manifold is homeomorphic to the three-sphere.*[20] Informally, the sphere is the only type of bounded three-dimensional space possible that contains no holes. Formulated a century ago this conjecture, unlike Fermat's Last Theorem but like Kepler's Conjecture, has the immediacy of a fundamental structural characterization. But unlike, say, Riemann's Hypothesis, there was no general presumption about which way it would go; groups worked through the 1980s and into the 1990s both to try to find a proof and to try to find a counterexample. With increasing effort expended, resolution of the conjecture had become more and more prized, and the Clay Institute in 2000 offered a million dollars for a confirmed proof, as one of several outstanding problems for the millenium.[21]

[19] What follows is taken from McLarty [14].

[20] See Collins [6] for a short, and Szpiro [18] for a book-length, account.

[21] See the website www.claymath.org/millennium/ for the problems and an account of their importance for mathematics.

Like Wiles's proof, Perelman's argument was actually to resolve a more general conjecture, in this case the classification of all three-manifolds according to William Thurston's *geometrization.* In 2002 and 2003 Perelman sketched his argument in three papers, amounting to 68 dense and terse pages, and gave a series of talks across the United States and Europe in 2003. Because of the centrality of geometrization, the argument began being vetted by the experts. Seminars and workshops worked their way through; sketches were filled in with detailed proof; and there were reorganized expositions.[22] All this increased the surveyability and the intelligibility, and the high-level configuration of ideas and methods together with Perelman's ability to convey verbally the local and global structures of the argument gradually led over a couple of years to a general acceptance of his proof based quite avowedly on authority. Eventually, three pairs of authors published accounts in 2006 detailing Perelman's proof: the Kleiner-Lott notes in 200 pages filled out the Perelman papers, like a Talmudic exegesis, to provide a proof of geometrization; the Cao-Zhu paper amounting to 326 pages also provided a proof of geometrization; and the Morgan-Tian monograph in 521 pages gave a careful, "self-contained" account of just the Poincaré Conjecture solution. In 2006, Perelman was awarded the Fields Medal,[23] and just March of this year, 2010, was awarded the million dollar Clay prize for the Poincaré Conjecture.

We tuck in here another example, though it does not have to do as much with proof surveyability but effectiveness. Also in 2006, Terence Tao was awarded the Fields Medal for his solutions of problems over work in a broad range of areas. His best-known result has been his 2004 joint result with Ben Green: *There are arbitrarily long arithmetical progressions of prime numbers.* They brought together a cocktail of sophisticated mathematical ideas to solve a very old problem, and the result *is* surveyable, but there is another issue. The Green-Tao Theorem was another non-constructive existence assertion, in that it provided no algorithmic means of providing an arithmetical progression of specified length.[24] In fact, before they had established this theorem of modern mathematics in 2004, 23 was the length of the longest known arithmetic progression of primes, and as of 12 April 2010, it is 26, which as of the end of 2012 still holds the record. On one side there is the steady work of improving specific computational procedures—in effect new mathematical arguments and proofs—and on the other, there are the enormously sophisticated methods of modern mathematics—both focused on an age old issue about the prime

[22]See the website http://www.math.lsa.umich.edu/ lott/ricciflow/perelman.html.

[23]The Fields Medal is the most prestigious award given to a mathematician, with several awarded every four years at the International Congress of Mathematicians. Perelman was the first to refuse to accept the medal.

[24]In fact, the other well-known result of Klaus Roth, of 1952, that any set of positive integers of positive density has arithmetic progressions of length three, was the beginning of results toward the Green-Tao result.

numbers. On the latter side, mathematicians are moving ahead toward an old conjecture of Paul Erdős, that if a set A of natural numbers has the property that $\Sigma_{n \in A} 1/n = \infty$, then A contains arbitrarily long arithmetical progressions of prime numbers. One has to look for a new kind of proof, and there is even talk about undecidability, the complexity of definition of A becoming an issue.

The weight of these various examples shows how far away mathematics now is from being comprehended by any formal notio of proof and any theory of mathematical knowledge, and how the limits of human intelligibility are being put to the test. A complicated proof is like the Grand Tour: We are first given an itinerary and told that the connections can be made by straightforward means. The capitals Paris and Rome loom large, and in their terms we start to find our way about. We are soon taken in by Florence, seduced by Venice. The connections become shorter, and we explore the lesser churches and monuments. History begins to play an increasingly important role, and we put together a picture for ourselves of how it all fits. In our further travels we see more and more of the smaller towns and villages, but we also return refreshed to the large capitals. We make friends and look up relatives.

The argumentation is more important than the theorem; it is the journey, not the destination, that counts.

References

[1] Kenneth Appel and Wolfgang Haken. Every planer map is four colorable. *Illinois Journal of Mathematics*, 21:439–657, 1977.

[2] Kenneth Appel and Wolfgang Haken. *Every Planer Map is Four-Colorable.* American Mathematical Society, Providence, 1989.

[3] Michael Aschbacher. The status of the classification of the finite simple groups. *Notices of the American Mathematical Society*, 51, 2004.

[4] George S. Boolos. Constructing Cantorian counterexamples. *Journal of Philosophical Logic*, 26:237–239, 1997.

[5] Nicolas Bourbaki. *Eléments de Mathématique. I. Théorie des Ensembles. Chapter III: Ensembles ordonnés, cardinaux, nombres entiers.* Hermann, Paris, 1956.

[6] Graham P. Collins. The shapes of space. *Scientific American*, 291:94–103, 2004.

[7] John W. Dawson. Why do mathematicians re-prove theorems? *Philosophia Mathematica*, 14:269–286, 2006.

[8] Walter Feit and John Thompson. Stability of groups of odd order. *Pacific Journal of Mathematics*, 13:775–1029, 1964.

[9] Timothy Gowers. *Mathematics, A Very Short Introduction.* Oxford University Press, Oxford, 2002.

[10] Thomas C. Hales. Cannonballs and honeycombs. *Notices of the American Mathematical Society*, 47:440–449, 2000.

[11] Thomas C. Hales. Some algorithms arising in the proof of the Kepler conjecture. In *Discrete and Computational Geometry*, Algorithms and Combinatorics 25, pages 489–507, Berlin, 2003. Springer.

[12] Thomas C. Hales. A proof of the kepler conjecture. *Annals of Mathematics*, 162:1065–1185, 2005.

[13] Akihiro Kanamori. The mathematical import of Zermelo's well-ordering theorem. *The Bulletin of Symbolic Logic*, 3:281–311, 1997.

[14] Colin McLarty. What does it take to prove Fermat's Last Theorem? Goethendieck and the logic of number theory. *The Bulletin of Symbolic Logic*, 2010.

[15] Yehuda Rav. Why do we prove theorems? *Philosophia Mathematica*, 7:5–41, 1999.

[16] Neil Robertson, Daniel P. Sanders, Paul Seymour, and Robin Thomas. A new proof of the Fcour-Color Theorem. *Electronic Research Announcements of the American Mathematical Society*, 2:17–25, 1996.

[17] George G. Szpiro. *Kepler's Conjecture.* John Wiley & Sons, Hoboken, 2003.

[18] George G. Szpiro. *Poincaré's Prize: The Hundred-Year Quest to Solve One of Math's Greatest Puzzles.* Penguin Group, New York, 2007.

[19] Alfred Tarski. On well-ordered subsets of any set. *Fundamenta Mathematicae*, 32:176–183, 1939. Reprinted in [20, vol. 2, pp.551–558].

[20] Alfred Tarski. *Collected Papers.* Birhkäuser, Basel, 1986. Steven R. Givant and Ralph N. McKenzie (editors).

[21] Friedrich Waismann. *Wittgenstein and the Vienna Circle.* Basil Blackwell, 1979. Edited by Brain McGuinness.

[22] Andrew Wiles. Modular elliptic curves and Fermat's Last Theorem. *Annals of Mathematics*, 141:443–551, 1995.

[23] Ernst Zermelo. Beweis, dass jede Menge wohlgeordnet werden kann (Aus einem an Herrn Hilbert gerichteten Briefe. *Mathematische Annalen*, 59:514–516, 1904. Reprinted and translated with commentary in Zermelo [25, pp.80–119].

[24] Ernst Zermelo. Untersuchungen über die grundlagen der mengenlehre I. *Mathematische Annalen*, 65:261–281, 1908. Reprinted and translated with commentary in Zermelo [25, pp.160–229].

[25] Ernst Zermelo. *Collected Works*. Springer, 2010. Edited by Heinz-Dieter Ebbinghaus and Akihiro Kanamori.

8 Aspect-Perception and the History of Mathematics

In broad strokes, mathematics is a massive yet multifarious edifice and mode of thinking based on networks of *conceptual constructions*. With its richness, variety and complexity, any discussion of the nature of mathematics cannot but account for these networks through its evolution in history and practice. What is of most import is the emergence of knowledge, and the carriers of mathematical knowledge are *proofs*, more generally arguments and procedures, as embedded in larger contexts. One does not really get to *know* a proposition, but rather a proof, the complex of argument taken altogether as a conceptual construction. Propositions, or rather their prose statements, gain or absorb their sense from the proofs made on their behalf, and yet proofs can achieve an autonomous status beyond their initial contexts. Proofs are not merely stratagems or strategies; they and thus their evolution are what carries forth mathematical knowledge.

Especially with this emphasis on proofs, aspect-perception—*seeing* an aspect, seeing something *as* something, seeing something *in* something—emerges as a schematic for or approach to what and how we know, and this for quite substantial mathematics. There are sometimes many proofs for a single statement, and a proof argument can cover many statements. Proofs can have a commonality, itself a proof; proofs can be seen as the same under new light; and disparate proofs can be correlated, this correlation itself amounting to a proof. Less malleably, statements themselves can be seen as the same in one way different in another, this bolstered by their proofs.

With this, aspect-perception counsels the history of mathematics, taken in two neighboring senses. There is the history, the patient accounting of people and their mathematical accomplishments over time, and there is the mathematics, evolutionary analysis of results and proofs over various contexts. In both, there would seem to be the novel or the surprising. Whether or not there is creativity involved, according to one measure or another, analysis through aspects fosters understanding of the byways of mathematics.

In what follows, the first section briefly describes and elaborates aspect-perception with an eye to mathematics. Then in each of the succeeding two sections, substantial topics are presented that particularly draw out and show aspect-perception at work.

Republished from Brendan Harrington, Dominic Shaw and Michael Beaney (editors), *Aspect Perception After Wittgenstein: Seeing-As and Novelty*, Routledge, New York, 2018, pp.109-130, with permission from Tayler & Francis Group LLC.

Aspects of this paper were presented at 2013 seminars at Carnegie Mellon University and at the University of Helsinki; many thanks to the organizers for having provided the opportunity. The paper has greatly benefitted from discussions with Juliet Floyd.

1 Aspect-Perception

Aspect-perception is a sort of meta-concept, one collecting a range of very different experiences mediating between seeing and thinking. Outwardly simple to instantiate, but inwardly of intrinsic difficulty, it defies easy reckoning but, once seen, it invites extension, application, and articulation. For the discussion and scrutiny of mathematics, it serves to elaborate and to focus aspect-perception in certain directions and with certain emphases. And for this, it serves to proceed through a deliberate arrangement of some *loci classici* for aspect-perception in the writings of Ludwig Wittgenstein.

Aspect-perception was a recurring motif for Wittgenstein in his discussions of perception, language, and mathematics. His later writings especially are filled with remarks, some ambitious and others elliptical, gnawing on a variety of phenomena of aspect-perception. It was already at work in his *Notebooks 1914-16*, and his 1921 *Tractatus*, with its *Bildhaftigkeit*, has at 5.5423 the Necker cube:

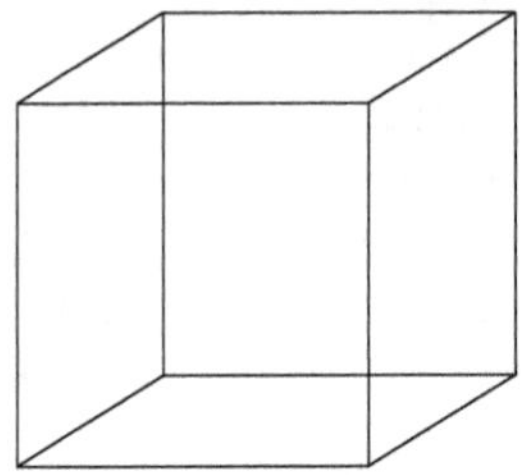

The figure can be seen in two ways as a cube, the left square in the front or the right square in the front. "For we really see two different facts." Among the points that were made here: A symbol, serving to articulate a truth, involves a projection by us into a space of possibility, hence multifarious relationships, not merely the interpretation of a sign. This early juncture in Wittgenstein correlates with aspects of symbolization in mathematics: as explicit in succeeding sections, different modes of organization to be brought out for purposes of proof can be carried by one geometric diagram or one algebraic equation.

In his 1934 *Brown Book* (cf. [1958, II,§16]), Wittgenstein discusses "seeing it *as a face*" in the "picture-face", a circular figure with four dashes inside, and in a picture puzzle, where "what at first sight appears as 'mere dashes' later appears as a face". "And in this way 'seeing dashes as a face' does not involve a comparison between a group of dashes and a real human face." We are taking or interpreting the dashes as a face. This middle juncture in Wittgenstein, at an aspect more conceptual than visual, correlates with aspects of contextual imposition in mathematics: As exhibited by the "commonality" (**) in §2 below, there can be a structure or a proof, logical yet lean, whose "physiognomy" can be newly seen by being placed in a rich conceptual or historical context.

In Wittgenstein's mature *Philosophical Investigations* [1953] aspect-perception comes to the fore in Part II, Section xi.[1] "I contemplate a face, and then suddenly notice its likeness to another. I *see* that it has not changed; and yet I see it differently. I call this experience [*Erfahrung*] 'noticing an aspect'."(p.193c) Wittgenstein works through an investigation of various schematic figures, particularly the Jastrow duck-rabbit (p.194e), a figure which can be seen either as a duck with its beak to the left or a rabbit with its ears to the left. With this, he draws out the distinction between "the 'continuous seeing' of an aspect" (seeing, with immediacy; later, 'regarding-as'), and "the 'dawning' of an aspect" (sudden recognition). With "continuous seeing" he navigates a subtle middle road between being caused by the figure to perceive, and the imposition of a subjective, private inner experience, undermining both as explanations of the phenomenon of seeing-as. Here follows a sequential arrangement of quotations, citing page and paragraph:

> 198c: The concept of a representation of what is seen, like that of a copy, is very elastic, and so *together with it* the concept of what is seen. The two are intimately connected. (Which is *not* to say that they are alike.)
>
> 199b: If you search in a figure (1) for another figure (2), and then find it, you see (1) in a new way. Not only can you give a new kind of description of it, but noticing the second figure was a new visual experience.
>
> 200f: When it looks as if there were no room for such a form between other ones you have to look for it in another dimension. If there is no room here, there *is* room in another dimension. [Example of imaginary numbers for the real numbers follows.]
>
> 201b: ... the aspects in a change of aspects are those ones which the figure might sometimes have *permanently* in a picture.
>
> 203e: 'The phenomenon is at first surprising, but a physiological explanation of it will certainly be found.'— Our problem is not a causal one but a conceptual one.
>
> 204g: Here it is *difficult* to see that what is at issue is the fixing of concepts. A *concept* forces itself on one.
>
> 208d: One *kind* of aspect might be called 'aspects of organization'. When the aspect changes parts of the picture go together which before did not.
>
> 212a: ... what I perceive in the dawning of an aspect is not a property of the object, but an internal relation between it and other objects.

For a visual context of some complexity pointing us toward the appreciation of aspects in mathematics, one can consider the Cubist paintings of Pablo

[1] Part II is renamed "Philosophy of Psychology—a Fragment" in the recent edition [2009] of *Philosophical Investigations.*

Picasso and Georges Braque. These elicit aspects of aspect-perception emphasized by Wittgenstein like seeing faces and objects from various perspectives; continuous seeing; dawning of an aspect; aspects there to be seen; possible blindness to an aspect for a fully competent person; and making sense of bringing such a person to see the aspect. Moreover, there is shifting of aspects beyond complementary pairs, as several aspects can be kept in play at once, some superposing on others, some internal to others, some at an intersection of others, and, with the painters' willful obstructionism, some petering out at borders and some incompatible with others in varying degrees.

With an eye to mathematics, we can locate aspect-perception among broad philosophical abstractions in the following ways:

> Aspect-perception is not (merely) *psychological* or *empirical*, but substantially *logical*, in working in spaces of logical possibility.
>
> Aspect-perception is not in the service of *conventionalism* and is not (only) about *language*.
>
> Aspect-perception, while having to do with *fact* and *truth*, is orthogonal to *value*.
>
> Aspect-perception can figure as a mode of *analysis* of concepts and states of affairs.
>
> Aspect-perception maintains *objectivity*, as aspects are there to be seen, but through a multifarious conception involving *modality*.

What, then, is the place and import of aspect-perception in mathematics and its history? The above points situate aspect-perception between seeing and thinking as logical—and so having to do with truth and objectivity—and not about convention or value and as possibly participating in analysis. In these various ways, aspect-perception can be seen to be fitting and indeed inherent in mathematical activity. At the very least, aspect-perception provides language, and so a way of thinking, for discussing and analyzing concepts, proofs, and procedures—how they are different or the same, how they can be compared or correlated. More substantially, since mathematics is a multifarious edifice of conceptual constructions, attention to aspects itself promotes seeing, seeing anew, and gaining insights. This is particularly so in connection with how we gauge *simplicity*, how we account for *surprise*, and how we come to *understand* mathematics.

On these last points, earlier remarks circa 1939 of Wittgenstein from *Remarks on the Foundations of Mathematics* have particular resonance. Part III starts with a discussion of proof, beginning: " 'A mathematical proof must be surveyable [*übersichtlich*].' Only a structure whose reproduction is an easy task is called a 'proof'." Aspect-perception casts light here, since seeing an argument organized in a specific way, e.g. through the projection of another or as figuring in a larger context, can lead to (the dawning of) a perception of simplicity and thereby newly found perspicuity. In this way, aspect-perception shows the limits of logic conceived to be (merely) a sequence of local implications.

In Part I, Appendix II, Wittgenstein discusses the surprising in mathematics. The first specific situation he considered is when a long algebraic expression is seen shrunk into a compact form and where being surprised shows (§2) "a phenomenon of failure to command a clear view [*übersehen*] and of the change of aspect of a seen complex."

> For one surely has this surprise only when one does not yet know the way. Not when one has the whole of it before one's eyes. ... The surprise and the interest, then come, so to speak, from the outside.

After the dawning of the aspect, there is no surprise, and what then remains of the surprise is the idea of seeing the logical space of possibilities. Wittgenstein subsequently wrote (§4):

> 'There's no mystery here!'—but then how can we have so much as believed that there was one?—Well, I have retraced the path over and over again and over and over again been surprised; and I never had the idea that here one can *understand* something.—So 'There's no mystery here!' means 'Just look about you!'

Though only elliptical, Wittgenstein here is suggesting that understanding, especially of novelty, as coming into play with the seeing and taking in of aspects.

Aspect-perception and mathematics have further useful involvements. Aspect-perception is an intrinsically difficult meta-concept in and through which to find one's way, and by going into the precise, structured setting of mathematics one can better gauge and reflect on its shades and shadows. One can draw out aspects and deploy them to make deliberate conceptual arrangements for communicating mathematics. And aspect-perception provides opportunity to bring in large historical and mathematical issues of context and method, and to widen the interpretive portal to ancient mathematics.

In the succeeding sections, we show aspect-perception at work in mathematics by going successively through two topics, chosen in part for their differing features to illuminate the breadth of aspects. §2 takes up the classical and conceptual issue of the irrationality of square roots, bringing out aspects geometric and algebraic, ancient and modern. §3 sets out a circularity in the development of the calculus having to do with the derivative of the sine function, retraces features of the concept in ancient mathematics, and considers possible ways out of the circularity, thus drawing out new aspects. A substantial point to keep in mind is that these topics are conceptually complex; aspect-perception can work and, indeed, is of considerable interest at higher levels of mathematics.[2] In each of the sections is presented a "new" result "found" by the author, but one sees that creativity is belied to a substantial extent by context.

2 Irrationality of Square Roots

For a whole number n which is not a square number ($4 = 2^2$, $9 = 3^2$, $16 = 4^2, \ldots$), its square root $\sqrt{n}$ is irrational, i.e. not a ratio of two whole numbers. This section attends to this irrationality; we shall see that its various aspects are far-ranging over time and mathematical context, but perhaps surprisingly, have a commonality. The irrationality can itself be viewed as an aspect of $\sqrt{n}$ separate from the seeing of it as or for its calculation as in Old Babylonian mathematics *c.*1800 BCE, an aspect embedded in conceptualizations about the nature of number and of mathematical proof as first seen in Greek mathematics. In particular, the irrationality of $\sqrt{2}$ was a pivotal result of Greek geometry established in the latter 5th century BCE. This result played an important role in expanding Greek concepts of quantity, and for contextually discussing $\sqrt{2}$ as well as the general $\sqrt{n}$, it is worth setting out, however briefly, aspects of quantity as then and now conceived.

For the Greeks, a *number* is a collection of *units*, what we denote today by $2, 3, 4, \ldots$ with less connotation of order than of cardinality. Numbers can be added and multiplied. A *magnitude* is a line, a (planar) region, a surface, a solid, or an angle. Magnitudes of the same kind can be added (e.g. region to region) and multiplied to get magnitudes of a another kind (e.g. line to line to get a region). Respecting this understanding, we will deploy modern notation with its algebraic aspect, this itself partly to communicate further mathematical sense. For example, Proposition I 47 of Euclid's *Elements*, the Pythagorean Theorem, states rhetorically "In right-angled triangles the square on the side opposite the right angle equals the sum of the squares on the sides containing the right angle" with "square" *qua* region. We will simply write the arithmetical $a^2 + b^2 = c^2$ where a, b are (the lengths of) the legs of a right triangle and c (the length of) the hypotenuse.

A *ratio* is a comparison between two numbers or two magnitudes of the same kind (e.g. region to region). Having ratio 2 to 3 we might today write as a relation $2\!:\!3$ or a quantity $\frac{2}{3}$, with the first being closer to the Greek concept. There is a careful historiographical tradition promoting the first, but we will nonetheless deliberately deploy the latter in what follows. There are several aspects to be understood here: the fractional notation itself can be read as the Greek ratio; it can be read as part of a modern numerical-algebraic construal; and finally, the two faces are assertively to be seen as coherent.

[2]This belies objections at times lodged against Wittgenstein that he only raised philosophical issues of pertinence to very simple mathematics.

A *proportion* is an equality of two ratios. We deploy the modern = as if for numerical quantities, this again having several aspects to be understood: it can be read as the Greek proportion; it can be read as an identity of two numerical quantities; and finally, the two faces are assertively to be seen as coherent. In what follows, the notation itself is thus to convey a breadth of aspect as well as a change of aspect, something not always made explicit.

Two magnitudes are *incommensurable* if there is no "unit" magnitude of which both are multiples. While we today objectify $\sqrt{n}$ as a (real) number, that $\sqrt{n}$ is irrational is also to convey, in what follows, a Greek geometric sense: a square containing n square units has a side which is incommensurable with the unit. The pivotal result that $\sqrt{2}$ is irrational was for the Greeks that the side s and diagonal d of a square are incommensurable: $d^2 = s^2 + s^2 = 2s^2$ by the Pythagorean Theorem, and the ratio $\frac{d}{s}(= \sqrt{2})$ is not a ratio of numbers. To say that this result triggered a *Grundlagenkrise* would be an exaggeration, but it undoubtedly stimulated both the development of ratio and proportion for general magnitudes in geometry and a rigorization of the elements and means of proof.

One of the compelling results of the broader context was just the generalization that $\sqrt{n}$ for non-square numbers n is irrational, sometimes called Theaetetus's Theorem. *Theaetetus* (*c*.369 BCE) is, of course, the great Platonic dialogue on epistemology. Socrates takes young Theaetetus (*c*.417-369 BCE) on a journey from knowledge as perception, to knowledge as true judgment, to knowledge as true judgment with *logos* (an account), and, in a remarkable circle, returns to perception: how can there even be knowledge of the first syllable SO of "Socrates"—is it a simple or a complex? Early in the dialogue (147c-148d), Theaetetus suggested conceptual clarification vis-à-vis square roots. He first noted that the elder geometer Theodorus (of Cyrene, *c*.465-398 BCE) had proved by diagrams the irrationality of $\sqrt{n}$ for non-square n up to 17. Then dividing the numbers into the "square" and the "oblong", he observed that they can be distinguished according to whether their square roots are numbers or irrational. In view of this and derivative commentary, Theaetetus has in varying degrees been credited with much of the content of the arithmetical book VII of Euclid's *Elements* and of Book X, the meditation on incommensurablilty and by far the longest book.

The avenues and byways of the Theodorus result and the Theaetetus generalization have been much discussed from both the historical and mathematical perspective.[3] In what follows we point out aspects there to be seen that coordinate across time and technique, and to this purpose we first review proofs for the irrationality of $\sqrt{2}$.

[3]See [Knorr, 1975] and [Fowler, 1999] for extended historical reconstructions based on different approaches, and see [Conway and Shipman, 2013] for the most recent mathematical tour.

The argument most often given today is algebraic, about $\sqrt{2}$: Assume that $\sqrt{2} = \frac{a}{b}$ for (whole) numbers a and b, so that, squaring, $a^2 = 2b^2$. a^2 is thus even, and so, consequently, is a, say $a = 2c$. But then, $4c^2 = 2b^2$ and so $2c^2 = b^2$. b^2 is thus even, and so consequently, is b, say $b = 2d$. But then, $\frac{a}{b} = \frac{c}{d}$. Now this reduction to a ratio of smaller numbers cannot be repeated forever (infinite regress), or had we started with the least possibility for b we would have a contradiction (*reductio ad absurdum*).

Cast geometrically in terms of the side and the diagonal of squares within squares, this is plausibly the earliest proof, found by the "Pythagoreans" in the first deductive theory, the even vs. odd (even times even is even, odd times odd is odd, and so forth). The proof is diagrammatically suggested in Plato's *Meno* 82b-85b, and, as an example of reasoning *per impossible*, in Aristotle's *Prior Analytics* I 23.

Another proof proceeds directly on a square, the features conveyable in the diagram showing half of a square with side s and diagonal d.

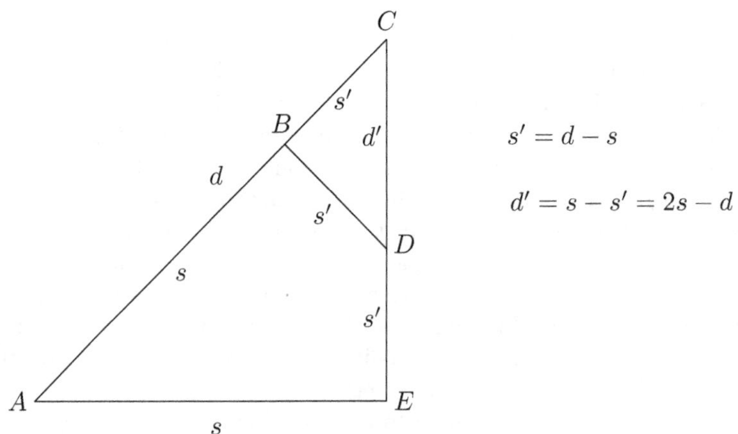

On the diagonal, a length s is laid off getting AB, and then a perpendicular BD is constructed. The three s's consequently indicate equal line segments, as can be seen using a series of what can be deduced to be isosceles triangles: the triangle ABE (formed by introducing line segment BE), the triangle BDE, and the triangle BCD. Now triangle BCD is also half of a square, with side s' and diagonal d' given in terms s and d as above. So, if s and d were commensurable, then so would be s' and d'. But this reduction cannot be repeated forever (infinite regress), or had we started out with commensurability with s and d being the smallest possible multiples of a unit, we would have a contradiction (*reductio*).

From the algebraic aspect, one sees that the components of a ratio have been made smaller:

$$\frac{d}{s} = \frac{d'}{s'} = \frac{2s-d}{d-s}, \quad \text{where } 0 < d-s < s\,.$$

This proof correlates with the Greek process of *anthyphairesis*, the Euclidean algorithm for line segments, whereby one works toward a common unit for two magnitudes by iteratively subtracting off the smaller from the larger. Because of this, the proof or something similar has been thought by some historians to be the earliest proof of incommensurability.[4] The proof appeared as a simple approach to irrationality in the secondary literature as early as in [Rademacher and Toeplitz, 1930, p.23] and recently, with the simple diagram as above, in [Apostol, 2000].

Proceeding to $\sqrt{n}$, Knorr [1975, chap.VI] worked out various diagrammatic versions of the $\sqrt{2}$ even-odd proof as possible reconstructions for Theodorus's $\sqrt{n}$ result n up to 17, and Fowler [1999, 10.3] provided various anthyphairetic proofs up to 19. The following proof of the general Theaetetus result appears to be new; at least it does not seem to appear put just so in the historical and mathematical literature.

Assume that $\frac{a}{b} = \sqrt{n}$. Laying off copies of b on a, the anthyphairetic "division algorithm", let $a = qb + r$ in algebraic terms with "quotient" q and "remainder" r where $0 < r < b$ ($r = 0$ would imply that $\frac{a}{b}$ is a number and n a square). Consider the following diagram generalizing the previous one for $\sqrt{2}$.

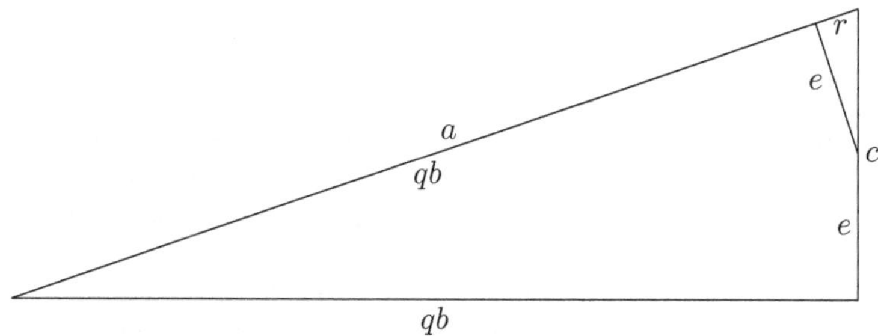

On the hypotenuse of length a, a length qb has been laid off, and so the two es are in fact the same as in the $\sqrt{2}$ case. This time, we appeal to *similarity*;

[4]See [Knorr, 1975, chap.II,sect.II] for a critical analysis. Knorr [1998] late in his life maintained, however, that a specific diagrammatic rendition of the proof was the original one.

the two triangles are seen to have pairwise the same angles, and so we have the proportion

$$\frac{a}{c} = \frac{c-e}{r}, \quad \text{or} \quad a = \frac{c^2 - ce}{r}.$$

There is a factor of b on both sides of the latter: $a = b\sqrt{n}$; $c^2 = a^2 - q^2b^2 = b^2n - q^2b^2$ by the Pythagorean Theorem; and $ce = qbr$, since $\frac{qb}{c} = \frac{e}{r}$ again by similarity. Reducing by b,

$$\sqrt{n} = \frac{bn - q^2b - qr}{r}, \quad \text{where } 0 < r < b,$$

so that the ratio $\frac{a}{b} = \sqrt{n}$ has been reduced to a ratio of smaller numbers. But as before, this reduction cannot be repeated forever (infinite regress) or had we started out with commensurability with smallest possible multiples, we would have a contradiction (*reductio*).

Since $r = a - qb$ and $qa = q^2b + qr$, one sees again that from the algebraic aspect, the components of a ratio have been made smaller:

$$(*) \qquad \sqrt{n} = \frac{bn - qa}{a - qb}, \text{ where } 0 < a - qb < b.$$

The author found this proof, and there is an initial sense of surprise in that through all the centuries there seems no record of a proof put just so. Is this creative? Novel? One should be loath to speculate in general for mathematics, but this is not an atypical episode in its progress. Perhaps there is surprise at first, but there quickly comes understanding by viewing aspects that are really there to be seen. With the $\sqrt{2}$ anthyphairetic proof given above as precedent, one is led to such a proof in order to account for the *generality* of Theaetetus's Theorem and his having been alleged to have had a proof. It could in fact have been the original proof; its use of the division algorithm and ratios is within the resources that were presumably available already to the elder Theodorus. One notices an aspect of generality emerging in context, like a face out of a picture puzzle.

Be that as it may, historians in their ruminations have attributed proofs to Theaetetus that can be drawn out from propositions in the arithmetical books VII and VIII of the *Elements*, the first having been attributed to Theaetetus himself in his efforts to rigorize his theorem.[5] Scanning these books, there are several propositions that lead readily to Theaetetus's Theorem:

According to book VII, "a *number* is a multitude composed of units" (Definition 2); "a number *is part of* [divides] a number, the less of the greater, when

[5] cf. [Knorr, 1975, chap.VII].

it measures the greater" (Definition 3); and "numbers *relatively prime* are those which are measured by a unit alone as a common measure" (Definition 12). Assuming that $\frac{a}{b} = \sqrt{n}$, so that $a^2 = nb^2$, each of the following propositions about numbers readily implies that n is a square:

(VII 27) If r and s are relatively prime, then r^2 and s^2 are relatively prime. (Assume that a and b are the least possible, so that they are relatively prime. As b^2 divides a^2, by the proposition b^2 must be the unit. Hence b must be the unit, and so n is a square.)

(VIII 14) If r^2 divides s^2, then r divides s. (It follows that b divides a and so n is a square. This proposition is not used elsewhere in the *Elements* and seems earmarked for Theaetetus's Theorem.)

(VIII 22) If r, s, t are in continued proportion (i.e. $\frac{r}{s} = \frac{s}{t}$) and r is a square, then t is a square. (Since $\frac{a^2}{nb} = \frac{nb}{n}$ and a^2 is a square, n is a square.)

Having allowed the positive conclusion that n after all could be a square, only the argument from VII 27 still depends on least choices for a and b. However, all the propositions depend on the much used VII 20, which is *about* least choices:

$$\text{If } \frac{a}{b} = \frac{c}{d}$$

and a and b are *least* possibilities for this ratio, then a divides c and b divides d. The proof of VII 20 given in the *Elements* seems patently incorrect. Here is a correct proof e.g. that b must divide d: Assume to the contrary that $d = qb + r$ with the division algorithm, where $0 < r < b$. $\frac{a}{b} = \frac{qa}{qb}$ (VII 17) and this together with $\frac{a}{b} = \frac{c}{d}$ implies $\frac{a}{b} = \frac{c-qa}{d-qb}$ (VII 12). But this contradicts the leastness assumption as $d - qb = r < b$.

VII 20 itself leads quickly to Theaetetus's Theorem:

$$\text{Assume } \frac{a}{b} = \sqrt{n}, \quad \text{so that } \frac{a}{b} = \frac{n}{\sqrt{n}} = \frac{nb}{a}.$$

Then if b is the least possibility for this ratio, then b divides a and so n is square.

These proofs of Theaetetus's Theorem drawn from the *Elements* are arithmetical and veer toward *reductio* formulations, while the "new" proof given earlier is diagrammatic and more suggestive of infinite regress. Is there, after all, a commonality of aspect? Yes, it is there to be seen, but somewhat hidden. It is seen through a simple algebraic proof of Theaetetus's Theorem using a scaling factor, which is mysterious at first:

If $\frac{a}{b} = \sqrt{n}$ and there is a number q such that $q < \sqrt{n} < q+1$, so that $qb < a < (q+1)b$, we then have the algebraic reduction

$$(**) \qquad \frac{a}{b} = \frac{a(\sqrt{n}-q)}{b(\sqrt{n}-q)} = \frac{bn-qa}{a-qb}, \quad \text{where } 0 < a-qb < b\,.$$

This q is just the q of the division algorithm $a = qb + r$ of the diagrammatic proof, and $(**)$ is a rendering of the $(*)$ after that proof. As for the arithmetical proof, the ratio reduction is there, but only as part of the *correct* proof of VII 20 given above, at the use of VII 12. These aspects of various propositions and proofs are all there to be seen in a kind of whirl, the interconnections leading to understanding.

Today, prime numbers and the Fundamental Theorem of Arithmetic, that every number has a unique factorization into prime numbers, are basic to number theory, and it is a simple exercise in counting prime factors to establish Theaetetus's Theorem. However, for over two millennia until Gauss the new simplicity afforded by the fundamental theorem was not readily attainable. There have recently been several accounts of the irrationality of $\sqrt[k]{n}$ *ab initio* that exhibit a minimum of resources though without conveying historical resonance, and these ultimately turn on $(**)$, what was there to be seen.[6]

Richard Dedekind in his 1872 *Stetigkeit und irrationale Zahlen*, the foundational essay in which he formulated the real numbers in terms of Dedekind cuts, provided (IV) what has been considered a short and interesting proof of the irrationality of $\sqrt{n}$ for non-square n: Assume that $\frac{a}{b} = \sqrt{n}$ with b the least possibility, and $q < \sqrt{n} < q+1$. Then algebraically

$$(***) \qquad (bn-qa)^2 - n(a-qb)^2 = (q^2-n)(a^2-nb^2)$$

However, $a^2 - nb^2$ is zero by assumption and so is the left side, and hence

$$\frac{bn-qa}{a-qb} = \sqrt{n} \quad \text{where } 0 < a-qb < b\,,$$

contrary to the leastness of b.

Again the scaling ratio of $(**)$ has emerged, but how had Dedekind gotten to it? During this time, Dedekind was steeped in algebraic number theory, particularly with his introduction of ideals. The *ring* $\mathbb{Z}[\sqrt{n}]$ consists of $x + y\sqrt{n}$ where x and y are integers, and the ring has a *norm* given by $N(x + y\sqrt{n}) = (x + y\sqrt{n})(x - y\sqrt{n}) = x^2 - ny^2$. The norm of a product is the product of the

[6] See [Beigel, 1991] and references therein. [Conway and Guy, 1996, p.185] conveys a proof in terms of fractional parts, which again amounts to $(**)$.

norms—Brahmagupta's identity, first discovered by the 7th century ce Indian mathematician. In these terms, $(***)$ above is just expressing

$$N((-q+\sqrt{n})\cdot(a+b\sqrt{n})) = N(-q+\sqrt{n})\cdot N(a+b\sqrt{n})\,.$$

The appearance of the scaling factor $\sqrt{n}-q$ of $(**)$ is motivated here in terms of norm reconstruing distance. Also, in this wider context of algebraic structures, it is well-known that unique factorization into "prime" elements may not hold, and so there is a *logical* reason to favor the Dedekind approach to the irrationality.

That $(**)$ emerges as a commonality in proofs of Theaetetus's Theorem is itself a notable aspect. Although indicating a proof on its own, $(**)$ remains thin and mysterious in juxtaposition with the ostensible significance of the result, both historical and mathematical. One sees more sides and angles, whether about number, discovery or proof, in the other proofs—embedded as they are in larger ways of thinking—and these aspects garner mathematical understanding of the proofs and related propositions.

3 Derivative of Sine

In this section, a basic circularity in textbook developments of calculus is brought to the fore, and this logical node is related to the ancient determination of the area of the circle. How to progressively get past this node is considered, the several ways bringing out different aspects of analysis, parametrization, and conceptualization. There is quite a lot of mathematical and historical complexity here, but this is requisite for bringing out the subtleties of aspect-perception in this case, especially of seeing something *as* something and *in* something.

The calculus of Newton and Leibniz revolutionized mathematics in the 17th century, with dramatically new methods and procedures that solved age-old problems and stimulated remarkable scientific advances. At the heart is a bifurcation into the differential calculus, which investigated instantaneous rates of change, like velocity and acceleration, and the integral calculus, which systematized total size or value, like areas and volumes. And the Fundamental Theorem of Calculus brought the two together as opposite sides of the same coin.

In modern standardized accounts, the differential calculus is developed with the notion of limit. Functions working on the real numbers are differentiated, i.e. corresponding functions, their derivatives, are determined that are to characterize their rates of change. The differentiation of the trigonometric functions is a consequential part of elementary differential calculus. The process can be reduced to determining that the derivative of the sine function is the cosine function, and this devolves, fortunately, to the determination of the derivative of the sine function evaluated at 0. This amounts to showing

$$(*) \qquad \lim_{\theta \to 0} \frac{\sin \theta}{\theta} = 1\,,$$

that the limit as θ approaches 0 of the ratio of $\sin \theta$ (the sine of θ) to θ is 1. This is the first interesting limit presented in calculus courses, bringing together angles and lengths. How is it proved?

In all textbooks of calculus save for a vanishing few, a geometric argument is invoked with the accompanying diagram.

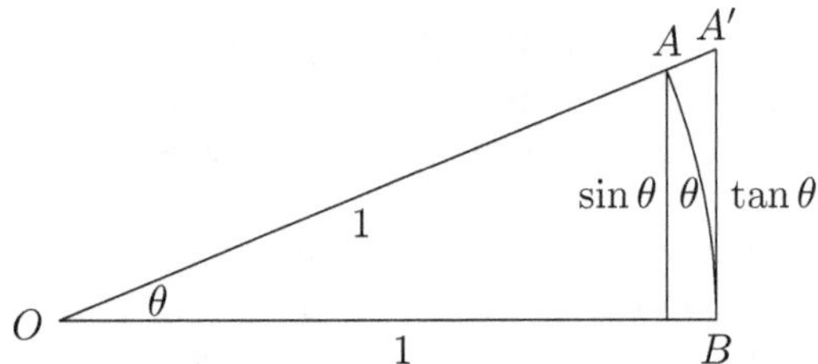

Consider the arc AB subtended by (a small) angle θ on the unit circle, the circle of radius 1, with center O. The altitude dropped from A has length $\sin \theta$, "opposite over hypotenuse", for the angle θ; the length of the circular arc AB is θ, the (radian) measure of the angle (with 2π for one complete revolution); and the length of $A'B$ is $\tan \theta$ (the tangent of θ), "opposite over adjacent". Once

$$(**) \qquad \sin \theta < \theta < \tan \theta$$

is established, pursuing the algebraic aspect and dividing through by $\sin \theta$ and then taking reciprocals yields

$$1 > \frac{\sin \theta}{\theta} > \frac{\sin \theta}{\tan \theta} = \cos \theta\,,$$

and since $\cos \theta$ (the cosine of θ), "adjacent over hypotenuse", approaches 1 as θ approaches 0, $(*)$ follows.

In geometric aspect, the first inequality of $(**)$ as a comparison of lengths is visually evident, but the second is less so. One can, however, proceed with areas: The area of triangle OAB (formed by introducing line segment AB) is $\frac{1}{2} \sin \theta$, "half the base times the height"; the area of circular sector OAB is $\frac{\theta}{2}$, since the ratio of this area to the area π of the unit circle is proportional to the ratio of θ to the circumference 2π; and $\frac{1}{2} \tan \theta$ is the area of triangle $OA'B$. With the figures subsumed one to the next, $(**)$ follows by comparison of areas.

But this is a circular argument! It relies on the area of the unit circle being π where 2π is the circumference, but the proof of this would have to entail taking a limit like $(*)$, or at least the comparison of lengths $(**)$ as in the diagram. And underlying this, what after all *is* the length of a curve, like an arc? We can elaborate on, and better see, this issue by looking at the determination of the area of a circle in Greek mathematics.[7]

Archimedes in his treatise *Measurement of a Circle* famously established that the area of a circle of radius r is equal to the area of a right triangle with sides r and the circumference. With this latter area being $\frac{1}{2} \cdot r \cdot (2\pi r)$, we today pursue the algebraic aspect and state the area as πr^2. Archimedes briefly sketched the argument in *Measurement* in terms of the right triangle; it is more directly articulated through Propositions 3-6 of his *On the Sphere and Cylinder I*:

The method was to inscribe regular n-gons (polygons with n equal sides) in the circle of radius r; to circumscribe with such; and then to take a limit as n gets larger and larger via the Eudoxan method of exhaustion. The accompanying diagram taken from *On the Sphere* pictures a sliver of the argument:

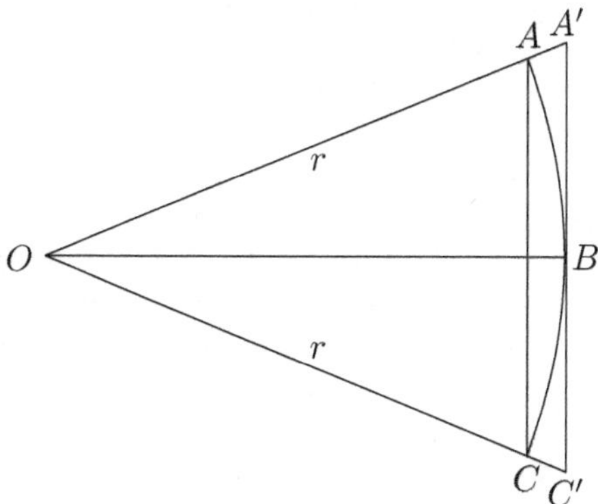

The triangle OAC is one of the n triangles making up an inscribing n-gon; the circular arc AC is $\frac{1}{n}$ of the circumference; and the triangle $OA'C'$ is one of the n triangles making up a circumscribing n-gon. This figure is just a coupling of the previous figure scaled to radius r with its mirror image, and so in (modern, radian) measure we would have:

The angle AOC is $\frac{2\pi}{n}$, and so with half of this as the angle $\theta = \frac{\pi}{n}$ of the previous figure, the line segment AC has length twice $r \sin \frac{\pi}{n}$, or $2r \sin \frac{\pi}{n}$. Similarly, the line segment $A'C'$ has length $2r \tan \frac{\pi}{n}$. Finally, the arc AC has length $\frac{2\pi r}{n}$, based on the circumference being $2\pi r$.

[7] This circularity was pointed by [Richman, 1993] and, in the context of ancient mathematics, by [Seidenberg, 1972]. Both pursue the trail in ancient mathematics at some length.

Archimedes in effect used the version

$$2r \sin \frac{\pi}{n} < \frac{2\pi r}{n} < 2r \tan \frac{\pi}{n}$$

of (**) to show, for the Eudoxan exhaustion—we would now say for the taking of a limit—that the perimeter of the inscribing and circumscribing polygons approximate the circumference from below and from above. This appeal to (**) is not itself justified in *Measure*, but the first two "assumptions" in the preface to *On the Sphere* serve: (1) The shortest distance between two points is that of the straight line connecting them (so $\sin\theta < \theta$); and (2) For two curves convex in the same direction and joining the same points, the one that contains the other has the greater length (so[8] $\theta < \tan\theta$).

With such assumptions, Archimedes had set out the conditions for how his early predecessors had constructed arc length. Archimedes' work evidently built on a pre-Euclidean tradition of geometric constructions,[9] in which an important motif had been how to *rectify* a curve, i.e. render it as a straight line. Indeed, Archimedes stated and conceptualized his area theorem as one sublimating the circumference as a straight line, the side of a triangle; he *could not* in any case have stated the area as πr^2, the Greek geometric multiplication having to do with areas of rectangles and not generally allowed for magnitudes.

These aspects of area and length are illuminated by Euclid's *Elements* XII 2: *Circles are to one another as the squares on their diameters.* This had been applied over a century before by Hippocrates of Chios in his "quadrature of the lunes", and Euclid managed a proof in his system with a paradigmatic use of the method of exhaustion that borders on Archimedes' later use. Commentators have pointed out how XII 2 falls short of getting to the actual ratio π, but in thinking through the aspects here, Euclid *could not* have gone further. In his rigorization over his predecessors, Euclid had famously restricted geometric constructions to straightedge and compass, and he had no way of rectifying a curve and so of formulating the ratio π. For Euclid, and Greek theoretical mathematics, area was an essentially simpler concept, from the point of view of proof, than length (for curves); area could be worked through congruent figures, and there were no beginning, "common notions" for length. Comparison of areas with one figure subsumed in another is simpler than comparison of length (of curves), and XII 2 epitomizes how far one can go with the first. Archimedes went further to the actual ratio π, for 20 centuries called "Archimedes' constant", but this depended on his assumptions (1) and (2).

Especially with this historical background uncovering aspects of length and area for the circle, one can arguably take (**) as logically immediate as part

[8]For this, one imagines in the first diagram a mirror image of the figure put atop it, say with a new point C corresponding to the old B. Comparing the arc CAB to the path $CA'B$ with (2), one gets $2\theta < 2\tan\theta$.

[9]cf. [Knorr, 1986].

of the concept of length. We today have a mathematical concept of rectifiable curves, a concept based on small straight chords approximating small arcs so that polygonal paths approximate curves. Seeing the concept of length from this aspect, Archimedes' assumptions (1) and (2) are more complicated in theory than (**) itself. Moreover, there is little explanatory value in proving (*), as it is presupposed in the definition of arc length.

The logical question remains whether we can avoid the theft of assuming what we want and move forward to the derivative of the sine with honest toil. There are several ways, each illuminating further aspects of how we are to take analysis and definition, from arc length to the sine function itself.

(a) Taking area as basic, *define* the measure of an angle itself in terms of the area of the subtended sector. To scale for radian measure, *define the measure of an angle to be twice the area of the sector it subtends in the unit circle.* Then by comparison of areas, $\frac{1}{2}\sin\theta < \frac{1}{2}\theta < \frac{1}{2}\tan\theta$, and (**) is immediate. With this, one could deduce à la Archimedes that the area of the unit circle is π where 2π is the circumference, and so the measure for one complete revolution is 2π, confirming that we do indeed have the standard radian measure. Defining angle measure in this way may be a pedagogical or curricular shortcoming, but the shift in logical aspect is quickly re-coordinated and moreover resonates with how the conceptualization of area is simpler than that of length.

G.H. Hardy's classic *A Course in Pure Mathematics* [1908, §§158, 217] and Tom Apostol's calculus textbook [1961, §1.38] are singular in pointing out the logical difficulty of defining the measure of an angle in terms of an unrigorized notion of arc length, and in advocating the definition of the measure of an angle in terms of area. [Hardy, 1908] advocated several approaches to the development of the trigonometric functions, two of which are conglomerated in (c) below. [Apostol, 1961] is the rare calculus textbook of recent memory that does not proceed circularly; it develops the integral calculus prior to the differential calculus, defines area as a definite integral, and only later defines length for rectifiable curves.[10]

(b) First define the length of a rectifiable curve in the usual way as an integral. Then, get the derivative of sine using methods of calculus:

Let $x = \sin\theta$ and $y = \cos\theta$, so that with Pythagorean relation $y = \sqrt{1-x^2}$,

$$\frac{dy}{dx} = \frac{-x}{\sqrt{1-x^2}}\,.$$

Anticipating the use of the length integral, note that

[10]The calculus textbook [Spivak, 2008, III.15] defines the sine and cosine functions as does [Apostol, 1961] and is not circular, but on the other hand it also (problem 27) gives the circular approach to the derivative of the sine as "traditional".

$$\sqrt{1 + (\frac{dy}{dx})^2} = \sqrt{1 + \frac{x^2}{1 - x^2}} = \frac{1}{\sqrt{1 - x^2}}\,.$$

Since, according to the parametrization, θ is the length of the arc from 0 to x,

$$\theta = \int_0^x \frac{1}{\sqrt{1 - t^2}} dt\,.$$

By the Fundamental Theorem of Calculus,

$$\frac{d\theta}{dx} = \frac{1}{\sqrt{1 - x^2}}\,,$$

so that by the Inverse Function (or Chain) Rule,

$$\frac{dx}{d\theta} = \frac{1}{\frac{d\theta}{dx}} = \sqrt{1 - x^2}\,,$$

or in terms of θ,

$$\frac{d}{d\theta} \sin\theta = \cos\theta\,.$$

One can similarly get $\frac{d}{d\theta}\cos\theta = -\sin\theta$.

This approach underscores how length can be readily comprehended with infinitesimal analysis and how the derivative of the sine being cosine can be rigorously established by a judicious ordering of the development of calculus. Importantly, the approach depends on the Fundamental Theorem, which in turn depends on the conceptualization of area as a definite integral. In this logical aspect, too, area is thus to be conceptually subsumed first. There would be a pedagogical or curricular shortcoming here as well, this time with the derivative of the sine popping out somewhat mysteriously.

The author found this non-circular proof that the derivative of the sine is the cosine, and could not find this approach in the literature. Is this a new theorem? Is it creative? Novel? Here, a logical gap was filled with familiar methods. The proof can be given as a student exercise, once a direction is set and markers laid. The task set would be to outline, with an astute ordering of the topics, a logical development of the calculus through to the trigonometric functions. If the Fundamental Theorem and the length of a rectifiable curve as an integral are put first, then the above route becomes available to the derivatives of the trigonometric functions. This logical aspect of the derivative of sine was there to be found, emerging with enough structure.

(c) Taking seriously the study of the trigonometric functions as part of *mathematical analysis*—the rigorous investigation of functions on the real and complex numbers through limits, differentiation, integration, and infinite series—*define* the sine function as an infinite series. One can follow a historical track as follows:

Let $f(x)$ by a function defined by the integral in (b):

$$f(x) = \int_0^x \frac{1}{\sqrt{1-t^2}} dt$$

Newton knew this to be, as in (b), the *inverse* of the sine function: $f(x) = \theta$ exactly when $\sin\theta = x$. He had come early on to the general binomial series, and so in his 1669 *De analysi* (cf. [Newton, 1968, pp.233ff])) he expanded the integrand $\frac{1}{\sqrt{1-t^2}}$ as an infinite series; integrated it term by term; and then, applying a key technique for inverting series term by term, determined the infinite series for sine:

$$S(x) = x - \frac{x^3}{3!} + \frac{x^5}{5!} - \frac{x^7}{7!} + \dots$$

One can, however, take this *ab initio* as simply a function to investigate. Term-by-term differentiation yields

$$C(x) = 1 - \frac{x^2}{2!} + \frac{x^4}{4!} - \frac{x^6}{6!} + \dots$$

Hence, by series manipulation, $S^2(x) + C^2(x) = 1$, which sets the stage as the Pythagorean Theorem for the unit circle. Next, define π as a parameter, the least positive x such that $S(x) = 0$.

With these, one works out that as x goes from 0 to π, the point $(S(x), C(x))$ in the coordinate plane traverses half of the unit circle and, through the length formula, that the circumference of the unit circle is 2π. Hence, taking $\sin x = S(x)$ and $\cos x = C(x)$ retrieves the familiar trigonometric functions and their properties, as well as the derivative of the sine being cosine.[11]

This approach draws out how the trigonometric functions can be developed separately and autonomously in the framework of mathematical analysis. The coordination with the classical study of the circle and its measurement then has a considerable aspectual variance: One can take the geometry of the circle as the main motivation; one can bring out interactions between the geometric and the analytic; or one can even avoid geometric "intuitions", a thematic feature of analysis into the 19th century.

[11]See the classics [Landau, 1934, chap.16] and [Knopp, 1921, §24] for details on this development.

Lest the analytic approach to the trigonometric functions still seems arcane or historically Whiggish, consider the function $g(x) = \int_0^x \frac{1}{\sqrt{1-t^4}} dt$, where the "2" of the previous integral has been replaced by "4". This "lemniscatic integral" arose as the length of the "lemniscate of Bernoulli" at the end of the 17th century, and it was the first integral which defied the Leibnizian program of finding equivalent expressions in terms of "known" functions (algebraic, trigonometric, or exponential functions and their inverses). At the end of the 18th century, the young Gauss focused on the *inverse* of the function $g(x)$ and found its crucial property of *double periodicity.* By 1827 the young Abel had also studied the inverse function, and in 1829 Jacobi wrote a treatise on the subject, from which such functions came to be known as *elliptic functions*, the integrals *elliptic integrals*, and the curves they parametrize *elliptic curves.* Thence, elliptic functions have played a large, unifying role in number theory, algebra, and geometry as they were extended into the complex plane. On that score, by 1857 Riemann had shown that the complex parametrizations are on a *torus*, a "doughnut", with double periodicity an intrinsic feature. This is how the geometric figure of the torus came to be of central import in modern mathematics—the arc of discovery going in the opposite direction from the circle to the integral $f(x) = \int_0^x \frac{1}{\sqrt{1-t^2}} dt$.

There is a broad matter of aspect to be reckoned with here, and, as a matter of fact, throughout this topic as well as the previous. Taking mathematics as based on networks of conceptual constructions, one sees through aspects various historical and logical progressions. Simply seeing a certain "face" on a topic in mathematics is to make connections with familiar contexts and modes of thinking, and this leads to the sometimes sudden dawning of logical connections. That aspects are logical thus has a further dimension in mathematics, that there are webs of logical connections. We can be provoked to seeing logical sequencings of results and themes, all there to be found. As we look and see, we can develop and reconstruct mathematics. In this, aspect-perception counsels the history of mathematics and draws forth understanding of it, that is, its truths as embedded in its proofs.

Stepping back further from the two topics presented in this article, one may surmise that many pieces of mathematics can be so presented as pictures at an exhibition of mathematics, with aspects and aspect-perception helping to get one about. Between seeing and thinking, aspect-perception is logical and can participate in the analysis of concepts. In the discussion of the irrationality of $\sqrt{n}$, we saw conceptions of number themselves at play, the diagrammatic geometry of the Greeks stimulating a remarkable advance, various arithmetical manipulations of number domesticating the irrationality, and the play of recent systematizations reinforcing a commonality. In the discussion of the derivative of the sine function, we saw a basic limit issue of calculus swirling with the ancient determination of the area of a circle, the involvement of Greeks conceptualizations of area and length, different approaches to establishing a rigorous progression, and how older concepts can be transmuted in a broad new context. Venturing some generalizing remarks, across mathematics there are many angles, faces, and views, and the noticing, continuous seeing, and dawning of many aspects. Especially in mathematics however, aspect-perception is not just about conventions or language. Rather, aspects get at objectivity from a range of perspectives, and thus collectively track and convey necessity, generality, and truth.

References

[Apostol, 1961] Tom M. Apostol. *Calculus, volume 1.* Blaisdell, New York, 1961.

[Apostol, 2000] Tom M. Apostol. Irrationality of the square root of two—a geometric proof. *The American Mathematical Monthly*, 107:841–842, 2000.

[Beigel, 1991] Richard Beigel. Irrationality without number theory. *The American Mathematical Monthly*, 98:332–335, 1991.

[Conway and Guy, 1996] John H. Conway and Richard K. Guy. *The Book of Numbers.* Springer-Verlag, New York, 1996.

[Conway and Shipman, 2013] John H. Conway and Joseph Shipman. Extreme proofs I: The irrationality of $\sqrt{2}$. *Mathematical Intelligencer*, 35(3):2–7, 2013.

[Fowler, 1999] David H. Fowler. *The Mathematics of Plato's Academy: A New Reconstruction.* Clarendon Press, Oxford, 1999. Second edition.

[Hardy, 1908] G. H. Hardy. *A Course in Pure Mathematics.* Cambridge at the University Press, Cambridge, 1908.

[Knopp, 1921] Konrad Knopp. *Theorie und Awendung der unendlichen Reihen.* Julius Springer, Berlin, 1921. Translated as *The Theory and Applications of Infinite Series*, Dover Publications, 1990.

[Knorr, 1975] Wilbur R. Knorr. *The Evolution of the Euclidean Elements.* D. Reidel Publishing Company, Dordrecht, 1975.

[Knorr, 1986] Wilbur R. Knorr. *The Ancient Tradition of Geometric Problems.* Birkhäuser, Boston, 1986.

[Knorr, 1998] Wilbur R. Knorr. "Rational diameters" and the discovery of incommensurability. *American Mathematical Monthly*, 105(5):421–429, 1998.

[Landau, 1934] Edmund Landau. *Einführung in die Differentialrechnung und Integralrechnung.* Nordhoff, Groningen, 1934. Translated as *Differential and Integral Calculus*, Chelsea Publishing Company, New York, 1951.

[Newton, 1968] Isaac Newton. De analysi per aequationes numero terminorum infinitas. In D.T. Whiteside, editor, *The Mathematical Papers of Isaac Newton, volume II 1667-1670*, pages 206–276. Cambridge University Press, Cambridge, 1968.

[Rademacher and Toeplitz, 1930] Hans Rademacher and Otto Toeplitz. *Von Zahlen und Figuren: Proben mathematischen Denkens für Liebhaber der Mathematik.* Julius Springer, Berlin, 1930. Latest version in English translation *The Enjoyment of Mathematics: Selections from Mathematics for the Amateur*, Dover, 1990.

[Richman, 1993] Fred Richman. A circular argument. *The College Mathematics Journal*, 24:160–162, 1993.

[Seidenberg, 1972] Abraham Seidenberg. On the area of a semi-circle. *Archive for History of Exact Sciences*, 9:171–211, 1972.

[Spivak, 2008] Michael Spivak. *Calculus.* Publish-or-Perish Press, Houston, 2008. Fourth edition.

[Wittgenstein, 1953] Ludwig Wittgenstein. *Philosophical Investigations.* Basil Blackwell, Oxford, 1953. Translated by G.E.M. Anscombe.

[Wittgenstein, 1958] Ludwig Wittgenstein. *The Blue and Brown Books.* Basil Blackwell, Oxford, 1958.

[Wittgenstein, 2009] Ludwig Wittgenstein. *Philosophical Investigations.* Wiley-Blackwell, Oxford, 2009. Translated by G.E.M. Anscombe, P.M.S. Hacker and Joachim Schulte.

9 Putnam's Constructivization Argument

Abstract. We revisit Putnam's constructivization argument from his *Models and Reality*, part of his model-theoretic argument against metaphysical realism. We set out how it was initially put, the commentary and criticisms, and how it can be specifically seen and cast, respecting its underlying logic and in light of Putnam's contributions to mathematical logic.

Hilary Putnam's constructivization argument, involving the axiom of constructibility $V = L$ of set theory, is at the cusp of mathematics and philosophy, being the most mathematically pronounced argument that he has put in the service of philosophical advocacy. In his shift in the mid-1970's to his internal realism, the argument appeared in his 1980 *Models and Reality* [15], as a "digression". Nonetheless, with subsequent commentary and criticisms it became considered a substantive piece of what has come to be called Putnam's "model-theoretic argument against metaphysical realism". What follows is a mainly mathematical meditation on the constructivization argument: how it was initially put, the commentary and criticisms, and how it can be specifically seen and cast respecting its underlying logic and in light of Putnam's mathematical work.

Putnam's contributions to mathematical logic—his work in recursion theory, on Hilbert's 10th Problem, on constructible reals and the ramified analytic hierarchy—are mainly from his early years. Whether mathematical results can or should be deployed to support philosophical positions at all, Putnam's subsequent deployment of model-theoretic arguments against an uncompromising realism was a novel and remarkable move.

At the outset, it should be said that we will not directly illuminate how the constructivization argument integrates with Putnam's broad philosophical stance at the time. For one thing, it argues only against a realist concept of set. Rather, we will bring out the flow of Putnam's thinking as he put his mathematical experience to work and how in its byways the constructivization argument actually worked.

In what follows, §1 reviews the constructivization argument, as presented in [15]. §2 describes the to and fro of commentary and criticisms of it in the literature. §3 takes a deeper look at the constructivization argument—the mathematical context, the inner logic, and the specific ways in which it can be taken. §4 coordinates the various criticisms, and in the process, consolidates the mathematical issues about the constructivization argument.

1 The Constructivization Argument

Putnam began his [15] with introductory remarks and then paragraphs headlined "The philosophical problem". He briefly recalled the Skolem-paradox ar-

Republished from Geoffrey Hillman and Roy T. Cook (editors), *Hilary Putnam on Logic and Mathematics*, Springer, Berlin, 2018, pp.235-247, with permission from Springer Nature.

gument about having unintended interpretations of set theory in which nondenumerable sets are "in reality" denumerable. He then specifically recalled the Downward Löwenheim-Skolem Theorem, according to which models can have countable elementary submodels. He pointed out that by the Skolem-paradox argument, "even a *formalization of total science* (if one could construct such a thing), or even a *formalization of all our beliefs* (whether they count as 'science' or not), could not rule out denumerable interpretations." With this showing that "theoretical constraints" "cannot fix the interpretation of the notion *set* in the 'intended' way", he proceeded to argue that even "operational constraints" cannot either. With the Downward Löwenheim-Skolem Theorem, "we can find a countable submodel of the 'standard model' (if there is such a thing)" that also preserves all the information the operational constraints provide. The philosophical problem that then emerges is that if axiomatic set theory does not capture the intuitive notion of set, then "understanding" might; but "understanding" cannot come to more than "*the way we use our language*"; yet even "*the total use of the language* (operational plus theoretical constraints) does not fix a unique interpretation."

In the next paragraphs of [15], headlined "An epistemological/logical digression", Putnam presented his constructivization argument, which amplifies the above argument with respect to constructibility. He briefly discussed Gödel's axiom $V = L$, that the set-theoretic universe V coincides with Gödel's universe L of constructible sets, and soon continued:

> [Gödel's] later view was that '$V = L$' is *really* false, even though it is consistent with set theory, if set theory itself is consistent.
>
> Gödel's intuition is widely shared among working set theorists. But does this 'intuition' make sense?
>
> Let MAG be a countable set of physical magnitudes which includes all magnitudes that sentient beings in this physical universe can actually measure (it certainly seems plausible that we cannot hope to measure more than a countable number of physical magnitudes). Let OP be the 'correct' assignment of values; that is, the assignment which assigns to each member of MAG the value that that magnitude actually has at each rational space-time point. Then all the information 'operational constraints' might give us (and in fact, infinitely more) is coded into OP.
>
> One technical term: an ω*-model* for a set theory is a model in which the *natural numbers* are ordered as they are 'supposed to be'; that is, the sequence of 'natural' numbers of the model is an ω-sequence.
>
> Now for a small theorem.[2] [[2] Barwise has proved the much stronger theorem that every countable model of ZF has a proper end extension which is a model of ZF + $V = L$ (in *Infinitary methods in the model theory of set*

theory, published in ***Logic Colloquium '69***). The theorem in the text was proved by me before 1963.]

THEOREM: *ZF plus $V = L$ has an ω-model which contains any given countable set of real numbers.*

Taking a countable set of reals as routinely coded by a single real, Putnam proceeded to provide an informal proof of his theorem using the Downward Löwenheim-Skolem Theorem to get a countable elementary submodel of L and then applying the Shoenfield Absoluteness Lemma. He noted in passing that "What makes [his] theorem startling" is that while a nonconstructible real cannot be in a β-model of $V = L$, it *can* be in an ω-model.

Putnam continued:

> Now, suppose we formalize *the entire language of science* within the set theory ZF + $V = L$. Any model for ZF which contains an abstract set isomorphic to OP can be extended to a model for this formalized language of science which is *standard with respect to* OP—hence, even if OP is nonconstructible 'in reality', we can find a model *for the entire language of science* which satisfies *everything is constructible* and which assigns the correct values to all the physical magnitudes in MAG at all rational space-time points.
>
> The claim Gödel makes is that '$V = L$' is false 'in reality'. But what on earth can this mean? It must mean, at the very least, that in the case just envisaged, the model we have described in which '$V = L$' holds would not be *the intended model.* But why not? It satisfies all theoretical constraints; and we have gone to great length to make sure it satisfies all operational constraints as well.

Putnam concluded this section:

> What the above argument shows is that if the 'intended interpretation' is fixed only by theoretical plus operational constraints, then if '$V \neq L$' does not follow from those theoretical constraints—if we do not *decide* to make $V = L$ true or to make $V = L$ false—then there will be 'intended models' in which $V = L$ is *true.* If I am right, then 'the relativity of set-theoretic notions' extends to a *relativity of the truth value of '$V = L$'* (and, by similar augments, of the axiom of choice and the continuum hypothesis as well).

2 Commentary and Criticisms

Putnam's advocacy of internal realism through articles starting in the later 1970s generated a philosophical literature both extensive and sundry. A focus was on his "model-theoretic argument against metaphysical realism", and eventually the most mathematically pronounced argument, the constructivization argument, itself came under sustained scrutiny in the literature, starting in the later 1990s. What is of particular interest is the extent to which a mathematically-based argument for a philosophical stance has elicited a range of responses about the mathematics and its applicability. In what follows, we review in chronological order the to and fro of commentary and criticisms.

Shapiro [17], on second-order logic and mathematical practice, briefly attended (p.724) to the constructivization argument. From a standard fact that he cites, if the set isomorphic to OP is nonconstructible then Putnam's final model of ZF $+ V = L$ containing the set would not have a (really) well-founded membership relation. But for Shapiro, "one can surely claim that the well-foundedness of the membership relation is a 'theoretical constraint' on (intended) models of set theory."

Levin [13] mounted a detailed critique of the constructivization argument, in terms of the semantics of first-order logic. On its face a response to Putnam on reference and constructibility, it seems a tissue of conflations about constants and terms and their interpretations model to model. The argument devolves to what OP is, its role, its coding as a real number, and whether that real is constructible—all this riddled with confusions and missing the thrust of Putnam's argument.

Velleman [19] reviewed Levin [13] and vetted it along the lines above. At the beginning, Velleman pointed out that Putnam's theorem (as stated in the first quoted passage of §1) cannot be provable in ZFC as it implies the consistency of ZFC. (By Gödel's Second Incompleteness Theorem, no theory, unless inconsistent, can establish its own consistency.) "[T]here must be a mistake in Putnam's proof"; the mistake is that "the Löwenheim-Skolem theorem is only applicable to sets, not to proper classes such as L"; and: "For example, the proof can be fixed by adding the hypothesis that there is an inaccessible cardinal κ, and then applying the Löwenheim-Skolem theorem to the set L_κ rather that to L."

Dümont [10] undertook a "detailed reconstruction" of Putnam's "model-theoretic argument(s)", and ultimately concluded that he "fails to give convincing arguments for rejecting mathematical or metaphysical realism". While mainly concentrating on Putnam's Skolemization argument, Dümont did attend, briefly, to the constructivization argument. Following his overall tack, he took Putnam to have failed to give a convincing answer to the realist who replies (p.348-9) to "the fact that V = L does not follow from the theoretical and operational constraints": "After all the theoretical and operational constraints have their source in *our* theoretical and empirical investigations and of course our faculties are limited. So our inability to fix one intended model only reflects our restricted access to the independently existing set-theoretical universe."

Bays [3] mounted a broad critique of the constructivization argument, both its mathematics and its philosophy. He argued first that "a key step in Putnam's argument rests on a mathematical mistake", discussing its philosophical ramifications; second, that "even if Putnam could get his mathematics to work, his argument would still fail on purely philosophical grounds", and third, that "Putnam's mathematical mistakes and his philosophical mistakes are surprisingly closely related".

As Velleman [19] had done, Bays indicated (p.366f) that Putnam's proof of his theorem (as stated in the first quoted passage of §1) is mistaken, as the Downward Löwenheim-Skolem Theorem cannot be applied to L, a proper class, and that the proof can be patched up e.g. by assuming that there is an inaccessible cardinal. Bays, however, argued (p.339f) that such patch-ups involving additional assumptions "do very little toward salvaging [Putnam's] overall philosophical argument". If in ZFC + XYZ one establishes that there is a model of ZFC + $V = L$, then XYZ would be part of the theoretical constraints yet would not hold in the model. The problem is "intrinsic" (because of Gödel's Second Incompleteness Theorem).

After criticizing Putnam's argument on philosophical grounds, Bays at the end made a connection between the mathematical and the philosophical. Putnam is not being fair to the realist, as (p.349):

> When the realist tries to 'stand back' from his set theory to talk about that theory's interpretation—to specify, for instance, that this interpretation must be transitive, or well founded, or satisfy second-order ZFC—Putnam accuses him of 'begging the question.' Although Putnam's own model-theoretic talk should be viewed as talk *about* set theory, the realist's talk must be viewed as talk *within* set theory.

Gaifman [11], on non-standard models in a broader perspective, brought up the constructivization argument, pointing out that "Putnam's proof contains a mathematical error" and that one needs an "additional assumption" to be believed by the realist. With this granted, Gaifman went on, favorably: "if s [coding OP] is not in L, the [final] model is not well-founded, but this makes no difference; we can carry out all our physical measurements, while assuming that $V = L$".

Gaifman proceeded to point out how a realist can appreciate the investigation of various structures, e.g. in which "false" propositions hold. He objected to Putnam's approach of treating "the problem as one that should be decided by appeal to general pragmatic criteria [operational constraints] and some blurry ideal of rationality [theoretical constraints]".

Bellotti [5] examined the constructivization argument and critiques thus far. Getting to Bays [3], Bellotti opined (p.404-5) that his charge that Putnam made a mathematical mistake "seems unfair, since Putnam is not clear about the theory in which he is working". *Contra* Bays, Bellotti argued that effecting Putnam's argument with an additional assumption, e.g. having an inaccessible cardinal, does not weaken Putnam's philosophical point. Such an additional assumption can be taken to be part of "our best theory of the world", and "Putnam *can* obtain a final model which satisfies the necessary assumption". On the other hand, Bellotti agreed with what Bays [3] had at the end (quoted above), that Putnam is not fair to his opponents, in that "he does not allow them what he allows himself", e.g. arbitrating what is an intended model. Following Shapiro [17], Bellotti focused on the ill-foundedness of the final model (p.408):

> ... Putnam's models for nonconstructible reals are so definitely unintended (they are not well-founded, although they 'believe' themselves to be such) to lose much of their disquieting character for any philosophical reflection on unintended models of set theory.

In a reply to Bellotti [5], Bays [4] mainly reaffirmed his [3] position. Going on at considerable length, he nuanced and finessed, one specific point *contra* Bellotti: Putnam is working and needs to work in a fixed theory. Bays newly opined (p.133f), taking account of recent criticism, that "the issues involved in the other parts of [Putnam's] argument are more fundamental", with "the big conceptual questions " being "intended" vs. "unintended" models, "standard" vs. "non-standard" models, and the role of second-order logic. At the end, Bays concluded:

> ... I think it's still important to focus some of our attention on the purely mathematical problems in Putnam's argument. It's not that they are the *only* problems in this argument or even that they're the *deepest* problems in this argument; it's that they're the problems which are most closely connected to the things which make this argument philosophically interesting."

Finally, Button [8], in an account framed by Bays' criticisms, set out the details and imperatives of the "metamathematics of Putnam's model-theoretic arguments". Button discussed at length what he took to be some related mathematics, e.g. the Completeness Theorem and weak set theories, anticipating concerns and reactions of the metaphysical realist. He did forward a simple overall line of argument, that although "Bays' challenge poses considerable problems for the constructivization argument", "it has no impact at all on the Skolemization or the permutation arguments".[1] For these two arguments, only a conditional is needed: 'if there is a model at all, there is an unintended one".

[1]The permutation argument is another model-theoretic argument from Putnam [15]; any theory with a model has multiple distinct yet isomorphic models given by permuting elements, and so there is a fundamental semantic indeterminacy.

3 Constructivization Revisited

Spurred by the commentary and criticisms, we here take a deeper look at Putnam's constructivization argument—the text, the mathematical context, the underlying logic, and the specific ways, in the end, in which it can be taken. Putnam's footnote just before his theorem (cf. in the first quoted passage of §1), though not discussed by any of the commentators, provides textual evidence, with its two items each serving as points of beginning in what follows.

How did Putnam actually conceive of and render his argument? In the footnote, he wrote that he had proved the theorem to be deployed before 1963. We look at the historical context here to get an appropriate construal of his theorem.

Putnam's 1963 [14] was a short yet seminal paper on constructible sets of integers.[2] In it, Putnam established, with ω_1^L being the least uncountable ordinal in the sense of L:

(∗) There is an ordinal $\alpha < \omega_1^L$ such that
there is no set of integers in $L_{\alpha+1} - L_\alpha$.

Gödel, of course, had established that $L_{\omega_1^L} = \bigcup_{\alpha<\omega_1^L} L_\alpha$ contains every constructible set of integers, thereby establishing the relative consistency of the Continuum Hypothesis. Putnam's theorem revealed that sets of integers are not steadily constructed up the L_α hierarchy, with his proof of (∗) actually showing that there are arbitrarily large $\alpha < \omega_1^L$ such that there is no set of integers in $L_{\alpha+1} - L_\alpha$. Putnam [14] also showed that by the Shoenfield Absoluteness Lemma, which had just appeared in Shoenfield [18], for any Δ_2^1 ordinal γ, there is an $\alpha < \omega_1^L$ such that there is no set of integers in $L_{\alpha+\gamma} - L_\alpha$.[3] This early and astute use of the Lemma is consonant with its use in Putnam's proof of his theorem for his constructivization argument.

Putnam's (∗) stimulated the dissertation work of his student Boolos on the recursion-theoretic analysis of the constructible sets of integers, this leading to their [7]. According to Jensen in his classic [12], p.230: "To my knowledge, the first to study the fine structure of L for its own sake was Hilary Put[nam] who, together with his pupil George Bool[o]s first proved some of the results in §3."

How Putnam [14] proceeded with the proof of (∗) anticipated his later constructivization argument. At the outset, he credited Cohen with the method; on his way to forcing, Cohen [9] had shown that there is a minimal ∈-model of set theory, and to do this he closed off $\{0, 1, 2, \ldots, \omega\}$ under set-theoretic operations and most crucially, under the instances of the Replacement Schema, assuming that this is possible and appealing to the Löwenheim-Skolem Theorem to get the countability of the resulting model.

[2]Sets of integers are routinely identifiable with, and called, reals, but we stick with the thematic trajectory here for a while.

[3]In fact, this holds, by a straightforward modification of his argument, for *any* $\gamma < \omega_1^L$.

Recasting this, Putnam argued for $(*)$ by initially appealing to the Downward Löwenheim-Skolem Theorem to get a countable elementary submodel of L, and proceeding to its transitive isomorph $\langle M, \in\rangle$, so that $M = L_\gamma$ for some countable γ. He then pointed out that there is no set of integers in $L_{\omega_1+1} - L_{\omega_1}$ by Gödel, and hence by elementarity that there is an ordinal $\alpha \in M$ such that there is no set of integers in $L_{\alpha+1} - L_\alpha$.

With VB (von Neumann-Bernays) being his working set theory, Putnam next pointed out: "...essentially the preceding argument can be formalized in VB. Of course, we cannot construct a model of *all* of VB in VB and also prove that it is a model." He then described formalizing argument in $\langle L_{\omega_1+2}, \in\rangle$ instead.

Finally, all this relativizes to L, and so there is an $\alpha < \omega_1^L$ such that there is no set of integers in $L_{\alpha+1} - L_\alpha$.

Now to Putnam's theorem (as stated in the first quoted passage of §1) for the constructivization argument, essentially:

$(**)$ For any real s, there is an ω-model of ZF $+ V = L$ containing s.

The first point is that this cannot be a theorem of ZF, simply because of its asserting the existence of a model of ZF and hence the consistency of ZF. This would have been clear to anyone versed in mathematical logic as Putnam certainly was, and his [14] remarks on VB above bears this out. *Putnam did not make a mathematical mistake in stating the theorem*, for surely he did not intend to state a theorem of ZF.

Proceeding to Putnam's proof of $(**)$ as given in [15], one next sees the connection to his proof of $(*)$, described above. It is quite so, as commentators have observed, that using the Downward Löwenheim-Skolem Theorem on L requires additional resources beyond ZFC. This could be said also of Cohen's [9] proof and of Putnam's [14] argument. However, one sees in these argumentations from the early 1960s that they were proceeding informally to get at the fact of the matter. Putnam [14] understood that there is the Cohen minimal model conditionally "if there is any well-founded model" (p.269), and noted that his argument for $(*)$ with the Downward Löwenheim-Skolem Theorem can be carried out in $\langle L_{\omega_1+2}, \in\rangle$, as a model of full set theory is not required.

Similarly, *Putnam's $(**)$ is a theorem of informal mathematics*, stating a fact of the matter to be accepted by the metaphysical realist. His proof, getting quickly to the crucial use of Shoenfield Absoluteness, was meant, it would seem, to provide sufficient deductive ballast to usher the realist to the truth of $(**)$. If one does insist on a ZFC theorem, then the following is appropriate for an appeal to Shoenfield Absoluteness:

(1) If for any real s there is an $\in$-model of ZF containing s, then
for any real s there is an ω-model of ZF $+ V = L$ containing s.

Given a constructible real s, there is by the hypothesis an $\in$-model M of ZF containing s and hence (by Cohen's argument!) there is such a model of form $\langle L_\gamma, \in\rangle$. Hence, the Π^1_2 statement formalizing "$\forall s \exists\, \omega$-model of ZF $+ V = L$ containing s" is satisfied in L, and the result follows by Shoenfield Absoluteness.

That (1) is a *conditional* assertion in ZFC leads to a pivotal point about Putnam's model-theoretic arguments. Both his Skolemization and constructivization arguments are rhetorically in the form of a *reductio*, and the underlying logic can be carried by the conditional: if there is a model at all, then there is an unintended one. This being said, one can see what Putnam would have had in mind for the mathematics to be invoked by looking again at his footnote just before his theorem (cf. the first quoted passage of §1).

Putnam began the footnote with "Barwise has proved the much stronger theorem", that:

(2) Every countable model of ZF has a proper end extension which is a model of ZF $+ V = L$.

If $\langle M, E\rangle$ and $\langle N, E'\rangle$ are models of ZF, then the second is an *extension* of the first if $M \subseteq N$ and the membership relation E' extends the membership relation E; moreover, it is an *end extension* if for any $a \in M$ and $b \in N$, $b\, E'\, a$ implies that $b \in M$, i.e. elements of M have no new members in N. Barwise's theorem was a culmination of both the investigation of end extensions of models of set theory and the application of infinitary logic to the construction of models of set theory. His [1] proof can be described as a complex application of the Barwise Compactness Theorem and the Shoenfield Absoluteness Lemma; the proof, rendered in the elegant terms of "admissible covers", appears as the last in his book [2]. Barwise's theorem is evidently a strong "upward" Löweinheim-Skolem Theorem, in that one gets an end extension that also satisfies $V = L$. The analogy extends to having a sort of Skolem paradox for models of set theory, with any countable model of ZF being extendible to a canonically slimmest kind of model. This thematically suggests a role in the constructivization argument.

With Putnam in the footnote writing of Barwise's theorem as "much stronger" than his, we take the tack of deploying Barwise's theorem itself, rather than Putnam's, to effect a specification of the constructivization argument:

Both the Downward Löweinheim-Skolem Theorem and Barwise's theorem are conditional theorems of ZFC. With the former, having a (set) model of ZF that contains an abstract set isomorphic to OP amounts to having a countable such model. With the latter, having a countable model of ZF having an abstract set isomorphic to OP amounts to having such a model that also satisfies $V = L$. Thus, we have a ZFC rendition of Putnam's "Any model of ZF which contains an abstract set isomorphic to OP can be extended to a model for this formalized language of science which is *standard with respect to OP*" (cf. the second quoted passage of §1). Of course, the theorems used provide a close relationship between the resulting models and the initial one.

Putnam's Skolemization argument really turns on assuming that there is some model of set theory compatible with theoretical and operational constraints, and then showing that there is a countable one. Its logical structure analogous, his constructivization argument turns on assuming that there is some model of set theory compatible with theoretical and operational constraints, and then showing that there is one satisfying $V = L$. The first implication deployed the Downward Löweinheim-Skolem Theorem, and the second can be effected with Barwise's "upward" theorem.

Putnam [15] had deployed ω-models in order to preserve the sense of OP coded as a real. For a specification of his argument just turning on ω-models, one can argue as above with the following immediate corollary of Barwise's theorem:

(3) If there is a countable $\in$-model of ZF containing a real s, then there is an ω-model of ZF + $V = L$ containing s.

With (1) schematized as $\forall s\varphi \longrightarrow \forall s\psi$, (3) is seen as the stronger version $\forall x(\varphi \longrightarrow \psi)$.

In summary, Putnam's constructivization argument was directed against a realist concept of set. An "epistemological/logical digression" as he put it, it has the rhetorical form of his Skolemization argument, that if there is a model at all, then there is an unintended one. Putnam simply pointed to a mathematical fact of the matter (**) for his argument, but if the realist insists, one can present a conditional ZFC theorem (1) to him. In fact, there are stronger ZFC results, e.g. (2) and (3), that can be invoked. The constructivization argument has various aspects, various ways of putting it and of taking it. However, its overall philosophical thrust and import would not seem to depend on its underlying mathematics. Several results and theorems can be cited or invoked, each perhaps toning the argument in different directions, but not affecting its overall philosophical arc.

4 Critical Coordination

In this final section, we coordinate various criticisms (§2) that have been made in the literature of the constructivization argument, and in the process, consolidate mathematical issues about the argument beyond what was brought out in §3. In the broad, Putnam famously attempted to use model theory, i.e. mathematics, to draw metaphysical conclusions. The particular, constructivization argument, depending on a mathematical contingency new at the time, became surprisingly pivotal in the philosophical literature decades later. Mathematics having a precision, there were specifics that could be aired and argued, and with some more confident than others about the mathematics, commentators generated a fine-grained mesh of interpretation and assessment. While adjudication is often not be the order of the day for philosophical arguments, those involving mathematical results can arguably be illuminated by seeing how they turn on or can be taken according to the mathematics. In what follows, we initially follow a simplified dialectical arc.

Bays [3, 4] has been the most persistent and uncompromising in his criticism of Putnam's constructivization argument. Caught up in the mathematics, Bays urged repeatedly that Putnam's proof of his theorem is "mistaken" and maintained that there is an "intrinsic" problem here because of the Second Incompleteness Theorem, and then that the overall argument, in rhetorically pursuing such paths, is compromised.

Taken to an extreme, if no theorems asserting the existence of models are to be allowed at all, then Putnam's argument would collapse through vacuity. This simple *reductio* could not be what Bays was pursuing; he did acknowledge, for both Putnam's Skolemization and constructivization arguments, that one is assuming first that there is a model and then getting an unintended one. However, there is an ostensible asymmetry in how Bays proceeded:

Putnam had buttressed his Skolemization argument with the Downward Löwenheim-Skolem Theorem applied to "the standard model (if there is one)" which could formally be the proper class V of all sets, to get at a fact of the matter for the realist. If one insists on a ZFC theorem, then one can appeal to the Downward Löwenheim-Skolem Theorem for sets, starting from a set model and getting a countable one. Bays acknowledged the Skolemization argument in passing, this conditional avenue to getting an unintended model from a given model presumably being operative in Putnam's argument.

For the constructivization argument, Putnam had used the Downward Löwenheim-Skolem Theorem on L to get at a fact of the matter. Bays objected to this as such and did not pursue the underlying conditionality. However, if one insists on a ZFC theorem, (1) or (2), explicit in Putnam's footnote, could have been invoked, as discussed in §3.

Putnam's constructivization argument, it would seem, has a certain sense and an overall thrust. Its components can be addressed and debated variously, with its mathematical underpinnings renderable, e.g. with (2). Bays, in focusing on Putnam's "mathematical mistake"—and moreover treating it as symptomatic of Putnam's philosophical mistakes in general—seems to have fastened onto a relatively inconsequential eddy and distorted the overall flow of argumentation.

Bellotti [5], in arguing against Bays' [3] contention that Putnam had made a mathematical mistake in ZFC, first pointed out that Putnam had not specified his working theory. Belotti then became focused on possible extensions that could serve as that theory, to be part of "our best theory of the world".

One additional assumption beyond ZFC sufficient for Putnam's theorem that Bellotti mentioned is akin to the antecedent of (1) (cf. §3). Taking the conditional (1), one can stay in ZFC as the working theory while advancing the constructivization argument rhetorically against the realist.

At the end, Bellotti [5] affirmed (p.407) his "most serious objection to Putnam", that the final models for nonconstructible reals are "definitely unintended", having an ill-founded membership relation. On this Bellotti followed Shapiro [17], who had approached the issue from the perspective of second-order logic. The set-theoretic Axiom of Foundation asserts that the membership relation is well-founded, and if one is working in second-order logic, the axiom would indeed require any model to have a (really) well-founded membership relation. *Contra* Shapiro, one sees, however, that Putnam was working in first-order logic. Indeed, his Skolemization argument would not even get off the ground in second-order logic, as the Löwenheim-Skolem Theorem would not hold. One can pursue this sort of *reductio* to vacuity of course, but it would be by changing the very ground of the argument. *Contra* Bellotti, if one stays in first-order logic but requires intended models to be well-founded, this imposition of a second-order condition still goes against the very tenor of the constructivization argument. Theoretical and operational constraints are to be seen at work inside the final model, and (real) well-foundedness is something one only sees from outside the model.

Button [8] did point out, *contra* Bays, that Putnam's Skolemization argument turned on the conditional: if there is a model at all, then there is an unintended one. Button also pointed out how the Completeness Theorem, provable in a weak set theory, can carry this conditional, while Putnam had appealed to the Downward Löwenheim-Skolem Theorem. As part of his extended analysis, he could have seenxf Putnam's footnote in his constructivization argument, from which it becomes evident how it too turns on the conditional, which can be carried by Barwise's theorem (2).

Separate from this to and fro, Gaifman [11] interestingly waded through the mathematics of the constructivization argument, landing on a different shore. After acknowledging that Putnam's theorem can be based on some additional assumption(s) to be granted by the realist, Gaifman also pointed out that whether the final model is well-founded or not makes no difference—only what holds in the model, like $V = L$, is substantive to the argument. After this however, Gaifman took basic issue with ever launching such an argument, in view of the viability of non-standard models for the realist.

Stepping back, one sees that the mathematics of Putnam's constructivization argument has been chewed over variously, with the §3 articulation very much possible to hold up as a conditional challenge to the realist. Looking past the mathematics, the commentators, including Bays, went on to address substantive issues about how further to put and take the constructivization argument and determine the extent to which it is philosophically effective. These turn mainly on possible skeptical responses and where the realist stands dialectically in relation to the argument's components and what and how its moves are to be accepted. Be that as it may, however, the mathematics does stand as an interesting and robust part of the argument that Putnam put into play.

References

[1] Jon Barwise. Infinitary methods in the model theory of set theory. In Robin O. Gandy and C.M.E. Yates, editors, *Logic Colloquium '69*, volume 61 of *Studies in Logic and the Foundations of Mathematics*, pages 53–66. North-Holland, Amsterdam, 1971.

[2] Jon Barwise. *Admissible Sets and Structures. An Approach to Definability Theory*. Springer-Verlag, Berlin, 1975.

[3] Timothy Bays. On Putnam and his models. *The Journal of Philosophy*, 98:331–350, 2001.

[4] Timothy Bays. More on Putnam's models: A reply to Bellotti. *Erkenntnis*, 67:119–135, 2007.

[5] Luca Bellotti. Putnam and constructibility. *Erkenntnis*, 62:395–409, 2005.

[6] Paul Benacerraf and Hilary Putnam, editors. *Philosophy of Mathematics*. Cambridge University Press, Cambridge, 1983. Second edition.

[7] George S. Boolos and Hilary Putnam. Degrees of unsolvability of constructible sets of integers. *The Journal of Symbolic Logic*, 33:497–513, 1968.

[8] Tim Button. The metamathematics of Putnam's model-theoretic arguments. *Erkenntnis*, 74:321–349, 2011.

[9] Paul J. Cohen. A minimal model for set theory. *Bulletin of the American Mathematical Society*, 69:537–540, 1963.

[10] Jürgen Dümont. Putnam's model-theoretic argument(s). A detailed reconstruction. *Journal for General Philosophy of Science*, 30:341–364, 1999.

[11] Haim Gaifman. Non-standard models in a broader perspective. In Ali Enayat and Roman Kossak, editors, *Non-Standard Models of Arithmetic and Set Theory*, volume 361 of *Contemporary Mathematics*, pages 1–22. American Mathematical Society, Providence, 2004.

[12] Ronald B. Jensen. The fine structure of the constructible hierarchy. *Annals of Mathematical Logic*, 4:229–308, 1972.

[13] Michael Levin. Putnam on reference and constructible sets. *British Journal for the Philosophy of Science*, 48:55–67, 1997.

[14] Hilary Putnam. A note on constructible sets of integers. *Notre Dame Journal of Formal Logic*, 4:270–273, 1963.

[15] Hilary Putnam. Models and reality. *The Journal of Symbolic Logic*, 45:464–482, 1980. Delivered as a presidential address to the Association of Symbolic Logic in 1977. Reprinted in [16, p.1-25] and [6, p.421-444].

[16] Hilary Putnam. *Realism and Reason: Philosophical Papers Volume 3.* Cambridge University Press, Cambridge, 1983.

[17] Stewart Shapiro. Second-order languages and mathematical practice. *The Journal of Symbolic Logic*, 50:714–742, 1985.

[18] Joseph R. Shoenfield. The problem of predicativity. In Yehoshua Bar-Hillel, E.I.J. Poznanski, Michael O. Rabin, and Abraham Robinson, editors, *Essays on the Foundations of Mathematics*, pages 132–139. Magnes Press, 1961.

[19] Daniel Velleman. MR1439801, Review of Levin, 'Putnam on reference and constructible sets'. *Mathematical Reviews*, 98c:1364, 1998.

10 Kreisel and Wittgenstein

Georg Kreisel (15 September 1923 – 1 March 2015) was a formidable mathematical logician during a formative period when the subject was becoming a sophisticated field at the crossing of mathematics and logic. Both with his technical sophistication for his time and his dialectical engagement with mandates, aspirations and goals, he inspired wide-ranging investigation in the metamathematics of constructivity, proof theory and generalized recursion theory. Kreisel's mathematics and interactions with colleagues and students have been memorably described in *Kreiseliana* ([Odifreddi, 1996]). At a different level of interpersonal conceptual interaction, Kreisel during his life time had extended engagement with two celebrated logicians, the mathematical Kurt Gödel and the philosophical Ludwig Wittgenstein. About Gödel, with modern mathematical logic palpably emanating from his work, Kreisel has reflected and written over a wide mathematical landscape. About Wittgenstein on the other hand, with an early personal connection established Kreisel would return as if with an anxiety of influence to their ways of thinking about logic and mathematics, ever in a sort of dialectic interplay. In what follows we draw this out through his published essays—and one letter—both to elicit aspects of influence in his own terms and to set out a picture of Kreisel's evolving thinking about logic and mathematics in comparative relief.[1]

As a conceit, we divide Kreisel's engagements with Wittgenstein into the "early", "middle", and "later" Kreisel, and account for each in successive sections. §1 has the "early" Kreisel directly interacting with Wittgenstein in the 1940s and initial work on constructive content of proofs. §2 has the "middle" Kreisel reviewing Wittgenstein's writings appearing in the 1950s. And §3 has the "later" Kreisel, returning in the 1970s and 1980s to Wittgenstein again, in the fullness of time and logical experience.

Throughout, we detected—or conceptualized—subtle forth-and-back phenomena, for which we adapt the Greek term "chiasmus", a figure of speech for a reverse return, as in the trivial "never let a kiss fool you or a fool kiss you".[2] The meaning of this term will accrue to new depth through its use in this account to refer to broader and broader reversals.

1 Early Kreisel

At the intersection of generations, Kreisel as a young man had direct interactions with Wittgenstein in his last decade of life. Kreisel matriculated at Trinity

Republished from Paul Weingartner and Hans-Peter Leeb (editors), *Kreisel's Interests: On the Foundations of Logic and Mathematics*, College Publications, London, 2020, pp.1-32.

[1]Most of the essays appear, varyingly updated and in translation, in the helpful collection [Kreisel, 1990a]. Our quotations, when in translation, are drawn from this collection.

[2]I owe the use of this term to my colleague Jeffrey Mehlman's in his remarkable [2010, §7].

College, Cambridge, where he received a B.A. in 1944 and an M.A. in 1947. In between, he was in war service as Experimental Officer for the British Admiralty 1943-46, and afterwards, he held an academic position at the University of Reading starting in 1949. According to Kreisel [1958b, p.157], "I knew Wittgenstein from 1942 to his death. We spent a lot of time together talking about the foundations of mathematics, at a stage when I had read nothing on it other than the usual *Schundliteratur*." Indeed, they again had regular conversations in 1942, when we can fairly surmise that the 18-year old Kreisel would have been impressionable and receptive about the foundations of mathematics. For 1943-45, however, their generally separate whereabouts would have precluded much engagement. During 1946-1947, after the war, they had regular discussions on the philosophy of mathematics, although Wittgenstein had not written very much on the subject for two years.[3] At that time, Kreisel wrote his first paper in mathematical logic, [Kreisel, 1950]. From 1948 on, they would only have had intermittent contact, as Wittgenstein had resigned his professorship in 1947 and Kreisel took up his academic position at Reading in 1949. By the end of 1949, Kreisel had submitted for publication his [1951] and [1952a], the first papers on his "unwinding" of proofs. Wittgenstein was diagnosed with prostate cancer in 1949 and died in 1951. In what follows, we make what we can of the "early" Kreisel of the 1942 and 1946-7 conversations, our perception refracted through his published reminiscences.

Nearly half a century afterwards, Kreisel [1989a] provided "recollections and thoughts" about his 1942 conversations with Wittgenstein.[4] Early paragraphs typify the tone (p.131):

> I was eighteen when I got to know Wittgenstein in early 1942. Since my school days I had had those interests in foundations that force themselves on beginners when they read Euclid's *Elements* (which was then still done at school in England), or later when they are introduced to the differential calculus. I spoke with my 'supervisor', the mathematician Besicovitch. He sent me to a philosophy tutor in our College (Trinity), John Wisdom, at the time one of the few disciples of Wittgenstein. Wittgenstein was just then giving a seminar on the foundations of mathematics. I attended the meetings, but found the (often described and, for my taste, bad) theatre rather comic.
>
> Quite soon Wittgenstein invited me for walks and conversations. This was not entirely odd, since in his (and my) eyes I had at least one advantage over the other participants in the seminar: I did not study philosophy. Be that as it may, in his company (*à deux*) I had what in current jargon is called an especially positive *Lebensgefühl.*

Kreisel soon went on (p.133): "One day Wittgenstein suggested that we take a look at Hardy's [*A Course of*] *Pure Mathematics* together. This introduction

[3] See [Monk, 1990, p.499].

[4] What we quote from [Kreisel, 1989a] is taken from the English translation in [Kreisel, 1990a, chap.9]. [Kreisel, 1978c] provides a shorter account of the 1942 engagements with Wittgenstein.

to differential and integral calculus was a classic at the time, and, at least in England, very highly regarded." Kreisel thence put the book in a mathematical and historical context, mentioning that Wittgenstein "had only distaste" for it—"something in the style, and perhaps also in the content, was liable to have got in the way"—and opining that the "foundational ideal" in Hardy was passé and to be supplanted by Bourbaki. Kreisel then recalled (p.136):

> In the first few conversations about Hardy's book, Wittgenstein discussed everything thoroughly and memorably. The conversations were brisk and relaxed; never more than two proofs per conversation, never more than half an hour. Then one switched to another topic. After a few conversations the joint readings came to an end, even more informally than they had begun. It was, by then, clear that one could muddle through in the same manner.

As a matter of fact, Wittgenstein in his 1932-3 "Philosophy for Mathematicians" course had already read out passages from Hardy's book and worked through many examples.[5] What Kreisel writes coheres with Wittgenstein having made annotations in 1942 to his copy of the eighth, 1941 edition of Hardy's book.[6]

Just before these remarks, Kreisel had given a telling example from the conversations (p.135): "If $y = f(x)$ is (the equation of) a curve continuous in the interval $0 \leq x \leq 1$ and such that $f(0) < 0$ and $f(1) > 0$, then f intersects the x-axis. The job was to compute, from the proof (in Hardy) a point of intersection." This of course is the Intermediate Value Theorem, the classical example of a "pure existence" assertion. In a footnote, Kreisel elaborates: "The proof runs as follows. If $f(\frac{1}{2}) = 0$, let $x_0 = \frac{1}{2}$. Otherwise, consider the interval $\frac{1}{2} \leq x \leq 1$ if $f(\frac{1}{2}) < 0$, and the interval $0 \leq x \leq \frac{1}{2}$ if $f(\frac{1}{2}) > 0$, and start again. This so-called bisection procedure determines an x_0 such that $f(x_0) = 0$."[7]

Kreisel mentioned "constructive content" and how "...in the conversations one looked for suitable additional data". He elaborated elsewhere ([1978c, p.79]):

> Wittgenstein wanted to regard this proof as a *first step*, and restrict it by saying: the proof only gives an applicable method when the relevant decision (whether $f(\frac{1}{2})$ is equal to, greater than, or less than 0) can be done effectively (e.g. if f is a polynomial with algebraic coefficients).
>
> I still find Wittgenstein's suggestion (of a certain restriction) agreeable: *satisfaisant pour l'esprit.* But it is certainly not useful (since the restriction

[5] cf. [Wittgenstein, 1979].

[6] See [Floyd and Mühlhölzer, 2019] for accounts and interpretations of these annotations.

[7] This proof is a binary version of the original [Cauchy, 1821, note III] proof of the Intermediate Value Theorem, and there is a historical resonance here. Being a pure existence assertion, the formulation and proof of the Intermediate Value Theorem by Cauchy and [Bolzano, 1817] was a significant juncture in the development of mathematical analysis. Their arguments would not be rigorous without a background theory of real numbers as later provided e.g. by [Dedekind, 1872]. The glossy Dedekind-cut proof found in Hardy (§101) is embedded in that theory, and Wittgenstein raised issues about the extensionalist point of view generally —cf. [Floyd and Mühlhölzer, 2019].

> is hardly ever satisfied). A variant ([Kreisel, 1952b]) is *much* more useful: it applies when the restriction is only approximately satisfied, i.e. when one is able to decide not necessarily at $x = \frac{1}{2}$ itself, but sufficiently close to it (e.g. in the case of recursive analytic functions on $[0, 1]$).

Wittgenstein's suggestion here—what Kreisel finds "agreeable"—is quite astute, resonant with the Intermediate Value Theorem not being intuitionistically admissible. There are continuous functions for which it is not intuitionistically possible to decide for their values whether they are equal to, greater than, or less than 0.

The forth-and-back in the quotation about Wittgenstein's agreeable suggestion and then its lack of usefulness is a local *chiasmus* of some significance. Kreisel is best known today, of course, for pioneering the study of the constructive content of proofs and the metamathematics of constructivity. In recollections ([Kreisel, 1989a, p.131]) "still exceptionally vivid, though perhaps rose-colored", he is emphasizing in self-presentation the constructive content. The 18-year old Kreisel may fairly be said to have been launched into his life-long work by these early conversations with Wittgenstein. Kreisel subsequently wrote (p.136): "After the war I had a chance to go into mathematical logic in more detail; in particular, into consistency [WF] proofs. Instead of pursuing Hilbert's aim of eliminating dubious doubts about the usual methods of mathematics a more compelling application (better: interpretation) of those proofs occurred to me. Once again, the issue was a kind of constructive content; not, however, for items in some mathematical textbook, but for all derivations in some current formal systems." This was the direction of Kreisel's initial, and incisive, work in mathematical logic published in [Kreisel, 1950], of which more below.

On Wittgenstein's side, through 1942 he was actually working as a hospital dispensary porter in London toward the end of the Blitz, coming up to Cambridge on alternate weekends to deliver lectures on the foundations of mathematics (and presumably meeting with Kreisel then).[8] During this period, he penned remarks that would be compiled into Parts IV-VII of the *Remarks on the Foundations of Mathematics.*[9] Part V of the *Remarks* has an extensive discussion of non-constructive existence proofs and Dedekind cuts—Hardy's approach to the reals.

Kreisel (p.137) went on to write that Wittgenstein lent him a copy of *The Blue Book* at the beginning of summer 1942 and that he returned it by its end. *The Blue Book* was a text that Wittgenstein had dictated for his 1933-4 "Philosophy for Mathematicians" course and of which only a few copies were maintained. In *The Blue Book* Wittgenstein first brought forth the textures of

[8] cf. [Monk, 1990, chap.21,esp. p.443], [Wittgenstein, 1993].

[9] cf. [Monk, 1990, p.438]. The part numbers given for the *Remarks* are evidently for its second edition [Wittgenstein, 1978].

meaning and language that would be elaborated in the *Philosophical Investigations*, like "language games" and their understanding through "training" toward the beginning and what to make of "I am in pain" with respect to the "I" at the end. Notably, in the face of this Kreisel only mentioned raising a "malaise" with Wittgenstein about his notion of "family resemblances of concepts". Invested in mathematics, Kreisel gave as an example the mathematical concept of group with its subcategories, mentioning a latter-day motto of his, "*relatively* few distinctions for *relatively broad* domains of experience". He could be said to have sidestepped Wittgenstein's main thrust, as exemplified by his example of "game", where various games have family resemblances but there is no property joining all instances, and the generality may be open-ended and evolving.

At this point, we record a passage from the [Monk, 1990] biography, a part of which has been passed along several times about Kreisel vis-à-vis Wittgenstein:

> In 1944—when Kreisel was still only twenty-one—Wittgenstein shocked Rhees by declaring Kreisel to be the most able philosopher he had ever met who was also a mathematician. 'More able than Ramsey?' Rhees asked. 'Ramsey?!' replied Wittgenstein. 'Ramsey was a mathematician!'

Wittgenstein was steadily drawn to mathematicians for conversation and intellectual stimulation. In the early 1940s, he would have found interaction with Kreisel in the next generation newly stimulating.

The post-war, 1946-7 conversations may have been extensive and far-ranging, but we can only make something of two published recollections of Kreisel. The first is about style, from [Kreisel, 1978b, n.2]:

> The matter of jargon, or style, came up often in my conversations with W (from 1942 to his death in 1951). For example, once after W had invited F.J. Dyson, who at the time [1946–] had rooms in College next to W's, to discuss foundations. Dyson had said he did not wish to 'discuss' anything because *what* W had to say was not different from anything everybody was saying anyway, but he wanted to hear *how* W put it. W spoke to me of the occasion, agreeing very much with what Dyson had said, but finding Dyson's jargon a bit 'odd'. On another occasion, W said: Science is O.K.; if only it weren't so grey.

This resonates with what Kreisel wrote at the end of [1989a], that "The expository style (of Wittgenstein's conversations, where 'expository' would not apply to discussions) was at any rate for me much more successful", and "Wittgenstein's favorite quotation: *Le style, c'est l'homme*". Beginning with Wittgenstein's "distaste" for the style of Hardy's book, one can venture that the young Kreisel imbibed a sensibility to "style" so construed, this later seeping into his mathematical approach and writing.

The other recollection involves consistency proofs and the unprovability of consistency. As mentioned above, from [Kreisel, 1989a, p.136] one has "After the war I had a chance to go into mathematical logic in more detail; in particular, into consistency proofs", and at that time he had done the work to be published in [Kreisel, 1950]. From that publication, we can gather that he had begun by assimilating the 1939 *Grundlagen der Mathematik II* of Hilbert and Bernays.[10] Kreisel wrote in [1983a, pp.300f] about what would have been from 1946-7:

> A few days after receiving several short, reasonable explanations of Gödel's incompleteness proofs Wittgenstein opined full of enthusiasm that Gödel must be an exceptionally original mathematician, since he deduced arithmetical theorems from such banal—meaning: metamathematical—properties like *WF* [consistency]. In Wittgenstein's opinion Gödel had discovered an absolutely new method of proof.
>
> ...
>
> What he meant was that the metamathematical interpretation (made possible by the arithmetization of metamathematical concepts) makes the relevant arithmetical theorems immediately evident. This can be compared to the geometric interpretation of algebraic formulas, such as $ax^2 + ay^2 + bx + cy + d = 0$, from which it becomes obvious that two such equations cannot have more than two common roots (x, y), since two circles can intersect in at most two points.

There is ample evidence that Wittgenstein had already become aware of some of the ins and outs of Gödel's incompleteness theorem a decade earlier in 1937, when Turing's work came out.[11] What Kreisel is drawing attention to is Wittgenstein's apprehension of a "new method of proof", the metamathematical interpretation making the relevant arithmetical theorems "immediately evident".

Kreisel is known to have lectured on "Mathematical Logic" at the Moral Sciences Club on 27 February 1947, with Wittgenstein chairing.[12] The subject was presumably on the work to be published in [Kreisel, 1950].

Kreisel in that [1950] deftly provided "constructive content" to the Gödel incompleteness theorem, first exhibiting the sensitivity to recursiveness that would be a hallmark of his subsequent work. Drawing out recursive aspects of the Hilbert-Bernays 1939 *Grundlagen der Mathematik II* proof of Gödel's second incompleteness theorem, Kreisel established, in modern terms, that the Skolemized form of Gödel-Bernays set theory has no recursive model, exhibiting as a corollary a formula of first-order logic which has a model but no recursive model. Discussing at the end the definability of predicates through diagonalization, Kreisel provided the following telling, footnote 4:

---

[10]Kreisel elsewhere in [1987, p.395] wrote of "consistency proofs (which I had learnt in 1942 from Hilbert-Bernays Vol. 2)". This may have been, especially in the sense of first acquaintance, but the tenor of various other recollections would suggest first full assimilation later.

[11]cf. [Floyd, 2001]. *Remarks on the Foundations of Mathematics* [1956], Part I, drawn from 1937 manuscripts, has Wittgenstein ruminating over Gödel's proof of the incompleteness theorem.

[12]cf. [Wittgenstein, 1993, p.355].

> A great deal has been written since Poincaré on diagonal definitions occurring in a system of definitions. A very neat way of putting the point is due to Prof. Wittgenstein:
>
> Suppose we have a sequence of rules for writing down rows of 0 and 1, suppose the pth rule, the diagonal definition, say: write 0 at the nth place (of the pth row) if and only if the nth rule tells you to write 1 (at the nth place of the nth row); and write 1 if and only if the nth rule tells you to write 0. Then, for the pth place, the pth rule says: write nothing!
>
> Similarly, suppose the qth rule says: write at the nth place what the nth rule tells you to write at the nth place of the nth row. Then for the qth place, the qth rules says: write what you write!

Kreisel is acknowledging the rule-following versions of the Gödelian contrary, as well as the Turing direct, diagonalization arguments as given by Wittgenstein in conversation. As Kreisel moved forward with his "unwinding" [1951, 1952a] for constructive content of known proofs, this marks a closure point for the formative time of his direct engagement with Wittgenstein. Significantly, Kreisel will return to this footnote in the fullness of time and with a altered perspective, as will be discussed in §3.

On Wittgenstein's side, he with a change of aspect wrote in 1947 about Turing and rules ([Wittgenstein, 1980, §1096]):

> Turing's 'machines'. These machines are humans who calculate. And one might express what he says also in the form of games. And the interesting games would be such as brought one via certain rules to nonsensical instructions. I am thinking of games like the "racing game". One has received the order "Go on the same way" when this makes no sense, say because one has got into a circle. For that order makes sense only in certain positions. (Watson.)
>
> A variant of Cantor's diagonal proof:
>
> Let $N = F(k, n)$ be the form of the law for the development of decimal fractions. N is the nth decimal place of the kth development. The diagonal law then is $N = F(n, n) =$ Def $F'(n)$. To prove that $F'(n)$ cannot be one of the rules $F(k, n)$.
>
> Assume it is the 100th. Then the formation rule of $F'(1)$ runs $F(1, 1)$, of $F'(2)$ $F(2, 2)$ etc. But the rule for the formation of the 100th place of $F'(n)$ will run $F(100, 100)$; that is, it tells us only that the hundredth place is supposed to be equal to itself, and so for $n = 100$ it is *not* a rule.
>
> [I have namely always had the feeling that the Cantor proof did two things, while appearing to do only one.]
>
> The rule of the game runs "Do the same as . . ."—and in the special case it becomes "Do the same as you are doing".

This intensional, "rule" version of Turing's undecidability argument showing that the diagonal rule cannot be among the listed rules[13] corroborates Kreisel's footnote.

[13] [Floyd, 2012] calls this "Wittgenstein's diagonal argument" and analyzes it in great detail with respect to Turing's 1936 paper.

2 Middle Kreisel

In 1953, Wittgenstein's literary executors Elizabeth Anscombe and Rush Rhees published *Philosophical Investigations* [1953], what would become Wittgenstein's main legacy, out of manuscripts intended for publication. In 1956, the executors and G.H. von Wright published *Remarks on the Foundations of Mathematics* [1956], out of sporadic, working manuscripts from 1937-1944. And in 1958, Rush Rhees published *The Blue and Brown Books* [1958], two crafted texts from 1933-1935 sparsely circulated but never intended for publication. Kreisel, well into his career publishing five papers a year in mathematical logic and having met Gödel in Princeton, took it upon himself to provide extensive reviews of both the 1956 and 1958 publications. Let us proceed to this "middle" Kreisel with respect to Wittgenstein. Beyond our focus on Kreisel, it is of interest to take account of these reviews as part of the initial reception of Wittgenstein's works, especially in light of the considerable scholarship now attendant to this corpus.

Kreisel in his review [1958b] of the *Remarks on the Foundations of Mathematics* (RFM) took the compilation as presenting Wittgenstein's philosophy of mathematics, and contributed to setting a negative tone for its interpretation for quite some time. It is to be remembered, first of all, that RFM consists of unpolished, ruminating remarks never intended for publication and exhibit an evolution of thought and focus. Something of this as well as residual positivities for Kreisel were conveyed by him at the end of his review in a "Personal Note", which reveals an anxiety of influence:

> I knew Wittgenstein from 1942 to his death. We spent a lot of time together talking about the foundations of mathematics, at a stage when I had read nothing on it other than the usual *Schundliteratur*. I realise now from this book that the topics raised were far from the center of his interest though he never let me suspect it.
>
> What remains to me of the agreeable illusions produced by the discussions of this period is, perhaps, this: every significant piece of mathematics has a solid mathematical core (p.142, 16), and if we look honestly we shall see it. That is why Hilbert-Bernays vol. II, and particularly Herbrand's theorem satisfied me: it separates out the combinatorial (quantifier-free) part of a proof (in predicate logic) which is specific to the particular case, from the 'logical' steps at the end. Certain interpretations of arithmetic and analysis have a similar appeal for me. I realise that there are other points of view, but for the branches of mathematics just mentioned, I still see the mathematical core in the combinatorial or constructive aspect of the proof.
>
> I did not enjoy reading the present book. Of course I do not know what I should have thought of it fifteen years ago; now it seems to be a surprisingly insignificant product of a sparkling mind.

Whether Kreisel was personally miffed or not, Wittgenstein scholarship has shown that Wittgenstein often did not discuss directly with students and others at the time what was at "the center of his interest". The "agreeable illusions" is chiasmatic, as Kreisel by this time had incisively pursued "the combinatorial (quantifier free) part of a proof" in [1951, 1952a] and moreover had shifted the focus of consistency proofs onto such parts in [1958a].

As to the concluding "insignificant product of a sparkling mind", this would become quoted, but evidently the "product" is the literary executors', concocted out of varying working manuscripts.

Kreisel begins his review by discussing Wittgenstein's "general philosophy" as a sophisticated empiricism sensitive to the ways of language. Kreisel considers Wittgenstein's starting point to be (p.138): "he is not prepared to use the notions of mathematical object and mathematical truth as tools in philosophy." But Kreisel does not consider as convincing Wittgenstein's arguments against them, and (p.137) "his reduction to rules of language". For Kreisel, "the real objection to these notions is that, at any rate as far as I know, there does not exist a single significant development in philosophy based on them." With this pragmatic pronouncement, he simply skirts the depths of Wittgenstein's grapplings in RFM with the objectivity of rule-following. Kreisel's only allusion to this is in a footnote (p.138):

> ... it should be noted that Wittgenstein argues against a notion of mathematical object (presumably: substance), ... but, at least in places ... not against the objectivity of mathematics, through his recognition of formal facts.

Having ferreted this out of Wittgenstein, Kreisel himself would later become known for the dictum, "the objectivity of mathematics over the existence of mathematical objects".[14]

Kreisel next gets to Wittgenstein on proof. While a large part of RFM is devoted to aspects of proof, Kreisel here focuses on proof as related to theorem and, later in the review, on the equivalence of proofs (see below). Kreisel takes up as two themes that "A theorem is a rule of language and the proof tells us how to use the rule", and "The meaning of a theorem is determined only after the proof".[15] Kreisel discusses the various ways Wittgenstein approaches these themes at some length, but then deliberately reverses proof and theorem (p.140):

> Quite generally, it is simply not true that proof is primary and theorem derived, that only the proof determines the content of a theorem. In fact, Wittgenstein is wrong in saying that generally we change our way of looking at a theorem during the proof (p.122, 30), but equally often we change our way of looking at the proof as a result of restating the theorem; ...

[14]For example, [Putnam, 1975, p.70]: "The question of realism, as Kreisel long ago put it, is the question of the objectivity of mathematics and not the question of the existence of mathematical objects."

[15]Kreisel (p.136) refers to RFM II §39 for the first and RFM II §31 and III §30 for the second.

Kreisel will maintain this in his thinking as a *chiasmus*, elaborated with examples, but one can see it as a sort of surface reversal which can be subsumed into the greater depths of Wittgenstein's thinking.

First and foremost, Wittgenstein in RFM is seeing mathematics as a multifarious edifice of procedures and conceptual constructions, one for which proofs and methods of deduction as embedded in practice are crucial. Kreisel, in flattening the situation to a dichotomy between proof and theorem, and then shifting the weight back to theorem, eschews the complexity of interplay and moreover actually reinforces the importance of argument and construction. While Wittgenstein emphasizes how a proof accrues to the meaning of a theorem both by newly delineating its interplay of concepts and by providing procedural means for its further application, Kreisel emphasizes that (p.141) "*a theorem becomes an assertion about the actual structure of its own proof*"—which while focusing on theorem is in line with Wittgenstein's thinking. Kreisel's other way of shifting from proof to theorem is to emphasize that a proof yields new theorems, e.g. about structures.[16] Again, this is in accord with Wittgenstein's thinking, according to which a proof as procedure and becoming method is autonomous and would prove perforce various theorems.

Second, Kreisel continues from the above displayed passage with (p.140):

> e.g. if we are accustomed to the principle of proof that the totality of all subsets of a set is itself a set, we may reject it when it is pointed out to us that it is only valid for the notion of a combinatorial set and not, e.g. for the notion of a set as a rule of construction.

Pursuant of this—or with it as an anticipation—Kreisel in a later, critical part of the review, on "Higher Mathematics", writes (p.153): "Wittgenstein says (p.58, 6) that it was the diagonal argument which gave sense to the assertion that the set of all sequences (of natural numbers) is not enumerable." After describing the diagonal argument and posing it as a "definition" of non-enumerability, Kreisel then wrote (p.153): "What is wrong here? Well, after all there was a paradox, Skolem's paradox, which puzzled people. The mistake is to think that the diagonal argument applies *only* to the set of all sequences ..." Kreisel's allusion to Skolem's paradox, in purported line with the above displayed quote, is a local *chiasmus* in itself—about proof, theorem, and now the set of all sequences. Contrary to what Kreisel said about the diagonal argument being applicable in only one situation, Wittgenstein on the cited page

[16] Even much later, [Kreisel, 1983a, p.297] supports this versus Wittgenstein though with an oddly drawn example: "A caveman conjectures that $a^2 - b^2 = (a+b)(a-b)$ is valid for all even integers. Of course he is right. But the proof shows that the theorem has nothing to do with the distinctions of even and odd, integer or fraction. Therefore one formulates (the more general theorem for *arbitrary commutative rings*. This notion is determined by those few properties of the even integers which enter in the proof of $a^2 = b^2 = (a+b)(a-b)$. The more general theorem is more appropriate to the proof; in short: it is more meaningful."

had ruminated about "the diagonal procedure" in its various aspects, and wrote, rather, that "it gives sense to the mathematical proposition that the number so-and-so is different from all those of the system". A few pages earlier (pp.55f), he had discussed the diagonal procedure as a method, e.g. of transcending the algebraic numbers, and had expressed skepticism about the "idea" that the real numbers are not enumerable. Kreisel's simple gloss is seen to be overshadowed by Wittgenstein's wide-ranging remarks on the diagonal procedure as proof.[17]

The rest, and most, of the review concerns the "philosophy of mathematics". Kreisel had taken as Wittgenstein's conclusion in "general philosophy" (p.137):

> He regarded the traditional aims of philosophy, in particular of crude empiricism, as unattainable. He objected to a mathematical foundation of mathematics because the concepts used in the foundation are not sufficiently different from the [mathematical] concepts described (p. 171, 13) and, he thought (p. 177) that there are no mathematical solutions to his problems. He said the aim of a philosophy of mathematics should consist in a clarification of its grammar ...

For Kreisel, (p.143) "I do not accept his conclusions since I do no think that they are fruitful for further research." Again a pragmatic pronouncement, and after rejecting on these grounds Wittgenstein's main thrust, the "clarification of its grammar" as a matter of mathematical activity, Kreisel proceeds, over several pages, to counter Wittgenstein's negativity about foundations with the fruitfulness of contemporary investigations of set theory and of constructivity. On the latter, Kreisel is discerning about the differences between intuitionism and finitism, and here he does take Wittgenstein as making contributions to finitist investigations.

Having cast light from his direction on foundations, Kreisel in the concluding pages of the review returns to proof—the focus of Wittgenstein's "foundational" concerns—as newly to be considered in the wider context. While Kreisel had earlier chiasmatically shifted the weight to the range of theorems that a proof can prove, he gets here to the range of proofs and Wittgenstein's interest in characterizing the equivalence of proofs and how they might be compared. Kreisel writes that Wittgenstein (p.151) "does attempt to find a characterization of a very general sort by basing a comparison of proofs on the application, or, as he puts it (p. 155, 46) on what I can do with it." Though he finds limitations to this, Kreisel in support raises non-constructive existence proofs and "what we can do" with them—which is allusive to his own researches along these lines and their inspiration in his early conversations with Wittgenstein. Wittgenstein in RFM had ruminated over Gödel's proof of the incompleteness

[17]Notably, Kreisel in his next review [1960], to be discussed below, went to the extent of providing a "Correction" to the present review, allowing that Wittgenstein's (p.251) "remarks can be given a little more sense if an intensional notion of function (*rule* of calculation) is considered", and then giving three viable meanings of "enumeration". This resonates with how Wittgenstein was exploring the use of the diagonal argument and [Kreisel, 1950, n.4] as discussed at the end of §1; we will return to this at the end of §3, about a *mea culpa*.

theorem, mainly about its ostensible play with truth, provability, and consistency. Taking his arguments as "wild", Kreisel strikes a positive path through Gödel's arithmetization-of-syntax argument, delineating that (p.154) "all that one needs of the concept of truth is $\mathcal{R}$ or $\neg\mathcal{R}$." Wittgenstein, generally, raised issues around consistency as a formal concept, with respect to proofs and contradictions. Kreisel insisted on the fruitfulness, writing (p.156): "proofs of consistency and, more generally, of independence yield, perhaps, a better control over a calculus than anything else."

In his review [1958b] of RFM, aside from taking up objectivity vs. objects Kreisel mainly addressed what he regarded as challenges posed by the text concerning proof and foundations of mathematics as per meaning and knowledge. Variously flattening aspects, he set out contrasting viewpoints of proof vs. theorem, of the fruitfulness of foundational investigations, and even of specifics of the diagonal argument, the incompleteness theorem, and consistency. In this, he elaborated and promoted constructive aspects of proof.

Kreisel's review [1960] of *The Blue and Brown Books* can be seen as complementary, in that the text deals more centrally with language, and so what should be addressed is set in the seas of language rather than the precisification of mathematics. It will be remembered (cf. §1) that Wittgenstein lent Kreisel a copy of *The Blue Book* in the summer of 1942. The books first advance the method that would serve to buttress the mature *Philosophical Investigations.* In brief, Wittgenstein heralds the notion of a "language game" to shed light on the foundations of logic: the method utilizes simplified snapshots of portions of human language use to clarify meaning, understanding, and thinking. For concepts and categories, there is an exploration of the limits of reductive possibility, to be seen in the plasticities of language. For Kreisel, (p.240) "...quite natural developments of Wittgenstein's considerations may be formulated as a *reduction to the concrete*; for want of a better term I shall call it semi-behaviourism (with respect to mental acts) or semi-nomimalism (with respect to abstract objects)." This encapsulated interpretation is what Kreisel will discursively discuss in the review, and at the end of the review is a telling summary (p.251):

> As to content, the ideas of the book seem to be most relevant to the discipline which studies what is concrete (and whose exact delineation is yet to be evolved). On the positive side there are descriptions of little noticed phenomena (phenomenology) and reductions to concrete terms of many situations that are in the first place viewed abstractly. As described above a wider sense of 'reduction' is appropriate than is used in crude positivism or nominalism. This work shows convincingly a natural tendency of being unnecessarily abstract. On the negative side, we have Wittgenstein's theoretical positions; on analysis, there are seen to be cogent consequences of philosophical doctrines, which, roughly speaking, overestimate what can be done in concrete terms. Since the former

> seem to be easily refuted they are used in *reductio ad absurdum* arguments applied to the latter.
>
> As an introduction to the significant problems [of] traditional philosophy the books are deplorable.[2]
>
> [2]This is largely based on a personal reaction. I believe that early contact with Wittgenstein's outlook has hindered rather than helped me to establish a fruitful perspective on philosophy as a discipline in its own right, and not merely for example as methodology of highly developed sciences. ...

The last sentence and its footnote are darker still than what Kreisel had written in that "Personal Note" at the end of the RFM review [1958b], quoted at the beginning of this section.

In the body of the review, Kreisel rounds out his contentions about Wittgenstein's "reduction to the concrete" with (p.240) "some illustrations taken from the philosophy of mathematics." At first, Kreisel is broadly affirmative about how Wittgenstein describes (p.241) "often surprisingly successfully, situations which are normally considered to involve just those mental acts and abstract objects which he eliminates." Kreisel relates this to how (p.242) "detailed investigations in the foundations of mathematics"—of which he writes tellingly in a footnote "My own in this direction have certainly been influenced by the view of Wittgenstein's work here described"—"have revealed a similar situation with respect to a nominalist (finitist, or, more generally, predicative) elimination of such abstract objects as the totality of natural numbers or of functions." Kreisel points out how for a wide class of proofs Herbrand's theorem provides "an elimination in a quite precise and natural sense" and similarly, "in a large part of analysis, quantification over all real numbers can be eliminated". Concluding about Wittgenstein's "practice of philosophy", (p.242) "Both his examples and the studies in the foundations of mathematics show clearly that *we have a general tendency to describe language* and, in particular, mathematical practice, *by means of concepts whose level of abstraction is higher than the minimum actually needed.*"

In the extended Remark following, Kreisel significantly pulls back by suggesting that what he had earlier written (p.243) "may be too logically biased and even altogether pragmatic." Instead, "we may look at these books, particularly *The Brown Book*, as a contribution to the study of *what is concrete, of what is (immediately)* given." On this he brings in (p.243) "the theoretical question of the existence of sense-data" and Wittgenstein's "seeing X as Y". With the latter, Kreisel is astute enough to bring out something that would be central to Wittgenstein's later thinking, though by calling them "phenomenological studies" he diminishes their logical import. Kreisel pronounces (pp.243f): "though even in his later book *Philosophical Investigations* these phenomenological studies have not gone far enough to establish a discipline, the later work is incomparably better in this respect than the books under review."

Proceeding, Kreisel next considers Wittgenstein's "theoretical positions", which he takes to be (p.244):

> ... (i) negative assertions on what cannot be said (or: is not), such as what is common or essential to those cases which he describes as families of concepts, (ii) assertions on what should be accepted as a decisive criterion (equality or difference in) meaning, such as the actual use of a term, (iii) the identification of metaphysical distinctions with grammatical ones.

Addressing (i), Kreisel takes Wittgenstein as objecting (p.244) "(a) generally, to the introduction of an (abstract) object common to all instances of a general term, (b) to the assumption that a general term always corresponds to a (single?—presumably: well-defined) property." Addressing these, Kreisel again resorts to mathematical illustrations. For (a), he points out that properties of rotations in the plane and multiplication of complex numbers can be commonly derived from the group axioms, and while there is a distinction in the two applications (p.245) "it's a distinction without a difference" and "the distinction is not vivid". For (b), Kreisel alludes to "mechanical procedure" à la Turing, and notes that (p.246) "It seems very natural that one is not instantaneously convinced of correct characterisations even if the arguments are good on reflection." Finally, as to what is essential to a concept, Kreisel points to the great deal of clarity gained "by the rather surprising discovery that relatively few abstract structures were essential to the proofs in the greater part of current mathematics."

By remaining in the concrete and curtailed formulations of mathematics, Kreisel is reducing away from Wittgenstein's main thrust in *The Blue Book* about the contexts and ostensible workings of language and meaning. Wittgenstein [1958, p.17], discussing "our craving for generality", pointed out "We are inclined to think that there must be something in common to all games, say, and this common property is the justification for applying the general term 'game' to the various games; whereas games form a *family* the members of which have family likenesses", these overlapping in various ways. In a different direction (p.18), of "the man who has learnt to understand a general term, say, the term 'leaf'," "We say that he sees what is in common to all these leaves; and this is true if we mean what he can on being asked tell us certain features or properties which they have in common. But we are inclined to think that the general idea of a leaf is something like a visual image, but one which only contains what is common to all leaves", there being no such visual image. Finally, Wittgenstein somewhat anticipates the analytical and reductive approach that Kreisel is taking, with (p.18):

> Philosophers constantly see the method of science before their eyes, and are irresistibly tempted to ask and answer questions in the way science does. This tendency is the real source of metaphysics, and leads the philosopher into complete darkness. I want to say here that it can never be our job to reduce anything to anything, or to explain anything. Philosophy really *is* 'purely descriptive'.

Addressing (ii) of the penultimate displayed quote, Kreisel continues to take a reductive, scientific approach (p.247): "As far as *actual use* of words is concerned," it "may refer to the words spoken" or "it may also mean the *real* role of the word (as Wittgenstein puts it) undistorted by the vagaries of linguistic expression." It will become increasingly understood that Wittgenstein generally meant, rather, the use in a broad sense in our ordinary language. More attendant to the "real role", Kreisel opines "... in the cases of the eliminations of abstract terms ... there seems no doubt about the actual use ... "But in other cases the whole problem is thrown back to what is conceived as the real role"— on this referring to his discussion of "non-constructive" in the foundations of mathematics in his RFM review.

Addressing (iii), Kreisel first recalls that in his RFM review, he (p.247) "also questioned the value of the 'reduction' of metaphysics to grammar." Here, he refers to "syntactic" and "truth under the given interpretation" in mathematical logic, and opines, "I see no evidence that the grammatical distinctions which are to replace (problematic) metaphysical ones, are going to be described by means of less problematic concepts." This *is* a valid point, especially in answer to the temptation to take schematic formalization as elucidation of the large domains of truth and language. For Kreisel: "... the reference to grammar is deceptive for two additional reasons: First, ... one does not usually consider such questions as 'what is a noun' in a theoretical way .... Second, while it is apt to speak of a grammatical role of a word in a language, the difficulty of formulating this seems to be of an entirely different order from school grammar ..." Wittgenstein's use of the term "grammar" may indeed be deceptive at first, but it will become increasingly understood that he was taking it not as some sort of syntactic classification, but rather a tying of meaning to rules, of uses of general semantic types as these are correlated with syntactic categories in utterance and use.

Stepping back, one can fairly get the feeling that Kreisel in his review did not come to terms with Wittgenstein's frontal engagement with language and meaning. The *Philosophical Investigations* had come out in 1953, and the editor of *The Blue and Brown Books* had subtitled it *Preliminary Studies for the 'Philosophical Investigations'*. Nonetheless, Kreisel insisted on pursuing a path akin to the one taken in his RFM review, of making reductive logical pronouncements and alluding to logical-mathematical examples—managing, along the way, to make positive remarks about the elimination of abstract objects e.g. through Herbrand's theorem. In the large, Wittgenstein had begun to explore the seas of language, its waves to and fro, when reduction does not work to get at meaning.

3 Later Kreisel

In the fullness of time, after having pursued and stimulated avenues of research in constructive mathematics and proof theory and having had a substantive engagement with Gödel, Kreisel in several publications came again to engage with the words and ways of Wittgenstein. Latterly meditating on these in dialectical interplay with his own work and experience, Kreisel exhibited in style and tone a new, if commemorative, acknowledgement. This "later" Kreisel we pursue through his publications in chronological order, now to the further purpose of setting out his latter-day evolving thinking about logic and mathematics.

[Kreisel, 1976a], "Der unheilvolle Einbruch der Logik in die Mathematik", appeared among a collection of essays on Wittgenstein in honor of G.H. von Wright. The title is from *Remarks on the Foundations of Mathematics* IV 24, "The disastrous invasion of logic into mathematics". Kreisel takes this up as a theme of RFM—this in itself evincing a new positivity about that work—and proceeds to articulate his own thinking along these lines in light of contemporary developments.

Kreisel at the beginning cogently summarizes his line of thought (p.166):

> The aspects (of proofs and rules) which are regarded as basic in (1) current—somewhat pretentious—*logic*, are not only *different* from those which are essential in (2) current *mathematical practice* (which almost goes without saying), but actually *harmful* for a study of (2). The reason is that those basic questions of 'principle', concerning the validity of principles of proof and definition, appear more glamorous than the genuinely useful problems concerning current mathematical practice, and thereby divert attention from the latter. The 'practice' referred to in (2) includes not only applications inside or outside mathematics, but also facts of experience concerning mathematical reasoning: which (combinatorial) configurations and (abstract) ideas we handle easily.

Then setting out toward elaboration, Kreisel instinctively retrenches (p.168): "... at least in my own case, the quotation has not been of direct, not even of heuristic use. I have known it for nearly 20 years, and stressed its plausibility in my—otherwise rather negative—review of RFM. The brutal fact is that the quotation does not contain the remotest hint of how (the pretentious) logical analysis is to be replaced, that is *which concepts should be used in the analysis of proofs in the place of the 'basic' concepts of proof theory* and which questions should be asked in place of the 'principal' problems of proof theory ..." But later, "... the *value of Wittgenstein's quotation* (for me) can perhaps be summarized as follows: It is incisive and memorable, and so makes the reader familiar with a certain aim. If sometime later this aim is approximated, the reader is likely to take a closer look instead of moving on, breathlessly, to the next 'interesting' possibility."

Focusing on proofs and rules, Kreisel begins with the stark (p.169): "Proof theory is, in my opinion, a particularly crass example of that pretentious logic which was mentioned in the summary of this article ... The claims of proof theory to have uncovered the true, in particular, formal nature of mathematical reasoning surpass in pretentiousness the claims of most traditional philosophers." This is a bit of *chiasmus*, a reversal toward Wittgenstein, in that Kreisel had himself proceeded in collaborative work in proof theory during this period with something of such "claims" as incentive.

Be that as it may, according to Kreisel, "Wittgenstein's critique of proof theory and its principal problems (for example in the *Remarks*) is wildly exaggerated, and therefore quite unconvincing." (p.170) "Worse still, Wittgenstein's own attempts to characterize what is essential in proofs aren't much better (than Hilbert's)." First, he "stresses that proofs *create*—or at least use!—*new concepts*." Yet "the brutal fact remains that, somewhere or other, *propositions concerning these new concepts have to be proved too.*" And second, he stressed that "proofs must be graspable and memorable ... and visualizable if we mean literal seeing of some spatio-temporal configuration ..." "But all this is clearly secondary, as long as there are (genuine) doubts about the *principles of proof* that are used."

On this last, Kreisel makes an autobiographical remark revealing something of influence. He had a "long hesitation before studying the idea of simplicity or 'graspability' (*Übersehrbarkeit*) of proofs" (pp.173f):

> I just wasn't confident about finding a sensible measure in any direct way. First, I tried my hand at analyzing *simplicity* of principles of proof ..., by means of socalled autonomous progressions. Granted that these attempts were pretty faithful to the intended meaning, I soon came to this conviction: if the analyses are (even only) approximately right then those intended principles are just of little intrinsic interest ... So instead I went back to more traditional questions about proofs, in particular, infinite proofs in intelligently chosen languages with infinitely long expressions, and, above all, intuitionistic logic. ... What I overlooked was the witless way in which proofs entered! No recondite properties of proofs were involved, no relations between proofs or between proofs and other objects, nothing except their 'logical' aspects which occurs to us without any experience in mathematics at all! In short, nothing but the hackneyed business: The proofs establishes its conclusion (in particular a logic-free conclusion in the intuitionistic case).

Kreisel continued (p.174): "But, at last, I had become ... convinced that questions of validity are by no means theoretically senseless ... but that they are unrewarding at the present time." "At this stage it was natural to move so to speak to an opposite extreme, in particular, opposite to Hilbert's proof theory: I went about looking for methods of proof and properties of proof which are *trivial for proof theory*, but *essential for mathematical practice* ... to be analyzed by appropriate mathematical measures of *complexity*."

On this, Kreisel gives two extended examples, the first being *explicit definitions* (p.174):[18]

[18][Kreisel, 1977] elaborates along some lines, and in particular has a longer subsuming

> ... we think of explicit definitions as *introducing* new concepts, the definition being usually supplemented by a list of properties (of the new concept), which are proved by the use of the explicit definition. As is well-known, this way of introducing a new concept is trivial for Hilbert's proof theory, because such concepts are in an obvious way *eliminable*. On the other hand, for mathematical practice they are not only useful, but as it were typical—at least for modern mathematics, which is dominated by the *axiomatic method*. This proceeds as follows. A structure is defined explicitly in set theoretic or number theoretic terms, and then is shown to be, say, a unitary group: the axioms for unitary groups then constitute the supplementary 'list of properties' (of the structure or concept) mentioned above. The choice of such properties—or, as one says—of the proper *cadre*—is often the key to solve mathematical problems.

For Kreisel, his student Richard Statman in his 1974 dissertation made (p.175) "impressive progress by means of a suitable measure of complexity which is *relevant in a large number of cases*, in particular, for analyzing the role of explicit definitions."

The second example (p.177) "concerns a more subtle 'invasion' by logic, namely a somewhat exaggerated idea of the role of socalled logical languages, for example, of predicate logic of first order", the exaggeration to be considered concerning "the ideal form of a (mathematical) proposition". On this, Kreisel focuses on real closed fields. After mentioning Sturm's work on determining the number of zeros of a polynomial in an interval and noting that *effective* decisions can be made when the coefficients are algebraic,[19] Kreisel thence brought in, of course, Tarski and the decidability of the first-order theory of real closed fields as a generalization. On this though, Kreisel opined (p.178) "The trouble began when people started to get interested in the efficiency of decision procedures ...", and "... assumed that the 'ideal form' of 'the' decision problem for real closed ordered fields should deal with all formulas of the first-order language (of fields). They found so-called *upper and lower* bounds, namely $2^{2^{cn}}$ and 2^{cn} respectively, where n is the length of the formula." (p.179) "... the most obvious conclusion from the lower bounds is simply this: the full first-order language is *not* appropriate! And one would look for a *subclass* of that language [that has] a truly efficient procedure ... ". In the contemporaneous [1976b], Kreisel made proposals along these lines, and in [1982] worked out details for an application of Herbrand's Theorem for Σ_2 formulas.

After discussing related issues in budding computer science, Kreisel wraps up with "questions of 'principle' " and (p.186):

> I find it hard to have confidence in our whole 'critical' philosophical tradition, with it paradoxes, its dramatic claims either to see profound errors in our ordinary views or profound misconceptions in 2000 year old questions. It

account (pp.120ff) of explicit definitions.

[19]Notably, this harkens back to that 1942 conversation with Wittgenstein (cf. §1) on "constructive content" of the Intermediate Value Theorem and Kreisel's effectivization in [Kreisel, 1952a].

> all sounds like a paranoid's paradise, and forgets the most striking fact of intellectual experience: how our thoughts seem to adapt themselves to the objects concerned, as we study them and get familiar with them (in a detached way) and how, with this familiarity comes the judgment need to distinguish between plausible and implausible theories, substantial and superficial contributions.

After having elaborated in his own way about "the disastrous invasion of logic into mathematics", Kreisel here seems to come around to Wittgenstein's grounding faith in familiarity and the importance of our adaptability in coming to judgment.

[Kreisel, 1978b], "Lectures on the Foundations of Mathematics", is ostensibly a review of the compilation [Wittgenstein, 1976] of lectures notes for 1939 lectures put together by Cora Diamond. As if setting the stage, Kreisel quickly sketched Wittgenstein's progress *ab initio*, mentioning that (pp.98f) "W found that quite elementary mathematics provided excellent illustrations of weaknesses of traditional foundations, t.f. for short"—this incidentally setting a contrastive, positive tone from his RFM review.[20] But then, Kreisel shifts the purpose (p.99): "The main aim of this review is to restate the complaints of W and Bourbaki about t.f., with due regard for the discoveries of mathematical logic ... By and large, at least in the reviewer's view, the discoveries of logic support the principal complaints." With this Kreisel virtually ignores [Wittgenstein, 1976], writing of it dismissively that (n.[2]) it "does not even record what W said in the lectures, but what a bunch of students thought he had said", and referring to it only on one page (p.107).

Kreisel takes the principal target of W and Bourbaki to be (p.99) "the *formal-deductive presentation* of mathematics in a *universal system*". But while "Bourbaki simply record their impression (of set-theoretic foundations)", Kreisel writes of Wittgenstein that (pp.99f) "... W attempts to convert fundamentalists by 'deflating' the notions and thus the so-called fundamental problems of t.f., stated in terms of those notions. In W's words, he wants to *show the fly the way out of the fly bottle.* He does this with much ingenuity and patience, and some overkill."

Proceeding to "complaints", Kreisel gets to (p.101): "... the general complaint (of W and Bourbaki) is that t.f. may be *poor philosophy*, in the broader popular sense of 'philosophy', specifically, if in practice the general aims of foundations are better served by alternatives, for example, by ordinary careful scientific research and exposition." Taking as "principal complaint: better current ideas than t.f.", Kreisel discusses how both W and Bourbaki emphasize that "the *choice of explicit definitions* is incomparably more significant than the glamorous preoccupations of t.f., not only for discovery, a 'mathematical' affair,

[20] cf. §1.

but also for intelligibility, a principal factor in reliability."[21] Finally, Kreisel addresses (p.102) "specific complaints about some glamor issues of t.f.". The first is "the matter of *contradictions* as in the paradoxes, or their absence, consistency, as in Hilbert's program." "W had a particularly strong aversion", whereas "at least by implication, Bourbaki was unimpressed". The second example is "higher (infinite) cardinals". As in his RFM review, Kreisel connects this to the diagonal construction, but now mentions favorably how Wittgenstein "preferred to use the construction in the context of rules", recalling Wittgenstein's formulation as given in [Kreisel, 1950, n.4].[22]

Throughout Kreisel's discussion of "complaints", there is in contrast to his RFM review a softer attitude toward Wittgenstein. This continues into Kreisel's acknowlegement (p.103) of "W's advice"—what mainly he draws from the book ostensibly under review—that "when confronted ... by a philosophical problem about (mathematical) notions or proofs, we should see what we *do* with them, how we *use* them."[23] Kreisel concludes his "review" by "balancing the account on the positive side of t.f." He opined that to Wittgenstein the weaknesses of t.f. mattered less than the (pretentious) style, but proceeded to set out several examples—two from Gödel—for how such stylistic urgings may signal possibilities for progress.

[Kreisel, 1978a], "The motto of 'Philosophical Investigations' and the philosophy of proofs and rules," ostensibly takes up that motto, "All progress looks bigger than it is" interpreted as (p.13) "the ratio of *actual progress* (as judged by mature reflection) to *apparent progress* (measured by expectations after a few initial successes) is generally poor." With this as underlying thrust, Kreisel proceeds to elaborate on Wittgenstein's "family resemblances of concepts" and "principal pedagogic aim for philosophy", and discusses, in an extended appendix, "proofs and rules" to draw in recent logical experience. With this, Kreisel hovers closest to Wittgenstein's major work, *Philosophical Investigations.*

Kreisel starts by laying some groundwork about (p.15) "General features of traditional philosophy, and some of their implications": (a) "... traditional notions occur to us when we know very little." "... when we know very little, we tend to see superficial, abstract features of objects. And when we do see specific features we often cannot say very well—cannot 'define' in familiar terms—what we see." (b) "When we know very little, the main intellectual tools available are

[21] [Kreisel, 1976a] elaborated on explicit definitions, as described in our account of it above. Significantly, Kreisel there wrote (p.175): "As far as I know, Wittgenstein himself never stressed the role of explicit definitions particularly." Now, he is accrediting to Wittgenstein how he stressed the choice of explicit definitions.

[22] cf. end of §1.

[23] This recalls Wittgenstein's attempt to compare proofs according to "what we can do with them" in RFM, as already discussed by Kreisel in his RFM review (cf. §1).

a sense of coherence and, more generally, introspection." (c) "When we know very little compared to the scope of a question, we are often bad at guessing even remotely the methods needed for a satisfactory answer though often we recognize such an answer immediately when we see one." Kreisel peppers (b) and (c) with historical examples involving Galileo, Plato, Aristotle, Newton, and Cauchy.

Kreisel then focuses on (p.17) "Family resemblances of concepts", and, in connection, "the discovery of definitions". It will be remembered[24] that in 1942 Wittgenstein lent young Kreisel a copy of *The Blue Book*, and that, remarkably, Kreisel only reacted about family resemblances, and that with reference to the group concept. Here too, of the many themes of *Philosophical Investigations*, Kreisel concentrates on family resemblances, now with a remarkably literal twist (p.19): "As I see it (now), Wittgenstein's slogan of 'family resemblances' reminds one of a class of phenomena where the limitations of the traditional style are exceptionally vivid, and hence instructive. I mean the phenomena of *literal* family resemblances, say of the Hapsburgs or the Bourbons [sic]. What can we realistically expect from any definition of such a family resemblance, say, in the style of analytical philosophy?" This focus on literal family resemblances amounts to a local *chiasmus* moving in reverse to Wittgenstein's conceptualization of aspectual similarities and analogues. The tenor of *Philosophical Investigations* is to pursue aspects and work against definitions of family resemblances in terms of biological causes or necessary and sufficient conditions. Be that as it may, Kreisel proceeds to several points (p.19):

(a) "The first thing to expect is, probably, a *genuine theory* of literal family resemblances or some kind of practical mastery. As appears almost certain now, molecular biology is the appropriate tool here." With this scientific coordination of a question emerging "when we know very little", and especially with Kreisel soon following up with "We cannot expect to find a common element in ordinary experience", one can see Kreisel as proceeding orthogonally to Wittgenstein by looking for a genetic reduction. (b) "A second use to expect from a definition would be for the study of our actual *process of recognizing a family resemblance.* At least here, Kreisel is sensing the importance of what Wittgenstein wrote of as "seeing as" and "the dawning of an aspect". (c) (p.21) "An imaginative (clever) definition in this style, in terms of familiar things, may well be useful for *stimulating*—not the actual process of recognition of family of resemblances, but some of its useful results." Again taking a scientific approach, Kreisel mentions as an example of this kind of stimulation "*logical validity* in terms of *derivability*, say by Frege's rules".

Lastly, Kreisel attends to what he terms (p.22) "intimate pedagogy", what he took to be (p.14) "Wittgenstein's principal pedagogic aim for philosophy". Kreisel takes a particular tack (p.22): "Suppose we have come to the conclusion

[24]cf. §1.

that some given notion, for example, one of those grand traditional notions, has to do with a family resemblance .... Of course, we do not assume that such a conclusion, even if sound, can be conveyed convincingly, especially to individuals with very limited experience. We ask the pedagogic question: What can be done?" Emphasizing the need for discretion—not to make "grand" claims—and that "precise formalization" can be instructive, Kreisel proceeds to two examples, Tarski's truth definitions and Gödel's incompleteness theorem. Of the latter (p.24) "It unquestionably refutes the idea that, in mathematics, abstract notions are merely used as a *façon de parler*. Hilbert expressed this idea explicitly and precisely in his consistency programme. A more direct formulation of the idea, which is equally easy to make precise, is that a proof by use of abstract notions of a theorem stated in elementary form, can be straightforwardly converted into an elementary proof." Incidentally, Kreisel soon wrote revealingly (p.25): "*Digression* for readers who have seen my (constipated and fumbling) review in [[Kreisel, 1958b]] of *Remarks on the Foundations of Mathematics*. To me the single most disturbing (and most surprising) defect of those *Remarks* was and remains Wittgenstein's own fumbling."

As in his other articles concerning Wittgenstein, Kreisel insists on taking a scientific approach, and here, in an extended appendix, he further focuses on logic and mathematics to draw subtle distinctions about proofs and rules that round out his contentions. In particular, he has (p.27) "the novel twist of using notions from Brouwer's intuitionist foundations to examine a natural analogue of Church's thesis." While having taken on the motto from the *Philosophical Investigations*, Kreisel interestingly and chiasmatically proceeds away from its broad concerns towards the nuances of progress in logic and mathematics.

[Kreisel, 1983a], "Einige Erläuterungen zu Wittgensteins Kummer mit Hilbert und Gödel", starts out "I was very astonished by the *Remarks on the Foundations of Mathematics* when they came out, especially by those on Gödel's incompleteness theorems, for reasons that I can state precisely only now ..." The article, in a recapitulative way, engages Wittgenstein's views on consistency and incompleteness with a palatably seasoned appreciation.

Initially, Kreisel adopts and adapts Wittgenstein's (p.296) "proofs easy to take in and remember". In RFM, Wittgenstein had importantly discussed how mathematical proofs are to be "easy to take in and remember [*überschaubar und einprägsam*]" and "perspicuous [*übersichtlich*]". Kreisel declares that "... one of the main concerns of mathematics is to provide general guidelines for proofs to be easy to take in and remember." The guidelines are for what he analyzes into two parts as follows: "For usually one starts from a long, opaque proof and dissects it—with intuition—into a few lemmas, that is to say into a structure *easy to take in*. In this process one tries to formulate (or, if necessary to reformulate) the lemmas in such a way that the properties used in their proofs are easily assimilated by the memory, so that they are *easy to remember*." Kreisel frames this with elements from [Kreisel, 1976a] (discussed above), especially the appeal to properties that occur frequently and their axiomatic analysis for perspicuity. He then sets out (p.297): "Now we are ready to apply some of Wittgenstein's favorite slogans to the axiomatic analysis of proofs, e.g. the relatively original: the proof constructs (i.e., in the proof one discovers) new concepts, or the very popular one around 1930: only the proof gives meaning to the theorem that it proves." (It will be remembered that Kreisel in his [1958b, pp.140f] review of RFM had worked chiasmatically against these slogans.) Kreisel proceeds to give "two (entirely elementary) examples",[25] these evidently in the spirit of the elementary RFM examples.

Proceeding to Hilbert's program and consistency, Kreisel declaims (p.298): "Like many others around 1930, Wittgenstein was decidedly enthusiastic about the main component of Hilbert's program: formalization." Yet on two points Wittgenstein was critical: "Firstly, ... he thought it not fruitful to consider *all* calculations of a 'calculus. Put differently: formal provability (even by limited means) without regard to ease to take in and remember seemed to him a bad idealization." "Secondly, he was disturbed by Hilbert's exaggerated claims for the importance of consistency." On this last, Kreisel puts Wittgenstein in company with Brouwer and Russell as also "very critical", and mentions Gödel and Gentzen's criticism that "consistency at best guarantees the validity of universal theorems ..., whereas in practice one is rather more interested in existence theorems."

Considering next the shift from provability to proofs, Kreisel writes (p.300): "Since completeness and incompleteness only relate to provability, and have nothing to do with the structure of proofs, they lose their central role." But then, "What happens to incompleteness *proofs* when incompleteness itself loses its 'fundamental' significance? A normal person remembers the good advice: we have nothing to fear but fear itself. In other words, such proofs have more meaningful consequences ...." On this Kreisel relates an anecdote from the 1940s, the last of his quoted in §1, with how, with the incompleteness proofs, "In Wittgenstein's opinion, Gödel had discovered an absolutely new method of proof."

Kreisel ends with "Wittgenstein's expectations" (pp.301f):

> Above all the *Remarks* were meant to stimulate the reader to have his own thoughts; especially those readers who had already come close to Wittgenstein's thoughts. ...
>
> This expectation was confirmed by my own experience. When they came out, the *Remarks* did not help me at all. Since the end of the sixties I myself had started to consider structural properties of proofs. After a lecture in 1973 in which I presented these ideas and their development (also by Statman), Nagel drew my attention to the fact that these tendencies (certainly not the details)

[25]The first was given in footnote 16.

> reminded him of Wittgenstein's *Remarks.* I was absolutely unaware of this connection before then. But I am entirely aware of the additional confidence in my own thoughts that I derived afterwards from leafing through, e.g., Wittgenstein's *Zettel.* Added to this was a certain pleasure, at his skillful formulations and my reformulations of his less skillful ones.

That said, Kreisel retrenches with softer versions of criticisms from his RFM review: how Wittgenstein's specific examples were not fruitful for Kreisel; how he has no use for Wittgenstein's "fussing with *clarity* and *clarification*"; and Wittgenstein's "often erroneous contraposition of *clarification of existing knowledge* and *new constructions*". Nonetheless, throughout [Kreisel, 1983a] there is steady, serious engagement with Wittgensteinian incentives in RFM.

Kreisel's last articles concerning Wittgenstein are variously elliptical, reminiscent, or outright expressionistic. [Kreisel, 1983b] is a quick review of Kripke's 1982 book, *Wittgenstein: On Rules and Private Language*, a review that amounts to a series of chiasmatic remarks putting things in a series of different nutshells. [Kreisel, 1989a] is a collection of "recollections and thoughts" about conversations with Wittgenstein, from which we have already drawn in §1. And [Kreisel, 1989b], *Zu Wittgensteins Sensibilität*, written for a festschrift, is a remarkably expressionistic series of wide-ranging aphorisms, quips, repartees, and things that came to mind—but nevertheless an article that fully affirms Kreisel's deep engagement with Wittgenstein.

As a way of affirming and accentuating an overall *chiasmus* for Kreisel, his eventual reversal in attitude about RFM after his negative review [Kreisel, 1958b], we consider passages from a appendix to a long letter [1990b] that Kreisel wrote to Grigori Mints in 1990. First, from p.24:

> In a sense I might be said to have made fun of Wittgenstein in a review I wrote in the 50s of his *Remarks on the foundations of mathematics* (although this description certainly does not fit the way I felt about that volume nor about the review). I had made a mistake, which I noticed some 20 years later,* and have referred to it many times. But let me repeat it here, since you may not have taken in these references.
>
> *Main Mistake.* I did not look at the preface, where the editors say in the clearest possible terms that they had found a box full of notes by Wittgenstein, and that *they* had selected what, to *them, seemed most extraordinary.* N.B. I knew those editors! So, if I had looked at the preface this passage would have been an *immediate warning*: what is most extraordinary (=remarkable) to *them* was almost bound to be either wrong [?] or even an aberration.

With the * he references [Kreisel, 1979], a brief review of the second edition [Wittgenstein, 1978] of RFM that does not mention any mistake. In the preface to RFM [Wittgenstein, 1956], the editors nowhere state that they "had found a box full of notes", but do state (p.viie) "...what is here published is a *selection* from more extensive manuscripts".

Later on, Kreisel wrote (pp.25f):

> *Consequences of the main mistake.* Actually, in the last paragraph of the review (in the 50s) I said explicitly that I simply did not recognize in the *Remarks on the foundations of mathematics* what I had remembered minimally from my conversations with Wittgenstein. Fittingly (at least from my view of the world), I ignored what I remembered of Wittgenstein, and read the volume as a foil to my then current interests, mentioned above: What, if anything, does it say that is in conflict with—tacitly, the mere coherence of—the foundational tradition?
>
> *STAGGERING OVERSIGHT* on my part. I myself had put on record (in [[Kreisel, 1950]], somewhere in a highly visible footnote)—published during Wittgenstein's lifetime!—Wittgenstein's perfectly good understanding of Gödel's incompleteness theorem; tacitly, in the mid 40s, after I had explained it to him in $< \frac{1}{2}$ hour in WORDS CONGENIAL TO HIM. In accordance with his habit he recorded the explanation in his own words, incidentally stumbling thereby on—what later came to be called—Henkin's problem.

Henkin's problem is [Henkin, 1952]: "If Σ is any formal system adequate for recursive number theory, a formula (having a certain integer q as its Gödel number) can be constructed which expresses the proposition that the formula with Gödel number q is provable in Σ. Is this formula provable or independent of Σ?" [Kreisel, 1953] discussed an approach to this problem, and then [Löb, 1955] established provability for more general formulas and under minimal conditions on Σ, the result now known, of course, as Löb's Theorem.[26]

The overall *chiasmus* working its way through the previous and the current sections is *first*, the critical attitude Kreisel took to the *Remarks on the Foundations of Mathematics* in his review [Kreisel, 1958b] as particularly seen in his negative remarks about Wittgenstein's purported construals of the diagonal argument and the first incompleteness theorem, and *second*, a gradual working back, as traced in this section, to a nuanced assessment and appreciation in his articles in the 1970s and 1980s, particularly in [Kreisel, 1983a], ostensibly in tandem with the evolution of his own thinking and experience. The letter cements the *chiasmus* further, working various angles of mistakes and oversights. In particular, Kreisel not only records the *mea culpa* of his not recalling having mentioned Wittgenstein's rule-following version of the diagonal argument in first logic paper [Kreisel, 1950],[27] but credits Wittgenstein for actually formulating Henkin's problem, a problem he himself later worked on.

Stepping back and taking it all in from the beginning, one sees the "early" Kreisel as stirred to his lifelong engagement with constructivity and proof by conversations with Wittgenstein, particularly with the "combinatorial core" of consistency proofs. One sees the "middle" Kreisel with an anxiety of influence reacting negatively in his reviews of Wittgenstein publications, flattening his

[26]With *Bew* the provability predicate and $\ulcorner\varphi\urcorner$ the Gödel number of φ, Löb's Theorem asserts that for adequate Σ, if $\Sigma \vdash Bew(\ulcorner\varphi\urcorner) \longrightarrow \varphi$, then $\Sigma \vdash \varphi$. Henkin had asked whether $\Sigma \vdash \varphi$ or not in the special, fixed-point case when *is* φ.

[27]cf. end of §1.

work on language and insisting on the fruitfulness of research into constructivity and even set theory. Finally, one sees the "later" Kreisel in published essays interestingly integrating his latter-day, seasoned outlook on logic and mathematics with remembrances of the words and ways of Wittgenstein. Proceeding in dialectical engagement, Kreisel growingly acknowledges Wittgenstein as at least providing a conceptual context. But while aspiring to encompass Wittgenstein's broad ways of thinking about language, Kreisel would ultimately remain within the compass of logic and mathematics as set out by RFM. There, the engagement was enlivened by an appreciation of mathematical practice as the place to look for the important structural properties of proof; the role of explicit definitions in that regard; and the importance of proofs "easy to take in and remember". This rounds an arrow of influence, those early conversations with Wittgenstein having stimulated Kreisel to constructivity, logic, and proof.

References

[Bolzano, 1817] Bernard Bolzano. *Rein analytischer Beweis der Lehrsatzes, daß zwischen je zwey Werthen, die ein entgegengesetztes Resultat gewähren, wenigstens eine reelle Wurzel der Gleichung liege.* Gottliebe Hasse, Prague, 1817. Translation by Steve Russ in [Ewald, 1996], pages 225-248.

[Cauchy, 1821] Augustin-Louis Cauchy. *Cours d'analyse de l'École royale polytechnique.* Imprimérie royale, Paris, 1821.

[Dedekind, 1872] Richard Dedekind. *Stetigkeit und irrationale Zahlen.* F. Vieweg, Braunschweig, 1872. Translated in [Ewald, 1996], pages 765-779.

[Ewald, 1996] William Ewald. *From Kant to Hilbert.* Clarendon Press, Oxford, 1996.

[Floyd and Mühlhölzer, 2019] Juliet Floyd and Felix Mühlhölzer. Wittgenstein's Annotations to Hardy's *A Course in Pure Mathematics*, 2019. To appear.

[Floyd, 2001] Juliet Floyd. Prose versus proof: Wittgenstein on Gödel, Tarski and truth. *Philosophia Mathematica*, 9:280–307, 2001.

[Floyd, 2012] Juliet Floyd. Wittgenstein's diagonal argument: A variation on Cantor and Turing. In *Epistemology versus Ontology*, pages 24–55. Springer, Dordrecht, 2012.

[Henkin, 1952] Leon Henkin. A problem concerning provability. *The Journal of Symbolic Logic*, 17:160, 1952.

[Kreisel, 1950] Georg Kreisel. Note on arithmetic models for consistent formulae of the predicate calculus. *Fundamenta Mathematicae*, 37:265–285, 1950.

[Kreisel, 1951] Georg Kreisel. On the interpretation of non-finitist proofs—part I. *The Journal of Symbolic Logic*, 16:265–285, 1951.

[Kreisel, 1952a] Georg Kreisel. On the interpretation of non-finitist proofs. Part II. Interpretations of number theory. Applications. *The Journal of Symbolic Logic*, 17:43–58, 1952.

[Kreisel, 1952b] Georg Kreisel. Some elementary inequalities. *Indagationes Mathematicae*, 14:334–338, 1952.

[Kreisel, 1953] Georg Kreisel. On a problem of Henkin's. *Proceedings of the Koninklijke Nederlandse Akademie van Wetenschappen, series A*, 56:405–406, 1953.

[Kreisel, 1958a] Georg Kreisel. Mathematical significance of consistency proofs. *The Journal of Symbolic Logic*, 23:255–182, 1958.

[Kreisel, 1958b] Georg Kreisel. Wittgenstein's *Remarks on the Foundations of Mathematics. British Journal for the Philosophy of Science*, 9:135–158, 1958.

[Kreisel, 1960] Georg Kreisel. Wittgenstein's theory and practice of philosophy. *British Journal for the Philosophy of Science*, 11:238–251, 1960. Mainly a review of *The Blue and Brown Books.*

[Kreisel, 1976a] Georg Kreisel. "Der unheilvolle Einbruch der Logik in die Mathematik". *Acta Philosophica Fennica*, 28:166–187, 1976.

[Kreisel, 1976b] Georg Kreisel. What have we learnt from Hilbert's second problem? In Felix E. Browder, editor, *Mathematical Developments arising from Hilbert problems*, volume 28 of *Proceedings of Symposia in Pure Mathematics*, pages 93–130, Providence, 1976. American Mathematical Society.

[Kreisel, 1977] Georg Kreisel. On the kind of data needed for a theory of proofs. In Robin O. Gandy and J. Martin E. Hyland, editors, *Logic Colloquium 76*, volume 87 of *Studies in Logic and the Foundations of Mathematics*, pages 111–128. North-Holland, Amsterdam, 1977.

[Kreisel, 1978a] Georg Kreisel. The motto of 'Philosophical Investigations' and the philosophy of proofs and rules. *Grazer Philosophische Studien*, 6:13–38, 1978.

[Kreisel, 1978b] Georg Kreisel. *Wittgenstein's Lectures on the Foundations of Mathematics, Cambridge 1939. Bulletin of the American Mathematical Society*, 84:79–90, 1978. Reprinted in Stuart Shanker, editor, *Ludwig Wittgenstein: Critical Assessments*, 1986, Croom-Helm, London.

[Kreisel, 1978c] Georg Kreisel. Zu Wittgensteins Gesprächen und Vorlesungen über die Grundlagen der Mathematik. In E. Leinfellner, H. Berghel, and A. Hübner, editors, *Proceedings of the Second International Wittgenstein Symposium*, pages 79–81, Vienna, 1978.

[Kreisel, 1979] Georg Kreisel. *Remarks on the Foundations of Mathematics. American Scientist*, 67:619, 1979. Review of the second, 1978 edition.

[Kreisel, 1982] Georg Kreisel. Finiteness theorems in arithmetic: An application of Herbrand's theorem for Σ_2 formulas. In Jacques Stern, editor, *Proceedings of the Herbrand Symposium*, pages 39–55. North-Holland, Amsterdam, 1982.

[Kreisel, 1983a] Georg Kreisel. Einige Erläuterungen zu Wittgensteins Kummer mit Hilbert und Gödel. In *Epistemology and Philosophy of Science. Proceedings of the 7th International Wittgenstein Symposium*, pages 295–303, Vienna, 1983. Hölder-Pichler-Tempsky Verlag.

[Kreisel, 1983b] Georg Kreisel. Saul Kripke, *Wittgenstein on Rules and Private Language. Canadian Philosophical Reviews*, 3:287–289, 1983.

[Kreisel, 1987] Georg Kreisel. Proof theory: some personal recollections, 1987. in [Takeuti, 1987], pages 395-405.

[Kreisel, 1989a] Georg Kreisel. Zu Einigen Gesprächen mit Wittgenstein: Erinnerungen und Gedanken. In *Wittgenstein: Biographie-Philosophie-Praxis, Catalogue for the Exposition at the Wiener Secession, 13 September-29 October 1989*, pages 131–143, Vienna, 1989.

[Kreisel, 1989b] Georg Kreisel. Zu Wittgensteins Sensibilität. In W. Gombocz, H. Rutte, and W. Sauer, editors, *Traditionen und Perspektiven der analytischen Philosophie, Festschrift für Rudolf Haller*, pages 203–223, Vienna, 1989. Hölder-Pichler-Tempsky Verlag.

[Kreisel, 1990a] Georg Kreisel. *About Logic and Logicians: A Palimpsest of Essays by Georg Kreisel*, 1990. A two-volume unpublished collection of essays by Kreisel, selected and arranged by Piergiorgio Odifreddi.

[Kreisel, 1990b] Georg Kreisel. Appendix to a letter to Grigori Mints, 1990. Kreisel made the letter available e.g. to Mathieu Marion.

[Löb, 1955] Michael H. Löb. Solution of a problem of Leon Henkin. *The Journal of Symbolic Logic*, 20:115–118, 1955.

[Mehlman, 2010] Jeffrey Mehlman. *Adventures in the French Trade: Fragments Toward a Life*. Stanford University Press, Redwood City, 2010.

[Monk, 1990] Ray Monk. *Ludwig Wittgenstein: The Duty of Genius.* The Free Press, Macmillan, New York, 1990.

[Odifreddi, 1996] Piergiorgio Odifreddi, editor. *Kreiseliana. About and Around Georg Kreisel.* AK Peters, Wellesley, 1996.

[Putnam, 1975] Hilary Putnam. What is mathematical truth? In *Mathematics, Matter and Method. Philosophical Papers, Volume I*, pages 60–78. Cambridge University Press, Cambridge, 1975.

[Takeuti, 1987] Gaisi Takeuti. *Proof Theory.* North-Holland, Amsterdam, 1987.

[Wittgenstein, 1953] Ludwig Wittgenstein. *Philosophical Investigations.* Basil Blackwell, Oxford, 1953. Edited by G.E.M. Anscombe and R. Rhees and translated by G.E.M. Anscombe.

[Wittgenstein, 1956] Ludwig Wittgenstein. *Remarks on the Foundations of Mathematics.* Basil Blackwell, Oxford, 1956. Edited by G.H. von Wright, R. Rhees, G.E.M. Anscombe and translated by G.E.M. Anscombe. First edition.

[Wittgenstein, 1958] Ludwig Wittgenstein. *The Blue and Brown Books: Preliminary Studies for the 'Philosophical Investigations'.* Basil Blackwell, 1958. Edited by Rush Rhees.

[Wittgenstein, 1976] Ludwig Wittgenstein. *Wittgenstein's Lectures on the Foundations of Mathematics.* Cornell University Press, Ithaca, 1976. Edited by Cora Diamond from the notes of R.G. Bosanquet, Norman Malcolm, Rush Rhees, and Yorick Smythies. Second edition, University of Chicago, Chicago 1989.

[Wittgenstein, 1978] Ludwig Wittgenstein. *Remarks on the Foundations of Mathematics.* Basil Blackwell, Oxford, 1978. Edited by G.H. von Wright, R. Rhees, G.E.M. Anscombe, translated by G.E.M. Anscombe. Second edition with additions.

[Wittgenstein, 1979] Ludwig Wittgenstein. *Wittgenstein's Lectures, Cambridge, 1932-35: From the Notes of Alice Ambrose and Margaret Macdonald.* University of Chicago Press, Chicago, 1979.

[Wittgenstein, 1980] Ludwig Wittgenstein. *Remarks on the Philosophy of Psychology, vol. I.* Chicago University Press, Chicago, 1980. Edited by Elizabeth Anscombe and Georg Henrik von Wright.

[Wittgenstein, 1993] Ludwig Wittgenstein. *Philosophical Occasions, 1912-1915.* Hackett Publishing Company, Indianapolis, 1993. Edited by James C. Klagge and Alfred Nordmann.

Part III

Mathematics

11 Strong Axioms of Infinity and Elementary Embeddings (with Robert M. Solovay and William N. Reinhardt)

This is the expository paper on strong axioms of infinity and elementary embeddings originally to have been authored by Reinhardt and Solovay. It has been owed for some time and already cited with some frequency in the most recent set theoretical literature. However, for various reasons the paper did not appear in print for several years. The impetus for actual publication came from a series of lectures on the subject by Kanamori (Cambridge, 1975) and a set of notes circulated thereafter. Thus, although this present exposition in a detailed reworking of these notes, the basic conceptual framework was first developed by Reinhardt and Solovay some years ago. One factor which turns this delay in publication to advantage is that a more comprehensive view of the concepts discussed is now possible with the experience of the last few years, particularly in view of recent consistency results and also consequences in the presence of the axiom of determinacy. A projected sequel by Solovay to this paper will deal further with these considerations.

One of the most notable characteristics of the axiom of infinity is that its truth implies its independence of the other axioms. This, of course, is because the (infinite) set of hereditarily finite sets forms a model of the other axioms, in which there is no infinite set. Clearly, accepting an assertion whose truth implies its independence of given axioms requires the acceptance of new axioms. It is not surprising that the axiom of infinity should have this character (one would expect to have to adopt it as an axiom anyway), and moreover one would expect the existence of larger and larger cardinalities to have such character, as indeed it has. The procedures for generating cardinals studied by Mahlo [29] provided a notable example. It is remarkable that the new consequences of the corresponding (generalized) axioms of infinity also include arithmetic statements: this application of Gödel's second theorem is by now quite familiar. It is also remarkable that certain properties of cardinals which were originally introduced with little thought of "size" considerations should turn out to have this same character of implying their own independence (see, for example, Ulam [44] and Hanf [6]).

The monumental paper Keisler–Tarski [12] examined in detail three classes of "large cardinals": weakly compact cardinals, measurable cardinals, and strongly compact cardinals. Taking these as typical examples, it is reasonable to say (though we are in no way establishing an absolute criterion) that a property is of "large cardinal" character if it has the following two consequences:

(i) the existence of a cardinal which (at least in some inner model) is essentially "larger" than inaccessible cardinals and other "smaller" large cardinals (in the sense that it is a fixed point of reasonable thinning procedures, like

Republished from *Annals of Mathematical Logic* 13 (1978), pp.73-116, with permission from Elsevier.

Mahlo's, beginning from these cardinals);

(ii) a discernible new strength in set theory, not only in the provability of more formal statements (like Con(ZFC), etc.) but also in the existence of a richer structure on the cumulative hierarchy itself (for example, new combinatorial properties).

We observe in the preceding the emergence of interesting (and somewhat unexpected) mathematical connections among size, combinatorial properties, and syntactic strength.

It is well known that below each measurable cardinal there are many weakly compact cardinals. In fact, the experience of the last few years indicates that weak compactness is relatively weak, and many interesting train stops lie on the way from measurability to weak compactness (see Devlin [1] for a comprehensive survey). On the other hand, though strong compactness implies the consistency of the existence of any specified number of measurable cardinals, it is now known that the least measurable cardinal can be strongly compact (see Section 4 for a discussion). It is the purpose of this paper to consider even stronger large cardinal properties, and to investigate their various interrelationships, as well as the effects of their presence on the cumulative hierarchy of sets.

The circumstance that some mathematical problems give rise (unexpectedly) to large cardinal properties raises the question of adopting new axioms. One possibility, which seems a bit like cheating, is to "solve" the problem by adopting its solution, as an axiom. Another approach (suggested in the paper [12]) is to attempt to bypass the question by regarding all results showing that $P(\kappa)$ is a large cardinal property (which of course show that $\neg P(\kappa)$ has strong closure properties) as partial results in the direction of showing $\forall\kappa\neg P(\kappa)$. If, however, $\exists\kappa P(\kappa)$ should be true and have important consequences, this may appear somewhat futile, as comparison with the paradigm case of the axiom of infinity suggests. A third approach is to attempt to formulate new strong axioms of infinity. Ultimately, since this paper is an exposition of mathematics, the issue of whether the large cardinal properties we investigate are to be considered axiomatic or problematic can be left to one side. We do, however, wish to discuss briefly the problem of formulating strong axioms of infinity.

The whole question of what intuitive and set theoretical considerations should lead to the formulation of strong axioms of infinity is rather complicated and merits a systematic analysis (which we do not attempt here). There is some discussion of this in Wang [45, ch.VI] especially p.189, which gives a descriptive classification (due to Gödel) of the considerations which have so far led to such axioms. We remark that series of axioms such as $T_0 = \mathrm{ZF}, \ldots, T_n, T_{n+1} = T_n + \mathrm{Con}(T_n), \ldots$, or those of Mahlo, appear "endless" in that it always seems possible to use the same guiding idea to get yet stronger principles (although it is not clear how to express this precisely). This even seems a desirable characteristic in hierarchies of axioms: if they are given as an r.e. sequence, they must be incomplete, so we would hope that the guiding idea would continue further. The procedures we consider have this "endless" character up to a point, where a result of Kunen sets a delimitation to one kind of *prima facie* natural extension.

We can discern at least four motivating principles behind the large cardinal properties we formulate.

(i) *Generalization.* For instance, it is in many ways quite reasonable to attribute certain properties of ω to uncountable cardinals as well, and these considerations can yield the measurable and strongly compact cardinals. Also, in considerations involving measurable cardinals, natural strengthenings of closure properties on ultrapowers yield the supercompact cardinals (see Sections 1, 2).

(ii) *Reflection.* The ordinary Reflection Principle in set theory invites various generalizations, for instance the Π^n_m-indescribability of cardinals. In one approach more relevant to our context, what is involved is a formulation of various reflection properties Ω, the class of all ordinals, intuitively ought to have (formalized in an extended language), the antithetical realization that Ω ought to be essentially indescribable in set theory, and thus the synthesis in the conclusion that there must already be some cardinal at which these properties obtain. Note that this in itself is a reflection argument. Extendible cardinals especially can be motivated in this way (see Section 5 and Reinhardt [40]).

(iii) *Resemblance.* This is closely related to (ii). Because of reflection considerations and, generally speaking, because the cumulative hierarchy is neutrally defined in terms of just the power set and union operations, it is reasonable to suppose that there are $\langle V_\alpha, \in\rangle$'s which resemble each other. The next conceptual step is to say that there are elementary embeddings $\langle V_\alpha, \in\rangle \to \langle V_\beta, \in\rangle$. Since this argument can just as well be cast in terms of $\langle V_{f(\alpha)}, \in, X(\alpha)\rangle$'s, where $f(\alpha)$ and $X(\alpha)$ are uniformly definable from α, the elementary embeddings may well turn out not to be the identity. Strong axioms like λ-extendibility (see Section 5) or Vopěnka's Principle (see Section 6) can be motivated in this way.

(iv) *Restriction.* Known assertions can be weakened to gain more information and sharpen implications. Ramsey and Rowbottom cardinals can be considered to follow in this way from measurable cardinals, and the axioms of Sections 7-8 can be viewed as introducing a spectrum of perhaps consistent axioms arising from Kunen's inconsistency result (1.12).

In all these approaches, the recurring feature of the various postulations is the notion of *elementary embedding*, and this paper is organized around this main theme.

Let us once again say that this paper is an exposition of mathematics.

We consider that many of the methods and technical relationships that we encounter are not without some mathematical elegance. Thus, hopefully our exposition will add an esthetic element to other incentives for considering strong axioms of infinity, and this is by no means a factor to be underrated in the investigation of new mathematical concepts.

Our set theory is ZFC. However, the mathematics in this paper is not strictly formalizable in ZFC, since we discuss elementary embeddings of the whole set theoretical universe. Kelley–Morse (KM) is adequate, as satisfaction can be expressed there, but Bernays–Gödel (BG) is often sufficient for many purposes. Ultimately, most implications can be formalized in ZFC, since either the elementary embedding involved can be regarded as restricted to some set, or only one formula instance of the elementary schema need be used. For a method of formalizing elementary embeddings of V in ZFC through a system of approximations, see Gaifman [4, IV]; this formalization is adequate to take care of most of the embeddings we consider.

Much of our notation is standard, but we do mention the following: the letters $\alpha, \beta, \gamma, \ldots$ denote ordinals whereas $\kappa, \lambda, \mu, \ldots$ are reserved for cardinals. V_α is the collection of sets of rank $< \alpha$. θ^M denotes the usual relativization of a formula θ to a class M. If x is a set, $|x|$ is its cardinality, $\mathcal{P}x$ is its power set, and $\mathcal{P}_\kappa x = \{y \in \mathcal{P}x \mid |y| < \kappa\}$. If also $x \subseteq \Omega$, $\bar{x}$ denotes its order type in the natural ordering. The identity function with the domain appropriate to the context is denoted by id. Finally, $\square$ signals the end of a proof.

If I is a set, an ultrafilter $\mathcal{U}$ over I is a maximal filter on the Boolean algebra $\mathcal{P}I$. $\mathcal{U}$ is *uniform* iff whenever $X \in \mathcal{U}$, $|X| = |I|$; *non-principal* iff $\bigcap \mathcal{U} = \emptyset$; and κ*-complete* iff whenever $T \subseteq \mathcal{U}$ and $|T| < \kappa$, $\bigcap T \in \mathcal{U}$. A cardinal $\kappa > \omega$ is *measurable* iff there is a non-principal, κ-complete ultrafilter over κ. In context, "for almost every x" means for x in a set in the ultrafilter involved.

We would like to thank the referee for simplifications in regard to 1.14, 3.2, 3.3, 4.8, and for suggesting the formulation of 2.6 in order to make 2.7 and 5.11 clearer.

1 Elementary Embeddings and Ultrapowers

Elementary embeddings with domain either the whole set theoretical universe V or just some initial segment V_α play a basic thematic role in this paper. In this initial section we quickly review basic techniques and establish some of our notation and terminology by working through the paradigm case, which can be considered a natural way of motivating measurable cardinals. Also, we consider a result of Kunen which will establish an upper limit to our further efforts.

When we investigate elementary embeddings j of V into some inner model M, it is convenient to have the situation implicit in the notation $j: V \to M$, which we also take to include the assertion that j is not the identity function.

That j is an *elementary embedding* means that it preserves all relations definable in the language of set theory: if $x_1, \ldots, x_n$ are sets, and $\theta(v_1, \ldots, v_n)$ is a formula and $F(v_1, \ldots, v_n)$ a term in the language of set theory, then

$$\theta(x_1, \ldots, x_n) \quad \text{iff} \quad \theta^M(j(x_1), \ldots, j(x_n))$$

and

$$j(F(v_1, \ldots, v_n)) = F(j(x_1), \ldots, j(x_n)).$$

By an *inner model* we mean a transitive $\in$-model of ZFC containing all the ordinals. There is a formula $\mathrm{Inn}(M)$ and finitely many axioms $\varphi_0, \ldots, \varphi_n$ of ZFC so that for any class M,

$$\vdash_{\mathrm{ZFC}} (\varphi_0 \ \& \cdots \& \ \varphi_n)^M \ \& \ \bigcup M \subseteq M \ \& \ \Omega \subseteq M \longleftrightarrow \mathrm{Inn}(M)$$

and if φ is any theorem of ZFC without free variables,

$$(*) \qquad \vdash_{\mathrm{ZFC}} \mathrm{Inn}(M) \to \varphi^M .$$

For example, $\mathrm{Inn}(M)$ can assert that M is transitive, contains all the ordinals, and that, in M, the sequence $\langle V_\alpha^M \mid \alpha \in \Omega \rangle$ is definable. (Here, the V_α^M's satisfy, in M, the usual definition for the V_α's.) Since all instances of ZF axiom schema are needed in the proofs of the theorems $(*)$, this by no means implies that ZFC is finitely axiomatizable. We have:

$$\vdash_{\mathrm{ZFC}} \mathrm{Inn}(V), \quad \vdash_{\mathrm{ZFC}} \mathrm{Inn}(L), \quad \vdash_{\mathrm{ZFC}} \mathrm{Inn}(\mathrm{HOD}).$$

(HOD is the class of hereditarily ordinal definable sets, cf. Myhill–Scott [36].)

Assume now that $j \colon V \to M$. We will frequently use the preservation schema for j without comment, leaving the reader to see that the relations and functions involved are set theoretic. For example, in 1.1 we will use $j(\mathrm{rank}(x)) = \mathrm{rank}(j(x))$, and in 1.2 both

$$j(X_\alpha) = (j(X))_{j(\alpha)} = (j(X))_\alpha \quad \text{when} \quad j(\alpha) = \alpha,$$

and

$$j(\bigcap\{X_\alpha \mid \alpha < \gamma\}) = \bigcap\{(j(X))_\alpha \mid \alpha < j(\gamma)\}.$$

In the latter case, one must of course realize that X and γ are the free variables.

Proposition 1.1.

(i) *For every* α, $j(\alpha) \geq \alpha$.

(ii) j *moves some ordinal.*

Proof. (i) By transfinite induction. For (ii), let x be of least rank so that $j(x) \neq x$. Then if $\delta = \text{rank}(x)$, $j(\delta) > \delta$. Otherwise, if $y \in j(x)$, $\text{rank}(y) < \text{rank}(j(x)) = j(\text{rank}(x)) = \delta$, so that $j(y) = y$ and $y \in x$. But also, $y \in x$ implies $j(y) = y \in j(x)$. Hence, $j(x) = x$, which is a contradiction. □

Now let δ be the least ordinal moved by j. We say that δ is the *critical point* of the embedding j. A model theorist might be quick to see

Theorem 1.2. *Let $\mathcal{U}$ be defined by*

$$X \in \mathcal{U} \text{ iff } X \subseteq \delta \ \& \ \delta \in j(X).$$

Then $\mathcal{U}$ is a non-principal δ-complete ultrafilter over δ, and hence, δ is a measurable cardinal.

Proof. $\delta \in \mathcal{U}$, but $\alpha < \delta$ implies $\{\alpha\} \notin \mathcal{U}$ since $j(\{a\}) = \{a\}$. Also, $X \in \mathcal{U}$ iff $\delta \in j(X)$ iff $\delta \notin j(\delta - X)$ iff $\delta - X \notin \mathcal{U}$.

Finally, if $\gamma < \delta$ and $\{X_\alpha \mid \alpha < \gamma\} \subseteq \mathcal{U}$, $\delta \in \bigcap\{j(X_\alpha) \mid \alpha < \gamma\}$. But $j(\gamma) = \gamma$, so $\bigcap\{j(X_\alpha) \mid \alpha < \gamma\} = j(\bigcap\{X_\alpha \mid \alpha < \gamma\})$, i.e. $\bigcap\{X_\alpha \mid \alpha < \gamma\} \in \mathcal{U}$. □

Thus, elementary embeddings already give rise to measurable cardinals, and so these will be the "smallest" of the large cardinals to be considered. It might be worthwhile to note that direct arguments using j show that δ must be large.

(a) Since δ is obviously a limit ordinal, $V_\delta \models \mathrm{Z}$.

(b) By the argument of 1.1(ii), if $x \in V_\delta$, $j(x) = x$. Hence, $y \subseteq V_\delta$ implies $y = j(\delta) \cap V_\delta \in M$.

(c) If $F\colon V_\delta \to V_\delta$ and $x \in V_\delta$, then $F''x \in V_\delta$. This is so since $F = j(F) \mid V_\delta$ and $j(F)''j(x) = j(F''x) \subseteq V_\delta \in V_{j(\delta)}$, and thus $F''x \in V_\delta$ by elementarity. Hence, $V_\delta \models \mathrm{ZFC}$ and $V_{\delta+1} \models \mathrm{BG}$.

(d) If $x, y \in V_{\delta+1}$ and $V_{\delta+1} \models \theta(x,y)$ and θ is first-order, then for some $\alpha < \delta$, $V_{\alpha+1} \models \theta(x \cap V_\alpha, y \cap V_\alpha)$: using (b), we have $V_{\delta+1} \subseteq M$, so $V_{\delta+1} \models \theta(x,y)$ in M, so that in M there is an $\alpha < j(\delta)$ with $V_{\alpha+1} \models \theta(j(x) \cap V_\alpha, j(y) \cap V_\alpha)$. Thus, there is an $\alpha < \delta$ with $V_{\alpha+1} \models \theta(x \cap V_\alpha, y \cap V_\alpha)$ by elementarity. It follows that V_δ satisfies Bernays' schema, i.e. δ is second-order indescribable.

Since we will shortly get a converse to 1.2 (that is, if κ is a measurable cardinal, there is a $j\colon V \to M$ with critical point κ), the foregoing quickly establish the standard facts on the size of measurable cardinals.

To get that converse, we will take an ultrapower of V. So first, let us recall the general process with $\mathcal{D}$ an arbitrary ultrafilter over some index set I. As usual, define for f, g functions with domain I,

$$f \sim_{\mathcal{D}} g \text{ iff } \{i \in I \mid f(i) = g(i)\} \in \mathcal{D}.$$

$\sim_{\mathcal{D}}$ is an equivalence relation with each equivalence class a proper class. In order to form the ultrapower, we need to have equivalence types $[f]$ such that $f \sim_{\mathcal{D}} g$ iff $[f] = [g]$. It is customary to define $[f]$ as the equivalence class of f; in our case these are proper classes and this is inconvenient (since we prefer to stay within the language of ZFC). We may instead use the following device of Scott.

Definition 1.3. (Scott) If f is a function with domain I,

$$S_{\mathcal{D}}(f) = \{g \mid g \sim_{\mathcal{D}} f \;\&\; \forall h(h \sim_{\mathcal{D}} f \rightarrow \operatorname{rank}(g) \leq \operatorname{rank}(h)\}$$

That is, $S_{\mathcal{D}}(f)$ is the *collection of those $g \sim_{\mathcal{D}} f$ of least possible rank.* $S_{\mathcal{D}}(f)$ is a set, so we can define

$$V^I/\mathcal{D} = \{S_{\mathcal{D}}(f) \mid f \colon I \rightarrow V\},$$

and as a membership relation,

$$S_{\mathcal{D}}(f)\, E_{\mathcal{D}}\, S_{\mathcal{D}}(g) \text{ iff } \{i \in I \mid f(i) \in g(i)\} \in \mathcal{D}.$$

Thus, $\langle V^I/\mathcal{D}, E_{\mathcal{D}}\rangle$ is a (class) structure definable within the language of set theory. The following is basic, and proved by induction on length of formulas.

Theorem 1.4. (Łoś) *If $\theta(v_1, \ldots, v_n)$ is a formula of set theory and $f_1, \ldots, f_n$ are functions: $I \rightarrow V$, then*

$$\langle V^I/\mathcal{D}, E_{\mathcal{D}}\rangle \models \theta(S_{\mathcal{D}}(f_1), \ldots, S_{\mathcal{D}}(f_n)) \text{ iff } \{i \in I \mid \theta(f_1(i), \ldots, f_n(i))\} \in \mathcal{D}.$$

If $\mathcal{D}$ is a λ-complete ultrafilter, then this theorem extends to set theoretical formulas in the language $L_{\lambda\lambda}$. (Recall that $L_{\kappa\lambda}$ is the infinitary language allowing conjunctions of $< \kappa$ formulas and quantifications over $< \lambda$ variables.) By the theorem, there is a canonical elementary embedding $e_{\mathcal{D}} \colon V \rightarrow V^I/\mathcal{D}$ defined by

$$e_{\mathcal{D}}(x) = S_{\mathcal{D}}(\langle x \mid i \in I\rangle),$$

i.e. $e_{\mathcal{D}}(x)$ is the equivalence class of the constant function on I with value x.

We now assume that $\mathcal{D}$ is both *non-principal* and *ω_1-complete.* The following two propositions are basic tools:

Proposition 1.5. (Mostowski [35]) *Suppose $\langle A, E\rangle$ is a (possibly proper class) structure so that E is a binary relation on A, and*

(i) *E is well-founded,*

(ii) *E is extensional, i.e. if $a, b \in A$ and for any $x \in A$, xEa iff xEb, then $a = b$,*

(iii) *$\{x \mid xEa\}$ is a set for each $a \in A$.*

Then there is a unique isomorphism $h \colon \langle A, E\rangle \leftrightarrow \langle M, \in\rangle$ to a transitive $\in$-structure M, called the transitive collapse.

Proof. Define h by recursion on E by $h(x) = \{h(y) \mid yEx\}$. □

Proposition 1.6. *$E_{\mathcal{D}}$ on $V^I/\mathcal{D}$ is well-founded.*

Proof. If $\cdots S_{\mathcal{D}}(f_2)\, E_{\mathcal{D}}\, S_{\mathcal{D}}(f_1)\, E_{\mathcal{D}}\, S_{\mathcal{D}}(f_0)$, $x \in \bigcap_{n\in\omega}\{i \in I \mid f_{n+1}(i) \in f_n(i)\}$ implies $\cdots f_2(x) \in f_1(x) \in f_0(x)$, which is contradictory. □

In view of 1.4, the preceding proposition is essentially the observation that "E is well-founded" is expressible in $L_{\omega_1\omega_1}$. These results show that there is a canonical isomorphism $h_{\mathcal{D}} \colon \langle V^I/\mathcal{D}, E_{\mathcal{D}}\rangle \leftrightarrow \langle M_{\mathcal{D}}, \in\rangle$, where $M_{\mathcal{D}}$ is a transitive class. We have immediately from Łoś's Theorem that $\vdash_{\mathrm{ZFC}} \mathrm{Inn}(M_{\mathcal{D}})$. Setting $j_{\mathcal{D}} = h_{\mathcal{D}} \circ e_{\mathcal{D}}$ so that $j_{\mathcal{D}}$ is elementary we will use the typical notation

$$j_{\mathcal{D}} \colon V \to M_{\mathcal{D}} \approx V^I/\mathcal{D}$$

to depict the situation. Also, for any $f \colon I \to V$, we let

$$[f]_{\mathcal{D}} = h_{\mathcal{D}}(S_{\mathcal{D}}(f)),$$

i.e. $[f]_{\mathcal{D}}$ is the transitive collapse of the ultrapower equivalence class of f. $\mathcal{D}$ as a subscript will often be dropped in these and similar situations when it is clear from the context. Recalling some comments before 1.3, notice that $f \sim_{\mathcal{D}} g$ iff $[f]_{\mathcal{D}} = [g]_{\mathcal{D}}$. Consequently, $[f]_{\mathcal{D}}$ could initially have played the role of $S_{\mathcal{D}}(f)$ in the construction of well-founded ultrapowers, in which case $V^I/\mathcal{D}$ is immediately identified with $M_{\mathcal{D}}$.

The following proposition is very useful.

Proposition 1.7. *With $\mathcal{D}$ as above, $j = j_{\mathcal{D}}$, $M = M_{\mathcal{D}}$, etc.*

(i) *if $j''x \in M$, and $y \subseteq M$ is such that $|y| \leq |x|$, then $y \in M$,*

(ii) *$j''(|I|^+) \notin M$.*

Proof. For (i), let $y = \{[t_a] \mid a \in x\} \subseteq M$, and define $T \colon j''x \to y$ by $T(j(a)) = [t_a]$. Since T enumerates y, it suffices to show that $T \in M$. So, we need a g so that $[g] = T$, i.e.

(a) domain $([g]) = j''x$.

(b) for $a \in x$, $[g](j(a)) = [t_a]$.

Let $[f] = j''x$. By Łoś's Theorem (1.4) if for each $i \in I$ we set domain$(g(i)) = f(i)$, and $g(i)(a) = t_\alpha(i)$ for each $a \in$ domain$(g(i))$, then clearly g is as required.

For (ii), assume that $j''(|I|^+) = [f] \in M$. If $A = \{i \mid |f(i)| \leq |I|\} \in \mathcal{D}$, since $|I|^+$ is regular, there is an $\alpha \in |I|^+ - \bigcup\{f(i) \mid i \in A\}$. But then $j(\alpha) \notin [f]$. If $B = \{i \in I \mid |f(i)| > |I|\} \in \mathcal{D}$, define h on B by induction on some well-ordering $\leq_I$ of I so that

$$h(i) \in f(i) - \{h(j) \mid j <_I i \;\&\; j \in B\}.$$

Then $[h] \in [f]$, yet h is not constant on any set in $\mathcal{D}$, as $\mathcal{D}$ is non-principal. Hence, in either case we get a contradiction from the assumption that $j''(|I|^+) = [f]$. □

If $|x| = \lambda$ in 1.7(i), the result can simply be stated as: M is *closed under λ-sequences*. In the future, we will also use ${}^{\lambda}M \subseteq M$ to denote this. 1.7(ii) puts an upper limit on the closure of M.

We now consider $\mathcal{U}$ a non-principal κ-complete ultrafilter over κ a measurable cardinal, and $j\colon V \to M \approx V^\kappa/\mathcal{U}$. The next theorem deals with this situation, and its (i) completes the converse to 1.2.

Theorem 1.8.

(i) *κ is the critical point of j.*

(ii) *${}^{\kappa}M \subseteq M$ but ${}^{\kappa^+}M \not\subseteq M$.*

(iii) *$\kappa < 2^\kappa < j(\kappa) < (2^\kappa)^+$.*

(iv) *$\mathcal{U} \notin M$.*

Proof. For (i), we first prove by induction on α that if $\alpha < \kappa$, $j(\alpha) = \alpha$. Suppose, towards a contradiction, that $j(\alpha) > \alpha$. Let $[f] = \alpha$. Then $\{\beta \mid f(\beta) < \alpha\} \in \mathcal{U}$, so that by the κ-completeness of $\mathcal{U}$, there is $\xi < \alpha$ so that $\{\beta \mid f(\beta) = \xi\}$. But then, $[f] = j(\xi) = \xi$, which is absurd.

Next, if id: $\kappa \to \kappa$ is the identity function on κ, $\kappa \leq [\mathrm{id}] < j(\kappa)$ since for each $\delta < \kappa$, $\{\alpha \mid \delta < \alpha < \kappa\} \in \mathcal{U}$. Thus κ is the critical point of j.

For (ii), use 1.7 and the fact that $j''\kappa = \kappa \in M$.

To show (iii), first note that $j(\kappa) =$ order type of $\{[f]_\mathcal{U} \mid f \in {}^\kappa\kappa\}$, so that $j(\kappa) < (2^\kappa)^+$. Also, in M, $j(\kappa)$ is measurable, hence strongly inaccessible, so that $(2^\kappa)^M < j(\kappa)$. But by (ii), $\mathcal{P}(\kappa) = \mathcal{P}^M(\kappa)$ so that $2^\kappa \leq (2^\kappa)^M$, since $M \subseteq V$.

For (iv), assume that $\mathcal{U} \in M$. Since ${}^\kappa\kappa = ({}^\kappa\kappa)^M$, in M one can evaluate $j(\kappa)$, so that $j(\kappa) < ((2^\kappa)^+)^M$ as in the previous paragraph. But this contradicts the strong inaccessibility of $j(\kappa) \in M$. □

The following corollary has been, of course, much strengthened in recent years by the work of Gaifman, Rowbottom, Silver and others.

Corollary 1.9. (Scott [41]) *If there is a measurable cardinal,* $V \neq L$.

Proof. Assume $V = L$. Then the M as above is an inner model satisfying the axiom of constructibility. Hence, $M = L$. But $V \neq M$ by 1.8 (iv), a contradiction. □

This corollary has as an easy consequence the fact that there is no elementary embedding $j\colon V \to L$. The situation with HOD is unclear. But, how about an elementary embedding of V into V itself? Kunen showed that this is not possible in ZFC. His proof uses a simple case of a combinatorial result of Erdős and Hajnal. But first, for the reader's interest a short proof of the general case due to Galvin and Prikry is presented. The result is concerned with the so-called Jónsson's problem (see Devlin [1] for details), and shows that if we allow an infinitary operation, there are Jónsson algebras of every infinite cardinality.

Definition 1.10. For any set x, a function f is called *ω-Jónsson* over x iff $f\colon {}^{\omega}x \to x$ and whenever $y \subseteq x$ with $|y| = |x|$, $f''{}^{\omega}y = x$.

Theorem 1.11. (Erdős–Hajnal [2]) *For every infinite cardinal λ, there is an ω-Jónsson function over λ.*

Proof. (Galvin–Prikry [5]) For the special case $\lambda = \omega$, there is a simple inductive argument: Let $\{\langle X_\alpha, n_\alpha\rangle \mid \alpha < 2^\omega\}$ enumerate $[\omega]^\omega \times \omega$. (Recall that $[\omega]^\omega$ is the collection of infinite subsets of ω.) By induction on $\alpha < 2^\omega$ pick $s_\alpha \in {}^{\omega}X_\alpha$ so that $s_\alpha \neq s_\beta$ for $\beta < \alpha$ and set $f(s_\alpha) = n_\alpha$. Then any extension of f to all of ${}^{\omega}\omega$ is ω-Jónsson over ω.

Suppose now that $\lambda > \omega$. Let $\mathcal{I}$ be any maximal collection of subsets of λ so that members of $\mathcal{I}$ have order type ω and are mutually almost disjoint. By the special case above, we can assume that for each $x \in \mathcal{I}$ there is a function f_x ω-Jónsson over x. Define now a function $g\colon {}^{\omega}\lambda \to \lambda$ by:

$$g(s) = \begin{cases} f_x(s) & \text{if the range of } s \text{ is infinite and} \\ & \quad s \in {}^{\omega}x \text{ for some } x \in \mathcal{I}, \\ 0 & \text{otherwise.} \end{cases}$$

Then g is well-defined.

It suffices to find an $A \subseteq \lambda$ so that $|A| = \lambda$, yet $B \subseteq A$ and $|B| = \lambda$ implies $g''{}^{\omega}B \supseteq A$, for then an ω-Jónsson function over A can easily be derived from g. If no such A exists, there are sets $\lambda \supseteq A_0 \supseteq A_1 \supseteq A_2 \cdots$ each of cardinality λ, and $a_n \in A_n - A_{n+1}$ so that $a_n \notin g''{}^{\omega}A_{n+1}$. If $y = \{a_n \mid n \in \omega\}$, by maximality of $\mathcal{I}$ there is an $x \in \mathcal{I}$ so that $x \cap y$ is infinite. Let $x \cap y = \{a_{n_0}, a_{n_1}, \ldots\}$, where $n_0 < n_1 < \cdots$. Now $t = \{a_{n_1}, a_{n_2}, \ldots\} \subseteq A_{n_1}$. But by definition of f_x, there is an $s \in {}^\omega t$ so that $g(s) = a_{n_0}$. Hence, $a_{n_0} \in g''{}^\omega A_{n_1}$, which is a contradiction. □

Theorem 1.12. (Kunen [16]) *There is no non-trivial elementary embedding of V into itself.*

Proof. Argue by contradiction and suppose $j \colon V \to V$ with critical point κ. Set $\lambda = \sup\{j''\kappa \mid n \in \omega\}$, where $j^0(\kappa) = \kappa$ and $j^{n+1}(\kappa) = j(j^n(\kappa))$. Note that $j(\lambda) = \sup\{j^{n+1}(\kappa) \mid n \in \omega\} = \lambda$.

Now let f be ω-Jónsson over λ; then $j(f)$ is also ω-Jónsson over λ. Consider the set $X = j''\lambda$. Since $|X| = \lambda$, let $x \in {}^\omega X$ so that $j(f)(x) = \kappa$. But if $x(n) = j(\alpha_n)$ for $n \in \omega$, $x = j(y)$ where $y \in {}^\omega\lambda$ and $y(n) = \alpha_n$. Hence, $\kappa = j(f)(j(y)) = j(f(y))$, contradicting the fact that κ is not in the range of j. □

As Kunen remarks, since the λ in the above proof is a strong limit cardinal of cofinaiity ω, the argument of the special case of 1.11 suffices to produce an ω-Jónsson function over κ. (The full result of 1.11 will be used in our forthcoming 3.3 and 3.4.) Note also that more generally, if $j \colon V \to M$ and λ is defined in the same way, then there is actually a subset of λ not in M. This way of formulating Kunen's result is formalizable in ZFC through the stratagem of Gaifman [4]; the notion $j \colon V \to V$ is not. Since AC was used in the proof of 1.11, we may still ask

Open Question 1.13. In ZF (without AC) can there be a non-trivial elementary embedding of the universe into itself?

Kunen's result will limit our efforts in that we cannot embed the universe into too "fat" an inner model. Pending an answer to 1.13, one can perhaps best view this fact as a structural limitation imposed on V by the Axiom of Choice.

Finally, we note the following generalization of 1.8(iv) concerning ultrapowers. This was known before Kunen's result (1.12), and established the special case that no $j \colon V \to V$ could be the result of taking an ultrapower.

Proposition 1.14. *D is a non-principal, ω_1-complete ultrafilter over some cardinal ν, then $\mathcal{D} \notin M_{\mathcal{D}}$.*

Proof. Let $M = M_{\mathcal{D}}$ and $j = j_{\mathcal{D}}$ and assume that $\mathcal{D} \in M$. It follows that $\mathcal{P}\nu \subseteq M$ and ${}^\nu\nu \subseteq \mathcal{P}(\nu \times \nu) \subseteq M$. Note that for any ordinal α, $j(\alpha) =$ the order type of $\{[f]_{\mathcal{D}} \mid f \in {}^\nu\alpha\}$. Thus, $j''\nu \in M$, since $j''\nu$ is just the collection of such order types for α ranging over ordinals below ν, and can be properly defined in M as $\mathcal{D} \in M$ and ${}^{\nu}\nu = ({}^{\nu}\nu)^M$.

By 1.7(i), it follows that M is closed under ν-sequences, and in particular ${}^{\nu}\nu^+ = ({}^{\nu}\nu^+)^M$. The argument in the previous paragraph can now be used again to show that $j''(\nu^+) \in M$, thereby contradicting 1.7(ii). $\square$

2 Supercompactness

Though there can be no elementary embedding $j\colon V \to V$, we noted that if $j_{\mathcal{U}}\colon V \to M_{\mathcal{U}}$ arises from a non-principal κ-complete ultrafilter over κ a measurable cardinal, then $M_{\mathcal{U}}$ is not even closed under κ^+-sequences. The following is an intermediary notion, and seems the proper generalization of measurability.

Definition 2.1. If $\kappa \leq \lambda$, κ is λ-*supercompact* iff there is an elementary embedding $j\colon V \to M$ so that:

(i) j has critical point κ and $j(\kappa) > \lambda$,
(ii) ${}^{\lambda}M \subseteq M$.

κ is *supercompact* iff κ is λ-supercompact for all $\lambda \geq \kappa$.

It follows from (ii) that M contains all sets hereditarily of cardinal $\leq \lambda$. Note that from Section 1, κ is κ-supercompact iff κ is measurable. It will be shown shortly that if κ is 2^κ-supercompact, then κ is actually the κth measurable cardinal (see 3.5). Kunen noticed that (i) in the above definition can be replaced by simply "j has critical point κ", since for some integer n, $j^n(\kappa) > \lambda$—else we would again get the contradiction of 1.12. A further argument is needed to show that the nth iterate of j embeds V into an inner model closed under λ-sequences (see [16]).

2.2. In the situation of 2.1, since ${}^{\lambda}M \subseteq M$, the embedding j immediately suggests considering the following ultrafilter:

$$X \in \mathcal{U} \text{ iff } j''\lambda \in j(X)\,.$$

Note that since $|j''\lambda| = \lambda < j(\kappa)$ in M, $j''\lambda \in \mathcal{P}_{j(\kappa)}j(\lambda)^M$. Hence, $\mathcal{P}_\kappa\lambda \in \mathcal{U}$, and we can consider this set to be the underlying index set for $\mathcal{U}$. $\mathcal{U}$ has the following properties:

(i) $\mathcal{U}$ is a κ-complete ultrafilter.
(ii) $\mathcal{U}$ is non-principal, and for any $\alpha \in \lambda$, $\{x \in \mathcal{P}_\kappa\lambda \mid \alpha \in x\} \in \mathcal{U}$,
(iii) If f is a function defined on a set in $\mathcal{U}$ so that $\{x \in \mathcal{P}_\kappa\lambda \mid f(x) \in x\} \in \mathcal{U}$, then there is an $\alpha \in \lambda$ so $\{x \in \mathcal{P}_\kappa\lambda \mid f(x) = \alpha\} \in \mathcal{U}$.

By (i) and (ii) if $y \in \mathcal{P}_\kappa\lambda$, $\{x \mid y \subseteq x\} \in \mathcal{U}$. For a proof of (iii), note that if $j(f)(j''\lambda) \in j''\lambda$, then $j(f)(j''\lambda) = j(a)$ for some $a \in \lambda$.

Definition 2.3. If $\kappa \leq \lambda$, an ultrafilter over $\mathcal{P}_\kappa\lambda$ is *normal* iff it satisfies (i), (ii) and (iii) as above. More generally, an ultrafilter $\mathcal{U}$ over $\mathcal{P}_\kappa I$, where I is a set, is *normal* iff it satisfies (i), (ii) and (iii) with λ replaced by I. Finally, without reference to κ, an ultrafilter $\mathcal{U}$ over $\mathcal{P}I$ (i.e. $\mathcal{U} \subseteq \mathcal{PP}I$) is *normal* iff it satisfies (ii) and (iii) with λ replaced by I.

If $f\colon I \to \lambda$ is bijective, then f induces a bijection between normal ultrafilters over $\mathcal{P}_\kappa I$ and those over $\mathcal{P}_\kappa\lambda$, so for most purposes, it suffices to consider ultrafilters over sets of form $\mathcal{P}_\kappa\lambda$.

Note that an ω_1-complete ultrafilter $\mathcal{U}$ over $\mathcal{P}I$ is normal iff $[\mathrm{id}] = j_{\mathcal{U}}''I$, where id: $\mathcal{P}I \to \mathcal{P}I$ is the identity map. This easy but central fact will be used repeatedly throughout the rest of this paper.

Just as in Section 1, having produced an ultrafilter from an embedding, one can hope to reverse the process by taking an ultrapower. So, let $\mathcal{U}$ over $\mathcal{P}_\kappa\lambda$ be normal and consider the canonical $j\colon V \to M \approx V^{\mathcal{P}_\kappa\lambda}/\mathcal{U}$. Then:

(i) ${}^\lambda M \subseteq M$. Use 1.7(i) and the fact that $[\mathrm{id}]= j''\lambda$.

(ii) κ is the critical point of j and $j(\kappa) > \lambda$. We have $\{x \mid |x| < \kappa\} \in \mathcal{U}$, so $|[\mathrm{id}]| = |j''\lambda| < j(\kappa)$. But $|j''\lambda| = \lambda$ in M, since M is closed under λ-sequences.

We have shown

Theorem 2.4. *If $\kappa \leq \lambda$, the following are equivalent:*

(i) *κ is λ-supercompact.*

(ii) *There is a normal ultrafilter over $\mathcal{P}_\kappa\lambda$.*

Note the following reversibility: if $\mathcal{U}$ is normal over $\mathcal{P}_\kappa\lambda$, $j_{\mathcal{U}}\colon V \to M_{\mathcal{U}}$, and $\mathcal{U}'$ is defined from $j_{\mathcal{U}}$ as in 2.2, then $\mathcal{U}' = \mathcal{U}$.

Though normal ultrafilters have been defined generally over index sets $\mathcal{P}\lambda$, we have seen how sets of form $\mathcal{P}_\kappa\lambda$ naturally come into play. In fact, if $\mathcal{U}$ is an ω_1-complete normal ultrafilter over $\mathcal{P}\lambda$, then for some $\mu < \lambda$, $\mathcal{P}_\mu\lambda \in \mathcal{U}$:

Let $j\colon V \to M$ be the corresponding embedding with critical point κ. Since ${}^\lambda M \subseteq M$ by normality, Kunen's argument shows that for some integer n, we must have $j^n(\kappa) < \lambda \leq j^{n+1}(\kappa)$. If $\lambda < j^{n+1}(\kappa)$, set $\mu = j^n(\kappa)$ so that $\mu < \lambda$. If $\lambda = j^{n+1}(\kappa)$, note that a simple argument inducing on the $j^i(\kappa)$'s and using ${}^\lambda M \subseteq M$ shows that λ must be inaccessible in V. Thus, if we set $\mu = (j^n(\kappa))^+$ in this case, we have $\mu < \lambda$. Finally in either case, note that $|[\mathrm{id}]| = |j''\lambda| = \lambda < j(\mu)$, so that $\{x \subseteq \lambda \mid |x| < \mu\} \in \mathcal{U}$, i.e. $\mathcal{P}_\mu\lambda \in \mathcal{U}$.

We now show that if κ is supercompact and $\theta \geq \kappa$, then V_θ can be expressed as an ultraproduct of V_γ's with $\gamma < \kappa$.

Theorem 2.5. *Let $\theta \geq \kappa$, and assume that κ is $|V_\theta|$-supercompact. Let $\mathcal{U}$ be a normal ultrafilter over $\mathcal{P}_\kappa V_\theta$, and $j\colon V \to M$ the corresponding embedding. For convenience, set $X = \mathcal{P}_\kappa V_\theta$. Suppose that in M, $\theta = [\langle \theta_x \mid x \in X\rangle]$. Then*

(i) *For almost every x in X, x is an elementary submodel of V_α, and the transitive collapse gives an isomorphism $\pi_x\colon x \cong V_{\theta_x}$.*

(ii) *Set $j_x = \pi_x^{-1}$. Then $j|V_\theta = [\langle j_x \mid x \in X\rangle]$.*

(iii) *There is an isomorphism*

$$(*) \qquad V_{\theta+1} \cong \prod_{x\in X} V_{\theta_x+1}/\mathcal{U}$$

which can be explicated as follows: if $y \in V_\theta$, $y = [\langle \pi_x(y) \mid x \in X \;\&\; y \in x\rangle]$. If $y \in V_{\theta+1}$, $y = [\langle \pi_x{}''(y \cap x) \mid x \in X\rangle]$.

Proof. $j''V_\theta$ is an elementary submodel of $V^M_{f(\theta)}$, and the transitization map $\pi\colon j''V_\theta \cong V_\theta$ has inverse $j|V_\theta$. We also have that $[\mathrm{id}] = j''V_\theta$ in M, and that $j|V_\theta$ and π are both in M as they are hereditarily of cardinality $\leq |V_\theta|$. From these facts and Łoś's Theorem (1.4), (i) and (ii) of the theorem now follow.

For (iii), note that the right side of $(*)$ is clearly isomorphic to $V^M_{\theta+1}$. However, using the fact that M is closed under $|V_\theta|$-sequences, it is easy to prove by induction on $\alpha \leq \theta+1$ that $V_\alpha = V^M_\alpha$. This establishes $(*)$.

Now let $y \in V_\theta$, and suppose $y = [\langle y_x \mid x \in X\rangle]$. Then $y_x \in V_{\theta_x}$ for almost every $x \in X$. Moreover, $j(y) = [(\langle j_x(y_x) \mid x \in X\rangle]$. But by definition, $j(y) = [\langle y \mid x \in X\rangle]$. Hence, for almost every $x \in X$, we have $j_x(y_x) = y$, i.e. $\pi_x(y) = y_x$, which was to be proved.

Finally, let $y \in V_{\theta+1}$. We wish to show that $y = [\langle \pi_x{}''(y \cap x) \mid x \in X\rangle]$. For this, it suffices to prove that $y = \pi''(j(y) \cap j''V_\theta)$. But $j(y) \cap j''V_\theta = j''y$, and $\pi(j(z)) = z$ for $z \in V_\theta$. So our claim is evident. □

Note that since, in the notation of 2.5, $j(\kappa) > |V_\theta| \geq \theta$, we have that $\theta_x < \kappa$ for almost every x.

Let us say that a property $P(x)$ is *local* iff it has the form $\exists\delta(V_\delta \models \psi(x))$. The ultraproduct representation makes it evident that if $P(x)$ is local and for some γ, $P(\gamma)$ holds, then if κ is supercompact, $P(\gamma)$ holds for some $\gamma < \kappa$. This is worth some elaboration.

Definition 2.6. Recall the Lévy hierarchy of formulas (see Lévy [20], Definition 1). For any transitive M, say Σ_n (resp. Π_n) *relativizes down to* M iff whenever $P(x)$ is Σ_n (resp. Π_n), if $a \in M$ and $P(a)$ holds, then $M \models P(a)$.

Σ_n relativizes down to M iff Π_{n+1} does. According to [20] Theorem 36, if $|V_\theta| = \theta$, Σ_1 (and hence Π_2) relativizes down to V_θ. Moreover, it is easy to construct a sentence Φ so that $V_\theta \models \Phi$ iff $|V_\theta| = \theta$. With these points in mind, suppose now that $P(x)$ is Σ_2, say $\exists y Q(x,y)$ where Q is Π_1. Then,

$$P(x) \text{ iff } \exists\beta[V_\beta \models (\Phi \,\&\, \exists y Q(x,y)] .$$

Thus, any Σ_2 property is local (as we have defined this notion just before 2.6). Conversely, it is well known that "x is a V_α" is Π_1, and hence a local property is one given by a Σ_2 formula.

Theorem 2.7. *If κ is supercompact, Σ_2 (and hence Π_3) relativize down to V_κ.*

Proof. Suppose $P(x)$ is $\exists y Q(x,y)$ where Q is Π_1. Let $a \in V_\kappa$ so that $P(\alpha)$ holds and fix b so that $Q(a,b)$ holds. By supercompactness, let $j \colon V \to M$ with critical point κ so that $b \in M \cap V_{j(\kappa)}$. Note that $j(a) = a$. Thus, $(V_{j(\kappa)} \models P(a))^M$, so $V_\kappa \models P(\alpha)$. $\square$

Observe that 2.6 is optimal, since the Σ_3 sentence "There is a supercompact cardinal" fails in V_κ if κ is the least supercompact cardinal. (Cf. 5.8 below, or note that "x is not supercompact" is Σ_2 and apply 2.6.)

It is pertinent here to discuss the question of cardinal powers in the context of large cardinals. Silver [42] showed that in $L[\mathcal{U}]$, where $\mathcal{U}$ is a normal ultrafilter over a measurable cardinal, the GCH holds, and hence: Con(ZFC & there is a measurable cardinal) implies Con(ZFC & there is a measurable cardinal & GCH). Kunen [17] then showed that Con(ZFC & there is a measurable cardinal κ & $2^\kappa > \kappa^+$) implies $\forall\alpha$ Con(ZFC & there are α measurable cardinals). This surprising result certainly indicated that strong assumptions would be necessary to get a model with a measurable cardinal κ so that $2^\kappa > \kappa^+$.

It was Silver who first found such a model: he showed that if κ is supercompact in the ground model, there is a forcing extension in which $2^\kappa > \kappa^+$ and κ is still measurable. (A more precise formulation of Silver's result is possible in the terminology of Section 5: If κ is $\eta+\delta+1$-extendible in the ground model, there is a forcing extension in which $2^\kappa \geq \aleph_{\kappa+\delta}$ and κ is still η-extendible if $\eta > 0$, or measurable if η was 0.) Combining this result with one further extension using Prikry forcing (to change the cofinality of a measurable cardinal to ω) yielded the first example of a singular strong limit cardinal κ so that $2^\kappa > \kappa^+$. Magidor [26] then proved the following relative consistency result, using in part Silver's result. Given a cardinal with a sufficient degree of supercompactness in the ground model, there is a forcing extension in which $2^{\aleph_\omega} = \aleph_{\omega+2}$ and for every integer n, $2^{\aleph_n} \leq \aleph_{n+2}$. Recent results of Jensen [8] and Mitchell seem to indicate that one cannot expect to weaken the initial large cardinal assumption by very much. See also the end of Section 7 for a result of Magidor with a stronger conclusion, but also starting with the stronger assumption of hugeness.

Silver's method of iterated forcing, first used to get the result mentioned in the previous paragraph, is often called Backward Easton Forcing. It is applicable to a variety of problems in set theory, and in particular can be used to show that the presence of large cardinals has little effect on the cardinal powers of regular cardinals. For example, Menas [32] shows that if κ is supercompact in the ground model and Φ is a function on regular cardinals so that

(i) $\lambda \leq \mu$ implies $\Phi(\lambda) \leq \Phi(\mu)$,
(ii) $\mathrm{cf}(\Phi(\lambda)) > \lambda$, and
(iii) Φ is local (as previously defined, i.e. Σ_2),

then there is a Backward Easton forcing extension in which cardinals are preserved, $2^\lambda = \Phi(\lambda)$ for every regular λ, and κ is still supercompact.

When a cardinal κ is supercompact, other than the technical result that there are κ measurable cardinals below κ (see 3.5), little is known about the behavior of the set theoretical universe below κ which does not already follow from the measurability of κ.

Concerning the behavior of the universe above κ, several interesting facts are known. Some of these already follow from the weaker assumption of strong compactness (see Section 4), but one exception is the fact that the second-order Löwenheim-Skolem Theorem holds for structures with underlying set of cardinality at least κ. Magidor [22] shows that, in an appropriate sense, we need the strength of supercompactness in this case.

In Section 4, the main consistency results involving supercompactness and strong compactness are stated. Section 5 contains several results on supercompactness in the context of extendibility, and another characterization (5.7). To conclude this section, we mention that there are combinatorial characterizations of supercompact cardinals (Magidor [24]), in terms of concepts first formulated by Jech [7]. Also, Prikry [39] has recently formulated a concept of *real-supercompactness* in analogy to real-valued measurability, and observed that if it is consistent that a supercompact cardinal exists, then it is consistent that 2^ω is real-supercompact. He also showed that several consequences of supercompactness that we will discuss shortly (see 4.6, 4.7) also follow from real-supercompactness.

3 Normal Ultrafilters

With the introduction of normal ultrafilters in the previous section, we now take time to investigate them in some detail. First, some technicalities; recall that if x is a set of ordinals, $\bar{x}$ denotes its order type.

Proposition 3.1. *If $\mathcal{U}$ is normal over $\mathcal{P}_\kappa\lambda$* and $\alpha \leq \lambda$,

(i) $j_{\mathcal{U}}''\alpha = [\langle x \cap \alpha \mid x \in \mathcal{P}_\kappa\lambda\rangle]_{\mathcal{U}}$.
(ii) $\alpha = [\langle \overline{x \cap \alpha} \mid x \in \mathcal{P}_\kappa\lambda\rangle]_{\mathcal{U}}$.

Proof. $j''\alpha = j''\lambda \cap j(\alpha)$ and $\overline{j''\alpha} = \alpha$. □

Proposition 3.2. *If $\mathcal{U}$ is normal over $\mathcal{P}_\kappa\lambda$, M_λ is actually closed under $\lambda^{<\kappa}$-sequences.*

Proof. Let $j\colon V \to M_{\mathcal{U}} \approx V^{\mathcal{P}_\kappa\lambda}/\mathcal{U}$. By 1.7(i), it suffices to show that $j''(\mathcal{P}_\kappa\lambda) \in M_{\mathcal{U}}$, since $\lambda^{<\kappa} = |\mathcal{P}_\kappa\lambda|$. However, $j''(\mathcal{P}_\kappa\lambda) = \mathcal{P}_\kappa(j''\lambda) = (\mathcal{P}_\kappa(j''\lambda))^{M_{\mathcal{U}}} \in M_{\mathcal{U}}$. (Here, the first equality holds as $j(x) = j''x$ for any $x \in \mathcal{P}_\kappa\lambda$, and the second, as $M_{\mathcal{U}}$ is closed under λ-sequences and $j''\lambda \in M_{\mathcal{U}}$.) □

We next present another structural result about normal ultrafilters, which has already appeared in Solovay [43]. First, a preliminary observation.

Proposition 3.3. *Let $\kappa \leq \lambda$ and G be a ω-Jónsson function over λ. If $\mathcal{U}$ is a normal ultrafiilter over $\mathcal{P}_\kappa\lambda$,*

$$\{x \mid G|''^{\omega}x \text{ is } \omega\text{-Jónsson over } x\} \in \mathcal{U}\,.$$

Remarks. G exists by 1.11. Our definition of a function being ω-Jónsson is slightly stronger than the one used in [43], but there is little difference in the manipulations.

Proof of 3.3. By Łoś's Theorem (1.4), it suffices to show $j(G)|^{\omega}j''\lambda$ is ω-Jónsson over $j''\lambda$. (Here, we need not distinguish between V and M, as M is closed under λ-sequences.) So, suppose $X \subseteq j''\lambda$ and $|X| = |j''\lambda| = \lambda$. If $Y = j^{-1}(X)$, since G is ω-Jónsson over λ, we have $G''^{\omega}Y = \lambda$. So, given any $\alpha < \lambda$, let $s \in {}^{\omega}Y$ such that $G(s) = \alpha$. Then $j(\alpha) = j(G(s)) = j(G)(j(s))$, and $j(s) \in {}^{\omega}X$. Thus, we have shown $j(G)''^{\omega}X = \lambda$, which was to be proved. □

The following is Theorem 2 of [43].

Theorem 3.4. *Suppose $\kappa \leq \lambda$ are regular cardinals and $\mathcal{U}$ is a normal ultrafilter over $\mathcal{P}_\kappa\lambda$. If $F\colon \mathcal{P}\lambda \to \lambda$ is defined by $F(x) = \sup(x)$, then there is an $X \in \mathcal{U}$ so that $F|X$ is one-to-one.*

Proof. Set $X = \{x \in \mathcal{P}_\kappa\lambda \mid \bar{x}$ is inaccessible, x is closed under ω-sequences, and $G|^{\omega}x$ is ω-Jónsson is over $x\}$. By 3.1, 3.3, and further use of Łoś's Theorem (1.4), $X \in \mathcal{U}$. We show that X has the required property.

Assume that $x, y \in X$, and $\sup(x) = \gamma = \sup(y)$. As $\bar{x}, \bar{y}$ are inaccessible and x, y are both closed under ω-sequences, a simple argument shows that $x \cap y$

is cofinal in γ, and hence $|x| = |x \cap y| = |y|$. Thus, by the ω-Jónsson property, $x = G''^{\omega}(x \cap y) = y$. $\square$

It is not hard to see that if $Y \in \mathcal{U}$, $\bigcup \mathcal{P}x \,|\, x \in Y\} = \mathcal{P}_\kappa\lambda$. Thus, by 3.4 we can conclude that if κ is supercompact and $\kappa \leq \lambda$ is regular, then $\lambda^{<\kappa} = \lambda$. This fact already follows from the strong compactness of κ (Solovay [43])—see Section 4 for further remarks.

Recall now that the term "normal" is already known in another context: if κ is measurable, a non-principal κ-complete ultrafilter $\mathcal{U}$ over κ is normal iff in $M_{\mathcal{U}}$, $\kappa = [\mathrm{id}]$. This, of course, just means that whenever $f \in {}^{\kappa}\kappa$ is such that $\{\alpha < \kappa \,|\, f(\alpha) < \alpha\} \in \mathcal{U}$, there is a $\gamma < \kappa$ so that $\{\alpha < \kappa \,|\, f(\alpha) = \gamma\} \in \mathcal{U}$. Normal ultrafilters turn out to be rather special, but one can always get them from elementary embeddings. In fact, it is easily shown that the ultrafilter of 1.2 is normal.

We now have a notion of normality in two senses, but in fact, there is a one-to-one correspondence between normal ultrafilters over $\mathcal{P}_\kappa\kappa$ and normal ultrafilters over κ. If $\mathcal{V}$ is normal over κ, $\mathcal{U} = \{X \subseteq \mathcal{P}_\kappa\kappa \,|\, X \cap \kappa \in \mathcal{V}\}$ is normal over $\mathcal{P}_\kappa\kappa$. Conversely, if $\mathcal{U}$ over $\mathcal{P}_\kappa\kappa$ is normal, $\kappa \in \mathcal{U}$. If not, then $\{x \in \mathcal{P}_\kappa\kappa \,|\, x \text{ is not an ordinal}\} \in \mathcal{U}$. For such x, let $f(x) \in x$ be so that $f(x)$ is the least above some ordinal not in x. By normality, $[f] = j(\gamma)$ for some $\gamma < \kappa$, but this contradicts $\{x \in \mathcal{P}_\kappa\kappa \,|\, \gamma \subseteq x\} \in \mathcal{U}$.

Determining the number of normal ultrafilters possible over a measurable cardinal has turned out to be an interesting problem. Kunen [15] showed that in $L[\mathcal{U}]$, the universe constructed from a normal ultrafilter $\mathcal{U}$ over κ, $\mathcal{U} \cap L[\mathcal{U}]$ is the only normal ultrafilter over κ. Kunen and Paris [19] showed that if κ is measurable in the ground model, there is a forcing extension in which κ carries the maximal number of normal ultrafilters, i.e. 2^{2^κ}. Then Mitchell [33] more recently showed that if κ is 2^κ-supercompact and τ is $\leq \kappa$ or one of the terms κ^+ or κ^{++}, there is an inner model in which κ is measurable and carries exactly τ normal ultrafilters, It would still be desirable to get Mitchell's relative consistency results starting from just the measurability of κ.

In Mitchell's model with exactly two normal ultrafilters over κ, one contains the set $\{\alpha < \kappa \,|\, \alpha \text{ is measurable}\}$ and the other does not. In this regard, consider the following two propositions.

Proposition 3.5. *If κ is 2^κ-supercompact, there is a normal ultrafilter $\mathcal{U}$ over κ so that $\{\alpha < \kappa \,|\, \alpha \text{ is measurable}\} \in \mathcal{U}$. Hence, 2^κ-supercompactness is already enough to assure that κ is the κth measurable cardinal.*

Proof. Let $j\colon V \to M$ with critical point κ, so that M is closed under 2^κ-sequences. If $\mathcal{U}$ defined by

$$x \in \mathcal{U} \quad \text{iff} \quad X \subseteq \kappa \;\&\; \kappa \in j(X)\,,$$

$\mathcal{U}$ is normal over κ, as before. But since M is closed under 2^κ-sequences, it is not hard to see that every ultrafilter over κ is a member of M. Hence, κ is measurable in M, i.e. $\{\alpha < \kappa \mid \alpha \text{ is measurable}\} \in \mathcal{U}$ by the definition of $\mathcal{U}$. □

Proposition 3.6. *Any measurable cardinal κ carries a normal ultrafilter so that* $\{\alpha < \kappa \mid \alpha \text{ is not measurable}\} \in \mathcal{U}$.

Proof. By induction. Let $\mathcal{V}$ be a normal ultrafilter over κ. Set $T = \{\alpha < \kappa \mid \alpha \text{ is measurable}\}$. If $T \notin \mathcal{V}$, the theorem certainly holds for κ. So assume $T \in \mathcal{V}$. For each $\alpha \in T$, by induction hypothesis let $\mathcal{U}_\alpha$ be normal over α so that $\{\beta < \alpha \mid \beta \text{ is not measurable }\} \in \mathcal{U}_\alpha$. Define $\mathcal{U}$ over κ by

$$X \in \mathcal{U} \text{ iff } \{\alpha < \kappa \mid X \cap \alpha \in \mathcal{U}_\alpha\} \in \mathcal{V}.$$

It is not hard to check that $\mathcal{U}$ is a normal ultrafilter over κ so that $\{\alpha < \kappa \mid \alpha \text{ is not measurable}\} \in \mathcal{U}$. □

The two propositions show that if κ is 2^κ-supercompact, there are at least two normal ultrafilters over κ. In fact, there are 2^{2^κ} normal ultrafilters over κ, and this result is a special case of a general result on the number of normal ultrafilters over $\mathcal{P}_\kappa\lambda$. To prove the main theorem (3.8) from which this will follow, we develop some technical machinery of independent interest.

Consider the following situation. $\kappa \le \lambda < \mu$ and there is a normal uitrafilter U over $\mathcal{P}_\kappa\lambda$. For $X \subseteq \mathcal{P}_\kappa\mu$ if we let $X|\lambda = \{x \cap \lambda \mid x \in X\}$ and set

$$\mathcal{U}|\lambda = \{X|\lambda \mid X \in \mathcal{U}\},$$

it is not hard to see that $\mathcal{U}|\lambda$ is a normal ultrafilter over $\mathcal{P}_\kappa\lambda$. In fact,

$$X \in \mathcal{U}|\lambda \text{ iff } X \subseteq \mathcal{P}_\kappa\lambda \;\&\; j_{\mathcal{U}}''\lambda \in j_{\mathcal{U}}(X).$$

Consider the following diagram:

$$\begin{array}{ccc} V & \xrightarrow{j_0} & M_0 \approx V^{\mathcal{P}_\kappa\lambda}/\mathcal{U}|\lambda \\ & \searrow^{j_1} & \downarrow k \\ & & M_1 \approx V^{\mathcal{P}_\kappa\mu}/\mathcal{U} \end{array}$$

where k is defined by

$$k([\langle f(x) \mid x \in \mathcal{P}_\kappa\lambda\rangle]_{\mathcal{U}|\lambda} = [\langle f(x \cap \lambda) \mid x \in \mathcal{P}_\kappa\mu\rangle]_{\mathcal{U}}.$$

Proposition 3.7. *k is an elementary embedding and the diagram commutes.*

Proof. Straightforward. $\square$

For those familiar with the Rudin-Keisler ordering on ultrafilters, note that $\mathcal{U}|\lambda$ is just $h_*(\mathcal{U})$, where $h\colon \mathcal{P}_\kappa\mu \to \mathcal{P}_\kappa\lambda$ is defined by $h(x) = x \cap \lambda$, and k is the associated elementary embedding.

Concerning the action of k, first note that if $\alpha \leq \lambda$, $k(\alpha) = \alpha$ since

$$k(\alpha) = k([\langle \overline{x \cap \alpha} \,|\, x \in \mathcal{P}_\kappa\lambda\rangle]) = [\langle \overline{x \cap \lambda \cap \alpha} \,|\, x \in \mathcal{P}_\kappa\mu\rangle] = \alpha\,.$$

Next, we show $k(x) = x$ for $x \in \mathcal{P}_\kappa\lambda$. Let $\gamma < \kappa$, and suppose $h\colon \gamma \to x$ is surjective. Then $h \in M_0$. Since $k(\gamma) = \gamma$, $k(h)$ is a function with domain γ. For $\xi < \gamma$, $k(h)(\xi) = k(h(\xi)) = h(\xi)$ by the preceding paragraph. So $k(h) = h$, whence $k(x) = \text{range}(k(h)) = \text{range}(h) = x$.

Next, since $k(\mathcal{P}_\kappa\lambda) = \mathcal{P}_\kappa\lambda^{M_1} = \mathcal{P}_\kappa\lambda$, an argument using elementarity and the preceding paragraph shows that $k(X) = X$ for $X \in \mathcal{P}\mathcal{P}_\kappa\lambda$.

Finally, if $X \in \mathcal{P}\mathcal{P}\mathcal{P}_\kappa\lambda \cap M_0$, then $k(X) = X$, since $\mathcal{P}\mathcal{P}_\kappa\lambda \subseteq M_0$ by 3.2. Using these results, a straightforward argument (Menas [30], 2.6 and preceding) shows that if $\mu \geq 2^{\lambda^{<\kappa}}$, the least ordinal moved by k is $(2^{\lambda^{<\kappa}})^{M_0}$. However, this particular fact will not be used in this paper.

We are now is a position to prove the main result of this section.

Theorem 3.8. *κ is $2^{\lambda^{<\kappa}}$-supercompact, $X \in \mathcal{P}\mathcal{P}\mathcal{P}_\kappa\lambda$ implies that there is a $\mathcal{V}$ normal over $\mathcal{P}_\kappa\lambda$ so that $X \in M_\mathcal{V}$.*

Proof. Let $\mathcal{U}$ be normal over $\mathcal{P}_\kappa(2^{\lambda^{<\kappa}})$ and consider the diagram:

$$\begin{array}{ccc} V & \xrightarrow{j_0} & M_0 \approx V^{\mathcal{P}_\kappa\lambda}/\mathcal{U}|\lambda \\ & \searrow^{j_1} & \downarrow k \\ & & M_1 \approx V^{\mathcal{P}_\kappa(2^{\lambda^{<\kappa}})}/\mathcal{U} \end{array}$$

Argue by contradiction. Let $\varphi(X,\kappa,\lambda)$ iff $X \in \mathcal{P}\mathcal{P}\mathcal{P}_\kappa\lambda$ and for any $\mathcal{V}$ normal over $\mathcal{P}_\kappa\lambda$, $X \notin \mathcal{P}\mathcal{P}\mathcal{P}_\kappa\lambda \cap M_\mathcal{V}$. Suppose $\exists X\varphi(X,\kappa,\lambda)$. Then $M_1 \models \exists X\varphi(X,\kappa,\lambda)$. This is so since every normal $\mathcal{V}$ over $\mathcal{P}_\kappa\lambda$ is in M_1, and $\mathcal{P}\mathcal{P}\mathcal{P}_\kappa\lambda \cap M_\mathcal{V}$ is correctly "computed" in M_1: every function $\mathcal{P}_\kappa\lambda \to V_\kappa$ is in M_1 and $\mathcal{P}\mathcal{P}\mathcal{P}_\kappa\lambda \cap M_\mathcal{V} \subseteq (V_{\lambda+4})^{M_\mathcal{V}} \subseteq (V_{j_\mathcal{V}(\kappa)})^{M_\mathcal{V}} = j_\mathcal{V}(V_\kappa)$.

Recall now the properties of k discussed just before the theorem. Since $k(\kappa) = \kappa$ and $k(\lambda) = \lambda$, $M_0 \models \varphi(X,\kappa,\lambda)$. Let $X_0 \in M_0$ so that $M_0 \models \varphi(X_0,\kappa,\lambda)$. $X_0 \in \mathcal{P}\mathcal{P}\mathcal{P}_\kappa\lambda$, so $k(X_0) = X_0$. Hence, $M_1 \models \varphi(X_0,\kappa,\lambda)$. But this contradicts $X_0 \in M_0 = M_{\mathcal{U}|\lambda}$ and the definition of φ. $\square$

The following improves the result of Magidor [23].

Corollary 3.9. *If κ is $2^{\lambda^{<\kappa}}$-supercompact, there are $2^{2^{\lambda^{<\kappa}}}$ normal ultrafilters over $\mathcal{P}_\kappa\lambda$. Hence, if κ is 2^κ-supercompact, there are 2^{2^κ} normal ultrafilters over K.*

Proof. If $\mathcal{U}$ is normal over $\mathcal{P}_\kappa\lambda$, note first that $|\mathcal{PPP}_\kappa\lambda \cap M_\mathcal{V}| < (2^{\lambda^{<\kappa}})^+$. This is so because

$$(2^{\lambda^{<\kappa}})^{M_\mathcal{V}} < j_\mathcal{V}(\kappa) < (2^{\lambda^{<\kappa}})^+,$$

the first inequality because $j_\mathcal{V}(\kappa)$ is inaccessible in $M_\mathcal{V}$ and the second, a trivial upper bound (see the analogous 1.8(iii)). Hence, since by 3.8,

$$\mathcal{PPP}_\kappa\lambda = \bigcup\{\mathcal{PPP}_\kappa\lambda \cap M_\mathcal{V} \mid \mathcal{V} \text{ is normal over } \mathcal{P}_\kappa\lambda\},$$

there must be

$$|\mathcal{PPP}_\kappa\lambda| = 2^{2^{\lambda^{<\kappa}}}$$

normal ultrafilters over $\mathcal{P}_\kappa\lambda$. □

By examining the proofs of 3.8 and 3.9, one can check that if κ is 2^κ-supercompact there are 2^{2^κ} normal ultrafilters over κ containing the set $\{\alpha < \kappa \mid \alpha \text{ is measurable}\}$. But paradoxically, the following question is still open.

Open Question 3.10. If κ is 2^κ-supercompact, is it provable that there is more than one normal ultrafilter over κ containing the set $\{\alpha < \kappa \mid \alpha \text{ is not measurable}\}$?

We can also prove something about the following ordering defined on ultrafilters.

Definition 3.11. If $\mathcal{U}$ and $\mathcal{V}$ are ω_1-complete ultrafilters, $\mathcal{U} \lhd \mathcal{V}$ iff $\mathcal{U} \in M_\mathcal{V}$.

$\lhd$ on normal ultrafilters over fixed $\mathcal{P}_\kappa\lambda$ is a well-founded partial ordering (by a generalization of 1.1 of Mitchell [33]). Notice that if $\mathcal{V}$ is normal over $\mathcal{P}_\kappa\lambda$, there are at most $2^{\lambda^{<\kappa}}$ normal ultrafilters over $\mathcal{P}_\kappa\lambda$ which are $\lhd$ predecessors of $\mathcal{V}$, by the first fact used in the proof of 3.9.

Corollary 3.12. *If κ is $2^{\lambda^{<\kappa}}$-supercompact, there is $\lhd$ chain of normal ultrafilters over $\mathcal{P}_\kappa\lambda$ of length $(2^{\lambda^{<\kappa}})^+$. Hence, if κ is 2^κ-supercompact, there is a $\lhd$ chain of normal ultrafilters over κ of length $(2^\kappa)^+$.*

Proof. Given at most $2^{\lambda^{<\kappa}}$ normal ultrafilters, one can code them as some $X \in \mathcal{PPP}_\kappa\lambda$. Thus, 3.8 can be used to find some normal $\mathcal{V}$ over $\mathcal{P}_\kappa\kappa$ so that $X \in M_\mathcal{V}$. □

For an application of the order $\lhd$, see Magidor [27]. He shows that if $\mu < \kappa$ are regular, a $\lhd$ chain of order type κ of normal ultrafilters over κ can be used to define a partial ordering such that forcing with it preserves all cardinals but changes the cofinality of μ. This, of course, complements Prikry forcing.

To conclude this section, we mention that, analogous to Rowbottom's partition result for normal ultrafilters over a measurable cardinal, there is the following partition property: say that a normal ultrafilter $\mathcal{U}$ over $\mathcal{P}_\kappa\lambda$ has the *partition property* iff whenever $f\colon [\mathcal{P}_\kappa\lambda]^2 \to 2$, there is an $X \in \mathcal{U}$ and an $i < 2$ so that $x, y \in X$, $x \subseteq y$, and $x \neq y$ imply that $f(\{x, y\}) = i$. Menas established a characterization of this property and showed that 3.8 holds with the additional requirement that the normal ultrafilters over $\mathcal{P}_\kappa\lambda$ each satisfy the partition property. However, Solovay has shown that if $\kappa < \lambda$, κ is λ-supercompact and λ is measurable, then there is a normal ultrafilter over $\mathcal{P}_\kappa\lambda$ without the partition property. Also, Kunen proved that the least $\mu > \kappa$ so that there is a normal ultrafilter over $\mathcal{P}_\kappa\mu$ without the partition property is Π^2_1-indescribable and strictly less than the least ineffable cardinal greater than κ. See Menas [31] for the details.

4 Strong compactness

The concept of strong compactness is discussed in Keisler–Tarski [12] and historically was motivated by efforts to generalize the Compactness Theorem of lower predicate calculus to infinitary languages $\mathcal{L}_{\kappa\kappa}$. Supercompactness was conceived partly in order to ostensibly strengthen the definition of strong compactness in a desirable manner, but Solovay conjectured that the two concepts coincide. Though this conjecture stood for some time, it is now known to be false (see 4.4 and after).

In view of our thematic approach, in this section we consider strong compactness as formulated in terms of elementary ernbeddings and ultrafilters. The connection with $\mathcal{L}_{\kappa\kappa}$ and other equivalent formulations are given in a variety of sources.

Definition 4.1. If $\kappa \leq \lambda$ and $\mathcal{U}$ is an ultrafilter over $\mathcal{P}_\kappa\lambda$, then $\mathcal{U}$ is *fine* iff $\mathcal{U}$ is κ-complete and for each $\alpha < \lambda$, $\{x \in \mathcal{P}_\kappa\lambda \mid \alpha \in x\} \in \mathcal{U}$.

Thus, the definition leaves out clause (iii) of the definition of normality (2.3).

Definition 4.2. If $\kappa \leq \lambda$, κ is *λ-compact* iff there is a fine ultrafilter over $\mathcal{P}_\kappa\lambda$. κ is *strongly compact* iff κ is λ-compact for all $\lambda \geq \kappa$.

Trivially, if κ is λ-superpcompact, κ is λ-compact, and κ is measurable iff κ is κ-compact. If $\kappa \leq \lambda < \mu$ and κ is μ-compact, and $\mathcal{U}$ is a fine ultrafilter over

$\mathcal{P}_\kappa\mu$, then $\mathcal{U}|\lambda$ is a fine ultrafilter over $\mathcal{P}_\kappa\lambda$, so that κ is λ-compact. The reader is cautioned that there is a different, equally natural notion of λ-compactness often seen in the literature.

We proceed immediately to the characterization:

Theorem 4.3. *If $\kappa \leq \lambda$, the following are equivalent:*

(i) *κ is λ-compact.*

(ii) *There is a $j\colon V \to M$ with critical point κ so that: $X \subseteq M$ and $|X| \leq \lambda$ implies there is a $Y \in M$ so that $X \subseteq Y$ and $M \models |Y| < j(\kappa)$.*

(iii) *If $\mathcal{F}$ is any κ-complete filter over an index set I so that $\mathcal{F}$ is generated by $\leq \lambda$ sets, then $\mathcal{F}$ can be extended to a κ-complete ultrafilter over I.*

Proof. (i) $\to$ (ii). Let $\mathcal{U}$ be fine over $\mathcal{P}_\kappa\lambda$, and consider $j\colon V \to M \approx V^{\mathcal{P}_\kappa\lambda}/\mathcal{U}$. If $X = \{[f_\alpha] \mid \alpha < \lambda\} \subseteq M$, set $G(x) = \{f_\alpha(x) \mid \alpha \in x\}$ and $Y = [G]$.

(ii) $\to$ (iii). Suppose $\mathcal{F}$ is as hypothesized, and generated by elements of $T \subseteq \mathcal{P}I$, where $|T| \leq \lambda$. By *(ii)* let $Y \supseteq j''T$ so that $Y \in M$ and $M \models |Y| < j(\kappa)$. In M, $j(\mathcal{F})$ is a $j(\kappa)$-complete filter and $j(\mathcal{F}) \cap Y$ is a subset of cardinality $< j(\kappa)$. Hence, there is a $c \in M$ so that $c \in \bigcap(j(\mathcal{F}) \cap Y)$. Set $X \in \mathcal{U}$ iff $X \subseteq \mathcal{I}$ & $c \in j(X)$. Then $\mathcal{U}$ is a κcomplete ultrafilter which extends $\mathcal{F}$.

(iii) $\to$ (i). Extend the κ-complete filter over $\mathcal{P}_\kappa\lambda$ generated by the sets $\{x \in \mathcal{P}_\kappa\lambda \mid \alpha \in x\}$ for $\alpha < \lambda$ to a κ-complete ultrafilter. □

4.3(ii) thus shows the weakness of λ-compactness; with λ-supercompactness one can always assert that $X = Y$. We mention here that Ketonen [14] has another characterization: if $\kappa \leq \lambda$ are regular, κ is λ-compact iff for every regular μ so that $\kappa \leq \mu \leq \lambda$ there is a uniform κ-complete ultratilter over μ. For a further discussion of fine ultrafilters, normal ultrafilters, and connections involving the Rudin-Keisler ordering, consult Sections 2.1-2.3 of Menas [30].

The following result indicates that strong compactness and supercompactness are not the same concept.

Proposition 4.4. (Menas [30]) (i) *If κ measurable and a limit of strongly compact cardinals, then κ is strongly compact.*

(ii) *if κ is the least cardinal as in* (i)*, then κ is not 2^κ-supercompact.*

Proof. For (i), let $\mathcal{U}$ let be a non-principal κ-complete ultrafilter over κ so that $A = \{\alpha < \kappa \mid \alpha \text{ is strongly compact}\} \in \mathcal{U}$. If $\lambda \geq \kappa$, for $\alpha \in A$ let $\mathcal{U}_\alpha$ be fine over $\mathcal{P}_\alpha\lambda$. Define $\mathcal{V}_\lambda$ by

$$X \in \mathcal{V}_\lambda \text{ iff } X \subseteq \mathcal{P}_\kappa\lambda \;\&\; \{\alpha \mid X \cap \mathcal{P}_\alpha\lambda \in \mathcal{U}_\alpha\} \in \mathcal{U}.$$

Then $\mathcal{V}_\lambda$ is fine over $\mathcal{P}_\kappa\lambda$.

For (ii), argue by contradiction, and suppose κ were 2^κ-supercompact. Let $j: V \to M$ with critical point κ so that M is closed under 2^κ-sequences. By definition of κ and elementarity, we have in M that $j(\kappa)$ is the least measurable cardinal which is a limit of strongly compact cardinals. But M is closed under 2^κ-sequences so that κ is measurable in M, and also, if $\alpha < \kappa$ is strongly compact, $j(\alpha) = \alpha$ is strongly compact in the sense of M. Hence, in M, κ is also measurable and a limit of strongly compact cardinals, which is a contradiction. □

It is a consequence of the existence of an extendible cardinal (by an easy strengthening of 5.9 of the next section) that there are many cardinals as in 4.4 (i). Menas was able to establish the following result.

(a) (Menas [30]) Con(ZFC & there is a measurable cardinal that is the limit of strongly compact cardinals) implies Con(ZFC & there is exactly one strongly compact cardinal κ, and κ is not even κ^+-supercompact),

The following are results of Magidor.

(b) (Magidor [25]) Con(ZFC & there is a supercompact cardinal) implies Con (ZFC & there is a supercompact cardinal with no strongly compact cardinals below it).

(c) (Magidor [25]) Con(ZFC & there is a strongly compact cardinal) implies Con(ZFC & there is a strongly compact cardinal with no measurable cardinals below it).

Though Kunen [15] has shown that the existence of a strongly compact cardinal implies the existence of inner models with any specified number of measurable cardinals, the results (a), (b) and (c) indicate that strong compactness is a rather pathological concept in the hierarchy of large cardinals. Perhaps it should ultimately be regarded as a generalization of weak compactness in the same spirit that supercompactness is a generalization of measurability.

In connection with these considerations, recall that in the previous section we showed that if κ is supercompact there are many normal ultrafilters over κ. However, the following is still open.

Open Question 4.5. If κ is strongly compact, can it be proved that there is more than one normal ultrafilter over κ?

4.6. Concerning the effect of the existence of a strongly compact cardinal κ on the behavior of the set theoretical universe, Solovay [43] has proved that if $\lambda \geq \kappa$ and λ is a singular strong limit cardinal, then $2^\lambda = \lambda^+$. As noted by Paris, Prikry and probably others, this result now follows from the easier result of [43] that if $\lambda \geq \kappa$ and λ is regular, then $\lambda^{<\kappa} = \lambda$, and Silver's recent solution to most cases of the singular cardinals problem:

Assume $\lambda \geq \kappa$ and λ is a singular strong limit cardinal. If $\mathrm{cf}(\lambda) < \kappa$, it is immediate that $2^\lambda = \lambda^{\mathrm{cf}(\lambda)} \leq (\lambda^+)^{\mathrm{cf}(\lambda)} = \lambda^+$. But if $\mathrm{cf}(\lambda) \geq \kappa$, $S = \{\alpha < \lambda \mid \alpha$ is a singular strong limit cardinal of cofinality $< \kappa\}$ is a stationary subset of λ and $\alpha \in S$ implies $2^\alpha = \alpha^+$ by the previous sentence. Silver's result states that if μ is a singular cardinal of uncountable cofinality so that those $\alpha < \mu$ with $2^\alpha = \alpha^+$ forms a stationary subset of μ, then $2^\mu = \mu^+$. Thus, we can conclude $2^\lambda = \lambda^+$.

However, we note that the further results in [43] on powers of cardinals cannot ostensibly be simplified in this way.

It is also proved in [43] that if $\kappa \leq \lambda$ and κ is λ^+-supercompact, then Jensen's combinatorial principle $\Box_\lambda$ fails. (see Jensen [9] for the result that if $V = L$, $\Box_\mu$ holds for every infinite cardinal μ). Since then, Gregory proved that the failure of $\Box_\lambda$ already follows from the λ^+-compactness of κ. His proof used the notion of a λ-free Abelian group, but a direct combinatorial proof is possible. Following a comment of Kunen, we present the ideas involved in general context.

For regular $\lambda > \omega$, let us call the following principle E_λ: There is a set $S \subseteq \{\alpha < \lambda \mid \mathrm{cf}(\alpha) = \omega\}$ stationary in λ so that for all limit ordinals $\xi < \lambda$, $S \cap \xi$ is not stationary in ξ. Jensen [9] proved that if $V = L$, then E_λ fails just in case λ is weakly compact cardinal. The following proposition is well-known.

Proposition 4.7. *Suppose that $\lambda > \omega$ and $\Box_\lambda$ holds. Then E_{λ^+} holds, and in fact, whenever T a stationary subset of λ, there is a stationary $S \subseteq T$ so that for any limit ordinal $\xi < \lambda^+$, $S \cap \xi$ is not stationary in ξ.*

Proof. Let us first recall the principle $\Box_\lambda$: There is a sequence $\langle C_\xi \mid \xi < \lambda^+\rangle$ so that for any $\xi < \lambda^+$, we have

(i) $C_\xi \subseteq \xi$, and if ξ is a limit ordinal, C_ξ is closed and unbounded in ξ.
(ii) The order type of C_ξ is $\leq \lambda$.
(iii) If γ is a limit point of C_ξ, then $C_\gamma = C_\xi \cap \lambda$.

Suppose now that T is a stationary subset of λ^+. Without loss of generality, we can assume that T consists of limit ordinals. For $\alpha \leq \lambda$, set $S_\alpha = \{\xi \in T \mid C_\xi$ has order type $\alpha\}$. Then, as the S_α's partition T into λ parts by property (ii), there is some $\alpha_0 \leq \lambda$ so that $S = S_{\alpha_0}$ is still stationary in λ^+. We now claim that for any limit ordinal $\xi < \lambda^+$, $S \cap \xi$ is not stationary in ξ. There are three cases:

(a) $\mathrm{cf}(\xi) = \omega$. Then as S consists of limit ordinals, S is disjoint from any sequence of type ω of successor ordinals, cofinal in ξ.

(b) $\mathrm{cf}(\xi) > \omega$ and C_ξ has order type $\leq \alpha_0$. Let $\overline{C_\xi}$ consist of the limit points of C_ξ. Then as $\mathrm{cf}(\xi) > \omega$, $\overline{C_\xi}$ is closed and unbounded in ξ, and $(S \cap \xi) \cap \overline{C_\xi} = \emptyset$ by property (iii).

(c) $\mathrm{cf}(\xi) > \omega$ and C_ξ has order type $> \alpha_0$. Let $\gamma \in C_\xi$ so that $C_\xi \cap \gamma$ is of type α_0. Then if $\overline{C_\xi}$ is defined as in (b), we have by property (iii) that $(S \cap \xi) \cap (\overline{C_\xi} - (\gamma + 1)) = \emptyset$.

Thus, the claim is proved, and the proposition follows. $\square$

The following theorem establishes the connection to large cardinals.

Theorem 4.8. *If λ is regular and there is a uniform, ω_1-complete ultrafilter over λ, then E_λ fails.*

Proof. Let $\mathcal{U}$ be uniform, ω_1-complete over λ. Let $f\colon \lambda \to \lambda$ so that $\sup\{j_{\mathcal{U}}(\alpha) \mid \alpha < \lambda\} = [f]_{\mathcal{U}}$. Such an f exists since the supremum in question is $\leq$ [id]. Set

$$\mathcal{V} = \{X \subseteq \lambda \mid f^{-1}(X) \in U\}.$$

It is not hard to see that $\mathcal{V}$ is a uniform, ω_1-complete ultrafilter over λ, which is *weakly normal* in the following sense: If $g\colon \lambda \to \lambda$ so that $\{\alpha < \lambda \mid g(\alpha) < \alpha\} \in \mathcal{V}$, then for some $\delta < \lambda$, $\{\alpha < \lambda \mid g(\alpha) < \delta\} \in \mathcal{V}$.

Let us first prove a preliminary

Fact. $\{\alpha < \lambda \mid \mathrm{cf}(\alpha) > \omega\} \in \mathcal{V}$.

Otherwise, for almost every α let $\langle \alpha_n \mid n \in \omega\rangle$ be cofinal in α. By the weak normality of $\mathcal{V}$, for every $n \in \omega$ there is a $\delta_n < \lambda$ so that $X_n = \{\alpha < \lambda \mid \alpha_n < \delta_n\} \in \mathcal{V}$. The $Y = \bigcap X_n \in \mathcal{V}$, but if $\delta = \sup \delta_n < \lambda$, notice that $\alpha \in Y$ implies $\alpha \leq \eta$, contradicting the fact that $\mathcal{V}$ is uniform. The Fact is thus proved.

We now proceed with the main proof. Let S be a stationary subset of λ so that $S \subseteq \{\alpha \mid \mathrm{cf}(\alpha) = \omega\}$. To show that E_λ fails, we establish in fact that

$$(*) \qquad \{\xi \mid S \cap \xi \text{ is stationary in } \xi\} \in \mathcal{V}.$$

Let us suppose to the contrary, and derive a contradiction. Then for almost every ξ, there is a closed and unbounded $C_\xi \subseteq \xi$ so that $S \cap C_\xi = \emptyset$. Define

$$C = \{\alpha < \lambda \mid \{\xi < \lambda \mid \alpha \in C_\xi\} \in \mathcal{V}\}.$$

Because $\mathcal{V}$ is ω_1-complete and as almost every ξ has cofinality $> \omega$ by the Fact, C is ω-closed: given $\alpha \in C$ for $n \in \omega$, $\sup \alpha_n \in C$. Also, we claim that C is an unbounded subset of λ. To show this, let $\beta_0 < \lambda$ be arbitrary. Using the weak normality of $\mathcal{V}$, for each integer n we can define functions $f_n\colon \lambda \to \lambda$ and ordinals β_n by induction so that

(a) $A_n = \{\xi < \lambda \mid f_n(\xi) \in C_\xi \ \& \ f_n(\xi) > \beta_n\} \in \mathcal{V}$, and
(b) $B_n = \{\xi < \lambda \mid f_n(\xi) < \beta_{n+1}\} \in \mathcal{V}$.

Since $\mathcal{V}$ is ω_1-complete, $X = \bigcap(A_n \cap B_n) \in \mathcal{V}$. But then if $\beta = \sup \beta_n$, it is not hard to show using the set X (and the Fact proved earlier) that for almost every ξ, $\beta \in C_\xi$.

We have just established that C is ω-closed and unbounded in X. Since S is a stationary subset of $\{\alpha < \lambda \mid \mathrm{cf}(\alpha) = \omega\}$, we must have $C \cap S \neq \emptyset$. Having arrived at this contradiction, we have thus established $(*)$ and the theorem. □

The following is now immediate from 4.7 and 4.8.

Corollary 4.9. (Gregory) *If κ is λ^+-compact, then $\square_\lambda$ fails.*

We remark that the main idea in the proof of 4.8 is due to Jensen, Prikry and Silver, and is stated (somewhat obscurely) in Theorem 20 of Prikry [38]. Let E^κ_λ for regular $\kappa < \lambda$ be like E_λ but with the required stationary $S \subseteq \{\alpha \mid \mathrm{cf}(\alpha) = \kappa\}$. The method of proof actually shows (see [38]) that if there is a uniform ultrafilter over λ which is κ-descendingly complete, then E^κ_λ fails. Prikry had to assume some form of the GCH to get a weakly normal ultrafilter related to such a $\mathcal{U}$, but it is now known that no such hypothesis is needed (see Theorem 2.5 of [10]).

The referee has outlined a proof of 4.8 (which similarly generalizes for E^κ_λ) that does not use weak normality. Suppose that $S \subseteq \lambda$ witnesses E_λ. For $\alpha \in S$ let $\langle \gamma^n_\alpha \mid n \in \omega\rangle$ be cofinal in α. Call a function $f\colon S \to \omega$ a *disjointer* iff whenever $\alpha \neq \beta$, $m > f(\alpha)$ and $n > f(\beta)$ implies $\gamma^m_\alpha \neq \gamma^n_\beta$. By the regressive function lemma, there is no disjointer for S. However, one can show by induction $\xi < \lambda$ that there are disjointers f_ξ for $S \cap \xi$, using the fact that $S \cap \xi$ is not stationary in ξ. If $\mathcal{U}$ were uniform, ω_1-complete over λ, one can obtain a disjointer for S (and hence a contradiction) by taking the ultraproduct of the f_ξ.

5 Extendible Cardinals

We now consider an axiom which implies the existence of many supercompact cardinals. The notion of *extendibility* is motivated in Reinhardt [40] by considerations involving strong principles of reflection and resemblance formalized in an extended theory which allows transfinite levels of higher type objects over the set theoretical universe V. Essentially, Cantor's Ω, the class of all ordinals, is hypothesized to be extendible in this context. With the natural reflection down into the realm of sets, we have the concept of an extendible cardinal. (As Reinhardt [40] points out, however, this sort of internal formalization within V rather begs the question if we want to discuss fundamental issues about the nature of V and Ω.)

More simplistically, recall that Kunen's Theorem (1.12) showed that one cannot embed V into too fat an inner model. As a natural weakening one can instead consider embedding initial segments of the universe into larger initial segments, $j\colon V_\alpha \to V_\beta$ where $\alpha \leq \beta$. (As before, implicit in this notation is the assertion that j is not the identity.) This approach may be conceptually helpful, but the exact form of the following definition owes its origin to the considerations of [40].

Definition 5.1. If $\eta > 0$ κ is *η-extendible* iff there is a ζ and a $j\colon V_{\kappa+\eta} \to V_\zeta$ with critical point κ, where $\kappa+\eta < j(\kappa) < \zeta$. κ is *extendible* iff κ is η-extendible for every $\eta > 0$.

Remarks. (i) Since $\eta \geq \kappa \cdot \kappa$ implies $\kappa + \eta = \eta$, the exact form of the above definition is distinctive only for small η.

(ii) When $\eta < \kappa$, it is not hard to see that $\zeta = j(\kappa) + \eta$. This fact will be used without further reference. Note that in such cases and especially for η an integer, it is clear that η-extendibility is just a postulate of resemblance : With $j\colon V_{\kappa+\eta} \to V_{j(\kappa)+\eta}$, V_κ and $V_{j(\kappa)}$ are indistinguishable as far as $(\eta+1)$th order properties are concerned.

(iii) If κ is η-extendible and $0 < \delta < \eta$, then κ is δ-extendible. Since the term V_α is definable from α, if ζ and j are as in 5.1 for the η-extendibility of κ, then $j|V_{\kappa+\delta}\colon V_{\kappa+\delta} \to V_{j(\kappa+\delta)}$ is also an elementary embedding, as

$$\begin{aligned} V_{j(\kappa+\delta)} \models \varphi(x) &\longleftrightarrow V_{\kappa+\eta} \models (V_{\kappa+\delta} \models \varphi(x)) \\ &\longleftrightarrow V_\zeta \models (V_{j(\kappa+\delta)} \models \varphi(j(x))) \\ &\longleftrightarrow V_{j(\kappa+\delta)} \models \varphi(j(x))\,. \end{aligned}$$

That restrictions of elementary embeddings in this way are also elementary will be assumed in what follows.

(iv) For the related concept of *complete η-extendibility* and some results concerning it, see Gaifman [4].

(v) The requirement $j(\kappa) > \kappa + \eta$ in 5.1 can be regarded as a natural one, reminiscent of the definition 2.1 of λ-supercompactness. It is a useful, but not stringent, condition; in fact, if it were ever the case that $j(\kappa) \leq \kappa + \eta$, we would have a much stronger axiom ($A_2(\kappa)$ of Section 8). In this connection, we remark that the definition of extendibility given in [40] contains an equivocation (pointed out by Wang) between statements E_0 and E in 6.2c, which leaves Axiom 6.3 unclear. We take this opportunity to resolve this ambiguity: what was intended was E, rather than E_0 (this corresponds to including the condition $j(\kappa) > \kappa + \eta$ in Definition 5.1). This should not affect the ensuing discussion in [40], since it is argued there that the additional condition is natural, though not forced by the guiding idea. (In 6.3 of [40], it is suggested that the critical step in arriving at Kunen's contradiction is the treatment of $V_{\kappa+\eta}$ as a universe in itself, which moreover is absolute in a strong sense (in our formulation, this amounts to setting $\kappa + \eta = \zeta$, rather than anything in the guiding idea behind extendibility as expressed in E or E_0.) The following result shows that (full) extendibility as a concept is not affected, in any case.

Proposition 5.2. *κ is extendible* iff *for every $\delta > \kappa$, there is a ζ and a $j\colon V_\delta \to V_\zeta$ with critical point κ.*

Proof. The forward direction is immediate. For the converse, it suffices to show from the hypothesis that if $\eta \geq \kappa \cdot \kappa$, then κ is η-extendible.

Given such an η, first get an auxiliary ordinal $\gamma > \eta$ so that

(a) $\mathrm{cf}(\gamma) = \omega_1$.

(b) whenever there is a $k\colon V_\eta \to V_\zeta$ with critical point κ so that $k(\kappa) < \gamma$, there is such a k (with the same value for $k(\kappa)$) for some ζ so that $\zeta < \gamma$.

We can use a reflection argument to get (b), and (a) can easily be arranged.

By hypothesis, let $j\colon V_\gamma \to V_\rho$ with critical point κ. Set $\kappa_0 = \kappa$, and for $n \in \omega$, $\kappa_{n+1} = j(\kappa_n)$ whenever possible (i.e. whenever $\kappa_n < \gamma$). If κ_n is defined for each integer n, then $\sup\{\kappa_n \mid n \in \omega\} < \gamma$ since $\mathrm{cf}(\gamma) = \omega_1$. But then, we can now get a contradiction using Kunen's argument (1.12). Thus, it follows that there is an n so that $\kappa_n < \gamma \leq \kappa_{n+1}$.

To conclude the proof, it suffices to establish $P(m)$ for every $m < n+1$ by induction on m, where $P(m)$ states: there is a ζ and an $i\colon V_\eta \to V_\zeta$ with critical point κ so that $i(\kappa) = \kappa_{m+1}$. This is so, since η-extendibility would follow from $P(n)$ and $\kappa_n < \gamma \leq \kappa_{n+1}$.

Define $\tilde{j} = j|V_\eta$. Then $\tilde{j}\colon V_\eta \to V_{j(\eta)}$ with critical point κ so that $\tilde{j}(\kappa) = \kappa_1$. Hence, we have $P(0)$. Now assume $P(m)$, where $m < n$. Then, because $\kappa_{m+1} < \gamma$, by the property (b) of γ, there is an $i\colon V_\eta \to V_\zeta$ for some $\zeta < \gamma$, with critical point κ so that $i(\kappa) = \kappa_{m+1}$. Thus, by the elementarity of j, we have that in V_ρ (and hence in V) there is an $\tilde{i}\colon V_{j(\eta)} \to V_{j(\zeta)}$ with critical point $j(\kappa)$ so that $\tilde{i}(j(\kappa)) = j(\kappa_{m+1}) = \kappa_{m+2}$. Recalling the definition of $\tilde{j}$ above, we can now conclude that $\tilde{i} \circ \tilde{j}\colon V_\eta \to V_{j(\zeta)}$ with critical point κ so that $\tilde{i} \circ \tilde{j} = \kappa_{m+2}$. Hence, $P(m+l)$ holds, and the proof is complete. $\square$

If κ is supercompact, it is consistent that there is no strongly inaccessible cardinal $> \kappa$, since if there were one, we can cut off the universe at the least one and still get have a model of set theory in which κ is supercompact. However, suppose that κ is even 1-extendible, with $j\colon V_{\kappa+1} \to V_{j(\kappa)+1}$. Then by elementarity $j(\kappa)$ is inaccessible in $V_{j(\kappa)+1}$, and hence in V. Similarly, if κ is 2-extendible with $j\colon V_{\kappa+2} \to V_{j(\kappa)+2}$, by elementarity $j(\kappa)$ is measurable in

$V_{j(\kappa)+2}$ and hence in V. Thus, the extendibility of a cardinal κ implies the existence of large cardinals $> \kappa$.

These considerations begin to show how strongly the existence of an extendible cardinal affects the higher levels of the cumulative hierarchy, and why λ-extendibility cannot be formulated, as λ-supercompactness can, merely in terms of the existence of certain ultrafilters. They also point to the close relationship between extendibility and principles of reflection and resernblance. See the end of this section for an elaboration in terms of the Lévy hierarchy of formulas.

We now proceed to investigate extendibility, particularly in connection with supercompactness. Ultimately, we will establish that any extendible cardinal κ is supercompact and is the κth supercompact cardinal, and that the least supercompact cardinal is not even 1-extendible. Note first that 1-extendibility is already quite strong.

Proposition 5.3. *If κ is 1-extendible, then κ is measurable and there is a normal ultrafilter $\mathcal{U}$ over κ so that* $\{\alpha < \kappa \mid \alpha \text{ is measurable}\} \in \mathcal{U}$.

Proof. Let $j\colon V_{\kappa+1} \to V_{j(\kappa)+1}$ with critical point κ. Then $\mathcal{U}$ defined by $X \in \mathcal{U}$ iff $X \subseteq \kappa \ \&\ \kappa \in j(X)$ is normal over κ. Certainly $\mathcal{U} \in V_{j(\kappa)+1}$, so $V_{j(\kappa)+1} \models \kappa$ is measurable, i.e. $\{\alpha < \kappa \mid \alpha \text{ is measurable}\} \in \mathcal{U}$. □

We now proceed to establish some connections between degrees of supercompactness and degrees of extendibility.

Proposition 5.4. *If κ is $|V_{\kappa+\eta}|$-supercompact and $\eta < \kappa$, there is a normal ultrafilter $\mathcal{U}$ over κ so that* $\{\alpha < \kappa \mid \alpha \text{ is } \eta\text{-extendible}\} \in \mathcal{U}$.

Proof. Let $j\colon V \to M$ be as $|V_{\kappa+\eta}|$-supercompactness. Then $V_{\kappa+\eta} \in M$, since $V_{\kappa+\eta}$ is hereditarily of cardinality $\leq |V_{\kappa+\eta}|$. Similarly, if we set $e = j|V_{\kappa+\eta}$, then $e \in M$. Now in M, $e\colon V_{\kappa+\eta} \to j(V_{\kappa+\eta})$ is an elementary embedding with critical point κ and $e(\kappa) > \kappa+\eta$, since this is true in V. Thus, κ is η-extendible in M. If we let $\mathcal{U}$ be the usual normal ultrafilter over κ corresponding to j, then it follows that $\{\alpha < \kappa \mid \alpha \text{ is } \eta\text{-extendible}\} \in \mathcal{U}$. □

Thus, supercompactness implies the existence of many cardinals with some degree of extendibility. One conjecture which expresses confidence in even 2-extendibility is the following.

Open Question 5.5. Does Con(ZFC & there is a 2-extendible cardinal) imply Con(ZFC & there is a strongly compact cardinal)?

Of course, by 5,4 an affirmative answer to this question would imply that the consistency strength of strong compactness is very weak compared to that of supercompactness.

The following proposition reverses the process of 5.4.

Proposition 5.6. *If κ is η-extendible and $\delta+1<\eta$, then κ is $|V_{\kappa+\delta}|$-supercompact. Hence, if κ is extendible, then it is supercompact.*

Proof. Suppose $j\colon V_{\kappa+\eta}\to V_\zeta$ is as in η-extendibility. Since $j(\kappa)$ is really inaccessible and $j(\kappa)>\kappa+\delta$, $j(\kappa)>|V_{\kappa+\delta}|$. Hence, since $\delta+1<\eta$ so that $\mathcal{P}\mathcal{P}V_{\kappa+\delta}\subseteq V_{\kappa+\eta}$, we can define a normal ultrafilter over $\mathcal{P}_\kappa V_{\kappa+\delta}$ as usual:

$$X\in\mathcal{U}\quad\text{iff}\quad j''V_{\kappa+\delta}\in j(X)\ . \qquad \square$$

Incidentally, the methods of 5.4 and 5.6 yield another characterization of supercompactness, which was also noticed by Magidor [22].

Theorem 5.7. *κ is supercompact iff for every $\eta>\kappa$ there is an $\alpha<\kappa$ and a $j\colon V_\alpha\to V_\eta$ with critical point γ so that $j(\gamma)=\kappa$.*

Proof. For the forward direction, fix $\eta>\kappa$ and let $j\colon V\to M$ be as in the $|V_\eta|$-supercompactness of κ. Then just as in 5.4, $j|V_\eta\colon V_\eta\to V^M_{j(\eta)}$ is an elementary embedding which is in M. Thus, M models the following: "There is an $\alpha<j(\kappa)$ and an elementary embedding $e\colon V_\alpha\to V_{j(\eta)}$ with a critical point γ such that $e(\gamma)=j(\kappa)$." The result now follows from the elementarity of j.

For the converse, fix $\eta>\kappa$ and let $j\colon V_{\alpha+\omega}\to V_{\eta+\omega}$ for some $\alpha<\kappa$, with critical point γ so that $j(\gamma)=\kappa$. As in 5.6, since $\mathcal{P}\mathcal{P}_\gamma\alpha\subseteq V_{\alpha+\omega}$, j determines a normal ultrafilter $\mathcal{U}$ over $\mathcal{P}_\gamma\alpha$. But $\mathcal{U}\in V_{\alpha+\omega}$ and so $j(\mathcal{U})$ is a normal ultrafilter over $\mathcal{P}_{j(\gamma)}j(\alpha)=\mathcal{P}_\kappa\eta$. $\square$

The following proposition on supercompactness will be used in the next theorem, but is interesting in its own right.

Proposition 5.8. *If κ is α-supercompact for every $\alpha<\lambda$ and λ is supercompact, then κ is supercompact.*

Proof. Let $\mu\geq\lambda$. We must get a normal ultrafilter over $\mathcal{P}_\kappa\mu$. For each $x\in\mathcal{P}_\lambda\mu$ so that $|x|\geq\kappa$ let $\mathcal{U}_x$ be normal over $\mathcal{P}_\kappa x$. Such $\mathcal{U}_x$ exist since $|x|<\lambda$. Let $\mathcal{V}$ be normal over $\mathcal{P}_\lambda\mu$, and define $\mathcal{U}$ over $\mathcal{P}_\kappa\mu$ by

$$X\in\mathcal{U}\quad\text{iff}\quad\{x\in\mathcal{P}_\lambda\mu\,|\,\mathcal{P}_\kappa x\cap X\in\mathcal{U}_x\}\in\mathcal{V}\,.$$

Since $j_{\mathcal{U}_x}{}''x=[\mathrm{id}]_{\mathcal{U}_x}$, it is not hard to see that $\mathcal{U}$ is a normal ultrafilter over $\mathcal{P}_\kappa\mu$. $\square$

Theorem 5.9. *If κ is supercompact and* 1*-extendible, then there is a normal ultrafilter $\mathcal{U}$ so that* $\{\alpha < \kappa \mid \alpha \text{ is supercompact}\} \in \mathcal{U}$. *Hence, the least supercompact cardinal is not* 1*-extendible.*

Proof. Let $j\colon V_{\kappa+1} \to V_{j(\kappa)+1}$ be as in 1-extendibility, and let $\mathcal{U}$ be the usual normal ultrafilter over κ corresponding to j, as in 5.3. Now $V_{j(\kappa)+1} \models \kappa$ is δ-supercompact for all $\delta < j(\kappa)$, since $j(\kappa)$ is inaccessible. Hence, $A = \{\alpha < \kappa \mid \alpha$ is δ-supercompact for all $\delta < \kappa\} \in \mathcal{U}$. But by the previous proposition, $\alpha \in A$ implies α is supercompact. □

Note that if κ is extendible, $\gamma_0 = \kappa + 1$, and $\gamma_{n+1} =$ least ordinal ζ so that there is a $j\colon V_{\gamma_n} \to V_\zeta$ as in γ_n-extendibility, then if $\gamma = \sup\{\gamma_n \mid n \in \omega\}$, we have $V_\gamma \models \kappa$ is extendible. However, V_γ may not model ZFC. In contrast, consider the following.

Proposition 5.10. *If $\kappa < \lambda$, κ is extendible and λ is supercompact, then* $V_\lambda \models \kappa$ is extendible.

Proof. Suppose that $\kappa \cdot \kappa < \alpha < \lambda$. We must show that $V_\lambda \models \kappa$ is α-extendible. Since κ is α-extendible in V, there is a $j\colon V_\alpha \to V_\beta$ with critical point κ, so that $j(\kappa) > \alpha$. If $\beta < \lambda$, we are done, so assume $\beta \geq \lambda$.

Let $k\colon V \to M$ be as in the $|V_\beta|$-supercompactness of λ. Then $M \models j\colon V_\alpha \to V_\beta$ with critical point κ, and $j(\kappa) > \alpha$ and $\beta < k(\lambda)$. So, by elementarity, in V there is a $\delta < \lambda$, and $\bar{j}\colon V_\alpha \to V_\beta$ with critical point κ, and $\bar{j}(\kappa) > \alpha$. This, $V_\lambda \models \kappa$ is α-extendible. □

One can also prove 5.10 using the fact that the property "κ is extendible" is Π_3 in the Lévy hierarchy and so reflects down to V_κ for λ supercompact, by 2.7.

We thus see that the extendibility of a cardinal κ can already be comprehended in V_λ where λ is a supercompact cardinal $> \kappa$. This shows in particular that it is consistent to assume there are no supercompact cardinals above an extendible cardinal. Perhaps, 5.10 may serve to allay suspicions about extendibility, which might arise from the fact that it has as a consequence the existence of proper classes of various large cardinals.

In terms of the Lévy hierarchy, this last fact about extendibility can be expressed as follows. Any local property (i.e. Σ_2, see Section 2) of an extendible cardinal κ holds for a proper class of cardinals. The example of supercompactness shows that one cannot prove in ZFC that every Π_2 property of an extendible cardinal κ holds for some $\lambda > \kappa$. Finally, it follows readily from 5.10 that "κ is the least extendible cardinal" is Π_3 (being equivalent to "κ is extendible & $\forall \mu < \kappa(V_\kappa \models \mu$ is not extendible)")—and certainly holds for no $\lambda > \kappa$.

Theorem 5.11. *If κ is extendible, then Σ_3 (and hence Π_4) relativize down to* V_κ. (Recall 2.6 for this notion.)

Proof. Actually, we only use the fact that there are arbitrarily large inaccessibles $\lambda > \kappa$ with $V_\kappa \prec V_\lambda$. Suppose $P(x)$ is $\exists y Q(x, y)$ where Q is Π_2. Let $a \in V_\kappa$ so that $P(a)$ holds, and fix b such that $Q(a, b)$. Let $\lambda > \kappa$ be inaccessible so that $b \in V_\lambda$ and $V_\kappa \prec V_\lambda$. Since, by a remark just after 2.6, Π_2 relativizes down to V_λ, we have $V_\lambda \models Q(a, b)$. Thus, $V_\lambda \models P(a)$, and so $V_\kappa \models P(a)$. □

Note that the Σ_4 sentence "There is an extendible cardinal" is false in V_λ if λ is the least extendible cardinal (by 5.6 and 5.10). Thus, Π_4 in 5.11 is optimal.

6 Vopěnka's Principle

We next consider an axiom of a different character both from supercompactess and from extendibility. Bearing in mind our theme of elementary embedding and considerations of resemblance, the motivation behind the following statement is evident, especially in the context of model theory.

> Vopěnka's Principle. Given a proper class of (set) structures of the same type, there exists one that can be elementarily embedded in another.

This concept was also considered independently by Keisler. It may not be immediately clear that Vopěnka's Principle is a very strong axiom of infinity at all, but we shall prove that the principle implies the existence of many extendible cardinals in a strong sense. In Section 8, it will be shown that the principle actually has a natural place in a hierarchy of strong axioms of infinity.

Vopěnka's Principle definitely cannot be formulated in ZFC, and in this section, we will freely use quantification and comprehension over classes. However, all the manipulations can be carried out in ZFC within some V_κ where κ is inaccessible (indeed, this is how A_8 of Section 8 is stated). Formulated in this way, note that Vopěnka's Principle is a second order statement about V_κ, whereas even measurability is third order. Indeed, one significant way in which Vopěnka's Principle differs from our previous axioms is that it does not merely assert the existence of a large cardinal with higher order properties, but provides a framework in which many such cardinals can be shown to exist.

We now assume Vopěnka's Principle throughout this section. The approach here is reminiscent of Lévy [21] in that a natural filter is developed and used as a tool. Recall that Ω is the class of all ordinals. Call a sequence of structures $\langle \mathcal{M}_\alpha \mid \alpha \in \Omega \rangle$ a *natural sequence* iff each $\mathcal{M}_\alpha$ is of the same fixed type and specifically of form $\langle V_{\xi_\alpha}; \in, \{\alpha\}, A_\alpha \rangle$ where A_α codes a finite number of relations and $\alpha < \beta$ implies $\alpha < \xi_\alpha \leq \xi_\beta$. The specification of $\{\alpha\} \in \mathcal{M}_\alpha$ insures that whenever $\alpha < \beta$ and $j \colon \mathcal{M}_\alpha \to \mathcal{M}_\beta$ is elementary, $j(\alpha) = \beta$.

Definition 6.1. If $X \subseteq \Omega$ is a class, X is *enforced* by a natural sequence $\langle \mathcal{M}_\alpha \mid \in \Omega \rangle$ whenever $\alpha < \beta$ and $j_\alpha \colon \mathcal{M}_\alpha \to \mathcal{M}_\beta$, the critical point of j is a member of X. X is *enforceable* iff X is enforced by some natural sequence.

Proposition 6.2. *The enforceable classes form a proper filter over* Ω.

Proof. Clearly, if X is enforceable and $X \subseteq Y \subseteq \Omega$, then Y is enforceable. $\emptyset$ is not enforceable since we are assuming Vopěnka's Principle. Suppose now that X and Y are enforced by $\langle \mathcal{M}_\alpha \mid \alpha \in \Omega \rangle$ and $\langle \mathcal{N}_\alpha \mid \alpha \in \Omega \rangle$ respectively. Set $\mathcal{A}_\alpha = \langle V_{\xi_\alpha}; \in, \{\alpha\}, \langle \mathcal{M}_\alpha, \mathcal{N}_\alpha \rangle \rangle$ where V_{ξ_α} is the union of the underlying sets of $\mathcal{M}_\alpha$ and $\mathcal{N}_\alpha$. Then $X \cap Y$ is enforced by $\langle \mathcal{A}_\alpha \mid \alpha \in \Omega \rangle$: If $j \colon \mathcal{A}_\alpha \to \mathcal{A}_\beta$ with critical point κ, $j|\mathcal{M}_\alpha$ and $j|\mathcal{N}_\alpha$ both have critical point κ, i.e. $\kappa \in X \cap Y$. $\square$

Proposition 6.3. *Every closed unbounded subclass of* Ω *is enforceable.*

Proof. Suppose $C \subseteq \Omega$ is closed unbounded. For each ordinal α let γ_α be the least limit point of C greater than α, and set

$$\mathcal{M}_\alpha = \langle V_{\gamma_\alpha}; \in, \{\alpha\}, C \cap \gamma_\alpha \rangle .$$

We show that C is enforced by $\langle \mathcal{M}_\alpha \mid \alpha \in \Omega \rangle$. Suppose $j \colon \mathcal{M}_\alpha \to \mathcal{M}_\beta$ with critical point κ, and assume $\kappa \notin C$. Then $\rho = \sup(C \cap \kappa) < \kappa$ and if η is the least element of C greater than ρ, $\kappa < \eta < \gamma_\alpha$ since γ_α is a limit point of C. As η is definable from ρ in $\mathcal{M}_\alpha$ and $j(\rho) = \rho$, $j(\eta) = \eta$. But, as usual, $\lambda = \sup\{j^n(\kappa) \mid n \in \omega\}$ is the least ordinal greater than κ fixed by j, so that $\lambda \leq \eta$. We can now derive a contradiction by Kunen's argument (1.12), since γ_α is a limit ordinal $> \eta$. Thus, $\kappa \in C$. $\square$

In fact, the enforceable classes form a normal, Ω-complete filter over Ω; see Kanamori [11]. This paper discusses strong versions of Vopěnka's Principle related to n-hugeness (see Section 7 for this concept), which are analogous to the n-subtle and n-ineffable cardinals studied by Baumgartner.

Proposition 6.4. $\{\alpha \in \Omega \mid \alpha \text{ is extendible}\}$ *is enforceable.*

Proof. Define $F \colon \Omega \to \Omega$ by

$$F(\alpha) = \begin{cases} \alpha & \text{if } \alpha \text{ is extendible,} \\ \alpha + \beta & \beta \text{ is the least so that } \alpha \text{ is not} \\ & \beta\text{-extendible otherwise.} \end{cases}$$

If $C = \{\delta \mid F|\delta \colon \delta \to \delta\}$, C is closed unbounded. Define $\langle \mathcal{M}_\alpha \mid \alpha \in \Omega \rangle$ for this C, just as in the previous proposition.

It suffices to show that if $j \colon \mathcal{M}_\alpha \to \mathcal{M}_\beta$ with critical point κ, then κ is extendible. If not, let $F(\kappa) = \rho > \kappa$. Since $\kappa < \gamma_\alpha$ and $\gamma_\alpha \in C$, we have $\rho < \gamma_\alpha$ by definition of C. So, $j|V_\rho \colon V_\rho \to V_{j(\rho)}$ is elementary with critical point κ. Finally, by the proof of the previous proposition, $\kappa \in C$ and so $j(\kappa) \in C$ (recall that $C \cap \gamma_\alpha$ is a specified relation in $\mathcal{M}_\alpha$). Hence, $\kappa < j(\kappa)$ implies $\rho < j(\kappa)$. This also implies $\rho < j(\rho)$. If $\rho = \kappa + \beta$, all these facts show that κ was β-extendible after all, which contradicts the definition of F. □

Proposition 6.5. *If X is enforceable and $Y = \{\alpha \mid \alpha$ is measurable and for some normal ultrafilter $\mathcal{U}$ over α, $X \cap \alpha \in \mathcal{U}\}$ then Y is also enforceable.*

Proof. For each ordinal α, set

$$\mathcal{M}_\alpha = \langle V_{\alpha+\omega}; \in, \{\alpha\}, \{X \cap \alpha\} \rangle .$$

It suffices to show that if $j \colon \mathcal{M}_\alpha \to \mathcal{M}_\beta$ with critical point κ and $\kappa \in X$, then $\kappa \in Y$. Let $\mathcal{U}$ be the normal ultrafilter over κ corresponding to j. We have

$$j(X \cap \kappa) = j(X \cap \alpha \cap \kappa) = j(X \cap \alpha) \cap j(\kappa) = X \cap \beta \cap j(\kappa) = X \cap j(\kappa) .$$

But $\kappa \in X \cap j(\kappa)$, so $X \cap \kappa \in \mathcal{U}$. □

Theorem 6.6. *Assuming Vopěnka's Principle, the class of extendible cardinals κ which carry a normal ultrafilter containing $\{\alpha < \kappa \mid \alpha$ is extendible$\}$ is stationary in Ω.*

Proof. See 6.3, 6.4, and 6.5. □

All in all, Vopěnka's Principle seems to have an unbridled strength, and the relative ease with which strong consequences can be derived from it may lead one to be rather suspicious of the principle. However, it is the weakest in the hierarchical list of axioms we will consider in Sections 7-8.

In the remainder of this section, we present two alternative characterizations of Vopěnka's Principle. In addition to their intrinsic interest, these results will be useful in the discussion of Backward Easton forcing in a projected sequel to this paper.

Let A be a class. We are going to relativize the notions of supercompactness and extendibility to the class A.

Definition 6.7. A cardinal κ is *A-extendible* iff for every $\alpha > \kappa$, there is an elementary embedding

$$j \colon \langle V_\alpha; \in, A \cap V_\alpha \rangle \to \langle V_\beta; \in, A \cap V_\beta \rangle$$

with κ the critical point of j and $\alpha < j(\kappa) < \beta$.

A cardinal κ is *A-supercompact* iff for every $\eta > \kappa$ there is an $\alpha < \kappa$ and an elementary embedding

$$j\colon \langle V_\alpha; \in, A \cap V_\alpha\rangle \to \langle V_\eta; \in, A \cap V_\eta\rangle$$

with critical point γ such that $j(\gamma) = \kappa$ (cf. 5.7).

We: can also give a characterization in terms of normal ultrafilters. Let $\mathcal{U}$ be a normal ultrafilter over $\mathcal{P}_\kappa V_\eta$. Recall (2.5) that if $\eta = [\langle \eta_x \mid x \in \mathcal{P}_\kappa V_\eta\rangle]$, for almost all $x \in \mathcal{P}_\kappa V_\eta$ there is an elementary embedding $j_x\colon V_{\eta_x} \to V_\eta$ with $x = j_x{}''V_{\eta_x}$.

We say that $\mathcal{U}$ is *A-normal* iff for almost all x,

$$j_x\colon \langle V_{\eta_x}; \in, A \cap V_{\eta_x}\rangle \to \langle V_\eta; \in, A \cap V_\eta\rangle$$

is elementary. Evidently $\mathcal{U}$ is A-normal iff $\mathcal{U}$ is $(A \cap V_\eta)$-normal.

Proposition 6.8. *κ is A-supercompact* iff *for every $\eta \geq \kappa$, there is an A-normal ultrafilter over $\mathcal{P}_\kappa V_\eta$.*

Proof. We show that if κ is A-supercompact and $\eta \geq \kappa$, there is an A-normal ultrafilter over $\mathcal{P}_\kappa V_\eta$. The other implication is left to the reader (cf. the proof of 5.7.)

By the A-supercompactness of κ, let $\alpha < \kappa$ and

$$j\colon \langle V_{\alpha+\omega}; \in, A \cap V_{\alpha+\omega}\rangle \to \langle V_{\eta+\omega}; \in, A \cap V_{\eta+\omega}\rangle$$

with critical point γ such that $j(\gamma) = \kappa$. Note that $j(A \cap V_\alpha) = A \cap V_\eta$.

Define $\mathcal{U}$ over $\mathcal{P}_\gamma V_\alpha$ by $X \in \mathcal{U}$ iff $j''V_\alpha \in j(X)$. Then, as usual, $\mathcal{U}$ is a normal ultrafilter over $\mathcal{P}_\gamma V_\alpha$. Let $\alpha = [\langle \alpha_x \mid x \in \mathcal{P}_\gamma V_\alpha\rangle]_{\mathcal{U}}$, and recall 2.5(i) in what follows. For almost every x, $\langle x; \in, A \cap x\rangle$ is an elementary submodel of $\langle V_\alpha; \in, A \cap V_\alpha\rangle$. Thus, $\mathcal{U}$ is A-normal iff the transitive collapse $\pi_x\colon x \cong V_{\alpha_x}$ "preserves A" for almost every x iff the transitive collapse $\pi\colon j''V_\alpha \cong V_\alpha$ "preserves A". But this last formulation is evident since $\pi(j(x)) = x$ for $x \in V_\alpha$, and j "preserves A".

Thus, $\mathcal{U}$ is an $(A \cap V_\alpha)$-normal ultraflter over $\mathcal{P}_\gamma V_\alpha$. But then $j(\mathcal{U})$ is a $j(A \cap V_\alpha)$-normal ultrafilter over $\mathcal{P}_\kappa V_\eta$. Since $j(A \cap V_\alpha) = A \cap V_\eta$, $j(\mathcal{U})$ is the desired A-normal ultrafilter over $\mathcal{P}_\kappa V_\eta$. □

Note that if $A = \{\gamma\}$, then γ is not A-supercompact or A-extendible. However,

Theorem 6.9. *The following are equivalent:*

(1) *Vopěnka's Principle,*

(2) *For every class A, there is an A-extendible cardinal,*

(3) *For every class A, there is an A-supercompact cardinal.*

We remark that our proof will show in fact that if (1) holds, the classes of A-extendible cardinals and A-supercompact cardinals are enforceable.

Proof. The proof of 6.4 adapts to show that (1) $\to$ (2). Also, the proof of 6.8 can be used to show that every A-extendible cardinal is A-supercompact. Thus (2) $\to$ (3).

Finally, let $\langle \mathcal{M}_\xi \mid \xi \in \Omega \rangle$ be as in the statement of Vopěnka's Principle. Let $A = \{\langle \xi, \mathcal{M}_\xi \rangle \mid \xi \in \Omega\}$. Let κ be A-supercompact. Let $\theta > \kappa$ be such that:

(i) $\xi < \theta \to \mathcal{M}_\xi \in V_\theta$, and

(ii) $|V_\theta| = \theta$.

Let $\alpha < \kappa$, and

$$j\colon \langle V_\alpha; \in, A \cap V_\alpha \rangle \to \langle V_\theta; \in, A \cap V_\theta \rangle$$

be an elementary embedding that maps its critical point γ onto κ. Then j induces an elementary embedding of $\mathcal{M}_\gamma$ into $\mathcal{M}_\kappa$. □

7 On the Verge of Inconsistency

Having examined several axioms increasing in strength and motivated with different but definite plausibility arguments in mind, we now take a more pragmatic approach. Kunen's result (1.12) sets an upper bound to our efforts in an essential way, but it is still of interest to see what weaker principles can possibly be retained without inconsistency in ZFC. In this and the next section we work downward through weaker and weaker axioms that suggest themselves, are at least as strong as Vopěnka's Principle, but are not directly ruled out by Kunen's argument.

Tacit in this section is the assumption that if j is some elementary embedding with critical point κ, then $\kappa_0 = \kappa$, and for each integer n, $\kappa_{n+1} = j(\kappa_n)$ if κ_n is still in the domain of j, and $\kappa_\omega = \sup\{\kappa_n \mid n \in \omega\}$, again, if definable at all. (Of course, the κ_n's depend on j, but the j being discussed should be clear from the context.)

If there were a $j\colon V_\alpha \to V_\beta$ with $\kappa_\omega \leq \alpha$, note that we must have $\alpha < \kappa_\omega + 2$. This is so since what is needed to get Kunen's contradiction is that a function ω-Jónsson over κ_ω, i.e. a certain function: ${}^\omega\kappa_\omega \to \kappa_\omega$, be the domain of j—but all such functions are in $V_{\kappa_\omega+2}$. When a function ω-Jónsson over κ_ω is in the domain of an elementary embedding as in, say, $j\colon V \to M$, then by Kunen's argument we must have $V_{\kappa_\omega+1} \not\subseteq M$. However, we can still consider the following statements.

11. There is a $j\colon V_{\kappa_\omega+1} \to V_{\kappa_\omega+1}$.

I2. There is a $j\colon V \to M$ with $V_{\kappa_\omega+1} \subseteq M$.

13. There is a $j\colon V_{\kappa_\omega} \to V_{\kappa_\omega}$.

Notice that in I1 we have specified the range of j to be included in $V_{\kappa_\omega+1}$, but this is true since $j(\kappa_\omega) = \kappa_\omega$; similarly for I3. In fact, I1 and I3 are the only possible forms that an axiom of the type "there is a non-trivial elementary embedding of some V_α into itself" can take.

The proof of the next proposition uses iteration and limit ultrapower techniques.

Proposition 7.1. (Gaifman) I1 *implies* I2.

Proof. See IV.8 of Gaifman [4]. □

Proposition 7.2. 12 *implies* 13.

Proof. If j is as in I2, then $j(V_{\kappa_\omega}) = V^M_{\kappa_\omega} = V_{\kappa_\omega}$, so that

$$j|V_{\kappa_\omega}\colon V_{\kappa_\omega} \to V_{\kappa_\omega}\,. \qquad \square$$

Next, it is natural to consider postulations with weaker closure requirements on the range of the embedding.

Definition 7.3. If n is an integer, κ *n-huge* iff there is a $j\colon V \to M$ with critical point κ so that ${}^{\kappa_n}M \subseteq M$. κ is *huge* (Kunen) iff κ is 1-huge.

Note that κ is 0-huge iff κ is measurable, and κ n-huge implies κ_n is inaccessible in V itself. It is interesting to note that, reminiscent of λ-supercompactness, a characterization exists in terms of the existence of certain ultrafilters.

Theorem 7.4. *κ is n-huge iff there is a κ-complete normal ultrafilter $\mathcal{U}$ over some $\mathcal{P}\lambda$, and cardinals $\kappa = \lambda_0 < \lambda_1 < \cdots < \lambda_n = \lambda$ so that for each $i < n$, $\{x \mid \overline{x \cap \lambda_{i+1}} = \lambda_i\} \in \mathcal{U}$.*

Proof. If $: V \to M$ as in n-hugemess, define $\mathcal{U}$ over $\mathcal{P}\kappa_n$ by

$$X \in \mathcal{U} \ \text{ iff } \ j''\kappa_n \in j(X)\,.$$

Then as in 2.2, we can show that $\mathcal{U}$ is normal and κ-complete. Also, note that

$$\overline{j''\kappa_n \cap j(\kappa_i)} = \overline{j''\kappa_i} = \kappa_i \ \text{ for } \ 0 \le i \le n\,,$$

so that we can set $\kappa_i = \lambda_i$.

Conversely, take j, M the ultrapower as usual. Then, as in 2.4, $[\mathrm{id}] = j''\lambda$ and M is closed under λ-sequences. Also, as in 3.1, we have that for $0 \leq i < n$,

$$j(\lambda_i) = [\langle \overline{x \cup \lambda_{i+1}} \,|\, x \in \mathcal{P}\lambda \rangle] = \lambda_{i+1}\,. \qquad \square$$

Theorem 7.5. *If* I3 *or* κ *is* $n+1$*-huge, then there is a normal utrafilter* $\mathcal{U}$ *over* κ *so that* $\{\alpha < \kappa \,|\, \alpha \text{ is } n\text{-huge}\} \in \mathcal{U}$.

Proof. Suppose, for example, that $j\colon V \to M$ as in the $n+1$-hugeness of κ. Since ${}^{\kappa_{n+1}}M \subseteq M$, M certainly contains the ultrafilter described in 7.4 for the n-hugeness of κ, arising from j. Hence, $M \models \kappa$ is n-huge, and we can take $\mathcal{U}$ to be the (usual) normal ultrafilter over κ corresponding to j. $\square$

It seems likely that I1, I2 and I3 are all inconsistent since they appear to differ from the proposition proved inconsistent by Kunen only in an inessential technical way. The axioms asserting the existence of n-huge cardinals, for $n > 1$, seem (to our unpracticed eyes) essentially equivalent in plausibility: far more plausible that I3, but far less plausible than say extendibility.

Kunen's work [18] relates 1-hugeness (i.e. hugeness) to the theory of saturated ideals. He shows that Con(ZFC & there is a huge cardinal) implies Con(ZFC & there is a countably complete, ω_2-saturated ideal on ω_1 containing all the singletons). Kunen also indicates a heuristic argument suggesting that the consistency of something slightly weaker than hugeness ($\mathrm{A}_3(\kappa)$ of the next section) should follow from the consistency of the existence of such a non-trivial ω_2-saturated ideal on ω_1.

More recently, Laver has announced a refinement of Kunen's argument to get an ideal over ω_1 with an even stronger property, which has as a consequence the following polarized partition relation:

$$\begin{pmatrix} \aleph_2 \\ \aleph_1 \end{pmatrix} \to \begin{pmatrix} \aleph_1 \\ \aleph_1 \end{pmatrix}_{\aleph_0}$$

Since the GCH can also be arranged in this particular case, this answers Problem 27 of Erdős–Hajnal [3] in the negative (at least if we assume the consistency of the existence of a huge cardinal!). Prikry [37] had previously shown that there is a combinatorial principle which implies, in a strong sense, that the partition relation does not hold, and that this principle can be made to hold by forcing. Jensen then showed that this principle holds in L.

An ultrafilter $\mathcal{U}$ is called (κ, λ)*-regular* iff there are λ sets in $\mathcal{U}$ any κ of them having empty intersection. This concept was formulated by Keisler in the context of model theory some time ago. Recently, it has been shown that the existence of, for instance, a uniform ultrafilter over ω_1 which is not (ω, ω_1)-regular leads to consequences of large cardinal character (see Kanamori [10] and Ketonen [13]). In this context, we state the following result of Magidor [28], proved by using a variation of the Kunen [18] argument: Con(ZFC & there is a huge cardinal) implies Con(ZFC & there is a uniform ultrafilter over ω_2 not (ω, ω_2)-regular). At present, there is no other known way to get uniform ultrafilters with any degree of irregularity over any ω_n.

Finally. the most recent relative consistency result involving hugeness is the following, due once again to Magidor [26]: Con(ZFC & there is a huge cardinal with a supercompact cardinal below it) implies Con(ZFC & $2^{\aleph_\omega} = \aleph_{\omega+2}$, yet for every integer n, $2^{\aleph_n} = \aleph_{n+1}$). This, of course, solves the so-called singular cardinals problem at $\aleph_\omega$. Previously, we had remarked (end of Section 2) that Magidor had found a model in which $2^{\aleph_\omega} = \aleph_{\omega+2}$ and $\aleph_\omega$ is a strong limit cardinal, assuming the consistency strength of the existence of a cardinal with sufficient degree of supercompactness. To get the exactitude of the GCH actually holding below $\aleph_\omega$, Magidor found it necessary to start from his stronger assumption.

It is tempting to, speculate on the further relevance of huge cardinals in considerations involving the lower orders of the cumulative hierarchy. After all, it is such empirical evidence which gained for measurability a certain respectability, if not acceptance.

8 Below Huge

This sectlon contains the rest of the new axioms to be considered in this paper. They are indeed to fill in the gap between the concept of hugeness and the relatively weak one of extendibility with a spectrum of statements. Though we are thus continuing to take a pragmatic approach, hopefully these further axioms will prove interesting in their own right. At least, their motivations should be clear in the context of this paper. By a *natural model* of KM (Kelley–Morse) we mean one of form $V_{\kappa+1}$, where κ is inaccessible and elements of V_κ are to be the "sets". The axioms are as follows.

$A_1(\kappa)$. There is a $j\colon V \to M$ with critical point κ, so that ${}^{j(\kappa)}M \subseteq M$. ($\kappa$ is huge.)

$A_2(\kappa)$. There is a $j\colon V_\alpha \to V_\beta$ with critical point κ, so that $j(\kappa) \leq \alpha$.

$A_3(\kappa)$. There is a $j\colon V \to M$ with critical point κ, so that ${}^{\lambda}M \subseteq M$ for every $\lambda < j(\kappa)$.

$A_4(\kappa)$. There is a $\lambda > \kappa$ and normal ultrafilter $\mathcal{U}$ over $\mathcal{P}_\kappa\lambda$ so that if $M \approx V^{\mathcal{P}_\kappa\lambda}/\mathcal{U}$ and $f \in {}^{\kappa}\kappa$, then $M \models j(f)(\kappa) < \lambda$.

$\mathrm{A}_5(\kappa)$. There is a normal ultrafilter $\mathcal{U}$ over κ with the following property: Suppose $\langle \mathcal{M}_\xi \mid \xi < \kappa \rangle$ is a sequence of structures of the same type with each $\mathcal{M}_\xi \in V_\kappa$. Then for some $X \in \mathcal{U}$, whenever $\alpha, \beta \in X$ and $\alpha < \beta$, there is an elementary embedding $\mathcal{M}_\alpha \to \mathcal{M}_\beta$ which fixes any element of V_α in its domain and moves α if α is in the domain.

$\mathrm{A}_6(\kappa)$. $V_{\kappa+1}$ is a natural model of KM and Vopěnka's Principle: Given a proper class of (set) structures of the same type, there exists one that can be elementarily embedded into another.

$\mathrm{A}_6^*(\kappa)$. $V_{\kappa+1}$ is a natural model of KM and the following: There is a stationary class S so that for any integer n and $\alpha_0 < \alpha_1 < \cdots < \alpha_n < \beta_0 < \beta_1 < \cdots < \beta_n$ all in S, there is a $j\colon V_{\alpha_n} \to V_{\beta_n}$ with critical point α_0 and $j(\alpha_i) = \beta_i$ for $0 \le i < n$.

$\mathrm{A}_7(\kappa)$. κ is extendible and carries a normal ultrafilter containing $\{\alpha < \kappa \mid \alpha$ is extendible$\}$ as a member.

It is convenient in this section to use the following terminology: if $\psi(\cdot)$ and $\varphi(\cdot)$ are both formulas with one free variable, say that $\psi(\kappa$ *strongly implies* $\varphi(\kappa)$ iff $\psi(\kappa)$ implies $\varphi(\kappa)$ and also that there is a normal ultrafilter over κ containing $\{\alpha < \kappa \mid \varphi(\alpha)\}$ as a member. We shall prove that $\exists\kappa A_1(\kappa)$ implies $\exists\kappa A_2(\kappa)$ in a strong sense, $\mathrm{A}_i(\kappa)$ strongly implies $\mathrm{A}_{i+1}(\kappa)$ for $1 < i < 5$, and that $\mathrm{A}_6(\kappa)$ and $\mathrm{A}_6^*(\kappa)$ are both strongly implied by $\mathrm{A}_5(\kappa)$. In Section 6, we showed that $\mathrm{A}_6(\kappa)$ implies $\mathrm{A}_7(\kappa)$ in a suitably strong sense; in this section, we show that $\mathrm{A}_6^*(\kappa)$ similarly implies $\mathrm{A}_7(\kappa)$ in a strong sense. However, the exact relationship between $\mathrm{A}_6(\kappa)$ and $\mathrm{A}_6^*(\kappa)$ is as yet unclear. The rest of this section is devoted to the task of establishing these implications.

In terms of relative consistency, we are dealing with very strong principles. However, it should be pointed out that, except for $\mathrm{A}_7(\kappa)$, all of these various assertions about κ, as well as the notion of n-hugeness, are local properties, and so do not even imply that κ is supercompact. By 2.6, if any one of these properties holds for some cardinal at all, then it holds for a cardinal less than the least supercompact cardinal. On the other hand, as remarked by Morgenstern [34], a straightforward application of work of Magidor shows that, for example, it is consistent for the least huge cardinal to be larger than the least strongly compact cardinal.

Theorem 8.1. $\mathrm{A}_1(\kappa)$ *implies that there is a normal ultrafilter* $\mathcal{U}$ *over* κ *so that* $\{\alpha < \kappa \mid \mathrm{A}_2(\alpha)\} \in \mathcal{U}$.

Proof. Let $j\colon V \to M$ show that κ is huge. As in previous arguments, $V_{j(\kappa)} \in M$ and $j|V_{j(\kappa)} \in M$. Hence, as $j|V_{j(\kappa)}\colon V_{j(\kappa)} \to j(V_{j(\kappa)})$, this is also true in M. Thus, it follows that if $\mathcal{U}$ the normal ultrafilter over κ corresponding to

j, $\{\alpha < \kappa \mid$ there are β, γ, and e so that $e\colon V_\beta \to V_\gamma$ with critical point α and $e(\alpha) \leq \beta\} \in \mathcal{U}$. But from this, the result follows. □

For the passage from A_2 to A_3 some auxiliary notions are needed. If $\kappa < \lambda$, say that a sequence $\langle \mathcal{U}_\eta \mid \kappa \leq \eta < \lambda\rangle$ is *coherent* iff each $\mathcal{U}_\eta$ is a normal ultrafilter over $\mathcal{P}_\kappa\eta$ and $\kappa \leq \eta \leq \zeta < \lambda$ implies $\mathcal{U}_\zeta|\eta = \mathcal{U}_\eta$. If $\langle \mathcal{U}_\eta \mid \kappa \leq \eta < \lambda\rangle$ is coherent, for convenience let $M_\eta = M_{\mathcal{U}_\eta}$, and $j_\eta\colon V \to M_\eta$ the canonical elementary embedding. Recall that by 3.7, there exist elementary embeddings $k_{\eta\zeta}\colon M_\eta \to M_\zeta$ for $\kappa \leq \eta \leq \zeta < \lambda$ so that $k_{\eta\zeta}|\eta$ is the identity. In this notation, consider the following statement.

$A_3^*(\kappa)$. There is a strongly inaccessible cardinal $\lambda > \kappa$ and a coherent sequence $\langle \mathcal{U}_\eta \mid \kappa \leq \eta < \lambda\rangle$ so that if $\kappa \leq \eta < \lambda$ and $\eta \leq \rho < j_\eta(\kappa)$, Then there is a ζ so that $\eta \leq \zeta < \lambda$ and $\kappa_{\eta\zeta}(\rho) = \zeta$.

$A_3^*(\kappa)$ is a technical statement that we show is actually equivalent to $A_3(\kappa)$, and then we prove that $A_2(\kappa)$ strongly implies $A_3^*(\kappa)$. Ostensibly, $A_3^*(\kappa)$ involves proper classes j_η and $k_{\eta\zeta}$, but actually it can be considered a statement in $j_\eta|V_\lambda$ and $k_{\eta\zeta}|V_\lambda$. Hence, $A_3^*(\kappa)$ can be expressed in the form $\exists\lambda > \kappa(V_{\lambda+1} \models \varphi(\kappa))$ for some suitable set-theoretical formula φ. This aspect of $A_3^*(\kappa)$ becomes significant in 8.3.

8.2. Theorem. $A_3(\kappa)$ *is equivalent to* $A_3^*(\kappa)$.

Proof. Suppose first that $A_3(\kappa)$ and let $j\colon V \to M$ with critical point κ, so that M is closed under $< j(\kappa)$-sequences. If we set $\lambda = j(\kappa)$, then λ is inaccessible. For $\kappa \leq \eta < \lambda$ be the (usual) normal ultrafilter over $\mathcal{P}_\kappa\eta$ corresponding to j. Then it is not hard to see that $\langle \mathcal{U}_\eta \mid \kappa \leq \eta < \lambda\rangle$ is coherent.

Assume now $\kappa \leq \eta < \lambda$, $f\colon \mathcal{P}_\kappa\eta \to \kappa$, and $\eta \leq [f]_{\mathcal{U}_\eta} < j_\eta(\kappa)$. (We are using the notation developed for coherent sequences just after 8.1.) Then by definition of $\mathcal{U}_\eta$, $\eta = \overline{j''\eta} \leq j(f)(j''\eta) < j(\kappa) = \lambda$. Let $\zeta = j(f)(j''\eta)$. But then, $\overline{j''\zeta} = \zeta = j(f)(j''\zeta \cap j(\eta))$ and so $\{x \in \mathcal{P}_\kappa\zeta \mid \bar{x} = f(x \cap \eta)\} \in \mathcal{U}_\xi$, i.e. $k_{\eta\zeta} = ([f]_{\mathcal{U}_\eta}) = \zeta$ by the definition of $k_{\eta\zeta}$ (see 3.7 and before). Thus, $A_3^*(\kappa)$ holds.

Suppose now that $A_3^*(\kappa)$. Continuing to use the notation developed for coherent sequences, note that $\langle \mathcal{M}_\eta, k_{\eta\zeta} \mid \kappa \leq \eta \leq \zeta < \lambda\rangle$ forms a directed system. Since λ is regular, the direct limit of this system is well-founded, so let $\tilde{M}$ be its transitive collapse. There are canonical embeddings $k_\eta\colon M_\eta \to \tilde{M}$ and $\tilde{j}\colon V \to \tilde{M}$, where for $\kappa \leq \eta < \lambda$, $\tilde{j} = k_\eta \circ j_\eta$.

It suffices to show that $\tilde{M}$ is closed under $< \lambda$-sequences, and that $\tilde{j}(\kappa) = \lambda$. First, note that if $s = \{x_\alpha \mid \alpha < \gamma\} \subseteq \tilde{M}$ where $\gamma < \lambda$, then by regularity of λ there is an $\eta < \lambda$ so that $t = \{y_\alpha \mid \alpha < \gamma\} \subseteq M_\eta$, where for $\alpha < \lambda$, $k_\eta(y_\alpha) = x_\alpha$. We can certainly assume that $\gamma < \eta$ so that $t \in M_\eta$ and hence $k_\eta(t) = s \in \tilde{M}$.

To show that $\tilde{j}(\kappa) = \lambda$, note first that if $\eta < \lambda$, since $\eta < j_\eta(\kappa)$,

$$\eta \leq k_\eta(\eta) < k_\eta(j_\eta(\kappa)) = \tilde{j}(\kappa) .$$

Hence, $\tilde{j}(\kappa) \geq \lambda$. Conversely, suppose $\rho < \tilde{j}(\kappa)$. There is an η and a $\rho' < j_\eta(\kappa)$ so that $k_\eta(\rho') = \rho$. The case $\rho' < \eta$: since $k_{\eta\zeta}$ fixes ρ' for $\eta \leq \xi < \lambda$, $\rho = k_\eta(\rho') = \rho' < j_\eta(\kappa)$. The case $\eta \leq \rho'$: by $\mathrm{A}_3^*(\kappa)$ there is a ζ so that $\eta \leq \zeta < \lambda$ and $k_{\eta\zeta}(\rho') = \xi$; thus, whenever $\zeta \leq \xi < \lambda$, we have $k_{\eta\zeta}(\rho') = \zeta$, so that $\rho = k_\eta(\rho') = \zeta < j_\zeta(\kappa)$. In either case, there is some $\delta < \lambda$ so that $\rho < j_\delta(\kappa)$, and hence

$$\rho < j_\delta(\kappa) < (2^{\delta^{<\kappa}}) < \lambda$$

as λ is inaccessible. Thus, $\tilde{j}(\kappa) = \lambda$.

The proof is complete. □

Theorem 8.3. $\mathrm{A}_2(\kappa)$ *strongly implies* $\mathrm{A}_3^*(\kappa)$.

Proof. Suppose $j\colon V_\alpha \to V_\beta$ as for $\mathrm{A}_2(\kappa)$. Since $\lambda = j(\kappa)$ is inaccessible and $j(\kappa) \leq \alpha$, by using an argument entirely analogous to the first part of 8.2, it can be shown that $\mathrm{A}_3^*(\kappa)$ holds—note that j need only be defined on V_λ. It now follows that $\mathrm{A}_3^*(\kappa)$ holds within V_β (recall the observation made just before 8.2). Thus, if $\mathcal{U}$ is the normal ultrafilter over κ corresponding to j, $\{\alpha < \kappa \mid \mathrm{A}_3^*(\alpha)\} \in \mathcal{U}$. □

Theorem 8.4. $\mathrm{A}_3(\kappa)$ *strongly implies* $\mathrm{A}_4(\kappa)$.

Proof. Let $j\colon V \to M$ with critical point κ so that M is closed under $< j(\kappa)$-sequences. Since $j(\kappa)$ is inaccessible and $f \in {}^\kappa\kappa$ implies $j(f)(\kappa) < j(\kappa)$, let μ be such that $\sup\{j(f)(\kappa) \mid f \in {}^\kappa\kappa\} < \mu < j(\kappa)$. Let $\mathcal{V}$ be the normal ultrafilter over $\mathcal{P}_\kappa\mu$ corresponding to j. Then

$$\begin{aligned} j_{\mathcal{V}}(f)(\kappa) < \mu \ \text{ iff } & \ \{x \mid f(\overline{x \cap \kappa}) < \bar{x}\} \in \mathcal{V} \\ \text{iff } & \ j(f)(\overline{j''\mu \cap j(\kappa)}) < \overline{j''\mu} \\ \text{iff } & \ j(f)(\kappa) < \mu . \end{aligned}$$

Thus, $\mathrm{A}_4(\kappa)$ holds, but then it also holds in M. That $\mathrm{A}_3(\kappa)$ strongly implies $\mathrm{A}_4(\kappa)$ follows, with $\mathcal{U}$ the normal ultrafilter over κ corresponding to j. □

Theorem 8.5. $\mathrm{A}_4(\kappa)$ *strongly implies* $\mathrm{A}_5(\kappa)$.

Proof. Let $j\colon V \to M$ be determined by a normal ultrafilter over $\mathcal{P}_\kappa\lambda$, as provided by $\mathrm{A}_4(\kappa)$. Let $\mathcal{U}$ be normal over κ corresponding to j. We first show that $\mathrm{A}_5(\kappa)$ is satisfied with this $\mathcal{U}$.

Suppose $\langle \mathcal{M}_\xi \,|\, \xi < \kappa \rangle$ are structures of the same type so that each $\mathcal{M}_\xi \in V_\kappa$. We may assume that the underlying set of $\mathcal{M}_\xi$ has the form V_η for η a limit ordinal $> \xi$, and that $\in$ is a predicate of $\mathcal{M}_\xi$. (If not, replace $\mathcal{M}_\xi$ by $\langle V_\eta; \in, \{\mathcal{M}_\xi\}\rangle$ where η is the least limit ordinal greater than $\max\{\xi, \operatorname{rank}(\mathcal{M}_\xi)\}$.)

Claim. If $\alpha < \kappa$ and $X_\alpha = \{\xi \,|$ there is an elementary embedding $\mathcal{M}_\alpha \to \mathcal{M}_\xi$ with critical point $\alpha\}$, $T = \{\alpha < \kappa \,|\, X_\alpha \in \mathcal{U}\} \in \mathcal{U}$.

To show the claim, first set

$$j(\langle \mathcal{M}_\xi \,|\, \xi < \kappa \rangle) = \langle \mathcal{M}'_\xi \,|\, \xi < j(\kappa)\rangle$$
$$j(\langle \mathcal{M}'_\xi \,|\, \xi < j(\kappa)\rangle) = \langle \mathcal{M}''_\xi \,|\, \xi < j^2(\kappa)\rangle$$
$$j(\langle X_\alpha \,|\, \alpha < \kappa \rangle) = \langle X'_\kappa \,|\, \alpha < j(\kappa)\rangle$$

Note that $\xi < \kappa$ implies $\mathcal{M}_\xi = \mathcal{M}'_\xi$. If $\alpha < \kappa$, $X_\alpha \in \mathcal{U}$ iff $\mathcal{M}_\alpha$ is elementarily embeddable into $\mathcal{M}_\kappa$ with critical point α. So if $\alpha < j(\kappa)$, $X'_\alpha \in j(\mathcal{U})$ iff in M, $\mathcal{M}'_\alpha$ is elementarily embeddable into $\mathcal{M}''_{j(\kappa)}$ with critical point α. is elementarily embeddable into $\mathcal{M}'_\kappa$ with critical point α. Hence, $T \in \mathcal{U}$ iff $\kappa \in j(T)$ iff $X'_\kappa \in j(\mathcal{U})$ iff in M, $\mathcal{M}'_\kappa$ is elementarily embeddable into $\mathcal{M}''_{j(\kappa)}$ with critical point κ.

Now $j|\mathcal{M}'_\kappa \colon \mathcal{M}'_\kappa \to \mathcal{M}''_{j(\kappa)} = j(\mathcal{M}'_\kappa)$ is elementary, so it suffices to show that $j|\mathcal{M}'_\kappa \in M$. Define $f \in {}^\kappa\kappa$ by $f(\xi) = |\mathcal{M}_\xi|$. By the $\mathrm{A}_4(\kappa)$ property of j, $j(f)(\kappa) < \lambda$. Thus, since M is closed under λ-sequences and $j|\mathcal{M}'_\kappa$ is just a set of ordered pairs of elements of M of cardinality $|\mathcal{M}'_\kappa| = j(f)(\kappa)$, $j|\mathcal{M}'_\kappa \in M$. The claim is proved.

Having shown that $T \in \mathcal{U}$, by the diagonal intersection property of normal ultrafilters we have that $\{\alpha \in T \,|\, \beta < \alpha \;\&\; \beta \in T \text{ implies } \alpha \in X_\beta\} \in \mathcal{U}$. This set satisfies $\mathrm{A}_4(\kappa)$ for $\langle \mathcal{M}_\xi \,|\, \xi < \kappa\rangle$.

Finally, another application of the $\mathrm{A}_4(\kappa)$ property of j shows that $2^\kappa \le (2^\kappa)^M < \lambda$, so that our $\mathcal{U} \in M$. Thus $M \models \mathrm{A}_5(\kappa)$, and so $\{\alpha < \kappa \,|\, \mathrm{A}_5(\alpha)\} \in \mathcal{U}$. □

Theorem 8.6. $\mathrm{A}_5(\kappa)$ *strongly implies* $\mathrm{A}_6(\kappa)$.

Proof. $\mathrm{A}_6(\kappa)$ is immediate from $\mathrm{A}_5(\kappa)$. But if $j \colon V \to M$ corresponds to any normal ultrafilter $\mathcal{U}$ over κ, then $M \models \mathrm{A}_6(\kappa)$, as $V_{\kappa+1} \subseteq M$. Thus, $\{\alpha < \kappa \,|\, \mathrm{A}_6(\alpha)\} \in \mathcal{U}$. □

Theorem 8.7. $\mathrm{A}_5(\kappa)$ *strongly implies* $\mathrm{A}^*_6(\kappa)$.

Proof. It suffices to show $\mathrm{A}^*_6(\kappa)$, for the result would then follow as in the proof of 8.6. Let $\mathcal{U}$ be as in $\mathrm{A}_5(\kappa)$. It is sufficient to find an $X_n \in \mathcal{U}$ satisfying the condition for the S of $A_6^*(\kappa)$ for a fixed n, as we can then take a countable intersection. Define a function $F\colon [\kappa]^{2n+2} \to 2$ by

$$F(\langle \alpha_0, \ldots, \alpha_n, \beta_0, \ldots, \beta_n \rangle) = 0 \text{ iff } \alpha_0 < \cdots < \alpha_n < \beta_0 < \cdots < \beta_n$$

and there is a $j\colon V_{\alpha_n} \to V_{\beta_n}$ with critical point α_0 so that $j(\alpha_i) = \beta_i$ for $i < n$. By the partition property for normal ultrafilters, there is a $Y \in \mathcal{U}$ and an $m < 2$ so that $F''[Y]^{2n+2} = m$. It suffices to get a contradiction from the assumption that $m = 1$.

For α in such a Y, let $\alpha^1, \ldots, \alpha^n$ be the first n members of Y after α, and set

$$\mathcal{M}_\alpha = \langle V_\alpha; \in, \{\alpha\}, \{\alpha^1\}, \ldots, \{\alpha^{n-1}\} \rangle.$$

By the hypothesis $A_5(\kappa)$, there is an $X \in \mathcal{U}$ so that $\alpha, \beta \in X$ and $\alpha < \beta$ implies there is a $k\colon \mathcal{M}_\alpha \to \mathcal{M}_\beta$ with critical point α. But this would be contradictory for any $\alpha, \beta \in X \cap Y$ so that $\beta > \alpha^n$. □

Theorem 8.8. *Suppose* $A_6^*(\kappa)$ *and let* S *be a stationary class as provided. Then*

(i) *If* $\alpha, \beta \in S$ *and* $\alpha < \beta$, *the inclusion map* $V_\alpha \to V_\beta$ *is elementary,*
(ii) $\alpha \in S$ *implies* $V_\alpha \prec V_\kappa$.
(iii) $\alpha \in S$ *implies* $V_\kappa \models A_7(\alpha)$.

Proof. For (i), let $\alpha < \beta < \gamma_1 < \gamma_2 < \gamma_3$ all in S. Let $j\colon V_{\gamma_1} \to V_{\gamma_3}$ with critical point α and $j(\alpha) = \gamma_2$; and let $k\colon V_{\gamma_1} \to V_{\gamma_3}$ with critical point β and $k(\beta) = \gamma_2$. Now $j|V_\alpha\colon V_\alpha \to V_{\gamma_2}$ and $k|V_\beta\colon V_\beta \to V_{\gamma_2}$ are elementary, and both are identities. Hence, $V_\alpha \prec V_{\gamma_2}$ and $V_\beta \prec V_{\gamma_2}$, so that $V_\alpha \prec V_\beta$.

(ii) follows from (i) by union of elementary classes.

To show (iii), work within V_κ. Note first that each element of S is an extendible cardinal. Also, if $\alpha < \gamma_1 < \gamma_2 < \gamma_3$ all in S, let $j\colon V_{\gamma_1} \to V_{\gamma_3}$ with critical point α and $j(\alpha) = \gamma_2$. By 5.10, V_{γ_3} is extendible. Hence, if $\mathcal{U}$ is the normal ultrafilter over corresponding to j, $\{\beta < \alpha \mid \beta \text{ is extendible}\} \in \mathcal{U}$. □

Open Question 8.9. What is the relationship between A_6^* and A_6?

References

[1] Keith J. Devlin. Some weak versions of large cardinal axioms. *Ann. Math. Logic*, 5:291–325, 1972/73.

[2] Erdős, Paul and András Hajnal. On a problem of B. Jónsson. *Bull. Acad. Polon. Sci. Sér. Sci. Math. Astronom. Phys.*, 14:19–23, 1966.

[3] Erdős, Paul and András Hajnal. Unsolved problems in set theory. In *Axiomatic Set Theory (Proc. Sympos. Pure Math., Vol. XIII, Part I, Univ. California, Los Angeles, Calif., 1967)*, pages 17–48. Amer. Math. Soc., Providence, R.I., 1971.

[4] Haim Gaifman. Elementary embeddings of models of set-theory and certain subtheories. In *Axiomatic set theory (Proc. Sympos. Pure Math., Vol. XIII, Part II, Univ. California, Los Angeles, Calif., 1967)*, pages 33–101, 1974.

[5] Fred Galvin and Karel Prikry. Infinitary Jonsson algebras and partition relations. *Algebra Universalis*, 6(3):367–376, 1976.

[6] William P. Hanf. Incompactness in languages with infinitely long expressions. *Fund. Math.*, 53:309–324, 1963/64.

[7] Thomas J. Jech. Some combinatorial problems concerning uncountable cardinals. *Ann. Math. Logic*, 5:165–198, 1972/73.

[8] Ronald B. Jensen. The Core Model. Privately circulated.

[9] Ronald B. Jensen. The fine structure of the constructible hierarchy. *Ann. Math. Logic*, 4:229–308; erratum, ibid. 4 (1972), 443, 1972. With a section by Jack Silver.

[10] Akihiro Kanamori. Weakly normal filters and irregular ultrafilters. *Trans. Amer. Math. Soc.*, 220:393–399, 1976.

[11] Akihiro Kanamori. On Vopěnka's and related principles. In *Logic Colloquium '77 (Proc. Conf., Wrocław, 1977)*, volume 96 of *Stud. Logic Foundations Math.*, pages 145–153. North-Holland, Amsterdam-New York, 1978.

[12] H. Jerome Keisler and Alfred Tarski. From accessible to inaccessible cardinals. Results holding for all accessible cardinal numbers and the problem of their extension to inaccessible ones. *Fund. Math.*, 53:225–308, 1963/64.

[13] Jussi Ketonen. Strong compactness and other cardinal sins. *Ann. Math. Logic*, 5:47–76, 1972/73.

[14] Jussi Ketonen. Nonregular ultrafilters and large cardinals. *Trans. Amer. Math. Soc.*, 224:61–73, 1976.

[15] Kenneth Kunen. Some applications of iterated ultrapowers in set theory. *Ann. Math. Logic*, 1:179–227, 1970.

[16] Kenneth Kunen. Elementary embeddings and infinitary combinatorics. *J. Symbolic Logic*, 36:407–413, 1971.

[17] Kenneth Kunen. On the GCH at measurable cardinals. In *Logic Colloquium '69 (Proc. Summer School and Colloq., Manchester, 1969)*, pages 107–110. North-Holland, Amsterdam, 1971.

[18] Kenneth Kunen. Saturated ideals. *J. Symbolic Logic*, 43(1):65–76, 1978.

[19] Kenneth Kunen and Jeffrey B. Paris. Boolean extensions and measurable cardinals. *Ann. Math. Logic*, 2(4):359–377, 1970/71.

[20] Azriel Lévy. A hierarchy of formulas in set theory. *Mem. Amer. Math. Soc.*, 57:76, 1965.

[21] Azriel Lévy. The sizes of the indescribable cardinals. In *Axiomatic Set Theory (Proc. Sympos. Pure Math., Vol. XIII, Part I, Univ. California, Los Angeles, Calif., 1967)*, pages 205–218. Amer. Math. Soc., Providence, R.I., 1971.

[22] Menachem Magidor. On the role of supercompact and extendible cardinals in logic. *Israel J. Math.*, 10:147–157, 1971.

[23] Menachem Magidor. There are many normal ultrafilters corresponding to a supercompact cardinal. *Israel J. Math.*, 9:186–192, 1971.

[24] Menachem Magidor. Combinatorial characterization of supercompact cardinals. *Proc. Amer. Math. Soc.*, 42:279–285, 1974.

[25] Menachem Magidor. How large is the first strongly compact cardinal? or A study on identity crises. *Ann. Math. Logic*, 10(1):33–57, 1976.

[26] Menachem Magidor. On the singular cardinals problem. I. *Israel J. Math.*, 28(1-2):1–31, 1977.

[27] Menachem Magidor. Changing cofinality of cardinals. *Fund. Math.*, 99(1):61–71, 1978.

[28] Menachem Magidor. On the existence of nonregular ultrafilters and the cardinality of ultrapowers. *Trans. Amer. Math. Soc.*, 249(1):97–111, 1979.

[29] Paul Mahlo. Über lineare transfinite Mengen. *Ber. Verhandl. Sachs. Akad. Wiss. Leipzig, Math.-Naturw. Kl.*, 63:187–225, 1911.

[30] Telis K. Menas. On strong compactness and supercompactness. *Ann. Math. Logic*, 7:327–359, 1974/75.

[31] Telis K. Menas. A combinatorial property of $p_k\lambda$. *J. Symbolic Logic*, 41(1):225–234, 1976.

[32] Telis K. Menas. Consistency results concerning supercompactness. *Trans. Amer. Math. Soc.*, 223:61–91, 1976.

[33] William J. Mitchell. Sets constructible from sequences of ultrafilters. *J. Symbolic Logic*, 39:57–66, 1974.

[34] Carl F. Morgenstern. On the ordering of certain large cardinals. *J. Symbolic Logic*, 44(4):563–565, 1979.

[35] Andrzej Mostowski. An undecidable mathematical statement. *Fund. Math.*, 36:143–164, 1949.

[36] John Myhill and Dana Scott. Ordinal definability. In *Axiomatic Set Theory (Proc. Sympos. Pure Math., Vol. XIII, Part I, Univ. California, Los Angeles, Calif., (1967)*, pages 271–278. Amer. Math. Soc., Providence, R.I., 1971.

[37] Karel Prikry. On a problem of Erdős, Hajnal and Rado. *Discrete Math.*, 2:51–59, 1972.

[38] Karel Prikry. On descendingly complete ultrafilters. In *Cambridge Summer School in Mathematical Logic (1971)*, pages 459–488. Lecture Notes in Math., Vol. 337. American Mathematical Society, 1973.

[39] Karel Prikry. Ideals and powers of cardinals. *Bull. Amer. Math. Soc.*, 81(5):907–909, 1975.

[40] W. N. Reinhardt. Remarks on reflection principles, large cardinals, and elementary embeddings. In *Axiomatic set theory (Proc. Sympos. Pure Math., Vol. XIII, Part II, Univ. California, Los Angeles, Calif., 1967)*, pages 189–205, 1974.

[41] Dana Scott. Measurable cardinals and constructible sets. *Bull. Acad. Polon. Sci. Sér. Sci. Math. Astronom. Phys.*, 9:521–524, 1961.

[42] Jack Silver. The consistency of the GCH with the existence of a measurable cardinal. In *Axiomatic Set Theory (Proc. Sympos. Pure Math., Vol. XIII, Part I, Univ. California, Los Angeles, Calif., 1967)*, pages 391–395. Amer. Math. Soc., Providence, R.I., 1971.

[43] Robert M. Solovay. Strongly compact cardinals and the GCH. In *Proceedings of the Tarski Symposium (Proc. Sympos. Pure Math., Vol. XXV, Univ. California, Berkeley, Calif., 1971)*, pages 365–372, 1974.

[44] Stanislaw M. Ulam. Zur Masstheorie in der allgemeinen Mengenlehre. *Fundamenta Mathematicae*, 16:140–150, 1930.

[45] Hao Wang. *From Mathematics to Philosophy.* Humanities Press, New York, 1974.

12 On Gödel Incompleteness and Finite Combinatorics (with Kenneth McAloon)

Gödel's paper on formally undecidable propositions [3] raised the possibility that finite combinatorial theorems could be discovered which are independent of powerful axiomatic systems such as first-order Peano Arithmetic. An important advance was made by J. Paris in the late 1970's; building on joint work with L. Kirby, he used model-theoretic techniques to investigate arithmetic incompleteness and proved theorems of finite combinatorics which were unprovable in Peano Arithmetic [11]. The Paris-Harrington paper [13] gives a self-contained presentation of the proof that a straightforward variant of the familiar finite Ramsey Theorem is independent of Peano Arithmetic. In this paper, we consider a simple finite corollary of a theorem of infinite combinatorics of Erdős and Rado [1] and show it to be independent of Peano Arithmetic. This formulation avoids the Paris-Harrington notion of *relatively large* finite set and deals with a generalized notion of partition. This shift of focus also provides for simplifications in the proofs and directly yields a level-by-level analysis for subsystems of Peano Arithmetic analogous to that in [12].

We have tried to provide a treatment of the proof whose organization and brevity make it suitable for expository purposes. These results were first discussed in 1982, and almost all the details were worked out by a year later. We would like to thank Peter Clote for his later interest and involvement in this web of ideas.

1 Definitions and the Main Results

We begin by recalling Ramsey's Theorem. Let $[X]^n$ denote the collection of subsets of X of cardinality n. If X is a set of natural numbers and if f is a function with domain $[X]^n$, we write $f(x_1, \ldots, x_n)$ for $f(\{x_1, \ldots, x_n\})$ with the understanding that $x_1 < x_2 < \cdots < x_n$. In keeping with notation used in logic and in Ramsey theory, we identify each natural number n with the set of its predecessors: $n = \{0, 1, \ldots, n-1\}$. Also, we shall use N to denote the set of natural numbers as well as its cardinality. If n, k and r are either N or members of N, $X \to (k)^n_r$ means that whenever $f\colon [X]^n \to r$ there is $H \in [X]^k$ such that f is constant on $[H]^n$; in this case we say that H is *homogeneous* for f. Ramsey [14] established the Infinite Ramsey Theorem:

$$\text{For any } n, r \in N, \quad N \to (N)^n_r$$

as well as the Finite Ramsey Theorem:

$$\text{For any } n, r, k \in N, \text{ there is an } m \in N \text{ such that } m \to (k)^n_r\,.$$

Republished from *Annals of Pure and Applied Logic* 33 (1987), pp.23-41, with permission from Elsevier.

In 1950, Erdos and Rado [1] established a generalization of Ramsey's Theorem with no restrictions on the number of cells of the partition. If f has domain $[X]^n$, where $X \subseteq N$, we say that $H \subseteq X$ is *canonical for* f if there is a $v \subseteq n$ satisfying the following condition: if $s, t \in [H]^n$ are construed as increasing functions mapping n into X, then

$$f(s) = f(t) \leftrightarrow s \,|\, v = t \,|\, v\,.$$

In other words, $f(x_0, \ldots, x_{n-1}) = f(y_0, \ldots, y_{n-1}) \leftrightarrow x_i = y_i$ for all $i \in v$. We shall write $v = v(H)$ when v makes H canonical. By way of example, if $v(H) = \emptyset$, then H is homogeneous for f in the usual sense; and if $v(H) = n$, then f is injective on $[H]^n$.

Using the Infinite Ramsey Theorem, Erdős and Rado established:

Theorem 1.1 [1]. *For any $n \in N$, if $f\colon [N]^n \to N$, then there is an infinite $H \subseteq N$ canonical for f.*

The Infinite Ramsey Theorem in turn can be seen to be an immediate corollary of Theorem 1.1: because of the '$\leftrightarrow$' requirement in the definition of 'canonical', if the range of the partition f is finite, then $v = \emptyset$ is the only possible case.

A function $f\colon [X]^n \to N$, where $X \subseteq N$, is said to be *regressive* if $f(s) < \min(s)$ for all s such that $\min(s) > 0$. Of course, there cannot be large homogeneous sets for such functions in the usual sense, but there is a natural notion of homogeneity here: we say that a set $H \subseteq X$ is *min-homogeneous* for f if $\min(s) = \min(t)$ implies $f(s) = f(t)$; that is, if $f|[H^n]$ only depends on the minimum element. If $k, n \in N$, the notation $X \to (k)^n_{\text{reg}}$ means that whenever $f\colon [X]^n \to N$ is regressive, there is $H \in [X]^k$ min-homogeneous for f.

The following is a straightforward corollary to the Erdős-Rado result.

Corollary 1.2. *For any $n \in N$, $N \to (N)^n_{\text{reg}}$.*

Proof. The case $n = 1$ is trivial; so assume $n > 1$. Let $f\colon N^n \to N$ be regressive and let H be an infinite canonical set for f with $v = v(H)$. We claim that either $v = \emptyset$ or $v = \{0\}$. Suppose to the contrary that v contains some $i \neq 0$. If h is the least non-zero element of H, there would be arbitrarily many n-tuples from H with first element h which disagree on v and hence are mapped to different values less than h by f, a contradiction. Now, if $v = \emptyset$, f is homogeneous on $[H]^n$, and if $v = \{0\}$, f is min-homogeneous on $[H]^n$, because f must be injective according to the minimum element; that is, $f(s) = f(t) \leftrightarrow \min(s) = \min(t)$.

Now consider the proposition

(∗) For any $n, k \in N$, *there is an* $m \in N$ such that $m \to (k)^n_{\text{reg}}$.

There is a simple argument by contradiction which establishes (∗) from Corollary 1.2: if for some $n, k \in N$, there are regressive counterexamples $f_m \colon [m]^n \to m$ for every $m \in N$, use $f \colon N^{n-1} \to N$ defined by

$$f(x_1, \ldots, x_n) = f_{x_n}(x_0, \ldots, x_{n-1})$$

to derive a contradiction.

It is the proposition (∗) that we will prove independent of first-order Peano Arithmetic. The notion of regressive function comes from Set Theory and the combinatorics of regular cardinals. We will remark further on this connection later but here let us finish with the preliminaries necessary to state our main results.

Peano Arithmetic, abbreviated PA, is the first-order theory in the language containing $0, 1, +, \cdot, <$ axiomatized by the defining properties of these primitive notions together with the induction scheme for all formulas (allowing parameters). A function $f \colon N \to N$ is *provably recursive* in PA if $f(m) = n$ just in case PA $\vdash F(m, n)$ for some formula $F(x, y)$ which is Δ_1 in PA satisfying PA $\vdash \forall x \exists y F(x, y)$. Thus if f is a provably recursive function in PA, the fact that f is total is a consequence of PA. A function f is said to *eventually dominate* the function g iff $f(n) > g(n)$ for all but at most a finite number of integers $n \in N$.

Our main results are

Theorem A. *The assertion* (∗) *is not provable in* PA.

Theorem B. *The function* $\nu(n)$ = *the least* m *such that* $m \to (2n)^n_{\text{reg}}$ *is not provably total in* PA *and eventually dominates every provably recursive function of* PA.

Theorem C. *The function* $v_2(n)$ = *the least* m *such that* $m \to (n)^2_{\text{reg}}$ *eventually dominates all primitive recursive functions; its rate of growth is approximately equal to that of the Ackermann function.*

These results are, respectively, Corollaries 2.3, 2.4 and 4.6 below.

For comparison and for future reference, let us recall the Paris-Harrington variant of the Finite Ramsey Theorem which is also independent of PA, [13]. We say that $H \subseteq N$ is *relatively large* if H has at least as many elements as its minimum element, that is $|H| \geq \min(H)$. The notation $X \to_* (k)^n_r$

requires that the appropriate partitions have relatively large homogeneous sets of cardinality $\geq k$. Paris and Harrington showed that the proposition

(PH) For any $r, n, k \in N$, there is an $m \in N$ such that $m \rightarrow_* (k)^n_r$

is true but not provable in PA.

Our paper owes a great deal to the Ketonen-Solovay paper [4]. There, a direct combinatorial proof of the Paris-Harrington result is given. This approach uses a level-by-level analysis of the rate of growth of the functions involved in terms of the Grzegorczyk-Wainer hierarchy for the provably recursive functions of PA. Regressive functions are actually an integral part of [4] although homogeneity is still tied there to the notion of relatively large finite set. In an earlier manuscript version of [4] the analogies with combinatorial notions from the study of large cardinals in Set Theory were more explicitly developed. In Set Theory the key result on regressive functions is Fodor's Lemma [2]: if $f : \kappa \rightarrow \kappa$ is regressive and κ is a regular uncountable cardinal, then f is constant on a stationary subset of κ. The result that $\kappa \rightarrow (\kappa)^m_{\text{reg}}$ for measurable cardinals κ due to Rowbottom, [15]. The combinatorial theorem $(*)$ is thus a true miniaturization of combinatorics on transfinite cardinal numbers.

We next give a direct, finitistic proof that (PH) implies $(*)$. This result appears as Lemma 1.8 in [4]. It can be formalized in PA or in *Primitive Recursive Arithmetic* (PRA); this is the first-order theory in the language containing $0, 1, +, \cdot <$ and a function symbol for each primitive recursive function, axiomatized by the defining properties of the primitive notions, the recursion equations for the primitive recursive functions and the induction scheme only for quantifier-free formulas. Induction for Σ_0 formulas (i.e. allowing bounded quantifiers) is provable in PRA.

Theorem 1.3 [4]. (PH) *implies* $(*)$.

Proof. Given $n, k \in N$, first find $m \in N$ such that whenever $h : [m]^{n+1} \rightarrow 3$, there is $H \subseteq m$ homogeneous for h such that $|H| \geq \min(H)+n$ and $|H| \geq k+n$. This self-refinement of (PH) is a straightforward consequence of it, cf. [13, 2.14]. So given $f : [m]^n \rightarrow m$ regressive, define $g : [m]^{n+1} \rightarrow 3$ by

$$g(x_0, \ldots, x_n) = \begin{cases} 0 & \text{if } f(x_0, \ldots, x_{n-1}) = f(x_0, x_2, \ldots, x_n), \\ 1 & \text{if } f(x_0, \ldots, x_{n-1}) < f(x_0, x_2, \ldots, x_n), \\ 2 & \text{if } f(x_0, \ldots, x_{n-1}) > f(x_0, x_2, \ldots, x_n). \end{cases}$$

Let $H \subseteq m$ be homogeneous for g and satisfy $|H| \geq \min(H) + n$ and $|H| \geq k + n$. As f is regressive, the first condition on H insures that g is constantly 0 on $[H]^{n+1}$, else there would be too many different values of f below $\min(H)$. By the second condition on H, we can take H' to consist of the first k elements of H and have $|H - H'| \geq n$. We now argue that H' is min-homogeneous for f as follows: given $x_0 < x_1 < \cdots < x_{n-1}$ and $x_0 < y_1 < \cdots < y_{n-1}$ all from H', let $z_1 < \cdots < z_{n-1}$ be $n-1$ elements in H such that $z_1 > \max(x_{n-1}, y_{n-1})$. Then $f(x_0, x_1, \ldots, x_{n-1}) = f(x_0, x_2, \ldots, x_{n-1}, z_1) = f(x_0, x_3, \ldots, z_1, z_2) = \cdots = f(x_0, z_1, \ldots, z_{n-1})$, and similarly for $f(x_0, y_1, \ldots, y_{n-1})$. Thus $f(x_0, x_1, \ldots, x_{n-1}) = f(x_0, y_1, \ldots, y_{n-1})$.

The method of proof we will use in the next section to establish the independence of $(*)$ from PA is model-theoretic; it is thus in the tradition of [13], [11] and [5]. In [6], Kirby and Paris give an elegant independence result based on work of Goodstein and on [4]. Their result seems to require correlation with the function hierarchies and thus far has eluded a quick model-theoretic treatment.

In Section 3, we develop the combinatorics to carry through the [4] scheme for $(*)$, thereby eliminating the original bootstraps arguments from relatively large. Finally, in Section 4 we refine the previous arguments as well as discuss generalizations of $(*)$. Interestingly enough, the refinements provide a level-by-level analysis of subsystems of PA using indiscernibles.

2 Independence

The main purpose of this section is to provide a model-theoretic proof of the independence of $(*)$, which, stripped of exegesis, is remarkably brief. Harrington's idea of diagonal indiscemibles for reducing full induction to induction for Σ_0 formulas is still crucial, but by focusing on $(*)$ we can avoid the diverting combinatorics of [13] needed for procuring and spreading out the indiscemibles. The idea for the proof of the following proposition occurs in [9, p.406] and was also noticed by Laver.

Proposition 2.1. *Assume $(*)$. For any $e, k, n \in N$ and formulas $\psi_0, \ldots, \psi_e$ in the language of arithmetic in at most $n+1$ free variables, there is a set $H \in [N]^k$ which constitute diagonal indiscernibles for these formulas, i.e. given $c_0 < c_1 < \cdots < c_n$ and $c_0 < d_1 < \cdots < d_n$ all in H and any $p < c_0$, then*

$$\psi_i(p, c_1, \ldots, c_n) \leftrightarrow \psi_i(p, d_1, \ldots, d_n)$$

holds for each $i \leq e$.

Proof. We shall assume $k \geq 2n+1$ for technical reasons. Let $m \to (w)^{2n+1}_{\text{reg}}$, where $w \to (k+n)^{2n+1}_{e+2}$. Given $x_0 < \cdots < x_{2n} < m$, if there is an $i \leq e$ and a $p < x_0$ such that $\psi_i(p, x_1, \ldots, x_n)$ and $\psi_i(p, x_{n+1}, \ldots, x_{2n})$ have different truth values, then let $f(x_0, \ldots, x_{2n})$ be the least such p and $i(x_0, \ldots, x_{2n})$ the least such i. Otherwise, set $f(x_0, \ldots, x_{2n}) = 0$ and $i(x_0, \ldots, x_{2n}) = e+1$.

By hypothesis on m, there is a set $H_0 \in [m]^w$ which is min-homogeneous for f. Next, by hypothesis on w, there is an $H_1 \in [H_0]^{k+n}$ and a fixed $i \leq e+1$ such that $i = i(x_0, \ldots, x_{2n})$ for every $\{x_0, \ldots, x_{2n}\} \in [H_1]^{2n+1}$.

Suppose first that $i = e+1$. Then let $z_1 < \cdots < z_n$ be the last n elements of H_1, and $H = H_1 - \{z_1, \ldots, z_n\}$. Given any $c_0 < c_1 < \cdots < c_n$ and $c_0 < d_1 < \cdots < d_n$ all from H, for each $i \leq e$ and each $p < c_0$, $\psi_i(p, c_1, \ldots, c_n)$ must have the same truth value as $\psi_i(p, z_1, \ldots, z_n)$ and so also must $\psi_i(p, d_1, \ldots, d_n)$.

Thus, the argument would be complete if we can derive a contradiction from the assumption that $i \leq e$. To this end, let $x_0 < \cdots < x_{3n}$ be all from H_1. (Remember that we are assuming that $k \geq 2n+1$.) By the min-homogeneity of H_0, there is a fixed value $p < x_0$ for f on any $2n+1$ sequence starting with x_0. However, at least two of $\psi_i(p, x_1, \ldots, x_n)$, $\psi_i(p, x_{n+1}, \ldots, x_{2n})$, and $\psi_i(p, x_{2n+1}, \ldots, x_{3n})$ must have the same truth value, contradicting the choice of $i \leq e$.

Notice that this proposition is formalizable in any theory which has a truth predicate for the formulas concerned. We will only require it for Σ_0 formulas in what follows, so we can take PRA as the theory, since there is a primitive recursive truth predicate for the Σ_0 formulas. Of course, in a quick exposition, PRA can be replaced by PA.

Also, there is a proof of the proposition using only $(*)$ for exponent $n+1$ rather than $2n+1$, and this has implications in the level-by-level analysis of Paris, [12]—see Section 4.

We now turn to the standard sort of model-theoretic result leading to independence. Recall that any non-standard model M of enough of arithmetic has a proper initial segment which we can identify with N, the set of natural numbers. In general, if I is a proper initial segment of M, we write $I < M$. Also, if $a, b \in M$ and $a \in I < M$ yet $b \notin I$, we write $a < I < b$. As our I's will be closed under $+$ and $\cdot$, we will regard them as substructures of M. Finally, $[a, b]$ denotes as usual the closed interval $\{x \mid a \leq x \leq b\}$.

Theorem 2.2. *Suppose that* $M \models \mathrm{PRA} \ \& \ [a, b] \to (2c)^c_{\mathrm{reg}}$, *where* $c \in M - N$. *Then there is an* $I < M$ *with* $a < I < b$ *such that* $I \models \mathrm{PA}$.

Proof. We shall first get $\langle c_i \mid i \in N \rangle$, with $a < c_i < b$ for each i, which constitute diagonal indiscernibles for all the Σ_0 formulas, i.e., whenever ψ is Σ_0 in say $n+1$ free variables, and $i_0 < i_1 \cdots < i_n$ and $i_0 < j_1 < \cdots < j_n$, then

$$M \models \forall p < c_{i_0} (\psi(p, c_{i_1}, \ldots, c_{i_n}) \leftrightarrow \psi(p, c_{j_1}, \ldots, c_{j_n})) .$$

One way to do this is as follows: Let $\sigma(k)$ assert that there are at least k diagonal indiscernibles in the interval between a and b for the first k Σ_0

formulas (in some standard coding). By the proposition and the succeeding comment about PRA, it is easy to see that for each $k \in N$, $M \models \sigma(k)$:

In the notation of the lemma, it suffices to find a min-homogeneous set of size $\geq w$ for some regressive f on $[[a,b]]^{2n+1}$, where $w, n \in N$. Remembering that $c \in M - N$, we can define $\bar{f}$ on $[[a,b]]^c$ by $\bar{f}(s) = f(\text{first } 2n+1 \text{ elements of } s)$. By $[a,b] \to (2c)^c_{\text{reg}}$ there is a set $\bar{X}$ of size $2c$ min-homogeneous for $\bar{f}$. If we then let X consist of the first c elements of $\bar{X}$, then X is min-homogeneous for f—the last c members of $\bar{X}$ can be used to extend any $2n+1$-tuple from X to a c-tuple from $\bar{X}$.

Note that we can take $\sigma(k)$ to be primitive recursive, since it can be gotten from the primitive recursive truth predicate for Σ_0 formulas through bounded quantification. Hence, by 'overspill' (which in this case just asserts that N is not definable in $M \models$ PRA) there is a $t \in M - N$ such that $M \models \sigma(t)$. We can take $\langle c_i \mid i \in N \rangle$ to be the first N of the t indiscernibles provided.

We now let I be determined by $\langle c_1 \mid i \in N \rangle$, i.e., $I = \{x \in M \mid \exists i(x < c_i)\}$, and establish that $I \models$ PA:

First, note that if $i_0 < i_1 < i_2$, then for any $p < c_{i_0}$, $p + c_{i_1} = c_{i_2}$ would imply $p + c_{i_1} = c_j$ for any $j > i_2$ by diagonal indiscernibility, contradicting the distinctness of the c_i's. Hence, $c_{i_0} + c_{i_1} \leq c_{i_2}$, and I is *closed under addition.*

Next, suppose that $i_0 < i_1 < i_2$ then for some $p < i_{c_0}$,

$$(\#) \qquad p \cdot c_{i_1} < c_{i_2} \leq (p+1) \cdot c_{i_1}\,.$$

Then by adding c_{i_1} to both sides of the first inequality we get $(p+1) \cdot c_{i_1} < c_{i_1} + c_{i_2}$ and for any $j > i_2$, $c_{i_1} + c_{i_2} \leq c_j$ by the previous paragraph, so that $(p+1) \cdot c_{i_1} < c_j$. But this would contradict the diagonal indiscernibility applied to the second inequality of $(\#)$. Thus, there is no such p, and so $c_{i_0} \cdot c_{i_1} \leq c_{i_2}$ and I is *closed under multiplication.*

It remains to establish that I satisfies the Induction schema. Notice that if $p < c_{i_0}$ and ψ is any Σ_0 formula, then since Σ_0 formulas are absolute between $I < M$ and the c_i's are diagonal indiscernibles,

$$
\begin{aligned}
& I \models \exists x_1 \forall x_2 \cdots \exists x_n \psi(p, x_1, \ldots, x_n) \\
& \text{iff } \exists i_1{>}i_0 \forall i_2{>}i_1 \cdots \exists i_n{>}i_{n-1} I \models \exists x_1{<}c_{i_1} \forall x_2{<}c_{i_2} \cdots \exists x_n{<}c_{i_n} \psi(p, x_1, \ldots, x_n) \\
& \text{iff } \exists i_1{>}i_0 \forall i_2{>}i_1 \cdots \exists i_n{>}i_{n-1} M \models \exists x_1{<}c_{i_1} \forall x_2{<}c_{i_2} \cdots \exists x_n{<}c_{i_n} \psi(p, x_1, \ldots, x_n) \\
& \text{iff } M \models \exists x_1{<}c_{i_0+1} \forall x_2{<}c_{i_0+2} \cdots \exists x_n{<}c_{i_0+n} \psi(p, x_1, \ldots, x_n) \\
& \text{iff } I \models \exists x_1{<}c_{i_0+1} \forall x_2{<}c_{i_0+2} \cdots \exists x_n{<}c_{i_0+n} \psi(p, x_1, \ldots, x_n)\,.
\end{aligned}
$$

Now to verify Induction, suppose that ϕ is any formula in parameters $p_0, \ldots, p_e$ and variable of induction x. It suffices to establish that if $\exists x \phi(p_0, \ldots, p_e, x)$, then there is a $<$-least such x. But if $\phi(p_0, \ldots, p_e, \bar{x})$, we can slightly modify ϕ by using a primitive recursive pairing function to construe $p_0, \ldots, p_e, \bar{x}$ as one

p and blocks of like quantifiers as one quantifier, find an i_0 such that $p < c_{i_0}$, and implement the above reduction to a Σ_0 formula. But M satisfies Induction for Σ_0 formulas (allowing parameters) since $M \models$ PRA, and since $I < M$, so does I. Hence, $I \models$ PA.

(In terms of the Indicator Theory of Paris and Kirby, the proof of 2.2 shows that

$$Y(a,b) = \max c([a,b] \to (2c)^c_{\text{reg}})$$

is an Indicator for models of PA.)

Let $\Pi_n(N)$ be the true Π_n sentences of Arithmetic. We can include $\Pi_1(N)$ in the independence result:

Corollary 2.3. (*) *is not provable in* PA $+ \Pi_1(N)$.

Proof. Let $M \models \text{PA} + \Pi_1(N)$, and a in $M - N$. We can suppose that there is a least b such that $M \models [a,b] \to (2a)^a_{\text{reg}}$. By the Theorem, there is an $I < M$ such that $a < I < b$ and $I \models$ PA. But Π_1 sentences persist downward, so $I \models \Pi_1(N)$. Finally, $I \models \neg\exists x([a,x] \to (2a)^a_{\text{reg}})$, since the coding of subsets of any $x \in I$ is certainly absolute between I and M.

The above corollary establishes our Theorem A. We now turn to Theorem B. We first argue by contradiction to show that the map $\bar{\nu}(n) =$ the least $m([n,m] \to (2n)^n_{\text{reg}})$ eventually dominates every provably recursive function of PA. So suppose that $\bar{\nu}(n) \leq g(n)$ on an infinite set $D \subseteq N$ for some provably recursive g. By an ultrapower construction with D in the ultrafilter, there is a non-standard elementary extension $M > N$ with infinite a in M such that $M \models \bar{\nu}(a) \leq g(a)$. By the construction in the proof of Corollary 2.3, there is $I < M$ such that $a < I < g(a)$ and $I \models$ PA. But then $I \models \neg\exists y[y = g(a)]$ contradicting the assumption that g is provably recursive in PA.

If we now define ν by $\nu(n) =$ the least m such that $m \to (2n)^n_{\text{reg}}$, then it is easy to correlate $\bar{\nu}$ to ν. For example, $\bar{\nu}(k) \leq \nu(2k)$ for any k: If $m \to (4k)^{2k}_{\text{reg}}$, then for any regressive $f\colon [[k,m]]^k \to m$, define regressive $\bar{f}\colon [m]^{2k} \to m$ by $\bar{f}(s) = f(\text{first k elements of } s)$ if $\min(s) \geq k$, and $= 0$ otherwise. If $\bar{H} \in [m]^{4k}$ is min-homogeneous for $\bar{f}$, since $[k,m] \cap \bar{H}$ has at least $3k$ elements, let H be the first $2k$ of these. Then H is min-homogeneous for f, since the last k members of $\bar{H}$ can be used to extend any k-tuple from H to a $2k$-tuple from $\bar{H}$.

It is not difficult to show that if $\nu(2n)$ dominates every provably recursive function of PA, so does $\nu(n)$. We have thus established:

Corollary 2.4. *The function $\nu(n) =$ the least m such that $m \to (2n)^n_{\text{reg}}$ is not provably total in* PA *and eventually dominates every provably recursive function of* PA.

We can now go on to establish as in [7] or [13] that, in PA $(*)$ is actually equivalent to the 1-Consistency of PA, i.e., the statement "PA together with the Π_1-theory of the universe is consistent". Put yet another way, $(*)$ is equivalent to the Gödel statement Con(PA $+ T_1$), where T_1 is the set of Π_1-sentences true according to some standard complete Π_1-formula. Since (PH) is equivalent to this principle, we have:

Corollary 2.5. (PH) *and* $(*)$ *are equivalent (and this is formalizable in* PA*).*

Very recently (March, 1985), Paris has established this corollary directly by clever combinatorial means. (See also the remarks after 4.7.) Of the several mathematical propositions now known to be equivalent to the 1-Consistency of PA, our proofs might argue for $(*)$ as leading most directly to independence.

3 Combinatorics

The emphasis of the previous section was on a short, global proof of the independence of $(*)$. We now turn to further combinatorial consequences of $(*)$, primarily in order to provide those adjustments to Ketonen and Solovay [4] needed when only $(*)$ is assumed. Assuming a known functional hierarchy result about the functions provably recursive in PA, the elegant paper [4] establishes the independence of (PH) by entirely combinatorial means, that is to say by means directly formalizable in PRA without appeal to higher principles such as Compactness. In brief, they first recall the Grzegorczyk-Wainer Hierarchy $\langle F_\alpha \mid \alpha < \epsilon_0 \rangle$ (where ϵ_0 is the least ordinal ϵ such that $\epsilon^\omega = \epsilon$ in ordinal exponentiation), a hierarchy of functions $F_\alpha \colon N \to N$ which, under the ordering of eventual dominance, is increasing in α and cofinal in the class of provably recursive functions. They then establish from (PH) that for any F_α, there is an $f \colon [N]^n \to N$ for some $n \in N$ such that any relatively large homogeneous set H for f has the property that $x < y$ both in H implies $F_\alpha(x) \leq y$. Actually, they establish a level-by-level correlation: there is such an $f \colon [N]^n \to N$ for F_α just in case $\alpha < \gamma_{n-2}$ (where $\langle \gamma_i \mid i \in N \rangle$ is inductively defined by $\gamma_0 = \omega$ and $\gamma_{i+1} = \omega^{\gamma_i}$, so that $\epsilon_0 = \sup \gamma_i$). Finally, they show that this directly implies that Ramsey functions in the context of (PH), for example $\sigma(n) =$ least $m(m \to_* (n+1)^n_n)$, eventually dominate any F_α, and thus that (PH) is not provable in PA. (Going beyond the results of [13], they complete their *tour de force* by providing careful upper bounds for $\sigma(n)$. However, we will not deal with this aspect here.)

The following combinatorial propositions highlight the arguments needed to adopt the [4] scheme to $(*)$. Routine applications of the Finite Ramsey Theorem are involved here as well as in [4], but recalling that that theorem is provable in PRA, everything will be formalizable in PRA.

(For the rest of the section, our notation implicitly assumes that our min-homogeneous sets H are finite, when we need to adjust them by eliminating a $\max(H)$. This anticipates the [4] application, and no such adjustments are necessary for infinite H.)

Proposition 3.1. *There is a (primitive recursive) function $p\colon N \to N$ such that: For any $n, e \in N$, whenever $f_i\colon [N]^n \to N$ is regressive for each $i \leq e$, there is a $\rho\colon [N]^{n+1} \to N$ regressive such that: If $\bar{H}$ is min-homogeneous for ρ, then $H = \bar{H} - (p(e) \cup \{\max(\bar{H})\})$ is min-homogeneous for each f.*

The proof is an adaptation of the Harrington idea in [13], Lemmas 2.7 and 2.8, to which the following lemmas correspond:

Lemma 3.2. *If $f\colon [N]^n \to N$ is regressive, then $H \subseteq N$ is min-homogeneous for f iff every $u \subseteq H$ of cardinality $n+1$ is min-homogeneous for f.*

Proof. If $H \subseteq N$ is not min-homogeneous, let $x_0 < \cdots < x_{n-1}$ be the lexicographically least sequence drawn from H such that there are $x_0 < y_1 < \cdots < y_{n-1}$ all from H with $f(x_0.x_1, \ldots, x_{n-1}) \neq f(x_0, y_1, \ldots, y_{n-1})$, where we can take $\langle y_1, \ldots, y_{n-1}\rangle$ to be the lexicographically least with this property. If i is the least such that $x_i \neq y_i$, then it is not difficult to see that $f(x_0, \ldots, x_{n-1}) = f(x_0, \ldots, x_i, y_i, y_{i+1}, \ldots, y_{n-2})$ and thus $u = \{x_0, \ldots x_i, y_i, y_{i+1}, \ldots, y_{n-1}\}$ is not min-homogeneous for f.

For any $x \in N$, let $\log(x) =$ least $d \in N$ such that $2^d \geq x$. Then any $y < x$ can be represented as $(y_0, \ldots, y_{\log(x)-1})_2$ in binary notation, with each $y_i < 2$. Notice that $2\log(x) + 1 \leq x$ for every $x \geq 7$. (So 7 plays a peculiar role in the present context for a reason entirely different than in [13, p.1137]. The following lemma allows us to press down the values of a regressive function further, at the cost of increasing the exponent.

Lemma 3.3. *If $f\colon [N]^n \to N$ is regressive, there is an $\bar{f}\colon [N]^{n+1} \to N$ regressive such that:*

(i) $\bar{f}(s) < 2\log(\min(s)) + 1$, *and*

(ii) *if $\bar{H}$ is min-homogeneous for $\bar{f}$, then $H = \bar{H} - (7 \cup \{\max(\bar{H})\})$ is min-homogeneous for f.*

Proof. Write $f(s) = (y_0(s), \ldots, y_{d-1}(s))_2$, where $d = \log(\min(s))$. Define $\bar{f}$ on

$[N]^{n-1}$ by:

$$\bar{f}(x_0,\ldots,x_n) = \begin{cases} 0 & \text{if either } x_0 < 7, \text{ or } \{x_0,\ldots,x_n\} \\ & \text{is min-homogeneous for } f, \\ 2i + y_i(x_0,\ldots,x_{n-1}) + 1 & \text{otherwise, where } i < \log(x_0) \text{ is} \\ & \text{the least such that } \{x_0,\ldots,x_n\} \\ & \text{is not min-homogeneous for } y_i. \end{cases}$$

Then $\bar{f}$ is regressive and satisfies (i).

To verify (ii), suppose that $\bar{H}$ is min-homogeneous for $\bar{f}$ and H is as described. If $\bar{f} \mid [H]^{n+1} = \{0\}$, then we are done by the previous lemma. Suppose on the contrary that there are $x_0 < \cdots < x_n$ all in H such that $\bar{f}(x_0,\ldots,x_n) = 2i + y_i(x_0,\ldots,x_{n-1}) + 1$. Given any $s,t \in [\{x_0,\ldots,x_n\}]^n$ with $\min(s) = \min(t) = x_0$, note that $\bar{f}(s \cup \{\max(\bar{H})\}) = \bar{f}(x_0,\ldots,x_n) = \bar{f}(t \cup \{\max(\bar{H})\})$ by min-homogeneity. But then, $y_i(s) = y_i(t)$, to that $\{x_0,\ldots,x_n\}$ was min-homogeneous for y_i after all—a contradiction.

Proof of Proposition 3.1. Note that, given any $e \in N$, for sufficiently large $x \in N$, $(2\log(x)+1)^{e+1} \leq x$; let $p(e)$ be the least such x. We shall verify the proposition using this p. So, suppose that n, e and f_i for $i \leq e$ are as given. To each f_i associate $\bar{f}_i$ as in the previous lemma, and define ρ on $[N]^{n+1}$ by:

$$\rho(s) = \begin{cases} \langle \bar{f}_0(s),\ldots,\bar{f}_e(s)\rangle & \text{if } \min(s) \geq p(e), \\ 0 & \text{otherwise.} \end{cases}$$

By the definition of p, ρ can be coded as a regressive function. Since $p(e) \geq 7$ for any e, the proof is now complete because of the previous lemma.

The next important juncture in [4] is Lemma 1.8, the verification of $(*)$ for relatively large homogeneous sets and some straightforward generalizations of it. We discussed in Section 1 how (PH) implies $(*)$, and the generalizations are inductively derivable from $(*)$ alone. Actually, [4] in Lemma 1.9 requires a self-refinement of $(*)$ in which the values on the min-homogeneous set are non-decreasing; the following proposition accomplishes the task directly from $(*)$:

Proposition 3.4. *$f\colon [N]^n \to N$ is regressive, there are $\sigma_1\colon [N]^{n+1} \to N$ regressive and $\sigma_2\colon [N]^{n+1} \to 2$ such that: If $H \subseteq N$ of cardinality $> n+1$ is both* *min-homogeneous for* σ_1 *and homogeneous for* σ_2*, then* $H - \{\max(H)\}$ *is min-homogeneous for* f *and if* $s, t \in [H]^n$ *with* $\min(s) < \min(t)$*, then* $f(s) \leq f(t)$.

Proof. Define $\sigma_1 \colon [N]^{n+1} \to N$ by:

$$\sigma_1(x_0, \ldots, x_n) = \min(f(x_0, \ldots, x_{n-1}), f(x_1, \ldots, x_n)),$$

and $\sigma_2 \colon [N]^{n+1} \to 2$ by:

$$\sigma_2(x_0, \ldots, x_n) = \begin{cases} 0 & \text{if } f(x_0, \ldots, x_{n-1}) \leq f(x_1, \ldots, x_n), \\ 1 & \text{otherwise.} \end{cases}$$

Now let H be as hypothesized, and suppose first that σ_2 is constantly 0 on $[H]^{n+1}$. By using $\max(H)$ as the last argument in applications of σ_1, it is straightforward to see that the conclusions of the Proposition are satisfied.

Assume to the contrary that σ_2 is constantly 1 on $[H]^{n+1}$. Let $x_0 < \cdots < x_{n+1}$ be $n+2$ elements from H. Then $f(x_0, \ldots, x_{n-1}) > f(x_1, \ldots, x_n) > f(x_2, \ldots, x_{n+1})$ by two applications of σ_2, so that $\sigma_1(x_0, \ldots, x_n) = f(x_1, \ldots, x_n)$ and $\sigma_1(x_0, x_2, \ldots, x_{n+1}) = f(x_2, \ldots, x_{n+1})$. But this contradicts the min-homogeneity of σ_1 on H, and the proof is complete.

[4] now proceeds to establish the results about the Grzegorczyk-Wainer Hierarchy alluded to earlier. They rely on an inductive bootstraps argument based on relatively large homogeneous sets, but we describe how this can be avoided. To be concrete, we establish the analogue to their prototype result for F_ω (their Theorem 1.9); the corresponding juncture in the general case (their Theorem 3.5) can be handled similarly.

For any function $f \colon N \to N$, let $F^1(x) = F(x)$, and inductively, $F^{n+1}(x) = F(F^n(x))$. The first $\omega + 1$ functions in the Grzegorczyk-Wainer Hierarchy are:

$$F_0(x) = x + 1; \quad F_{n+1}(x) = F_n^{x+1}(x); \quad \text{and} \quad F_\omega(x) = F_x(x).$$

It can be shown by induction that if $m \leq n$ and $x \leq y$, then $F_m(x) \leq F_n(y)$, a fact assumed below.

Proposition 3.5. *There are functions* $\tau_1 \colon [N]^2 \to N$ *regressive,* $\tau_2 \colon [N]^2 \to N$ *regressive, and* $\tau_3 \colon [N]^2 \to 2$ *so that the following is true: Suppose* $H \subseteq N$ *of cardinality* > 2 *is: (a) min-homogeneous for* τ_1 *and has the property that if* $s, t \in [H]^2$ *with* $\min(s) < \min(t)$*, then* $\tau_1(s) \leq \tau_1(t)$*; (b) min-homogeneous for* τ_2*; and (c) homogeneous for* τ_3*. Then for any* $x < y$ *both in* H*,* $F_\omega(x) \leq y$.

Proof. Define $\tau_1\colon [N]^2 \to N$ by:

$$\tau_1(x,y) = \begin{cases} 0 & \text{if } F_\omega(x) \leq y, \\ e-1 & \text{otherwise, where } e \text{ is the least such that } y < F_e(x). \end{cases}$$

Notice that $0 < e \leq x$ here, since F_0 is the successor function and $F_\omega(x) = F_x(x)$. Thus, τ_1 is a well-defined regressive function.

Next, define $\tau_2\colon [N]^2 \to N$ by:

$$\tau_2(x,y) = \begin{cases} 0 & \text{if } F_\omega(x) \leq y, \\ k-1 & \text{otherwise, where } F^k_{\tau_1(x,y)}(x) \leq y < F^{k+1}_{\tau_1(x,y)}(x). \end{cases}$$

τ_2 is also a well-defined regressive function, since $0 < k \leq x$ by an appeal to the definition of the F_n's.

Finally, define $\tau_3\colon N^2 \to 2$ by:

$$\tau_3(x,y) = \begin{cases} 0 & \text{if } F_\omega(x) \leq y, \\ 1 & \text{otherwise.} \end{cases}$$

Suppose now that H is as hypothesized. To conclude the argument, we shall establish that τ_3 is constantly 0 on H. Assume to the contrary, and let $x < y < z$ all be from H. If $e-1 = \tau_1(x,y)$ and $k-1 = \tau_2(x,y)$, then

$$F^k_{e-1}(x) \leq y < z < F^{k+1}_{e-1}(x)$$

by min-homogeneity. However, the leftmost inequality implies $F^{k+1}_{e-1}(x) \leq F_{e-1}(y)$ since the F_n's are non-decreasing. Also, $e-1 \leq \tau_1(y,z)$ by condition (a) of the Proposition on H, so that

$$F_{e-1}(y) \leq F_{\tau_1(y,z)}(y) \leq z\,.$$

Thus, we have arrived at the contradiction $z < z$.

With these arguments in hand, one can reorganize the [4] scheme in several ways to establish that functions like $\nu(n) =$ least $m(m \to (2n)^n_{\text{reg}})$ eventually dominate every F_α with $\alpha < \epsilon_0$. We should point out a weakness of our Propositions as they now stand: If we trace the exponent n needed to procure one regressive function to combine the two functions τ_1 and τ_2 of Proposition 3.5, then first by Proposition 3.4 we need a $\sigma_1\colon [N]^3 \to N$ taking $\tau_1 = f$ in that proposition, and then we must combine τ_2 and σ_1 by Proposition 3.1, to finally get one function: $[N]^4 \to N$. But by analogy with the level-by-level correlation in [4] between n and $\alpha < \gamma_{n-2}$ alluded to at the beginning of the section, we ought to be able to get the exponent down from 4 to 3. The following proposition provides the necessary augmentation to Proposition 3.1, by showing that we can incorporate one function: $[N]^{n+1} \to N$ into the proceedings.

Proposition 3.6. *Let* $p\colon N \to N$ *be as in Proposition* 3.1. *For any* $n, e \in N$, *whenever* $f_i\colon [N]^n \to N$ *is regressive for each* $i \le e$, *and* $f\colon [N]^{n+1} \to N$ *is regressive, there are* $\rho_1\colon [N]^{n+1} \to N$ *regressive and* $\rho_2\colon [N]^{n+1} \to 2$ *such that if* $\bar{H}$ *is both min-homogeneous for* ρ_1 *and homogeneous for* ρ_2, *then* $H = \bar{H} - (p(e) \cup \{\max(\bar{H})\})$ *is min-homogeneous for each* f_i *with* $i \le e$ *and for* f.

Proof. The idea is extend the proof of Proposition 3.1 by taking advantage of the fact that in Lemma 3.3, if H is sufficiently large and min-homogeneous for $\bar{f}$, then $\bar{f}$ on $[H]^{n+1}$ is in fact constantly 0. So, for $i \le e$, let $\bar{f}_i$ correspond to f_i as before, and define $\rho_2\colon [N]^{n+1} \to 2$ by:

$$\rho_2(s) = \begin{cases} 0 & \text{if } \bar{f}_i(s) \neq 0 \text{ for some } i \le e, \\ 1 & \text{otherwise.} \end{cases}$$

Then define $\rho_1\colon [N]^{n+1} \to N$ regressive by

$$\rho_1(s) = \begin{cases} \langle \bar{f}_0(s), \ldots, \bar{f}_e(s) \rangle & \text{if } \rho_2(s) = 0 \text{ and } \min(s) \ge p(e), \\ f(s) & \text{otherwise.} \end{cases}$$

As before, ρ_1 can be coded as a regressive function.

Suppose now that $\bar{H}$ is as hypothesized, and H is as defined from $\bar{H}$. If ρ_2 on $[H]^{n+1}$ were constantly 0, we can derive a contradiction as in the proof of Lemma 3.3. Thus, ρ_1 on $[H]^{n+1}$ must be constantly 1, and so the proof is complete.

It can now be checked in detail that (the idea of the proof of) this proposition can be used to prove results for $(*)$ fully analogous to [4].

4 Refinement and Generalization

In this concluding section, we discuss first technical improvements for previous propositions which have a consequence about subsystems of PA, and then a generalization of $(*)$ based on the growth rate of functions.

Let $\mathrm{I}\Sigma_n$ be the subsystem of PA consisting of the defining properties of the primitive notions together with the induction schema restricted to Σ_n formulas. Thus, $\mathrm{I}\Sigma_1$ already subsumes PRA. Let $(\mathrm{PH})_n$ denote the restriction of (PH) to fixed exponent n, and $(*)_n$ the restriction of $(*)$ to fixed exponent n. Paris in his meticulous paper [12] ramifies his model-theoretic analysis of PA by providing a strong level-by-level correlation of $\mathrm{I}\Sigma_n$ for $n > 0$ with several propositions, including $(\mathrm{PH})_{n+1}$. Although the idea of diagonal indiscernibles may seem tailored for a global proof, we show how it can be used in a level-by-level analysis with the focus on $(*)_n$. Such a possibility was considered by Hajek, and perhaps others.

First, we outline a result of Clote which will correlate the coming results with $(\mathrm{PH})_{n+1}$. The quick proof of Theorem 1.3 actually shows that $(\mathrm{PH})_{n+1}$ implies $(*)_n$. Clote noticed that by working through Mills' notion of arboricity [10], the result can be sharpened to what we shall later see is best possible.

Proposition 4.1 (Clote). *For $n > 0$, $(\mathrm{PH})_n$ implies $(*)_n$.*

Proof (outline). The following notions are due to Mills [10]. If $A \subseteq N$ and $f\colon N \to N$, an *$f(x)$-small-branching A-tree* is a tree T with field A such that aTb implies $a < b$, and any $a \in A$ has at most $f(a)$ immediate successors in T. Set $A^0 = A - \{\max(A)\}$. Then A is *0-fold c-$f(x)$-arboreal iff* $|A| \geq c+1$; and inductively, A is *$(n+1)$-fold c-$f(x)$-arboreal* iff for every $f(x)$-small-branching A-tree T, there is a path P through T such that P^0 is n-fold c-$f(x)$-arboreal. Finally, if $X \subseteq N$ and $F\colon [X]^{n+1} \to N$, say that $H \subseteq N$ is *prehomogeneous for* F iff $F \mid [H]^{n+1}$ only depends on the first n elements, i.e., $F(x_0, \ldots, x_{n-1}, a) = F(x_0, \ldots, x_{n-1}, b)$ for any $x_0 < \cdots < x_{n-1} < a < b$ all from H.

The argument of [10, Lemma 3.5] establishes the following: Suppose $c, n \geq 1$ and A is n-fold c-x^{2^x}-arboreal with $\min(A) > 0$. Suppose also that f with domain $[A]^{n+1}$ is regressive. Then, there is an $H \subseteq A$ such that H is prehomogeneous for f and H_0 is $(n-1)$-fold c-x^{2^x}-arboreal.

Thus, the following is an immediate corollary by induction, since prehomogeneous and min-homogeneous coincide at exponent 2: For any n, suppose $c \geq 1$ and A is n-fold $(c-1)$-x^{2^x}-arboreal with $\min(A) > 0$. Then, $A \to (c)^{n+1}_{\mathrm{reg}}$.

Finally, [10, Theorem 3.6] provides careful lower and upper bounds on the sizes of A such that $A \to_* (n+2)^{n+1}_c$ in terms of arboricity. It is immediate from this theorem that for any n, if A is such that $\min(A) \geq \max\{n^2, 2c\}$, $A \to_* (n+2)^{n+1}_c$ implies that A is n-fold $(c-n-1)$-x^{2^x}-arboreal, and thus that $A \to (c-n)^{n+1}_{\mathrm{reg}}$. Careful information is provided here which more than suffices to show that $(\mathrm{PH})_{n+1}$ implies $(*)_{n+1}$ for any n.

Paris [12, Theorem 3.6] established that in $\mathrm{I}\Sigma_1$, $(\mathrm{PH})_{n-1}$ is equivalent to the 1-Consistency of $\mathrm{I}\Sigma_n$ for $n \geq 1$. (Recalling the discussion at the end of Section 2, this means $\mathrm{Con}(\mathrm{I}\Sigma_n + T_1)$, where T_1 is the set of Π_1 sentences true according to some standard complete Π_1 formula.) With Proposition 4.1 in hand, we shall focus on $(*)_{n+1}$ and show how it can be used to generate indiscemibles to establish the 1-Consistency of $I\Sigma_n$.To do this, we first establish the conclusion of Proposition 2.1 assuming only $(*)_{n+1}$, improving the $(*)_{2n+1}$ of the short proof given. The following lemma should have a familiar ring:

Lemma 4.2. *There are three regressive functions η_1, η_2, η_3: $[N]^2 \to N$ such that whenever $\bar{H}$ is min-homogeneous for all of them, then $H = \bar{H}$ - (the last three elements of $\bar{H}$) has the property that $x < y$ both in H implies $x^x \leq y$.*

Proof. Define η_1, η_2, η_3: $[N]^2 \to N$ by:

$$\eta_1(x,y) = \begin{cases} 0 & \text{if } x + x \leq y \\ y - x & \text{otherwise,} \end{cases}$$

$$\eta_2(x,y) = \begin{cases} 0 & \text{if } x \cdot x \leq y \\ u & \text{otherwise, where } u \cdot x \leq y < (u+1) \cdot x, \end{cases}$$

$$\eta_3(x,y) = \begin{cases} 0 & \text{if } x^x \leq y \\ v & \text{otherwise, where } x^v \leq y < x^{v+1}, \end{cases}$$

Suppose that $\bar{H}$ is as hypothesized, and let $z_1 < z_2 < z_3$ be the last three elements of $\bar{H}$. If $x < y$ are both in $\bar{H} - \{z_3\}$, then since $\eta_1(x,y) = \eta_1(x, z_3)$, clearly we must have $\eta_1(x,y) = 0$. Hence, η_1 on $[\bar{H} - \{z_3\}]^2$ is constantly 0.

Next, assume that $x < y$ are both in $\bar{H} - \{z_2, z_3\}$ and $\eta_2(x,y) = u > 0$. Then

$$u \cdot x \leq z_2 < (u+1) \cdot x$$

by min-homogeneity, and $(u+1) \cdot x \leq x + y$ by adding x to both sides of the first inequality, which in turn is $< y + y \leq z_2$ by the previous paragraph. But this leads to the contradiction $z_2 < z_2$. Hence, on $[\bar{H} - \{z_2, z_3\}]^2$ is constantly 0.

Finally, we can iterate the argument to show that η_3 on $[\bar{H} - \{z_1, z_2, z_3\}]^2$ is constantly 0, and so the proof is complete.

This is all that we will need of a clearly inductive argument which proceeds through the classical Grzegorczyk Hierarchy, or equivalently, through $\langle F_n \mid n \in \omega \rangle$ of Section 3.

Here is the heralded improvement:

Proposition 4.3. *The conclusion of Proposition* 2.1 *already follows from* $(*)_{n+1}$.

Proof. Given the $e, k, n \in N$ and formulas $\psi_0, \dots, \psi_e$ in at most $n+1$ free variables, first define $q\colon N \to N$ by $q(x) =$ largest $d \in N$ such that $2^{e+1} \cdot d \leq x$, and then define $f\colon [N]^{n+1} \to N$ by:

$$f(x_0, \dots, x_n) = \langle \delta_{ip} \mid i \leq e \ \&\ p < q(x_0) \rangle$$

where

$$\delta_{ip} = \begin{cases} 0 & \text{if } \psi_i(p, x_1, \dots, x_n) \text{ is true,} \\ 1 & \text{otherwise.} \end{cases}$$

By definition of q, we can code f as a regressive function.

The idea now is to combine the functions of the previous lemma with f to get a min-homogeneous set spread out enough to accommodate q. However, a direct application of Proposition 3.6 would be restricted to $n \geq 2$, so we use an idea similar to the proof of that proposition. First define $g\colon [N]^2 \to 4$ from the functions of the previous lemma as follows:

$$g(x, y) = \begin{cases} 0 & \text{if } \eta_j(x, y) = 0 \text{ for } j = 1, 2, 3, \\ j & \text{otherwise, where } j \text{ is the least such that } \eta_j(x, y) \neq 0. \end{cases}$$

Then define $h\colon [N]^{n+1} \to N$ regressive by:

$$h(x_0, \dots, x_n) = \begin{cases} \eta_j(x_0, x_1) & \text{if } g(x_0, x_1) = j > 0, \\ f(x_0, \dots, x_n) & \text{otherwise.} \end{cases}$$

(We assume from here that $n > 0$, else the Proposition is trivial.)

By $(*)_{n+1}$, let H_0 of cardinality $2^{e+1} + 2k + 3$ be min-homogeneous for h and homogeneous for g. Let $z_1 < z_2 < z_3$ be the last three elements of H_0, and set $H_1 = H_0 - \{z_1, z_2, z_3\}$. If g on $[H]^2$ were constantly $j > 0$, then we can derive a contradiction as in the previous lemma. Thus, we can assume that $h = f$ on $[H_1]^{n+1}$, and $x < y$ both in H_1 implies $x^x \leq y$.

Now let H_2 consist of the last $2k$ elements of H_1. Then $x < y$ both in H_2 implies $2^{(e+1)x} \leq x^x \leq y$ since $2^{e+1} \leq x$, so that $x \leq q(y)$. Thus, if H_3 consists of every other element of H_2, then H_3 constitutes the desired k diagonal indiscernibles since H_2 is min-homogeneous for f.

Now the corresponding model-theoretic result:

Theorem 4.4. *Suppose that $n > 0$ and $M \models$ PRA & $[a, b] \to (c)^{n+1}_{\text{reg}}$, where $c \in M - N$. Then there is an $I < M$ with $a < I < b$ such that $I \models \mathrm{I}\Sigma_n$.*

Proof. We proceed as in the proof of Theorem 2.2. First, we can get diagonal indiscemibles $\langle c_i \mid i \in N \rangle$ in the interval $[a, b]$ for all Σ_0 formulas but in at most $n+1$ free variables, using Proposition 4.3. Let $I < M$ be determined by the c_i's, and note that I is closed under addition and multiplication. (The proof in Theorem 2.2 works only for $n \geq 2$, but we could cite Proposition 4.3 where we spread out the indiscernibles directly.) Finally, the verification of the Induction schema for Σ_n formulas proceeds as before, since we only need n alterations of quantifiers.

Corollary 4.5. *If* $n > 0$, $(*)_{n+1}$ *is not provable in* $\mathrm{I}\Sigma_n + \Pi_1(N)$.

Proof. As for Corollary 2.3.

Proceeding as with $(*)$, we can observe that

$$Y(a, b) = \max c([a, b] \to (c)^{n+1}_{\text{reg}})$$

is an indicator for models of $\mathrm{I}\Sigma_n$, and that in PRA, $(*)_{n+1}$ implies $\mathrm{Con}(\mathrm{I}\Sigma_n + T_1)$, the 1-Consistency of $\mathrm{I}\Sigma_n$.

We have the following corollaries:

Corollary 4.6. *The function* $\nu_2(n) =$ *the least* m *such that* $m \to (n)^2_{\text{reg}}$ *eventually dominates every primitive recursive function; its rate of growth is approximately that of the Ackermann function.*

Corollary 4.7. *If* $n > 0$, *the following are equivalent in* $\mathrm{I}\Sigma_1$*:*

(a) $(\mathrm{PH})_{n+1}$,
(b) $(*)_{n+1}$,
(c) $\mathrm{Con}(\mathrm{I}\Sigma_n + T_1)$.

Corollary 4.6 is our Theorem C; the proof of its first part is entirely analogous to that of Corollary 2.4. The proof of the second part is part of the proof that in PRA, $(*)_{n+1}$ implies $\mathrm{Con}(\mathrm{I}\Sigma_n + T_1)$. Corollary 4.7 follows from the result of Paris [12] mentioned earlier, that $(\mathrm{PH})_{n+1}$ is equivalent to $\mathrm{Con}(\mathrm{I}\Sigma_n + T_1)$. Our argument with indiscemibles may be more direct than Paris' argument from $(\mathrm{PH})_{n+1}$. However, at present we see no way to establish $\mathrm{Con}(\mathrm{I}\Sigma_n + T_1)$ implies $(*)_{n+1}$ other than to go through his argument using concepts from [4] and developing the [4] scheme for $(*)$. Very recently (March 1985), Paris has provided a clever combinatorial argument to show that $(*)_{n+1}$ implies $(\mathrm{PH})_{n+1}$ directly.

We now discuss a simple way to extend $(*)$ based on the growth rate of functions, in the spirit of [8]. Given any $F \colon N \to N$, $n \in N$, and $X \subseteq N$, say that a function $f\colon [X]^n \to N$ is *F-regressive* if $f(s) < F(\min(s))$ for all s such that $F(\min(s)) > 0$. For $n, k, m \in N$, $m \to (k)^n_{F-\mathrm{reg}}$ means that whenever $f\colon [m]^n \to N$ is F-regressive, there is an $H \in [m]^k$ min-homogeneous for f. Consider now the propositions

$(*)_F$ For any $n, k \in N$, there is an $m \in N$ such that $m \to (k)^n_{F-\mathrm{reg}}$.

Thus, $(*)$ is just the special case when F is taken to be the identity function, and for any F, $(*)_F$ follows from the Erdős-Rado Theorem 1.1 by the same sort of argument as for $(*)$. The corresponding generalization of (PH) discussed in [8] results from replacing relatively large by $|H| \geq F(\min(H))$.

The following characterization makes clear how $(*)_F$ can be incorporated into the known contexts:

Proposition 4.8. *For any increasing function $F\colon N \to N$, $(*)_F$ is equivalent to: For any $n, k \in N$, there is an $m \in N$ such that whenever $f\colon [m]^n \to m$ is regressive, there is an $H \in [m]^k$ which is min-homogeneous for f and has the property that $x < y$ both in H implies $F(x) \leq y$.*

Proof. In the forward direction, define $g\colon [N]^2 \to N$ by:

$$g(x, y) = \begin{cases} 0 & \text{if } F(x) \leq y, \\ y & \text{otherwise.} \end{cases}$$

g is clearly F-regressive, and whenever H finite of cardinality > 2 is min-homogeneous for g, then g on $[H - \{\max(H)\}]^2$ is constantly 0. By $(*)_F$ we can find an m sufficiently large such that: given any $f\colon [m]^n \to m$ regressive, we can combine it with $g \,|\, [m]^2$ by a version of Proposition 3.1, and find a min-homogeneous set for both f and g of cardinality k.

For the converse, the argument of [13, Lemma 2.14] can be used. Define $h\colon N \to N$ by $h(x) =$ largest y such that $F(y) \leq x$ (and 0 if there is no such y). Now given $n, k \in N$, let m be as provided, and suppose that $f\colon [m]^n \to N$ is F-regressive. Define $\bar{f}$ on $[m]^n$ by

$$\bar{f}(x_0, \ldots, x_{n-1}) = \begin{cases} f(h(x_0), \ldots, h(x_{n-1})) & \text{if } h(x_i) \neq h(x_j) \text{ for } 0 \leq i < j < n, \\ 0 & \text{otherwise.} \end{cases}$$

Since this value is $< F(h(x_0)) \leq x_0$ whenever $F(h(x_0)) \neq 0$, $\bar{f}$ is a regressive function. So, let $\bar{H} \in [m]^k$ be min-homogeneous for $\bar{f}$, satisfying: $x < y$ both in $\bar{H}$ implies $F(x) \leq y$. Consider $H = \{h(x) \,|\, x \in \bar{H}\}$; the last condition on $\bar{H}$ guarantees that h is one-to-one on $\bar{H}$, so that $|H| = k$. H is clearly min-homogeneous for f.

We can now proceed as in [8] to formulate for each $n \in N$ a function G_n based on some complete Π_n formula so that over PA, $(*)_{G_n}$ is equivalent to the n-Consistency of PA, i.e. Con(PA + T_n) where T_n is the set of Π_n-sentences true according to a standard complete Π_n-formula. Moreover, one can continue this process into the transfinite as in [8] with a hierarchy of faster and faster growing functions.

We conclude with some remarks on regressive partition relations and the Reverse Mathematics program of Friedman, Simpson and others. It has been observed that over the base theory RCA_0 (Recursive Comprehension Axiom), the system ACA_0 (Arithmetic Comprehension Axiom) is equivalent to the system axiomatized by the principle $N \to (N)^3_2$. It is not known whether the superscript 3 can be replaced by 2. This is the so-called "3-2 problem", which has the following recursion-theoretic formulation: Is there a recursive map $f\colon [N]^2 \to 2$ such that for any infinite homogeneous H, $0' \leq_T H$? Peter Clote has observed that over RCA_0, the system axiomatized by $N \to [N]^2_{\mathrm{reg}}$ is equivalent to ACA_0. Thus the exponent can be lowered if regressive partitions are used.

References

[1] P. Erdös and R. Rado. A combinatorial theorem. *J. London Math. Soc.*, 25:249–255, 1950.

[2] Géza Fodor. Eine Bemerkung zur Theorie der regressiven Funktionen. *Acta Sci. Math. (Szeged)*, 17:139–142, 1956.

[3] Kurt Gödel. Über formal unentscheidbare Sätze der Principia Mathematica und verwandter Systeme I. *Monatsh. Math. Phys.*, 38(1):173–198, 1931.

[4] Jussi Ketonen and Robert Solovay. Rapidly growing Ramsey functions. *Ann. of Math. (2)*, 113(2):267–314, 1981.

[5] Laurence A. S. Kirby and Jeffrey B. Paris. Initial segments of models of Peano's axioms. In *Set theory and hierarchy theory, V (Proc. Third Conf., Bierutowice, 1976)*, pages 211–226. Lecture Notes in Math., Vol. 619, 1977.

[6] Laurence A. S. Kirby and Jeffrey B. Paris. Accessible independence results for Peano arithmetic. *Bull. London Math. Soc.*, 14(4):285–293, 1982.

[7] Kenneth McAloon. Les rapports entre la méthode des indicatrices et la méthode de Gödel pour obtenir des résultats d'indépendance. In *Models of arithmetic (Sem., Univ. Paris VII, Paris, 1977) (French)*, volume 73 of *Astérisque*, pages 31–39. Soc. Math. France, Paris, 1980.

[8] Kenneth McAloon. Progressions transfinies de théories axiomatiques, formes combinatoires du théorème d'incomplétude et fonctions récursives à croissance rapide. In *Models of arithmetic (Sem., Univ. Paris VII, Paris, 1977) (French)*, volume 73 of *Astérisque*, pages 41–58. Soc. Math. France, Paris, 1980.

[9] Kenneth McAloon. On the complexity of models of arithmetic. *J. Symbolic Logic*, 47(2):403–415, 1982.

[10] George H. Mills. A tree analysis of unprovable combinatorial statements. In *Model theory of algebra and arithmetic (Proc. Conf., Karpacz, 1979)*, volume 834 of *Lecture Notes in Math.*, pages 248–311. Springer, Berlin-New York, 1980.

[11] Jeffrey B. Paris. Some independence results for Peano arithmetic. *J. Symbolic Logic*, 43(4):725–731, 1978.

[12] Jeffrey B. Paris. A hierarchy of cuts in models of arithmetic. In *Model theory of algebra and arithmetic (Proc. Conf., Karpacz, 1979)*, volume 834 of *Lecture Notes in Math.*, pages 312–337. Springer, Berlin-New York, 1980.

[13] Jeffrey B. Paris and Leo Harrington. A mathematical incompleteness in Peano arithmetic. In *Handbook of mathematical logic*, volume 90 of *Stud. Logic Found. Math.*, pages 1133–1142. North-Holland, Amsterdam, 1977.

[14] Frank P. Ramsey. On a problem of formal logic. *Proc. London Math. Soc. (2)*, 30(4):264–286, 1929.

[15] Frederick Rowbottom. Some strong axioms of infinity incompatible with the axiom of constructibility. *Ann. Math. Logic*, 3(1):1–44, 1971.

13 Regressive Partition Relations, n-Subtle cardinals, and Borel Diagonalization

Abstract. We consider natural strengthenings of H. Friedman's Borel diagonalization propositions and characterize their consistency strengths in terms of the n-subtle cardinals. After providing a systematic survey of regressive partition relations and their use in recent independence results, we characterize n-subtlety in terms of such relations requiring only a finite homogeneous set, and then apply this characterization to extend previous arguments to handle the new Borel diagonalization propositions.

In previous papers [6,7] we showed how regressive partition relations provide a simplifying and unifying scheme for establishing the independence of the Paris-Harrington as well as the Friedman [3] propositions. In these contexts the more informative approach of using regressive partition relations to generate indiscemibles in models can replace the abstract diagonalization technique of Cantor and Gödel for substantiating transcendence. Friedman's proposition correlated with the n-Mahlo cardinals. Here we show how the regressive partition formulation leads directly to an extension that correlates with the n-subtle cardinals, far stronger in consistency strength. In Section 1 we provide a systematic survey of regressive partition relations, their use in independence results, and related open questions. In Section 2 we establish a regressive partition result about n-subtle cardinals, and finally in Section 3 we use it to motivate and characterize the aforementioned extension.

1 Regressive Partition Relations

Let X be a set of ordinals and $n \in \omega$. If f is a function with domain $[X]^n$, we write $f(\alpha_0, \ldots, \alpha_{n-1})$ for $f(\{\alpha_0, \ldots, \alpha_{n-1}\})$ with the understanding that $\alpha_0 < \cdots < \alpha_{n-1}$. Such a function is called *regressive* iff $f(\alpha_0, \ldots, \alpha_{n-1}) < \alpha_0$ whenever $\alpha_0 < \cdots < \alpha_{n-1}$ all belong to X and $\alpha_0 > 0$. There is a natural notion of homogeneity for such a function $f\colon Y \subseteq X$ is *min-homogeneous for* f iff whenever $\alpha_0 < \cdots < \alpha_{n-1}$ and $\beta_0 < \cdots < \beta_{n-1}$ all belong to Y, $\alpha_0 = \beta_0$ implies $f(\alpha_0, \ldots, \alpha_{n-1}) = f(\beta_0, \ldots, \beta_{n-1})$. In other words, f on an n-tuple from Y depends only on the first element. We write $X \to (\gamma)^n_{\text{reg}}$ iff whenever f on $[X]^n$ is regressive, there is a $Y \in [X]^\gamma$ min-homogeneous for f. If the usual partition relation emanating from Ramsey's Theorem can be viewed as a generalization of the Pigeon-Hole Principle, then the regressive partition relation can be regarded as a generalization of Fodor's Theorem on regressive functions on stationary sets. The relation is actually a special case of the canonical partition relation of Erdős-Rado [2]. The following is an immediate

Republished from *Annals of Pure and Applied Logic* 20 (1991), pp.65-77, with permission from Elsevier.

consequence of their canonical generalization of Ramsey's Theorem:

$$\text{For any } n \in \omega,\ \omega \to (\omega)^n_{\text{reg}}\,.$$

In Kanamori-McAloon [7] the direct "miniaturization"of this proposition,

(∗) For any $n, k \in \omega$ there is an $m \in \omega$ such that $m \to (k)^n_{\text{reg}}$,

is shown to be equivalent to the well-known Paris-Harrington [8] proposition and hence unprovable (in a strong sense) in Peano Arithmetic. In fact, it is shown that (∗) for fixed n is equivalent to Paris-Harrington for fixed n and hence unprovable in $I\Sigma_{n-1}$, induction restricted to Σ_{n-1} formulas. The transparent independence proofs in [7], which quickly provide indiscernibles for models, argue for the efficacy of regressive partitions in this context. We mention here two open questions:

1.1. Question. Is the following proposition independent of Peano Arithmetic? For any $n \in \omega$ there is an $m \in \omega$ such that $m \to (n+2)^n_{\text{reg}}$.

The $n+2$ here is the minimal value for a non-trivial partition relation and allows very little flexibility; the [7] independence proof ostensibly needs $2n$ in place of $n+2$. There is an analogous open question concerning the Paris-Harrington proposition.

In the context of the Friedman-Simpson Reverse Mathematics program, it has been observed that over the base theory RCA_0 (Recursive Comprehension Axiom), the system ACA_0 (Arithmetical Comprehension Axiom) is equivalent to the system axiomatized by $\omega \to (\omega)^3_2$. It is not known whether the superscript 3 can be replaced by 2. This is the so-called "3-2 Problem", another problem about minimal hypotheses for transcendence, and has the following recursion-theoretic formulation:

1.2. Question. Is there a recursive map $f\colon [\omega]^2 \to 2$ such that for any infinite homogeneous H, $0' \leq_T H$?

Clote has observed that over RCA_0 the system axiomatized by $\omega \to (\omega)^2_{\text{reg}}$ is equivalent to ACA_0, so that the exponent can be lowered if regressive partitions are used.

Turning to the infinite case, we already mentioned the Erdős-Rado result for ω. To get min-homogeneous sets of size $\kappa > \omega$ with exponent 2, $\lambda \to (\kappa)^2_{\text{reg}}$, simple arguments show that λ can be taken accessible from κ as in the familiar Erdős-Rado Theorem for ordinary partition relations. However, it turns out that for exponents ≥ 3 regressive partition relations provide characterizations of the n-Mahlo cardinals for $n \in \omega$, as was first established by Schmerl.

Recall that the n-Mahlo cardinals are the least large cardinals conceptually transcending inaccessibility: κ is 0-Mahlo iff κ is (strongly) inaccessible; and κ is $n+1$-Mahlo iff every closed unbounded subset of κ contains an n-Mahlo cardinal. The following was established by Schmerl in a different notation:

1.3. Theorem (Schmerl [9]). *The following are equivalent for cardinals $\kappa > \omega$ and $n \in \omega$:*

(a) κ is n-Mahlo.

(b) For any $m \in \omega$ and unbounded $X \subseteq \kappa$, $X \to (m)^{n+3}_{\text{reg}}$.

(c) For any unbounded $X \subseteq \kappa$, $X \to (n+5)^{n+3}_{\text{reg}}$.

Rather unexpectedly, a partition relation for $\kappa > \omega$ only requiring a finite homogeneous set characterizes a large cardinal. This idea is pursued in useful form for the n-subtle cardinals in Section 2.

The following theorem completed the characterization of regressive partition relations.

1.4. Theorem (Schmerl [9] for (b), Hajnal-Kanamori-Shelah [5] for (c)). *The following are equivalent for cardinals $\kappa > \omega$ and $0 < n < \omega$:*

(a) κ is n-Mahlo.

(b) For any $\gamma < \kappa$ and unbounded $X \subseteq \kappa$, $X \to (\gamma)^{n+2}_{\text{reg}}$.

(c) For any closed unbounded $C \subseteq \kappa$, $C \to (\omega)^{n+2}_{\text{reg}}$.

Although the partition relation is preserved upon increasing the set on the left, imposing conditions on all unbounded $X \subseteq \kappa$ enables one to have characterizations *at* κ. Keeping this in mind, Theorems 1.3 and 1.4 show how one works one's way up through the regressive partition relation for exponents $n \geq 3$: Getting non-trivial min-homogeneous sets of size $m < \omega$ for $n = 3$ requires inaccessibility. Suddenly, getting one of size ω for $n = 3$ requires a 1-Mahlo cardinal κ. Moreover, we can then get min-homogeneous sets of any size $< \kappa$ for $n = 3$, as well as of any size $< \omega$ for $n = 4$. Repeating the pattern, to get a min-homogeneous set of size ω for $n = 4$ requires a 2-Mahlo, and so forth.

We next discuss the interplay between these characterizations of n-Mahlo cardinals and a "Borel diagonalization" proposition of Friedman [3]. He formulated and investigated several rather concrete propositions about Borel measurable functions (and in later work about spaces of groups and the like, and finite propositions—see Stanley [10] and Friedman [4]) which turned out through clever coding to have remarkably strong consistency strengths in terms of large cardinal hypotheses in set theory.

To recapitulate some notation and concepts, let I be the unit interval of reals and $Q = {}^{\omega}I$ (the Hilbert cube) the set of countable sequences drawn from

I. If $n \in \omega$ and $y, z \in {}^nQ$, say that $y \sim z$ iff there is a permutation ρ of ω, which is the identity except at finitely many arguments, such that $y(i) \circ \rho = z(i)$ for each $i < n$. Let us say that a function F with domain ${}^nQ$ is *totally invariant* iff whenever $y, z \in {}^nQ$, and $y \sim z$, then $F(y) = F(z)$. A function G with domain $Q \times {}^nQ$ is *right-invariant* iff whenever $x \in Q$, $y, z \in {}^nQ$, and $y \sim z$, then $G(x, y) = G(x, z)$.

Friedman's proposition P from his [3] is $\forall n \in \omega\, P_n$, where

(P_n) Suppose $F\colon Q \times {}^nQ \to I$ is Borel and right-invariant. Then for any $m \in \omega$ there is a sequence $\langle x_k \mid k < m\rangle$ of distinct elements of Q such that: whenever $s < t_1 < \cdots < t_n < m$, $F(x_s, \langle x_{t_1}, \ldots, x_{t_n}\rangle)$ is the first coordinate of x_{s+1}.

Note the analogy between the conclusion and min-homogeneity. Friedman motivated P as a sequential generalization of a basic Borel diagonalization proposition that he established in ZFC:

> If $F\colon Q \to I$ is Borel and totally invariant, then there is an $x \in Q$ such that $F(x) \in$ the range of x.

This was in turn motivated by Cantor's original topological proof that I is not countable, which amounted to showing that "totally invariant" cannot be dropped from above. As Friedman emphasized, "Borel" can be replaced by "finitely Borel", i.e. of a finite rank in the Baire hierarchy, without affecting the strength of P and thus bringing it into the fold of "concrete" mathematics. In particular, unlike other propositions like Suslin's Hypothesis, P is absolute with respect to relativization to the constructible universe L.

Friedman established:

1.5. Theorem (Friedman [3]). *The following are equivalent:*

(a) P(even just for finitely Borel functions).

(b) For any $a \subseteq \omega$ and $n \in \omega$, there is an ω-model containing a of ZFC + $\exists\kappa(\kappa$ *is n-Mahlo*$)$.

In the forward direction, P_{n+4} is used with an appropriate right-invariant Borel function to generate a finite sequence of reals that corresponds to a set of indiscernible ordinals in a "min" sense in an ω-model of ZFC + $V = L$. The characterization 1.3 is then invoked to show that there is a n-Mahlo cardinal in the model. In the converse direction, given a Borel function F as hypothesized in P_n, one works with a countable ω-model containing an $a \subseteq \omega$ coding F of ZFC + $\exists\kappa(\kappa$ is n-Mahlo$)$. In the Levy collapse of an n-Mahlo cardinal κ, ordinals $< \kappa$ are associated with members of Q, and Theorem 1.3 is used with a function based on corresponding forcing terms to verify P_n.

Kanamori [6] refined the proof of Theorem 1.5 and developed more technical propositions $\overline{P}_n$ for $n \in \omega$ to provide near equivalences for a level-by-level analysis:

1.6. Theorem (Kanamori [6]). *For any* $n \in \omega$*:*

(a) if $\overline{P}_{n+2}$ *holds (even just for Borel functions of rank* < 3*), then for any* $a \subseteq \omega$ *there is an* ω*-model containing* a *of* ZFC $+ \exists\kappa(\kappa$ *is* n*-Mahlo).*

(b) If for any $a \subseteq \omega$ *there is an* ω*-model containing* a *of* ZFC $+ \exists\kappa\exists\delta > \kappa(\kappa$ *is* n*-Mahlo and* $L_\kappa[a] \prec L_\delta[a])$*, then* $\overline{P}_{n+2}$ *holds.*

This was motivated by a question of Friedman, reminiscent of Questions 1.1 and 1.2 in minimizing hypotheses, that remains unresolved:

1.7. Question. Is P_3 independent of ZFC?

P_2 may also be independent, with the overall scheme suggesting that it may entail the existence of an ω-model of ZFC $+ \exists\kappa(\kappa$ is inaccessible).

The refinement of Theorem 1.6 over Theorem 1.5 is based to a large extent on a succinct extension of Theorem 1.3:

1.8. Theorem (Kanamori [6]). *Suppose that* $n \in \omega$ *and* X *is a set of ordinals such that* $X \cap \omega = \emptyset$*. Then* $X \to (n+5)^{n+3}_{\text{reg}}$ *iff* $X \cap \kappa$ *is unbounded in* κ *for some* n*-Mahlo cardinal* κ*.*

The point is that the regressive partition relation for a single set X requiring only a finite min-homogeneous set entails the existence of n-Mahlo cardinals. $X \cap \omega = \emptyset$ corresponds to the $\kappa > \omega$ case in Theorem 1.3, avoiding the known cases $\leq \omega$. Theorem 1.8 inspired an analogous assertion about n-subtle cardinals, which is the crucial ingredient in the extension of P in Section 3.

2 n-Subtle Cardinals

The n-subtle cardinals were introduced by Baumgartner [1] as generalizations of the subtle cardinals, isolated by Jensen and Kunen in their investigation of combinatorial principles in L. Compatible with $V = L$, the cardinals chart the territory between the weakly compact cardinals and the existence of $0^\sharp$ in the hierarchy of large cardinal hypotheses in set theory. Through a combinatorial analysis of their incipient definitions, Baumgartner provided regressive partition characterizations to which Theorem 1.3 bears an evident relation.

For X a set of ordinals and $n \in \omega$, we write $X \to (\gamma)^n_{\text{regr}}$ iff whenever f on $[X]^n$ is regressive, there is a $Y \in [X]^\gamma$ homogeneous for f (in the usual sense). Requiring homogeneous rather than just min-homogeneous sets turns out to be a considerable strengthening. For present purposes, we can comprehend the n-subtle cardinals through the following characterization.

2.1. Theorem (Baumgartner [1]). *Suppose that* $0 < n < \omega$. *Then the following are equivalent for a cardinal* κ:

(a) κ *is* n*-subtle.*

(b) *For any closed unbounded* $C \subseteq \kappa$, $C \to (n+2)^{n+1}_{\mathrm{regr}}$.

(c) *For any closed unbounded* $C \subseteq \kappa$, *and* f *regressive on* $[C]^{n+1}$, *there is an inaccessible cardinal* $\lambda < \kappa$ *and an unbounded* $Y \subseteq \lambda$ *homogeneous for* f.

The following is the needed analogue of Theorem 1.8; although the proof is similar, we include it because of the subtle differences.

2.2. Theorem. *Suppose that* $0 < n < \omega$ *and* X *is a set of ordinals such that* $X \cap 2 = \emptyset$. *If* $X \to (n+2)^{n+1}_{\mathrm{regr}}$, *then* $X \cap \kappa$ *is unbounded in* κ *for some* n*-subtle cardinal* κ.

Unlike Theorem 1.8 this is not an equivalence, with X the set of successor ordinals below an n-subtle cardinal being a counterexample. The refinement to $X \cap 2 = \emptyset$ is the natural one in the present context, but the first lemma toward the theorem provides a more useful condition.

2.3. Lemma. *Under the assumptions of Theorem 2.2,* $X \to (\gamma)^n_{\mathrm{regr}}$ *iff* $X \sim \omega \to (\gamma)^n_{\mathrm{regr}}$.

Proof. In the non-trivial direction, suppose that f is regressive on $[X \sim \omega]^n$; we must find a $Y \in [X \sim \omega]^\gamma$ homogeneous for f. Define g on $[X]^n$ as follows:

$$f(s) \begin{cases} f(s) & \text{if } s \cap \omega = \emptyset, \\ \min(s) - 1 & \text{if } s \subseteq \omega, \text{ else} \\ 0 & \text{if } |s \cap \omega| \text{ is even,} \\ 1 & \text{if } |s \cap \omega| \text{ is odd.} \end{cases}$$

g is regressive since $X \cap 2 = \emptyset$, so by hypothesis there is an $Y \in [X]^\gamma$ homogeneous for g. We can, of course, assume that $\gamma > n$. If $Y \subseteq \omega$, we can easily derive a contradiction using the second clause of g. If $Y \cap \omega \neq \emptyset$ and $Y \sim \omega \neq \emptyset$, then we can easily derive a contradiction using the third and fourth clauses of g. Hence, $Y \cap \omega = \emptyset$, and we are done. □

The following lemma contains the crux of Theorem 2.2.

2.4. Lemma. *Suppose that* $2 \leq n < \omega$ *and for some limit ordinal* η, C *and* X *are subsets of* $\eta \sim \omega$ *with* C *closed unbounded. If* $C \not\to (\gamma)^n_{\mathrm{regr}}$ *and* $X \cap \xi \not\to (\gamma)^n_{\mathrm{regr}}$ *for every* $\xi < \eta$, *then* $X \not\to (\gamma)^n_{\mathrm{regr}}$.

Proof. We first handle the cases $n \geq 3$. Set $\overline{C} = C \cup \{\omega\}$. For each $\alpha \in X$ set $\psi(\alpha) = \sup(\overline{C} \cap (\alpha + 1))$, an element of $\overline{C}$ since C is closed unbounded and $\min(X) \geq \omega$. We first define the *type* of a member of $[X]^n$ according to $\overline{C}$ as follows: If $\alpha_0 < \cdots < \alpha_{n-1}$ all belong to X, let $\langle \xi_0, \ldots, \xi_k \rangle$ enumerate the set $\{\psi(\alpha_i) \mid i < n\}$ in increasing order, and set $r_i = |\{i \mid \psi(\alpha_i) = \xi_j\}|$ for $j \leq k$. Then the *type* of $\{\alpha_0, \ldots, \alpha_{n-1}\rangle$ is $\langle r_0, \ldots, r_k \rangle$, which we can assume through sequence coding is a natural number $\neq 1$.

Next, let g attest to $C \not\to (\gamma)^n_{\text{regr}}$ and g_ξ attest to $X \cap \xi \not\to (\gamma)^n_{\text{regr}}$ for $\xi < \eta$. Since $C, X \subseteq \eta \sim \omega$, we can assume through renumbering that the ranges of g and of the g_ξ's do not contain 1 or any number coding a type. Now define G on $[X]^n$ as follows:

$$G(\alpha_0, \ldots, \alpha_{n-1}) = \begin{cases} 1 & \text{if } \omega = \psi(\alpha_0) < \cdots < \psi(\alpha_{n-1}), \\ g(\psi(\alpha_0), \ldots, \psi(\alpha_{n-1})) & \text{if } \omega < \psi(\alpha_0) < \cdots < \psi(\alpha_{n-1}), \\ g_\xi(\alpha_0, \ldots, \alpha_{n-1}) & \text{if } \psi(\alpha_0) = \cdots = \psi(\alpha_{n-1}), \\ & \quad \text{where } \xi \text{ is the next element} \\ & \quad \text{of } C \text{ after } \psi(\alpha_1), \\ \text{type of } \{\alpha_0, \ldots, \alpha_{n-1}\} & \text{otherwise.} \end{cases}$$

G is regressive, so suppose that $Y \subseteq X$ is homogeneous for G. We can assume that Y has at least $n + 1$ elements.

Using $n \geq 3$ and the last clause of G, it is simple to see that ψ must be either constant or injective on Y. If ψ is constant on Y, then Y cannot have ordertype γ by the third clause of G. If ψ is injective on Y, then $\psi(\min(Y)) \neq \omega$ by the first clause of G. In this case, Y cannot have ordertype γ by the second clause of G. This completes the proof for it $n \geq 3$.

In the special case $n = 2$, for every infinite ordinal ρ let f^e_ρ be a bijection between ρ and the *even* ordinals less than ρ, and f^o_ρ a bijection between ρ and the *odd* ordinals $\neq 1$ less than ρ. Now define G on $[X]^2$ as before, but modulated by these functions:

$$G(\alpha_0, \alpha_1) = \begin{cases} 1 & \text{if } \omega = \psi(\alpha_0) < \psi(\alpha_1), \\ f^e_{\psi(\alpha_0)}(g(\psi(\alpha_0), \psi(\alpha_1))) & \text{if } \omega < \psi(\alpha_0) < \psi\alpha_1), \\ f^o_{\alpha_0}(g_\xi(\alpha_0, \alpha_1)) & \text{if } \psi(\alpha_0 = \psi(\alpha_1), \end{cases}$$

We can now argue as before, discerning cases by whether the constant value on the homogeneous set is even or odd. The proof is complete. □

Proof of Theorem 2.2. With the given hypotheses, let η be the least ordinal such that $X \cap \eta \to (n+2)^{n+1}_{\text{regr}}$. Then $\eta > \omega$ by a simple argument. By Lemma 2.3 we can assume that $X \cap \omega = \emptyset$. But then Lemma 2.4 implies that for any closed unbounded $C \subseteq \eta$, $C \to (n+2)^{n+1}_{\text{regr}}$, i.e. η is n-subtle. □

3 The Proposition *H*

The clear analogy between Theorems 1.8 and 2.2 and the formulation and analysis in [6] of $\overline{P}_n$ (which we did not bother to state here) leads to the following extension of P:

$$H \text{ is } \forall n \in \omega\, H_n,$$

where

(H_{n+1}) Suppose that $F_1\colon Q \times {}^nQ \to I$ and $F_2\colon Q \times Q \to I$ are Borel and right-invariant such that $F_1(x, y) \in$ the range of x for $x \in Q$ and $y \in {}^nQ$. Then for any $m \in \omega$ there is a sequence $\langle x_i \mid i < m\rangle$ of distinct elements of Q such that: (a) if $s < t < m$, then $F_2(x_s, x_t)$ is the first coordinate of x_{s+1}, and (b) $\langle x_i \mid i < m\rangle$ is homogeneous for F_1.

Here, of course, "homogeneous" means that $F(x_s, \langle x_{t_1}, \ldots, x_{t_n}\rangle)$ is independent of the choice of $s < t_1 < \cdots < t_n < m$. The lack of dependence even on s corresponds to the move to homogeneity from min-homogeneity, and the "choice function" condition $F_1(x, y) \in$ range of x corresponds to regressiveness. The diagonalization analogy between H_n and P_n is maintained in the use of F_2 and (a) which figure in the proof below; from an esthetic point of view, eliminating them is desirable, and may be possible with a more subtle analysis.

The following is the main theorem of the paper. Because of the analogy established between the n-Mahlo and n-subtle cardinals, the proof amounts to a modification of that of Theorem 1.5. Consequently, we only provide details on the amendments, based in the forward direction on the approach of [6].

3.1. Theorem. *The following are equivalent:*

(a) H (even just for Borel functions of rank 3).

(b) For any $a \subseteq \omega$ and $0 < n < \omega$, there is an ω-model containing a of ZFC *+ $\exists\kappa(\kappa$ is n-subtle$)$.*

Remark. A level-by-level analysis along the lines of 1.6 is presumably possible at the cost of developing more technical propositions akin to $\overline{P}_n$; a finer proof than the one below would then have to be developed.

Proof. *(a)* $\to$ *(b)*. As in [3], let $\mathcal{L}$ be the language of second-order arithmetic augmented by class variables for *subsets* of $\mathcal{P}(\omega)$ (but no quantifiers for these variables). Any $A \subseteq \mathcal{P}(\omega)$ is regarded as a structure for $\mathcal{L}$ in the natural way, with first-order variables ranging over members of A that happen to be integers, second-order variables ranging over members of A, and class variables ranging over arbitrary subsets of A. A formula ψ of $\mathcal{L}$ is Σ^1_k if it has $k-1$ alternations of second-order quantifiers beginning with an existential quantifier, followed by only bounded first-order quantifiers. For $x \subseteq \omega$ let $|x| = \{\{m \mid 2^n 3^m \in x\} \mid n \in \omega\} \subseteq \mathcal{P}(\omega)$. Modifying Friedman's notion of (n,k)-critical sequence, say that $\langle x_i \mid i < d \rangle$ for $d \in \omega$ is an *n-sublime* sequence iff each $x_i \subseteq \omega$ and:

(i) for all $s < t < d$ and all Σ^1_2 formulas ψ we have $x_s \in |x_t|$ and $\{j \in \omega \mid |x_t| \models \psi(j, x_s)\} \in |x_{s+1}|$, and

(ii) for all $t_1 < \cdots < t_n < d$, $u_1 < \cdots < u_n < d$, $a \in |x_{\min\{t_1,u_1\}}|$ and Σ^1_2 formulas ψ in a finite collection Ψ (described below), we have

$$|x_{t_{n+1}}| \models \psi(a, |x_{t_1}|, \ldots, |x_{t_n}|) \text{ iff } |x_{u_{n+1}}| \models \psi(a, |x_{u_1}|, \ldots, |x_{u_n}|).$$

Here, Ψ is a finite collection of Σ^1_2 formulas which can be determined *a priori*, so that the above indiscernibility property for these formulas suffices to push through the main argument below for $(a) \to (b)$ of the theorem. The use of Σ^1_2 formulas follows [6], and is the reason why we can restrict H to Borel functions of rank < 3.

3.2. Lemma. *If $n > 0$ and H_{n+1} holds, then for any $d \in \omega$ there is an n-sublime sequence of length d.*

Proof. To apply H_{n+1}, we make the natural switch from I to $\mathcal{P}(\omega)$. For $x \in {}^\omega\mathcal{P}(\omega)$ let $\overline{x} = \{2^n 3^m \mid m \in x(n)\}$ and $\mathrm{Rng}(x)$ be the range of x. For any formula ψ of $\mathcal{L}$ let $\#\psi$ denote its Gödel number in some fixed arithmetization.

Define $F_1\colon {}^\omega\mathcal{P}(\omega) \times {}^{n+1}({}^\omega\mathcal{P}(\omega)) \to \mathcal{P}(\omega)$ and $g\colon {}^{n+2}({}^\omega\mathcal{P}(\omega)) \to \omega \times 3$ as follows: Suppose that $x, x_1, \ldots, x_{n+1} \in {}^\omega\mathcal{P}(\omega)$.

Case I. There is a Σ^1_2 formula $\psi \in \Psi$ as in (ii) above and $a \in \mathrm{Rng}(x)$ such that either

(Ia)
$$\mathrm{Rng}(x_n) \models \psi(a, \mathrm{Rng}(x), \mathrm{Rng}(x_1), \ldots, \mathrm{Rng}(x_{n-1})) \text{ and}$$
$$\mathrm{Rng}(x_{n+1}) \models \neg\psi(a, \mathrm{Rng}(x_1), \ldots, \mathrm{Rng}(x_n)).$$

or

(Ib)
$$\mathrm{Rng}(x_n) \models \neg\psi(a, \mathrm{Rng}(x), \mathrm{Rng}(x_1), \ldots, \mathrm{Rng}(x_{n-1})) \text{ and}$$
$$\mathrm{Rng}(x_{n+1}) \models \psi(a, \mathrm{Rng}(x_1), \ldots, \mathrm{Rng}(x_n)).$$

Then let a_0 be such an a so that $a = x(n)$ with n minimal, and for this a let ψ_0 be such a ψ with $\#\psi$ minimal. Set

$$F_1(x, \langle x_1, \ldots, x_{n+1}\rangle) = a_0, \quad \text{and}$$

$$g(x, x_1, \ldots, x_{n+1}) = \begin{cases} \langle \#\psi_0, 1\rangle & \text{if (Ia) holds,} \\ \langle \#\psi_0, 2\rangle & \text{if (Ib) holds.} \end{cases}$$

Case II. There is no such ψ. Then set

$$F_1(x, \langle x_1, \ldots, x_{n+1} \rangle) = x(0), \quad \text{and}$$
$$g(x, x_1, \ldots, x_{n+1}) = \langle 0, 0 \rangle .$$

Next, define $F_2 \colon {}^{\omega}\mathcal{P}(\omega) \times {}^{\omega}\mathcal{P}(\omega) \to \mathcal{P}(\omega)$ just as in 3.1 of [6], as follows: Suppose that $x, y \in {}^{\omega}\mathcal{P}(\omega)$.

Case 1. There is a Σ^1_2 formula ϕ such that $\{j \in \omega \mid \mathrm{Rng}(y) \models \phi(j, \overline{x})\} \notin \mathrm{Rng}(y)$. Then let ϕ_0 be such a formula with $\#\phi$ minimal, and set $F_2(x, y) = \{j \in \omega \mid \mathrm{Rng}(y) \models \phi_0(j, \overline{x})\}$.

Case 2. There is no such ϕ. Then set $F_2(x, y) = \{\#\phi \mid \phi \text{ is } \Sigma^1_2 \text{ and } \mathrm{Rng}(y) \models \phi(\overline{x})\}$.

Suppose now that $d \in \omega$ is given, assuming $d \geq n + 4$ for non-triviality. F_1 and F_2 satisfy the hypotheses of H_{n+1} (after the switch from I to $\mathcal{P}(\omega)$), and since Ψ in (ii) of n-sublime is finite, the range of g is finite. Hence, we can first apply H_{n+1} to get a sequence sufficiently long so that, by an application of the Finite Ramsey Theorem we can extract a subsequence $\langle x_i \mid i < d \rangle$ satisfying the conclusions of H_{n+1} and so that g is constant on ascendingly indexed $n+2$-tuples drawn from the subsequence.

We can now show that $\langle \overline{x_i} \mid i < d \rangle$ is n-sublime. By the argument for 3.1 of [6] using F_2, clause (i) must be satisfied. It is easy to check that the constant value of g must be $\langle 0, 0 \rangle$, and hence by a simple indiscernibility argument clause (ii) must also be satisfied. This completes the proof of Lemma 3.2. □

The rest of the proof of $(a) \to (b)$ is just as in [3] and [6]. Starting with a sufficiently long $n+1$-sublime sequence, we can build an ω-model of ZFC with an initial segment of the x_i's in the sequence corresponding to "ordinals" in the model. By Theorem 2.2 and the indiscernibility property (ii) of n-sublime, we can then show that in the model there must be an n-subtle cardinal. It can be checked in the argument of [6] that, indeed, only finitely many Σ^1_2 formulas, which we had anticipated with the collection Ψ, need be involved in (ii) of sublime. Finally, for the precise statement of (b), given any $a \subseteq \omega$ it can be used as a parameter in the Σ^1_2 formulas in the definition of n-sublime so that it will be a member of $|x_1|$.

$(b) \to (a)$. In this direction we try to exhibit the main ideas by following [3] as closely as we can, foregoing the refinements of [6], for the benefit of the reader. In particular, we outline the argument with $\sim$ in the definition of "right-invariant" replaced by $\approx$, where $x_0 \approx x_1$ means that x_0 and x_1 have the same range. The distracting modifications for getting the result with $\sim$ are just as in [3].

Toward the verification of H_n ($n > 0$) and maintaining the switch from I to $\mathcal{P}(\omega)$, suppose that $F_1 \colon {}^\omega\mathcal{P}(\omega) \times {}^n({}^\omega\mathcal{P}(\omega)) \to \mathcal{P}(\omega)$ such that $F_1(x,y) \in$ the range of x and $F_2 \colon {}^\omega\mathcal{P}(\omega) \times {}^\omega\mathcal{P}(\omega) \to \mathcal{P}(\omega)$ such that $F_2(x,y) \in$ the range of x are both Borel and right-invariant. Let $a \subseteq \omega$ code Borel codes for F_1 and F_2, and let M be a countable ω-model containing a of ZFC + "κ is $n+1$-subtle". Let C be the "Levy collapse" forcing notion in M, consisting of finite partial functions $f \colon \kappa \times \omega \to V_\kappa^M$ such that $f(\alpha, i) \in V_\alpha^M$.

Suppose now that $G \subseteq C$ is generic over M. Define $\overline{G} \colon \kappa \times \omega \to V_\kappa^M$ by $\overline{G}(\alpha, i) = x$ iff $\exists f \in G(f(\alpha,i) = x)$. For $x \in M$, set $E(G,x) = \{k \in \omega \mid \exists f \in G(\langle k, f\rangle \in x)\}$. Finally, for limit ordinals $\delta < \kappa$ define $T(G,\delta) \in {}^\omega\mathcal{P}(\omega)$ by $T(G,\delta)(m) = E(G, \overline{G}(\delta, m))$.

The following is Lemma 5.12 of [3] and is established using right-invariance:

3.3. Lemma. *Suppose that $\delta < \delta_1 < \kappa$ are limit ordinals and $f \in C$. Then for any $k \in \omega$,*

$$f \models k \in F_2(T(\dot{G},\delta), T(\dot{G},\delta_1)) \text{ iff } f|((\delta+1)\times\omega) \models k \in F_2(T(\dot{G},\delta), T(\dot{G},\delta_1)).$$

The analogous assertion holds for F_1.

With the choice function condition $F_1(x,y) \in$ range of x, the values of F_1 are determined by even less of the given condition, and this opens the door to the application of $n+1$-subtlety. In what follows, we write $p \,||\, \phi$ for p decides ϕ. i.e. $p \Vdash \phi$ or $p \Vdash \neg\phi$.

3.4. Lemma. *Suppose that $\delta < \delta_1 < \cdots < \delta_n < \kappa$ are limit ordinals, $f \in C$, and $k \in \omega$. If*

$$f \;||\; k \in F_1(T(\dot{G},\delta), T(\dot{G},\delta_1), \ldots, T(\dot{G},\delta_n)),$$

then there is a $g \le f$ and a $\gamma < \delta$ such that

$$g|(\gamma \times \omega) \;||\; k \in F_1(T(\dot{G},\delta), T(\dot{G},\delta_1), \ldots, T(\dot{G},\delta_n)).$$

Proof. Since $F_1(x,y) \in$ range of x, let $g \le f$ be such that

$$g \Vdash z = T(\dot{G},\delta)(m) = F_1(T(\dot{G},\delta), T(\dot{G},\delta_1), \ldots, (T(\dot{G},\delta_n))$$

for some term z and $m \in \omega$. By definition of T and E, we can consider z to be definable from $G|(\gamma\times\omega)$ for some $\gamma < \delta$. We now show that $g|(\gamma\times\omega) \,||\, k \in F_1(T(\dot{G},\delta), T(\dot{G},\delta_1), \ldots, T(\dot{G},\delta_n))$.

Let $h \leq g|(\gamma{\times}\omega)$ be arbitrary, and set $j = \max\{i{+}1 \mid \exists\alpha(\langle\alpha, i\rangle \in \text{domain}(h))\}$. Define $\hat{g} \in C$ by

$$\hat{g}(\alpha, i) = \begin{cases} g(\alpha, i) & \text{if } \alpha < \gamma \text{ and } g(\alpha, i) \text{ is defined,} \\ g(\alpha, i+j) & \text{if } \alpha \geq \gamma \text{ and } g(\alpha, i) \text{ is defined,} \\ \text{undefined} & \text{otherwise.} \end{cases}$$

By an automorphism argument, it follows that

$$\hat{g} \Vdash z = T(\dot{G}, \delta)(m+j) = F_1(T(\dot{G}, \delta), (T(\dot{G}, \delta_1), \ldots, (T(\dot{G}, \delta_n)) .$$

But clearly $\hat{g}$ and h are compatible, so we are done. $\square$

Continuing now with the main argument, we work in M. Let $C = \{\alpha < \kappa \mid \alpha \text{ is a strong limit cardinal}\}$. Since $n+1$-subtle cardinals are inaccessible, C is a closed unbounded subset of κ. Define a function H on $[C]^{n+2}$ as follows: Suppose that $\delta_0 < \delta_1 < \cdots < \delta_{n+1}$ all belong to C.

Case I. There is an $f \in C$ and a $k \in \omega$ such that either

$$\text{(Ia)} \qquad \begin{aligned} f \Vdash k \in\ & F_1(T(\dot{G}, \delta_0), T(\dot{G}, \delta_1), \ldots, T(\dot{G}, \delta_n)) \\ & \sim F_1(T(\dot{G}, \delta_1), T(\dot{G}, \delta_2), \ldots, T(\dot{G}, \delta_{n+1})) \end{aligned}$$

or

$$\text{(Ib)} \qquad \begin{aligned} f \Vdash k \in\ & F_1(T(\dot{G}, \delta_1), T(\dot{G}, \delta_2), \ldots, T(\dot{G}, \delta_{n+1})) \\ & \sim F_1(T(\dot{G}, \delta_0), T(\dot{G}, \delta_1), \ldots, T(\dot{G}, \delta_n)) \end{aligned}$$

By Lemma 3.4 we can assume that for some $\gamma < \delta_0$,

$$f|(\gamma \times \omega) \mid\mid k \in F_1(T(\dot{G}, \delta_0), T(\dot{G}, \delta_1), \ldots, T((\dot{G}, \delta_n)) .$$

Taking f least possible in some fixed well-ordering and γ least possible for this f, set

$$H(\delta_0, \delta_1, \ldots, \delta_{n+1}) = \begin{cases} \langle f|(\gamma \times \omega), 1\rangle & \text{if (Ia) holds,} \\ \langle f|(\gamma \times \omega), 2\rangle & \text{if (Ia) holds.} \end{cases}$$

Case II. There is no such $f \in C$ and $k \in \omega$. Then set $H(\delta_0, \delta_1, \ldots, \delta_{n+1}) = \langle 0, 0\rangle$.

We can regard H as a regressive function on $[C]^{n+2}$ through coding, since the number of possibilities for f in *Case* I is small. Suppose that $Y \subseteq C$ is homogeneous for H. Then assuming that $|Y| \geq n+3$, *Case* I cannot occur on ascending $n+2$-tuples drawn from Y: If for some $\delta_0 < \cdots < \delta_{n+2}$ all belonging to Y and $0 < i \leq 2$,

$$H(\delta_0, \ldots, \delta_{n+1}) = \langle g, i \rangle = H(\delta_1, \ldots, \delta_{n+2}),$$

then g decides $k \in F_1(T(\dot{G}, \delta_1), \ldots, T(\dot{G}, \delta_{n+1}))$ in one way by the second equality, but by the first equality and definition, g is extendible to a condition that decides it in the *other* way. Hence, *Case* II occurs, and the argument of [3] can now be used to get a homogeneous sequence for F_1.

To further handle F_2, note that by Theorem 2.1(c) we can assume that Y is unbounded in some inaccessible cardinal λ. Hence, Theorem 1.3(b) is more than enough so that with the original [3] argument based on Lemma 3.3, we can extract arbitrarily long finite subsequences of Y min-homogeneous for a regressive function corresponding to F_2 and show that the full conclusion of H_{n+1} can be satisfied. This completes our (indication of) the proof of the main Theorem 3.1. □

We point out that Lemma 3.4 was needed to insure that H can be regarded as regressive on a *closed* unbounded set $\subseteq \kappa$ so that we can extract a homogeneous set by $n+1$-subtlety. The function corresponding to F_2 based on Lemma 3.3 need only be regressive on an unbounded set $\subseteq \lambda$ to extract a min-homogeneous set, so a simpler strategem is available—see the H_1^+ idea at the end of the proof of Theorem A in [6].

As in [6] it is possible to develop an infinitary version of H_n with an $F_1 \colon Q \times {}^{<\omega}Q \to I$ to get a principle with consistency strength at least that of ZFC + $\exists\kappa\forall n(\kappa$ is n-subtle$)$. However, getting an equiconsistency result seems difficult here.

References

[1] J. Baumgartner, Ineffability properties of cardinals I, in: A. Hajnal, R. Rado, and V. Sós, eds. Infinite and Finite Sets, Colloq. Math. Soc. Janos Boylai 10 (North-Holland, Amsterdam, 1975) 109-130.

[2] P. Erdős and R. Rado, A combinatorial theorem, J. London Math. Soc. 25 (1950) 249-255.

[3] H. Friedman, On the necessary use of abstract set theory, Adv. in Math. 41 (1981) 209-280.

[4] H. Friedman, On the necessary use of abstract set theory in finite mathematics, Adv. in Math. 60 (1986) 92-122.

[5] A. Hajnal, A. Kanamori, and S. Shelah, Regressive partition relations for infinite cardinals, Trans. Amer. Math. Soc. 299 (1987) 145-154.

[6] A. Kanamori, Regressive partitions and Borel diagonalization, J. Symbolic Logic 54 (1989) 540-552.

[7] A. Kanamori and K. McAloon, On Gödel incompleteness and finite combinatorics, Ann. Pure Appl. Logic 33 (1987) 23-41.

[8] J. Paris and L. Harrington, A mathematical incompleteness in Peano Arithmetic, in: J. Barwise, ed., Handbook of Mathematical Logic (North-Holland, Amsterdam, 1977) 1133-1142.

[9] J. Schmerl, A partition property characterizing cardinals hyperinaccessible of finite type, Trans. Amer. Math. Soc. 188 (1974) 281-291.

[10] L. Stanley, Borel diagonalization and abstract set theory, in: L.A. Harrington, M.D. Morley, A. Scedrov, and S.G. Simpson, eds., Harvey Friedman's Research on the Foundations of Mathematics (North-Holland, Amsterdam 1985) 11-86.

Part IV

Lives in Set Theory

14 Gödel and Set Theory

Kurt Gödel (1906–1978) with his work on the constructible universe L established the relative consistency of the Axiom of Choice (AC) and the Continuum Hypothesis (CH). More broadly, he ensured the ascendancy of first-order logic as the framework and a matter of method for set theory and secured the cumulative hierarchy view of the universe of sets. Gödel thereby transformed set theory and launched it with structured subject matter and specific methods of proof. In later years Gödel worked on a variety of set-theoretic constructions and speculated about how problems might be settled with new axioms. We here chronicle this development from the point of view of the evolution of set theory as a field of mathematics. Much has been written, of course, about Gödel's work in set theory, from textbook expositions to the introductory notes to his collected papers. The present account presents an integrated view of the historical and mathematical development as supported by his recently published lectures and correspondence. Beyond the surface of things we delve deeper into the mathematics. What emerges are the roots and anticipations in work of Russell and Hilbert, and most prominently the sustained motif of truth as formalizable in the "next higher system". We especially work at bringing out how transforming Gödel's work was for set theory. It is difficult now to see what conceptual and technical distance Gödel had to cover and how dramatic his re-orientation of set theory was. What he brought into set theory may nowadays seem easily explicated, but only because we have assimilated his work as integral to the subject. Much has also been written about Gödel's philosophical views about sets and his wider philosophical outlook, and while these may have larger significance, we keep the focus on the motivations and development of Gödel's actual mathematical constructions and contributions to set theory. Leaving his "concept of set" alone, we draw out how in fact he had strong mathematical instincts and initiatives, especially as seen in his last, 1970 attempt at the continuum problem.

1 From Truth to Set Theory

Gödel's advances in set theory can be seen as part of a steady intellectual development from his fundamental work on completeness and incompleteness. Two remarkably prescient passages in his early publications serve as our point of departure. His incompleteness paper [1931], submitted for publication 17 November 1930, had a footnote 48a:

Republished from *The Bulletin of Symbolic Logic* 13 (2007), pp.153-188, with permission from The Association for Symbolic Logic.

This is a much expanded version of a third of Floyd–Kanamori [2006] and a source for an invited address given at the annual meeting of the Association for Symbolic Logic held at Montreal in May 2006. The author expresses his gratitude to Juliet Floyd and John Dawson for helpful comments and suggestions.

> As will be shown in Part II of this paper, the true reason for the incompleteness inherent in all formal systems of mathematics is that the formation of ever higher types can be continued into the transfinite (cf. D. Hilbert, "Über das Unendliche", Math. Ann. 95, p. 184), while in any formal system at most denumerably many of them are available. For it can be shown that the undecidable propositions constructed here become decidable whenever appropriate higher types are added (for example, the type ω to the system P). An analogous situation prevails for the axiom system of set theory.

This passage has been made much of,[1] whereas the following has not. It appeared in a summary [1932b], dated 22 January 1931, of a talk on the incompleteness results given in Karl Menger's colloquium. Notably, matters in a footnote, perhaps an afterthought then, have now been expanded to take up fully one-third of an abstract on incompleteness:

> If we imagine that the system Z is successively enlarged by the introduction of variables for classes of numbers, classes of classes of numbers, and so forth, together with the corresponding comprehension axioms, we obtain a sequence (continuable into the transfinite) of formal systems that satisfy the assumptions mentioned above, and it turns out that the consistency (ω-consistency) of any of those systems is provable in all subsequent systems. Also, the undecidable propositions constructed for the proof of Theorem 1 [the Gödelian sentences] become decidable by the adjunction of higher types and the corresponding axioms; however, in the higher systems we can construct other undecidable propositions by the same procedure, and so forth. To be sure, all the propositions thus constructed are expressible in Z (hence are number-theoretic propositions); they are, however, not decidable in Z, but only in higher systems for example, in that of analysis. In case we adopt a type-free construction of mathematics, as is done in the axiom system of set theory, axioms of cardinality (that is, axiom postulating the existence of sets of ever higher cardinality) take the place of type extensions, and it follows that certain arithmetic propositions that are undecidable in Z become decidable by axioms of cardinality, for example, by the axiom that there exist sets whose cardinality is greater than every α_n, where $\alpha_0 = \aleph_0$, $\alpha_{n+1} = 2^{\alpha_n}$.

The salient points of these passages is that the addition of the next "higher type" to a formal system leads to newly provable propositions of the system;

[1] See e.g. Kreisel [1980: 183,195,197], a memoir on Gödel, and Feferman [1987], where the view advanced in the footnote is referred to as "Gödel's doctrine".

the iterative addition of higher types can be continued into the transfinite; and in set theory, new propositions become analogously provable from "axioms of cardinality". The transfinite heritage from Hilbert [1926], cited in footnote 48a, will be discussed in §5. Here we discuss the connections with the frameworks of types and of truth, which can be associated respectively with Bertrand Russell and Alfred Tarski.

Mathematical logic was emerging from the Russellian world of orders and types, and Gödel's work would reflect and transform Russell's initiatives. Russell's *ramified theory of types* is a scheme of logical definitions based on *orders* and *types* indexed by the natural numbers. Russell proceeded "intensionally"; he conceived this scheme as a classification of propositions based on the notion of *propositional function*, a notion not reducible to membership (extensionality). Proceeding however in modern fashion, we may say that the universe is to consist of *objects* stratified into disjoint types T_n, where T_0 consists of the *individuals*, $T_{n+1} \subseteq \{X \mid X \subseteq T_n\}$, and the types T_n for $n > 0$ are further ramified into orders O^i_n with $T_n = \bigcup_i O^i_n$. An object in O^i_n is to be defined either in terms of individuals or of objects in some fixed O^j_m for some $j < i$ and $m \leq n$, the definitions allowing for quantification only over O^j_m. This precludes Russell's Paradox and other "vicious circles", as objects can only consist of previous objects and are built up through definitions only referring to previous stages. However, in this system it is impossible to quantify over all objects in a type T_n, and this makes the formulation of numerous mathematical propositions at best cumbersome and at worst impossible. So Russell was led to introduce his axiom of *Reducibility*, which asserts that *for each object there is a predicative object having exactly the same constituents*, where an object is *predicative* if its order is the least greater than that of its constituents. This axiom in effect reduced consideration to individuals, predicative objects consisting of individuals, predicative objects consisting of predicative objects consisting of individuals, and so on—the *simple theory of types.*[2]

The above quoted Gödel passages can be considered a point of transition from type theory to set theory. The system P of footnote 48a is Gödel's streamlined version of Russell's theory of types built on the natural numbers as individuals, the system used in [1931]. The last sentence of the footnote calls to mind the other reference to set theory in that paper; Gödel [1931: 178] wrote of his comprehension axiom IV, foreshadowing his approach to set theory, "This axiom plays the role of [Russell's] axiom of reducibility (the comprehension axiom of set theory)." The system Z of the quoted [1932b] passage is already the more modern first-order Peano arithmetic, the system in which Gödel in his abstract described his incompleteness results. The passage envisages the introduction of higher-type variables, which would have the effect of re-establishing

[2]In substantial criticism based on how mathematics ought to be regarded as a "calculus of extensions", Frank Ramsey [1926] emphasized and advocated this reduction.

the system P, but as one proceeds to higher and higher types, that "all the [unprovable] propositions constructed are expressible in Z (hence are number-theoretic propositions)" is an important point about incompleteness. The last sentence of the [1932b] passage is Gödel's first remark on set theory of substance, and significantly, his example of an "axiom of cardinality" to take the place of type extensions is essentially the one that both Abraham Fraenkel [1922] and Thoralf Skolem [1923] had pointed out as unprovable in Ernst Zermelo's [1908] axiomatization of set theory and used by them to motivate the axiom of Replacement.

We next face head on the most significant underlying theme broached in our two quoted passages. Gödel's engagement with truth at this time, whether with conviction or caution,[3] could be viewed as his *entrée* into full-blown set theory. In later, specific terms, first-order satisfaction involves canvassing *arbitrary* variable assignments, and higher-order satisfaction requires, in effect, scanning all *arbitrary* subsets of a domain.

In the introduction to his dissertation on completeness Gödel [1929] had already made informal remarks about satisfaction, discussing the meaning of " 'A system of relations *satisfies* [*erfüllt*] a logical expression' (that is, the sentence obtained through substitution is true [*wahr*])." In a letter to Paul Bernays of 2 April 1931 Gödel[4] described how to define the unary predicate that picks out the Gödel numbers of the "correct" ["*richtig*"] sentences of first-order arithmetic. Gödel then remarked, as he would in similar vein several times in his career, "Simultaneously and independently of me (as I gathered from a conversation), Mr. Tarski developed the idea of defining the concept 'true proposition' in this way (for other purposes, to be sure)." Finally, Gödel emphasized the "decidability of the undecidable propositions in higher systems" specifically through the use of the truth predicate.

The semantic, recursive definition of the satisfaction relation, both first-order and higher-order, was first systematically formulated in set-theoretic terms by Tarski [1933][1935], to whom is usually attributed the undefinability of truth for a formal language within the language.[5] However, evident in Gödel's thinking was the *necessity* of a higher system to capture truth, and in fact Gödel maintained to Hao Wang [1996: 82] that he had come to the undefinability of arithmetical truth in arithmetic already in the summer of 1930.[6]

---

[3]Cf. Feferman [1984].

[4]See Gödel [2003a: 97ff].

[5]See Tarski [1933][1935] §5, theorem I; in a footnote Tarski wrote: "We owe the method used here to Gödel, who has employed it for other purposes in his recently published work, [Gödel [1931]] ...." See Feferman [1984], Murawski [1998], Krajewski [2004], and Woleński [2005] for more on Gödel and Tarski vis-á-vis truth.

[6]Wang [1996: 82] reports Gödel as having conveyed the following: Gödel began work on Hilbert's problem to establish the consistency of analysis in the summer of 1930. Gödel quickly distinguished two problems: to establish the consistency of analysis relative to number theory, and to establish the consistency of number theory relative to finitary number theory.

In a letter to Zermelo of 12 October 1931 Gödel[7] pointed out that the undefinability of truth leads to a quick proof of incompleteness: The class of provable formulas *is* definable and the class of true formulas is not, and so there must be a true but unprovable formula. Gödel also cited his [1931] footnote 48a, and this suggests that he himself invested it with much significance.[8]

Higher-order satisfaction is particularly relevant both for footnote 48a and the [1932b] abstract. Rudolf Carnap at this time was working on his *Logical Syntax of Language*, and in a manuscript attempted a definition of "analyticity" for a language that subsumed the theory of types. Working upward, he provided an adequate definition of truth for first-order arithmetic. In a letter to Carnap of 11 September 1932 Gödel[9] pointed out however that Carnap's attempted recursive definition for second-order formulas contained a circularity. Gödel wrote:

> ...this error may only be avoided by regarding the domain of the function variables not as the predicates of a definite language, but rather as all sets and relations whatever.[1] On the basis of this idea, in the second part of my work [[1931]] I will give a definition for "truth", and I am of the opinion that the matter may not be done otherwise ....
>
> [1]This doesn't necessarily involve a Platonistic standpoint, for I assert only that this definition (for "analytic") be carried out within a definite language in which one already has the concepts "set" and "relation".

The semantic definition of second-order truth requires "all sets and relations whatever" and must be carried out where one "already has the concepts 'set' and 'relation' ".[10]

A succeeding letter of 28 November 1932 from Gödel to Carnap elaborated on Gödel's footnote 48a.[11] Gödel never actually wrote a Part II to his [1931] and laconically admitted in the letter that such a sequel "exists only in the realm of ideas". Gödel then clarified how the addition of an infinite type ω

For the first problem, Gödel found that he had to rely on the concept of *truth* for number theory, not just the consistency of a formal system for it, and this soon led him to establish the undefinability of truth. The second problem led, of course, to the incompleteness theorems. Note here that Gödel had already focused on establishing *relative* consistency results.

[7]Cf. [2003a: 423ff].

[8]In a letter to Gödel of 21 September 1931 Zermelo (cf. Gödel [2003a: 420ff]) had actually given the argument for the undefinability of arithmetical truth in arithmetic, thinking that he had found a contradiction in Gödel [1931] whereas he had only conflated truth with provability. This followed the one meeting between Zermelo and Gödel, for which see Kanamori [2004: §7].

[9]Cf. [2003: 347].

[10]In a reply of 25 September 1932 to Gödel [2003: 351], Carnap seems somewhat the foil when he asked how this last is to be understood, and further: "Can you define the concept 'set' within a definite formalized semantics?"

[11]Cf. [2003: 355].

to the [1931] system P would render provable the unprovable propositions he had constructed — specifically since a truth definition can now be provided. Significantly, Gödel wrote however:

> ...the interest of this definition does not lie in a clarification of the concept 'analytic' since one employs in it the concepts 'arbitrary sets', etc., which are just as problematic. Rather I formulate it only for the following reason: with its help one can show that undecidable sentences become decidable in systems which ascend farther in the sequence of types.

The definition of truth is not itself clarificatory, but it does serve a *mathematical* end.

Tarski, of course, *did* put much store in his systematic definition of truth for formal languages, and Carnap would be much influenced by Tarski's work on truth. Despite their contrasting attitudes toward truth, Gödel's and Tarski's approaches had similarities. Tarski's [1933][1935] undefinability of truth result is couched in terms of languages having "infinite order", analogous to Gödel's [1931] system P having infinite types, and Gödel's infinite type ω is analogous to Tarski's "metalanguage". In a postscript in his [1935: 194,n.108], Tarski acknowledged Gödel's footnote 48a.

In a lecture [1933o] Gödel expanded on the themes of our quoted passages. He propounded the axiomatic set theory "as presented by Zermelo, Fraenkel, and von Neumann" as "a natural generalization of the [simple] theory of types, or rather, what becomes of the theory of types if certain superfluous restrictions are removed." First, instead of having separate types with sets of type $n+1$ consisting purely of sets of type n, sets can be *cumulative* in the sense that sets of type n can consist of sets of *all* lower types. If S_n is the collection of sets of type n, then: S_0 is the type of the individuals, and recursively, $S_{n+1} = S_n \cup \{X \mid X \subseteq S_n\}$. Second, the process can be continued into the transfinite, starting with the cumulation $S_\omega = \bigcup_n S_n$, proceeding through successor stages as before, and taking unions at limit stages. Gödel [1933o: 46] again credited Hilbert for opening the door to the formation of types beyond the finite types. As for how far this cumulative hierarchy of sets is to continue, the "first two or three types already suffice to define very large ordinals" ([1933o: 47]) which can then serve to index the process, and so on, in an "autonomous progression" in later terminology. In a prophetic remark for set theory and new axioms, Gödel observed: "We set out to find a formal system for mathematics and instead of that found an infinity of systems, and whichever system you choose out of this infinity, there is one more comprehensive, i.e., one whose axioms are stronger." Further echoing the quoted [1932b] passage Gödel [1933o: 48] noted that for any formal system S there is in fact an arithmetical proposition that cannot be proved in S, unless S is inconsistent. Moreover, if S is based on the theory of types, this arithmetical proposition becomes provable if to S is adjoined "the next higher type and the axioms concerning it."

Gödel's approach to set theory, with its emphasis on hierarchical truth, should be set into the context of the axiomatic development of the subject.[12] Zermelo [1908] had provided the initial axiomatization of "the set theory of Cantor and Dedekind", with characteristic axioms Separation, Infinity, Power Set, and of course, Choice. Work most substantially of John von Neumann [1923][1928] on ordinals led to the incorporation of Cantor's transfinite numbers as now the ordinals and the axiom schema of Replacement for the formalization of transfinite recursion. Von Neumann [1929] also formulated the axiom of Foundation, that every set is well-founded, and defined the *cumulative hierarchy* in his system via transfinite recursion: The axiom entails that the universe V of sets is globally structured through a stratification into cumulative "ranks" V_α, where with $\mathcal{P}(X) = \{Y \mid Y \subseteq X\}$ denoting the power set of X,

$$V_0 = \emptyset;\ V_{\alpha+1} = \mathcal{P}(V_\alpha);\ V_\delta = \bigcup_{\alpha<\delta} V_\alpha \text{ for limit ordinals } \delta\,;$$

and

$$V = \bigcup_\alpha V_\alpha\,.$$

Zermelo in his remarkable [1930] subsequently provided his final axiomatization of set theory, proceeding in a second-order context and incorporating both Replacement (which subsumes Separation) and Foundation. These axioms rounded out but also focused the notion of set, with the first providing the means for transfinite recursion and induction and the second making possible the application of those methods to get results about *all* sets. Gödel's coming work would itself amount to a full embrace of Replacement and Foundation but also first-order definability, which would vitalize the earlier initiative of Skolem [1923] to establish set theory on the basis of first-order logic.[13] The now standard axiomatization ZFC is essentially the first-order version of the Zermelo [1930] axiomatization, and ZF is ZFC without AC.

2 The Constructible Universe L

Set theory was launched on an independent course as a distinctive field of mathematics by Gödel's formulation of the class L of *constructible* sets through which he established the relative consistency of AC in mid-1935 and CH in mid-1937.[14] In his first announcement, communicated 9 November 1938, Gödel [1938] wrote:

[12] For a fuller account documenting the contributions of many, see Kanamori [1996].

[13] See §6 for a comparative analysis of the approaches of Zermelo and Gödel.

[14] See Dawson [1997: 108,122]; in one of Gödel's *Arbeitshefte* there is an indication that he established the relative consistency of CH in the night of 14-15 June 1937.

> "[The] 'constructible' sets are defined to be those sets which can be obtained by Russell's ramified hierarchy of types, if extended to include transfinite orders. The extension to transfinite orders has the consequence that the model satisfies the impredicative axioms of set theory, because an axiom of reducibility can be proved for sufficiently high orders."

This points to two major features of the construction of L:

(i) Gödel had refined the cumulative hierarchy of sets described in his lecture [1933o] to a hierarchy of *definable* sets which is analogous to the orders of Russell's *ramified* theory. Despite the broad trend in mathematical logic away from Russell's intensional intricacies and toward versions of the simple theory of types, Gödel had assimilated the ramified theory and its motivations as of consequence and now put the theory to a new use, infusing its intensional character into an extensional context.

(ii) Gödel continued the indexing of the hierarchy through *all* the ordinals as given beforehand to get a class model of set theory and thereby to achieve *relative* consistency results. His earlier [1933o: 47] idea of using large ordinals defined in low types for further indexing in a bootstrapping process would not suffice. That "an axiom of reducibility can be proved for sufficiently high orders" is an opaque allusion to how Russell's problematic axiom would be rectified in the consistency proof of CH (see §3) and more broadly to how the axiom of Replacement provided for new sets and enough ordinals.[15] Von Neumann's ordinals would be the spine for a thin hierarchy of sets, and this would be the key to both the AC and CH results.

In a brief account [1939a] Gödel informally presented L much as is done today: For any set x let $\mathrm{def}(x)$ denote the collection of subsets of x definable over $\langle x, \in\rangle$ via a first-order formula allowing parameters from x. Then define

$$L_0 = \emptyset;\ L_{\alpha+1} = \mathrm{def}(L_\alpha),\ L_\delta = \bigcup\nolimits_{\alpha<\delta} L_\alpha \text{ for limit ordinals } \delta;$$

and the *constructible universe*

$$L = \bigcup\nolimits_\alpha L_\alpha\,.$$

Toward the end Gödel [1939a: 31] pointed out that L "can be defined and its theory developed in the formal systems of set theory themselves." This is a

[15]Wang [1980: 129], reporting on conversations with Gödel in 1976, wrote how he "spoke of experimenting with more and more complex constructions [of ordinals for indexing] for some extended period somewhere between 1930 and 1935." Kreisel [1980: 193,196] wrote of Gödel's "reservations" about Replacement which initially held him back from considering all the ordinals as being given beforehand.

remarkable understatement of arguably the central feature of the construction of L:

(iii) L is a class definable in set theory via a transfinite recursion that could be based on the formalizability of def(x), the definability of definability. Gödel had not embraced the definition of truth as itself clarificatory,[16] but through his work he in effect drew it into mathematics to a new mathematical end. Though understated in Gödel's writing, his great achievement here as in arithmetic is the submergence of metamathematical notions into mathematics.

In the proof of the incompleteness theorem, Gödel had encoded provability — syntax — and played on the interplay between truth and definability. Gödel now encoded satisfaction — semantics — with the room offered by the transfinite indexing, making truth, now definable for levels, part of the formalism and part of the subject matter. In modern parlance, an *inner model* of ZFC is a transitive (definable) class containing all the ordinals such that, with membership and quantification restricted to it, the class satisfies each axiom of ZFC. Gödel in effect argued in ZF to show that L is an inner model of ZFC, and moreover that L satisfies CH. He thus established the relative consistency Con(ZF) implies Con(ZFC + CH). In what follows, we describe his proofs that L is an inner model of ZFC and in §3 that L satisfies CH.

In his sketch [1939a] Gödel simply argued for the ZFC axioms holding in L as evident from the construction, with the extent of the ordinals and the sets provided by def(x) sufficient to establish Replacement in L. Only at the end when he was attending to formalization did he allude to the central issue of relativization. For here and later, recall that for a formula φ and classes C and M, φ^M and C^M denote the relativizations to M of φ and C respectively, i.e. φ^M denotes φ but with the quantifiers restricted to the elements of M, and C^M denotes the class defined by the relativization to M of a defining formula for C. Gödel's [1939a] arguments for relative consistency amount to establishing φ^L as theorems of set theory for various φ starting with the axioms of set theory themselves, and could only work if $\text{def}^L(x) = \text{def}(x)$ for $x \in L$. This *absoluteness of first-order definability* is central to the proof if L is to be formally defined via the def(x) operation, but notably Gödel himself would never establish this absoluteness explicitly, preferring in his one rigorous published exposition of L to take an approach that avoids def(x) altogether.

In his monograph [1940], based on 1938 lectures, Gödel provided a specific, formal presentation of L in a class-set theory developed by Paul Bernays [1937], a theory based in turn on a theory of von Neumann [1925]. First, Gödel carried out a paradigmatic development of "abstract" set theory through the ordinals and cardinals with features that have now become common fare, like

[16]See the last displayed passage in §1.

his particular well-ordering of pairs of ordinals.[17] Gödel then used eight binary operations, producing new classes from old, to generate L set by set via transfinite recursion. This veritable "Gödel numbering" with ordinals bypassed the def(x) operation and made evident certain aspects of L. Since there is a direct, definable well-ordering of L, choice functions abound in L, and AC holds there.

Much of the analysis of L would have to be devoted to verifying Replacement or at least Separation there, this requiring an analysis of the first-order formalization of set properties. It has sometimes been casually asserted that Gödel [1940] through his eight operations provided a finite axiomatization of Separation, but this cannot be done. Through closure under the operations one does get Separation for *bounded* formulas, i.e. those formulas all of whose quantifiers can be rendered as $\forall x \in y$ and $\exists x \in y$. Gödel established using Replacement (in V) that for any set $x \subseteq L$, there is a $y \in L$ such that $x \subseteq y$ (9.63 of [1940]). He then established that a wide range of classes $C \subseteq L$ satisfy the condition that for any $x \in L$, $x \cap C \in L$, that C is "amenable" in later terminology. With this, he established σ^L for every axiom σ of ZFC, the relativized instances of Replacement being the most crucial to confirm.[18] Having bypassed def(x), this argumentation makes no appeal to absoluteness.

3 Consistency of the Continuum Hypothesis

Gödel's proof that L satisfies CH consisted of two separate parts. He established the implication $V = L \rightarrow$ CH, and, in order to apply this implication within L, the absoluteness $L^L = L$ to establish the desired $(\mathrm{CH})^L$. That $V = L \rightarrow$ CH established a connection between two quite non-absolute concepts, the power set and successor cardinality of an infinite set, and the absoluteness $L^L = L$ effected the requisite relativization. That $L^L = L$ had been asserted in his first announcement [1938], and follows directly from $\mathrm{def}^L(x) = \mathrm{def}(x)$ for $x \in L$, which was broached in the sketch [1939a]. In [1940], his approach to $L^L = L$ was rather through the evident absoluteness of the eight generating operations which in particular entailed that being a (von Neumann) ordinal is absolute and ensured the internal integrity of the generation of L. There is a nice resonance here with Gödel [1931], in that there he had catalogued a series of functions to be primitive recursive whereas now he catalogued a series of set-theoretic operations to be absolute — the submergence of provability (syntax) for arithmetic evolved to the submergence of definability (semantics) for set theory. The argument in fact shows that for any inner model M of ZFC, $L^M = L$. Decades later many inner models based on first-order definability would be investigated for which absoluteness

[17]In footnote 14 added in a 1951 printing of his [1940] Gödel (cf. [1990: 54]) even used the device later attributed to Dana Scott [1955] for reducing classes to sets by restricting to members of lowest rank.

[18]Jech [2002: §13] presents a modern version of Gödel's argument.

considerations would be pivotal, and Gödel had formulated the canonical inner model.

Gödel's argument for $V = L \to \mathrm{CH}$ rests, as he himself wrote in a brief summary [1939], on "a generalization of Skolem's method for constructing enumerable models." This was the first significant use of Skolem functions since Skolem's own [1920] to establish the Löwenheim-Skolem theorem. Gödel [1939a] specifically established:

(*) For infinite α, every constructible subset of L_α belongs to some L_β for a β of the same cardinality as α.

It is straightforward to show that for infinite α, L_α has the same cardinality as that of α. It follows from (*) that in L, the power set of L_ω is included in L_{ω_1}, and so CH follows. (Gödel emphasized the Generalized Continuum Hypothesis (GCH), that $2^{\aleph_\alpha} = \aleph_{\alpha+1}$ for all α, and $V = L \to \mathrm{GCH}$ follows by analogous reasoning.) Gödel [1939a] proved (*) for an $X \subseteq L_\alpha$ such that $X \in L$ by getting a set $M \subseteq L$ containing X and sufficiently many ordinals and definable sets so that M will be isomorphic to some L_β, the construction of M ensuring that β has the same cardinality as α. Gödel's approach to M, different from the usual approach taken nowadays, can be seen as proceeding through layers defined recursively, a new layer being defined via closure according to new Skolem functions and ordinals based on the preceding layer. This was indeed a "generalization of Skolem's method", being an *iterative* application of Skolem closures. M having been sufficiently bolstered, Gödel then confirmed that M is isomorphic with respect to $\in$ to some L_β, making the first use of the now familiar Mostowski transitive collapse.

Gödel in his monograph [1940], having proceeded without def(x), formally carried out his [1939a] argument in terms of his eight operations, and this had the effect of obscuring the Skolem definability and closure. There is, however, an economy of means that can be seen from Gödel [1940]: The arguments there demonstrated that absoluteness is not necessary to establish either that L is an inner model of ZFC or that $V = L \to \mathrm{CH}$; absoluteness is only necessary where it is intrinsic, to establish $L^L = L$.

Until the 1960s accounts of L dutifully followed Gödel's [1940] presentation, and papers generally in axiomatic set theory often used and referred to Gödel's specific listing and grouping of his class-set axioms. However, modern expositions of L proceed in ZFC with the direct formalization of def(x), first formulating satisfaction-in-a-structure and coding this in set theory. They then establish Replacement or Separation in L by appealing to an L analogue of the ZF Reflection Principle, drawn from Richard Montague [1961: 99] and Azriel Levy [1960: 234].[19] Moreover, they establish $V = L \to \mathrm{CH}$ via some version

[19]The principle asserts that for any (first-order) formula $\varphi(v_1, \ldots, v_n)$ in the free variables as displayed and any ordinal β, there is a limit ordinal $\alpha > \beta$ such that for any $x_1, \ldots, x_n \in V_\alpha$

of the *Condensation Lemma*: If δ is a limit ordinal and X is an elementary substructure of L_δ, then there is a β such that X is isomorphic to L_β. Instead of Gödel's hand-over-hand algebraic approach to get $(*)$, one incorporates the satisfaction-in-a-structure relation and takes at least a Σ_1-elementary substructure of an ambient L_δ in a uniform fashion using its Skolem functions. This higher-level approach is indicative of how the satisfaction relation has been assimilated into modern set theory but also of what Gödel's approach had to encompass.

One is left to speculate why, and perhaps to rue that, Gödel did not himself articulate a reflection principle for use in L or some version of the Condensation Lemma based on the model-theoretic satisfaction-in-a-structure relation. The requisite Skolem closure argument would have served as a motivating *entrée* into his [1939a] proof of CH in L. Moreover, this approach would have provided a thematic link to Gödel's later advocacy of the heuristic of reflection, described in §7. Finally, with satisfaction-in-a-structure becoming the basis of model theory after Tarski-Vaught [1957] and the ZF Reflection Principle emerging only through the infusion of model-theoretic methods into set theory around 1960, a fuller embrace by Gödel of the satisfaction relation might have accelerated the process. That infusion was stimulated by Tarski through his students, and this sets in new counterpoint Gödel's indirect engagement with truth and satisfaction.[20]

Gödel's fine grained [1940] approach made transparent the absoluteness of L without having to confront $\mathrm{def}(x)$, but it also obfuscated the intuitive underpinnings of definability and the historical motivations, and this may have hindered the understanding of L for years. On the other hand, once L became assimilated, Gödel's [1940] presentation would serve as the direct precursor for Ronald Jensen's [1972] potent and fruitful fine structure theory.

4 Descriptive Set Theory Results

In his first announcement [1938] Gödel listed together with the Axiom of Choice and the Generalized Continuum Hypothesis two other propositions that hold in L. These were propositions of descriptive set theory, the definability theory

we have $\varphi[x_1,\ldots,x_n]$ *iff* $\varphi^{V_\alpha}[x_1,\ldots,x_n]$, where again φ^M denotes the relativization of the formula φ to M. This principle is equivalent to Replacement and Infinity in the presence of the other ZF axioms.

[20] It is, however, notable that in a seminal paper, Tarski [1931] gave a precise, set-theoretic formulation of the concept of a set of reals being first-order definable in the structure $\langle \mathrm{Reals}, +, \times\rangle$ that bypassed formulating the concept of satisfaction in this structure. Rather, Tarski worked with Boolean combinations and geometric projections. Like Gödel, Tarski at the time worked to dispense with the metamathematical uderpinnings. With the development of mathematical logic, we now see results stated there as leading to the decidability of real closed fields via the elimination of quantifiers.

of the continuum.[21] To state them in modern terms, we first recall some terminology: With $\mathbb{R}$ the set of real numbers and considering $\mathbb{R}^n$ as a topological space in the usual way, suppose that $Y \subseteq \mathbb{R}^n$. Y is $\mathbf{\Sigma}^1_1$ (*analytic*) *iff* Y is the projection $pB = \{\langle x_1, \ldots, x_n\rangle \mid \exists y(\langle x_1, \ldots, x_n, y\rangle \in B)\}$ of a Borel subset B of $\mathbb{R}^{n+1}$. (Equivalently, Y is the image under a continuous function of a Borel subset of some $\mathbb{R}^k$.) Y is $\mathbf{\Pi}^1_1$ *iff* $\mathbb{R}^n - Y$ is $\mathbf{\Sigma}^1_1$. Y is $\mathbf{\Sigma}^1_2$ *iff* Y is the projection of a $\mathbf{\Pi}^1_1$ subset of $\mathbb{R}^{n+1}$. Y is $\mathbf{\Pi}^1_2$ *iff* $\mathbb{R}^n - Y$ is $\mathbf{\Sigma}^1_2$. Y is $\mathbf{\Delta}^1_2$ *iff* it is both $\mathbf{\Sigma}^1_2$ and $\mathbf{\Pi}^1_2$. Proceeding thus through finite indices we get the hierarchy of *projective* sets. A set of reals has the *perfect set property* if either it is countable or else has a perfect subset.[22] Gödel's propositions following from $V = L$ can be cast as follows:

(a) There is a $\mathbf{\Delta}^1_2$ set of reals which is not Lebesgue measurable.

(b) There is a $\mathbf{\Pi}^1_1$ set of reals which does not have the perfect set property.

It had been known from Luzin [1914] that every $\mathbf{\Sigma}^1_1$ set is Lebesgue measurable and has the perfect set property, and so (a) and (b) provided an explanation in terms of relative consistency about the lack of progress up the projective hierarchy.

Gödel never again mentioned (a) or (b) in print, and only in an endnote to a 1951 printing of his [1940] did he describe a relevant result. There, he pointed out that the inherent [1940] well-ordering of L when restricted to its reals is a $\mathbf{\Sigma}^1_2$ subset of $\mathbb{R}^2$, describing how generally to incorporate his [1940] development into the definability context of descriptive set theory. When every real is in L, this $\mathbf{\Sigma}^1_2$ well-ordering is $\mathbf{\Delta}^1_2$ and does not satisfy Fubini's Theorem for Lebesgue measurable subsets of the plane, and this is one way to confirm (a). (b) is most often derived indirectly; what may have been Gödel's original argument is given in Kanamori [2003: 170].

Correspondence with von Neumann casts some light here. In a letter to von Neumann of 12 September 1938 Gödel[23] pointed out: "The theorem on one-to-one continuous images of [$\mathbf{\Pi}^1_1$] sets, which we had discussed at our last meeting, turned out to be false (refuted by Mazurkiewicz in Fund[amenta Mathematicae] 10). ... I now even have some results in the opposite direction ..." What was at issue here were images under *one-to-one* continuous functions. Gödel had been working on ongoing mathematics and would use L to address a mathematical question by giving a negative consistency result as per the axioms of set theory — a new kind of impossibility result.

With the reconstrual of projections as continuous real functions, the $\mathbf{\Sigma}^1_2$ sets are exactly the sets that are the continuous images of $\mathbf{\Pi}^1_1$ sets. Noting

[21]See Moschovakis [1980] or Kanamori [2003] for the concepts and terminology of descriptive set theory. See Kanamori [1995] for the emergence of descriptive set theory.

[22]A set of reals is perfect if it is non-empty, closed, and has no isolated points.

[23]Cf. [2003a: 361].

that Sierpiński had asked whether the *one-to-one* continuous image of any $\mathbf{\Pi}^1_1$ set is again $\mathbf{\Pi}^1_1$, Mazurkiewicz [1927] had observed that this was not so by showing that the difference of any two $\mathbf{\Sigma}^1_1$ sets is the one-to-one continuous image of a $\mathbf{\Pi}^1_1$ set. In his letter to von Neumann, Gödel proceeded to announce the consistency of (b) and the version of (a) asserting that there is a non-measurable (and hence not $\mathbf{\Pi}^1_1$) one-to-one continuous image of a $\mathbf{\Pi}^1_1$ set. With the recasting of continuous functions as projections, his actual statement of (a) in [1938], communicated two months later on 9 November 1938, was in the stringent and telling form: There is a non-measurable set such that both it and its complement are *one-to-one* projections of $\mathbf{\Pi}^1_1$ subsets of $\mathbb{R}^2$. Gödel was focused on one-to-one images, and one can reconstruct this consistency result with L.[24]

In a letter to Gödel of 28 February 1939 von Neumann (Gödel [2003a: 363]) brought to his attention the paper of Motokiti Kondô [1939]. For $A, B \subseteq \mathbb{R}^2$, A is *uniformized* by B *iff* $B \subseteq A$ and $\forall x(\exists y(\langle x,y\rangle \in A) \leftrightarrow \exists! y(\langle x,y\rangle \in B))$. Kondô [1939] had established the culminating result of the early, classical period of descriptive set theory, that every $\mathbf{\Pi}^1_1$ subset of $\mathbb{R}^2$ can be uniformized by a $\mathbf{\Pi}^1_1$ set. With this it was immediate that *every* $\mathbf{\Sigma}^1_2$ set is the *one-to-one* continuous image of a $\mathbf{\Pi}^1_1$ set, and (a) as stated above is indeed equivalent to Gödel's original [1938] form. In a letter to von Neumann of 20 March 1939 Gödel[25] wrote: "The result of Kondô is of great interest to me and will definitely allow an important simplification in the consistency proof of [(a)] and [(b)] of the attached offprint."[26]

Gödel's results (a) and (b) can be put into a broad historical context. Cantor's early preoccupation was with sets of reals and the like, and substantially motivated by his CH he both developed the transfinite numbers and investigated topological properties of sets of reals. In particular, he established that the closed sets have the perfect set property and so "satisfy the CH" since perfect sets have the cardinality of the continuum. Zermelo developed *abstract* set theory, with $\in$ having no privileged interpretation and sets regulated and generated by axioms.[27] In the first decades of the 20th Century *descriptive* set theory carried forth the investigation of sets of reals through the Borel and analytic

[24]For example, one can apply the idea used in Kanamori [2003: 170].

[25]Cf. [2003a: 365].

[26]All this to and fro tends to undermine the eye-catching remark of Kreisel [1980: 197] that: "... according to Gödel's notes, not he, but S. Ulam, steeped in the Polish tradition of descriptive set theory, noticed that the definition of the well-ordering ... of subsets of ω was so simple that it supplied a non-measurable PCA [i.e. $\mathbf{\Sigma}^1_2$] set of real numbers ... ."

[27]See Kanamori [2004] for Zermelo and set theory. Concerning "abstract", Fraenkel in his text *Abstract Set Theory* [1953] distinguished between *abstract sets* (the nature of whose elements are not of concern) and *sets of points* (typically numbers). In the early years "general set theory" was also sometimes used with connotations similar to "abstract set theory", though Zermelo himself consistently used "general set theory" to refer to axiomatic set theory without Infinity. The latter-day Skolem [1962] was still entitled *Abstract Set Theory*.

sets into the projective sets, while in abstract set theory Cantor's transfinite numbers were incorporated into the axiomatic framework by von Neumann with his ordinals. Then formal definability was brought into descriptive set theory by Tarski [1931], which before his well-known paper [1933] on truth dealt with the concept of a first-order definable set of reals, and by Kuratowski-Tarski [1931] and Kuratowski [1931], which pursued the basic connection between existential number quantifiers and countable unions and between existential real quantifiers and projection and used these "logical symbols" to aid in the classification of sets in the Borel and projective hierarchies. Gödel in his monograph [1940: 3] developed "abstract" set theory, and in that 1951 endnote started *ab initio* to correlate definability in L with formal definability in descriptive set theory. Gödel's results (a) and (b) constitute the first real synthesis of abstract and descriptive set theory, in that the axiomatic framework is brought to bear on the investigation of definable sets of reals.

5 L through the Lectures

Gödel's posthumously published lectures [1939b] and [1940a] provide considerable insight into his motivations and development of L. Both Hilbert and Russell loom large in Gödel's lecture [1939b], given at Hilbert's Göttingen on 15 December 1939. Gödel recalled at length Hilbert's previous work [1926] on CH and cast his own as an analogical development, one leading however to the constructible sets as a model for set theory. Hilbert [1926] apparently thought that if he could show that from any given formalized putative disproof of CH, he could prove CH, then CH would have been established. At best, Hilbert's argument could only establish the relative *consistency* of CH; this was evident to Gödel, who unlike Hilbert saw the distinction between truth and consistency clearly and wrote [1939b: 129]: "the first to outline a *program* for a consistency proof of the continuum hypothesis was *Hilbert*." For Hilbert, any disproof of CH would have to make use of number-theoretic functions whose definitions in his system needed his ϵ-symbol, his well-known device for abstracting quantification. He thus set out to replace the use of such functions by functions defined instead by transfinite recursion through the countable ordinals and via recursively defined higher-type functionals. The influence of Russell's ramified hierarchy is discernible here both in the preoccupation with definability and with the introduction of a type hierarchy, albeit one extended into the transfinite. Finally, Hilbert's scheme rested on establishing a bijection between such definitions and the countable ordinals to establish CH.

Gödel started his description of L by recalling two main lemmas in Hilbert's argument and casting two main features of L in analogous fashion. Contrasting his approach with Hilbert's however, Gödel [1939b: 131] emphasized about L that "*the model ... is by no means finitary*; in other words, the transfinite and impredicative procedures of set theory enter into its definition in an essential way, and that is the reason why one obtains only a relative consistency proof [of CH] ..." Gödel then pointed out a crucial property of L to which there was no Hilbertian counterpart, that it has "a certain *invariance*" property, i.e. the absoluteness $L^L = L$. To motivate the model Gödel again referred to Russell's ramified theory of types. Gödel first described what amounts to the orders of that theory for the simple situation when the members of a countable collection of real numbers are taken as the individuals and new real numbers are successively defined via quantification over previously defined real numbers, and emphasized that the process can be continued into the transfinite. He [1939b: 135] then observed that this procedure can be applied to sets of real numbers and the like, as individuals, and moreover, that one can "intermix" the procedure for the real numbers with the procedure for sets of real numbers "by using in the definition of a real number quantifiers that refer to sets of real numbers, and similarly in still more complicated ways." Gödel called a *constructible* set "the most general [object] that can at all be obtained in this way, where the quantifiers may refer not only to sets of real numbers, but also to sets of sets of real numbers and so on, *ad transfinitum*, and where the indices of iteration ... can also be arbitrary transfinite ordinal numbers." Gödel [1939b: 137] considered that although this definition of constructible set might seem at first to be "unbearably complicated", "the *greatest generality yields,* as it so often does, at the same time the *greatest simplicity*." Gödel was picturing Russell's ramified theory of types by first disassociating the types from the orders, with the orders here given through definability and the types represented by real numbers, sets of real numbers, and so forth. Gödel's intermixing then amounted to a recapturing of the complexity of Russell's ramification, the extension of the hierarchy into the transfinite allowing for a new simplicity.

Gödel [1939b: 137] went on to describe the universe of set theory, "the objects of which set theory speaks", as falling into "a transfinite sequence of Russellian [simple] types" the cumulative hierarchy of sets that he had described in his [1933o]. He then formulated the constructible sets as an analogous hierarchy, the hierarchy of [1939a]. Giving priority to the ordinals, Gödel had introduced transfinite Russellian orders through definability, and the hierarchy of types was spread out across the orders. The jumble of the *Principia Mathematica* had been transfigured into the model L of the constructible universe. Gödel forthwith pointed out a salient difference between the V and the L hierarchies with respect to cardinality: Whereas $|V_{\alpha+1}| > |V_\alpha|$ because of the use of the power set operation, $|L_{\alpha+1}| = |L_\alpha| = |\alpha|$ for infinite α.

In a comment bringing out the intermixing of types and orders, Gödel [1939b: 141] pointed out that "there are sets *of lower type* that *can* only *be defined* with the help of *quantifiers for sets of higher type*." Constructible subsets of L_ω will first appear high in the L hierarchy; in terms of the [1933o: 48] remarks, sets of natural numbers will encode truth propositions about higher

L_α's. However, these cannot be arbitrarily high. Gödel [1939b: 143] announced the version of $(*)$ (cf. §3) for countable ordinals as the crux of the consistency proof of CH. He subsequently asserted that "this fundamental theorem constitutes the corrected core of the so-called Russellian axiom of reducibility." Thus, Gödel established another connection between L and Russell's ramified theory of types. But while Russell had to *postulate* his axiom of Reducibility for his finite orders, Gödel was able to *prove* an analogous form for his transfinite hierarchy, one that asserts that the types are delimited in the hierarchy of orders. Not only did Gödel resurrect the ramified theory with L, but his transfinite type extension rectified Russell's ill-fated axiom. Reflecting a remark from [1931] quoted in §1 about the axiom of Reducibility as "the comprehension axiom of set theory", Gödel wrote [1939b: 145]:

> This character of the fundamental theorem as an axiom of reducibility is also the reason why *the axioms of classical* mathematics hold for the model of the constructible sets. For after all, as Russell showed, the axioms of reducibility, infinity and choice are the only axioms of classical mathematics that do not have [a] tautological character. To be sure, one must observe that the axiom of reducibility appears in different mathematical systems under different names and in different forms, for example, in Zermelo's system of set theory as the axiom of separation, in Hilbert's systems in the form of recursion axioms, and so on.

This passage shows Gödel to be holding a remarkably synthetic, unitary view, viewing as he does Russell's axiom of Reducibility, Zermelo's Separation axiom, and Hilbert's [1926] recursion axioms all as one. Actually, $(*)$ as such is not necessary to establish that L is a model of set theory; it is sufficient that for any α, the constructible subsets of L_α all belong to some L_β and for this one only needs the full extent of the ordinals as bolstered by Replacement. That $(*)$ is sufficient but separate is acknowledged by Gödel when he next wrote: "Now the axiom of reducibility holds for the constructible sets on the basis of the fundamental theorem ..." Thus, it is more proper to regard Reducibility, Replacement, and the Reflection Principle (cf. end of §3) all as one, and the thrust of Gödel's comments on Reducibility are more in this direction.

Gödel in his lecture did not detail the proof of $L^L = L$, mentioning only that [1939b: 145] "an essential point in it is that the notion of ordinal number is absolute: that is, ordinal number in the model of the constructible sets means the same as ordinal number itself." He then launched into a detailed account of the proof of the "fundamental theorem", i.e. $(*)$ for countable ordinals, the proof being the one sketched in [1939a]. This lecture of Gödel's is a remarkably clear presentation of both the mathematical and historical development of L, and had it become widely accessible together with his [1940], it would no doubt have accelerated the assimilation of L.

Hilbert and Russell also figure prominently in a later lecture [1940a] on CH given on 15 November 1940 at Brown University, of which we mainly describe the new ground covered. Gödel began by announcing that he had "succeeded in giving the [consistency] proof a new shape which makes it somewhat similar" to Hilbert's [1926] attempt and proposed to sketch the new proof, considering it "perhaps the most perspicuous". First however, Gödel described the issues involved in general terms and reviewed the def(x) construction of L. Once again he emphasized that his argument for showing that CH holds in L proves an axiom of reducibility, this time putting more stress on Separation [1940a: 178]: "...it is not surprising that the axioms of set theory hold for the constructible sets, because the axiom of reducibility or its equivalents, e.g., Zermelo's Aussonderungsaxiom [Separation], is really the only essential axiom of set theory." Gödel then turned to his new approach and introduced the concept of a relation being "recursive of order α" for ordinals α. This concept is a generalization of the notion of definability, a generalization obtained by essentially interweaving the operation def(x) with a recursion scheme akin to Hilbert's for his [1926] hierarchy of functionals.[28] As Gödel [1940a: 180] wrote: "The difference between this notion of recursiveness and the one that Hilbert seems to have had in mind is chiefly that I allow quantifiers to occur in the definiens." This, of course, is a crucial difference, and having separated out arithmetical aspects of definability à la Hilbert, Gödel [1940a: 181] because of the quantifiers had to face head on "defining recursively the metamathematical notion of truth" à la Tarski. This 1940 juncture is arguably when Gödel came closest, having never written that part II to his [1931], to describing what could have been its contents:

> Now this metamathematical notion of truth, i.e., the class of numbers of truth propositions, can be defined by a method similar to the one which Tarski applied for the system of *Principia mathematica*. The point is to well-order all propositions of our domain in such a manner that the truth of each depends in a precisely describable manner on the truth of some of the preceding; this gives then the desired recursive definition.

Using the new concept of recursiveness – better, new concept of definability – Gödel gave a model of Russell's *Principia*, construed as his [1931] system P, in which CH holds. The types of this model were essentially coded versions of $L_{\omega_{n+1}} - L_{\omega_n}$. Echoing his [1931] footnote 48a, Gödel [1940: 184] subsequently wrote:

> You know every formal system is incomplete in the sense that it can be enlarged by new axioms which have approximately the same

[28] As analyzed by Solovay in his introductory notes (cf. Gödel [1995: 122]), for $\alpha > \omega$, a relation on α is recursive of order α exactly when it appears in $L_{\alpha\cdot\omega}$.

> degree of evidence as the original axioms. The most general way of accomplishing these enlargements is by adjoining higher types, e.g., the type ω for the system of *Principia mathematica*. But you will see that my proof goes through for systems of arbitrarily high type.

However high a transfinite type that one wanted to include, one can similarly establish the relative consistency of CH in the corresponding "inner model".

A coda, returning to truth: Years later, in a letter to Hao Wang of 7 March 1968 Gödel[29] wrote, in implicit criticism of Hilbert:

> ...there was a special obstacle which *really* made it *practically impossible* for constructivists to discover my consistency proof. It is the fact that the ramified hierarchy, which had been invented *expressly for constructive purposes,* had to be used in an *entirely nonconstructive way.* A similar remark applies to the concept of mathematical truth, where formalists considered formal demonstrability to be an *analysis* of the concept of mathematical truth and, therefore, were of course not in a position to *distinguish* the two.

Wang [1996: 250ff] described how Gödel in January 1972 retrospectively contrasted Hilbert's approach to CH with his.

6 Set Theory Transformed

Gödel with L brought into set theory a *method* of construction and argument and thereby affirmed several features of its axiomatic presentation. First, Gödel showed how first-order definability can be formalized and used in a transfinite recursive construction to establish striking new mathematical results. This significantly contributed to a lasting ascendancy for first-order logic, which beyond its *sufficiency* as a logical framework for mathematics was seen to have considerable *operational efficacy*. Gödel's construction moreover buttressed the incorporation of Replacement and Foundation into set theory. Replacement was immanent in the arbitrary extent of the ordinals for the indexing of L and in its formal construction via transfinite recursion. In his analysis of Russell's mathematical logic Gödel [1944: 147] again wrote about how with L he had proved an axiom of reducibility, and in fact that "...all impredicativities are reduced to one special kind, namely the existence of certain large ordinal numbers (or well-ordered sets) and the validity of recursive reasoning for them." As for Foundation, underlying the construction was the well-foundedness of sets. Gödel in a footnote to his account [1939a: fn12] wrote about his axiom A, i.e. $V = L$: "In order to give A an intuitive meaning, one has to understand by 'sets' all objects obtained by building up the simplified hierarchy of types on

[29] Cf. [2003a: 404].

an empty set of individuals (including types of arbitrary transfinite orders)." Gödel [1947: 518ff] later wrote:[30]

> ... there exists a satisfactory foundation of Cantor's set theory in its whole original extent, namely axiomatics of set theory, under which the logical system of *Principia mathematica* (in a suitable interpretation) may be subsumed.
>
> It might at first seem that the set-theoretical paradoxes would stand in the way of such an undertaking, but closer examination shows that they cause no trouble at all. They are a very serious problem, but not for Cantor's set theory. ... This concept of set ... according to which a set is anything obtainable from the integers (or some other well-defined objects) by iterated application of the operation "set of", and not something obtained by dividing the totality of all existing things into two categories, has never led to any antinomy whatsoever; that is, the perfectly "naïve" and uncritical working with this concept of set has so far proved completely self-consistent.

A new emphasis here is on the inherent consistency of the cumulative hierarchy stratification, which, to emphasize, is provided by the axioms, most saliently Foundation interacting with Replacement, Power Set, and Union.

The approaches of Gödel and of Zermelo [1930] (mentioned in §1) to set theory merit comparison with respect to the emergence of the cumulative hierarchy view, the focus on models of set theory, and subsequent influence.[31] Zermelo had first adopted Foundation, thereby promoting the cumulative hierarchy view of sets, and posited an endless procession of models of his axioms of form V_κ for inaccessible cardinals[32] κ with one model a set in the next. Both Zermelo and Gödel advocated direct transfinite reasoning, with Zermelo proceeding in an avowedly second-order axiomatic context and Gödel formalizing first-order definability in his transfinite extension of the theory of types. Gödel came close to Zermelo [1930] in his informal sketch [1939a] about L when he stated his relative consistency results in terms of the axioms of Zermelo [1908] as rendered in first-order logic and asserted that L_Ω, where Ω is "the first inaccessible number", is a model of Zermelo's axioms together with Replacement. Also, making his only explicit reference to Zermelo [1930], Gödel [1947: 520] later gave the existence of inaccessible cardinals as the simplest example of an axiom that asserts still further iterations of the 'set of' operation and can

[30] Here the footnotes to the text are excised.

[31] Kreisel [1980] draws this comparison for didactic purposes.

[32] An uncountable cardinal κ is *inaccessible* if κ is a regular cardinal, i.e. if $\alpha < \kappa$ and $F\colon \alpha \to \kappa$, then $\bigcup F``\alpha < \kappa$, and κ is a strong limit cardinal, i.e. if $\beta < \kappa$, then $2^\beta < \kappa$.

supplement the axioms of set theory without arbitrariness.[33] Beyond the imprint on Gödel himself, which could be regarded as significant, Zermelo [1930] seemed to have had little influence on the further development of set theory, presumably because of its second-order lens and its lack of rigorous detail and attention to relativism.[34] On the other hand, Gödel's work with L with its incisive analysis and use of first-order definability was readily recognized as a signal advance. Issues about consistency, truth, and definability were brought to the forefront, and the CH result established the mathematical importance of a hierarchical analysis. As the construction of L was gradually digested, the sense it promoted of a cumulative hierarchy reverberated to become the basic picture of the universe of sets.

How Gödel transformed set theory can be broadly cast as follows: On the larger stage, from the time of Cantor, sets began making their way into topology, algebra, and analysis so that by the time of Gödel, they were fairly entrenched in the structure and language of mathematics. But how were sets viewed among set *theorists*, those investigating sets as such? Before Gödel, the main concerns were what sets *are* and how sets and their axioms can serve as a reductive basis for mathematics. Even today, those preoccupied with ontology, questions of mathematical existence, focus mostly upon the set theory of the early period. After Gödel, the main concerns became what sets *do* and how set theory is to advance as an autonomous field of mathematics. The cumulative hierarchy picture was in place as subject matter, and the metamathematical methods of first-order logic mediated the subject. There was a decided shift toward epistemological questions, e.g. what can be proved about sets and on what basis.

7 Truth and New Axioms

A pivotal figure Gödel, what was his own stance? What he *said* would align him more with his predecessors, but what he *did* would lead to the development of methods and models. In a critical analysis [1944] of Russell's mathematical logic, a popular discussion [1947] of Cantor's continuum problem, and subsequent lectures and correspondence, Gödel articulated his philosophy of "conceptual realism" about mathematics. He espoused a staunchly objective "concept of set" according to which the axioms of set theory are true and are descriptive of an objective reality schematized by the cumulative hierarchy. Be that as it may, his actual mathematical work laid the groundwork for the development

[33]Gödel referenced Zermelo [1930] after writing: "[This] axiom, roughly speaking, means nothing else but that the totality of sets obtainable by exclusive use of the processes of formation of sets expressed in the other axioms forms again a set (and, therefore, a new basis for a further application of these processes)." This was just what Zermelo had emphasized; for Gödel there would also be the overlay of truth in the "next higher system".

[34]For the record, Kreisel [1980: 193] wrote that Zermelo's paper "made little impression" but adduced historically peculiar reasons.

of a range of models and axioms for set theory. Already in 1942 Gödel worked out for himself a possible model for the negation of AC in the framework of type theory.[35] In his steady intellectual development Gödel would continue to pursue the distinction between truth and provability into the higher reaches of set theory.

In oral, necessarily brief remarks at a conference Gödel [1946] made substantial mathematical suggestions that newly engaged truth in terms of absoluteness and with concepts involving the heuristic of *reflection.* Pursuing his "next higher system" theme Gödel explored possible absolute notions of demonstrability and definability, those not dependent on any particular formalism. For absolute demonstrability, Gödel again pointed out how formalisms can be transcended and the process iterated into the transfinite. And recalling his remarks about L, he pointed out that while no one formalism would embrace the entire process, "it could be described and collected in some non-constructible way". Gödel then charted new waters, with remarks having an anticipation in the [1932b] passage quoted in §1:

> In set theory, e.g., the successive extensions can most conveniently be represented by stronger and stronger axioms of infinity. It is certainly impossible to give a combinational and decidable characterization of what an axiom of infinity is; but there might exist, e.g. a characterization of the following sort: An axiom of infinity is a proposition which has a certain (decidable) formal structure and which in addition is true. Such a concept of demonstrability might have the required closure property, i.e., the following could be true: Any proof for a set-theoretic theorem in the next higher system above set theory (i.e., any proof involving the concept of truth which I just used) is replaceable by a proof from such an axiom of infinity. It is not impossible that for such a concept of demonstrability some completeness theorem would hold which would say that every proposition expressible in set theory is decidable from the present axioms plus some true assertion about the largeness of the universe of sets.

This a remarkably optimistic statement about the possibility of discovering new "true" axioms that will decide *every* set-theoretic proposition. The engagement with truth has introduced a new element, "strong axioms of infinity", and an argument by reflection: "Any proof for a set-theoretic theorem in the next higher system above set theory", i.e. if the satisfaction relation for V itself were available, "is replaceable by a proof from such an axiom of infinity." There is still an afterglow here from Russell's axiom of Reducibility as filtered through

[35]See e.g. Dawson [1997: 160].

Gödel's work. Reaching further back, there is more resonance with another notion of absoluteness, Cantor's of the absolutely infinite, or the Absolute.[36] Recast in terms of the cumulative hierarchy, the universe $V = \bigcup_\alpha V_\alpha$ cannot be comprehended, and so any particular property ascribable to it must already be ascribable to some rank V_α, some postulations becoming the strong axioms.

For absolute definability, Gödel pointed out that here also there is a transfinite hierarchy, one of "concepts of definability", and "it is not possible to collect together all these languages in one, as long as you have a finitistic concept of language." Whereas for demonstrability he had envisioned the use of strong axioms of infinity, for definability he turned to expanding the language by allowing constants for *every* ordinal. This is resonant with Gödel's formulation of L in that the main non-constructive feature is the indexing through the ordinals and their arbitrary extent is again brought to the fore and made use of. Gödel [1946: 3] made a crucial claim:

> By introducing the notion of truth for this whole transfinite language, i.e., by going over to the next language, you will obtain no new definable sets (although you will obtain new definable properties of sets).

The passages quoted in §1 and the construction of L had featured the introduction of higher types allowing for the definability of new satisfaction relations and hence new definable sets of lower type. Gödel saw that having the satisfaction relation for set theory for the enriched language with constants for every ordinal leads to a closure for definability, "no new definable sets", as separated from truth, "new definable properties of sets". Sets definable in the enriched language via the satisfaction relation are definable without it, and this reflection provides an absoluteness for definability.

Gödel's [1946] remarks would remain largely unknown in the succeeding two decades. John Myhill and Dana Scott in their [1971] carried out the development of the sets Gödel described, the *ordinal definable* sets. Gödel had at first described the constructible sets informally and shown that being constructible is itself formally definable in ZF; Gödel's claim above entails that being ordinal definable is likewise formally definable in ZF. This Myhill and Scott established with the ZF Reflection Principle, and this speaks to the road not taken by Gödel [1940] discussed at the end of §3. Moreover, as was anticipated by Gödel [1946: 4] the ordinal definable sets provided a new proof for the relative consistency of AC: HOD, the class of *hereditarily ordinal definable* sets is an inner model of ZFC. HOD has become an important feature of modern

[36] See Jané [1995] for the role of the absolute infinite in Cantor's conception of set. Wang [1996: 282ff] reported on how Gödel in 1975 acknowledged Cantor's "Absolute", particularly in connection with a set theory of Ackermann.

set theory, and important results about it have articulated Gödel's absolute definability motivation.[37]

In his article [1947] on Cantor's continuum problem Gödel put emphasis on how his philosophical outlook could be brought to bear on mathematical problems and effect mathematical programs. Of the three possibilities in axiomatic set theory, that CH could be demonstrable, disprovable, or undecidable, Gödel [1947: 519] regarded the third as the "most likely", and so advocated the search for a proof of the independence of CH, i.e. to establish Con(ZF) implies Con(ZFC + ¬CH) to complement his own relative consistency result with L. However, Gödel stressed that this would not "settle the question definitively" and turned to the possibility of new axioms. The axioms of set theory do not "form a system closed in itself", and so the "very concept of set on which they are based suggests their extension by new axioms which assert the existence of still further iterations of the operation 'set of'." Gödel then elaborated on the strong axioms of infinity he had alluded to in his [1946] by giving as examples the inaccessible cardinals (as mentioned in §6 in connection with Zermelo [1930]) and the Mahlo cardinals. These were entertained early in the development of set theory and are at the beginning of the modern hierarchy of *large cardinal hypotheses*, hypotheses that posit distinctive structure in the higher reaches of the cumulative hierarchy, most often by positing cardinals whose defining properties entail their inaccessibility from below in strong senses.[38]

Gödel pointed out two significant aspects of large cardinal hypotheses to which attention would be drawn many times in their development. First, in a new twist on the passages quoted in §1, each strong axiom of infinity "can, under the assumption of consistency, be shown to increase the number of decidable propositions even in the field of Diophantine equations." Large cardinal hypotheses establish the *consistency* of ZFC and stronger theories, and so even though they posit distinctive structure high in the cumulative hierarchy they lead to new simple, decidable propositions even about natural numbers.[39] Second, for the inaccessible and Mahlo cardinals and the like, the "undisprovability of the continuum hypothesis ... goes without change". These cardinals relativize to L, i.e. they retain their defining properties in L, and so the existence of these cardinals is consistent with CH.[40]

Gödel went on to speculate about possible strong axioms of infinity based on "hitherto unknown principles", and then, in a well-known passage, argued for new axioms just on extrinsic and pragmatic bases:

[37]Leaping forward, see Steel [1995].

[38]See Kanamori [2003] for the theory of large cardinals.

[39]The specific focus on Diophantine equations could already be seen in a lecture Gödel [193?], which anticipated now well-known work on Hilbert's 10th Problem.

[40]Actually, that inaccessible cardinals relativize to L was already noted in Gödel's first announcement [1938]. It would be a pivotal advance that not all large cardinals relativize to L (see below).

> ... even disregarding the intrinsic necessity of some new axiom, and even in case it had no intrinsic necessity at all, a decision about its truth is possible also in another way, namely, inductively by studying its 'success', that is, its fruitfulness in consequences and in particular in 'verifiable' consequences, i.e. consequences demonstrable without the new axiom, whose proofs by means of the new axiom, however, are considerably simpler and easier to discover, and make it possible to condense into one proof many different proofs. The axioms for the system of real numbers, rejected by the intuitionists, have in this sense been verified to some extent owing to the fact that analytical number theory frequently allows us to prove number-theoretical theorems which can subsequently be verified by elementary methods. A much higher degree of verification than that, however, is possible. There might exist axioms so abundant in their verifiable consequences, shedding so much light upon a whole discipline, and furnishing such powerful methods for solving given problems (and even solving them, as far as that is possible, in a constructivistic way) that quite irrespective of their intrinsic necessity they would have to be assumed at least in the same sense as any well established physical theory.

This advocacy of new axioms merely because of their "success" according to "fruitfulness of consequences" interestingly undercuts an avowedly realist position with a pragmatism that dilutes the force of "truth", but is resonant with subsequent investigations. Gödel [1947] concluded by forwarding the remarkable opinion that CH "will turn out to be wrong" since it has as paradoxical consequences the existence of "thin" (in various senses he articulated) sets of reals of the power of the continuum. These examples, one involving one-to-one continuous images, further emphasize how Gödel was aware of and influenced by the articulation of the continuum by the descriptive set theorists (cf. §4).

In 1963 Paul Cohen established the independences Con(ZF) implies Con (ZF + ¬AC) and Con(ZF) implies Con(ZFC + ¬CH), these being, of course, the inaugural examples of *forcing*, a remarkably general and flexible method for extending models of set theory. If Gödel's construction of L had launched set theory as a distinctive field of mathematics, then Cohen's method of forcing began its transformation into a modern, sophisticated one.

In a published revision [1964] of his [1947] Gödel took into account new developments, most notably Cohen's independence result for CH. As for large cardinals, in a new footnote 20 Gödel cited the emerging work on what are now known as the strongly compact, measurable, weakly compact, and indescribable cardinals, results which in particular showed that these cardinals are far larger in strong senses than the least inaccessible cardinal. Gödel mentioned in particular the pivotal result of Dana Scott [1961] that *if there is a measurable cardinal, then* $V \neq L$.[41] In an unpublished, 1966 revision of that footnote Gödel[42] argued that these "extremely strong axioms of infinity of an entirely new kind" are "supported by strong arguments from analogy, e.g., by the fact that they follow from the existence of generalizations of Stone's representation theorem to Boolean algebras with operations on infinitely many elements." He was evidently referring to the compact cardinals. This is the first appearance in his writing of the heuristic of *generalization* for motivating large cardinals. Recalling Cantor's unitary view of the transfinite as seamlessly extending the finite, some properties satisfied by $\aleph_0$ would be too accidental were they not ascribable to higher cardinals in an eternal recurrence.

In the tremendous expansion of set theory following the introduction of forcing, the theory of large cardinals developed a self-fueling momentum of its own and blossomed into a mainstream of set theory far overshadowing Gödel's early speculations. Nowhere would his words be acknowledged as having been a source of inspiration. On the other hand, an articulated and detailed hierarchy of large cardinal hypotheses was developed with the heuristics of reflection and generalization very much in play, and these hypotheses were shown to decide a wide range of strong set-theoretic propositions. Gödel's hopes that large cardinals could settle the continuum problem itself were dispelled by the observation of Levy–Solovay [1967], known by 1964, that small cardinality forcing notions preserve the defining properties of inaccessible large cardinals, so that CH is independent of their postulations. In a 1966 revision of his [1964] Gödel[43] himself implicitly acknowledged this. In a late, unpublished note [1972a] Gödel's advocacy of large cardinal hypotheses had two notable modulations. First, he speculated on their possible use to settle, not CH, but questions of "Goldbach type", i.e. Π^0_1 sentences of arithmetic. Second, Gödel pointed to what modern set theorists understand well:

> These principles show that ever more (and ever more complicated) axioms appear during the development of mathematics. For, in order only to understand the axioms of infinity, one must first have developed set theory to a considerable extent.

Extensive work through the 1970s and up to the present day has considerably strengthened the view that the emerging hierarchy of large cardinals

---

[41] Earlier, in a draft of a (presumably unsent) letter to Tarski of August 1961, Gödel [2003a: 273] had written: "I have heard it has been proved that there is no two valued denumerably additive measure for the first inacc. number. I still can't believe that this is true, but don't have the time to check it because I am working mainly on phil[osophy]. I understand the proof is based on some work of yours? You probably have heard of Scott's beautiful result that $V \neq L$ follows from the existence of any such measure for any set. I have not checked this proof either but the result does *not* surprise me."

[42] Cf. [1990: 260ff].

[43] Cf. [1990: 270].

provides *the* hierarchy of exhaustive principles against which all possible consistency strengths can be gauged, a kind of hierarchical completion of ZFC. First, the various hypotheses, though historically contingent, form a *linear* hierarchy with respect to relative consistency strength. Second, a wide range of strong statements arising in set theory and mathematics have been informatively bracketed in consistency strength between two large cardinal hypotheses. The stronger hypothesis implies that there is a forcing extension in which the statement holds; and if the statement holds, there is an L-like inner model satisfying the weaker hypothesis. *Equi*consistency results were established by refining proof ideas and weakening large cardinals to achieve optimal formulations. Throughout, in addition to their "intrinsic" significance, large cardinals amply exhibited "fruitfulness of consequences" by providing the context for quick proofs and illuminating methods, some later found not to require large cardinals at all. These developments have highlighted the contention that large cardinal hypotheses are not a matter of belief, but rather of method. Going far beyond the true and the false, large cardinals have provided the means for understanding strong statements of set theory and mathematics through relative consistency proofs.

Gödel's early advocacy of the search for new axioms can be seen as vindicated by these broad developments, although that vindication has been in much more subtle ways than he could have anticipated. In latter-day accounts, with modern set theory having reached a high degree of sophistication, there have been retrospective analyses that cast Gödel's sparse words across the vast modern landscape of large cardinal hypotheses, crediting them with enunciating "Gödel's program".[44]

Entering his sixties, mostly preoccupied with philosophy and health problems and despite his earlier advocacy of strong axioms of infinity, Gödel would draw on a distant mathematical initiative taken around the time of his birth to address the continuum problem anew.

8 Envoi

In a letter to Cohen of 22 January 1964 Gödel,[45] in connection with possible new uses of forcing, wrote:

> Once the continuum hypothesis is dropped the key problem concerning the structure of the continuum, in my opinion, is what Hausdorff calls the "Pantachie Problem",[1] i.e., the question of whether there exists a set of sequences of integers of power $\aleph_1$ which for any given sequence of integers contains one majoring it from a certain point on. Hausdorff evidently was trying to solve this problem af-

[44]See Kennedy–van Atten [2004], Koellner [2006], and Hauser [2006] for Gödel's program.
[45]Cf. [2003a: 383ff].

firmatively (see [Hausdorff [1907]] and [Hausdorff [1909]]). I was always suspecting that, in contrast to the continuum hypothesis, this proposition is correct and perhaps even demonstrable from the axioms of set theory. Moreover I have a feeling that, if your method does not yield a proof of independence here, it may lead to a proof of this proposition. At any rate it should be possible to prove the compatibility of the "Pantachie Hypothese" with $2^{\aleph_0} > \aleph_1$.

[1]In German the problem is frequently called "Problem der Wachstumsordnungen". Perhaps there exists some standard English expression for it, too.

In a letter to Stanisław Ulam of 10 February 1964 Gödel,[46] after praising Cohen's work, wrote similarly about the "Pantachie Problem". What Gödel was describing properly has to do with the "Scale Problem" of Hausdorff [1907: 152]. ${}^{\omega}\omega$, the set of functions from ω to ω, can be partially ordered according to: $f <^* g$, f is *eventually dominated by* g, *iff* $\exists m \in \omega \forall n \in \omega(m \leq n \rightarrow f(n) < g(n))$. A *$\kappa$-scale* is a subset of ${}^{\omega}\omega$ which according to $<^*$ is cofinal in ${}^{\omega}\omega$ and of ordertype κ. Without further elaboration, we shall extend these concepts in the expected way to other ordered sets besides ω. Hausdorff observed that CH implies that there is an ω_1-scale, and opined that the existence of an ω_1-scale is of significance independently of CH. This is echoed by Gödel in the above passage, but what he was "suspecting" there has an ironic twist.

It soon became known that in Cohen's *original* model for $\neg$CH, i.e. the one resulting from adding many Cohen reals, there is no ω_1-scale. On the other hand, if one adds many (Solovay) random reals to a model of CH, then any ω_1-scale in the ground model remains one in the generic extension.[47] Thus, the existence of ω_1-scales, like CH, comes under the purview of forcing and is independent of ZFC.

Because of its broader involvement in Gödel's later speculations, we review Hausdorff's work on pantachies as such. Most of Hausdorff [1907] is devoted to the analysis of pantachies and the main section V is entitled "On Pantachie Types".[48] The term "pantachie" derives from its initial use by Paul Du Bois-Reymond [1880] to denote everywhere dense subsets of the continuum and then to various notions connected with his work on rates of growth of real-valued functions and on infinitesimals.[49] Hausdorff redefined "pantachie" as

[46]Cf. [2003a: 298].

[47]Stephen Hechler in his dissertation of 1967 (cf. Hechler [1974]) introduced dominating reals, and by iterating his forcing established the general assertion that if in the sense of the ground model, κ and λ are cardinals of uncountable cofinality such that $2^{\aleph_0} \leq \kappa$ and $\lambda \leq \kappa$, then there is a cardinal-preserving generic extension in which $2^{\aleph_0} = \kappa$ and there is a λ-scale.

[48]See Plotkin [2005] for a penetrating analysis of Hausdorff's work on pantachies and more generally ordered sets, work remarkable for its depth and early appearance.

[49]At the end of his [1880] Du Bois-Reymond maintained that he rather than Cantor had come first to the concept of a dense subset of the continuum. In his book [1882] Du Bois-Reymond explained that his adjective 'pantachish' derives from the Greek words

a subset of ${}^{\omega}\mathbb{R}$ *maximal* with respect to being linearly ordered by the eventual dominance ordering, and a further refinement let to scales on ${}^{\omega}\omega$. This anticipated Hausdorff's later work on maximal principles, principles equivalent to the Axiom of Choice. For an ordered set $\langle X, <\rangle$, a (κ, λ^*)-*gap* is a set $\{x_\alpha \mid \alpha < \kappa\} \cup \{y_\alpha \mid \alpha < \lambda\} \subseteq X$ such that $x_\alpha < x_\beta < y_\gamma < y_\delta$ for $\alpha < \beta < \kappa$ and $\delta < \gamma < \lambda$, yet there is no $z \in X$ such that $x_\alpha < z < y_\gamma$ for $\alpha < \kappa$ and $\gamma < \lambda$. Pantachies were easily seen to have no countable cofinal or coinitial subset and no (ω, ω^*)-gaps. Regarding pantachies as higher order continua, it was natural to consider whether there could be (ω_1, ω_1^*)-gaps, their absence being a principle of higher-order continuity. Hausdorff established that with CH all pantachies are isomorphic and have (ω_1, ω_1^*)-gaps. In his [1909] he subsequently established that there is a pantachie with an (ω_1, ω_1^*)-gap without appeal to CH, and this recast from ${}^{\omega}\mathbb{R}$ to ${}^{\omega}\omega$ (cf. Hausdorff [1936]) was to become well-known in modern set theory as an "indestructible" ZFC gap, one that cannot be filled with any forcing that preserves $\aleph_1$. Hausdorff [1907: 151] asked in the concluding "The Pantachie Problem" subsection whether there could be a pantachie with no (ω_1, ω_1^*)-gaps. Strikingly, he [1907: 128] had shown earlier that if there were such a pantachie, then $2^{\aleph_0} = 2^{\aleph_1}$ and hence $\neg$CH. This was the first time that a question in ongoing mathematics had entailed the denial of CH.

In the late 1960s Gödel was mostly preoccupied with philosophy; through association with a new generation of set theorists he also kept abreast of the burgeoning developments in the subject. Yet, going his own way and struck by the plausibility of Hausdorff's old formulations, Gödel in 1970 proposed "orders of growth" axioms for deciding the value of $2^{\aleph_0}$ in two handwritten notes [1970a][1970b].[50]

In [1970a], entitled *Some considerations leading to the probable conclusion that the true power of the continuum is* $\aleph_2$, Gödel claimed to establish $2^{\aleph_0} = \aleph_2$ from the following axioms:

(1) For every $n \in \omega$, there is an ω_{n+1}-scale on ${}^{\omega_n}\omega_n$.

(2) In addition, for every $n \in \omega$, the set of all initial segments of all the functions in the ω_{n+1}-scale on ${}^{\omega_n}\omega_n$ has cardinality ω_n.

(3) There is a pantachie with every well-ordered increasing or decreasing descending subset having length at most ω_1.

πανταχῆ, πανταχοῦ for "everywhere". For real functions increasing without bound, Du Bois-Reymond had considered an ordering where $f < g$, $f \sim g$, or $f > g$ according to whether $\lim_{x \to \infty} f(x)/g(x)$ is zero, finite but not zero, or $+\infty$. He had advocated considering those f, g with $f \sim g$ as representing the same "order of infinity" and ranking these orders according to $<$. But of course, there are f, g incomparable according to Du Bois-Reymond's scheme, and on this basis Hausdorff [1907: 107] proclaimed that "*the infinitary pantachie* in the sense of Du Bois-Reymond *does not exist.*"

[50] See Solovay's introductory note in Gödel [1995: 405ff] and Brendle-Larson-Todorčević [∞] for extensive mathematical analyses.

(4) In addition, the pantachie has no (ω_1, ω_1^*)-gaps.

To modern eyes, there is an affecting, quixotic grandeur to this reaching back to primordial beginnings of set theory to charge the windmill once again. Gödel's only use of (4) was to apply Hausdorff's conclusion that CH fails, and then he argued that (1)-(3) implies that $2^{\aleph_0} \leq \aleph_2$. However, Martin pointed out that the argument does not work, and Solovay (cf. [Gödel [1995]: 412ff]) elaborated, showing how by adding many random reals it is consistent to have (1)-(3) and the continuum arbitrarily large. On the other hand, Brendle-Larson-Todorčević [∞] showed that there is a substantial part of Gödel's argument that does work to establish $2^{\aleph_0} \leq \aleph_2$ from propositions closely related to (1)-(3).

Gödel [1970a] took his axioms (1) and (2) to entail for all $m < n < \omega$ the existence of ω_{n+1}-scales on ${}^{\omega_n}\omega_m$ such that the set of initial segments of all the functions involved has cardinality ω_n.[51] In his attempted proof, he appealed to such a scale for $n = 2$ and $m = 1$. In fact, the existence of such a scale for $n = 1$ and $m = 0$ already implies CH, and this was the thrust of his [1970b], entitled *A proof of Cantor's continuum hypothesis from a highly plausible axiom about orders of growth.* At its end, Gödel wrote:

> It seems to me this argument gives *much* more likelihood to the truth of Cantor's continuum hypothesis than any counterargument set up to now gave to its falsehood, and it has at any rate the virtue of deriving the power of the set of *all* functions $\omega \longrightarrow \omega$ from that of certain *very* special sets of these functions.

A few years later, in a letter to Abraham Robinson of 20 March 1974 Gödel[52] wrote:

> Hausdorff proved that the existence of a 'continuous' system of orders of growth (i.e. one where every decreasing ω_1-sequence of closed intervals has a non-empty intersection) is incompatible with Cantor's Continuum Hypothesis. Surprisingly the same is true even for a 'dense' system, i.e., one where every decreasing ω_1 sequence of closed intervals, *all of which are larger than some fixed interval* I, has a non[-]empty intersection. I think many mathematicians will consider this to be a strong argument against the Continuum Hypothesis.

Here, the 'continuous' is clear, that there are no (ω_1, ω_1^*)-gaps, but 'dense' is not. Robinson was fatally inflicted with pancreatic cancer and died three weeks

[51] In an unsent letter to Tarski Gödel [1995: 424] soon disavowed this entailment.

[52] Cf. [2003a: 204].

after the date of this letter, on 11 April 1974.[53]

Wang [1996: 89] reported on how Gödel in 1976, two years before his own death, made the following observations:

> The continuum hypothesis may be true, or at least the power of the continuum may be no greater than aleph-two, but the generalized continuum hypothesis is definitely wrong.
>
> I have written up [some material on] the continuum hypothesis and some other propositions. Originally I thought [I had proved] that the power of the continuum is no greater than aleph-two, but there is a lacuna [in the proof]. I still believe the proposition to be true; even the continuum hypothesis may be true.

What are we to make of all this? In his [1947] Gödel had written with authority about the continuum problem, opining that CH would be shown independent, averring that it is actually false particularly because of its implausible consequences for the continuum, and suggesting that new strong axioms of infinity could settle the matter. With the revitalization of set theory after Cohen and perhaps partly spurred by the 1964 Levy-Solovay observation that large cardinal hypotheses have no direct effect on CH, Gödel pursued his rekindled interest in the very old initiatives of Hausdorff and formulated "orders of growth" axioms to inform the continuum problem anew. In this Gödel exhibited a remarkable fluidity, siding with his axioms and letting the mathematics attend to CH, come what may. In the end Gödel's strong mathematical instincts manifested themselves, and with the continuum problem still looming large and despite his "concept of set" and his once-held enthusiasm for large cardinals, he brought in old mathematical ideas from a different quarter and tried to push forward new mathematics. As set theory was to develop after Gödel, there would be a circling back, with deep and penetrating arguments from strong large cardinal hypotheses that, after all, lead to $2^{\aleph_0} = \aleph_2$.[54]

---

[53]It is striking to see Gödel offer comfort to a dying colleague by sharing a piece of mathematics with him. Earlier in the letter Gödel had written: "As you know I have unorthodox views about many things. Two of them would apply here: 1. I don't believe that any medical prognosis is 100% certain, 2. The assertion that our ego consists of protein molecules seems to me one of the most ridiculous every made. I hope you are sharing at least the second opinion with me." Exactly 18 years before to the day, Gödel on 20 March 1956 wrote to his friend von Neumann, dying of bone cancer (Gödel [2003a: 373]): "I hope and wish that your condition will soon improve further and that the latest achievements of medicine may, if possible, effect a complete cure." Gödel then went on to raise a mathematical issue, giving the first known formulation of the now well-known P = NP problem of computer science (cf. Hartmanis [1989]). There is something quite affecting, almost wry, in Gödel's conviction that mathematics is to trump everything.

[54]See Bekkali [1991], based on of lectures of Todorčević, for the results that the Perfect Forcing Axiom or Stationary Reflection at $\aleph_2$ implies $2^{\aleph_0} = \aleph_2$. See Woodin [1999] for the result that ψ_{AC} implies $2^{\aleph_0} = \aleph_2$.

References

1991 MOHAMED BEKKALI, *Topics in Set Theory*, Lecture Notes in Mathematics #1476, Springer-Verlag, New York.

1937 PAUL BERNAYS, A system of axiomatic set theory. Part I, *The Journal of Symbolic Logic 2*, 65–77; reprinted in: Gert H. Müller (editor), *Sets and Classes*, North-Holland, Amsterdam, 1–13.

∞ JÖRG BRENDLE, PAUL LARSON, and STEVO TODORČEVIĆ, Rectangular axioms, perfect set properties and decomposition, to appear.

1965 MARTIN DAVIS (editor), *The Undecidable: Basic papers on undecidable propositions, unsolvable problems and computable functions*, Raven Press, Hewlett, New York.

1997 JOHN W. DAWSON, Jr., *Logical Dilemmas: The Life and Work of Kurt Gödel*, A K Peters, Wellesley.

1880 PAUL DU BOIS-REYMOND, Der Beweis des Fundamentalsatzes der Integralrechnung: $\int_a^b F'(x)\,dx = F(b) - F(a)$, *Mathematisches Annalen 16*, 115–128.

1882 PAUL DU BOIS-REYMOND, *Die allgemeine Funktionentheorie I*, Lampp, Tübingen.

1984 SOLOMON FEFERMAN, Kurt Gödel: Conviction and caution, *Philosophia Naturalis 21*, 546–562; reprinted in [1998] below, 150–164.

1987 SOLOMON FEFERMAN, Infinity in mathematics: Is Cantor necessary? (Conclusion), in G. Toraldo di Francia (editor), *L'infinito nella Scienza*, Instituto della Enciclopedia Italiana, Roma, 151–209; reprinted in [1998] below, particularly 229–248.

1998 SOLOMON FEFERMAN, *In the Light of Logic*, Oxford University Press, New York.

2006 JULIET FLOYD and AKIHIRO KANAMORI, How Gödel transformed set theory, *Notices of the American Mathematical Society 53*, 417–425.

1922 ABRAHAM FRAENKEL, Zu den Grundlagen der Cantor-Zermeloschen Mengenlehre, *Mathematische Annalen 86*, 230–237.

1953 ABRAHAM FRAENKEL, *Abstract Set Theory*, North Holland, Amsterdam.

1929 KURT GÖDEL, *Über die Vollständigkeit der Logikkalküls* (doctoral dissertation, University of Vienna), reprinted and translated in [1986] below, 60–101.

1931 KURT GÖDEL, Über formal unentscheidbare Sätze der *Principia Mathematica* und verwandter Systeme I, *Monatshefte für Mathematik und Physik 38*, 173–198; reprinted and translated with minor emendations by the author in [1986] below, 144–195.

1932b KURT GÖDEL, Über Vollständigkeit und Widerspruchsfreiheit, *Ergebnisse eines mathematischen Kolloquiums 3*, 12–13; text and translation in [1995] below, 234–237.

1933o KURT GÖDEL, The present situation in the foundations of mathematics, handwritten text for an invited lecture, in [1995] below, 45–53, and the page references are to these.

1938 KURT GÖDEL, The consistency of the Axiom of Choice and of the Generalized Continuum Hypothesis, *Proceedings of the National Academy of Sciences U.S.A. 24*, 556–557; reprinted in [1990] below, 26–27.

1939 KURT GÖDEL, The consistency of the generalized continuum hypothesis, *Bulletin of the American Mathematical Society 45*, 93; reprinted in [1990] below, 27.

1939a KURT GÖDEL, Consistency-proof for the Generalized Continuum Hypothesis, *Proceedings of the National Academy of Sciences U.S.A. 25*, 220–224; reprinted in [1990] below, 28–32.

1939b KURT GÖDEL, Vortrag Göttingen; text and translation in [1995] below, 126–155, and the page references are to these.

193? KURT GÖDEL, Untitled lecture, in [1995] below, 164–175.

1940 KURT GÖDEL, *The Consistency of the Axiom of Choice and of the Generalized Continuum Hypothesis with the Axioms of Set Theory,* Annals of Mathematics Studies #3, Princeton University Press, Princeton; reprinted in [1990] below, 33–101.

1940a KURT GÖDEL, Lecture [on the] consistency [of the] continuum hypothesis (Brown University) in [1995] below, 175–185, and the page references are to these.

1944 KURT GÖDEL, Russell's mathematical logic, in Paul A. Schilpp (editor), *The Philosophy of Bertrand Russell,* The Library of Living Philosophers, vol. 5, Northwestern University, Evanston 123–153; reprinted in Gödel [1990] below, 119–141.

1947 KURT GÖDEL, What is Cantor's Continuum Problem? *American Mathematical Monthly 54*, 515–525; errata *55*(1948), 151; reprinted in [1990] below, 176–187; see also [1964] below.

1964 KURT GÖDEL, revised and expanded version of [1947] in: Paul Benacerraf and Hilary Putnam (editors), *Philosophy of Mathematics. Selected Readings*, Prentice Hall, Englewood Cliffs, 258–273; this version reprinted with emendations by the author in [1990] below, 254–270.

1970a KURT GÖDEL, Some considerations leading to the probable conclusion that the true power of the continuum is $\aleph_2$, handwritten document, in Gödel [1995] below, 420–422.

1970b KURT GÖDEL, A proof of Cantor's continuum hypothesis from a highly plausible axiom about orders of growth, handwritten document, in Gödel [1995] below, 422–423.

1972a KURT GÖDEL, Some remarks on the undecidability results, in Gödel [1990] below, 305–306.

1986 KURT GÖDEL, Solomon Feferman (editor-in-chief), *Collected Works, Volume I: Publications 1929–1936,* Oxford University Press, New York.

1990 KURT GÖDEL, Solomon Feferman (editor-in-chief), *Collected Works, Volume II: Publications 1938–1974,* Oxford University Press, New York.

1995 KURT GÖDEL, Solomon Feferman (editor-in-chief), *Collected Works, Volume III: Unpublished Essays and Lectures,* Oxford University Press, New York.

2003 KURT GÖDEL, Solomon Feferman and John W. Dawson, Jr. (editors-in-chief), *Collected Works, Volume IV: Correspondence A-G*, Clarendon Press, Oxford.

2003a KURT GÖDEL, Solomon Feferman and John W. Dawson, Jr. (editors-in-chief), *Collected Works, Volume V: Correspondence H-Z*, Clarendon Press, Oxford.

1989 JURIS HARTMANIS, Gödel, von Neumann, and the P = NP problem, *Bulletin of the European Association for Theoretical Computer Science 38*, 101–107.

1907 FELIX HAUSDORFF, Untersuchungen über Ordnungstypen, IV, V, *Berichte über die Verhandlungen der Königlich Sächsischen Gesellschaft der Wissenschaften zu Leipzig,*
Mathematisch-Physische Klasse 59, 84–159; translated in Plotkin [2005], 113–171.

1909 FELIX HAUSDORFF, Die Graduierung nach dem Endverlauf, *Abhandlungen der Königlich Sächsischen Gesellschaft der Wissenschaften zu Leipzig, Mathematisch-Physische Klasse 31*, 295–334; translated in Plotkin [2005], 271–301.

1936 FELIX HAUSDORFF, Summen von $\aleph_1$ Mengen, *Fundamenta Mathematicae 26*, 241–255; translated in Plotkin [2005], 305–316.

2006 KAI HAUSER, Gödel's Program Revisited, Part I: The Turn to Phenomenology, *The Bulletin of Symbolic Logic 21*, 529-590.

1974 STEPHEN M. HECHLER, On the existence of certain cofinal subsets of ${}^{\omega}\omega$, in: Thomas J. Jech (editor), *Axiomatic set theory*, Proceedings of symposia in pure mathematics, vol. 13, part 2, American Mathematical Society, Providence.

1926 DAVID HILBERT, Über das Unendliche, *Mathematische Annalen 95*, 161–190; translated into French by André Weil in *Acta Mathematica 48* (1926), 91-122; translated in van Heijenoort [1967], 367-392.

1995 IGNACIO JANÉ, The role of the absolute infinite in Cantor's conception of set, *Erkenntnis 42*, 375–402.

2002 THOMAS JECH, *Set Theory*, third millennium edition, revised and expanded, Springer, Berlin.

1972 RONALD B. JENSEN, The fine structure of the constructible hierarchy, *Annals of Mathematical Logic 4*, 229–308.

1995 AKIHIRO KANAMORI, The emergence of descriptive set theory, in: Jaakko Hintikka (editor), *From Dedekind to Gödel: Essays on the Development of the Foundations of Mathematics*, Synthese Library volume 251, Kluwer, Dordrecht, 241–262.

1996 AKIHIRO KANAMORI, The mathematical development of set theory from Cantor to Cohen, *The Bulletin of Symbolic Logic 2*, 1–71.

2003 AKIHIRO KANAMORI, *The Higher Infinite,* second edition, Springer-Verlag, Heidelberg.

2004 AKIHIRO KANAMORI, Zermelo and set theory, *The Bulletin of Symbolic Logic 10*, 487–553.

2004 JULIETTE C. KENNEDY and MARK VAN ATTEN, Gödel's modernism: On set-theoretic incompleteness, *Graduate Faculty Philosophy Journal 25*, 289–349.

2006 PETER KOELLNER, On the question of absolute undecidability, *Philosophia Mathematica.*

1939 MOTOKITI KONDÔ, Sur l'uniformisation des complémentaires analytiques et les ensembles projectifs de la seconde classe, *Japanese Journal of Mathematics 15*, 197-230.

1931 KAZIMIERZ KURATOWSKI, Evaluation de la classe Borélienne ou projective d'un ensemble de points à l'aide des symboles logiques, *Fundamenta Mathematicae 17*, 249-272.

1931 KAZIMIERZ KURATOWSKI and ALFRED TARSKI, Les opérations logiques et les ensembles projectifs, *Fundamenta Mathematicae 17*, 240-248; reprinted in Tarski [1986] vol. 1, 551–559; translated in Tarski [1983], 143–151.

2004 STANISŁAW KRAJEWSKI, Gödel on Tarski, *Annals of Pure and Applied Logic 127*, 303–323.

1980 GEORG KREISEL, Kurt Gödel, 28 April 1906 – 14 January 1978. *Biographical Memoirs of the Fellows of the Royal Society 26*, 149-224; Corrections *27* (1981), 697 and *28* (1982), 718.

1960 AZRIEL LEVY, Axiom schemata of strong infinity in axiomatic set theory, *Pacific Journal of Mathematics 10*, 223–238; reprinted in *Mengenlehre*, Wissensschaftliche Buchgesellschaft Darmstadt, 1979, 238-253.

1967 AZRIEL LEVY and ROBERT SOLOVAY, Measurable cardinals and the continuum hypothesis, *Israel Journal of Mathematics 5*, 234-248.

1927 STEFAN MAZURKIEWICZ, Sur une propriété des ensembles $C(A)$, *Fundamenta Mathematicae 10*, 172–174.

1961 RICHARD M. MONTAGUE, *Fraenkel's addition to the axioms of Zermelo*, in: Yehoshua Bar-Hillel, E.I.J. Poznanski, Michael O. Rabin, and Abraham Robinson (editors), *Essays on the Foundations of Mathematics*, dedicated to Professor A.A. Fraenkel on his 70th Birthday, Magnes Press, Jerusalem, 91–114.

1980 YIANNIS N. MOSCHOVAKIS, *Descriptive Set Theory*, North-Holland, Amsterdam.

1998 ROMAN MURAWSKI, Undefinability of truth. The Problem of priority: Tarski vs Gödel, *History and Philosophy of Logic 19*, 153–160.

1971 JOHN R. MYHILL and DANA S. SCOTT, Ordinal definability, in: Dana S. Scott (editor), *Axiomatic Set Theory*, Proceedings of Symposia in Pure Mathematics vol. 13, part 1, American Mathematical Society, Providence, 271–278.

2005 JACOB M. PLOTKIN (editor), *Hausdorff on Ordered Sets*, American Mathematical Society, Providence.

1925 FRANK P. RAMSEY, The foundations of mathematics, *Proceedings of the London Mathematical Society 25*, 338-384.

1955 DANA S. SCOTT, Definitions by abstraction in axiomatic set theory, *Bulletin of the American Mathematical Society 61*, 442.

1961 DANA S. SCOTT, Measurable cardinals and constructible sets, *Bulletin de l'Académie Polonaise des Sciences, Série des Sciences Mathématiques, Astronomiques et Physiques 9*, 521–524.

1920 THORALF SKOLEM, Logisch-kombinatorische Untersuchungen über die Erfüllbarkeit oder Beweisbarkeit mathematischer Sätze nebst einem Theoreme über dichte Mengen. Videnskaps-selskapets Skrifter, I. Matematisk-Naturvidenskabelig Klass (1920, #4), 1-36. Reprinted in [1970] below, 103–136. Partially translated in van Heijenoort [1967], 252–263.

1923 THORALF SKOLEM, Einige Bemerkungen zur axiomatischen Begründung der Mengenlehre, in: *Matematikerkongressen i Helsingfors den 4–7 Juli 1922, Den femte skandinaviska matematikerkongressen, Redogörelse*, Helsinki, Akademiska-Bokhandeln, 217-232; reprinted in [1970] below, 137–152; translated in van Heijenoort [1967], 290–301.

1962 THORALF SKOLEM, *Abstract Set Theory*, Notre Dame Mathematical Lecture Notes #8, Notre Dame.

1970 THORALF SKOLEM, Jens E. Fenstad (editor), *Selected Works in Logic*, Universitetsforlaget, Oslo.

1995 JOHN R. STEEL, $\mathrm{HOD}^{L(R)}$ is a core model below Θ, *The Bulletin of Symbolic Logic 1*, 75–84.

1931 ALFRED TARSKI, Sur les ensembles définissables de nombres réels, *Fundamenta Mathematicae 17*, 210–239; reprinted in Tarski [1986] below, vol. 1, 517–548; translated in Tarski [1983] below, 110–142.

1933 ALFRED TARSKI, Pojęcie prawdy w językach nauk dedukcyjnych (The concept of truth in the languages of deductive sciences), *Prace Towarzystwa Naukowego Warszawskiego, Wydział III, Nauk Matematyczno-fizycznych*

(Travaux de la Société des Sciences et des Lettres de Varsovie, Classe III, Sciences Mathématiques et Physiques) #34.

1935 ALFRED TARSKI, Der Wahrheitsbegriff in den formalisierten Sprachen, German translation of [1933] with a postscript, *Studia Philosophica 1*, 261–405; reprinted in [1986] below, vol. 2, 51–198; translated in [1983] below, 152–278.

1983 ALFRED TARSKI, *Logic, Semantics, Metamathematics. Papers from 1923 to 1938*, translations by J.H. Woodger, second edition, Hackett, Indianapolis.

1986 ALFRED TARSKI, *Collected Papers*, Steven R. Givant and Ralph N. McKenzie (editors), Birkhäuser, Basel.

1957 ALFRED TARSKI and ROBERT L. VAUGHT, Arithmetical extensions of relational systems, *Compositio Mathematica 13*, 81–102; reprinted in Tarski [1986], vol. 3, 653–674.

1967 JEAN VAN HEIJENOORT (editor), *From Frege to Gödel. A Source Book in Mathematical Logic, 1879–1931,* Cambridge, Harvard University Press 1967.

1923 JOHN VON NEUMANN, Zur Einführung der transfiniten Zahlen, *Acta Litterarum ac Scientiarum Regiae Universitatis Hungaricae Francisco-Josephinae (Szeged), sectio scientiarum mathematicarum, 1*, 199–208; reprinted in [1961] below, vol. 1, 24–33; translated in van Heijenoort [1967], 346–354.

1925 JOHN VON NEUMANN, Eine Axiomatisierung der Mengenlehre, *Journal für die reine und angewandte Mathematik 154*, 219–240; Berichtigung ibid. *155*, 128; reprinted in [1961] below, 34–56; translated in van Heijenoort [1967], 393–413.

1928 JOHN VON NEUMANN, Über die Definition durch transfinite Induktion und verwandte Fragen der allgemeinen Mengenlehre, *Mathematische Annalen 99*, 373–391; reprinted in [1961] below, vol. 1, 320–338.

1929 JOHN VON NEUMANN, Über eine Widerspruchfreiheitsfrage in der axiomatischen Mengenlehre, *Journal für die reine und angewandte Mathematik 160*, 227–241; reprinted in [1961] below, 494–508.

1961 JOHN VON NEUMANN, Abraham H. Taub (editor), *John von Neumann. Collected Works,* Vol. 1; Pergamon Press, New York.

1981 HAO WANG, *Popular Lectures on Mathematical Logic*, Van Nostrand Reinhold, New York.

1996 HAO WANG, *A Logical Journey. From Gödel to Philosophy*, The MIT Press, Cambridge.

2005 JAN WOLEŃSKI, Gödel, Tarski and Truth, *Revue International de Philosophie 59#4*, 459–490.

1999 W. HUGH WOODIN, *The Axiom of Determinacy, Forcing Axioms, and the Non-Stationary Ideal*, DeGruyter Series in Logic and Its Applications, volume 1.

1908 ERNST ZERMELO, Untersuchungen über die Grundlagen der Mengenlehre I, *Mathematische Annalen 65*, 261–281; translated in van Heijenoort [1967], 199–215.

1930 ERNST ZERMELO, Über Grenzzahlen und Mengenbereiche: Neue Untersuchungen über die Grundlagen der Mengenlehre, *Fundamenta Mathematicae 16*, 29–47.

1932 ERNST ZERMELO, Über Stufen der Quantifikation und die Logik des Unendlichen, *Jahresbericht der deutschen Mathematiker-Vereinigung (Angelegenheiten) 41*, 85–88.

15 Levy and Set Theory

Azriel Levy (1934–) did fundamental work in set theory when it was transmuting into a modern, sophisticated field of mathematics, a formative period of over a decade straddling Cohen's 1963 founding of forcing. The terms "Levy collapse", "Levy hierarchy", and "Levy absoluteness" will live on in set theory, and his technique of relative constructibility and connections established between forcing and definability will continue to be basic to the subject. Levy came into his prime at what was also a formative time for the State of Israel and has been a pivotal figure between generations in the flowering of mathematical logic at the Hebrew University of Jerusalem.[1] There was initially Abraham Fraenkel, and then Yehoshua Bar-Hillel, Abraham Robinson, and Michael Rabin. With Levy subsequently established at the university in the 1960's there would then be Saharon Shelah, Levy's student and current university president Menachem Magidor, and Ehud Hrushovski, all together with a stream of students who would achieve worldwide prominence in set theory.

What follows is a detailed account and analysis of Levy's work and contributions to set theory. Levy's work has featured several broad, interconnected themes coincident with those in the pioneering work of Fraenkel: Axiom of Choice independences using urelements, the investigation of axioms and the comparative strengths of set theories, and the study of formal definability. To these Levy brought in concerted uses of model-theoretic reflection arguments and the method of forcing, and he thereby played an important role in raising the level of set-theoretic investigation through metamathematical means to a new height of sophistication.

Already in his M.Sc. work under Fraenkel's supervision, Levy [1958, 1958a] used urelement models to establish independences for several definitions of finiteness. Levy's ground-breaking 1958 Hebrew University dissertation, *Contributions to the metamathematics of set theory* (in Hebrew with an English summary), under the joint supervision of Fraenkel and Robinson featured work in three directions: relative constructibility, reflection principles, and Ackermann's set theory. Each of these is discussed in turn in the first three sections. §4 discusses Levy's hierarchy of formulas of set theory and his well-known absoluteness result. This work as well as the further development of the thesis topics were undertaken when Levy was a Sloan postdoctoral fellow at the Massachussetts Institute of Technology 1958-1959 and a visiting assistant professor at the University of California at Berkeley 1959-1961. At Berkeley Levy through shared interests associated particularly with Robert Vaught. From 1961 Levy had academic positions at the Hebrew University, eventually becoming professor of mathematics there while continuing to have active visiting positions in

Republished from *Annals of Pure and Applied Logic* 140 (2006), pp.233-252, with permission from Elsevier.

[1]See Bentwich [1961] for a history of the Hebrew University to 1960.

the United States.

In the early 1960's Levy focused on independence results, first with Fraenkel's urelement approach as developed by Mostowski, and as soon as it appeared, with Cohen's method of forcing. These are discussed in sections §5 and §6, and in §7 and §8 Levy's work on definability using forcing is presented. The latter section is focused on the Levy collapse of an inaccessible cardinal, and in §9 his further work on large cardinals, measurable and indescribable cardinals, is described. Levy put capstones to his work in different directions in the late 1960's, and these are taken up in §10. There remains an envoi.

In such an account as this it is perhaps inevitable that the more basic results are accorded more of an airing and the subsequent developments are summarily sketched, giving the impression that the latter are routine emanations. This is far from the case, but on the other hand a detailed analysis of basic results becomes a natural undertaking when discussing Levy's work, since one sees and wants to bring out how they have become part and parcel of our understanding and investigation of set theory.

1 Relative Constructibility

Set theory was launched on an independent course as a distinctive field of mathematics by Gödel's work [1938, 1939] on the inner model L of *constructible* sets. Not only did this work establish the relative consistency of the Axiom of Choice (AC) and the Generalized Continuum Hypothesis (GCH), but it promoted a new relativism about the notion of set as mediated by first-order logic, which beyond its sufficiency as a logical framework for mathematics was seen to have considerable operational efficacy. Gödel's work however stood as an isolated monument for quite a number of years, the world war having a negative impact on mathematical progress. In retrospect, another inhibitory factor may have been the formal presentation of L in Gödel's monograph [1940]. There he pointedly avoided the use of the satisfaction predicate and, following John von Neumann's lead, used a class theory to cast, in effect, definability. This presentation of L, however, tended to obfuscate the metamathematical ideas, especially the reflection Skolem closure argument for the GCH. Levy's work in general would serve to encourage the use and exhibit the efficacy of metamathematical methods in set theory.

In the next generation, András Hajnal, Joseph Shoenfield and Levy came to generalize Gödel's construction in order to establish conditional independence results. Their presentations would be couched in the formalism of Gödel [1940], but the metamathematical ideas would soon become clear and accessible. Hajnal [1956, 1961] in his Hungarian dissertation essentially formulated for a given set A the *constructible closure* $L(A)$, the smallest inner model M of ZF such that $A \in M$. To summarize, for any structure N and subset y of its domain, y is *definable over* N *iff* there is a (first-order) formula $\psi(v_0, v_1, v_2, \ldots, v_n)$ in the free variables as displayed and $a_1, a_2, \ldots, a_n$ in the domain of N such that

$$y = \{z \mid N \models \psi[z, a_1, \ldots, a_n]\},$$

where $N \models \psi[z, a_1, \ldots, a_n]$ asserts that the formula ψ is satisfied in N when v_0 is interpreted as z and each v_i is interpreted as a_i for $i \geq 1$. For any set x now define

$$\mathrm{def}(x) = \{y \subseteq x \mid y \text{ is definable over } \langle x, \in \rangle\}.$$

Finally, with $\mathrm{tc}(x)$ denoting the transitive closure of set x, define:

$$L_0(A) = \mathrm{tc}(\{A\});\ L_{\alpha+1}(A) = \mathrm{def}(L_\alpha(A));\ L_\delta(A) = \textstyle\bigcup_{\alpha<\delta} L_\alpha(A) \text{ for limit } \delta > 0$$

and

$$L(A) = \textstyle\bigcup_\alpha L_\alpha(A)\ .$$

$L(\emptyset)$ is just Gödel's L, and the general construction starts instead with $L_0(A)$ as an "urelement" basis. Although $L(A)$ is indeed an inner model of ZF, unless $\mathrm{tc}(\{A\})$ has a well-ordering in $L(A)$, $L(A)$ does not satisfy the Axiom of Choice. This general situation was not broached by Hajnal, who used sets A of ordinals to establish the first relative consistency results about cardinal arithmetic after Gödel. Hajnal showed that in $L(A)$, if $A \subseteq \kappa^+$, then for every cardinal $\lambda \geq \kappa$, $2^\lambda = \lambda^+$, the case $\lambda = \kappa$ requiring some refinement of Gödel's original argument. In particular, if the Continuum Hypothesis (CH) fails, one can use an $A \subseteq \omega_2$ coding $\aleph_2$ sets of natural numbers and injections: $\alpha \to \omega_1$ for every $\alpha < \omega_2$ (so that $\omega_2^{L(A)} = \omega_2$ and $(2^{\aleph_0} = 2^{\aleph_1} = \omega_2)^{L(A)}$) to establish Con(ZFC + $\neg$CH) implies Con(ZFC + $\neg$CH + $\forall \alpha \geq 1 (2^{\aleph_\alpha} = \aleph_{\alpha+1})$), or more dramatically, if one can prove $2^{\aleph_0} \neq \aleph_2$ or $2^{\aleph_0} \neq 2^{\aleph_1}$, then one can prove the Continuum Hypothesis.

Levy [1957, 1960a] on the other hand developed for a given class A the inner model $L[A]$ of sets *constructible relative to* A, i.e. the smallest inner model M such that for every $x \in M$, $A \cap x \in M$. While $L(A)$ realizes the algebraic idea of building up a model starting from a basis of generators, $L[A]$ realizes the idea of building up a model using A construed as a predicate. Let

$$\mathrm{def}^A(x) = \{y \subseteq x \mid y \text{ is definable over } \langle x, \in, A \cap x \rangle\}\ ,$$

incorporating $A \cap x$ as a unary relation for definitions. Now define

$$L_0[A] = \emptyset\ ;\ L_{\alpha+1}[A] = \mathrm{def}^A(L_\alpha[A])\ ;\ L_\delta[A] = \textstyle\bigcup_{\alpha<\delta} L_\alpha[A] \text{ for limit } \delta > 0\ ;$$

and

$$L[A] = \textstyle\bigcup_\alpha L_\alpha[A]\ .$$

For sets A, unlike for $L(A)$ what remains of A is only $A \cap L[A] \in L[A]$, so that for example $L[\mathbb{R}] = L$ for the reals $\mathbb{R}$. $L[A]$ is more constructive since knowledge of A is incorporated through the hierarchy of definitions, and like L, $L[A]$ satisfies the Axiom of Choice for every A. Levy pointed out an important absoluteness, that with $\overline{A} = A \cap L[A]$, $L[A] = L[\overline{A}]$.

Shoenfield [1957, 1959] had separately sketched a special case of this construction to establish partial results toward the conditional independence Con (ZFC + $V \neq L$) implies Con(ZFC + GCH + $V \neq L$). Levy [1960a] then established this result by refining Shoenfield's argument to show that the full GCH can be proved in an appropriate $L[A]$.

The work of Hajnal and of Levy on constructibility elicited some interest in the correspondence between Paul Bernays and Gödel ([2003: 151-5]).[2] Although differing in their formal presentations, since Hajnal and Levy both worked with sets A of ordinals so that $L[A] = L(A)$, distinctions would surface only later. Because of its intrinsic absoluteness, Levy's construction $L[A]$ would become basic for the inner model theory of large cardinals. Hajnal's construction $L(A)$ has also become basic, particularly with the constructible closure of the reals $L(\mathbb{R})$ becoming the focal inner model for the Axiom of Determinacy.

2 Reflection Principles

Reflection has been an abiding motif in set theory, with its first appearance in a proof in Gödel's of the GCH in L. Gödel himself saw roots in Russell's Axiom of Reducibility and in Zermelo's Axiom of Separation, writing [1940a: 178] that: "...since an axiom of reducibility holds for constructible sets it is not surprising that the axioms of set theory hold for the constructible sets, because the axiom of reducibility or its equivalents, e.g., Zermelo's Aussonderungsaxiom is really the only essential axiom of set theory." A heuristic appearing early on for reflection was that any particular property attributable to the class of all ordinals, since its extent is not characterizable, should already be attributable to some cardinal. This heuristic was at work in the early postulations of large cardinals, cardinals that posit structure in the higher reaches of the cumulative hierarchy and prescribe their own transcendence over smaller cardinals. The weakest of the now standard large cardinals are the *inaccessible* cardinals, those uncountable regular cardinals κ such if $\alpha < \kappa$, then $2^\alpha < \kappa$, so that in the rank hierarchy V_κ models ZFC.

The formalization of reflection properties was one of the early developments of model-theoretic initiatives in set theory. With the basic concepts and methods of model theory being developed by Tarski and his students at Berkeley, Richard Montague [1961] in his 1957 Berkeley dissertation had studied reflec-

[2] In later correspondence, Bernays (Gödel [2003: 199]) referred Gödel to a review of Levy [1957] by Shepherdson [1959], but amusingly this review is mistitled and is actually a review of another paper drawn from Levy's dissertation, Levy [1958b].

tion properties in set theory and had shown that Replacement is not finitely axiomatizable over Zermelo set theory in a strong sense. Levy [1960b, 1960c] then exploited the model-theoretic methods to establish the broader significance of reflection principles and the close involvement of the Mahlo hierarchy of large cardinals.

The ZF *Reflection Principle*, drawn from Montague [1961: 99] and Levy [1960b: 234], asserts that for any (first-order) formula $\varphi(v_1, \ldots, v_n)$ in the free variables as displayed and any ordinal β, there is a limit ordinal $\alpha > \beta$ such that for any $x_1, \ldots, x_n \in V_\alpha$,

$$\varphi[x_1, \ldots, x_n] \;\; \textit{iff} \;\; \varphi^{V_\alpha}[x_1, \ldots, x_n] \, ,$$

where as usual φ^M denotes the relativization of the formula φ to M. The idea is to carry out a Skolem closure argument with the collection of subformulas of φ. Montague showed that the principle holds in ZF, and Levy showed that it is actually equivalent to the Replacement Schema together with the Axiom of Infinity in the presence of the other axioms of ZF. Through this work the ZF Reflection Principle has become well-known as making explicit how reflection is intrinsic to the ZF system.

Levy [1960b] took the ZF Reflection Principle as motivation for stronger reflection principles. The first in his hierarchy asserts that for any formula $\varphi(v_1, \ldots, v_n)$, there is an inaccessible cardinal α such that for any $x_1, \ldots, x_n \in V_\alpha$,

$$\varphi[x_1, \ldots, x_n] \;\; \textit{iff} \;\; \varphi^{V_\alpha}[x_1, \ldots, x_n] \, .$$

Levy showed that this principle is equivalent to the assertion that the class of inaccessible cardinals is definably stationary, i.e. every definable closed unbounded class of ordinals contains an inaccessible cardinal. Paul Mahlo [1911] [1912] [1913] had studied what are now known as the *weakly Mahlo* cardinals, those regular cardinals κ such that the set of smaller regular cardinals is stationary in κ, i.e. every closed unbounded subset of κ contains a regular cardinal. Levy's work thus established an evident connection between Mahlo's cardinals and structural principles about sets. Levy recast Mahlo's concept by replacing regular cardinals by inaccessible cardinals. On the other hand, whereas Mahlo had entertained arbitrary closed unbounded subsets, Levy's principle is restricted to definable closed unbounded classes. Be that as it may, it would be through Levy's work that Mahlo's cardinals would come into use in modern set theory cast as the *strongly Mahlo* cardinals, those regular cardinals κ such that the set of smaller *inaccessible* cardinals is stationary in κ.[3]

Levy proceeded to develop a hierarchy of reflection principles, the next principle being the one above with "inaccessible" replaced by "strongly Mahlo".

[3] See Kanamori [2005: §§1, 6].

Mahlo himself had developed a hierarchy of his cardinals, and Levy's work recast it as reflecting reflection: A reflection scheme is first formulated and is then itself reflected. In this way, Levy showed how the iterative formalization of reflection illuminates Mahlo's original scheme, formulated a half-century before.

Levy also substantiated how various reflection principles have proof-theoretic transcendence over each other. He had formulated the following, drawing on his dissertation work (cf. Levy [1958b]): For theories $T_0 \subseteq T$ in the same language and subsuming enough of arithmetic to encode formal consistency, T is *essentially reflexive* over T_0 if for any sentence σ, $T \vdash \sigma \rightarrow \text{Con}(T_0 + \sigma)$. This is an elegant formulation of the transcendence of one theory over another; note that no consistent extension of T is finitely axiomatizable over T_0, since for any σ, if $T_0 + \sigma$ were to extend T, we would have $T_0 + \sigma \vdash \text{Con}(T_0 + \sigma)$. Montague [1961] had shown in effect that ZF is essentially reflexive over Z, Zermelo set theory. Levy [1960c] considered "partial" reflection principles weaker than the ZF Reflection Principle and studied [1962] their minimal models of form V_α; Levy-Vaught [1961] showed that these partial principles are also essentially reflexive over Z. In [1960d] Levy showed that each of the strong reflection principles in his [1960b] hierarchy à la Mahlo is essentially reflexive over the previous. Moreover, Levy [1960d] showed that between any two of these reflection principles there is a whole spectrum of theories each essentially reflexive over the previous.

In further ramifications, to a volume dedicated to Fraenkel on the occasion of his 70th birthday Levy contributed a paper [1961b] that compared the Axiom of Choice with its global form, i.e. there is a class choice function for all sets, and showed that the set consequences of the global form follows from his reflection principle down to inaccessible cardinals. The elder Bernays [1961] also contributed a paper to that volume, one in which, inspired by Levy's work on reflection, he developed reflection principles based on second-order formulas which were seen to subsume all of Levy's principles.

The ZF Reflection Principle was foreshadowed in Gödel's remarks [1946]; he there introduced the ordinal-definable sets, and to develop their theory requires reflection in some form (cf. Gödel [1990: 146]). In his expository article on Cantor's Continuum Problem, Gödel [1947: 521] mentioned the Mahlo cardinals in connection with the proposal to search for new large cardinal axioms that would settle the Continuum Hypothesis. Bernays cited the paper Levy [1960b] in a letter of 12 October 1961 to Gödel (Gödel [2003: 196ff]), and Gödel noted in a letter of 13 August 1965 to Cohen (Gödel [2003: 385ff]), in a discussion about evidence for inaccessible cardinals, that "Levy's principle might be considered more convincing than analogy [with the integers]." What presumably impressed Gödel was how reflection, a persistent heuristic in his own work, had been newly marshaled to account for Mahlo's cardinals. Finally, Gödel wrote in a letter of 7 July 1967 to Robinson (Gödel [2003a: 195]):

> ...I perhaps stimulated work in set theory by my epistemological attitude toward it, and by giving some indications as to the further developments, in my opinion, to be expected and to be aimed at. I did not, however, give any clues as to how these aims were to be attained. This has become possible only due to the entirely new ideas, primarily of Paul J. Cohen and, in the area of axioms of infinity, of the Tarski school and of Azriel Levy.

3 Ackermann's Set Theory

Ackermann [1956] formulated a distinctive axiomatic theory of sets and classes, and this theory quickly came under the scrutiny of Levy whose extended analysis did a great deal to bring it into the fold of the standard ZF axiomatization. Much of the analysis was already present in Levy's dissertation and was subsequently extended in Levy [1959] and Levy-Vaught [1961]. Gödel wrote to Bernays in a letter of 30 September 1958 (Gödel [2003: 155]): "Of the results announced in the introduction to Levy's dissertation, the most interesting seems to me to be that on Ackermann's system of set theory. That really looks very surprising."

Ackermann's theory A is a first-order theory that can be cast as follows: There is one binary relation $\in$ for membership and one constant V; the objects of the theory are to be referred to as classes, and members of V as sets. The axioms of A are the universal closures of:

(1) Extensionality: $\forall z(z \in x \leftrightarrow z \in y) \longrightarrow x = y$.

(2) Comprehension: For each formula ψ not involving t, $\exists t \forall z(z \in t \longleftrightarrow z \in V \wedge \psi)$.

(3) Heredity: $x \in V \wedge (t \in x \vee t \subseteq x) \longrightarrow t \in V$.

(4) Ackermann's Schema: For each formula ψ in free variables $x_1, \ldots, x_n, z$ and having no occurrence of V, $x_1, \ldots, x_n \in V \wedge \forall z(\psi \rightarrow z \in V) \longrightarrow$
$\exists t \in V \forall z(z \in t \leftrightarrow \psi)$.

This last, a comprehension schema for sets, is characteristic of Ackermann's system. It forestalls Russell's Paradox, and its motivation was to allow set formation through properties independent of the whole extension of the set concept and thus to be considered sufficiently definite and delimited.

Ackermann [1956] himself argued that every axiom of ZF, when relativized to V, can be proved in A. However, Levy [1959] found a mistake in Ackermann's proof of the Replacement Schema, and whether Replacement can be derived from Ackermann's Schema would remain an issue for some time. Toward a closer correlation with ZF, Levy came to the idea of working with

A*: A together with the Axiom of Foundation relativized to V.

As for ZF, Foundation focuses the sets with a stratification into a cumulative hierarchy. Levy [1959] showed that, leaving aside the question of Replacement,

A* establishes substantial reflection principles. On the other hand, he also showed through a sustained axiomatic analysis that for a sentence σ of set theory (so without V) : *If σ relativized to V is provable in* A*, *then σ is provable in* ZF. The thrust of this work was to show that Ackermann's Schema can be assimilated into ZF—this is presumably what Gödel found surprising—and that ZF and A* have almost the same theorems for sets.

Levy-Vaught [1961] later observed by an inner model argument that, as for ZF and Foundation, if A is consistent, then so is A*. They then went on to confirm that the addition of Foundation to A was substantive; they showed that Ackermann's Schema is equivalent to a reflection principle in the presence of the other A^* axioms, and that A* establishes the existence of $\{V\}$ and the power classes $\mathcal{P}(V)$, $\mathcal{P}(\mathcal{P}(V))$, and so forth.

Years later, returning to the original issue about Replacement, William Reinhardt [1970] in his 1967 Berkeley dissertation under the supervision of Vaught built on Levy-Vaught [1961] to establish for A* what Ackermann could not establish for A: *Every axiom of* ZF, *when relativized to V, can be proved in* A*. Thus, A* and ZF do have exactly the same theorems for sets. Reinhardt also developed a theory of natural models of A*; these are connected to the indescribable cardinals (see §9) and led to further large cardinal postulations.[4]

4 Levy Hierarchy and Absoluteness

In his first work in a distinctive direction from his dissertation, Levy in [1959a], and much later in full exposition [1965b], formulated the now standard hierarchy of first-order formulas of the language of set theory. He showed that the hierarchy provides the scaffolding for an efficacious analysis of logical complexity, getting to a substantial absoluteness result that cast reflection in a new light.

For formulating his hierarchy Levy struck on the key idea of discounting *bounded* quantifiers, those that can be rendered as $\forall v \in w$ or $\exists v \in w$, an idea perhaps novel at the time in set theory but now subsumed into its modern sensibilities. There was an antecedent in the discounting of the bounded numerical quantifiers $\forall k < n$ and $\exists k < n$ in Stephen Kleene's [1943] formulation of the arithmetical hierarchy over the recursive predicates, but the motivations were rather different, and Levy had to make a conceptual leap because of the arbitrariness of sets.

In brief, a formula of set theory is Σ_0 and Π_0 in the Levy hierarchy if its only quantifiers are bounded. Recursively, a formula is Σ_{n+1} if it is of the form $\exists v_1 \ldots \exists v_k \varphi$ where φ is Π_n, and Π_{n+1} if it is of the form $\forall v_1 \ldots \forall v_k \varphi$ where φ is Σ_n. The classification of definable concepts in this hierarchy depends on the governing theory. For a set theory T, a formula φ is Σ_n^T *iff* for some Σ_n formula

[4]See Kanamori [2005: §23].

$\varphi', T \vdash \varphi \leftrightarrow \varphi'$; and similarly for Π_n^T. Σ_n^{ZF} and Π_n^{ZF} formulas are equivalent to formulas with blocks of like quantifiers contracted into one through applications of the Pairing Axiom. Also, bounded quantification does not add to complexity in ZF: If φ is Σ_n^{ZF} (respectively, Π_n^{ZF}), then so is $\exists v \in w\varphi$ and $\forall v \in w\varphi$. This depends on Replacement to "push" the bounded quantifiers to the right and is a crucial point about the Levy hierarchy. Finally, that Σ_0^T formulas are wide-ranging yet absolute for transitive models of weak set theories T has become a basic feature of the semantic analysis of set theory.

Levy [1959a] pointed out that his hierarchy is proper in ZF, i.e. there are formulas in $\Pi_n^{\mathrm{ZF}} - \Sigma_n^{\mathrm{ZF}}$ and in $\Sigma_n^{\mathrm{ZF}} - \Pi_n^{\mathrm{ZF}}$ for every $n > 0$, and that ZF establishes the consistency, for any particular n, of Zermelo set theory plus Replacement restricted to Σ_n formulas. Levy [1964c, 1965] worked out for each $n > 0$ a Σ_n (respectively, Π_n) satisfaction formula for the Σ_n (respectively, Π_n) formulas and thereby got careful hierarchy results. The antecedent was the Σ_n^0 universal predicate for the Σ_n^0 predicates in the Kleene arithmetical hierarchy, built directly on the normal forms of recursive predicates. Levy laid out satisfaction sequences à la Tarski level-by-level, once again drawing metamathematical methods into set theory.

The main advance of Levy [1959a] was a now well-known and basic absoluteness result. Shoenfield [1961] had established an absoluteness result seminal for modern descriptive set theory; he showed that, as we now say, every $\mathbf{\Sigma}_2^1$ set of reals is ω_1-Suslin in a constructible way, and concluded in particular that every (lightface) Σ_2^1 set of natural numbers is in L. As detailed in [1965b] Levy wove in the Shoenfield idea to establish in ZF together with the Axiom of Dependent Choices (DC) that any sentence (without parameters) Σ_1 in his hierarchy, if holding in V, also holds in L. More formally, we have the *Shoenfield-Levy Absoluteness Lemma*:

$$\text{For any } \Sigma_1 \text{ sentence } \sigma,\ \mathrm{ZF} + \mathrm{DC}\ \vdash\ \sigma \leftrightarrow \sigma^L.$$

Levy readily concluded that any Σ_1 or Π_1 theorem of ZF + $V = L$ is already a theorem of ZF + DC, so that any uses of e.g. GCH in a proof of a such sentence can be eliminated. Levy [1959a] actually pointed out that L can be replaced by a countable L_γ fixed for all σ, so that any Σ_1 sentence is absolute for every transitive $M \supseteq L_\gamma$.[5] Levy's proof, starting with a Σ_1 sentence σ, first appealed to DC to reflect down from the universe to a countable transitive model of σ. He then got a countable L_α modeling σ by applying Shoenfield's main idea of relying on the absoluteness of well-foundedness, i.e. the equivalence of no infinite descending chains and the existence of a ranking. One can view the Shoenfield and Levy absoluteness results as two sides of the same coin, one in the context of descriptive set theory and the other in the context of general set

[5]In modern terms, γ can be taken to be the least stable ordinal, where δ is *stable iff* $L_\delta \prec_1 L$, i.e. L_δ and L satisfy the same Σ_1 formulas with parameters from L_δ.

theory, with either one readily leading to the other.

The Shoenfield-Levy Absoluteness Lemma can be seen as an effective refinement of the ZF Reflection Principle that reflects a restricted sentence down to some countable L_γ, and as such it would find wide use in effective contexts like admissible set theory. Even just Levy's initial reflection down, in effect into the domain of hereditarily countable sets, would become basic to admissible set theory as the *Levy Absoluteness Principle.* In his book on admissible set theory Barwise [1975: 77] wrote: "One of the main features of this book (at least from our point of view) is the systematic use of the Levy Absoluteness Principle to simplify results by reducing them to the countable case."

5 Independence with Urelements

From the beginning Levy had a steady interest in the independence of choice principles and in the pre-Cohen era established penetrating results based on the Fraenkel-Mostowski method. To establish the independence of AC, Fraenkel had come to the fecund idea of starting with urelements, objects without members yet distinct from each other; building a model of set theory by closing off under set-theoretic operations; and exploiting automorphisms of the model induced by permutations of the urelements. Fraenkel [1922a] in one construction started with urelements $A = \{a_n \mid n \in \omega\}$ and considered a generated model in which for any set x there is a finite $s \subseteq A$ with the following property: x is fixed by any automorphism of the model induced by a permutation of A that fixes each member of s and at most interchanges pairs within the cells $\{a_{2n}, a_{2n+1}\}$ for $n \in \omega$. There can then be no choice function for the countable set of pairs $\{\{a_{2n}, a_{2n+1}\} \mid n \in \omega\}$ in the model, since for any purported such function one can take some a_{2n}, a_{2n+1} not in its support and apply a permutation interchanging them.

Andrzej Mostowski [1939] developed Fraenkel's constructions by imposing algebraic initiatives. First, the set of urelements can be structured e.g. with an ordering; second, a model is built based on invariance with respect to a specified *group* of permutations, group in the algebraic sense with respect to composition; and third, *supports* of sets are closely analyzed, a support of a set to be a set of urelements such that if a permutation fixes each, the induced automorphism also fixes the set. Mostowski in particular established that there is a model in which AC fails but the *Ordering Principle* (OP) holds, where:

(OP) Every set can be linearly ordered.

He began with a countable set of urelements ordered in the ordertype of the rationals; built a model based on the group of order-preserving permutations; established that every set has a $\subseteq$-least finite support; and showed by these means that the set of urelements cannot be well-ordered yet the model itself has a class linear ordering. Levy's initial work [1958, 1958a, 1960] in this direction was directly based on Mostowski's model for OP.

Coming into his own, Levy [1962a] constructed a model in which C_n holds for every natural number n yet $C_{<\aleph_0}$ fails, where

(C_κ) Every set consisting of sets of cardinality $\leq \kappa$ has a choice function

($C_{<\kappa}$) Every set consisting of sets of cardinality $< \kappa$ has a choice function

and moreover the *Axiom of Multiple Choice* (MC) holds, where

(MC) For any set x there is a function f on x such that
for any non-empty $y \in x$, $f(y)$ is a non-empty finite subset of y.

Levy began with a countable set of urelements A partitioned as $\bigcup_{k\in\omega} P_k$, where $P_k = \{a_1^k, \ldots, a_{p_k}^k\}$ with p_k the kth prime. Let π_k be that permutation of A fixing every member of $A - P_k$ and such that $\pi_k(a_i^k) = a_{i+1}^k$ for $1 \leq i < p_k$ and $\pi(a_{p_k}^k) = a_1^k$. Levy then took the group of permutations generated by the π_k's and generated a model based on finite supports. As in the Fraenkel model described above, the set $\{P_k \mid k \in \omega\}$ does not have a choice function so $C_{<\aleph_0}$ fails. That MC holds Levy affirmed with an argument also applicable to the Fraenkel model. Finally, the specifics of Levy's model came into play when he showed with algebraic arguments about cyclic permutations that C_n holds for every natural number n. That OP implies $C_{<\aleph_0}$ is simple to see, so that OP must fail in Levy's model. Hans Läuchli [1964] built another model in which $C_{<\aleph_0}$ holds yet OP fails.

Notably, it was later observed that in ZF, MC actually implies AC.[6] Thus, Levy's work shows that having urelements can separate these principles. Levy [1963] subsequently applied his [1962a] model in a considerable analysis of some graph-theoretic propositions studied by Mycielski. Also, Levy [1965] developed transfinite versions of his algebraic methods and argued e.g. that $C_{<\kappa}$ does not imply C_κ for limit alephs κ.[7] However, this seems to be the single instance when Levy was proved wrong, but even the error stimulated results.[8]

Levy [1964] established independence results for various choice principles indexed by alephs κ (cf. Jech [1973: 119ff]):

[6] See Jech [1973: 133]; the observation was first made by David Pincus in his 1969 Harvard dissertation. Working in the post-Cohen era, Pincus's results there (cf. Pincus [1972]) also showed how to "transfer" Levy's result to get the consistency relative to ZF of C_n holds for every natural number n and $C_{<\aleph_0}$ fails.

[7] Levy [1965] was a summary of Fraenkel-Mostowski methods given at a 1963 symposium; in that summary (p.225) Levy pointed out the open problem of whether $2 \cdot m = m$ for all infinite cardinals m implies AC, and eventually his student Gershon Sageev [1975] established that it does not.

[8] Paul Howard [1973] later established that in every Fraenkel-Mostowski model, $C_{<\aleph_0}$ already implies "C_∞", i.e. that every set consisting of well-orderable sets has a choice function. Pincus [1974] and Sageev [1975] independently established the ZF independence of $C_{\aleph_0}$ from $C_{<\aleph_0}$.

(DC_κ) Suppose that x is a set and r a binary relation such that for every $\alpha < \kappa$ and $s\colon \alpha \to x$ there is a $y \in x$ satisfying $s\ r\ y$. Then there is a function $f\colon \kappa \to x$ such that $f|\alpha\ r\ f(\alpha)$ for every $\alpha < \kappa$.

(AC_κ) Every set x with $|x| = \kappa$ has a choice function.

(W_κ) Every set x is comparable with κ, i.e. $|x| \leq \kappa$ or $\kappa \leq |x|$.

With DC_κ Levy generalized the Axiom of Dependent Choices, which is $DC_{\aleph_0}$. W_κ generalizes the proposition that every infinite set has a countable subset, which is $W_{\aleph_0}$.[9] DC_κ implies both AC_κ and W_κ. Levy was expanding on work of Mostowski [1948], who showed that there is a model satisfying $\neg AC_{\aleph_1} + DC_{\aleph_0}$ (and, as noticed later, $\neg W_{\aleph_1}$).

After drawing implications among these principles for various κ, Levy established several independences. He constructed a basic model by starting with λ urelements, considering all permutations, and working with supports of cardinality $< \lambda$. For a singular cardinal, taking $\lambda = \aleph_{\omega_1}$ as a typicality one gets $\neg AC_{\aleph_1} + DC_{\aleph_0} + \forall\kappa < \aleph_{\omega_1}(W_\kappa) + \neg W_{\aleph_{\omega_1}}$. For a successor cardinal, taking $\lambda = \aleph_1$ one gets:

$$\forall\kappa(AC_\kappa) + DC_{\aleph_0} + \neg W_{\aleph_1} \text{ (and so } \neg DC_{\aleph_1}).$$

In particular, Well-ordered Choice $\forall\kappa(AC_\kappa)$ does not imply $DC_{\aleph_1}$. Jensen [1967] later established the surprising result $\forall\kappa(AC_\kappa)$ implies $DC_{\aleph_0}$, that Well-ordered Choice actually implies Dependent Choices.

Levy constructed an interesting, second model, assuming $2^{\aleph_0} = \aleph_1$ and starting with a set of urelements ordered in the ordertype of the reals. He used then the group of order-preserving permutations and supports generated by the "Dedekind cuts" $(-\infty, r)$ to get:

$$\forall\kappa(AC_\kappa) + \neg DC_{\aleph_1} + W_{\aleph_1} + \neg W_{\aleph_2}.$$

Thus, $DC_{\aleph_1}$ and $W_{\aleph_1}$ were separated.

Already in the work on conditional independence results via relative constructibility there was an air of anticipation about possible independences from ZF. This became palatable in the work on choice principles by Fraenkel-Mostowski methods with, e.g. Levy [1964: 145] writing: "Even the independence of the axiom of choice itself is still an open problem for systems of set theory which do not admit urelements or non-founded sets. Thus we can hope, for the time being, to prove the above mentioned independence results only for a set theory which admits either urelements or non-founded sets." Of course, independence with respect to ZF was what was really wanted, and this would come about in a dramatic turn of events.

[9] Actually, Levy [1964] worked with a more involved proposition $H(\kappa)$, which has the property that W_κ is equivalent to $H(\kappa)$ together with $\forall\lambda < \kappa(W_\lambda)$.

6 ZF Independence

Paul Cohen [1963,1964] in 1963 established the independence of AC from ZF and the independence of the CH from ZFC. That is, complementing Gödel's relative consistency results with L Cohen established that Con(ZF) implies Con(ZF + ¬AC) and that Con(ZFC) implies Con(ZFC + ¬CH). These were, of course, the inaugural examples of *forcing*, a remarkably general and flexible method, with strong intuitive underpinnings, for extending models of set theory. If Gödel's construction of L had launched set theory as a distinctive field of mathematics, then Cohen's method of forcing began its transformation into a modern, sophisticated one. Cohen's particular achievement lies in devising a concrete procedure for extending well-founded models of set theory in a minimal fashion to well-founded models of set theory with new properties but without altering the ordinals. Set theory had undergone a sea-change, and beyond simply how the subject was enriched, it is difficult to convey the strangeness of it.[10]

Cohen circulated a preprint [1963] in April and soon started to give talks on his results. Already evident in the preprint were two features of forcing that resonated with Levy's work: Cohen relied on relative constructibility to extend a model of $V = L$ to one satisfying GCH + $V \neq L$, and Cohen used in effect a reflection argument to show that his beginning with a countable $\in$-model of ZF was formally unnecessary for getting consistency results relative to ZF. Levy first heard the details of Cohen's results at a July model theory conference at Berkeley, and later that summer fully assimilated forcing working with Solomon Feferman. Feferman had been extensively consulted by Cohen at Stanford when he was coming up with forcing, and Feferman [1965, 1965a] was the first after Cohen to establish results by forcing. In quick succession several abstracts appeared in the *Notices of the American Mathematical Society*: Levy [1963a], Feferman-Levy [1963b], Levy [1963c], and Levy [1963d], all received September 3, 1963 and soon after, Halpern-Levy [1964a], which was actually noted as received August 29, 1963. With this work Levy became the first after Cohen himself to exploit forcing in a sustained fashion to establish a series of significant results. We attend to the AC independence results through the rest of this section; the others, on definability, were seeds for papers which are discussed in the next section. Throughout, we assume familiarity with forcing and give only scant descriptions of the models.

Cohen had shown how to "collapse" a cardinal, i.e. adjoin a generic bijection to a smaller ordinal. Feferman-Levy [1963b] started with a model of ZFC + GCH and adjoined separate generic bijections between $\aleph_n$ and $\aleph_0$ for every natural number n. In the resulting generated model the former $\aleph_\omega$ becomes the new $\aleph_1$ so that it is singular, and moreover the reals are a countable union

[10] See Moore [1988] on the origins of forcing and for what follows.

of countable sets. $AC_{\aleph_0}$ thus fails, and it is seen in a drastic way how that principle is necessary to develop Borel and Lebesgue measure. In Cohen's own exposition [1966: 143] the Feferman-Levy model was presented in some detail.

James Halpern [1961, 1964] in his 1962 Berkeley dissertation[11] had shown that the *Boolean Prime Ideal Theorem* (BPI) holds in the Mostowski [1939] model for OP, where:

(BPI) Every Boolean algebra has a prime ideal.

The significance here is that BPI had become a focal choice principle, one that implies OP. Levy [1963] had dealt with BPI, observing that it fails in his [1962a] model. Halpern's argument required a new sophistication, with a Ramsey-type partition theorem being brought into play.

When Cohen's work appeared, Halpern–Levy [1964a] showed that in the original Cohen model for the independence of the AC, OP and in fact the *Kinna-Wagner Selection Principle* (KW) holds, where:

(KW) For any set x there is a function f such that whenever $y \in x$ has at least two elements, $f(y)$ is a non-empty proper subset of y.

Kinna–Wagner [1955] had formulated KW, a weak form of MC which they showed equivalent to: For any set x there is an ordinal α and a bijection between x and a subset of $\mathcal{P}(\alpha)$. KW implies OP, and though Halpern–Levy [1964a] emphasized the new independence of AC from OP, there are two observations about their getting KW which are worth mentioning: As noted earlier, the stronger MC implies AC in ZF. Also, KW was effectively unattainable through Fraenkel-Mostowski methods, as Mostowski [1958] had observed that KW fails in every Fraenkel-Mostowski model in which the urelements are not well-orderable.

Cohen's original model for the independence of AC was the result of adjoining countably many Cohen reals and the set x consisting of these, so that x has no well-ordering in the resulting model. Halpern-Levy [1964a] in effect argued in the Cohen model with the Cohen reals acting like urelements. Cohen [1965: 40] moreover acknowledged the similarities between his AC independence result and the previous Fraenkel-Mostowski models. In any case, the revelatory Halpern-Levy work initiated the process of "transferring" consistency results with Fraenkel-Mostowski models to ZF consistency results via forcing by correlating urelements with generic sets.

For the BPI, there was no routine transfer of the Halpern [1964] independence of AC from BPI. Levy saw the need for a strengthened, "tree" Ramsey-type partition theorem to effect a ZF independence result. Halpern-Läuchli [1966] then duly established this result. Finally, Halpern-Levy [1971] by 1966

[11] Halpern [1964] wrote that his dissertation was written under the "supervision of" Levy; Levy visited Berkeley the academic years 1959-1961.

had established that in Cohen's original model, BPI holds. This Halpern-Läuchli-Levy collaboration established a new level of sophistication in effecting a transfer from the Fraenkel-Mostowski context to the Cohen one. Work at this level would soon be pursued by Jech, Sochor, Pincus, and others, and the Halpern-Läuchli partition theorem would lead to an important extension by Richard Laver [1984], one also applied to forcing.

7 ZF Definability

The abstracts Levy [1963a, 1963c, 1963d] had to do with formal definability, and the papers Levy [1965a, 1970] provided extended accounts in a context of appropriate generality for the proofs. Levy probed the limits of ZFC definability, establishing consistency results about definable sets of reals and well-orderings and in descriptive set theory.

Heralded by Levy [1963a], Levy [1965a] established the relative consistency of ZFC + GCH together with there being a non-constructible real yet every definable set is constructible. Here, "definable" meant the broad notion of hereditarily ordinal-definable. A set x is *ordinal-definable iff* there is a formula $\psi(v_0, \ldots, v_n)$ in the free variables as displayed and ordinals $\alpha_1, \ldots, \alpha_n$ such that $x = \{y \mid \psi[y, \alpha_1, \ldots, \alpha_n]\}$. A set x is *hereditarily ordinal definable iff* the transitive closure of $\{x\}$ is ordinal definable. The ordinal-definable sets were introduced by Gödel [1946] as mentioned in §2, and their theory was developed by John Myhill and Dana Scott by 1964 (cf. Myhill-Scott [1971]) with explicit appeal to the ZF Reflection Principle, with which one can replace the informal satisfaction of $x = \{y \mid \psi[y, \alpha_1, \ldots, \alpha_n]\}$ by: for some V_α and ordinals $\alpha_1, \ldots, \alpha_n \in V_\alpha$, $x = \{y \in V_\alpha \mid \psi^{V_\alpha}[y, \alpha_1, \ldots, \alpha_n]\}$. OD denotes the (thus definable) class of ordinal-definable sets, and HOD, the class of hereditarily ordinal-definable sets.

Feferman [1965, 1965a] had shown that in Cohen's model which is the result of starting from a model of $V = L$ and adjoining a Cohen real there is no set-theoretically definable well-ordering of the reals, i.e. no formula in two free variables that defines such a well-ordering. Levy [1965a] showed that the model actually satisfies $V \neq L = \text{HOD}$, so that in particular there is no definable well-ordering of the reals even if ground model parameters are allowed in the definition. The crux is that the partial order of conditions for adjoining a Cohen real is *homogeneous* in the sense that for any pair p, q of conditions there is an automorphism e of the partial order such that p and $e(q)$ are compatible. Hence, for a formula $\varphi(v_1, \ldots, v_n)$ of the forcing language and sets $x_1, \ldots, x_n$ in the ground model, any condition forces $\varphi(\check{x}_1, \ldots, \check{x}_n)$ exactly when all conditions do. That HOD $= L$ follows by induction on rank. Levy's appeal to homogeneity and automorphism, related to his earlier work with urelements and realigned by the work of Cohen and Feferman, would become a basic motif that connects forcing with definability.

In an eventual sequel, Levy [1970] worked out his main delimitative results. He first considered Cohen's model which is the result of starting with a model of $V = L$ and collapsing ω_1^L, i.e. adjoining a generic bijection between ω and ω_1^L. As before one has $V \neq L = \mathrm{HOD}$, but this now easily implies that *every* OD *well-ordering of reals is at most countable.* This confirmed an announcement in Levy [1963a].

With his next theorem Levy [1970] provided an important delimitation for descriptive set theory, confirming an announcement in Levy [1963c]. Classical descriptive set theory, in its probing into the first levels of the projective hierarchy, had pushed against the limits of axiomatic set theory.[12] Levy presumably had assimilated this work in large part from John Addison at Berkeley (cf. Addison [1959]), but in any case Levy quickly saw how to apply forcing to illuminate the central issue of uniformization. For binary relations A and B,

A is *uniformized* by B *iff*

$$B \subseteq A \ \wedge \ \forall x(\exists y(\langle x, y\rangle \in A) \longrightarrow \exists! y(\langle x, y\rangle \in B)) .$$

$\exists!$ abbreviates the formalizable "there exists exactly one", and so this asserts that A can be refined to a function B. That every relation can be uniformized is a restatement of the Axiom of Choice. A high point of classical descriptive set theory was the result of Motokiti Kondô [1937] that, in terms of the projective hierarchy, every $\mathbf{\Pi}^1_1$ relation on reals can uniformized by a $\mathbf{\Pi}^1_1$ relation. This implied via projection that every $\mathbf{\Sigma}^1_2$ relation on reals can be uniformized by a $\mathbf{\Sigma}^1_2$ relation, and by looking at complements that not every $\mathbf{\Pi}^1_2$ relation on reals can be uniformized by a $\mathbf{\Pi}^1_2$ relation. Whether every $\mathbf{\Pi}^1_2$ relation on reals can be uniformized by a projective relation had remained open. Bringing in axiomatics Addison [1959] established that assuming $V = L$, for any $n \geq 2$ every $\mathbf{\Sigma}^1_n$ relation on reals can be uniformized by a $\mathbf{\Sigma}^1_n$ relation, so that in particular every $\mathbf{\Pi}^1_2$ relation on reals, being $\mathbf{\Sigma}^1_3$, can be uniformized by a $\mathbf{\Sigma}^1_3$ relation. In contradistinction, Levy established the relative consistency of there being a $\mathbf{\Pi}^1_2$ on reals that cannot be uniformized by any projective relation.

Levy considered Cohen's model which is the result of starting with a model of $V = L$ and adjoining (a sequence of) ω_1^L Cohen reals. Taking the cue from ordinal-definability, say that a set x is *real-ordinal-definable iff* there is a formula $\psi(v_0, \ldots, v_n, v_{n+1})$ in the free variables displayed, a V_α, a real $r \in V_\alpha$, and ordinals $\alpha_1, \ldots, \alpha_n \in V_\alpha$ such that $x = \{y \in V_\alpha \mid \psi^{V_\alpha}[y, \alpha_1, \ldots, \alpha_n, r]\}$. That is, a real parameter is to be allowed in the definition. Levy considered the relation A on reals defined by: $\langle f, g\rangle \in A$ *iff* $g \notin L[f]$. This relation, formulated in terms of his notion of relative constructibility, is $\mathbf{\Pi}^1_2$ by an elaboration of an argument in Addison [1959]. Suppose now that B is a real-ordinal-definable

[12]See Kanamori [2005: §§12–14] for the background and basic concepts of descriptive set theory.

relation on reals. By a definability argument the real parameter r in the definition can be taken to be one coding countably many of the Cohen reals. There is then an s such that $\langle r, s\rangle \in A$. If however there were an s such that $\langle r, s\rangle \in B$, then s, like B, would be real-ordinal-definable with real parameter r. But then, a homogeneity argument shows that $s \in L[r]$. Consequently, no real-ordinal-definable relation, and consequently no projective relation, can uniformize A.

8 Levy Collapse

The theory of large cardinals was revitalized by pivotal results in the early 1960's, and with Cohen's forcing large cardinals would enter the mainstream of set theory by providing hypotheses and methods to analyze strong set-theoretic propositions. Levy's earlier work on reflection principles had established a central place for Mahlo cardinals; in the post-Cohen era Levy made basic contributions to the fast growing theory of large cardinals.

In the last of the 1963 abstracts Levy [1963d], Levy announced a result that depended on what we now call the Levy collapse. In general terms, for infinite regular cardinals $\lambda < \kappa$, $\mathrm{Col}(\lambda, \kappa)$ is the partial order for adjoining a κ-sequence of surjections $\lambda \to \alpha$ for $\alpha < \kappa$. If κ is inaccessible, then $\mathrm{Col}(\lambda, \kappa)$ has the κ-chain condition;[13] hence, κ becomes the successor λ^+ of λ in the generic extension. The forcing with $\mathrm{Col}(\lambda, \kappa)$ is then called a *Levy collapse* of κ to λ^+. Already in the first flush of forcing Levy [1963d, 1970] used the Levy collapse of an inaccessible cardinal to ω_1 to establish the relative consistency of:

(*) Every real-ordinal-definable well-ordering of reals is at most countable.

Forcing ω_1 to be countable had led to the consistency of every OD well-ordering of reals being at most countable; forcing every α below an inaccessible to be countable provides enough closure to achieve (*). Levy also considered the proposition:

(**) Every real-ordinal definable set of reals is either countable or of cardinality $2^{\aleph_0}$.

The Levy collapse of an inaccessible to ω_1 entails CH so that (**) is vacuous, but Levy showed that in any further extension where many Cohen reals adjoined (**) continues to hold, this already inherent in the early abstract [1963d].

In deploying an inaccessible cardinal Levy was a pioneer in taking the modern approach to large cardinals: They are not novel hypotheses burdened by

[13] Actually, that $\mathrm{Col}(\lambda, \kappa)$ has the κ-chain condition only requires, by a so-called delta-system argument, that $\alpha < \kappa$ implies that $\alpha^{<\lambda} < \kappa$. Full inaccessibility is typically required in other parts of an argument using the Levy collapse.

ontological commitment but are the repository of means for carrying out mathematical arguments. Cohen [1966: 147] acknowledged this use of an inaccessible cardinal. As pointed out by Levy [1970: 131-2], either (*) or (**) implies that the ω_1 (of the universe) is an inaccessible cardinal in the sense of L. So, Levy's work was party to the first instance of an important phenomenon in set theory, the derivation of equiconsistency results based on the complementary methods of forcing and inner models.[14] After this heady introduction the Levy collapse would become standard fare in the theory of large cardinals, the way to render a large cardinal accessible yet still with substantial properties to establish the relative consistency of strong combinatorial propositions low in the cumulative hierarchy.

Levy's model was used by Robert Solovay to establish a now famous relative consistency result. Solovay played a prominent role in the forging of forcing as a general method, and he above all in this period raised the level of sophistication of set theory across its breadth from forcing to large cardinals. Already in the spring of 1964 Solovay [1965a, 1970] exposed what standard of argument was possible when showing that if an inaccessible cardinal is Levy collapsed to ω_1, *every real-ordinal-definable set of reals is Lebesgue measurable*, and proceeding to the corresponding inner model HROD of the hereditarily real-ordinal-definable sets, that

$$\text{HROD} \models \text{DC} + \text{“Every set of reals is Lebesgue measurable”}.$$

That this model satisfies Dependent Choices bolsters it as a *bona fide* one for mathematical analysis. Solovay thus illuminated the classical measure problem of Henri Lebesgue with the modern technique of forcing. Solovay also showed that the reals in this HROD model have several other substantial properties, one being the perfect set property: *Every set of reals is countable or else has a perfect subset.* This refined Levy's result with (**) above, any perfect set of reals having cardinality $2^{\aleph_0}$, and established the equiconsistency of the perfect set property with the existence of an inaccessible cardinal.[15] For quite some years, it was speculated that an inaccessible cardinal can be avoided for getting all sets of reals to be Lebesgue measurable. However, in 1979 Shelah [1984] established that DC + "Every set of reals is Lebesgue measurable" implies that ω_1 is inaccessible in L, vindicating Solovay's use of the Levy collapse for the Lebesgue measurability result.

Solovay [1970: 2] announced a joint result with Levy which eventually appeared in Levy-Solovay [1972]. Levy and Solovay built on the structure uncovered by Solovay in the Levy collapse model to establish a further "regularity"

[14] See Kanamori [2005: 135ff].

[15] Solovay [1970: 45ff] wrote: "Our proof that every [real-ordinal-definable] subset of [the reals] is countable or contains a perfect subset is, essentially, a slight refinement of the following result of Levy [1970]: Every uncountable [real-ordinal-definable] subset of [the reals] has power $2^{\aleph_0}$." Compare Levy [1970: 140ff].

property about sets of reals, being the union of an $\aleph_1$-sequence of Borel sets. In the classical investigation of the projective hierarchy, though the second level $\mathbf{\Sigma}^1_2$ seemed complicated, Sierpiński [1925] had established that *every* $\mathbf{\Sigma}^1_2$ *set of reals is the union of* $\aleph_1$ *Borel sets.* Of course, if CH holds, then every set of reals is the union of $\aleph_1$ Borel sets, namely the singletons of its members. It came to light with the emergence of Martin's Axiom that ¬CH together the converse of the Sierpiński result, that *every union of* $\aleph_1$ *Borel sets is* $\mathbf{\Sigma}^1_2$, is relatively consistent (Martin-Solovay [1970: §3]).

Considering arbitrary sets of reals, Levy and Solovay first noted with a simple construction via transfinite recursion that ZFC + ¬CH implies that there is a set of reals which is not the union of any $\aleph_1$ Borel sets, and then went on to show that some substantial use of AC is necessary. As with Levy's (**), Solovay [1970] had shown that, with V_1 the Levy collapse model and V_2 a further extension where many Cohen reals are adjoined, the propositions about Lebesgue measurability and so forth for real-ordinal-definable sets that hold in V_1 continue to hold in V_2, thereby establishing the further relative consistency of having a large continuum. Levy-Solovay [1972] showed that in any such V_2, every real-ordinal-definable set of reals is the union of $\aleph_1$ Borel sets. Specifically, such a set of reals is the union of Borel sets coded in V_1, and with CH holding in V_1, there are only $\aleph_1$ such Borel sets. Thus, Levy and Solovay established the relative consistency of ZFC + ¬CH + "Every real-ordinal-definable sets of reals is the union of $\aleph_1$ Borel sets". Passing to the inner HROD model Levy and Solovay then established the relative consistency of DC + "Every well-ordering of reals is at most countable" + "Every set of reals is the union of $\aleph_1$ Borel sets". In this way the property of being the union of $\aleph_1$ Borel sets was adjoined to the regularity properties of sets of reals illuminated by the Levy collapse model.

9 Measurable and Indescribable Cardinals

In addition to their collaborative work on the Levy collapse model, Levy [1964b] and Solovay [1965] independently established a general result about large cardinals that would become much cited in connection with the Continuum Problem. As is well-known, Gödel [1947: 520] had speculated that large cardinal postulations might decide CH. He himself took CH to be false and made remarks amounting to the observation that those large cardinals consistent with $V = L$ cannot disprove CH. Addressing this issue, Levy-Solovay [1967] showed that measurable cardinals κ remain measurable in "mild" forcing extensions, those via partial orders of cardinality $< \kappa$. That this betokened what would become a widely applied observation, that inaccessible large cardinals retain their characteristic properties in mild forcing extensions, is often regarded as springing from Levy-Solovay [1967] though they themselves wrote that this was well-known for many large cardinals. At the time there was a particular point in that Scott [1961] had dramatically established that the existence of a measurable cardinal contradicts $V = L$, and Levy-Solovay [1967] pointed out that measurable cardinals, though loosened from the moorings of constructibility, cannot decide CH and other issues like Suslin's Hypothesis. In years to come, the growing success of the theory of large cardinals led to more and more allusions to "Gödel's Program". The Levy-Solovay result would be consistently cited as a watershed, if only to point to a delimitation to be superseded by other, more subtle invocations of large cardinals in connection with the Continuum Problem.

Levy's last major contribution to the theory of large cardinals dealt with natural extensions of his earlier reflection principles (cf. §2) set in a higher order context of Π^m_n and Σ^m_n formulas.[16] William Hanf and Scott in their abstract [1961] considered higher-order reflection properties for structures $\langle V_\kappa, \in, R\rangle$ where κ is a cardinal and $R \subseteq V_\kappa$ and thereby provided a hierarchical scheme for large cardinals. For Q either Π^m_n or Σ^m_n,

κ *is Q-indescribable iff* for any $R \subseteq V_\kappa$ and Q sentence φ such that $\langle V_\kappa, \in, R\rangle \models \varphi$, there is an $\alpha < \kappa$ such that $\langle V_\alpha, \in, R \cap V_\alpha\rangle \models \varphi$.

Including $R \subseteq V_\kappa$ suffices to bolster the concept to accommodate general relational structures; as V_κ is closed under pairing, the definition is equivalent to one where R is replaced by any finite number of finitary relations.[17] Hanf-Scott [1961] observed that, with π^m_n the least Π^m_n-indescribable cardinal and σ^m_n the least Σ^m_n-indescribable cardinal, for $m > 0$: $\pi^m_n < \pi^m_{n+1}$ and if $n > 0$, then π^m_n is not Σ^m_n-indescribable. They also pointed out that the Π^1_1-indescribable cardinals are exactly the weakly compact cardinals and that measurable cardinals are Π^2_1-indescribable. This provided probably the earliest proof that below a measurable cardinal there are many weakly compact cardinals. Vaught [1963] subsequently pointed out that below a measurable cardinal there is cardinal

[16]Higher-order languages have typed variables of every finite type (or order), quantifications of these, and beyond the atomic formulas specified by the language, $X \in Y$ and $X = Y$ for any typed variables X and Y. In the intended semantics, if D is the domain of a structure, type 1 variables play the usual role of first-order variables, type 2 variables range over $\mathcal{P}(D)$, and generally, type $i+1$ variables range over $\mathcal{P}^i(D)$ where $\mathcal{P}^i$ denotes i iterations of the power set operation. A formula is Π^m_n *iff* it starts with a block of universal quantifiers of type $m+1$ variables, followed by a block of existential quantifiers of type $m+1$ variables, and so forth with at most n blocks in all, followed afterward by a formula containing variables of type at most $m+1$ and quantified variables of type at most m. A formula is Σ^m_n *iff* it starts instead with existential quantifiers. Of course, formulas containing only type 1 variables can be construed as the usual first-order formulas. This classification of formulas is cumulative because of the "at most": any Π^m_n or Σ^m_n formula is also Π^r_s and Σ^r_s for any $r > m$, or $r = m$ and $s > n$.

[17]Hanf-Scott [1961] formulated their concept for inaccessible κ and with κ in place of V_κ, but the difference is inessential as the R can code V_κ.

Π^m_n-indescribable for every $m, n \in \omega$. The evident connection between Levy's earlier reflection principles and the indescribable cardinals had an interconnecting node in the work of Bernays [1961], who had extended Levy's principles by positing, in effect, the Π^1_n-indescribability for every $n \in \omega$ of the class of all ordinals.

Levy [1971a] carried out a systematic study of the sizes of indescribable cardinals, extending aspects of a combinatorial study of large cardinals in Keisler-Tarski [1964]. The starting point of Levy's approach was that various large cardinal properties are not only attributable to cardinals, but to their subsets. For $X \subseteq \kappa$ and Q either Π^m_n or Σ^m_n,

X is *Q-indescribable in κ iff* for any $R \subseteq V_\kappa$ and Q sentence φ such that $\langle V_\kappa, \in, R\rangle \models \varphi$, there is an $\alpha \in X$ such that $\langle V_\alpha, \in, R \cap V_\alpha\rangle \models \varphi$.

This leads to the consideration of $\{X \subseteq \kappa \mid \kappa - X \text{ is not } Q\text{-indescribable in } \kappa\}$, which when κ is Q-indescribable is a (proper) filter, the *Q-indescribable filter over κ*.[18]

Using universal satisfaction formulas Levy showed that these definable filters have a crucial property: *For $m, n > 0$ and Q either Π^m_n or Σ^m_n, the Q-indescribable filter over κ is normal.* Recall that a filter F over a cardinal κ is *normal iff* it is closed under diagonal intersections, i.e. whenever $\{X_\alpha \mid \alpha < \kappa\} \subseteq F$, $\{\xi < \kappa \mid \xi \in \bigcap_{\alpha<\xi} X_\alpha\} \in F$. The previously known normal filters were the closed unbounded filters over regular uncountable cardinals and normal ultrafilters found over a measurable cardinal. Levy further established that $\{\alpha < \kappa \mid \alpha \text{ is } P\text{-indescribable in } \alpha\}$ is in the Q-indescribable filter over κ, where: (a) P is Π^1_n and Q is Π^1_{n+1}; or (b) $m > 1$ and $n > 0$, P is Σ^m_n, and Q is Π^m_n; or (c) $m > 1$ and $n > 0$, P is Π^m_n, and Q is Σ^m_n. These various results showed in concert that, just as normal ultrafilters over a measurable cardinal provide intrinsic senses to how large measurable cardinals are as had been shown in Keisler-Tarski [1964], e.g. normal ultrafilters are closed under Mahlo's Operation, so too do indescribable cardinals have inherent transcendence over smaller cardinals, specifically those in the indescribable hierarchy itself. The technique of ascribing large cardinal properties of cardinals also to their subsets has become an important part of large cardinal theory and has been used in particular by James Baumgartner [1975] to establish important hierarchical results about the n-subtle and n-ineffable cardinals.

10 Capstones

In the later 1960's, Levy capped his investigations in various directions with papers reflective of previous themes and techniques but also distinctive in how

[18] Levy himself called the members of this filter *weakly Q-enforceable* at κ, but we follow the formulation in Kanamori [2005: §6].

they resolve basic issues in axiomatics.

Levy's last result applying the Fraenkel-Mostowski method concerned Cantor's very notion of cardinality. The problem of cardinal definability in set theory is how to assign to every set x a set $|x|$ such that for every x, y we have $|x| = |y|$ *iff* $x \approx y$, i.e. there is a bijection between x and y. Of course, with AC the initial (von Neumann) ordinals construed as Cantor's alephs serve as such $|x|$. But even without AC, one can use the "trick" of Scott [1955] to formulate $|x|$ as the set of sets of least rank bijective with x. Levy [1969], in a 1966 conference proceedings, established that relative to ZF, ZF – Foundation + "There is no set-theoretic term τ such that for every x, y we have $\tau(x) = \tau(y)$ *iff* $x \approx y$" is consistent in several strong senses.[19] In this distinctive setting without Foundation the interplay between urelement constructions and forcing is not pertinent. It had been known for over a decade that the Fraenkel-Mostowski method with urelements can be recast, following Specker [1957] and Mendelson [1956], in ZF – Foundation with sets $a = \{a\}$ in the role of urelements. Levy ultimately relied on this recasting, but worked directly with urelements and automorphisms. In one model he used $\aleph_\omega$ urelements and generated an inner model in which they become a proper class; in another, he proceeded similarly but started from Mostowski's model for OP.[20] With the appearance of the forcing method, one might have thought that Fraenkel-Mostowski methods would be superseded, but in the years to come, there would be a continuing cottage industry investigating Fraenkel-Mostowski models as intrinsically interesting constructions in their own right.[21]

Levy and Georg Kreisel in their [1968] provided a detailed exposition that established a central place for proof-theoretic reflection principles in the comparative investigation of theories. Levy brought together his work on the transcendence of theories through set-theoretic reflection principles, and the inimitable Kreisel, whose hand is evident in the sections with the many italicizations, brought to bear is his initiatives in the proof theory of arithmetic and analysis. The main unifying motif was the proof-theoretic *Uniform Reflection Principle*:

$$\text{(URP(S))} \qquad \forall p \forall n(\mathrm{Prov}_{\mathrm{S}}(p, s(\ulcorner\varphi\urcorner, n)) \longrightarrow \varphi(n)).$$

Intended for theories sufficient to carry out Gödel numbering, $\mathrm{Prov}_{\mathrm{S}}(x, y)$ is to assert that x is the Gödel number of a proof in the theory S of the formula with

[19]This was also done, in a strong sense, by Robert Gauntt [1967].

[20]Pincus [1974] later addressed the issue of cardinal *representatives*, i.e. having a set-theoretic term τ such that for every x, y we have $\tau(x) = \tau(y)$ *iff* $x \approx y$, and moreover $|x| \approx x$. Transferring from Mostowski's model for OP Pincus showed that relative to ZF it is consistent to have ZF + "There are no cardinal representatives". Rather surprisingly, he also showed by an iterated forcing argument that relative to ZF it is consistent to have ZF + $\neg$AC + "There are cardinal representatives".

[21]See Howard-Rubin [1998].

Gödel number y; $s(\ulcorner\varphi\urcorner, n)$ is the Gödel number of the sentence obtained by substituting the numeral of the natural number n for the one free variable of φ; and finally, the uniformity has to do with having the parametrization with the numerical variable n. URP(S) is an assertion of soundness; instantiating to a φ refutable in S, URP(S) implies that something is not provable and hence the formal consistency of S. As Kreisel-Levy [1968: §1] pointed out, URP(S) actually subsumes both the assertion of ω-consistency for S and a general form of induction.

Kreisel and Levy established a strong, general result about how URP(S) leads to transcendence over S in terms of quantifier complexity: *If* U *is a theory in the same language as* S *and* URP(S) *is a theorem of* U, *then for no set* Σ *consisting of sentences of bounded quantifier complexity are the theorems of* U *provable in* S + Σ. For set theories, Σ_n in the Levy hierarchy typifies a set of formulas of bounded complexity. Levy [1959a] had already announced a result that implied that for any natural number n, no consistent extension of ZF can be obtained by adjoining to Zermelo set theory any set of Σ_n sentences. As noted, various results from Levy's previous work [1958b, 1960b, 1960d], which had the above form except that the Σ was a finite set of sentences, could now be strengthened. The general result was moreover applicable to Peano Arithmetic, Second-Order Arithmetic (Analysis), and the like to show that these theories cannot be axiomatized over weaker theories using any set of axioms of bounded quantifier complexity.

In the last sections, Kreisel-Levy [1968] established direct connections between URP(S) and schemas of transfinite induction. By the classical work of Gerhard Gentzen [1943], Peano Arithmetic (PA) establishes the coded schema of transfinite induction up to any particular ordinal less than ϵ_0, the least ordinal α such that $\alpha^\omega = \alpha$ in ordinal arithmetic, yet PA does not establish the schema of transfinite induction up to ϵ_0 itself. Kreisel and Levy showed that over PA, URP(PA) is equivalent to transfinite induction up to ϵ_0, and that over Second-Order Arithmetic Z_1, URP(Z_1) is equivalent to transfinite induction up to ϵ_0. These results provided elegant characterizations that connect two formulations of the consistency of well-known theories.

Levy [1964c, 1971b] investigated the logical complexity, in terms of his hierarchy of formulas, of basic statements of set theory like AC, GCH, and $V = L$. His result on AC typifies the articulation and argumentation, made possible by forcing. The Axiom of Choice is evidently Π_2. Levy showed that AC is not Σ_2 in the following strong sense: *For any* Σ_2 *sentence* σ, *if* ZF $\vdash \sigma \rightarrow$ AC, *then* ZF $\vdash \neg\sigma$. The following is the argument in brief:

Suppose that σ is Σ_2, say $\exists x \forall y \varphi(x, y)$ with φ being Σ_0, and ZF $\vdash \sigma \rightarrow$ AC. The following can then be formalized to establish ZF $\vdash \neg\sigma$: Assume to the contrary that there is a set x_0 such that $\forall y \varphi(x_0, y)$. Let M be a transitive structure with $x_0 \in M$ and modeling enough of set theory to construct a transitive forcing extension N in which AC fails and the ZF axioms that went into a proof of $\sigma \to$ AC hold. Since $\forall y\varphi(x_0, y)$ holds (in the universe) and φ is Σ_0, $\forall y\varphi(x_0, y)$ and hence σ holds in N. But then, this contradicts having that ZF proof of $\sigma \to$ AC.

This argument not only featured forcing as a model-theoretic method within a proof but also forcing over uncountable structures, for the x_0 above could be arbitrary. Modern set theory would come to incorporate many such tailored uses of forcing, and Levy's application to definability was a remarkably early instance.

Levy [1969a] provided an analysis of Π_2 statements of set theory in a different direction, one that addresses effectivity in terms of witnessing terms. Suppose that $\forall x\exists y\chi(x, y)$ is Π_2, with $\chi(x, y)$ being Σ_0, and recall that $\mathrm{tc}(x)$ denotes the transitive closure of x. Levy established that if ZF $\vdash \forall x\exists y\chi(x, y)$, then there is a set-theoretic term $\tau(u)$ such that

$$\mathrm{ZF} \vdash \forall x(\exists \text{ finite } u \subseteq \mathrm{tc}(x) \ \wedge \ \chi(x, \tau(u))).$$

Note that one cannot do much better for the AC assertion $\forall x\exists y(x = \emptyset \vee y \in x)$. Levy also established that if ZFC $\vdash \forall x\exists y\chi(x, y)$, then there is a set-theoretic term $\tau(u)$ such that

$$\mathrm{ZF} \vdash \forall x\forall r(r \text{ is a well-ordering of } \mathrm{tc}(x) \longrightarrow \chi(x, \tau(r))).$$

In his final article [1974] Levy came full circle back to the bedrock of the main comprehension schemas of ZF to investigate their forms anew. He began with the ZF axioms as most often given, as bi-conditionals with e.g. $\forall u\exists y\forall x(x \in y \longleftrightarrow x \subseteq u)$ for the Power Set Axiom. He first addressed the issue of the parameters allowed in the Separation and Replacement schemas. Let S_0 denote the Separation schema restricted to formulas $\varphi(x)$ with one free variable x, $\forall u\exists y\forall x(x \in y \longleftrightarrow x \in u \wedge \varphi(x))$. By a clever coding argument Levy established the positive result that, over a weak subtheory (Extensionality, Pairs, Union, and a weak form of Power Set), S_0 implies full Separation. He established an analogous result for Replacement. Hence, those universal quantifications of parameters, distracting when learning or teaching set theory, are not formally necessary after all.

Levy's main, negative results addressed another issue of self-refinement in axiomatics and showed that his aforementioned positive result is reasonably sharp. In the presence of Separation the generative axioms are sometimes given parsimoniously in a weaker, conditional form, e.g. $\forall u\exists y\forall x(x \subseteq u \longrightarrow x \in y)$ for the Power Set Axiom. With his positive result, the set T consisting of the usual ZF axioms, but with the Separation schema replaced by S_0 and the Replacement schema replaced by the conditional version, is an axiomatization of ZF. Levy established that full Separation is not a consequence of T if the Power Set axiom is weakened to the conditional form. He also established the analogous results for the conditional version of Union and the conditional version of Pairing.

Levy established these delimitative results by, in effect, taking Cohen's original AC independence model and building appropriate submodels. In this he appealed to the Halpern-Levy [1971] work on the Cohen model, the work that effected the first substantial transfer from the Fraenkel-Mostowski context to the Cohen context. Levy's sophisticated results on the independence of Separation are a fitting coda, one that resonates with the work Fraenkel [1922], who long ago and far away, steeped in the Hilbertian axiomatic tradition, established the independence of Separation from Zermelo's other axioms.

11 Envoi

Looking back over Levy's researches in set theory, we see a steady and in fact increasing exploitation of model-theoretic reflection and the method of forcing to establish substantial results about definability and axiomatizations. Levy often saw and developed potentialities after an initial ground-breaking move made by others, and had a way of establishing a full context with systematic, magisterial results. With his work set theory reached a new plateau in the direction of understanding the scope and limits of formal expressibility and derivability. With this assimilated, set theory would move forward over a broad range from the analysis of fine structure to a wealth of objectifications and principles provided by large cardinal hypotheses, becoming infused with more and more combinatorial arguments as well as sophisticated techniques involving forcing and inner models.

Around 1970 Levy turned to the writing of books, works that would establish a broad standard of understanding about set theory. The classic *Foundations of Set Theory* by Fraenkel and Bar-Hillel [1958] had become outdated because of the many advances made in the 1960's, and so a "second revised edition" Fraenkel–Bar-Hillel–Levy [1973] was brought out. Fraenkel was by then deceased, and Levy in fact carried out an almost complete rewriting of the second chapter. One section was published separately as Levy [1976]. Throughout the discussion of the axiomatic foundations one sees how the subject has become more elucidated by Levy's own work.

Levy's distinctive book *Basic Set Theory* [1979], largely written when he was a visiting professor at Yale University 1971-1972 and at the University of California at Los Angeles 1976-1977, provided a systematic presentation of the broad swath of set theory between elementary beginnings and advanced topics. Levy deliberately set out the extent of set theory before the use of model-theoretic methods and forcing, working out the extensive combinatorial development in a classical setting as rigorized by axiomatic foundations. In a way, it is quite remarkable that Levy forestalled the inclusion of most of his own work by insisting on this middle way. On the other hand, the book is a singular achievement of detailed exposition about what there is in set theory up to the use of the satisfaction predicate. The account of trees is typical, on the one hand a bit idiosyncratic in dealing with generalities but on the other hand broaching an interesting concept, that of a *thin* tree: Trees that have been studied on uncountable cardinals $\kappa > \omega_1$ usually have the property that their αth level has cardinality 2^α, but Levy raised the issue about having cardinality at most $|\alpha|$. Throughout the book, the specialist is treated to a reckoning of the historical sources for the concepts. And the student finds full and patient treatment of topics on which other texts might leave one queasy, like the set theorist's view of the reals. Descriptive set theorists from early on converted from the traditional construal of the real numbers as the continuum to the function space ${}^\omega\omega$ of functions: $\omega \to \omega$, and Levy explained in extensive detail the topological and measure-theoretic connections among the various "real" spaces. What is perhaps most notable is the appendix to the book, where at last Levy's own rigorous and axiomatic approach to set theory casts a telling light. There, he allayed another source of queasiness in accounts of set theory by showing, in some of the most thoroughgoing arguments in the book, that introduced class terms can be formally eliminated in regression back to the primitive language of set theory.

In the development of a mathematical field, a modest turn of events sometimes has an unforseen effect and achieves a folkloric distinction. Shelah had considerably developed his general concept of *proper forcing* by the early 1980's, a concept amenable to iteration schemes and having remarkably wide applications. Shelah lectured at the Hebrew University on proper forcing, and Levy took systematic notes. These notes, refined and edited, eventually became the bulk of the first three chapters of the monograph Shelah [1982] and of the subsequent book Shelah [1998] as therein acknowledged. For a generation of set theorists the Levy exposition was the entré into proper forcing; once drawn in, the return of Shelah's inimitable hand in subsequent chapters led to new realizations.

In later years, Levy wrote two texts [1985, 1997] in Hebrew on mathematical logic. The writing of books is an important venture for the advancement of mathematical fields, and as one who has written a book knows well, it is a difficult and time-consuming undertaking, especially when one aspires to codify mathematical knowledge over a broad range and to make its many facets available to succeeding generations.

Turning at last to his teaching and administrative work at the Hebrew University, there is a remarkable legacy of renowned students, as listed below. Levy himself has served as the Dean of the Faculty of Science and the Chairman of Institute of Mathematics and Computer Science, among other positions both in the university and with the academic union. And, as mentioned before, his student Menachem Magidor is currently president of the university.

Soon after his 60th birthday, a 1996 issue of the *Archive for Mathematical Logic* (vol.35, no.5-6), was dedicated to Levy, with the following words in the dedication (p.279):

> Azriel, besides being the important mathematician he is, is also a unique human being. His friends know him as the epitome of wisdom. He can always be approached for good advice, which is given without any personal interest, but purely out of a desire to help. His contribution to the public are innumerable. In any capacity he has held—University administrator, in the educational system in Israel, as a member of usual important editorial boards—you could always rely on his common sense, wisdom and devotion. Azriel is a symbol of intellectual honesty and integrity. His former students will always remember him as a devoted and inspiring teacher.

To this we add the words of Ecclesiastes 9:7-9 (King James Version):

> Go thy way, eat thy bread with joy, and drink thy wine with a merry heart; for God new accepteth they works.
>
> Let thy garments be always white; and let thy head lack no ointment.
>
> Live joyfully with the wife whom thou lovest all the days . . . which he hath given thee under the sun . . .

Doctoral Students of Ariel Levy

Dov Gabbay, *Non-classical logics*, 1969.

Shlomo Vinner, *Some problems in first order predicate calculus with numerical quantifier*, 1971. Second supervisor: Haim Gaifman.

Gadi Moran, *Size direction games over the real line*, 1972.

Menachem Magidor, *On supercompact cardinals*, 1972.

Gershon Sageev, *An independence result concerning the axiom of choice*, 1973.

Moti Gitik, *All uncountable cardinals can be singular*, 1979. Second supervisor: Menachem Magidor.

Uri Abraham, *Isomorphisms of Aronszajn trees and forcing without the generalized continuum hypothesis*, 1979. Second supervisor: Saharon Shelah.

Aharon Beller, *Applications of Jensen's coding technique*, 1980.

Ami Litman, *Combinatorial characterization of definable properties in the constructible universe*, 1981. Second supervisor: Menachem Magidor.

Hannah Perl, *Teaching mathematics in high school with graphing calculators*, 2002.

Publications of Azriel Levy

1957 Indépendance conditionnelle de $V = L$ et d'axiomes qui se rattachent au système de M. Gödel, *Comptes Rendus Hebdomadaires des Séances de l'Académie des Sciences, Paris 245*, 1582-1583.

1958 A note on definitions of finiteness, *The Bulletin of the Research Council of Israel, Section F: Mathematics and Physics 7F*, 83-84.

1958a The independence of various definitions of finiteness, *Fundamenta Mathematicae 46*, 1-13.

1958b Comparison of subtheories, *Proceedings of the American Mathematical Society 9*, 942-945.

1959 On Ackermann's set theory, *The Journal of Symbolic Logic 24*, 154-166.

1959a A hierarchy of formulae of set theory (abstract), *Notices of the American Mathematical Society 6*, 826.

1960 On models of set theory with urelements, *Bulletin de l'Academie Polonaise des Sciences, Serie des Sciences Mathématiques, Astronomiques et Physiques 8*, 463-465.

1960a A generalization of Gödel's notion of constructibility, *The Journal of Symbolic Logic 25*, 147-155.

1960b Axiom schemata of strong infinity in axiomatic set theory, *Pacific Journal of Mathematics 10*, 223-238; reprinted in *Mengenlehre*, Wissensschaftliche Buchgesellschaft Darmstadt, 1979, 238-253.

1960c Principles of reflection in axiomatic set theory, *Fundamenta Mathematicae 49*, 1-10.

1960d On a spectrum of set theories, *Illinois Journal of Mathematics 4*, 413-424.

1961 (with Robert L. Vaught) Principles of partial reflection in the set theories of Zermelo and Ackermann, *Pacific Journal of Mathematics 11*, 1045-1062.

1961a Axiomatization of induced theories, *Proceedings of the American Mathematical Society 12*, 251-253.

1961b Comparing the axioms of local and universal choice, in: Yehoshua Bar-Hillel, E.I.J. Poznanski, Michael O. Rabin, and Abraham Robinson (editors), *Essays on the Foundations of Mathematics*, dedicated to Professor A.A. Fraenkel on his 70th Birthday, Magnes Press, Jerusalem, 83-90.

1962 On the principles of reflection in axiomatic set theory, in: Ernest Nagel, Patrick Suppes, and Alfred Tarski (editors), *Logic, Methodology and Philosophy of Science*, Proceedings of the 1960 International Congress, Stanford University Press, Stanford, 87-93.

1962a Axioms of multiple choice, *Fundamenta Mathematicae 50*, 475-483.

1963 Remarks on a paper by J. Mycielski, *Acta Mathematica Academiae Scientiarum Hungaricae 14*, 125-130.

1963a Independence results in set theory by Cohen's method I (abstract), *Notices of the American Mathematical Society 10*, 592-593.

1963b (with Solomon Feferman) Independence results in set theory by Cohen's method II (abstract), *Notices of the American Mathematical Society 10*, 593.

1963c Independence results in set theory by Cohen's method III (abstract), *Notices of the American Mathematical Society 10*, 593.

1963d Independence results in set theory by Cohen's method IV (abstract), *Notices of the American Mathematical Society 10*, 593.

1964 The interdependence of certain consequences of the axiom of choice, *Fundamenta Mathematicae 54*, 135-157.

1964a (with James D. Halpern) The ordering theorem does not imply the axiom of choice (abstract), *Notices of the American Mathematical Society 11*, 56.

1964b Measurable cardinals and the continuum hypothesis, *Notices of the American Mathematical Society 11*, 769-770.

1964c A hierarchy of formulas in set theory (abstract), *The Journal of Symbolic Logic 29*, 226-227.

1965 The Fraenkel-Mostowski method for independence proofs in set theory, in: John W. Addison Jr., Leon Henkin, and Alfred Tarski (editors), *The Theory of Models*, Proceedings of the 1963 international symposium at Berkeley, North-Holland, Amsterdam, 221-228.

1965a Definability in axiomatic set theory I, in: Yehoshua Bar-Hillel (editor), *Logic, Methodology and Philosophy of Science*, Proceedings of the 1964 International Congress at Jerusalem, North-Holland, Amsterdam, 127-151.

1965b A hierarchy of formulas in set theory, *Memoirs of the American Mathematical Society 57*, 76pp.

1967 (with Robert M. Solovay) Measurable cardinals and the continuum hypothesis, *Israel Journal of Mathematics 5*, 234-248.

1968 (with Georg Kreisel) Reflection principles and their use for establishing the complexity of axiomatic systems, *Zeitschrift für Mathematische Logik und Grundlagen der Mathematik 14*, 97-142.

1968a On Von Neumann's axiom system for set theory, *American Mathematical Monthly 75*, 762-763.

1969 The definability of cardinal numbers, in: Jack J. Bulloff, Thomas C. Holyoke, and Samuel W. Hahn (editors), *Foundations of Mathematics.* Symposium papers commemorating the sixtieth birthday of Kurt Gödel, Springer-Verlag, Berlin.

1969a The effectivity of existential statements in axiomatic set theory, *Information Sciences 1*, 119-130.

1970 Definability in axiomatic set theory II, in: Yehoshua Bar-Hillel (editor), *Mathematcal Logic and Foundations of Set Theory*, Proceedings of an international colloquium, 11-14 November 1968, Jerusalem, North-Holland, Amsterdam, 129-145.

1971 (with James D. Halpern) The Boolean prime ideal theorem does not imply the axiom of choice, in: Dana S. Scott (editor), *Axiomatic Set Theory*, Proceedings of Symposia in Pure Mathematics vol.13, part I, American Mathematical Society, Providence, 83-134.

1971a The sizes of the indescribable cardinals, in: Dana S. Scott (editor), *Axiomatic Set Theory*, Proceedings of Symposia in Pure Mathematics vol.13, part I, American Mathematical Society, Providence, 205-218.

1971b On the logical complexity of several axioms of set theory, in: Dana S. Scott (editor), *Axiomatic Set Theory*, Proceedings of Symposia in Pure Mathematics vol.13, part I, American Mathematical Society, Providence, 219-230.

1972 (with Robert M. Solovay), On the decomposition of sets of reals to Borel sets, *Annals of Mathematical Logic 5*, 1-19.

1973 (with Abraham A. Fraenkel and Yehoshua Bar-Hillel) *Foundations of Set Theory*, North-Holland, Amsterdam, 404 pp.

1974 Parameters in the comprehension axiom schemas of set theory, in: Leon Henkin *et al.* (editors), *Proceedings of the Tarski Symposium*, Proceedings of Symposia in Pure Mathematics vol. 25, American Mathematical Society, Providence, 309-324.

1976 The role of classes in set theory (an excerpt from [1973]), in: Gert H. Müller (editor), *Sets and Classes*, North-Holland, Amsterdam, 173-215.

1979 *Basic Set Theory*, Springer-Verlag, Berlin, 391 pp.; reprinted by Dover Publications, 2003.

1985 *Mathematical Logic* (in Hebrew), A Course of the Open University of Israel.

1997 *Mathematical Logic A* (in Hebrew), Academon Press.

Other References

1956 WILHELM ACKERMANN, Zur Axiomatik der Mengenlehre, *Mathematische Annalen 131*, 336–345.

1959 JOHN W. ADDISON Jr., Some consequences of the axiom of constructibility, *Fundamenta Mathematicae 46*, 337–357.

1975 JON BARWISE, *Admissible Sets and Structures*, Springer-Verlag, Berlin.

1975 JAMES E. BAUMGARTNER, Ineffability properties of cardinals I, in: András Hajnal, Richard Rado, and Vera T. Sós (editors), *Infinite and Finite Sets*, Colloquia Mathematica Societatis Janos Bolyai vol. 10, North-Holland, Amsterdam.

1961 NORMAN BENTWICH, *The Hebrew University of Jerusalem, 1918–1960*, Weidenfeld and Nicolson, London.

1961 PAUL BERNAYS, Zur Frage der Unendlichkeitsschemata in der axiomatischen Mengenlehre, in: Yehoshua Bar-Hillel, E.I.J. Poznanski, Michael O. Rabin, and Abraham Robinson (editors), *Essays on the Foundations of Mathematics*, dedicated to Professor A.A. Fraenkel on his 70th Birthday, Magnes Press, Jerusalem, 3–49.

1963 PAUL J. COHEN, The independence of the Axiom of Choice, mimeographed preprint, Stanford University.

1963a PAUL J. COHEN, The independence of the continuum hypothesis. I, *Proceedings of the National Academy of Sciences U.S.A. 50*, 1143–1148.

1964 PAUL J. COHEN, The independence of the continuum hypothesis. II, *Proceedings of the National Academy of Sciences U.S.A. 51*, 105–110.

1965 PAUL J. COHEN, Independence results in set theory, in: John W. Addison Jr., Leon Henkin, and Alfred Tarski (editors), *The Theory of Models*, Proceedings of the 1963 international symposium at Berkeley, North-Holland, Amsterdam, 39–54.

1966 PAUL J. COHEN, *Set Theory and the Continuum Hypothesis*, Benjamin, New York.

1965 SOLOMON FEFERMAN, *Some applications of the notions of forcing and generic sets* (summary), in: John W. Addison Jr., Leon Henkin, and Alfred Tarski (editors), *The Theory of Models*, Proceedings of the 1963 international symposium at Berkeley, North-Holland, Amsterdam, 89–95.

1965a SOLOMON FEFERMAN, Some applications of the notions of forcing and generic sets, *Fundamenta Mathematicae 56*, 325–345.

1922 ABRAHAM A. FRAENKEL, Zu den Grundlagen der Cantor-Zermeloschen Mengenlehre, *Mathematische Annalen 86*, 230–237.

1922a ABRAHAM A. FRAENKEL, Über den Begriff 'definit' und die Unabhängigkeit des Auswahlaxioms, *Sitzungsberichte der Preussischen Akademie der Wissenschaften, Physikalish-mathematische Klass*, 253–257; translated in van Heijenoort (editor), *From Frege to Gödel: a source book in mathematical logic, 1879-1931*, Harvard University Press, Cambridge 1967.

1958 ABRAHAM A. FRAENKEL and YEHOSHUA BAR-HILLEL, *Foundations of Set Theory*, North-Holland, Amsterdam, 415pp.

1967 ROBERT J. GAUNTT, Undefinability of cardinality, in: *Lectures notes prepared in connection with the Summer Institute on Axiomatic Set Theory held at University of California, Los Angeles*, IV-M.

1943 GERHARD GENTZEN, Beweisbarkeit und Unbeweisbarkeit van Anfangsfällen der transfiniten Induktion in der reinen Zahlentheorie, *Mathematische Annalen 119*, 140-161, translated in: M.E. Szabo (editor), *The Collected Papers of Gerhard Gentzen*, North-Holland, Amsterdam 1969, 287-308.

1938 KURT GÖDEL, The consistency of the axiom of choice and of the generalized continuum-hypothesis, *Proceedings of the National Academy of Sciences U.S.A. 24*, 556–557; reprinted in [1990] below, 26–27.

1939 KURT GÖDEL, Consistency-proof for the generalized continuum hypothesis, *Proceedings of the National Academy of Sciences U.S.A. 25*, 220–224; reprinted in [1990] below, 28–32.

1940 KURT GÖDEL, *The Consistency of the Axiom of Choice and of the Generalized Continuum Hypothesis with the Axioms of Set Theory*, Annals of Mathematics Studies #3, Princeton University Press, Princeton; reprinted in [1990] below, 33-101.

1940a KURT GÖDEL, Lecture [on the] consistency [of the] continuum hypothesis, Brown University, in [1995] below, 175–185, and the page references are to these.

1946 KURT GÖDEL, Remarks before the Princeton bicentennial conference on problems in mathematics, 1–4, in [1990] below, 150–153.

1947 KURT GÖDEL, What is Cantor's continuum problem?, *American Mathematical Society 54*, 515–525; errata *55*, 151; reprinted in [1990] below, 176–187.

1990 KURT GÖDEL, *Collected Works*, Vol. 2, Solomon Feferman (editor-in-chief), Oxford University Press, New York.

1995 KURT GÖDEL, *Collected Works.* Vol. 3, Solomon Feferman (editor-in-chief), Oxford University Press, New York.

2003 KURT GÖDEL, *Collected Works*, Vol. 4, Solomon Feferman and John W. Dawson, Jr. (editors-in-chief), Clarendon Press, Oxford.

2003a KURT GÖDEL, *Collected Works*, Vol. 5, Solomon Feferman and John W. Dawson, Jr. (editors-in-chief), Clarendon Press, Oxford.

1956 ANDRÁS HAJNAL, On a consistency theorem connected with the generalized continuum problem, *Zeitschrift für Mathematische Logik und Grundlagen der Mathematik 2*, 131–136.

1961 ANDRÁS HAJNAL, On a consistency theorem connected with the generalized continuum problem, *Acta Mathematica Academiae Scientiarum Hungaricae 12*, 321–376.

1961 JAMES D. HALPERN, The independence of the axiom of choice from the Boolean prime ideal theorem (abstract), *Notices of the American Mathematical Society 8*, 279–280.

1964 JAMES D. HALPERN, The independence of the axiom of choice from the Boolean prime ideal theorem, *Fundamenta Mathmaticae 55*, 57–66.

1966 JAMES D. HALPERN and HANS LÄUCHLI, A partition theorem, *Transactions of the American Mathematical society 124*, 360–367.

1961 WILLIAM P. HANF and DANA S. SCOTT, Classifying inaccessible cardinals (abstract), *Notices of the American Mathematical Society 8*, 445.

1973 PAUL E. HOWARD, Limitations of the Fraenkel-Mostowski method of independence proofs, *The Journal of Symbolic Logic 38*, 416–422.

1998 PAUL E. HOWARD and JEAN E. RUBIN, *Consequences of the Axiom of Choice*, Mathematical Surveys and Monographs 59, American Mathematical Society, Providence.

1973 THOMAS J. JECH, *The Axiom of Choice*, North-Holland, Amsterdam.

1967 RONALD B. JENSEN, Consistency results for ZF, *Notices of the American Mathematical Society 14*, 137.

2005 AKIHIRO KANAMORI, *The Higher Infinite*, corrected second edition, Springer, Berlin.

1964 H. JEROME KEISLER and ALFRED TARSKI, From accessible to inaccessible cardinals, *Fundamenta Mathematicae 53*, 225–308; errata *57* (1965), 119.

1955 W. KINNA and K. WAGNER, Über eine Abschwächung des Auswahlpostulates, *Fundamenta Mathematicae 42*, 75–82.

1943 STEPHEN C. KLEENE, Recursive predicates and quantifiers, *Transactions of the American Mathematical Society 53*, 41–73.

1937 MOTOKITI KONDÔ, L'uniformization des complémentaires analytiques, *Proceedings of the Imperial Academy of Japan 13*, 287–291.

1964 HANS LÄUCHLI, The independence of the ordering principle from a restricted axiom of choice, *Fundamenta Mathematicae 54*, 31–43.

1984 RICHARD LAVER, Products of infinitely many perfect trees, *Journal of the London Mathematical Society 29*, 385–396.

1911 PAUL MAHLO, Über lineare transfinite Mengen, *Berichte über die Verhandlungen der Königlich Sächsischen Gesellschaft der Wissenschaften zu Leipzig, Mathematische-Physische Klass 63*, 187-225.

1912 PAUL MAHLO, Zur Theorie und Anwendung der ρ_0-Zahlen, *Berichte über die Verhandlungen der Königlich Sächsischen Gesellschaft der Wissenschaften zu Leipzig, Mathematische-Physische Klass 64*, 108-112.

1913 PAUL MAHLO, Zur Theorie und Anwendung der ρ_0-Zahlen II, *Berichte über die Verhandlungen der Königlich Sächsischen Gesellschaft der Wissenschaften zu Leipzig, Mathematische-Physische Klass 65*, 268-282.

1970 DONALD A. MARTIN and ROBERT M. SOLOVAY, Internal Cohen extensions, *Annals of Mathematical Logic 2*, 143–178.

1956 ELLIOT MENDELSON, The independence of a weak axiom of choice, *The Journal of Symbolic Logic 21*, 350–366.

1961 RICHARD M. MONTAGUE, *Fraenkel's addition to the axioms of Zermelo*, in: Yehoshua Bar-Hillel, E.I.J. Poznanski, Michael O. Rabin, and Abraham Robinson (editors), *Essays on the Foundations of Mathematics*, dedicated to Professor A.A. Fraenkel on his 70th Birthday, Magnes Press, Jerusalem, 91–114.

1988 GREGORY H. MOORE, The origins of forcing, in: Frank R. Drake and John K. Truss (editors), *Logic Colloquium '86*, North-Holland, Amsterdam.

1939 ANDRZEJ M. MOSTOWSKI, Über die Unabhängigkeit des Wohlordnungssatzes vom Ordnungsprinzip, *Fundamenta Mathematicae 32*, 201–252.

1948 ANDRZEJ M. MOSTOWSKI, On the principle of dependent choices, *Fundamenta Mathematicae 35*, 127–130.

1958 ANDRZEJ M. MOSTOWSKI, On a problem of W. Kinna and K. Wagner, *Colloquium Mathematicum 6*, 207–208.

1971 JOHN MYHILL and DANA S. SCOTT, Ordinal definability, in: Dana S. Scott (editor), *Axiomatic Set Theory*, Proceedings of Symposia in Pure Mathematics vol.13, part I, American Mathematical Society, Providence, 271–278.

1972 DAVID PINCUS, Zermelo-Fraenkel consistency results by Fraenkel Mostowski methods, *The Journal of Symbolic Logic 37*, 721–743.

1974 DAVID PINCUS, Cardinal representatives, *Israel Journal of Mathematics 18*, 321–344.

1970 WILLIAM N. REINHARDT, Ackermann's set theory equals ZF, *Annals of Mathematical Logic 2*, 189–249.

1975 GERSHON SAGEEV, An independence result concerning the axiom of choice, *Annals of Mathematical Logic 8*, 1–184.

1955 DANA S. SCOTT, Definitions by abstraction in axiomatic set theory, *Bulletin of the American Mathematical Society 61*, 442.

1961 DANA S. SCOTT, Measurable cardinals and constructible sets, *Bulletin de l'Académie Polonaise des Sciences, Série des Sciences Mathématiques, Astronomiques et Physiques 9*, 521-524.

1982 SAHARON SHELAH, *Proper Forcing*, Lecture Notes in Mathematics 940, Springer-Verlag, Berlin.

1984 SAHARON SHELAH, Can you take Solovay's inaccessible away?, *Israel Journal of Mathematics 49*, 1–47.

1998 SAHARON SHELAH, *Proper and Improper Forcing*, Perspectives in Mathematical Logic, Springer, Berlin.

1959 JOHN C. SHEPHERDSON, purported review of Levy [1957], *The Journal of Symbolic Logic 26*, 226.

1957 JOSEPH R. SHOENFIELD, Constructible sets (abstract), in: *Summaries of talks, Summer Institute for Symbolic Logic, Cornell University, 1957*, Institute for Defense Analysis, Princeton, vol. 2, 214.

1959 JOSEPH R. SHOENFIELD, On the independence of the axiom of constructibility, *American Journal of Mathematics 81*, 537–540.

1961 JOSEPH R. SHOENFIELD, The problem of predicativity, in: Yehoshua Bar-Hillel, E.I.J. Poznanski, Michael O. Rabin, and Abraham Robinson (editors), *Essays on the Foundations of Mathematics*, dedicated to Professor A.A. Fraenkel on his 70th Birthday, Magnes Press, Jerusalem, 132–139.

1925 WACŁAW SIERPIŃSKI, Sur une classe d'ensembles, *Fundamenta Mathematicae 7*, 237-243.

1965 ROBERT M. SOLOVAY, Measurable cardinals and the continuum hypothesis (abstract), *Notices of the American Mathematical Society 12*, 132.

1965a ROBERT M. SOLOVAY, The measure problem (abstract), *Notices of the American Mathematical Society 12*, 217.

1970 ROBERT M. SOLOVAY, A model of set theory in which every set of reals is Lebesgue measurable, *Annals of Mathematics 92*, 1–56.

1957 ERNST SPECKER, Zur Axiomatik der Mengenlehre (Fundierungs- und Auswahlaxiome), *Zeitschrift für Mathematische Logik und Grundlagen der Mathematik 3*, 173–210.

1963 ROBERT L. VAUGHT, Indescribable cardinals (abstract), *Notices of the American Mathematical Society 10*, 126.

16 Cohen and Set Theory

Paul Joseph Cohen (1934–2007) in 1963 established the independence of the Axiom of Choice (AC) from ZF and the independence of the Continuum Hypothesis (CH) from ZFC. That is, he established that Con(ZF) implies Con(ZF + ¬AC) and Con(ZF) implies Con(ZFC + ¬CH). Already prominent as an analyst, Cohen had ventured into set theory with fresh eyes and an open-mindedness about possibilities. These results delimited ZF and ZFC in terms of the two fundamental issues at the beginnings of set theory. But beyond that, Cohen's proofs were the inaugural examples of a new technique, *forcing*, which was to become a remarkably general and flexible method for extending models of set theory. Forcing has strong intuitive underpinnings and reinforces the notion of set as given by the first-order ZF axioms with conspicuous uses of Replacement and Foundation. If Gödel's construction of L had launched set theory as a distinctive field of mathematics, then Cohen's forcing began its transformation into a modern, sophisticated one.

The extent and breadth of the expansion of set theory henceforth dwarfed all that came before, both in terms of the numbers of people involved and the results established. With clear intimations of a new and concrete way of building models, set theorists rushed in and with forcing were soon establishing a cornucopia of relative consistency results, truths in a wider sense, with some illuminating classical problems of mathematics. Soon, ZFC became quite unlike Euclidean geometry and much like group theory, with a wide range of models of set theory being investigated for their own sake. Set theory had undergone a sea-change, and with the subject so enriched, it is difficult to convey the strangeness of it.

How did forcing come about? How did it develop into a general method? What is the extent of Cohen's achievement and its relation to subsequent events? How did Cohen himself view his work and mathematics in general at the time and late in his life? These are related questions that we keep in mind in what follows and reprise at the end, with an emphasis on the mathematical themes and details and the historical progression insofar as it draws out the mathematical development.[1]

Republished from *The Bulletin of Symbolic Logic* 14 (2008), pp.351-378, with permission from The Association for Symbolic Logic.

This is the full text of an invited address given at the annual meeting of the Association for Symbolic Logic held at Irvine in March 2008. My thanks to Juliet Floyd, and for points of detail to Andreas Blass, Solomon Feferman, Alexander Kechris, Elliott Mendelson, and Dana Scott.

[1]Moore [43] is a detailed, historical account of the origins of forcing; Cohen [19] is an extended reminiscence of the discovery of forcing four decades later; and Albers–Alexanderson–Reid [1],pp.43-58 is a portrait of Cohen based on an extended interview with him. We shall be referring to these in what follows. Yandell [61],pp.59–83 is another account of Cohen and his work, largely based on [43], [1], and a phone conversation with Cohen.

1 Before

Cohen was and is considered an analyst who made a transforming contribution to set theory. Because forcing is such a singular phenomenon, Cohen has at times been regarded as a bit of a brash carpetbagger, an opportunist who was fortunate and brilliant enough to have made a decisive breakthrough. Be that as it may, we try here to paint a more discriminating picture and to draw out some connections and continuities, both historical and mathematical, from a time when mathematics was simpler and not so Balkanized.

Cohen became a graduate student in 1953, at the age 19, at the University of Chicago; received a master's degree a year later; and then his doctorate in 1958. This was a veritable golden age for the emerging university and its department of mathematics, with figures like Irving Kaplansky and Saunders MacLane making strong impressions on the young Cohen[2] among others like Shiing-Shen Chern, Irving Segal, and André Weil who were making crucial advances in their fields. As for foundations, Paul Halmos and MacLane had interests and Elliott Mendelson was an instructor for 1955-6.

From early on Cohen had developed what was to be an abiding interest in number theory, an area of mathematics which he found attractive because of the simplicity of its statements and complexity and ingenuity of its proofs.[3] Cohen recalled working on a famous problem in diophantine approximation issuing from work of Axel Thue and Carl Siegel, loosely speaking whether an algebraic number can have only finitely "good" rational approximations, until one day the number theorist Swinnerton-Dyer knocked on his door and told him that it had been solved by Klaus Roth; for this Roth received the Fields Medal in 1958.[4] Whether it was from this formative period or later, Cohen, as can be seen in his writings, developed a conviction about the centrality of number theory in mathematics. Significantly, Cohen by this time also developed a significant interest in logic, this issuing from decidability questions in number theory.[5] He recalled spending time with future logicians, especially William Howard, Raymond Smullyan, and Stanley Tennenbaum, in this way picking up a good deal of logic, and also reading Kleene's *Introduction to Metamathematics*.[6] Also, one of his office-mates was Michael Morley, a soon-to-be prominent model-theorist.[7] In any case, there was little in the way of possibilities for apprenticeship in number theory at that time in Chicago.

Moving to a different field, Cohen, in his dissertation, worked in harmonic analysis with the well-known Antoni Zygmund of the Polish mathematical

[2]See [19],p.1071.
[3]See [21].
[4]See [1],p.49,56.
[5]See [1],p.50.
[6]See [43],p.154 and [1],p.51.
[7]See [21].

school, who had a large following and developed what then came to be known as the Chicago School of Analysis. Zygmund was among the next generation just after those who founded the Polish school, and, following up one path of the genealogy tree, was himself a student of Aleksander Rajchman and Stefan Mazurkiewicz, the latter a student of Wacław Sierpiński.[8] The incipient historical connection with set theory extended to Cohen's dissertation, entitled *Topics in the Theory of Uniqueness of Trigonometrical Series* — on the very subject that Georg Cantor first established results in 1871 about derived sets using his "symbols of infinity" that eventually became the ordinal numbers. The main work of the dissertation would remain unpublished, with only a note [7] extracted which established an optimal form of Green's theorem. This initial publication of Cohen's did have a *mathematical* thread which we expand on forthwith.

There is one antecedent kind of mathematical construction that trickles from the the beginnings of set theory into topology and analysis and impinges on forcing — by way of category. Set theory was born on that day in December in 1873 when Cantor established that the continuum is not countable. In a proof in which the much later diagonal argument is arguably implicit, Cantor defined, given a countable sequence of reals, a sequence of nested intervals so that any real in their intersection will not be in the sequence. Twenty-five years later, with much the same proof idea, René Baire in his 1899 thesis established the Baire Category Theorem, which in a felicitous formulation for our context asserts that *the intersection of countably many dense open sets is dense.*[9]

The Baire Category Theorem would become widely applied by the Polish school and, through its dissemination of ideas, across topology and analysis. Applications of the theorem establish existence assertions in a topological space through successive approximations. Typically the process can start anywhere in the space and therefore produce a large class of witnesses. The arguments can have an involved preamble having to do with the underlying topology and the dense open sets but then can be remarkably short. On the one hand, the existence assertion is not always buttressed by an "explicit construction", but on the other hand, there is the conclusion that apparent "pathologies" are almost everywhere.

In 1931 Stefan Banach [3] provided new sense to Weierstrass's classical construction of a continuous nowhere-differentiable functions by devising a remarkably short proof via the Baire Category Theorem that in the space $C[0,1]$ of continuous functions on the unit interval with the uniform metric (i.e. using the sup norm) *the nowhere differentiable functions are co-meager.* With Banach's result achieving expository popularity, Cohen would presumably have been cognizant of it, especially in the imported Polish air of Chicago.

[8]Incidentally, Cohen himself was the son of Jewish immigrants from Poland.

[9]This holds in abstract terms in any complete metric or locally compact space.

In any case, in that initial publication [7] of Cohen's drawn from his dissertation, there is an explicit appeal to the Baire Category theorem to attend to an issue about differentiability in a manner similar to Banach's.

A final connection, rather tenuous, has to do with stable dynamical systems. After Stephen Smale did his work on the Poincaré Conjecture for which he would get the Fields Medal in 1966, the same year that Cohen would, Smale in the mid-1960s initiated a program to determine the stable dynamical systems. For a compact differentiable manifold M and the space of diffeomorphisms $M \longrightarrow M$ with the uniform metric (all this possibly elaborated with higher derivatives), Smale in a well-known paper [49] defined a property of diffeomorphisms to be *generic* if the set of diffeomorphisms satisfying it is co-meager. Regarding diffeomorphisms as dynamical systems (via their iteration) Smale set out to show that the generic properties determine the stable dynamical systems. Smale often applied the Baire Category Theorem in various ways, but on the other hand he eventually came to a successful formulation of what a stable dynamical system is and moreover showed it to be distinct from genericity.

Back to Cohen. Before official receipt of his Ph.D. Cohen took up an instructorship at the University of Rochester for the academic year 1957-8 and then an instructorship at the Massachusetts Institute of Technology for 1958-9. At MIT Cohen came into contact with logicians Azriel Levy and Hartley Rogers, and among others, especially John Nash. Sylvia Nasar's well-known biography of Nash, *A Beautiful Mind* [44], candidly describes Cohen and his interactions with Nash. She wrote of Cohen (p.237):

> He spoke several languages. He played the piano. His ambitions were seemingly unlimited and he spoke, from one moment to the next, of becoming a physicist, a composer, even a novelist. [Eli] Stein, who became a close friend of Cohen's, said: "What drives Cohen is that he's going to be better than any other guy. He's going solve the big problems. He looks down on mathematicians who do mathematics for the sake of making incremental improvements in the field."

Nash had several years before done the work in game theory that would eventually lead to the Nobel Prize and established his most important and difficult result on the embeddability of compact Riemannian manifolds into Euclidean space. Nasar described how Nash and Cohen had long and charged discussions about mathematical problems, especially the Riemann Hypothesis.[10]

Of concrete results, Cohen in 1958 had started to work on measures on locally compact abelian groups. Within a year he made a significant advance on a conjecture of Littlewood in Fourier analysis about the lower bound on

[10] See [44],p.238. It is said that this association contributed to Nash's first psychiatric commitment for schizophrenia in 1959.

a exponential sum[11] and with it characterized the idempotent measures on locally compact abelian groups. For this Cohen would in 1964 receive the prestigious Bôcher Memorial prize awarded by the American Mathematical Society "for a notable paper in analysis published during the preceding six years". The citations for this award are mostly about a body of esteemed work, but in Cohen's case it was indeed for "a notable paper", [8]. Cohen was able to made a considerable advance with concrete means, at one point (p.196) using a lemma about finite integers that could have been derived via the Finite Ramsey Theorem. Notably, in a subsequent paper [9] Cohen continued his emphasis on concrete means by providing a method for eliminating appeals to the Axiom of Choice from several known applications of Banach algebras to classical analysis.

Cohen spent the years 1959–61 at the Institute for Advanced Study at Princeton as a fellow and then became an assistant professor at Stanford University. Having put aside the Riemann Hypothesis and with his particular work on locally compact abelian groups having gone as far as it could, Cohen during this period turned to big problems of logic. Both at the Institute and at Stanford, he discussed issues with Solomon Feferman, and at Stanford to a lesser extent with Georg Kreisel. In 1961-2 Cohen focused on the consistency of analysis and even conducted a seminar on his work, but abandoned his approach when it failed to get past arithmetic. Even then, his assimilation of and admiration for the consistency work in proof theory would play an important role in his later mathematical thinking. At the end of 1962 Cohen moved on to the independence of AC.[12]

What was the state of set theory at that time? In the axiomatic tradition Gödel's relative consistency result for AC and CH through the inner model L of constructibility had stood as an isolated monument for quite a number of years. To the extent that axiomatics and inner models were investigated at all, the papers were couched in the exacting formalism of Gödel's 1940 monograph [30], even to the reverent citation of the axioms through their groupings. Starting in the mid-1950s however, new model-theoretic initiatives informed the situation, and with Alfred Tarski established at the University of California at Berkeley a large part of the development would take place there. In the new terms, Gödel's main, CH argument for L was better understood as a direct Skolem hull and elementary substructure argument, something that had been obscured by [30].

The emergence of the ultraproduct construction for providing a concrete, algebraic means of building models led to a revitalization of the theory of large cardinals. Building on the work of Jerome Keisler, Dana Scott [45] in 1961 took an ultrapower of the set-theoretic universe V itself to establish that

[11] This is not what is widely known as "Littlewood's conjecture", about lattice points in Diophantine approximation theory.

[12] See [43],p.155.

having a measurable cardinal contradicts Gödel's Axiom of Constructibility $V = L$. With the ultrapower set theory was brought to the point of entertaining elementary *embeddings* into well-founded models. It was soon to be transfigured by a new means for getting well-founded *extensions* of well-founded models. At the 1962 International Congress of Mathematicians at Stockholm, Scott presented his result about measurability and L, and Cohen [10], his work on idempotent measures on locally compact abelian groups. At the 1966 congress at Moscow, Cohen was awarded the Fields Medal for the independence of the Axiom of Choice and the Continuum Hypothesis.

2 The Minimal Model

Cohen's progress to his independence results would be by way of the minimal model of set theory. Importantly, overt proof-theoretic approaches would give way to increasingly semantic approaches. In the retrospective [19], Cohen recalled at length how he had come to forcing. As to why there had been little work on the problem of independence, Cohen adduced two reasons. The first pertained to the obtuseness of Gödel's monograph [30] as compared to his first announcement [29], and what Cohen wrote is interestingly revelatory (p.1086ff):

> ... although the first note of Gödel was a very good sketch of his results, the publication of the formal exposition in his usual fastidious style gave the impression of a very technical, even partially philosophical, result. Of course, it was a perfectly good mathematical result with a relatively straightforward proof. Let me give some impressions that I had obtained before actually reading the Princeton monograph but after a cursory inspection. Firstly, it did not actually construct a model, the traditional method, but gave a concept, namely constructibility, to construct an *inner* model. Secondly, it had an exaggerated emphasis on relatively minor points, in particular, the notion of *absoluteness*, which somehow seemed to be a new philosophical concept. From general impressions I had of the proof, there was finality to it, an impression that somehow Gödel had mathematicized a philosophical concept, i.e., constructibility, and there seemed no possibility of doing this again, especially because the negation of CH and AC were regarded as pathological.

As here, Cohen persistently regarded "philosophical" attitudes with their emphasis on concepts and proofs as having been detrimental and would stress how it was a "mathematical" approach with constructions and models that led to success. The other reason that Cohen adduced for the lack of progress on independence was that rumor had it that Gödel had partially solved the independence problem, at least for AC.

Cohen went on to describe his early, 1962 speculations about independence. He first focused on the independence of just AC as simpler and because some work had already been done on it, namely the Fraenkel–Mostowski model-building with urelements (atoms). He eventually came to several conclusions (p.1088):

> One, there is no device of the type of Frankel-Mostowski or similar "tricks" which would give the result. Two, one would have to eventually analyze all possible proofs in some way and show that there is an inductive procedure to show that no proof is bringing one substantially closer to having a method of choosing one element from each set. Third, although there would have to be a semantic analysis in some sense, eventually one would have to construct a *standard* model.

A *standard* model is also known as an $\in$-model, a set which together with the membership relation restricted to it is a model of set theory; it was always clear to Cohen that any well-founded model of set theory is isomorphic to a transitive standard model, and the assumption of transitivity for standard models is implicit in his work. We see here how Cohen focused on standard models and their concreteness from the beginning. Presumably with his proof-theoretic work on the consistency of analysis influencing him, Cohen initially worked on devising some kind of induction on length of proofs. He continued (p.1089):

> It seemed some kind of inductive hypothesis would work, whereby if I showed that no "progress" was made in a choice function up to a certain point, then the next step would also not make any progress. It was at this point that I realized the connection with the models, specifically standard models. Instead of thinking about proofs, I would think about the formulas that defined sets, these formulas might involve other sets previous defined, etc. So if one thinks about sets, one sees that the induction is on the rank, and I am assuming that every set is defined by a formula. At this point I decided to look at Gödel's monograph, and I realized that this is exactly what the definition of constructibility does.

Cohen then saw that Gödel's construction did not correspond to the kind of proof analysis that he had in mind. "Namely, it is not specifically tailored to the axioms of ZF, but gives a very generous definition of 'construction'." (p.1090) From previous passages, it becomes obvious that this "generosity" is the incorporation of all the ordinals as impredicatively given. Cohen therefore modified the construction, and came up with his first, precursory result in set theory, that there is a minimal model of ZF, i.e. among all standard models there is one that is isomorphically embedded in every other (which of course must have transitive form L_γ for some countable $\gamma < \omega_1$). This result in particular implied that taking inner models cannot establish independence, and Cohen was happy with it, as it represented the first concrete progress he had made.[13]

Cohen recalled that both Kreisel and Scott urged that this result be published and he proceeded to do so though he was astounded that such a simple result was apparently unknown. Only later did Cohen become aware that he had duplicated a result of John Shepherdson published a decade earlier.[14]

The Shepherdson [46] and Cohen [12] papers are a study in contrasts that speaks to the historical distance and the coming breakthrough. While both papers are devoted to establishing essentially the same result, the former takes 20 pages and latter only 4. Shepherdson labors in the Gödelian formalism with its careful laying out of axioms and propositions in first-order logic, while Cohen proceeds informally and draws on mathematical experience. Shepherdson works out the relativization of formulas, worries about absoluteness and comes down to the minimal model, while Cohen takes an algebraic closure. As he writes (p.537),

> We observe that the idea of a minimal collection of objects satisfying certain axioms is well known in mathematics, for example, in group theory one often considers the subgroup generated by a collection of elements and in measure theory we define the Borel sets as the smallest σ-algebra of sets containing the open sets.

Cohen constructs the minimal model through a recursive definition based on closing off under the ZF axioms. Proceeding in modern terms, for a set X, let $\Gamma(X)$ denote the collection of $\{x, y\}$ for $x, y \in X$ as well as $\bigcup x$, $\mathcal{P}(x) \cap X$, and $\{z \in X \mid \exists w \in x R(w, z)\}$ whenever $R(\,\cdot\,,\,\cdot\,)$ is a functional relation defined with quantifiers restricted to X. The latter, of course, arranges for Replacement. Let $T_0 = \omega + 1$, i.e. the set consisting of all the natural numbers and the set of all the natural numbers, and recursively, $T_\alpha = \Gamma(\bigcup_{\beta<\alpha} T_\beta)$. For $M = \bigcup_\alpha T_\alpha$ being a model of ZF Cohen merely refers to the Gödel monograph. As he pointed out, the only difference between L and his construction is that he does not demand that $X \in \Gamma(X)$ — so that Replacement is being addressed while forestalling all the ordinals from getting into M. Cohen then argues that M is minimal and by the Löwenheim-Skolem theorem that M is countable; we would just note today that the closure ordinal γ, i.e. the least α such that $T_{\alpha+1} = T_\alpha$, is countable. In its way, there is an elegance to this construction in its concreteness and simplicity. Cohen here was working in a context where

[13]See [19],p.1090.

[14]See [43],p.155.

there are standard models of ZF, and though this would become an issue in forcing, he would henceforth work his ideas on the concrete backdrop provided by such models.

3 Forcing

For his frontal assault on independence, Cohen concentrated first on starting with a countable (transitive) standard model M of ZF and adjoining just a set a of integers to get a minimal extension $M(a)$, with a firm decision made not to alter the ordinals. There is a remarkable audacity and hope here in his trying to do something so basic, given the high axiomatic tradition steeped in formal logic. Cohen continued his recollection ([19], p.1091):

> To test the intuition, one should try to adjoin to M an element which enjoys no "specific" property to M, i.e., something akin to a variable adjunction to a field. I called such an element a "generic" element. Now the problem is to make precise this notion of a generic element.

By the middle of April 1963, everything came together for Cohen. He engagingly described the moment (p.1092):

> There are certain moments in any mathematical discovery when the resolution of a problem takes place at such a subconscious level that, in retrospect, it seems impossible to dissect it and explain its origin. Rather, the entire idea presents itself at once, often perhaps in a vague form, but gradually becomes more precise. Since the entire new "model" $M(a)$ is constructed by transfinite induction on ordinals, the definition of what is meant by saying a is generic must also be given by a transfinite induction. Yet a, as a set of integers, occurs very early in the rank hierarchy of sets, so there can be no question of building a by means of a transfinite induction. The answer is this: *the set a will not be determined completely, yet properties of a will be completely determined on the basis of very incomplete information about a* [my emphasis]. I would like to pause and ask the reader to contemplate the seeming contradiction in the above. This idea as it presented itself to me, appeared so different from any normal way of thinking, that I felt it could have enormous consequences. On the other hand, it seemed to skirt the possibility of contradiction in a very perilous manner. Of course, a new generation has arisen who imbibe this idea with their first serious exposure to set theory, and for them, presumably, it does not have the mystical quality that it had for me when I first thought of it. How could one decide whether a statement about a is true,

before we have a? In a somewhat exaggerated sense, it seemed that I would have to examine the very meaning of truth and think about it in a new way.

More informally, Cohen recalled: "What made it so exciting to me was how ideas which at first seemed merely philosophical could actually be made into precise mathematics. I went up to Berkeley to see Dana Scott and run the proof past him. I was very, very excited."[15] At Stanford Cohen described his argument to Feferman, lectured on it, and still in April, circulated a manuscript entitled "The independence of the Axiom of Choice".[16] In it Cohen presented four results by extending a countable (transitive) standard model of $V = L$:

(1) The consistency of ZFC + GCH + $V \neq L$ by adjoining what is now known as a Cohen real.

(2) The consistency of ZFC + $\neg$CH by adjoining $\aleph_2$ Cohen reals. He could not conclude at this time that $2^{\aleph_0}$ is exactly $\aleph_2$.

(3) The consistency of ZF + "There is no well-ordering of the reals" by adjoining countably many Cohen reals, the set consisting of them, but no enumeration —"the basic Cohen model".

(4) The consistency of ZF + "there is a countable sequence of pairs of sets of reals without a choice function" by adjoining a sequence $\langle \{A_i, B_i\} \mid i \in \omega \rangle$ where the A_i and B_i each consist of countably many Cohen reals —"the second Cohen model".

The thrust of Cohen's constructions, what gave them a crucial operational clarity, was to start with a (transitive) standard model of ZF and extend it to another without altering the ordinals. Cohen did appreciate that starting with a standard model of ZF is formally more substantial than assuming merely the consistency of ZF, and he indicated, as a separate matter, a syntactic way to pare down his arguments into formal relative consistency statements of the type Con(ZF) $\rightarrow$ Con(ZF + $\neg$AC).

The components of Cohen's forcing scheme as he first conceived it for extending a (transitive) standard model M of ZF to a generic extension $M(a)$ were as follows:

(a) In M, a *ramified language* together with *ranked terms* to denote sets in the extension is developed for the purpose of approaching satisfaction in

[15]See [1],p.53. The first sentence here is noteworthy in juxtaposition with the first passage quoted in the previous section which had: "somehow Gödel had mathematicized a philosophical concept, i.e. constructibility, and there seemed no possibility of doing this again ...".

[16]See [43],p.156.

the extension. The ramified language has quantifiers $\forall_\alpha$ and $\exists_\alpha$ indexed by the ordinals of M; at this time Cohen took the terms to be certain F_α's, well-ordered in the style of Gödel's monograph and its eight generators for generating L.

(b) In M, a set of *conditions* each regarded as providing partial information toward an eventual generic object a is devised, ordered according to amount of information.

(c) In M, the *forcing relation* "p forces φ", between conditions p and formulas φ of the ramified language is defined to specify when the information of p secures assertion φ in the extension. At this time Cohen took formulas to be in prenex form and equality to be given by the Axiom of Extensionality.

(d) Assuming M is countable, a *complete sequence* of stronger and stronger conditions $p_0, p_1, p_2, \ldots$ is devised so that every formula or its negation is forced by some member of the sequence, and through this sequence a generic object a is arrived at having the desirable properties to establish independence. This is a Baire category argument carried out outside of M; the set of Cohen reals over M is co-meager.

(e) That the resulting $M(a)$ consisting of the interpretations of the terms as determined by a is a model of ZF is established, based on the complete sequence, the definability of the forcing relation in M, and M being a model of ZF.

The genie was out of the bottle. Despite his relative inexperience Cohen had gone a great distance and squarely addressed big problems of set theory. At this moment of impact his scheme immediately generated excitement and came under considerable scrutiny, with some fitful questions raised a measure of the upending in thinking that was being wrought. In the thick of the to and fro, Cohen gave his second lecture on forcing at Princeton on 3 May,[17] and soon afterward met with Gödel himself. Kreisel had written to Gödel already on 15 April about Cohen's work and encouraged Gödel to see Cohen, and Cohen himself had written him on 24 April.[18]

Cohen in a letter of 9 May to Gödel wrote:

> ...what I am trying to say is that only you, with your pre-eminent position in the field, can give the "stamp of approval" which I would so much desire. I hope very much that you can study the manuscript thoroughly and by next week-end be willing to discuss it in more detail. Perhaps, at that time you would consider communicating to the *Proceedings of the National Academy of Sciences* a short note.

[17]This lecture was arranged by Nerode, at Cohen's request.

[18]See [43],p.157 for more details here and on the interactions with Gödel described below.

Gödel after a few days did pronounce Cohen's work to be correct and moreover agreed to communicate it to the PNAS. Cohen in the letter had also written: "I feel under a great nervous strain"; Gödel in a letter of 20 June to Cohen on various issues ended with ([31], p.383):

> I hope you are not under some nervous strain which hampers you in your work. You have just achieved the most important progress in set theory since its axiomatization [so more important than Gödel's own work with L]. So you have every reason to be in high spirits.

Through the summer Cohen worked on his PNAS communication, fielding numerous suggestions from Gödel in correspondence, and it eventually appeared in two installments [11, 13] in the winter of 1963-4.

4 First Expansion

From the beginning the potentialities of forcing were evident, and it began to be applied forthwith and, in the process, reworked for wide applicability, the technique soon to become general method. For Cohen, having taken formulas to be in prenex form the recursive definition of the forcing relation rested on the unraveling of the quantifiers, and the crucial clause was: p forces $\forall_\alpha x\varphi(x)$ *iff* for no q stronger than p does q force $\neg\varphi(t)$ for any term t of rank less than α. In May or June Scott had come up with the forcing symbol $\Vdash$ and had decoupled negation from Cohen's clause with: $p \Vdash \neg\varphi$ *iff* for no q stronger than p does $q \Vdash \varphi$.[19] Thus, the forcing relation could be defined for all formulas, and this clarifies how it is only the negation clause with its interaction of conditions that separates forcing from satisfaction. On the one hand, this negation clause is indicated by Cohen's formulation, but on the other hand, without it the general method cannot be developed nor can forcing arguments be made, as they are today, about assertions at large.

Feferman [26, 25] was the first after Cohen to get results with forcing, these both in set theory and in second-order arithmetic. Feferman had been working on whether Σ^1_1 AC could be shown independent via forcing from Δ^1_1 Comprehension for subsystems of second-order arithmetic, the idea being that the hyperarithmetic sets satisfy the latter and one might forcibly adjoin a counterexample to the former. He could not solve this problem, but in his efforts he came to understand forcing well, and working through May and June he established several results.[20]

[19]See [43], p.160. Years later, Scott (in the foreword to [4]) recalled the emergence of the Kripke and Beth models for intuitionistic logic and wrote: "...after Cohen's original announcement, I pointed out the analogy with intuitionistic interpretations, and along these lines Cohen simplified his treatment of negation at my suggestion."

[20]See [43],p.160. Only many years later did John Steel [57] use forcing to show that Δ^1_1 Comprehension does not imply Σ^1_1 AC; although ramified languages would soon be bypassed

Feferman relied on Scott's negation clause and of necessity used, instead of Cohen's Gödelian F_α's, abstraction terms corresponding directly to definitions in the ramified language — another felicitous move for the general reworking of forcing. In set theory he showed that adjoining one Cohen real entails that there is no definable well-ordering of the reals, initiating the use of forcing to study definability in ZFC. He also established a new $\neg$AC consistency, that there is no non-principal ultrafilter over ω, in the forcing extension adjoining countably many Cohen reals but not the set consisting of them — "the Feferman model".

In second-order arithmetic Feferman developed, for Kleene's rendering of the hyperarithmetic sets as the ramified analytic hierarchy, a corresponding ramified language and, forcing with it, established that there are incomparable hyperdegrees. Clifford Spector [56] had first established this by measure-theoretic means. In retrospect, Feferman's argument can be seen as an application of the Baire Category Theorem to establish an outright result. As such, the result amounts to a category-theoretic analogue of Spector's, for it shows that the pairs $\langle a, b \rangle$ of reals that are hyperarithmetically incomparable is co-meager, just as Spector's argument had shown that it has measure one. One significant dividend of the specific definability of the forcing relation is that Feferman's incomparable hyperdegrees can be seen to be recursive in Kleene's $\mathcal{O}$.

With Scott's treatment of negation in hand Feferman also came up with *weak forcing* $p \Vdash^* \varphi$, given by: $p \Vdash \neg\neg\varphi$. $p \Vdash^* \varphi$ *iff* $M(a) \models \varphi$ for any generic a approximated by p, and so $p \Vdash^* \varphi$ and $\vdash \varphi \rightarrow \psi$ implies $p \Vdash^* \psi$. With this closer connection to semantics and deducibility, it would be through the lens of $\Vdash^*$ that forcing would henceforth be viewed.

Robert Solovay would above all epitomize this period of great expansion in set theory, with his mathematical sophistication and fundamental results about and with forcing, in large cardinals, and in descriptive set theory. Following initial graduate study, also at the University of Chicago, in differential topology Solovay focused his energies on set theory after attending Cohen's 3 May Princeton lecture, quickly absorbing forcing and making his first incursions. For his $\neg$CH model resulting from adjoining $\aleph_2$ Cohen reals, Cohen had come to the Delta-system Lemma *ab initio* and through the consequent countable chain condition deduced that cardinals are preserved, but he could not initially conclude that $2^{\aleph_0}$ is exactly $\aleph_2$. In June both Cohen and Solovay independently came to this conclusion[21] by appealing anew to the countable chain condition.

Solovay carried out the first exploration of possible spectra of powers of

in general forcing, Steel's forcing re-established the specific importance of ramification for a careful hierarchical analysis.

[21] See [43],p.159,161-2. Cohen in a letter of 14 June to Gödel so confirmed the size of the continuum and also described how cardinals can be collapsed.

regular cardinals, introducing closure properties on the set of conditions and establishing the consistency of a finite conjunction of possibilities for the $\aleph_n$'s. Solovay subsequently gave seminar talks on forcing at the Institute for Advanced Study, and there in late Fall 1963 William Easton [23, 24] established his now well-known global result on powers of regular cardinals with class forcing.

Solovay's final advance in 1963 was a matter of technique. In his bootstrapping into a new context Cohen had relied on a version of Gödel's CH argument for L to establish that the Power Set Axiom holds in generic extensions. Solovay gave the now-standard sort of argument using Replacement, confirming that one does not have to start with a ground model satisfying $V = L$.

Cohen gave his third lecture on forcing on 4 July (Independence Day!) at a conference on model theory held at Berkeley in early summer 1963. Feferman and Solovay also presented their respective results with forcing. Cohen's paper [14] for the proceedings, judging from the results of others mentioned, was presumably written sometime in early 1964. The account, in its time frame, is notable in several respects: Cohen (p.52) posed two problems: Is there a (necessarily ill-founded) model of set theory with an automorphism σ whose square σ^2 is the identity? Cohen [18] would himself eventually answer this question positively with forcing. Is it consistent to have ZF + Countable AC + "all sets of reals are Lebesgue measurable"? Cohen considered this a "very important problem", one "requiring perhaps some basic elaboration of the ideas of forcing and generic sets." Solovay [50, 52] would soon answer this question positively in 1964, as described below.

Finally, Cohen (pp.53-4) separately from his argumentation with standard models detailed a specific approach for establishing formal relative consistency. In his PNAS communication [13] Cohen had proceeded semantically with standard models of finitely many ZF axioms in some enumeration, arguing that to get a standard model of the first p ZF axioms together with $2^{\aleph_0} = \aleph_2$ it suffices to start with a standard model of the first $f(p)$ axioms, for an arithmetical function f. Cohen now returned, interestingly, to what he had broached in his April manuscript, a syntactic approach about the forcing relation and logical deduction. He first replaced his standard model with the entire universe and proceeded to describe a series of primitive recursive functions e.g. "There is a primitive recursive function which assigns to the number of each axiom of Z-F a proof in Z-F that all [conditions] P force that axiom." Cohen's analysis exhibited his proof-theoretic experience and anticipated the relative consistency argument via the later Boolean-valued approach to forcing.

Actually, forcing can be carried out over any, not necessarily standard, model — so that formal relative consistency is immediate. Cohen himself would confirm this a decade later in his last paper [18] with mathematical results. The situation there required that he start with an ill-founded model; the "ordinals" remained unaltered in the forcing, but he had to argue separately for Foundation in the extension. Whatever is the case, standard models evidently played a central role in the discovery of forcing, and the simplifications and intuitive underpinnings afforded by them were crucial factors in the development of forcing as general method.

Azriel Levy, visiting Berkeley, first heard the details of Cohen's results at the model theory conference, and later that summer fully assimilated forcing working with Feferman. In quick succession several abstracts appeared: Levy [38], Feferman–Levy [27], Levy [39, 40], and Halpern-Levy [34]. With this work Levy became the first after Cohen to exploit forcing in a sustained fashion to establish a series of significant results. These had to do with further AC independences and the limits of definability in set theory and freely exploited the idea of *collapsing* a cardinal, i.e. adjoining a generic bijection to a smaller ordinal, as first set forth by Cohen.[22]

Feferman–Levy [27] started with a model of ZFC + GCH and collapsed every $\aleph_n$ to $\aleph_0$ — the "Feferman-Levy model." The former $\aleph_\omega$ becomes the new $\aleph_1$ so that it is singular, and the reals are a countable union of countable sets so that Countable AC fails in a drastic fashion.

Halpern–Levy [34] represents a line of work in set theory for which Cohen's epochal advance provided at least a semblance of continuity. James Halpern in his 1962 Berkeley dissertation[23] had shown that in a Fraenkel–Mostowski model with urelements the Boolean Prime Ideal Theorem (BPI), that every Boolean algebra has a prime ideal, holds, bringing a Ramsey-type partition theorem into play. Halpern–Levy [34] observed that the Ordering Principle (OP), that every set can be linearly ordered, holds in the basic Cohen model for the failure of AC. BPI was becoming a prominent choice principle, one which implied OP, and Levy saw the need for a strengthened "tree" partition theorem to "transfer" Halpern's result to a ZF forcing result. Halpern and Läuchli [33] duly established this theorem, and with it, Halpern and Levy [35] established that BPI also holds in the basic Cohen model. This Halpern–Läuchli–Levy collaboration was an important step forward in an emerging cottage industry of transferring results from Fraenkel–Mostowski models to ZF consistency results via forcing by correlating urelements with generic sets.

The results of Levy's abstracts [38, 39, 40], as eventually presented in expanded form in [41, 42], had to do with the limits of definability when successively, ordinal parameters and then real parameters are allowed. Extending the initial observation of Feferman [26], Levy [38, 41] showed that in Cohen's first model adjoining a Cohen real to a model of $V = L$, there is no definable well-ordering of the reals even if ground-model parameters are allowed in the definition. For this Levy first exploited the important *homogeneity* of the Co-

[22]See [14],p.51.
[23]See [32].

hen forcing conditions. Levy [38, 41] further showed that collapsing $\aleph_1$ entails moreover that every well-ordering of the reals definable with ground-model parameters is countable. Levy [39, 42] next established a delimitative result for descriptive set theory, that if to a model of $V = L$ one adjoins $\aleph_1$ Cohen reals, there is a Π^1_2 relation which cannot be uniformized by any projective relation. The relation is $\langle f, g\rangle \in R$ *iff* $g \notin L[f]$, and in the model it cannot be uniformized by any relation definable allowing ordinal and real parameters. Finally, Levy [40, 42] extended his results by collapsing every cardinal below an inaccessible cardinal to $\aleph_0$, rendering the cardinal itself the new $\aleph_1$. This entails that in the resulting model every well-ordering of the reals definable with ordinal *and* real parameters is countable, every new real having appeared in an early part of the collapsing. The "Levy collapse", thus devised to get at the limits of definability, would become a basic component of the investigation of large cardinals with forcing.

Still in the summer of 1963, Tennenbaum [58] established the independence of Souslin's Hypothesis from ZFC. This hypothesis is a classical proposition equivalent to the combinatorial assertion that there are no "Souslin trees", and Tennenbaum generically adjoined a Souslin tree. In having addressed a question from 1920, Tennenbaum's result is notable for having been the first after Cohen's for illuminating an outstanding classical problem and without involving Cohen's ways of adjoining reals and collapsing cardinals.

These various results of 1963 amounted to the first cresting of the wave created by Cohen. In the next several years, forcing, confirmed in its potency and applicability, was widely disseminated, leading to a great expansion of set theory as a field of mathematics, and advances with forcing at a higher plane were achieved, in large part by Solovay.

5 After

Forcing, even in its first year, began to be disseminated through seminars and courses across a wide range of universities. Already in 1963, J. Barkley Rosser gave seminar talks on forcing at the University of Wisconsin at Madison, and Chen-Chung Chang and Anil Nerode had a seminar at the Institute for Advanced Study at which Solovay gave talks on forcing. In 1964, Karl Prikry lectured on forcing in a Warsaw seminar and Yiannis Moschovakis at Harvard. Courses were given on forcing by Feferman at Stanford and by Levy at the Hebrew University. And at Berkeley a group of graduate students self-organized into a working seminar, a group including soon-to-be prominent set theorists James Baumgartner, Richard Laver, and William Mitchell.[24]

Cohen himself gave a comprehensive course at Harvard in the Spring of 1965 which resulted in a monograph [15]. This monograph would have served

[24]See [43],pp.160-2.

as a fine introduction to mathematical logic at the time, and as such it exhibits Cohen's by then impressive command of the subject.[25] The first third of the monograph is given over to the Gödel Incompleteness Theorems and recursive functions; Cohen's early interest in number theory comes through, and the culminating result presented is what is now known as Tennenbaum's theorem, that no countable non-standard model of Peano Arithmetic can be recursively presented. The middle third of the monograph is an incisive account of axiomatic set theory through Gödel's work on L to the minimal model, an account shedding Gödel's formal style and pivoting to a modern presentation. The final third of the monograph is given over, of course, to forcing and independence results. Cohen incorporated the methodological changes made by others in 1963 and presents, or least cites, most of the results of that year.

On the subject of establishing formal relative consistency in view of his standard model assumption, Cohen outlined both his syntactic approach about how all conditions P force a statement, described in his [14], and the semantic approach about using standard models of only finitely many axioms, described in his [13]. As to the former, he wrote (pp.147–8):

> Although this point of view may seem like a rather tedious way of avoiding models, it should be mentioned that in our original approach to forcing this syntactical point of view was the dominant point of view, and models were later introduced as they appeared to simplify the exposition. The peculiar role of the countability of M is here entirely avoided.

This stands in contrast to Cohen's early emphasis on the minimal model and his reminiscences ([19], p.1092) quoted earlier about his discovery of forcing through the contemplation of adding a new set a of integers to a standard model. This tension between syntactic consistency and actual model building would play an important role in Cohen's thinking about mathematics, particularly his coming espousal of formalism while at the same time emphasizing the operational importance of working with mathematical objects. This tension would henceforth remain in set theory once forcing as method became embedded into its fabric.

In his conclusion, Cohen offered the following point of view about CH from an "idealist" perspective (p.151):

> A point of view which the author feels may eventually come to be accepted is that CH is *obviously* false. The main reason one accepts the Axiom of Infinity is probably that we feel it absurd

[25] Incidentally, among those whom Cohen credited for help in the preparation of the manuscript were David Pincus, Thomas Scanlon, and Jon Barwise, all who would make significant contributions to mathematical logic.

> to think that the process of adding only one set at a time can exhaust the entire universe. Similarly with the higher axioms of infinity. Now $\aleph_1$ is the set of countable ordinals and this is merely a special and the simplest way of generating a higher cardinal. The set C [the continuum] is, in contrast, generated by a totally new and more powerful principle, namely the Power Set Axiom. It is unreasonable to expect that any description of a larger cardinal which attempts to build up that cardinal from ideas deriving from the Replacement Axiom can ever reach C. Thus C is greater than $\aleph_n$, $\aleph_\omega$, $\aleph_\alpha$ where $\alpha = \aleph_\omega$ etc. This point of view regards C is an incredibly rich set given to us by one bold new axiom, which can never be approached by any piecemeal process of construction. Perhaps later generations will see the problem more clearly and express themselves more eloquently.

Turning to the new work, the main achievements with forcing in the years immediately after 1963 would be due by Solovay. As we recede from the vortex of Cohen's initial advance, we give only a synopsis of these results, one which however hardly does justice to their full significance for the future development of set theory. Working on that Lebesgue measure problem of Cohen's, Solovay [50, 52] in 1964 established a result remarkable for its early sophistication and revelatory of what standard of argument is possible with forcing: In the extension resulting from the Levy collapse of an inaccessible cardinal, the inner model consisting of sets definable with ordinal and real parameters satisfies: The Principle of Dependent Choices and "all sets of reals are Lebesgue measurable".[26] The proof used Solovay's concept of a *random real* and the homogeneity and universality of the Levy collapse, themes significantly augmenting those from Levy's own work and still relevant today in the investigation of regularity properties of sets of reals.

By June 1965, Solovay and Tennenbaum [54] established the relative consistency of Souslin's Hypothesis. Forcing with a Souslin tree as an ordered set of conditions "kills" the tree, and they systematically killed all Souslin trees in the first genuine argument by iterated forcing.

In 1966 Solovay [51, 53] established the equiconsistency of there being a measurable cardinal and $2^{\aleph_0}$ being real-valued measurable. In the forward direction he adjoined many random reals and applied the Radon-Nikodym theorem of analysis, and in the converse direction he drew out the important concepts of *saturated ideal* and *generic ultrapower*, concepts that would become basic to the integration of forcing and large cardinals.

The development of forcing as method went hand in hand with this procession of central results, even at a basic level before iteration and integration with

[26] See [43] for the vicissitudes in the development of Solovay's result.

large cardinals. As early as 1963 Solovay realized the Cohen's framework can encompass arbitrary partial orders with the theory developed in terms of their incompatible members and dense sets. Later, Solovay brought in *generic filters*, a concept accredited to Levy,[27] loosening genericity from having a complete sequence and hence the countability of the ground model. While in Cohen's work sets of conditions were evidently concocted to yield generic objects directly witnessing existence assertions, in Levy's and Solovay's work various sets of conditions and their relation to genericity assumed a separate significance. For this a general formulation in terms of partial orders became desirable and incumbent.

When working on his Lebesgue measurability result, complications in describing when a formula holds over a range of generic extensions led Solovay to the idea of assigning a value to a formula from a complete Boolean algebra. He initially assigned to each formula a pair of sets, consisting first of those conditions forcing the formula and second of those conditions forcing its negation. Solovay described this to Scott in September 1965, and Scott pointed out that Solovay's Boolean algebra was essentially the algebra of regular open sets of the underlying topology. Working independently, Solovay and Scott soon developed the idea of recasting forcing entirely in terms of *Boolean-valued models.*[28] This approach showed how Cohen's ramified languages can be replaced by a more direct induction on rank and made evident how a countable standard model was not needed. By establishing in ZFC that e.g. there is a complete Boolean algebra assigning $\neg$CH Boolean value one, a semantic construction was replaced by a syntactic one that directly secured relative consistency.

Scott popularized Boolean-valued models in lectures at a four-week set theory conference held in the summer of 1967 at the University of California at Los Angeles (UCLA). This was by all accounts one of those rare, highly exhilarating conferences featuring groundbreaking papers and lectures presented that both summarized the progress and focused the energy of a new field opening up, this in large part due to Cohen's work.

At the UCLA conference, Cohen [17] presented his own thoughts on the foundations of set theory. Taking the critical point to be the existence of infinite totalities, Cohen framed the issue in dichotomous terms as a choice between "Realism" and "Formalism". He repeated his [14] contention that from a Realist perspective the size of the continuum may come to be considered very large because of the potency of the power set. On the other hand, he opted for Formalism, a choice which for him carries with it as a heavy weight (p.13) "the admission that CH, perhaps the first significant question about uncount-

[27]See [52],n.5.

[28]See [43],p.163. It turned out that Petr Vopěnka [60] leading his Prague seminar had developed a similar approach, though his earlier papers did not have much impact partly because of an involved formalism.

able sets which can be asked, has no intrinsic meaning." In the subsequent discussion, Cohen did point out how among mathematicians "there is a natural tendency to replace discussion of methods and statements by discussion of suitable abstractions which are considered as 'objects'," and recalled how with the expansion of the concept of function Weierstrass's continuous nowhere differentiable function came to have as legitimate an existence as $\sin x$. There is a quiet resonance here between Cohen's coming down on the Formalist side and his syntactic approach to formal relative consistency, described earlier, in light of his operational use of standard models. At the end Cohen put forward the view that "we do set theory because we have an informal consistency proof for it", and gave a sketch of how one might reduce the complexity of a putative contradiction à la Gentzen in proof theory.

Also at that UCLA conference Joseph Shoenfield [47] gave lucid lectures on "unramified forcing". Shoenfield both advanced the general partial order–generic filter formulation and eliminated the ramified language as in the Boolean-valued approach. Shoenfield had terms for denoting sets in the extension that were remarkably simple and clearly brought out how the only hierarchical dependence can be on the inherent well-foundedness of sets. Boolean-valued models, with their elegant algebraic trappings and seemingly more complete information, had held the promise of being the right approach to independence results. However, the view of forcing as a way of actually extending models with conditions held the reservoir of mathematical sense and the promise of new discovery, and bolstered by Shoenfield's simple formulation set theorists were soon proceeding in terms of partial orders and generic filters.

The theory and method of forcing stabilized by the early 1970s as follows:

(1) Forcing is a matter of partial orders and generic filters much as presented in [47], and as a heritage from Boolean-valued models, $\mathbf{1}$ denotes weakest condition and the direction $p < q$ is for p having more information than q.[29]

(2) V is typically construed as the ground model; a partial order $P \in V$ as a "notion of forcing" is specified to a purpose; a generic filter $G \notin V$ is posited; and an extension $V[G]$ taken, its properties argued for based on the combinatorial properties of P.

(3) Boolean algebras underscore the setting and are sometimes a necessary augmentation; most importantly, arguments about sub-extensions generated by terms and embeddings of extensions are sometimes best or necessarily cast in terms of Boolean algebras.

Item (2) is both representative and symptomatic of an underlying transformation of attitudes in and about set theory in large part brought on by

[29] There is a persistent Israeli revisionism in the other direction, following Saharon Shelah.

the advent of forcing. Up to Gödel's work on L and just beyond, the focus of set theory was on V as *the* universe of sets and what sets *are*. Set theory had born a special ontological burden because it is a theory of extensions to which mathematics can be reduced. On the one hand, sets appear to be central to mathematics. On the other hand, there seems to be no strong or evident metaphysical or epistemological basis for sets. This apparent perplexity led to metaphysical recastings of sets and more subtle appropriations like the sole reliance on some prior logical or iterative conception of set. After Gödel, set theory began to shed this ontological burden and the main concerns became what sets *do* and how set theory is to advance as an autonomous field of mathematics. With Cohen there was an infusion of mathematical thinking and of method and a proliferation of models, must as in other modern, sophisticated fields of mathematics. Taking V as the ground model goes against the sense of V as the universe of all sets and "Tarski's undefinability of truth", but actually V has become a *schematic* letter for a ground model. This further drew out that in set theory as well as in mathematics generally, it is a matter of method, not ontology.

Forcing has been interestingly adapted in a category theory context which is a casting of set theory in intuitionistic logic. The basic discovery, jointly due to William Lawvere and Myles Tierney, was that forcing can be interpreted as the construction of a certain topos of sheaves. The internal logic of the topos of presheaves over a partially ordered set is essentially Cohen's forcing, while passing to the subtopos of sheaves of the double-negation Grothendieck topology gives weak forcing. In the first, 1970 paper on elementary topoi Lawvere [36] gave a brief indication of how Cohen's independence of CH would look from a topos point of view, and then Tierney [59] provided the details. Later in 1980 Peter Freyd [28] gave a direct topos-theoretic proof of the independence of AC. Although his models could have been obtained by standard set-theoretic methods, they look simpler from the topos point of view than from the set-theoretic point of view. The details of correlation were worked by Solovay in unpublished notes, and by Andreas Blass and Andre Scedrov [5].[30]

In subsequent years Cohen would publish only five more papers. The first [16] was on decision procedures; the second [18] was a throwback on forcing; written many years later, the third [19] was an extended reminiscence of the discovery of forcing to which we have already referred; the fourth [20] was a paper on Skolem in which he articulated his views on the limits of proof; and the last [21] was a reminiscence of his career and interactions with Gödel, covering in summary terms ground covered elsewhere, for a conference on the centenary of Gödel's birth. We deal successively with [16, 18, 20].

In work that came full circle back to his budding interest in decision procedures in his graduate days at Chicago, Cohen [16] developed a concrete decision

[30] My thanks to Andreas Blass for all the particulars of this paragraph.

procedure for the p-adic fields. In well-known work of the mid-1960s, James Ax and Simon Kochen [2] had used ultraproducts to get a complete, recursive set of axioms for p-adic number theory and therewith a decision procedure. Cohen proceeded more directly as in the Tarski decision procedure for real-closed fields to carry out an elimination of quantifiers. With his constructive approach, Cohen was able to get a primitive recursive decision procedure as well to easily isolate the properties of fields with valuations which are being used.

In [18] Cohen answered a question that he had himself posed in [14] by getting a model of ZF with an automorphism σ whose square σ^2 is the identity. Such a σ would have to fix the ordinals, and so would be the identity if AC were to hold — since models of AC are determined by their sets of ordinals. Hence, this was a new way of getting $\neg$AC. Also, no model of ZF with an automorphism can be well-founded. As a variation on method, Cohen was able to start with a ill-founded countable model of ZF and force the desired result in an extension without altering the ordinals. To get a σ as desired, he introduced certain generic sets along rank levels and constructed σ generically through the use of a particular complete sequence. As mentioned earlier, this confirmed as a methodological point that forcing can be carried out on ill-founded models.

In [20], a paper given at a conference on mathematical proof at the British Royal Society, Cohen discussed Thoralf Skolem, the limits of proof and formalization, and "the ultimate pessimism deriving from Skolem's views". In the process, Cohen brought in his own experiences and approaches and summed up his own thinking. In Skolem Cohen evidently saw a kindred spirit, one whose mathematical work and conclusions in an earlier time he viewed his own as complementing and extending. The work was of course the Lowenheim-Skolem theorem leading to the Skolem Paradox, from which Skolem drew that conclusion that first-order axiomatization in terms of sets cannot be a satisfactory foundation for mathematics. This "pessimism" Cohen will extend, to assert (p.2408) that "it is unreasonable to expect that any reasoning of the type we call rigorous mathematics can hope to resolve all but the tiniest fraction of possible mathematical questions."

Cohen proceeded to give his own gloss on developments from Frege through Hilbert to Gödel. Coming to Skolem's infusion of witnessing constants, Cohen wrote (p.2411):

> The fundamental discovery of Lowenheim-Skolem, which is undoubtedly the greatest discovery in pure logic, is that the invention (or introduction) of 'constants' as in predicate calculus, is equivalent to the construction of a 'model' for which the statements hold.

This of course has an underlying resonance with forcing. Cohen pointed out how "Skolem's work received amazingly little attention", and went to accord

Skolem a high place in a salient passage that brings out Cohen's own standpoint (p.2411):

> Skolem wrote in a beautiful, intuitive style, totally precise, yet more in the spirit of the rest of mathematics, unlike the fantastically pedantic style of Russell and Whitehead. Thus, Hilbert even posed as a problem the very result that Skolem had proved, and even Gödel, in his thesis where he proved what is known as the Completeness Theorem, does not seem to have appreciated what Skolem had done, although in a footnote he does acknowledge that 'an analogous procedure was used by Skolem'. A possible explanation lies in the fact that Skolem emphasized models, and was amazingly prescient in some of his remarks concerning independence proofs in set theory. A discussion of the priority question can be found in the notes to Gödel's Collected Works (Gödel 1986). Gödel was undoubtedly sincere in his belief that his proof was in some sense new, and in view of his monumental contributions I in no way wish to find fault with his account. What is interesting is how the more philosophical orientation of logicians of the time, even the great Hilbert, distorted their view of the field and its results.

Earlier, Cohen had quoted the well-known passage from Skolem [48],p.229 in which Skolem considered adjoining a set of natural numbers to a model of set theory in connection with the continuum problem.

On constructivity, Cohen notably gave as what he believed to be "the first example of a truly non-constructive proof in number theory" the Thue-Siegel-Roth result that he had been working on in his first graduate days at Chicago.

Discussing consistency questions, Cohen sketched the idea of Gentzen's consistency proof for Peano Arithmetic "in my own version which I intend to publish some day." (p.2414) Cohen had a few years earlier given talks about this version at Stanford and Berkeley.

Getting to "the ultimate frontier", set theory, Cohen more or less reworks his earlier [17] remarks on the foundations of set theory, mentioning (p.2416) that "Through the years I have sided more firmly with the formalist position." Despite this, Cohen arrived at the ultimate pessimism via a basic ambivalence (p.2417):

> ...Even if the formalist position is adopted, in actual thinking about mathematics one can have no intuition unless one assumes that models exist and that the structures are real.
>
> So, let me say that I will ascribe to Skolem a view, not explicitly stated by him, that there is a reality to mathematics, but axioms cannot describe it. Indeed one goes further and says that there is

> no reason to think that any axiom system can adequately describe it.

Cohen then returned to the bedrock of number theory and gave as an example the twin primes conjecture as beyond the reach of proof. "Is it not very likely that, simply as a random set of numbers, the primes do satisfy the hypothesis, but there is no logical law that implies this?"[31] How about higher axioms of infinity resolving more and more arithmetical statements? "There is no intuition as to why the consideration of the higher infinite should brings us closer to solving questions about primes." Cohen speculated whether statistical evidence will someday count as proof, and ended:

> In this pessimistic spirit, I may conclude by asking if we are witnessing the end of the era of pure proof, begun so gloriously by the Greeks. I hope that mathematics lives for a very long time, and that we do not reach that dead end for many generations to come.

6 Envoi

Within a decade after Cohen's discovery forcing became a systematic part of the burgeoning field of modern set theory. That generic extensions have the same ordinals and satisfy ZF and the basic connection between the ground model and extension through the forcing relation are now simply taken for granted as part of the underlying theory. Rather, the focus is on the connections between the combinatorial properties of the partial order of conditions and structural properties of the extension. Through a natural progression of mathematical development there has evolved a vast range of different notions of forcing as themselves paradigms of construction, a vast technology for iterated forcing, and a vast web of interactions among forcing properties and central propositions of set theory.

Forcing has thus come to play a crucial role in the transformation of set theory into a modern, sophisticated field of mathematics, one tremendously successful in the investigations of the continuum, transfinite combinatorics, and strong propositions and their consistency strength. In all these directions forcing became integral to the investigation and became part of their very sense, to the extent that issues about the method became central and postulations in its terms, "forcing axioms", became pivotal.

Cohen would not be party to any of these further developments. It could be said that he was in the end a problem solver rather than a system builder.

[31] See [20],p.2418. On the other hand, Chen Jingrun [6] in 1966 proved, toward Goldbach's conjecture, that every sufficiently large even number is the sum of two primes or the sum of a prime and a *semiprime*, i.e. a product of two primes. In the process he proved, toward the twin primes conjecture, that there are infinitely many primes p such that $p + 2$ is either a prime or a semiprime.

Whatever is the case, Cohen seemed to evince little interest in the many new models of set theory and the elaboration of forcing as method and returned to the bedrock of number theory with a specifically formalist attitude toward mathematics.

With forcing so expanded into the interstices of set theory and the method so extensively amended from the beginning, what is the "it" of Cohen's forcing and his individual achievement? Cohen discovered a concrete and widely applicable means of operationally extending a standard model of set theory to another without altering the ordinals. The central technical innovation was the definable forcing relation, through which satisfaction for the extension could be approached in the ground model. Cohen's achievement was thus to be able to secure properties of new sets without having all of their members in hand and more broadly, to separate and then interweave truth and existence.

How singular a phenomenon is Cohen's forcing? On the precursory side, there were the Beth and Kripke semantics for modal and intuitionistic logic[32] in which satisfaction is not fixed in just a one model. But here, there is no thought of securing the *existence* of a new model. More pointedly, there were constructions in recursion theory, particularly Spector's two-quantifier argument[33] for getting minimal Turing degrees. But here, there is no thought of securing *truth* for a full model.

Of other roads to independence not taken, one might have thought at the time of appealing to an omitting-types theorem for first-order logic. But how does one ensure the second-order property of well-foundedness, without which other problems ensue? With later work in infinitary logic one can secure well-founded extensions. But how does one secure the Power Set Axiom?

The force of forcing is that, while appearing dramatically on the mathematical stage, it is a basically simple, though remarkable powerful, method. In reminiscences Cohen wrote[34]

> ...it's somewhat curious that in a certain sense the continuum hypothesis and the axiom of choice are not really difficult problems — they don't involve technical complexity; nevertheless, at the time they were considered difficult. One might say in a humorous way that the attitude toward my proof was as follows. When it was first presented, some people thought it was wrong. Then it was thought to be extremely complicated. Then it was thought to be easy. But of course it *is* easy in the sense that there is a clear philosophical idea. There were technical points, you know, which bothered me, but basically it was not really an enormously involved combinatorial problem; it was a philosophical idea.

[32]See Dummett [22].
[33]See [55]; the argument is cast as a forcing argument in Lerman [37].
[34]See [1],p.58.

Without bearing the historical weight of an earlier turn to logic in set theory and proceeding from "ordinary mathematics" to algebraitize truth and existence together, Cohen was able to cut through to a construction that actuates a *new way of thinking.* In Kantian terms, Cohen provided an *organon*, an instrument for the generation of new knowledge.

References

[1] Donald J. Albers, Gerald L. Alexanderson, and Constance Reid, editors. *More Mathematical People.* Harcourt Brace Jovanovich, Boston, 1990.

[2] James Ax and Simon Kochen. Diophantine problems over local fields II. A complete set of axioms for p-adic number theory. *American Journal of Mathematics*, pages 631–648, 1965.

[3] Stefan Banach. Über die Barie'sche Kategorie gewisser Funktionenmengen. *Studia Mathematica*, 3:174, 1931.

[4] John L. Bell. *Boolean-Valued Models and Independence Proofs in Set Theory.* Oxford Logic Guides. Clarendon Press, Oxford, 1977. Second edition, 1985; third edition, 2005.

[5] Andreas R. Blass and Andre Scedrov. Freyd's models for the independence of the axiom of choice. *Memoirs of the American Mathematical Society*, 79(404):134pp., 1989.

[6] Jingrun Chen. On the presentation of a large even number as the sum of a prime and the product of at most two primes. *Kexue Tongbao (Chinese Science Bulletin)*, 17:385–388, 1966. In Chinese.

[7] Paul J. Cohen. On Green's theorem. *Proceedings of the American Mathematical Society*, 10:109–112, 1959.

[8] Paul J. Cohen. On a conjecture of Littlewood and idempotent measures. *American Journal of Mathematics*, 82:191–212, 1960.

[9] Paul J. Cohen. A note on constructive methods in Banach algebras. *Proceedings of the American Mathematical Society*, 12:159–163, 1961.

[10] Paul J. Cohen. Idempotent measures and homomorphisms of group algebras. In *Proceedings of the International Congress of Mathematicians (Stockholm, 1962)*, pages 331–336. Institut Mittag-Leffler, Djursholm, 1963.

[11] Paul J. Cohen. The independence of the Continuum Hypothesis I. *Proceedings of the National Academy of Sciences U.S.A.*, 50:1143–1148, 1963.

[12] Paul J. Cohen. A minimal model for set theory. *Bulletin of the American Mathematical Society*, 69:537–540, 1963.

[13] Paul J. Cohen. The independence of the Continuum Hypothesis II. *Proceedings of the National Academy of Sciences U.S.A.*, 51:105–110, 1964.

[14] Paul J. Cohen. Independence results in set theory. In John W. Addison Jr., Leon Henkin, and Alfred Tarski, editors, *The Theory of Models, Proceedings of the 1963 International Symposium at Berkeley*, Studies in Logic and the Foundations of Mathematics, pages 39–54. North-Holland, Amsterdam, 1965.

[15] Paul J. Cohen. *Set Theory and the Continuum Hypothesis.* W.A.Benjamin, New York, 1966.

[16] Paul J. Cohen. Decision procedures for real and p-adic fields. *Communications on Pure and Applied Mathematics*, pages 131–151, 1969.

[17] Paul J. Cohen. Comments on the foundations of set theory. In Dana S. Scott, editor, *Axiomatic Set Theory*, volume 13 of *Proceedings of Symposia in Pure Mathematics*, pages 9–15. American Mathematical Society, Providence, 1971.

[18] Paul J. Cohen. Automorphisms of set theory. In *Proceedings of the Tarski Symposium*, volume 25 of *Proceedings of Symposia in Pure Mathematics*, pages 325–330. American Mathematical Society, Providence, 1974.

[19] Paul J. Cohen. The discovery of forcing. *Rocky Mountain Journal of Mathematics*, 32(4):1071–1100, 2002.

[20] Paul J. Cohen. Skolem and pessimism about proof in mathematics. *Philosophical Transactions of The Royal Society*, 363:2407–2418, 2005.

[21] Paul J. Cohen. My interaction with Kurt Gödel: The man and his work. In *Horizons of Truth.* Cambridge University Press, Cambridge, 2009. In honor of the 100th annniversay year of Kurt Gödel (2006).

[22] Michael Dummett. *Elements of Intuitionism*, volume 2 of *Oxford Logic Guides.* Clarendon Press, Oxford, 1977.

[23] William B. Easton. Proper class of generic sets. *Notices of the American Mathematical Society*, 11:205, 1964. Abstract.

[24] William B. Easton. Powers of regular cardinals. *Annals of Mathematical Logic*, 1:139–178, 1970. Abridged version of Ph.D. thesis, Princeton University 1964.

[25] Solomon Feferman. Some applications of the notions of forcing and generic sets. *Fundamenta Mathematicae*, 56:325–345, 1965.

[26] Solomon Feferman. Some applications of the notions of forcing and generic sets (summary). In John W. Addison Jr., Leon Henkin, and Alfred Tarski, editors, *The Theory of Models, Proceedings of the 1963 International Symposium at Berkeley*, Studies in Logic and the Foundations of Mathematics, pages 89–95. North-Holland, Amsterdam, 1965.

[27] Solomon Feferman and Azriel Levy. Independence results in set theory by Cohen's method II. *Notices of the American Mathematical Society*, 10:593, 1963. Abstract.

[28] Peter J. Freyd. The axiom of choice. *Journal of Pure and Applied Algebra*, 19:103–125, 1980.

[29] Kurt F. Gödel. Consistency proof for the generalized continuum hypothesis. *Proceedings of the National Academy of Sciences U.S.A.*, 25:220–224, 1939.

[30] Kurt F. Gödel. *The Consistency of the Axiom of Choice and of the Generalized Continuum Hypothesis with the Axioms of Set Theory.* Number 3 in Annals of Mathematics Studies. Princeton University Press, Princeton, 1940.

[31] Kurt F. Gödel. *Collected Works, volume IV: Correspondence A–G.* Clarendon Press, Oxford, 2003. Edited by Solomon Feferman and John W. Dawson Jr.

[32] James D. Halpern. The independence of the Axiom of Choice from the Boolean Prime Ideal Theorem. *Fundamenta Mathematicae*, 55:57–66, 1964.

[33] James D. Halpern and Hans Läuchli. A partition theorem. *Transactions of the American Mathematical Society*, 124:360–367, 1966.

[34] James D. Halpern and Azriel Levy. The ordering theorem does not imply the axiom of choice. *Notices of the American Mathematical Society*, 11:56, 1964. Abstract.

[35] James D. Halpern and Azriel Levy. The Boolean prime ideal theorem does not imply the axiom of choice. In Dana S. Scott, editor, *Axiomatic Set Theory*, volume 13 of *Proceedings of Symposia in Pure Mathematics*, pages 83–134. American Mathematical Society, Providence, 1971.

[36] F. William Lawvere. Quantifiers and sheaves. In *Actes du Congrès International des Mathématiciens (Nice 1970)*, volume 1, pages 329–334. Gauthier-Villars, Paris, 1971.

[37] Manuel Lerman. *Degrees of Unsolvability*. Perspectives in Mathematical Logic. Springer-Verlag, Berlin, 1983.

[38] Azriel Levy. Independence results in set theory by Cohen's method I. *Notices of the American Mathematical Society*, 10:592–593, 1963. Abstract.

[39] Azriel Levy. Independence results in set theory by Cohen's method III. *Notices of the American Mathematical Society*, 10:593, 1963. Abstract.

[40] Azriel Levy. Independence results in set theory by Cohen's method IV. *Notices of the American Mathematical Society*, 10:593, 1963. Abstract.

[41] Azriel Levy. Definability in axiomatic set theory I. In Yehoshua Bar-Hillel, editor, *Logic, Methodology and Philosophy of Science. Proceedings of the 1964 International Congress at Jerusalem*, Studies in Logic and the Foundations of Mathematics, pages 127–151. North-Holland, Amsterdam, 1965.

[42] Azriel Levy. Definability in set theory II. In Yehoshua Bar-Hillel, editor, *Mathematical Logic and Foundations of Set Theory. Proceedings of an International Colloquium*, Studies in Logic and the Foundations of Mathematics, pages 129–145. North-Holland, Amsterdam, 1970.

[43] Gregory H. Moore. The origins of forcing. In Frank R. Drake and John K. Truss, editors, *Logic Colloquium '86*, Studies in Logic and the Foundations of Mathematics, pages 143–173. North-Holland, Amsterdam, 1988.

[44] Sylvia Nasar. *A Beautiful Mind*. Simon and Schuster, New York, 1998.

[45] Dana S. Scott. Measurable cardinals and constructible sets. *Bulletin de l'Académie Polonaise des Sciences, Série des Sciences Mathématiques, Astronomiques et Physiques*, 9:521–524, 1961.

[46] John C. Shepherdson. Inner models of set theory — part III. *The Journal of Symbolic Logic*, 18:145–167, 1953.

[47] Joseph R. Shoenfield. Unramified forcing. In Dana S. Scott, editor, *Axiomatic Set Theory*, volume 13 of *Proceedings of Symposia in Pure Mathematics*, pages 357–381. American Mathematical Society, Providence, 1971.

[48] Thoralf Skolem. Einige Bemerkungen zur axiomatischen Begründung der Mengenlehre. In *Matematikerkongressen i Helsingfors den 4–7 Juli 1922, Den femte skandinaviska matematikerkongressen, Redogörelse*, pages 217–232. Akademiska-Bokhandeln, Helsinki, 1923.

[49] Stephen Smale. Differentiable dynamical systems. *Bulletin of the American Mathematical Society*, 73:747–817, 1967.

[50] Robert M. Solovay. The measure problem. *Notices of the American Mathematical Society*, 12:217, 1965. Abstract.

[51] Robert M. Solovay. Real-valued measurable cardinals. *Notices of the American Mathematical Society*, 13:721, 1966.

[52] Robert M. Solovay. A model of set theory in which every set of reals is Lebesgue measurable. *Annals of Mathematics*, 92:1–56, 1970.

[53] Robert M. Solovay. Real-valued measurable cardinals. In Dana S. Scott, editor, *Axiomatic Set Theory*, volume 13 of *Proceedings of Symposia in Pure Mathematics*, pages 397–428. American Mathematical Society, Providence, 1971.

[54] Robert M. Solovay and Stanley Tennenbaum. Iterated Cohen extensions and Souslin's problem. *Annals of Mathematics*, 94:201–245, 1971.

[55] Clifford Spector. On degrees of recursive unsolvability. *Annals of Mathematics*, 64:581–592, 1956.

[56] Clifford Spector. Measure-theoretic construction of incomparable hyperdegrees. *The Journal of Symbolic Logic*, 23:280–288, 1958.

[57] John R. Steel. Forcing with tagged trees. *Annals of Mathematical Logic*, 15:55–74, 1978.

[58] Stanley Tennenbaum. Souslin's problem. *Proceedings of the National Academy of Sciences U.S.A.*, 59:60–63, 1968.

[59] Myles Tierney. Sheaf theory and the continuum hypothesis. In *Toposes, Algebraic Geometry and Logic*, volume 274 of *Lecture Notes in Mathematics*, pages 13–42. Springer, Berlin, 1972.

[60] Petr Vopěnka. The general theory of ∇-models. *Commentationes Mathematicae Universitatis Carolinae*, 8:145–170, 1967.

[61] Benjamin H. Yandell. *The Honors Class. Hilbert's Problems and their Solutions.* A K Peters, Natick, 2002.

17 Kunen and Set Theory

Abstract. This commemorative article provides an account of Kenneth Kunen's work in set theory.[1]

Kenneth Kunen (1943–) figured principally in the development of set theory in all the major directions, this during a formative period for the subject when it was transmuting to a sophisticated field of mathematics. In fact, several of Kunen's results and proofs still frame modern set theory or serve as standards to be reckoned with in its further progress. This is all the more notable as much of the work was done in a short run of about four years from his 1968 thesis. The work has an incisiveness and technical virtuosity remarkable for the time as well as a sense of maturity and definiteness. Typically, others may have started a train of conceptual constructions, but Kunen made great leaps forward advancing the subjects, almost heroically, to what would seem to be the limits achievable at the time. There is a sense of movement from topic to topic and then beyond to applications of set theory, with heights scaled and more heights beckoning.

In what follows, we chronicle Kunen's singular progress over the broad swath of set theory. The work of the years 1968-1972, especially, deserve full airing, and we recall the initiatives of the time as well as describe the ramifications of the advances made. As almost all of this work is in the mainstream of modern set theory, we only set the stage in a cursory way and recall the most immediately relevant concepts, relying in part on the readers' familiarity,[2] but then dwell on the particulars of how ideas and proofs became method. In this way we reaffirm Kunen's work in set theory as central to the subject and of lasting significance for its development.

1 Thesis

Kunen's 1968 thesis [43] was itself a notable landmark in the development of modern set theory, and in its breadth and depth the thesis conveyed a sense of broad reach for the emerging field. For discussing its role and significance, we quickly sketch, though in a deliberate way to our purpose, how set theory was beginning its transmutation into a field of mathematics with sophisticated methods.

With the emergence of the now basic ultraproduct construction in model theory, Dana Scott in 1961 took an ultrapower of the entire set-theoretic universe V to establish that having a measurable cardinal contradicts Gödel's

Republished from *Topology and its Applications* 158 (2011), pp.43-60, with permission from Elsevier.

[1]This paper was written whilst the author was a senior fellow at the Lichtenberg-Kolleg at the University of Göttingen. He expresses his gratitude for the productive conditions and support of the kolleg.

[2]See Jech [33] for the set-theoretic concepts and results, and Kanamori [38] for those involving large cardinals.

Axiom of Constructibility $V = L$. The ultrapower having brought set theory to the point of entertaining elementary *embeddings* into well-founded models, it was soon transfigured by a new means for getting well-founded *extensions* of well-founded models. This was *forcing*, of course, discovered by Cohen in 1963 and used by him to establish the independence of the Axiom of Choice (AC) and of the Continuum Hypothesis (CH). With clear intimations of a new, concrete, and flexible way of building models, many rushed into set theory, and, with forcing becoming method, were soon establishing a cornucopia of relative consistency results.

Robert Solovay epitomized this initial period of sudden expansion with his mathematical sophistication and many results over the expanse of set theory. In 1964, he established his now-famous result that if an inaccessible cardinal is Levy collapsed to make it ω_1, there is an inner model of ZF + DC in which all sets of reals are Lebesgue measurable.[3] Toward this result, he formulated a new generic real, the random real. In 1965, he worked out with Stanley Tennenbaum the iterated forcing proof of the consistency of Suslin's Hypothesis and therewith of Martin's Axiom (MA). In 1966, Solovay established the equi-consistency of the existence of a measurable cardinal and the real-valued measurability of $2^{\aleph_0}$, i.e. having a (countably additive) measure extending Lebesgue measure to all sets of reals. In the process, he worked out structural consequences of saturated ideals and generic ultrapowers.

While the "forcing king" was forging forth, the University of California at Berkeley became a hotbed of activity in set theory, with Jack Silver's 1966 thesis the initial high point, with its large cardinal analysis of L leading to the set of integers $0^{\#}$ (also isolated by Solovay). To the south, at the University of California at Los Angeles, a huge multi-week conference was held in the summer of 1967 that both summarized the progress already made and focused the energy of a new field opening up.

As a graduate student at Stanford University, Kunen [42] initially developed certain initiatives of Georg Kreisel about effectiveness and compactness of infinitary languages, as did Jon Barwise. Interest in this direction would persist through to a joint Barwise-Kunen study [5] of a cardinal characteristic of infinitary languages. At that 1967 UCLA conference, Kunen [47] presented work on a somewhat related topic, indescribability and the continuum. After the conference, however, Kunen had moved squarely into set theory, with his thesis of a year later having 10 items in its bibliography from the packaged proceedings of the conference.

That thesis, *Inaccessibility Properties of Cardinals* [43], of August 1968 and with Scott as advisor, consisted of two parts. Part I dealt with strong inaccessibility properties, mainly applications of iterated ultrapowers. Part II

[3]DC is the Axiom of Dependent Choice, a weak form of the Axiom of Choice sufficient for developing the classical theory of measure and category for the reals.

discussed weak inaccessibility properties, particularly of the continuum. The next section describes the work of the first part and its extensions, and here we continue with a description of the work of the second.

To the extent that Kunen's Part II was disseminated, it served as an early exemplar for the subsequent and wide-ranging analyses of the continuum assuming real-valued measurability and assuming Martin's Axiom. In an introductory section Kunen already stated an important result, a carefully formulated Π^2_1-indescribability for real-valued measurable cardinals. This result was fully aired by Solovay in his eventual real-valued measurability paper [90, §6]. In succeeding sections Kunen addressed three problems going in different directions:

(a) Is every subset of the plane in the σ-algebra generated by arbitrary rectangles (products of subsets of the line)?

(b) Is the Boolean algebra of all subsets of the reals modulo sets of cardinality less than the continuum weakly (ω, ω) distributive?

(c) Must every set of reals of cardinality less than the continuum be of Lebesgue measure 0 (or of first category)?

In addressing these questions, Kunen exhibited a fine and quick understanding of the contexts recently established by the consistency work of Solovay on real-valued measurability and on Martin's Axiom and a remarkable working knowledge of classical set-theoretic work on the continuum.

Question (a) was inspired by a presentation of Richard Mansfield (cf. his [74]) at the UCLA conference. Kunen drew on work from Hausdorff [31], nowadays well-known as the source of the "Hausdorff gap", to establish how MA leads to a positive conclusion in a structured sense. He then observed how his indescribability result for real-valued measurability leads to a negative conclusion. R. H. Bing, W. W. Bledsoe and R. Daniel Mauldin [8] subsequently established results about the number of steps needed, starting from the rectangles, to generate all the subsets.

Question (b) Kunen reformulated forthwith in terms of functions $f : \omega \to \omega$ under the eventual dominance ordering, as had already been done classically, and showed that both real-valued measurability and MA entail positive conclusions.

For question (c) Kunen first recalled Solovay's results that under MA any union of fewer than continuum many Lebesgue measure zero sets is again measure zero, and likewise for any union of first category sets. Kunen then went on to observe that MA implies that there is a (generalized) Luzin set, i.e. a set of cardinality the continuum whose intersection with any first category set is of smaller cardinality, and that MA implies that every set of cardinality less than the continuum has strong measure zero. A set of reals has *strong measure zero* if for any sequence $\langle a_i \mid i \in \omega \rangle$, there are open intervals I_i of length at most a_i for $i \in \omega$, whose union covers the set. (Kunen mentioned that Borel in 1919 conjectured that all sets of strong measure zero are countable; with this question recalled in the post-Cohen area, Richard Laver [69], with the first clear use of countable-support iterated forcing, famously established the consistency of Borel's conjecture.) Finally, Kunen showed that under real-valued measurability the answers to Question (c) are yes for category, and, applying his indescribability result (Σ^1_2-indescribability suffices), no for measure.

Kunen would make important contributions to the theory of MA itself; see Section 8. See Fremlin [21] and Fremlin [20] for compendium accounts of the mature theory for real-valued measurability and Martin's Axiom respectively, and Kunen's pioneering role situated in context.

2 Iterated Ultrapowers

The most sustained of Kunen's work in his early period was on applications of iterated ultrapowers, and this work was to become foundational for modern inner model theory. The infusion of forcing into set theory had induced a broad context extending beyond its applications and sustained by model-theoretic methods, a context which included central developments having their source in Scott's 1961 ultrapower result that measurable cardinals contradict $V = L$. Haim Gaifman [25, 26] invented *iterated ultrapowers* and established incisive results about and with the technique, and this work most immediately stimulated the results of Kunen [43, 45] on inner models of measurability.

For a normal ultrafilter U over a measurable cardinal κ, the inner model $L[U]$ of sets constructible relative to U is easily seen with $\overline{U} = U \cap L[U]$ to satisfy $L[U] \models$ "$\overline{U}$ is a normal ultrafilter". With no presumption that κ is measurable (in V) and taking $U \in L[U]$ from the beginning, call $\langle L[U], \in, U\rangle$ a *κ-model* iff $\langle L[U], \in, U\rangle \models$ "U is a normal ultrafilter over κ". Solovay observed that in a κ-model, GCH holds above κ by a version of Gödel's argument for L and that κ is the only measurable cardinal by a version of Scott's argument. Silver [84, 86] then established that the full GCH holds, thereby establishing the relative consistency of GCH and measurability; Silver's proof turned on a local structure $L_\alpha[U]$ being *acceptable* in the later parlance of inner model theory.

Kunen made Gaifman's technique of iterated ultrapowers integral to the subject of inner models of measurability. For a κ-model $\langle L[U], \in, U\rangle$, the ultrapower of $L[U]$ by U with corresponding elementary embedding j provides a $j(\kappa)$-model $\langle L[j(U)], \in, j(U)\rangle$, and this process can be repeated and the elementary embeddings composed. At limit stages, one can take the direct limit of models, which when well-founded can be identified with the transitive collapse. In terms of the stage thus set, Kunen established in his thesis [43] and subsequent paper [45]:

(a) [43, 45] For any κ, there is at most one κ-model.

(b) [45] For any κ-model and κ'-model with $\kappa < \kappa'$, the latter is an iterated ultrapower of the former.

These are the definitive structure results for inner models of measurability, results that argued forcefully for the coherence and consistency of the concept of measurability.

The argument for (b) is similar to that for (a), which we can recapitulate in terms of its essential components as follows: Suppose that $L[U]$ and $L[V]$ are both κ-models. By Gaifman's work, all the iterated ultrapowers of each are well-founded, i.e. $L[U]$ and $L[V]$ are *iterable.* Through a helpful representation of the iterates in terms of the successive critical points, for any regular $\nu > \kappa^+$, the νth iterated ultrapower of each works out, quite remarkably, to be $L[\mathcal{C}_\nu]$, where $\mathcal{C}_\nu$ is the closed unbounded filter over ν. In particular, *comparison* of $L[U]$ and $L[V]$ can be carried out by iterating them sufficiently many times. Next, with $j^U_{0\nu}$ the corresponding elementary embedding of $L[U]$ into $L[\mathcal{C}_\nu]$, if Λ is any proper subclass of the image $j^U_{0\nu}$"ON of the ordinals that includes all of κ, then the Skolem hull of Λ in $L[\mathcal{C}_\nu]$ is isomorphic to $L[U]$. The analogous result holds for $j^V_{0\nu}$ and $L[V]$. Finally, the intersection of $j^U_{0\nu}$"ON and $j^V_{0\nu}$"ON is in fact such a Λ—what we here call *stability*—and so $L[U] = L[V]$. Inner model theory would develop from Kunen's work into a mainstream of modern set theory, and the essential components of the above proof—iterability, comparison, stability—would remain at the heart of the subject (cf. the end of this section).

Beyond his structure results, Kunen in his [45] established a range of results about what happens in κ-models to ultrafilters, large cardinals, and so forth. As for generalizations, Kunen [43, 45] in 1971 showed that his structure results can be extended to inner models $L[\mathcal{U}]$ where $\mathcal{U}$ is a sequence of normal ultrafilters over various measurable cardinals whose length is less than the least of them. With this, he [43, 45] established that if there is a strongly compact cardinal, then there are inner models with arbitrarily many measurable cardinals. Subsequently in 1969, Kunen [48] was able to draw the same conclusion from the existence of a measurable cardinal κ satisfying $\kappa^+ < 2^\kappa$; this conclusion, which applied a combinatorial result of his student Jussi Ketonen about independent functions, provided the first inkling of the strength of this proposition about measurable κ, and popularized it as a focal one.[4]

From the beginning, Kunen had emphasized that iterated ultrapowers can be taken of an inner model M with respect to an ultrafilter U even if $U \notin M$, as

[4]One of the triumphs of modern set theory would be to gauge the consistency strength of having a measurable cardinal κ satisfying $\kappa^+ < 2^\kappa$. Moti Gitik [27, 28], building on work of Woodin and Mitchell, showed that this proposition is equi-consistent with Mitchell's $\exists\kappa(o(\kappa) = \kappa^{++})$.

long as U is an *M-ultrafilter*, i.e. U in addition to having M-related ultrafilter properties also satisfies an "amenability" condition for M. A crucial dividend was a characterization of the existence of $0^\#$ that secured its central importance in inner model theory. Motivated by work of Gaifman [25] on iterated ultrapowers and constructible sets, Silver in his thesis [85, 88] had investigated the generation of L with indiscernibles provided by large cardinals, and he and Solovay [89] had independently isolated the set of integers $0^\#$ as encoding the corresponding theory. With $0^\#$, any increasing shift of the Silver indiscernibles provides an elementary embedding $j\colon L \to L$. Kunen [43, Theorem 4.7] established conversely that any such embedding generates indiscernibles, so that $0^\#$ *exists iff there is a (non-identity) elementary embedding* $j\colon L \to L$. Starting with such an embedding, Kunen defined a corresponding ultrafilter U over the least ordinal moved and showed that U is an L-ultrafilter with which the iterated ultrapowers of L are well-founded. The successive images of the critical point were seen to be indiscernibles for L, giving $0^\#$. As inner model theory was to develop, this sort of sharp analysis would become a schematic cornerstone: the "sharp" of an inner model M would encapsulate transcendence over M, and the *non-rigidity* of M, that there is a (non-identity) elementary embedding $j\colon M \to M$, would provide equivalent structural sense.

Kunen's final application of iterated ultrapowers was to establish a notable result about an infinitary language generalization of constructibility. At that UCLA conference, Chen-Chung Chang [11] had presented for cardinals κ a class C^κ of sets constructible using the infinitary language $L_{\kappa\kappa}$ and observed various generalizations of the properties of $L\,(= C^\omega)$. Using iterated ultrapowers, Kunen [50] showed that unlike L, these models do not generally have intrinsic well-orderings: *If κ is regular and there are κ^+ measurable cardinals, then the Axiom of Choice fails in C^{κ^+}*. Toward this result Kunen established a striking lemma that revealed an unexpected global constraint on measurable cardinals: *For any ordinal ξ, the set of measurable cardinals κ for which there is a κ-complete ultrafilter over κ such that the corresponding ultrapower embedding moves ξ is finite.* Later, William Fleissner [14] provided a proof that does not use iterated ultrapowers.

It is a notable historical happenstance that the number of normal ultrafilters over a measurable cardinal would stimulate much of the early work in modern inner model theory. Scott had made normality central to the study of measurable cardinals through his ultrapower analysis. Kunen was evidently motivated in large part to consider κ-models in order to establish, which he did, that in any κ-model the defining normal ultrafilter is unique. Already in his thesis Kunen established with forcing that it is relatively consistent for a measurable cardinal κ to have the maximal number 2^{2^κ} of normal ultrafilters. Jeffrey Paris in his thesis [79] independently established this result as well as several of Kunen's results about κ-models. In a joint paper [62] written by

Paris, the consistency result was presented as well as some results about saturated ideals, and this early method of forcing that preserved large cardinals came to be called *Kunen-Paris forcing.*

At the end of their paper, Kunen and Paris pointedly asked whether it is consistent to have some intermediate number of normal ultrafilters between 1 and the maximal number 2^{2^κ}, e.g. 2. William Mitchell in 1972, just after completing his own pioneering, Berkeley thesis, considered this question and soon provided the first substantive extension of Kunen's inner model results. Little was known outright about normal ultrafilters, but Kunen [43] did observe that for any measurable cardinal κ, there is always a normal ultrafilter U over κ such that $\{\alpha < \kappa \mid \alpha \text{ is not measurable}\} \in U$. Thus, if there were, rather extravagantly, a normal ultrafilter W over κ such that $\{\alpha < \kappa \mid \alpha \text{ is measurable}\} \in W$, then U and W would be distinct. Taking this as a beginning point for an inner model construction, Mitchell [78] formulated what is now known as the *Mitchell order* $\triangleleft$.

For normal ultrafilters U and U' over κ, $U' \triangleleft U$ *iff* U' is in the ultrapower $\mathrm{Ult}(V, U)$ of the universe by U, i.e. there is an $f\colon \kappa \to V$ representing U' in the ultrapower, so that $\{\alpha < \kappa \mid f(\alpha) \text{ is a normal ultrafilter over } \alpha\} \in U$ and κ is already a limit of measurable cardinals. $U \triangleleft U$ always fails, and generally, $\triangleleft$ is a well-founded relation by a version of Scott's argument that measurable cardinals contradict $V = L$. Consequently, to each U can be recursively assigned a rank $o(U) = \sup\{o(U') + 1 \mid U' \triangleleft U\}$, and to a cardinal κ, the supremum $o(\kappa) = \sup\{o(U) + 1 \mid U \text{ is a normal ultrafilter over } \kappa\}$. By a cardinality argument, if $2^\kappa = \kappa^+$ then $o(\kappa) \leq \kappa^{++}$.

Mitchell [78] devised the concept of a *coherent* sequence of ultrafilters ("measures"), a doubly-indexed sequence that has just enough ultrafilters to witness the $\triangleleft$ relationships, and was able to establish uniqueness results for inner models $L[\mathcal{U}] \models$ "$\mathcal{U}$ is a coherent sequence of ultrafilters". In this first inner model extension of Kunen's work one sees the closest connection to the essential components of his argument for the uniqueness of κ-models. Mitchell affirmed that these $L[\mathcal{U}]$'s are *iterable* in that arbitrary iterated ultrapowers via ultrafilters in $\mathcal{U}$ and its successive images are always well-founded. He then effected a *comparison*, as any $L[\mathcal{U}_1]$ and $L[\mathcal{U}_2]$ have respective iterated ultrapowers $L[\mathcal{W}_1]$ and $L[\mathcal{W}_2]$ such that $\mathcal{W}_1$ is an initial segment of $\mathcal{W}_2$ or vice versa. This he achieved through a process of *coiteration* of least differences: At each stage, one finds the lexicographically least coordinate at which the current iterated ultrapowers of $L[\mathcal{U}_1]$ and $L[\mathcal{U}_2]$ differ and takes the respective ultrapowers by the differing ultrafilters; the difference is eliminated as ultrafilters never occur in their ultrapowers. Finally, Mitchell applied a generalization of Kunen's *stability* argument to establish e.g. that in $L[\mathcal{U}]$ the only normal ultrafilters over α for any α are those that occur in $\mathcal{U}$. For example, if one starts with a coherent sequence that corresponds to $o(\kappa) = 2$, there are exactly two normal ultrafilters over κ, differentiated as U and W were in the discussion above of Kunen's observation. As inner model theory would develop, coiteration would become embedded as basic for comparison and $\exists\kappa(o(\kappa) = \kappa^{++})$ would become a pivotal large cardinal proposition for gauging consistency strength.

The fine-structural "core" models for larger and larger cardinals would be generated by "mice", local versions of structures that incorporate some large cardinal structure, and always crucial would be their iterability. Iterability allows for the possibility of comparison, now more substantive in that iterates of mice may not coincide but one could be an initial segment of the same relative constructible hierarchy as the other. Iterability and comparison would thus guide the further development of inner model theory for getting canonical structures, and the need for new forms of iteration to get iterable mice and, after that, to effect comparison, would lead to new local and global structures and new procedures. Furthermore, Kunen's stability argument would also surface in new emanations. Loosely speaking, for either the Mitchell core model for coherent sequences of ultrafilters or the more general Steel core model for coherent sequences of extenders, an initial inner model K^c is defined which is "universal" in that it compares at least as long as any other similarly defined model. Then, certain elementary substructures of K^c are considered which are "thick" in that their transitizations are also universal. These thick substructures are Skolem hulls of certain classes of ordinals, which as in Kunen's original argument are actually fixed by all the relevant embeddings. Finally, the "true" core model K is recursively defined as a certain thick elementary submodel of K^c.[5]

Ironically, after four decades of inner model theory and what seems like a lifetime of experience, the Kunen-Paris question of the number of normal ultrafilters was revisited and seen, finally, not to be a matter of stronger hypotheses at all but of stronger methods. In 2009 Sy Friedman and Menachem Magidor [23] showed how to get all the possible values for the number of normal ultrafilters, starting with just one measurable cardinal κ and carrying out forcing that featured adjoining subsets of κ with perfect sets in the style of Sacks forcing.

3 Combinatorial Principles

In work quite different from his thesis and yet of fundamental importance for modern set theory, Kunen in 1969, together with Ronald Jensen, established the framing results about now basic combinatorial principles and related large cardinals in the constructible universe L. The consummate master of constructibility was to be Jensen, whose systematic analysis transformed the subject with the introduction of the fine structure theory for L. In 1968 Jensen [34] made his first breakthrough by showing that $V = L$ *implies the failure of*

[5]For details on inner model theory, see the Steel and Schimmerling chapters in the *Handbook of Set Theory* [18].

Suslin's Hypothesis, i.e. (there is a Suslin tree)^L, applying L for the first time after Gödel to establish a relative consistency about a classical proposition. Inspired by Jensen's construction, the ubiquitous Solovay established that $V = L$ *implies Kurepa's Hypothesis*, i.e. (there is a Kurepa tree)^L. The combinatorial features of L that enabled these constructions were soon isolated in two combinatorial principles for a regular cardinal κ, $\diamondsuit_\kappa$ ("diamond") and $\diamondsuit^+_\kappa$ ("diamond plus") respectively. To recall just the first:

($\diamondsuit_\kappa$) There is a sequence $\langle S_\alpha \mid \alpha < \kappa\rangle$ with $S_\alpha \subseteq \alpha$ such that for any $X \subseteq \kappa$, $\{\alpha < \kappa \mid X \cap \alpha = S_\alpha\}$ is stationary in κ .

$\diamondsuit_{\omega_1}$ implies that there is a Suslin tree, and $\diamondsuit^+_{\omega_1}$ implies that there is a Kurepa tree. In notes [35] written at Rockefeller University in 1969, Jensen presented his collaborative work with Kunen on these combinatorial principles and related new large cardinals. As the notes were scrupulous about assigning credit, one sees confirmed Kunen's substantial role in the development of a now basic part of set theory.

First, Kunen established that $\diamondsuit^+_\kappa$ *implies* $\diamondsuit_\kappa$, and this he did by establishing the equivalence of $\diamondsuit_\kappa$ with: ($\diamondsuit'_\kappa$) There is a sequence $\langle S_\alpha \mid \alpha < \kappa\rangle$ with $S_\alpha \subseteq P(\alpha)$ and $|S_\alpha| \leq |\alpha|$ such that for any $X \subseteq \kappa$, $\{\alpha < \kappa \mid X \cap \alpha \in S_\alpha\}$ is stationary in κ . This equivalence is still regarded as a notable combinatorial result, and it spawned the extended investigation of variations of $\diamondsuit_\kappa$.

Second, Kunen and Jensen independently formulated the following now well-known large cardinal concept, a $\forall\exists$ version of $\diamondsuit_\kappa$: A regular cardinal κ is *ineffable* iff for any sequence $\langle S_\alpha \mid \alpha < \kappa\rangle$ with $S_\alpha \subseteq \alpha$, there is an $X \subseteq \kappa$ stationary in κ such that for $\alpha < \beta$ both in X, $S_\beta \cap \alpha = S_\alpha$. Kunen forthwith situated ineffability by characterizing it in terms of a strong version of the partition property characterization of weak compactness and showing that ineffability implies Π^1_2-indescribability. He and Jensen showed that the least ineffable cardinal is smaller than the least κ satisfying $\kappa \longrightarrow (\omega)^{<\omega}$ and is larger than the least totally indescribable cardinal, i.e. one which is Π^n_m-indescribable for every $n, m \in \omega$. They also showed that if κ is ineffable, then (κ is ineffable)^L.

With respect to the motivating connection with combinatorial principles, Jensen and Kunen established that *if* κ *is ineffable, then* $\diamondsuit^+_\kappa$ *fails.* Jensen established the converse under the assumption $V = L$, and hence that *in* L, κ *is ineffable iff* $\diamondsuit^+_\kappa$ *fails.*

What about ineffability and $\diamondsuit_\kappa$? The following weak form of ineffability was isolated: A regular cardinal κ is *subtle* iff for every closed unbounded $C \subseteq \kappa$ and sequence $\langle S_\alpha \mid \alpha < \kappa\rangle$ with $S_\alpha \subseteq \alpha$, there are $\alpha < \beta$ both in C such that $S_\beta \cap \alpha = S_\alpha$. Kunen, in the last theorem of the notes [35], showed that *if* κ *is subtle, then* $\diamondsuit_\kappa$ *holds.*

Ineffable and subtle cardinals would become staple for the theory of large cardinals. In a far-reaching systematic investigation of generalizations of ineffable cardinals, the n-ineffable cardinals, Baumgartner [6] would uncover an elegant closely-knit hierarchy, one in fact that could be taken to have as a basis generalizations of the subtle cardinals, the n-subtle cardinals. Eventually, the existence of n-subtle cardinals would be characterized [37] in terms of having a finite homogeneous set for "regressive" partitions and by even more refined means [22] to establish that certain propositions of finitary mathematics have strong consistency strength.

4 The Inconsistency

In 1970 Kunen [46] established an ultimate delimitation for the entire hierarchy of large cardinals by showing that a *prime facie* extension of ideas leads to an outright inconsistency. With large cardinal hypotheses having become central in modern set theory for gauging consistency strength, "Kunen's inconsistency" will live on as an ultimate upper bound on the strength of propositions of set theory and so of mathematics.

In the late 1960s, Solovay and William Reinhardt, as a graduate student at Berkeley, were charting out hypotheses stronger than measurability based on the concept of elementary embedding. Solovay and Reinhardt independently formulated the concept of *supercompactness*, as based on postulating elementary embeddings $j: V \to M$ of the universe V into an inner model M. Assuming always that this notation commits to having j not the identity, it has a critical point crit(j), i.e. a least ordinal moved by j. Measurability had become focal through the embedding characterization, *κ is measurable iff κ is the critical point of an elementary embedding $j: V \to M$ for some inner model M*, in which case one can through the ultrapower construction assume additionally that ${}^{\kappa}M \subseteq M$, i.e. arbitrary κ-sequences drawn from M are again members of M. Solovay and Reinhardt formulated (cf. [92]) the global concept: κ is *supercompact* iff κ is γ-supercompact for every γ, where κ is *γ-supercompact* iff there is an elementary embedding $j: V \to M$ for some inner model M with crit$(j) = \kappa$, $\gamma < j(\kappa)$, and ${}^{\gamma}M \subseteq M$. The imposition of strong closure properties on M allowed for strong reflection arguments to take place, and supercompactness has achieved a central place in the large cardinal hierarchy both as the provenance of forcing relative consistency results and as a goal for the development of inner model theory.

At the end of his 1967 Berkeley thesis Reinhardt briefly considered the following postulation as an ultimate extension: There is an elementary embedding $j: V \to V$. Reinhardt's proposal led to a dramatic turn of events. After initial results aroused some suspicion, Kunen [46] established in ZFC: *There is no elementary embedding $j: V \to V$.*

This result delimited the whole large cardinal enterprise. It could have been that $j: V \to V$ would serve as the culmination of the guiding idea of closure conditions on target models of elementary embeddings; a new guiding idea in some orthogonal direction would have been exploited to formulate still stronger hypotheses; and so on. Rather, in quickly resolving the situation with an appropriately simple statement, Kunen's result sharply defined the context and showed that a completion of ZFC in a specific sense exists. The particular forms of the result were intriguing and unexpected, and although the original proof had an *ad hoc* flavor, what it established has not since been superseded by any stronger inconsistency result.

Kunen's original proof applied a result of Paul Erdős and András Hajnal [13] from the partition calculus and about algebras with infinitary operations. As pointed out by Kunen, for the particular case of their result applied, there was a simple recursive construction available, and Kunen's main argument was itself short and transparent. The argument led to two particular forms of the result:

(a) For any δ, there is no elementary embedding $j\colon V_{\delta+2} \to V_{\delta+2}$.

(b) If $j\colon V \to M$ is a elementary embedding into some inner model M and δ is the least ordinal above $\operatorname{crit}(j)$ such that $j(\delta) = \delta$, then $\{j(\alpha) \mid \alpha < \delta\} \notin M$.

The first provides a striking, local version, and the second points to a particular set that cannot be in the target model. For both, the δ can be given a concrete sense as emerging from the proof as follows: Setting $\kappa_0 = \operatorname{crit}(j)$ and $\kappa_{n+1} = j(\kappa_n)$, $\delta = \sup\{\kappa_n \mid n \in \omega\}$.

Several proofs of Kunen's result, as articulated by (a) and (b) above, have emerged, attesting both to the importance of the result and to its resilience as the point of transition to inconsistency. In the late 1980s, Hugh Woodin gave a proof based on splitting stationary sets, and Mikio Harada, on elementarity and closure.[6] In the mid-1990s, Jindřich Zapletal [97] provided a proof using a fundamental consequence of Saharon Shelah's pcf theory, the existence of "scales" on δ^+ for singular cardinals δ.

Arguments for Kunen's result have established limitative consequences in the theory of ideals and generic elementary embeddings, a subject we also broach in Section 7. Kunen's original argument shows e.g. there is no normal, fine, precipitous ideal over $[\aleph_{\omega+1}]^{\aleph_{\omega+1}}$, and Zapletal's, that for singular cardinals λ, there is no λ^+-saturated normal, fine ideal over $[\lambda]^\lambda$.[7]

The resilience of Kunen's transition to inconsistency has been affirmed in another, more substantive way, by the use in relative consistency results of, and

[6]See Kanamori [38, pp.319-322].

[7]For both results, see Section 6.2 of Foreman's chapter in the *Handbook of Set Theory* [18].

the development of a coherent theory for, hypotheses just short of the inconsistency. The relative consistency results were the first that were established for strong determinacy hypotheses, a subject to which we turn in Section 6. In 1978 Martin [75] established that if there is an "iterable" elementary embedding $j\colon V_\delta \to V_\delta$, then $\mathbf{\Pi}^1_2$-Determinacy holds, i.e. every $\mathbf{\Pi}^1_2$ set of reals is determined. Woodin then considered the following possibility, just short of Kunen's inconsistency:

(I0) There is an elementary embedding $j\colon L(V_{\delta+1}) \to L(V_{\delta+1})$ with $\mathrm{crit}(j) < \delta$.

Here, $L(V_{\delta+1})$ is the constructible closure of $V_{\delta+1}$, the least inner model of set theory containing all the sets of rank at most δ. In 1984 Woodin established that I0 implies $\mathrm{AD}^{L(\mathbb{R})}$, i.e. every set of reals in $L(\mathbb{R})$, the constructible closure of the reals, is determined. The Martin and Woodin results established an initial mooring for strong determinacy hypotheses in the hierarchy of large cardinals; as is well-known, by mid-1985 splendid, definitive results were established that secured equi-consistencies.

As to coherent theories for hypotheses just short of Kunen's inconsistency, Laver starting in the late 1980s investigated the collection of elementary embeddings $j\colon V_\delta \to V_\delta$ for a fixed δ as an algebra satisfying the "left distributive law", taking the cue from identities first applied by Martin in his $\mathbf{\Pi}^1_2$ result. Laver's results on a normal form and the solvability of a word problem spawned a cottage industry, one which through Patrick Dehornoy's infusion of the infinite braid group generated remarkable connections and initiatives.[8] After his initial success with I0, Woodin developed a detailed and coherent theory of $L(V_{\delta+1})$ under I0. (Laver [71, 72] summarized this theory and developed it further.) What became evident was a striking analogy between the theory of $L(V_{\delta+1})$ under I0 and the theory of $L(\mathbb{R})$ under $\mathrm{AD}^{L(\mathbb{R})}$. Woodin has developed this analogy in both directions and has lately speculated about connections with an ultimate inner model for large cardinals (cf. his [96]).

5 Ultrafilters

Kunen would be a mainstay of the Department of Mathematics at the University of Wisconsin, and one of his principal mathematical connections would be with the pioneering initiatives of Mary Ellen Rudin in set-theoretic topology. The first connection involved the Stone-Čech compactification $\beta\mathbb{N}$ of the discrete countable space $\mathbb{N}$, identifiable with the set of ultrafilters over ω topologized by taking as basic open sets $O_X = \{U \mid X \in U\}$ for $X \subseteq \omega$. In a groundbreaking paper, husband Walter Rudin [81] had early on established that $\mathbb{N}^* = \beta\mathbb{N} - \mathbb{N}$ (discarding the distinguishing principal ultrafilters) is not homogeneous under CH, since there are then distinguished points, the *p-points*—

[8] See Dehornoy's chapter in the *Handbook of Set Theory* [18].

those points for which the intersection of any countably many neighborhoods is again a neighborhood. Zdeněk Frolík [24] then established that $\mathbb{N}^*$ is not homogeneous in ZFC, using "integrating sums" of points. Mary Ellen Rudin [80] formatively focused the investigation of $\beta\mathbb{N}$ on partial orders, mainly two, growing out of this earlier work. In the 1969 Wisconsin thesis of David Booth (cf. his [9]), Kunen's first student, the two orders were named the *Rudin-Frolík ordering* and the *Rudin-Keisler ordering*, with the latter defined for ultrafilters in general as follows: For ultrafilters U over I and V over J, $U \leq V$ iff there is a function $f\colon J \to I$ such that for all $X \subseteq I$, $X \in U$ iff $f^{-1}(X) \in V$. (This natural projection ordering had also been considered by the model-theorist H. Jerome Keisler, another mainstay of the Wisconsin department.) Several results of Kunen appeared in [9], and in particular it was he who established (cf. Theorem 4.9) the now familiar characterization of the Rudin-Keisler minimal ultrafilters as the *Ramsey ultrafilters*, p-points of a special sort.

Separately in 1969, Kunen [44, 51] showed: *Assuming* CH, *(a) there are points in* $\mathbb{N}^*$ *which are not p-points nor the limits of any countable set,* and establishing a conjecture of Mary Ellen Rudin, *(b) there is a countable* $X \subseteq \mathbb{N}^*$ *such that each point in* X *is a limit point of* X*, yet not a limit of any discrete countable subset of* $\mathbb{N}^*$. In these ways Kunen amply showed that there are Rudin-Frolík minimal points which are not p-points. Investigations under CH and more generally MA would show that they lead to a host of such distinguishing features and a variegated structure for $\mathbb{N}^*$.

An important thrust of Kunen's work on ultrafilters, on the other hand, would be in the direction of the possibilities in just ZFC. Also in 1969, Kunen (cf. [44, 49]) established in ZFC, applying a classical, 1930s construction of an independent family of sets to enable a $2^{\aleph_0}$-length recursion, that *there are Rudin-Keisler incomparable ultrafilters over* ω. The proof showed in general that *for any cardinal* κ *there are* 2^κ *pairwise Rudin-Keisler incomparable, uniform ultrafilters over* κ.[9]

The general setting led to a further ZFC result. Keisler had formulated the concept of "good" ultrafilter, and had shown that for a cardinal $\lambda > \omega$, the λ-good, countably incomplete ultrafilters are exactly those that produce λ-saturated ultraproducts of structures for a language of cardinality less than λ. He was only able to show assuming $2^\kappa = \kappa^+$ that there are κ^+-good, countably incomplete ultrafilters over κ. Every countably incomplete ultrafilter is ω_1-good, and so the issue is only substantive for $\kappa > \omega$. Applying an elaboration of independent family of sets leading to an independent family of functions, Kunen [49] established in ZFC that *for any cardinal* κ*, there are* (2^κ) κ^+*-good, countably incomplete ultrafilters over* κ. In the 1974 book *Theory of Ultrafilters* [12], independent families of functions are given conspicuous treatment and Kunen's result on good ultrafilters is showcased as their "fundamental existence

[9]An ultrafilter over κ is *uniform* if every member has cardinality κ.

theorem".

In a similar vein, Kunen and Karel Prikry [64] established a result in ZFC about descendingly incomplete ultrafilters over uncountable cardinals, one that had previously required the Generalized Continuum Hypothesis (GCH), by appealing to some classical combinatorial constructions. This work too is aired in the aforementioned book.

In the direction of limitations of ZFC for $\beta\mathbb{N}$, already in 1971 Kunen [49, p.301][51] in effect had established: *If one adjoins at least* $(2^{\aleph_0})^+$ *random reals, then there are no Ramsey ultrafilters in the extension.*[10] This framed the stage for the issue of whether the more topologically relevant p-points always exist, when in 1977 Shelah (cf. [82, 95]) visiting Wisconsin duly established that *it is consistent that there are no p-points in* $\mathbb{N}^*$. Soon afterwards, Kunen focused attention on what arguably could have been the pivotal concept from the beginning by defining a *weak p-point* to be a point which is not the limit of any countable sequence, and establishing in ZFC (cf. [55]) that *there are weak p-points in* $\mathbb{N}^*$. This "effectively" settled the issue of homogeneity for $\mathbb{N}^*$ in ZFC by producing specific, distinguishing points. Kunen's proof was elegant in context; he relaxed the condition of being an ω_2-good ultrafilter (there are no such ultrafilters over ω) to an "ω_1-OK" ultrafilter, which are weak p-points, and then enhanced the independent function argument to establish that there are ω_1-OK ultrafilters. The investigation of $\beta\mathbb{N}$ has been carried forth in such vein, especially by Kunen, Jan van Mill and Alan Dow, but we leave off further discussion of this as situated in general topology.

While on the subject of ultrafilters, we tuck in some 1971 results of Kunen on a partition property for ultrafilters. In that year, Solovay and Telis Menas investigated the natural generalization of the well-known Rowbottom partition property for normal ultrafilters over κ to normal ultrafilters over $\mathcal{P}_\kappa\gamma$ for $\gamma \geq \kappa$ as given by the supercompactness of κ. Solovay initially showed under GCH that if κ is supercompact, then every normal ultrafilter over $\mathcal{P}_\kappa\gamma$ for small γs like κ^+, κ^{++}, etc. has the partition property. Menas (cf. [76]) showed that for any $\gamma \geq \kappa$ there are always the maximal possible number of normal ultrafilters over $\mathcal{P}_\kappa\gamma$ with the partition property, and moreover eliminated the GCH assumption from Solovay's results. On the other hand, Solovay had shown that the partition property does not always hold; e.g. if κ is supercompact and $\lambda > \kappa$ is measurable, then there is a normal ultrafilter over $\mathcal{P}_\kappa\lambda$ without the partition property. With his knowledge of indescribability and ineffability, Kunen (cf. Kunen-Pelletier [63]) improved Solovay's result in these directions: *If* κ *is supercompact and* $\lambda > \kappa$ *is ineffable, then there are stationarily many* $\gamma < \lambda$ *such that there is a normal ultrafilter over* $\mathcal{P}_\kappa\gamma$ *without the partition property, and moreover, the least such* γ *is* Π^2_1*-indescribable.* These results are

[10] Much later in Hart-Kunen [29], this result was technically improved with $\aleph_2$ replacing $(2^{\aleph_0})^+$.

still the delimitative results for the partition property today.

Kunen's familiarity with ultrafilters contributed substantially to his playing a crucial role in a novel setting without AC, as we describe in the next section.

6 Determinacy

Kunen had a spectacular run through the projective sets under the Axiom of Determinacy (AD) in the spring and summer of 1971. It was at the beginning of the formative period when the structure theory for the projective sets under determinacy hypotheses was being worked out, a theory that would come to be considered their "correct" theory. As with the combinatorial principles (cf. Section 3) Kunen entered the fray fully equal to the task and quickly worked at the forefront as a full-fledged pioneer in the subject. AD seemed to exude remarkable deductive power, and Kunen seemed to get at what was achievable in the initial foray, erecting conceptual constructions that would not be bettered for over a decade.

For $Y \subseteq \mathbb{R}^{k+1}$, the *projection of* Y is $pY = \{\langle x_1, ..., x_k\rangle \mid \exists y(\langle x_1, ..., x_k, y\rangle \in Y)\}$. With Suslin having essentially noted in 1917 that *a set of reals is analytic iff it is the projection of a Borel subset of* $\mathbb{R}^2$, the early descriptive set theorists had taken the geometric operation of projection to be basic and defined the projective sets to be those in a corresponding hierarchy, which in modern notation is as follows: For $A \subseteq \mathbb{R}^k$, A is $\mathbf{\Sigma}^1_1$ *iff* $A = pY$ for some Borel set $Y \subseteq \mathbb{R}^{k+1}$; A is $\mathbf{\Pi}^1_n$ *iff* $\mathbb{R}^k - A$ is $\mathbf{\Sigma}^1_n$; A is $\mathbf{\Sigma}^1_{n+1}$ *iff* $A = pY$ for some $\mathbf{\Pi}^1_n$ set $Y \subseteq \mathbb{R}^{k+1}$; and A is $\mathbf{\Delta}^1_n$ *iff* A is both $\mathbf{\Sigma}^1_n$ and $\mathbf{\Pi}^1_n$. However, the early descriptive set theorists could not make much headway in their structural investigation beyond the first level of the projective hierarchy, and this had to await the coming of a new paradigm, a new way of thinking.

For $A \subseteq {}^\omega\omega$, let $G(A)$ be the following *game*: There are two players I and II. I initially chooses $x(0) \in \omega$, then II chooses $x(1) \in \omega$, then I chooses $x(2) \in \omega$, then II chooses $x(3) \in \omega$, and so forth. With the resulting $x \in {}^\omega\omega$, I wins $G(A)$ *iff* $x \in A$, and otherwise II wins. A *strategy* is a function that tells a player what move to make given the sequence of previous moves, and a *winning strategy* is a strategy such that if a player plays according to it he always wins no matter what his opponent plays. A is *determined* if either I or II has a winning strategy in $G(A)$. With increasing interest in game-theoretic approaches, in 1962 Jan Mycielski and Hugo Steinhaus proposed the Axiom of Determinacy (AD), that *every* $A \subseteq {}^\omega\omega$ *is determined.* AD postulated in pure form a new kind of dichotomy, one that was seen through coding information to be applicable to a broad range of issues about sets of reals. AD contradicted AC, but from the beginning it was thought that AD could animate $L(\mathbb{R})$, the constructible closure of the reals $\mathbb{R}$ $(= {}^\omega\omega)$, with unfettered uses of AC relegated to the universe at large.

In 1967 two results brought determinacy to the foreground of set theory, one about the transfinite and the other about definable sets of reals. Solovay established that AD *implies that* ω_1 *is measurable*, injecting emerging large cardinal techniques and results into a novel setting without AC. David Blackwell provided a new, determinacy proof of a classical result of Kuratowski, that *the* $\mathbf{\Pi}^1_1$ *sets have the reduction property.* Martin in particular saw the potentialities in both directions and soon made incisive contributions to investigations with and of determinacy. Martin initially made a simple but crucial observation in the first direction, that *assuming* AD, *the filter over the Turing degrees generated by the Turing cones is an ultrafilter.* This provided another proof of Solovay's measurability result and allowed Solovay to establish in 1968 that AD *implies that* ω_2 *is measurable.*

The advances in the investigation of definable sets of reals generally under AD would be in terms of their analysis as projections of trees. For purposes of descriptive set theory, T is a *tree on* $\omega \times \kappa$ iff (a) T consists of pairs $\langle s, t\rangle$ where s is a finite sequence drawn from ω and t is a finite sequence of the same length drawn from κ, and (b) if $\langle s, t\rangle \in T$, s' is an initial segment of s and t' is a initial segment of t of the same length, then also $\langle s', t'\rangle \in T$. For such T, $[T]$ consists of pairs $\langle f, g\rangle$ corresponding to infinite branches, i.e. f and g are ω-sequences such that for any finite initial segment s of f and finite initial segment t of g of the same length, $\langle s, t\rangle \in T$. In modern terms with ${}^\omega\omega$ taken for the reals $\mathbb{R}$, $A \subseteq {}^\omega\omega$ is κ-*Suslin* iff there is a tree on $\omega \times \kappa$ such that $A = p[T] = \{f \mid \exists g(\langle f, g\rangle \in [T])\}$. $[T]$ is a closed set in the space of $\langle f, g\rangle$'s where $f \colon \omega \to \omega$ and $g \colon \omega \to \kappa$, and so otherwise complicated sets of reals, if shown to be κ-Suslin, are newly comprehended as projections of closed sets.

The analytic (i.e. $\mathbf{\Sigma}^1_1$) sets of reals are exactly the ω-Suslin sets. Membership in a $\mathbf{\Pi}^1_1$ set is thus characterizable as having *no* infinite branch through a tree on $\omega \times \omega$, and this well-foundedness can be converted set-theoretically into *having* an order-preserving map into ω_1, which amounts to having an infinite branch through a tree on $\omega \times \omega_1$. This witnessing possibility can be extended by an existential quantifier, and thus Joseph Shoenfield in 1961 established in ZF that *every* $\mathbf{\Sigma}^1_2$ *set is* ω_1-*Suslin.* Martin and Solovay analogously "dualized" $\omega \times \omega_1$ trees to get trees of order-preserving maps, maps on which a well-ordering could be imposed by using a measurable cardinal. Martin then refined this Martin-Solovay tree analysis to get a contextually optimal one using Silver-type indiscernibles. He thus established in ZF that *assuming the existence of (Silver-type) indiscernibles for* $L[a]$ *for every real* a, *every* $\mathbf{\Sigma}^1_3$ *set is* ω_ω-*Suslin* with trees having strong "homogeneity" properties, and as a remarkable contingency, that *assuming* AD, *for* $2 < n < \omega$, ω_n *is singular of cofinality* ω_2!

The structure theory of the projective sets under determinacy hypotheses, what brought the theory to prominence and what the classical descriptive set theorists had aspired for, was the inductive propagation of properties up the projective hierarchy. This propagation was not actually based directly on tree representations, but rather on related properties corresponding to determined games. Moschovakis isolated the *prewellordering property* and the central *scale property*, and established his *periodicity theorems*, from which the structure results flowed. Having defined $\Theta = \sup\{\xi \mid \text{there is a surjection: } {}^\omega\omega \to \xi\}$ to delineate the effect of AD on the transfinite, Moschovakis defined as definability analogues the *projective ordinals* $\boldsymbol{\delta}^1_n = \sup\{\xi \mid \text{there is a } \boldsymbol{\Delta}^1_n \text{ prewellordering of length } \xi\}$.[11] His work established the importance of these ordinals in Suslin representations; e.g. with a corresponding amount of determinacy, *for any* $n \in \omega$, *the* $\boldsymbol{\Sigma}^1_{2n+2}$ *sets are exactly the* $\boldsymbol{\delta}^1_{2n+1}$*-Suslin sets.*

Kunen's entrée into this subject was to establish in early 1971, independently with Martin, a basic theorem using scales, the Kunen-Martin Theorem of ZF + DC: *Every* κ*-Suslin well-founded relation on the reals has length less than* κ^+. This had as consequences some basic facts under AD about the projective ordinals: For each $n \in \omega$, $\boldsymbol{\delta}^1_{2n+2} = (\boldsymbol{\delta}^1_{2n+1})^+$, and, with Alexander Kechris having shown that the $\boldsymbol{\Sigma}^1_{2n+1}$ sets are exactly the λ_{2n+1}-Suslin sets for λ_{2n+1} a cardinal of cofinality ω, that $\boldsymbol{\delta}^1_{2n+1} = \lambda^+_{2n+1}$.

Martin had extended Solovay's measurability result for $\omega_1 (= \boldsymbol{\delta}^1_1)$ to show that *assuming* AD, *for odd* $n \in \omega$, $\boldsymbol{\delta}^1_n$ *is measurable.* Having arrived at the scene, Kunen provided a uniform proof that *assuming* AD, *for all* $n \in \omega$, $\boldsymbol{\delta}^1_n$ *is measurable.* At this point, Kunen had quickly provided the facts about the projective ordinals that completed their basic theory.

By the summer of 1971, Kunen had brought to bear his familiarity of ultrafilters to their study under AD. Generalizing the early uses of the Martin cone filter to establish measurability, Kunen observed that under AD, ω_1 is "strongly compact up to Θ": *Assuming* AD, *for any* $\lambda < \Theta$, *any* $\aleph_1$*-complete filter over* λ *can be extended to an* $\aleph_1$*-complete ultrafilter over* λ. In fact, any such filter is included in a Rudin-Keisler projection of the Martin cone filter. This in turn had the corollary: *Assuming* AD + DC, *for any* $\lambda < \Theta$, $\beta\lambda = \{U \mid U \text{ is an ultrafilter over } \lambda\}$ *is well-orderable.* Kunen thus pointed out the possibility that under AD there is a great deal of global structure to ultrafilters, and this stimulated the subsequent wide investigation of the structure theory and its applications.

Kunen's main contribution in determinacy was to be to their possible determination through the use of their ultrafilter theory to code the projective sets. With $\boldsymbol{\delta}^1_1 = \omega_1$ being a classical result, Martin established under AD that $\boldsymbol{\delta}^1_2 = \omega_2$. His aforementioned Suslin analysis of $\boldsymbol{\Sigma}^1_3$ sets established under AD that $\boldsymbol{\delta}^1_3 = \omega_{\omega+1}$, and thus the Kunen-Martin result entailed that $\boldsymbol{\delta}^1_4 = \omega_{\omega+2}$. But what is $\boldsymbol{\delta}^1_5$? In a few weeks Kunen provided the contextually optimal ultrafilter analysis up to $\boldsymbol{\delta}^1_3$ and laid out a program for the calculation of $\boldsymbol{\delta}^1_5$, a program that was to be entirely successful, but only more than a decade later.

[11] A prewellordering $\preceq$ is a well-ordering except that there could be distinct x, y such that $x \preceq y$ and $y \preceq x$.

Much of this theory, already extravagant with the infusion of measurability, was driven by infinite-exponent partition relations. $\kappa \longrightarrow (\kappa)^\lambda$ asserts that if the increasing functions from λ into κ are partitioned into two cells, then there is an $H \subseteq \kappa$ of cardinality κ such that all the increasing functions from λ into H are in one cell. The *strong partition property for* κ is the assertion $\kappa \longrightarrow (\kappa)^\kappa$ and the *weak partition property for* κ is the assertion $\forall\lambda < \kappa((\kappa \longrightarrow (\kappa)^\lambda)$. The connection with ultrafilters was mainly through an observation of Eugene Kleinberg. For $\lambda < \kappa$ both regular, let $\mathcal{C}_\kappa^\lambda$ denote the filter over κ generated by the λ-closed unbounded subsets, i.e. the filter generated by the closed unbounded filter $\mathcal{C}_\kappa$ together with the set $\{\xi < \kappa \mid \mathrm{cf}(\xi) = \lambda\}$. Kleinberg pointed out that in ZF: *If* $\lambda < \kappa$ *are both regular and* $\kappa \longrightarrow (\kappa)^{\lambda+\lambda}$, *then* $\mathcal{C}_\kappa^\lambda$ *is a normal ultrafilter over* κ.

Martin established, quite dramatically at the time, that *assuming* AD, ω_1 *has the strong partition property.* Solovay had actually shown that the closed unbounded filter $\mathcal{C}_{\omega_1}$ witnesses the measurability of ω_1, and moreover that the ultrapower ${}^{\omega_1}\omega_1/\mathcal{C}_{\omega_1}$ is (order-isomorphic) to ω_2 and has a "canonical" representation property. Martin and Paris in early 1971 applied this to lift ω_1 having the strong partition property to show: *Assuming* AD + DC, ω_2 *has the weak partition property and so there are exactly two normal ultrafilters over* ω_2, *namely* $\mathcal{C}_{\omega_2}^{\omega}$ *and* $\mathcal{C}_{\omega_2}^{\omega_1}$.

Kunen drew his immediate inspiration from the Martin-Paris work, and proceeded to characterize the ultrafilters over ω_2 under AD + DC as those (Rudin-Keisler) equivalent to a product of normal ultrafilters over ω_1 and ω_2, this work also showing that ω_2 does *not* have the strong partition property.

Pushing upward from this in a considerable refinement of the Martin-Paris work, Kunen analyzed ultrapowers by the product ultrafilters $(\mathcal{C}_{\omega_1})^n$ and exploited the strong partition property for ω_1 to get a revelatory representation of subsets of ω_ω as countable unions of "simple" sets, sets comprehendible through an analysis of ultrafilters over ω_ω. This provided a $\mathbf{\Delta}_3^1$ coding of the subsets of ω_ω, which in turn established that *assuming* AD + DC, $\boldsymbol{\delta}_3^1\,(= \omega_{\omega+1})$ *has the weak partition property and so there are exactly three normal ultrafilters over* $\boldsymbol{\delta}_3^1$, *namely* $\mathcal{C}_{\omega_{\omega+1}}^{\omega}$, $\mathcal{C}_{\omega_{\omega+1}}^{\omega_1}$, *and* $\mathcal{C}_{\omega_{\omega+1}}^{\omega_2}$.[12] Proceeding as before, Kunen characterized the ultrafilters over $\omega_{\omega+1}$ under AC + DC as equivalent to products of normal ultrafilters.

For tree representations, Kunen "dualized" the Martin-Solovay tree (mentioned above) to get representations for the $\mathbf{\Pi}_3^1$ sets and hence the $\mathbf{\Sigma}_4^1$ sets, applying ultrafilters over $\lambda < \omega_{\omega+1}$, and so *assuming* AD + DC, $\mathbf{\Sigma}_4^1$ *sets are* $\omega_{\omega+1}$-*Suslin* with trees having strong homogeneity properties as given by the weak partition property for $\omega_{\omega+1}$. Continuing, Kunen could again dualize to get representations for the $\mathbf{\Pi}_4^1$ sets and hence the $\mathbf{\Sigma}_5^1$ sets, but absent the strong

[12]See Solovay [91], which provides the details as he worked them out in lectures in 1976-1977.

partition property for $\omega_{\omega+1}$, could not get the requisite homogeneity properties to proceed further.

By the later 1970s, a complete structure theory for the projective sets was in place, a resilient edifice founded on determinacy with both strong buttresses and fine details—save for one lacuna, the determination of the projective ordinals. Kunen's 1971 work, communicated in short, pithy notes, remained the high point of the contextually optimal analysis, and the work's several aspects were given extended exposition in the latter 1970s in Kechris [39], Solovay [91], and Kechris [40]. Be that as it may, the ground lay fallow well into the 1980s.

In a veritable *tour de force*, Steve Jackson, a student of Martin, completed the determination of the projective ordinals. With the starting point a new analysis of normal ultrafilters over $\boldsymbol{\delta}^1_3$ by Martin that led to a putative lower bound for $\boldsymbol{\delta}^1_5$, Jackson determined $\boldsymbol{\delta}^1_5$ in his 1983 UCLA thesis. By 1985 Jackson had carried out the determination of all the projective ordinals, with the large part of the upper bound calculations presented in his formidable [32]. Define ordinals $E(n)$ for $n \in \omega$ by: $E(0) = 1$, and $E(n+1) = \omega^{E(n)}$ in ordinal exponentiation. Jackson established: *Assuming* AD + DC, *for* $n \in \omega$, $\boldsymbol{\delta}_{2n+3} = \omega_{E(2n+1)+1}$ *and* $\boldsymbol{\delta}^1_{2n+3}$ *has the strong partition property.*

Jackson's proof proceeded by induction, starting at the basis with the weak partition property for $\boldsymbol{\delta}^1_3$ and establishing in turn the upper bound $\boldsymbol{\delta}^1_5 \leq \omega_{E(3)+1}$; the strong partition property for $\boldsymbol{\delta}^1_3$ and thus the lower bound $\omega_{E(3)+1} \leq \boldsymbol{\delta}^1_5$; and the weak partition property for $\boldsymbol{\delta}^1_5$; and iteratively repeating this cycle. A crucial ingredient of the proof was the Kunen idea of representing sets as countable unions of simple sets, these comprehendible through an analysis of ultrafilters—these being the main complications toward developing tree representations with good homogeneity properties.[13]

7 Saturated Ideals

Kunen established results formative for the theory of saturated ideals, with one of the arguments, devised in 1972, becoming a bulwark of method for the modern theory of ideals and generic elementary embeddings. Saturated ideals, particularly $\aleph_1$-saturated ones related to measure, had already occurred in his thesis. His classic paper [53] on the subject, appearing relatively late, set out the various possibilities for saturated ideals and featured two elegant arguments which settled the remaining cases. Here we frame the context and, in turn, describe the workings of the arguments and, of one, the 1972 one, its subsequent reach.

Let I be an ideal over κ.[14] I is λ-*saturated* iff for $\{X_\alpha \mid \alpha < \lambda\} \subseteq P(\kappa) - I$

[13] See Sections 4 and 5 of Jackson's Chapter in the *Handbook of Set Theory* [18] for a schematically presented proof.

[14] That is, I is an ideal on the power set $P(\kappa)$, i.e. I is closed under the taking of subsets and of unions. In fact, we always assume that I contains all singletons and is κ-complete,

there are $\beta < \gamma < \lambda$ such that $X_\beta \cap X_\gamma \in P(\kappa) - I$ (i.e. the corresponding Boolean algebra has no antichains of cardinality λ).

Solovay's 1966 work on real-valued measurable cardinals had brought to the foreground the concepts of saturated ideal, generic ultrapower, and generic elementary embedding. For an ideal I over κ, forcing with the members of $P(\kappa) - I$ as conditions, and p stronger than q when $p - q \in I$, engenders an ultrafilter on the ground model power set $P(\kappa)$. With this, one can construct an ultrapower of the ground model in the generic extension and a corresponding elementary embedding. It turns out that the κ^+-saturation of the ideal ensures that this generic ultrapower is well-founded. Thus, a synthesis of forcing and ultrapowers is effected, and this raised enticing possibilities for having such large cardinal-type structure low in the cumulative hierarchy.

The first result of Kunen [53] addressed the relative consistency of having a κ-saturated ideal over an inaccessible cardinal κ in a non-trivial sense. Kunen-Paris [62] had established that starting with a measurable cardinal κ, there is a forcing extension in which $\kappa = 2^{\aleph_0}$ is a regular limit cardinal and there is a κ-saturated ideal over κ, yet no λ-saturated ideal over κ for any $\lambda < \kappa$. Answering a remaining question (cf. his [45, p.225]) Kunen [53] established: *If κ is a measurable cardinal, then there is a cardinal-preserving forcing extension in which κ is inaccessible but not measurable and there is a κ-saturated ideal over κ.* Kunen cleverly devised a forcing T that adjoins a κ-Suslin tree through a large cardinal κ in such a way that the forcing combined with the further forcing for shooting a cofinal branch through the tree is equivalent to having just added one Cohen subset of κ in the first place. For the actual model, Kunen deftly applied Silver's recently arrived at "reversed Easton" iterated forcing and master condition technique to get to a preliminary forcing extension in which κ is measurable and, moreover, adjoining a further Cohen subset of κ retains measurability. He then followed this up with his T to get his final model. κ is not measurable in this model as there is a κ-Suslin tree. However, by design there is a κ-c.c. forcing over this model which resurrects measurability and so in particular secures a κ-saturated ideal over κ. Finally, a simple lemma about κ-c.c. forcing shows that there was already a κ-saturated ideal in Kunen's model!

The second result of Kunen [53] addressed the relative consistency of having a κ^+-saturated ideal over a successor cardinal κ. By his work [45] with iterated ultrapowers, the consistency strength of having such an ideal was stronger than having a measurable cardinal. To implement his argument Kunen unabashedly appealed to the strongest embedding hypothesis to date for carrying out a relative consistency construction. A cardinal κ is *huge* iff there is an elementary embedding $j\colon V \to M$ for some inner model M with $\mathrm{crit}(j) = \kappa$ and ${}^{j(\kappa)}M \subseteq M$. Kunen established: *If κ is a huge cardinal, then there is a forcing extension in which $\kappa = \omega_1$ and there is an $\aleph_2$-saturated ideal over ω_1.*

i.e. I is closed under the taking of unions of fewer than κ members.

Huge cardinals are consistency-wise much stronger than supercompact cardinals and were latterly situated at the bottom of a hierarchy, the hierarchy of n-huge cardinals, that reach up to the strong hypotheses just short of Kunen's inconsistency (cf. Section 4). Supercompact cardinals, with their strong reflection properties, would become much applied in relative consistency results, but huge cardinals would remain a landmark only through Kunen's application for quite some time.

With a $j\colon V \to M$ with critical point κ, $\lambda = j(\kappa)$, and ${}^{\lambda}M \subseteq M$ as given by the hugeness of κ, Kunen collapsed κ to ω_1 and then collapsed λ to ω_2 in such a way so as to be able to define a saturated ideal. The first collapse was a specifically devised, iteratively constructed "universal" collapse P, and the second collapse was a "Silver collapse" S drawn from Silver's relative consistency result (cf. his [87]) for Chang's Conjecture. To show that the resulting forcing extension $V^{P*\dot{S}}$ is as desired, Kunen first used the devised universality of P to show that $P * \dot{S}$ is a subforcing of $j(P)$, and moreover, the rest of the forcing to get $j(P)$ has the λ-c.c., this requiring ${}^{\lambda}M \subseteq M$. He then showed that the Silver collapse allowed for a Silver master condition which enabled the lifting of the embedding j to $V^{P*\dot{S}}$ in the further, λ-c.c. extension. Finally, Kunen was able, as in his previous forcing argument but with some complications, to pull back a corresponding ultrafilter to show because of the λ-c.c. that there is already a λ-saturated ideal in $V^{P*\dot{S}}$.

From the late 1970s on, Kunen's argument, as variously elaborated and amended, would become a prominent tool for producing saturated ideals and other strong phenomena at accessible cardinals. Magidor [73] provided a variation on the theme that only required an "almost" huge cardinal, replacing the master condition with a "master sequence"; on the other hand, Kunen's model satisfies Chang's Conjecture whereas Magidor's does not. Foreman in his thesis (cf. [15]) established, starting from a 2-huge cardinal, the relative consistency of a three-cardinal version of Chang's Conjecture; this involved a considerable complication of the Kunen argument which successively collapsed three cardinals. Foreman [16] subsequently established that if there is a huge cardinal, then in a forcing extension there is a set model of ZFC satisfying "for all regular cardinals κ there is a κ^+-saturated ideal over κ"; this involved a delicate iteration of Kunen's forcing together with Radin forcing to have all the regular cardinals. Laver [70] used Kunen's argument with an "Eastonized" version of the Silver collapse to get a stronger saturation property.[15]

As for the proposition that there is an $\aleph_2$-saturated ideal over ω_1 itself, the importance of such ideals grew in the 1980s and with Kunen's result seen as setting an initial high bar for the stalking of its consistency strength, reflective and then definitive results were established. In 1984 Foreman, Magidor, and

[15]See Sections 7.6 through 7.13 of Foreman's chapter in the *Handbook of Set Theory* [18] for a systematic account of Kunen's work as thus variously elaborated and amended.

Shelah [19] established penetrating results that led to a new understanding of strong propositions and the possibilities with forcing. The focus was on a new, maximal forcing axiom, Martin's Maximum (MM), and they showed that *if there is a supercompact cardinal* κ*, there is a forcing extension in which* $\kappa = \omega_2$ *and* MM *holds.* They then established that MM *implies that the nonstationary ideal over* ω_1 *is* $\aleph_2$*-saturated.* Not only was the upper bound for the consistency strength of having an $\aleph_2$-saturated ideal over ω_1 considerably reduced from Kunen's huge cardinal, but for the first time the consistency of the nonstationary ideal over ω_1 being $\aleph_2$-saturated was established relative to large cardinals. Kunen had naturally enough collapsed a large cardinal to ω_1 in order to transmute strong properties of the cardinal into an $\aleph_2$-saturated ideal over ω_1, and this sort of direct connection had become the rule. The new discovery was that a collapse of a large cardinal to ω_2 instead can provide enough structure to secure such an ideal. In fact, Foreman, Magidor, and Shelah showed that even the usual Levy collapse of a supercompact cardinal to ω_2 engenders an $\aleph_2$-saturated ideal over ω_1. In terms of method, the central point is that the existence of sufficiently large cardinals implies the existence of substantial generic elementary embeddings with small critical points like ω_1.

Woodin in 1984 drew out what turned out to be a critical concept in this direction, that of a *Woodin cardinal*, and as is well-known, this concept turned out, remarkably, to play a central role in both establishing the consistency of determinacy hypotheses and in developing inner model theory. Reducing the consistency strength for saturated ideals, in 1985 Shelah [83] established: *If* κ *is a Woodin cardinal, then there is a forcing extension in which* $\kappa = \omega_2$ *and the nonstationary ideal over* ω_1 *is* $\aleph_2$*-saturated.* Finally, with the inner model theory brought up to this level, John Steel [93] established: *If there is an* $\aleph_2$*-saturated ideal over* ω_1 *and there is a measurable cardinal, then there is an inner model with a Woodin cardinal.* Thus, having an $\aleph_2$-saturated ideal over ω_1, first shown relatively consistent by Kunen, has essentially been gauged on the scale of large cardinals, and this at a Woodin cardinal.

As a final note, Kunen had observed by considering where the ordinals go in generic ultrapowers that *there is no uniform, (even only)* $\aleph_1$*-complete,* $\aleph_2$*-saturated ideal over any cardinal between* $\aleph_\omega$ *and* $\aleph_{\omega_1}$. Foreman [17] has recently generalized this and carried out a systematic study of such "forbidden intervals".

8 Martin's Axiom

From 1973 on, Kunen vigorously pursued research in set-theoretic topology (later, "general topology") and areas reaching into artificial intelligence, but of course, there would continue to be significant results in set theory. Martin's Axiom had been a focal presence in his thesis; in later years, Kunen framed the limits of this axiom, in terms of possibilities both for "gaps" and for when parametrized MA could first fail.

In 1975, Kunen [60] (cf. Baumgartner [7, Theorem 4.2]) provided an incisive analysis of possible "gaps" in $\mathcal{P}(\omega)$ under eventual inclusion $\subseteq^*$, i.e. $X \subseteq^* Y$ iff $|X - Y|$ is finite, which has important MA consequences. Considering pairs $\langle \mathcal{A}, \mathcal{B} \rangle$ with $\mathcal{A} = \langle X_\alpha \mid \alpha < \kappa \rangle$ and $\mathcal{B} = \langle Y_\gamma \mid \gamma < \lambda \rangle$ such that $\alpha < \beta < \kappa$ implies $X_\alpha \subseteq^* X_\beta$ and $\gamma < \eta < \lambda$ implies $Y_\eta \subseteq^* Y_\gamma$, *$\langle \mathcal{A}, \mathcal{B} \rangle$ is a (κ, λ^*)-gap* iff there is no $Z \subseteq \omega$ such that $\alpha < \kappa$ and $\gamma < \lambda$ implies $X_\alpha \subseteq^* Z \subseteq^* Y_\gamma$. Hausdorff [31] had famously constructed an (ω_1, ω_1^*)-gap which in the post-Cohen area was seen to be c.c.c.-indestructible, i.e. no interpolant Z can be adjoined by any c.c.c. forcing. There is a natural partial order $P(\mathcal{A}, \mathcal{B})$ with finite conditions for adjoining an interpolant Z, and Kunen showed: *(a) If $\langle \mathcal{A}, \mathcal{B} \rangle$ is not a (κ, λ^*)-gap, then $P(\mathcal{A}, \mathcal{B})$ has the c.c.c., and (b) If $\kappa = \lambda = \omega_1$, then there is a c.c.c. partial order that creates an uncountable antichain in $P(\mathcal{A}, \mathcal{B})$.* It follows that under MA, any $\langle \mathcal{A}, \mathcal{B} \rangle$ with $\kappa = \lambda = \omega_1$ is an (ω_1, ω_1^*)-gap: By (b) the very application of MA shows that the natural way of creating an interpolant does not have the c.c.c., and by (a) there can consequently be no interpolant at all.

Kunen used this pretty strategem to show in effect: *It is consistent to have* MA + $2^{\aleph_0} = \aleph_2$ + *"there are no (ω_1, ω_2^*)-, (ω_2, ω_1^*)- or (ω_2, ω_2^*)-gaps".* Starting with CH and $\diamondsuit_{\omega_2}$ for cofinality ω_1 ordinals, he carried out a finite-support c.c.c. iteration using the diamond sequence to anticipate possible (ω_1, ω_2^*)-, (ω_2, ω_1^*)- or (ω_2, ω_2^*)-gaps. These anticipations will have embedded in them (ω_1, ω_1^*)-gaps, which by the above strategem will henceforth remain unfilled through the iteration, and so could not have anticipated (ω_1, ω_2^*)-, (ω_2, ω_1^*)- or (ω_2, ω_2^*)-gaps after all! There is a nice historical resonance of Kunen's "self-denying" strategem with the very way that diamond principles were established in L. Kunen's analysis would be pursued in terms of partition relations in the important paper Abraham-Rubin-Shelah [1].

In 1980, Kunen (cf. [57]) refined the foregoing argument to make his final reckoning of MA, that parametrized MA can first fail at a singular cardinal: *If θ is a singular cardinal of cofinality ω_1, then there is a c.c.c. partial order forcing $2^{\aleph_0} > \theta$ and "θ is the least cardinal κ such that there is a c.c.c. partial order with a family of κ dense subsets for which there is no filter meeting them all".* Kunen now considered pairs $\langle \mathcal{A}, \mathcal{B} \rangle$ with $\mathcal{A}$ and $\mathcal{B}$ no longer necessarily having internal $\subseteq^*$ relationships, and was able to preserve a (θ, θ^*)-gap through the finite-support c.c.c. iteration as a counterexample to MA for meeting θ dense sets.

The Kunen [60] strategem would engagingly surface after three decades in Hart-Kunen [30, Theorem 5.9] for whether "Cantor trees" have the c.c.c. and just as for gaps, a consistency result with MA + $2^{\aleph_0} = \aleph_2$. Furthermore, in the related Kunen-Raghavan [65], which in method has a historical resonance with Kunen [57], another limit on possibilities is set by showing that there are models of MA with the continuum arbitrarily large in which there are "Gregory trees" and in which there are no such trees.

9 Envoi

When one thinks of modern set theory—that autonomous field of mathematics engaged in the continuing investigation of the transfinite numbers and definable sets of reals, employing remarkably elegant and sophisticated methods, and elucidating the consistency strength of strong propositions, indeed the strongest of mathematics—one thinks of Kunen's early work, the work of the years 1968-1972, as crucial for its development and fundamental for its articulation. However, Kunen's subsequent work in set theory, both the expository work and the extended collaborative work, has been considerable and far-ranging, and in its way substantive in its complementarity.

With his early accomplishments in set theory in place, Kunen within a decade provided several magisterial expositions that illuminated different aspects of the subject. For the 1977 *Handbook of Mathematical Logic* [4], edited by his colleague Barwise and the mother of all handbooks in logic, Kunen provided a chapter [52] on combinatorics. Going through the stuff of stationary sets, enumeration principles, trees, almost disjoint sets, partition calculus, and large cardinals, the chapter presented a remarkably integrated view of classical initiatives and modern developments. In 1980, Kunen's book [54] appeared and quickly became the standard for both the basics of the subject through definability as well as independence proofs through forcing. The presentation was precise and to the point, and articulation of methods, unburdened and accessible. Even to the notation, a generation would imbibe set theory through this careful account. For the 1984 *Handbook of Set-Theoretic Topology* [68] that he himself edited together with Jerry Vaughn, Kunen provided an account [56] of random and Cohen reals. Random reals are, and will always remain, a bit of a mystery for some, but with his command of measure algebras Kunen presented a coordinated, forcing account of both measure and category.

Very recently, in the fullness of time, Kunen has provided a text *The Foundations of Mathematics* [59]. With the sure hand of experience in mathematical logic and computer science, Kunen filled a niche by providing a readable beginning graduate level account of set theory, model theory, and recursion theory.

Kunen's research work after the early 1970s in what would be considered set theory proper was almost all with collaborators. In 1975 Kunen provided a chart about various consistency possibilities for cardinal sizes related to measure and category, and Arnold Miller [77] presented this chart as well as established some results for it. The "Kunen-Miller" chart would elaborate a large part of the later Cichoń diagram, a focal diagram for the burgeoning investigation of the possible orderings of cardinal invariants in the 1980s and 1990s.

Kunen-Tall [66] surveyed the landscape between Martin's Axiom and Suslin's

Hypothesis; their division of consequences of MA into "combinatorial" ones that readily imply $2^{\aleph_0} > \aleph_1$ and other "Suslin" type consequences would, with the first, anticipate extensive work on weak forms of MA and, with the second, anticipate Todorčević's [94] Open Coloring Axiom.

Carlson-Kunen-Miller [10] constructed a minimal degree that collapses ω_1. Kunen-Miller [61] and Keisler-Kunen-Leth-Miller [41] investigated descriptive set theory from the point of view of compact sets and over hyperfinite sets respectively. Hart-Kunen [29] studied weak measure extension axioms, axioms that posit that measures on σ-algebras can be extended to encompass a few more sets, and provided (§5) an illuminating random-graph proof of why adding $\aleph_2$ random reals leaves no Ramsey ultrafilters. Kunen-Tall [67] considered reals in elementary submodels of set theory.

In the new millennium, Baker-Kunen [2] generalized Kunen's construction of weak p-points to uniform ultrafilters over regular κ, and Baker-Kunen [3] promulgated a general approach in terms of Stone spaces of Boolean algebras and the unifying concept of a "hatpoint". Juhász-Kunen [36] explored a principle about elementary submodels which holds in models resulting from adding any number of Cohen reals to a ground model of GCH. Kunen [58] investigated a reflection property for compactness of spaces and in its terms characterized the least supercompact cardinal. Hart-Kunen [30] and Kunen-Raghavan [65] were mentioned at the end of the previous section.

The perusal of these later publications raises an overarching issue of subject and technique. Why, after all, are all of these set-theoretic and many others topological? A large body of Kunen's publications have to do with the investigation of general topological spaces using the sophisticated methods and instrumental postulations of modern set theory. But, we cannot say at what point technique begins or where it ends. We see worked in Kunen's hands a large subject, one with classical roots in the investigations of Sierpiński, Hausdorff, and others publishing in *Fundamenta Mathematicae* and conveyed in spirit in the second part of Kunen's thesis—now general topology perhaps, but imbued with modern set theory.

References

[1] Uri Abraham, Matatyahu Rubin, and Saharon Shelah. On the consistency of some partition theorems for continuous colorings, and the structure of $\aleph_1$-dense real order types. *Annals of Pure and Applied Logic*, 29:123–206, 1985.

[2] Joni Baker and Kenneth Kunen. Limits in the uniform ultrafilters. *Transactions of the American Mathematical Society*, 353:4083–4093, 2001.

[3] Joni Baker and Kenneth Kunen. Matrices and ultrafilters. In Miroslav Hušek and Jan van Mill, editors, *Recent Progress in General Topology II*, pages 59–81. North Holland, Amsterdam, 2002.

[4] K. Jon Barwise, editor. *Handbook of Mathematical Logic.* North-Holland, Amsterdam, 1977.

[5] K. Jon Barwise and Kenneth Kunen. Hanf numbers for fragments of L_{ω_1,ω_1}. *Israel Journal of Mathematics*, 10:407–413, 1971.

[6] James E. Baumgartner. Ineffability properties of cardinals I. In András Hajnal, Richard Rado, and Vera T. Sós, editors, *Infinite and Finite Sets*, volume 10 of *Colloquia Mathematica Societatis Janos Bolyai.* North-Holland, Amsterdam, 1975.

[7] James E. Baumgartner. Applications of the Proper Forcing Axiom. In Kenneth Kunen and Jerry E. Vaughn, editors, *Handbook of Set-Theoretic Topology*, pages 913–959. North-Holland, Amsterdam, 1984.

[8] R. H. Bing, W. W. Bledsoe, and R. Daniel Mauldin. Sets generated by rectangles. *Pacific Journal of Mathematics*, 51:27–36, 1974.

[9] David Booth. Ultrafilters on a countable set. *Annals of Mathematical Logic*, 2:1–24, 1970.

[10] Timothy Carlson, Kenneth Kunen, and Arnold W. Miller. A minimal degree which collapses ω_1. *The Journal of Symbolic Logic*, 49:298–300, 1984.

[11] Chen-Chung Chang. Sets constructible using $L_{\kappa\kappa}$. In Dana S. Scott, editor, *Axiomatic Set Theory*, volume 13(1) of *Proceedings of Symposia in Pure Mathematics*, pages 1–8. American Mathematical Society, Providence, 1971.

[12] W. Wistar Comfort and Stylianos Negrepontis. *The Theory of Ultrafilters.* Springer-Verlag, Berlin, 1974.

[13] Paul Erdős and András Hajnal. On a problem of B. Jónsson. *Bulletin de l'Académie Polonaise des Sciences, Série des Sciences Mathématiques, Astronomiques et Physiques*, 14:19–23, 1966.

[14] William G. Fleissner. Lemma on measurable cardinals. *Proceedings of the American Mathematical Society*, 49:517–518, 1975.

[15] Mattew Foreman. Large cardinals and strong model theoretic transfer properties. *Transactions of the American Mathematical Society*, 272(2):427–463, 1982.

[16] Matthew Foreman. More saturated ideals. In Alexander S. Kechris, D. Anthony Martin, and Yiannis N. Moschovakis, editors, *Cabal Seminar 79-81. Proceedings, Caltech-UCLA Logic Seminar 1979-81*, volume 1019 of *Lecture Notes in Mathematics*, pages 1–27. Springer-Verlag, Berlin, 1983.

[17] Matthew Foreman. Forbidden intervals. *The Journal of Symbolic Logic*, 74:1081–1099, 2009.

[18] Matthew Foreman and Akihiro Kanamori, editors. *Handbook of Set Theory*. Springer, Berlin, 2010.

[19] Matthew Foreman, Menachem Magidor, and Saharon Shelah. Martin's Maximum, saturated ideals, and nonregular ultrafilters. I. *Annals of Mathematics*, 127(1):1–47, 1988.

[20] David H. Fremlin. *Consequences of Martin's Axiom*, volume 84 of *Cambridge Tracts in Mathematics*. Cambridge University Press, Cambridge, 1984.

[21] David H. Fremlin. Real-valued-measurable cardinals. In Haim Judah, editor, *Set Theory of the Reals*, volume 6 of *Israel Mathematical Conference Proceedings*, pages 151–304. American Mathematical Society, Providence, 1993.

[22] Harvey M. Friedman. Subtle cardinals and linear orderings. *Annals of Pure and Applied Logic*, 107:1–34, 2001.

[23] Sy D. Friedman and Menachem Magidor. The number of normal ultrafilters. *The Journal of Symbolic Logic*, 74:1969–1080, 2009.

[24] Zdeněk Frolik. Sums of ultrafilters. *Bulletin of the American Mathematical Society*, 73:87–91, 1967.

[25] Haim Gaifman. Measurable cardinals and constructible sets (abstract). *Notices of the American Mathematical Society*, 11:771, 1964.

[26] Haim Gaifman. Elementary embeddings of models of set theory and certain subtheories. In Thomas J. Jech, editor, *Axiomatic Set Theory*, volume 13(2) of *Proceedings of Symposia in Pure Mathematics*, pages 33–101. American Mathematical Society, Providence R.I., 1974.

[27] Moti Gitik. The negation of the singular cardinals hypothesis from $o(\kappa) = \kappa^{++}$. *Annals of Pure and Applied Logic*, 43:209–234, 1989.

[28] Moti Gitik. The strength of the failure of the singular cardinals hypothesis. *Annals of Pure and Applied Logic*, 51:215–240, 1991.

[29] Joan E. Hart and Kenneth Kunen. Weak measure extension axioms. *Topology and Applications*, 85:219–246, 1998.

[30] Joan E. Hart and Kenneth Kunen. Inverse limits and function algebras. *Topology Proceedings*, 30:501–521, 2006.

[31] Felix Hausdorff. Summen von $\aleph_1$ Mengen. *Fundamenta Mathematicae*, 26:241–255, 1936.

[32] Steve Jackson. AD and the projective ordinals. In Alexander Kechris, D. Anthony Martin, and John R. Steel, editors, *Cabal Seminar 81-85. Proceedings, Caltech-UCLA Logic Seminar 1981-1985*, volume 1333 of *Lecture Notes in Mathematics*, pages 117–220. Springer-Verlag, Berlin, 1988.

[33] Thomas J. Jech. *Set Theory.* Springer, Heidelberg, 2003. Third millennium edition.

[34] Ronald B. Jensen. Souslin's Hypothesis is incompatible with $V = L$ (abstact). *Notices of the American Mathematical Society*, 15:935, 1968.

[35] Ronald B. Jensen. Some combinatorial properties of L and V, 1969. Available at http://www.mathematik.hu-berlin.de/~raesch/org/jensen.html.

[36] István Juhász and Kenneth Kunen. The power set of ω. Elementary submodels and weakenings of CH. *Fundamenta Mathematicae*, 170:257–265, 2001.

[37] Akihiro Kanamori. Regressive partition relations, n-subtle cardinals, and Borel diagonalization. *Annals of Pure and Applied Logic*, 52:65–77, 1991.

[38] Akihiro Kanamori. *The Higher Infinite. Large Cardinals in Set Theory from their Beginnings.* Springer, 2003. Second edition.

[39] Alexander S. Kechris. AD and projective ordinals. In Alexander Kechris and Yiannis Moschovakis, editors, *Cabal Seminar 76-77. Proceedings, Caltech-UCLA Logic Seminar 1976-1977*, volume 689 of *Lecture Notes in Mathematics*, pages 91–132. Springer-Verlag, Berlin, 1978.

[40] Alexander S. Kechris. Homogeneous trees and projective scales. In Alexander S. Kechris, D. Anthony Martin, and Yiannis N. Moschovakis, editors, *Cabal Seminar 77-79. Proceedings, Caltech-UCLA Logic Seminar 1977-79*, volume 839 of *Lecture Notes in Mathematics*, pages 33–73. Springer-Verlag, Berlin, 1981.

[41] H. Jerome Keisler, Kenneth Kunen, Arnold W. Miller, and Steven Leth. Descriptive set theory over hyperfinite sets. *The Journal of Symbolic Logic*, 54:1167–1180, 1989.

[42] Kenneth Kunen. Implicit definability and infinitary languages. *The Journal of Symbolic Logic*, 33:446–451, 1968.

[43] Kenneth Kunen. *Inaccessibility Properties of Cardinals*. PhD thesis, Stanford University, 1968.

[44] Kenneth Kunen. On the compactification of the integers. *Notices of the American Mathematical Society*, 17:299, 1970.

[45] Kenneth Kunen. Some applications of iterated ultrapowers in set theory. *Annals of Mathematical Logic*, 1:179–227, 1970.

[46] Kenneth Kunen. Elementary embeddings and infinitary combinatorics. *The Journal of Symbolic Logic*, 36:407–413, 1971.

[47] Kenneth Kunen. Indescribability and the continuum. In Dana S. Scott, editor, *Axiomatic Set Theory*, volume 13(1) of *Proceedings of Symposia in Pure Mathematics*, pages 199–203. American Mathematical Society, Providence R.I., 1971.

[48] Kenneth Kunen. On the GCH at measurable cardinals. In Robin O. Gandy, editor, *Logic Colloquium '69*, pages 107–110. North-Holland, Amsterdam, 1971.

[49] Kenneth Kunen. Ultrafilters and independent sets. *Transactions of the American Mathematical Society*, 172:299–306, 1972.

[50] Kenneth Kunen. A model for the negation of the axiom of choice. In Adrian R. D. Mathias and Hartley Rogers Jr., editors, *Cambridge Summer School in Mathematical Logic*, volume 337 of *Lecture Notes in Mathematics*, pages 489–494. Springer-Verlag, Berlin, 1973.

[51] Kenneth Kunen. Some points in $\beta\mathrm{N}$. *Mathematical Proceedings of the Cambridge Philosophical Society*, 80:385–398, 1976.

[52] Kenneth Kunen. Combinatorics. In K. Jon Barwise, editor, *Handbook of Mathematical Logic*, pages 371–403. North-Holland, Amsterdam, 1977.

[53] Kenneth Kunen. Saturated ideals. *The Journal of Symbolic Logic*, 43:65–76, 1978.

[54] Kenneth Kunen. *Set Theory. An Introduction to Independence Proofs*, volume 102 of *Studies in Logic and the Foundations of Mathematics*. North-Holland, Amsterdam, 1980.

[55] Kenneth Kunen. Weak p-points in $\mathbb{N}^*$. In Á Császár, editor, *Topology (Proceedings of the Fourth Colloquum), volume II*, volume 23 of *Colloquia Mathematica Societatis János Bolyai*, pages 741–749. North-Holland, Amsterdam, 1980.

[56] Kenneth Kunen. Random and Cohen reals. In Kenneth Kunen and Jerry E. Vaughn, editors, *Handbook of Set-Theoretic Topology*, pages 887–911. North-Holland, Amsterdam, 1984.

[57] Kenneth Kunen. Where MA first fails. *The Journal of Symbolic Logic*, 53:429–433, 1988.

[58] Kenneth Kunen. Compact spaces, compact cardinals, and elementary submodels. *Topology and its Applications*, 130:99–109, 2003.

[59] Kenneth Kunen. *The Foundations of Mathematics*, volume 19 of *Studies in Logic*. College Publications, 2009.

[60] Kenneth Kunen. (κ, λ)-gaps under MA, August 1975. Unpublished.

[61] Kenneth Kunen and Arnold W. Miller. Borel and projective sets from the point of view of compact sets. *Mathematical Proceedings of the Cambirdge Philosophical Society*, 94:399–409, 1983.

[62] Kenneth Kunen and Jeffrey B. Paris. Boolean extensions and measurable cardinals. *Annals of Mathematical Logic*, 2:359–377, 1971.

[63] Kenneth Kunen and Donald H. Pelletier. On a combinatorial property of Menas related to the partition property for measures on supercompact cardinals. *The Journal of Symbolic Logic*, 48:475–481, 1983.

[64] Kenneth Kunen and Karel L. Prikry. On descendingly incomplete ultrafilters. *The Journal of Symbolic Logic*, 36:650–652, 1971.

[65] Kenneth Kunen and Dilip Raghavan. Gregory trees, the continuum, and Martin's Axiom. *The Journal of Symbolic Logic*, 74:712–720, 2009.

[66] Kenneth Kunen and Franklin D. Tall. Between Martin's Axiom and Suslin's Hypothesis. *Fundamenta Mathematicae*, 102:173–181, 1979.

[67] Kenneth Kunen and Franklin D. Tall. The real line in elementary submodels of set theory. *The Journal of Symbolic Logic*, 65:683–691, 2000.

[68] Kenneth Kunen and Jerry E. Vaughn, editors. *Handbook of Set-Theoretic Topology*. North-Holland, Amsterdam, 1984.

[69] Richard Laver. On the consistency of Borel's conjecture. *Acta Mathematica*, 137:151–169, 1976.

[70] Richard Laver. An $(\aleph_2, \aleph_2, \aleph_0)$-saturated ideal on ω_1. In Dirk van Dalen, Daniel Lascar, and Timothy J. Smiley, editors, *Logic Colloquium '80 (Prague, 1980)*, volume 108 of *Studies in Logic and the Foundations of Mathematics*, pages 173–180. North-Holland, Amsterdam, 1982.

[71] Richard Laver. Reflection of elementary embedding axioms on the $L[V_{\lambda+1}]$ hierarchy. *Annals of Pure and Applied Logic*, 107:227–235, 2001.

[72] Richard Laver. On very large cardinals. In *Paul Erdős and his Mathematics II*, volume 11 of *Bolyai Society Mathematical Studies*, pages 453–469. Springer, Berlin, 2002.

[73] Menachem Magidor. On the existence of nonregular ultrafilters and the cardinality of ultrapowers. *Transactions of the American Mathematical Society*, 249(1):109–134, 1979.

[74] Richard Mansfield. The solution to one of Ulam's problems concerning analytic sets. II. *Proceedings of the American Mathematical Society*, 26:539–540, 1970.

[75] D. Anthony Martin. Infinite games. In Olli Lehto, editor, *Proceedings of the International Congress of Mathematicians, Helsinki 1978, volume 1*, pages 269–273. Academica Scientiarum Fennica, Helsinki, 1980.

[76] Telis K. Menas. A combinatorial property of $p_\kappa\lambda$. *The Journal of Symbolic Logic*, 41:225–234, 1976.

[77] Arnold W. Miller. Some properties of measure and category. *Transactions of the American Mathematical Society*, 266:93–114, 1981.

[78] William J. Mitchell. Sets constructible from sequences of ultrafilters. *The Journal of Symbolic Logic*, 39:57–66, 1974.

[79] Jeffrey B. Paris. *Boolean Extensions and Large Cardinals.* PhD thesis, University of Manchester, 1969.

[80] Mary Ellen Rudin. Partial orders on the types in βN. *Transactions of the American Mathematical Society*, 155:353–362, 1971.

[81] Walter Rudin. Homogeneity problems in the theory of Čech compactifications. *Duke Mathematical Journal*, 23:409–420, 1955.

[82] Saharon Shelah. *Proper Forcing*, volume 940 of *Lecture Notes in Mathematics.* Springer-Verlag, Berlin, 1982.

[83] Saharon Shelah. Iterated forcing and normal ideals on ω_1. *Israel Journal of Mathematics*, 60:345–380, 1987.

[84] Jack H. Silver. The consistency of the generalized continuum hypothesis with the existence of a measurable cardinal (abstract). *Notices of the American Mathematical Society*, 13:721, 1966.

[85] Jack H. Silver. *Some Applications of Model Theory in Set Theory.* PhD thesis, University of California at Berkeley, 1966.

[86] Jack H. Silver. The consistency of the generalized continuum hypothesis with the existencre of a measurable cardinal. In Dana S. Scott, editor, *Axiomatic Set Theory*, volume 13(1) of *Proceedings of Symposia in Pure Mathematics*, pages 391–395. American Mathematical Society, Providence R.I., 1971.

[87] Jack H. Silver. The independence of Kurepa's conjecture and two-cardinal conjectures in model theory. In Dana S. Scott, editor, *Axiomatic Set Theory*, volume 13(1) of *Proceedings of Symposia in Pure Mathematics*, pages 383–390. American Mathematical Society, Providence R.I., 1971.

[88] Jack H. Silver. Some applications of model theory in set theory. *Annals of Mathematical Logic*, 3:45–110, 1971.

[89] Robert M. Solovay. A nonconstructible Δ^1_3 set of integers. *Transactions of the American Mathematical Society*, 127:50–75, 1967.

[90] Robert M. Solovay. Real-valued measurable cardinals. In Dana S. Scott, editor, *Axiomatic Set Theory*, volume 13(1) of *Proceedings of Symposia in Pure Mathematics*, pages 397–428. American Mathematical Society, Providence R.I., 1971.

[91] Robert M. Solovay. A Δ^1_3 coding of subsets of ω_ω. In Alexander Kechris and Yiannis Moschovakis, editors, *Cabal Seminar 76-77. Proceedings, Caltech-UCLA Logic Seminar 1976-1977*, volume 689 of *Lecture Notes in Mathematics*, pages 133–150. Springer-Verlag, Berlin, 1978.

[92] Robert M. Solovay, William N. Reinhardt, and Akihiro Kanamori. Strong axioms of infinity and elementary embeddings. *Annals of Mathematical Logic*, 13:73–116, 1978.

[93] John R. Steel. *The Core Model Iterability Problem*, volume 8 of *Lecture Notes in Logic.* Springer-Verlag, Berlin, 1996.

[94] Stevo Todorčević. *Partition Problems in Topology*, volume 84 of *Contemporary Mathematics.* American Mathematical Society, Providence R.I., 1989.

[95] Edward L. Wimmers. The Shelah P-point independence theorem. *Israel Journal of Mathematics*, 43:28–48, 1982.

[96] W. Hugh Woodin. Suitable extender sequences, 2009.

[97] Jindřich Zapletal. A new proof of Kunen's inconsistency. *Proceedings of the American Mathematical Society*, 124(7):2203–2204, 1996.

18 Laver and Set theory

Richard Joseph Laver (20 October 1942 – 19 September 2012) was a set theorist of remarkable breadth and depth, and his tragic death from Parkinson's disease a month shy of his 70th birthday occasions a commemorative and celebratory account of his mathematical work, work of an individual stamp having considerable significance, worth, and impact. Laver established substantial results over a broad range in set theory from those having the gravitas of resolving classical conjectures through those about an algebra of elementary embeddings that opened up a new subject. There would be crisp observations as well, like the one, toward the end of his life, that the ground model is actually definable in any generic extension. Not only have many of his results as facts become central and pivotal for set theory, but they have often featured penetrating methods or conceptualizations with potentialities that were quickly recognized and exploited in the development of the subject as a field of mathematics.

In what follows, we discuss Laver's work in chronological order, bringing out the historical contexts, the mathematical significance, and the impact on set theory. Because of his breadth, this account can also be construed as a mountain hike across heights of set theory in the period of his professional life. There is depth as well, as we detail with references the earlier, concurrent, and succeeding work.

Laver became a graduate student at the University of California at Berkeley in the mid-1960s, just when Cohen's forcing was becoming known, elaborated and applied. This was an expansive period for set theory with a new generation of mathematicians entering the field, and Berkeley particularly was a hotbed of activity. Laver and fellow graduate students James Baumgartner and William Mitchell, in their salad days, energetically assimilated a great deal of forcing and its possibilities for engaging problems new and old, all later to become prominent mathematicians. Particularly influential was Fred Galvin, who as a post-doctoral fellow there brought in issues about order types and combinatorics. In this milieu, the young Laver in his 1969 thesis, written under the supervision of Ralph McKenzie, exhibited a deep historical and mathematical understanding when he affirmed a longstanding combinatorial conjecture with penetrating argumentation. §1 discusses Laver's work on Fraïssé's Conjecture and subsequent developments, both in his and others' hands.

For the two academic years 1969-71, Laver was a post-doctoral fellow at the University of Bristol, and there he quickly developed further interests, e.g. on consistency results about partition relations from the then *au courant* Martin's Axiom. §3 at the beginning discusses this, as well as his pursuit in the next several years of saturated ideals and their partition relation consequences.

For the two academic years 1971-3, Laver was an acting assistant professor at the University of California at Los Angeles; for Fall 1973 he was a research

Republished from *Archive for Mathematical Logic* 55 (2016), pp.133-164, with permission from Springer Nature BV.

associate there; and then for Spring 1974 he was a research associate back at Berkeley. During this time, fully engaged with forcing, Laver established the consistency of another classical conjecture, again revitalizing a subject but also stimulating a considerable development of forcing as method. §2 discusses Laver's work on the Borel Conjecture as well as the new methods and results in its wake.

By 1974, Laver was comfortably ensconced at the University of Colorado at Boulder, there to pursue set theory, as well as his passion for mountain climbing, across a broad range. He was Assistant Professor 1974-7, Associate Professor 1977-80, and Professor from 1980 on; and there was prominent faculty in mathematical logic, consisting of Jerome Malitz, Donald Monk, Jan Mycielski, William Reinhardt, and Walter Taylor. Laver not only developed his theory of saturated ideals as set out in §3, but into the 1980s established a series of pivotal or consolidating results in diverse directions. §4 describes this work: indestructibility of supercompact cardinals; functions $\omega \to \omega$ under eventual dominance; the $\aleph_2$-Suslin Hypothesis; nonregular ultrafilters; and products of infinitely many trees.

In the mid-1980s, Laver initiated a distinctive investigation of elementary embeddings as given by very strong large cardinal hypotheses. Remarkably, this led to the freeness of an algebra of embeddings and the solvability of its word problem, and stimulated a veritable cottage industry at this intersection of set theory and algebra. Moving on, Laver clarified the situation with even stronger embedding hypotheses, eventually coming full circle to something basic to be seen anew, that the ground model is definable in any generic extension. This is described in the last, §5.

In the preparation of this account, several chapters of [Kanamori *et al.*, 2012], especially Jean Larson's, proved to be helpful, as well as her compiled presentation of Laver's work at Luminy, September 2012. Just to appropriately fix some recurring terminology, a *tree* is a partially ordered set with a minimum element such that the predecessors of any element are well-ordered; the *αth level* of a tree is the set of elements whose predecessors have order type α; and the *height* of a tree is the least α such that the αth level is empty. A forcing poset has the *κ-c.c. (κ-chain condition)* when every antichain (subset consisting of pairwise incompatible elements) has size less than κ, and a forcing poset has the *c.c.c. (countable chain condition)* if it has the $\aleph_1$-c.c.

1 Fraïssé's Conjecture

Laver [1][2] in his doctoral work famously established Fraïssé's Conjecture, a basic-sounding statement about countable linear orderings that turned out to require a substantial proof. We here first reach back to recover the historical roots, then describe how the proof put its methods at center stage, and finally, recount how the proof itself became a focus for analysis and for further application.

Cantor at the beginnings of set theory had developed the ordinal numbers [*Anzalen*], later taking them as order types of well-orderings, and in his mature *Beiträge* presentation [1895] also broached the order types of linear orderings. He (§§9-11) characterized the order types θ of the real numbers and η of the rational numbers, the latter as the type of the countable dense linear ordering without endpoints. With this as a beginning, while the transfinite numbers have become incorporated into set theory as the (von Neumann) ordinals, there remained an *indifference to identification* for linear order types as primordial constructs about order, as one moved variously from canonical representatives to equivalence classes according to order isomorphism or just taking them as *une façon de parler* about orderings.

The first to elaborate the transfinite after Cantor was Hausdorff, and in a series of articles he enveloped Cantor's ordinal and cardinal numbers in a rich structure festooned with linear orderings. Well-known today are the "Hausdorff gaps", but also salient is that he had characterized the scattered [*zerstreut*] linear order types, those that do not have embedded in them the dense order type η. Hausdorff [1908, §§10-11] showed that for regular $\aleph_\alpha$, the scattered types of cardinality $< \aleph_\alpha$ are generated by starting with 2 and regular ω_ξ and their converse order types ω_ξ^* for $\xi < \alpha$, and closing off under the taking of sums $\Sigma_{i \in \varphi} \varphi_i$, the order type resulting from replacing each i in its place in φ by φ_i. With this understanding, scattered order types can be ranked into a *hierarchy*.

The study of linear order types under order-preserving embeddings would seem a basic and accessible undertaking, but there was little scrutiny until the 1940s. Ostensibly unaware of Hausdorff's work, Ben Dushnik and Edwin Miller [1940] and Wacław Sierpiński [1946, 1950], in new groundbreaking work, exploited order completeness to develop uncountable types embedded in the real numbers that exhibit various structural properties. Then in 1947 Roland Fraïssé, now best known for the Ehrenfeucht-Fraïssé games and Fraïssé limits, pointed to basic issues for *countable* order types in four conjectures. For types φ and ψ, write $\varphi \leq \psi$ *iff* there is an (injective) order-preserving embedding of φ into ψ and $\varphi < \psi$ *iff* $\varphi \leq \psi$ yet $\psi \not\leq \varphi$. Fraïssé's [1948] first conjecture, at first surprising, was that there is no infinite $<$-descending sequence of countable types. Laver would affirm this, but in a strong sense as brought out by the emerging theory and the eventual method of proof.

A general notion applicable to classes ordered by embeddability, Q with a $\leq_Q$ understood is *quasi-ordered* if $\leq_Q$ is a reflexive, transitive relation on Q. Reducing with the equivalence relation $q \equiv r$ *iff* $q \leq_Q r$ and $r \leq_Q q$, one would get a corresponding relation on the equivalence classes which is anti-symmetric and hence a partial ordering; the preference however is to develop a theory doing without this, so as to be able to work directly with members of

Q.[1] Q is *well-quasi-ordered* (wqo) if for any $f\colon \omega \to Q$, there are $i < j < \omega$ such that $f(i) \leq_Q f(j)$. Graham Higman [1952] came to wqo via a finite basis property and made the observation, simple with Ramsey's Theorem, that Q is wqo *iff* (a) Q is *well-founded*, i.e. there are no infinite $<_Q$-descending sequences (where $q <_Q r$ *iff* $q \leq_Q r$ yet $r \not\leq_Q q$), and (b) there are no infinite *antichains*, i.e. sets of pairwise $\leq_Q$-incomparable elements. For a Q quasi-ordered by $\leq_Q$, the subsets of Q can be correspondingly quasi-ordered by: $X \leq Y$ *iff* $\forall x \in X \exists y \in Y (x \leq_Q y)$. With this, Higman established that if Q is wqo, then so are the *finite* subsets of Q. In his 1954 dissertation Joseph Kruskal [1960] also came to well-quasi-ordering, coining the term, and settled a conjecture about trees: For trees T_1 and T_2, $T_1 \leq T_2$ *iff* T_1 is homeomorphically embeddable into T_2, i.e. there is an injective $f\colon T_1 \to T_2$ satisfying $f(x \wedge y) = f(x) \wedge f(y)$, where $\wedge$ indicates the greatest lower bound. Kruskal established that the *finite* trees are wqo.

Pondering the delimitations to the finite, particularly that there had emerged a simple example of a wqo Q whose full power set $\mathcal{P}(Q)$ is not wqo, Crispin Nash-Williams [1965] came up with what soon became a pivotal notion. Identifying subsets of ω with their increasing enumerations, say that a set B of non-empty finite subsets of ω is a *block* if every infinite subset of ω has an initial segment in B. For non-empty finite subsets s, t of ω, write $s \lhd t$ *iff* there is a $k < \min(t)$ such that s is a proper initial segment of $\{k\} \cup t$. Finally, Q with $\leq_Q$ is *better-quasi-ordered* (bqo) if for any block B and function $f\colon B \to Q$, there are $s \lhd t$ both in B such that $f(s) \leq_Q f(t)$. bqo implies wqo, since $\{\{i\} \mid i \in \omega\}$ is a block and $\{i\} \lhd \{j\}$ *iff* $i < j$, and this already points to how bqo might be a useful strengthening in structured situations. Nash-Williams observed that if Q is bqo then so is $\mathcal{P}(Q)$, and established that the infinite trees of height at most ω are bqo.

With this past as prologue, Laver [1][2] in 1968 dramatically established Fraïssé's Conjecture in the strong form: *the countable linear order types are bqo.* Of course, it suffices to consider only the scattered countable types, since any countable type is embeddable into the dense type η. In a remarkably synthetic proof, Laver worked up a hierarchical analysis building on the Hausdorff characterization of scattered types; develop a labeled tree version of Nash-Williams's tree theorem; and established a main preservation theorem, Q bqo $\longrightarrow$ $Q^{\mathcal{M}}$ bqo, the latter consisting of Q-labeled ordered types in a class $\mathcal{M}$. Actually, Laver established his result for the large class $\mathcal{M}$ of *σ-scattered* order types, countable unions of scattered types, working up a specific hierarchy for these devised by Fred Galvin.

Laver's result, both in affirming that the countable linear order types have

[1] A quasi-order is also termed a *pre-order*, and in iterated forcing, to the theory of which Laver would make an important contribution (cf. §2), one also prefers to work with pre-orders of conditions rather than equivalence classes of conditions.

the basic wqo connecting property and being affirmed with a structurally synthetic and penetrating proof, would stand as a monument, not the least because of a clear and mature presentation in [2]. wqo and bqo were brought to the foreground; the result was applied and analyzed; and aspects and adaptations of both statement and proof would be investigated. Laver himself [3][7][12] developed the theory in several directions.

In [3], Laver proceeded to a decomposition theorem for order types. As with ordinals, an order type φ is *additively indecomposable* (AI) *iff* whenever it is construed as a sum $\psi + \theta$, then $\varphi \leq \psi$ or $\varphi \leq \theta$. Work in [2] had shown that any scattered order type is a finite sum of AI types and that the AI scattered types can be generated via "regular unbounded sums". Generalizing homeomorphic embedding to a many-one version, Laver established a tree representation for AI scattered types as a decomposition theorem, and then drew the striking conclusion that *for σ-scattered φ, there is an $n \in \omega$ such that for any finite partition of φ, φ is embeddable into a union of at most n parts.* In [7], Laver furthered the wqo theory of finite trees; work there was later applied by [Kozen, 1988] to establish a notable finite model property. Finally in [12], early in submission but late in appearance, Laver made his ultimate statement on bqo. He first provided a lucid, self-contained account of bqo theory through to Nash-Williams's subtle "forerunning" technique. A tree is *scattered* if the complete binary tree is not embeddable into it, and it is *σ-scattered* if it is a countable union of scattered, downward-closed subtrees. As a consequence of a general preservation result about labeled trees, Laver established: *the σ-scattered trees are bqo.* Evidently stimulated by this work, Saharon Shelah [1982] investigated a bqo theory for uncountable cardinals based on whenever $f\colon \kappa \to Q$ there are $i < j < \kappa$ such that $f(i) \leq_Q f(j)$, discovering new parametrized concepts and a large cardinal connection.

"Fraïssé's Conjecture", taken to be the (proven) proposition that countable linear orders are wqo, would newly become a focus in the 1990s with respect to the reverse mathematics of provability in subsystems of second-order arithmetic.[2] Richard Shore [1993] established that the countable *well*-orderings being wqo already entails the system ATR_0. Since the latter implies that any two countable well-orderings are comparable, there is thus an equivalence. Antonio Montalbán [2005] proved that every hyperarithmetic linear order is mutually embeddable with a recursive one and [2006] showed that Fraïssé's Conjecture is equivalent (over the weak theory RCA_0) to various propositions about linear orders under embeddability, making it a "robust" theory. However, whether Fraïssé's Conjecture is actually equivalent to ATR_0 is a longstanding problem of reverse mathematics, with e.g. [Marcone and Montalbán, 2009] providing a partial result. The proposition, basic and under new scrutiny, still has the one proof that has proved resilient, the proof of Laver [2] going through the hierar-

[2]See [Marcone, 2005] for a survey of the reverse mathematics of wqo and bqo theory.

chy of scattered countable order types and actually establishing bqo through a preservation theorem for labeled order types.

Into the 21st Century, there would finally be progress about possibilities for extending Laver's result into non-σ-scattered order types and trees. Laver [12] had mentioned that Aronszajn trees (cf. §4.3) are not wqo assuming Ronald Jensen's principle $\diamondsuit$ and raised the possibility of a relative consistency result. This speculation would stand for decades until in 2000 Stevo Todorcevic [2007] showed that no, there are $2^{\aleph_1}$ Aronszajn trees pairwise incomparable under (just) injective order-preserving embeddability. Recently, on the other hand, Carlos Martinez-Ranero [2011] established that under the Proper Forcing Axiom (PFA), Aronszajn lines *are* bqo. Aronszajn lines are just the linearizations of Aronszajn trees, so this is a contradistinctive result. Under PFA, Justin Moore [2009] showed that there is a universal Aronszajn line, a line into which every Aronszajn line is embeddable, and starting with this analogue of the dense type η, Martinez-Canero proceeded to adapt the Laver proof. Generally speaking, a range of recent results have shown PFA to provide an appropriately rich context for the investigation of general, uncountable linear order types; [Ishiu and Moore, 2009] even discussed the possibility that the Laver result about σ-scattered order types, newly apprehended as prescient as to how far one can go, is sharp in the sense that it cannot be reasonably extended to a larger class of order types.

2 Borel Conjecture

Following on his Fraïssé's Conjecture success, Laver [5][8] by 1973, while at the University of California at Los Angeles, had established another pivotal result with an even earlier classical provenance and more methodological significance, the consistency of "the Borel Conjecture". A subset X of the unit interval of reals has *strong measure zero* (Laver's term) *iff* for any sequence $\langle \epsilon_n \mid n \in \omega \rangle$ of positive reals there is a sequence $\langle I_n \mid n \in \omega \rangle$ of intervals with the length of I_n at most ϵ_n such that $X \subseteq \bigcup_n I_n$. Laver established with iterated forcing the relative consistency of $2^{\aleph_0} = \aleph_2$ + "Every strong measure zero set is countable". We again reach back to recover the historical roots and describe how the proof put its methods at center stage, and then how both result and method stimulated further developments.

At the turn of the 20th Century, Borel axiomatically developed his notion of measure, getting at those sets obtainable by starting with the intervals and closing off under complementation and countable union and assigning corresponding measures. Lebesgue then developed his extension of Borel measure, which in retrospect can be formulated in simple set-theoretic terms: A set of reals is *null iff* it is a subset of a Borel set of measure zero, and a set is *Lebesgue measurable iff* it has a null symmetric difference with some Borel set, in which case its Borel measure is assigned. With null sets having an amorphous feel,

Borel [1919] studied them constructively in terms of rates of convergence of decreasing measures of open covers, getting to the strong measure zero sets. Actually, he only mentioned them elliptically, writing that they would have to be countable but that he did not possess an "entirely satisfactory proof".[3] Borel would have seen that no uncountable closed set of reals can have strong measure zero, and so, that no uncountable Borel set can have strong measure zero. More broadly, a perfect set (a non-empty closed set with no isolated points), though it can be null,[4] is seen not to have strong measure zero. So, it could have been deduced by then that no uncountable analytic set, having a perfect subset, can have strong measure zero. While all this might have lent an air of plausibility to strong measure zero sets having to be countable, it was also known by then that the Continuum Hypothesis (CH) implies the existence of a *Luzin set*, an uncountable set having countable intersection with any meager set. A Luzin set can be straightforwardly seen to have strong measure zero, and so Borel presumably could not have possessed a "satisfactory proof".

In the 1930s strong measure zero sets, termed Wacław Sierpiński's "sets with Property C", were newly considered among various special sets of reals formulated topologically.[5] Abram Besicovitch came to strong measure zero sets in a characterization result, and he provided, in terms of his "concentrated sets", a further articulated version of CH implying the existence of an uncountable strong measure zero set. Then Sierpiński and Fritz Rothberger, both in 1939 papers, articulated the first of the now many cardinal invariants of the continuum, the bounding number. (A family F of functions: $\omega \to \omega$ is *unbounded* if for any $g\colon \omega \to \omega$ there is an f in F such that $\{n \mid g(n) \leq f(n)\}$ is infinite, and the *bounding number* b is the least cardinality of such a family.) Their results about special sets established that (without CH but just) $b = \aleph_1$ implies the existence of an uncountable strong measure zero set. Strong measure zero sets having emerged as a focal notion, there was however little further progress, with Rothberger [1952] retrospectively declaring "the principal problem" to be whether there are uncountable such sets.[6]

Whatever the historical imperatives, two decades later Laver [5][8] duly established the relative consistency of "the Borel Conjecture", that all strong measure zero sets can be countable. Cohen, of course, had transformed set theory in 1963 by introducing forcing, and in the succeeding decades there were broad advances made through the new method involving the development

[3][Borel, 1919, p.123]: "Un ensemble énumérable a une mesure asymptotique inférieure à toute série donnée a l'avance; la réciproque me paraît exacte, mai je n'en possède pas de démonstration entièrement satisfaisante."

[4]The Cantor ternary set, defined by Cantor in [1883], is of course an example.

[5]cf. [Steprāns, 2012, pp.92-102] for a historical account.

[6][Rothberger, 1952, p.111] "...the principal problem, viz., to prove with the axiom of choice only (without any other hypothesis) the existence of a non-denumerable set of property C, this problem remains open."

both of different forcings and of forcing techniques. Laver's result featured both a new forcing, for adding a *Laver real*, and a new technique, adding reals at each stage in a *countable support iteration*.

For adding a Laver real, a condition is a tree of natural numbers with a finite trunk and all subsequent nodes having infinitely many immediate successors. A condition is stronger than another if the former is a subtree, and the longer and longer trunks union to a new, generic real: $\omega \to \omega$. Thus a Laver condition is a structured version of the basic Cohen condition, which corresponds to just having the trunk, and that structuring revises the Sacks condition, in which one requires that every node has an eventual successor with two immediate successors. Already, a Laver real is seen to be a dominating real, i.e. for any given ground model $g\colon \omega \to \omega$ a Laver condition beyond the trunk can be pruned to always take on values larger than those given by g. Thus, the necessity of making the bounding number $\mathfrak{b}$ large is addressed. More subtly, Laver conditions exert enough infinitary control to assure that for any uncountable set X of reals in the ground model and with f being the Laver real, there is no sequence $\langle I_n \mid n \in \omega \rangle$ of intervals in the extension with length $I_n < \frac{1}{f(n)}$ such that $X \subseteq \bigcup_n I_n$.

Laver proceeded from a model of CH and adjoined Laver reals iteratively in an iteration of length ω_2. The iteration was with countable support, i.e. a condition at the αth stage is a vector of condition names at earlier stages, with at most countably many of them being non-trivial. This allowed for a tree "fusion" argument across the iteration that determined more and more of the names as actual conditions and so showed that e.g. for any countable subset of the ground model in the extension, there is a countable set in the ground model that covers it. Consequently, ω_1 is preserved in the iteration and so also the $\aleph_2$-c.c., so that all cardinals are preserved and $2^{\aleph_0} = \aleph_2$ in the extension. Specifically for the adjoining of Laver reals, Laver crowned the argument as follows:

Suppose that X is an $\aleph_1$ size set of reals in the extension. Then it had already occurred at an earlier stage by the chain condition, and so at that stage the next Laver real provides a counterexample to X having strong measure zero. But then, there is enough control through the subsequent iteration with the "fusion" apparatus to ensure that X still will not have strong measure zero.

Laver's result and paper [8] proved to be a turning point for iterated forcing as method. Initially, the concrete presentation of iteration as a quasi-order of conditions that are vectors of forcing names for local conditions was itself revelatory. Previous multiple forcing results like the consistency of Martin's Axiom had been cast in the formidable setting of Boolean algebras. Henceforth, there would be a grateful return to Cohen's original heuristic of conditions approximating a generic object, with the particular advantage in iterated forcing of seeing the dynamic interaction with forcing names, specifically names for later conditions. More centrally, Laver's structural results about countable support iteration established a scaffolding for proceeding that would become standard fare. While the consistency of Martin's Axiom had been established with the finite support iteration of c.c.c. forcings, the new regimen admitted other forcings and yet preserved much of the underlying structure of the ground model.

Several years later Baumgartner and Laver [16] elaborated the countable support iteration of Sacks forcing, and with it established consistency results about selective ultrafilters as well as about higher Aronszajn trees (cf. §4.3). They established: *If κ is weakly compact and κ Sacks reals are adjoined iteratively with countable support, then in the resulting forcing extension $\kappa = \omega_2$ and there are no $\aleph_2$-Aronszajn trees.* Groundbreaking for higher Aronszajn trees, that they could be no $\aleph_2$-Aronszajn trees had first been pointed out by Jack Silver as a consequence of forcing developed by Mitchell (cf. [1972, p.41]) and significantly, that forcing was the initial instance of a countable support iteration. However, it worked in a more involved way with forcing names, and the Baumgartner-Laver approach with the Laver scaffolding made the result more accessible.

By 1978 Baumgartner had axiomatically generalized the iterative addition of reals with countable support with his "Axiom A" forcing, and in an influential account [1983] set out iterated forcing and Axiom A in an incisive manner. Moreover, he specifically worked through the consistency of the Borel Conjecture by iteratively adjoining Mathias reals with countable support, a possible alternate approach to the result pointed out by Laver [8, p.168]. All this would retrospectively have a precursory air, as Shelah in 1978 established a general, subsuming framework with his *proper forcing.* With its schematic approach based on countable elementary substructures, proper forcing realized the potentialities of Laver's initial work and brought forcing to a new plateau. Notably, a combinatorial property of Laver forcing, "the Laver property", was shown to be of importance and preserved through the iteration of proper forcings.[7]

As for Laver reals and Laver's specific [8] model, Arnold Miller [1980] showed that in that model there are no q-point ultrafilters, answering a question of the author. Later, in the emerging investigation of cardinal invariants, Laver forcing would become *the* forcing "associated" with the bounding number b,[8] in that it is the forcing that increases b while fixing the cardinal invariants not immediately dependent on it. [Judah and Shelah, 1990] exhibited this with the Laver [8] model.

And as for the Borel Conjecture itself, the young Hugh Woodin showed in 1981 that adjoining any number of random reals to Laver's model preserves the Borel Conjecture, thereby establishing the consistency of the conjecture with

[7]cf. [Bartoszyński and Judah, 1995, 6.3.E].
[8]cf. [Bartoszyński and Judah, 1995, 7.3.D].

the continuum being arbitrarily large. The sort of consistency result that Laver had achieved has become seen to have a limitative aspect in that countable support iteration precludes values for the continuum being larger than $\aleph_2$, and at least for the Borel Conjecture a way was found to further increase the size of the continuum. [Judah *et al.*, 1990] provided systematic iterated forcing ways for establishing the Borel Conjecture with the continuum arbitrarily large.

3 Partition Relations and Saturated Ideals

Before he established the consistency of Borel's conjecture, Laver, while at the University of Bristol (1969-1971), had established [6] relative consistency results about partition relations low in the cumulative hierarchy. Through the decade to follow, he enriched the theory of saturated ideals in substantial part to get at further partition properties. This work is of considerable significance, in that large cardinal hypotheses and infinite combinatorics were first brought together in a sustained manner.

In the well-known Erdős-Rado partition calculus, the simplest case of the ordinal partition relation is $\alpha \longrightarrow (\beta)^2_2$, the proposition that for any partition $[\alpha]^2 = P_0 \cup P_1$ of the 2-element subsets of α into two cells P_0 and P_1, there is a subset of α of order type β all of whose 2-element subsets are in the same cell. The unbalanced relation $\alpha \longrightarrow (\beta, \gamma)^2$ is the proposition that for any partition $[\alpha]^2 = P_0 \cup P_1$, either there is a subset of α order type β all of whose 2-element subsets are in P_0 or there is a subset of α of order type γ all of whose 2-elements subsets are in P_1. Ramsey's seminal 1930 theorem amounts to $\omega \longrightarrow (\omega)^2_2$, and sufficiently strong large cardinal properties for a cardinal κ imply $\kappa \longrightarrow (\kappa)^2_2$, which characterizes the weak compactness of κ. Laver early on focused on the possibilities of getting the just weaker $\kappa \longrightarrow (\kappa, \alpha)^2$ for small, accessible κ and a range of $\alpha < \kappa$.

In groundbreaking work, Laver [6] showed that Martin's Axiom (MA) has consequences for partition relations of this sort for $\kappa \leq 2^{\aleph_0}$. Laver was the first to establish relative consistency results, rather than outright theorems of ZFC, about partition relations for accessible cardinals. Granted, Karel Prikry's [1972] work was important in this direction in establishing a negation of a partition relation consistent, particularly as he did this by forcing a significant combinatorial principle that would subsequently be shown to hold in L. Notably, Erdős bemoaned how the partition calculus would now have to acknowledge relative consistency results. Laver's work, in first applying MA, was also pioneering in adumbration of arguments for the central theorem of Baumgartner and András Hajnal [1973], that $\omega_1 \longrightarrow (\alpha)^2_2$ for every $\alpha < \omega_1$, a ZFC theorem whose proof involved appeal to MA and absoluteness. As for the stronger, unbalanced relation, the young Stevo Todorcevic [1983] by 1981 established the consistency of MA $+ 2^{\aleph_0} = \aleph_2$ together with $\omega_1 \longrightarrow (\omega_1, \alpha)^2$ for every $\alpha < \omega_1$.

By 1976, Laver saw how saturated ideals in a strong form can drive the argumentation to establish unbalanced partition relations for cardinals. Briefly, I is a *κ-ideal iff* it is an ideal over κ (a family of subsets of κ closed under the taking of subsets and unions) which is non-trivial (it contains $\{\alpha\}$ for every $\alpha < \kappa$ but not κ) and κ-complete (it is closed under the taking of unions of fewer than κ of its members). Members of a κ-ideal are "small" in the sense given by I, and mindful of this, such an ideal is *λ-saturated iff* for any λ subsets of κ not in I there are two whose intersection is still not in I. Following the founding work of Robert Solovay on saturated ideals in the 1960s, they have become central to the theory of large cardinals primarily because they can carry strong consistency strength yet appear low in the cumulative hierarchy. κ is a measurable cardinal, as usually formulated, just in case there is a 2-saturated κ-ideal, and e.g. if κ Cohen reals are adjoined, then in the resulting forcing extension: $\kappa = 2^{\aleph_0}$ and there is an $\aleph_1$-saturated κ-ideal. Conversely, if there is a κ^+-saturated κ-ideal for some κ, then in the inner model relatively constructed from such an ideal, κ is a measurable cardinal.

In a first, parametric elaboration of saturation, Laver formulated the following property: A κ-ideal I is *(λ, μ, ν)-saturated iff* every family of λ subsets of κ not in I has a subfamily of size μ such that any ν of its members has still has intersection not in I. In particular, a κ-ideal is λ-saturated *iff* it is $(\lambda, 2, 2)$-saturated. In the abstract [14], for a 1976 meeting, Laver announced results subsequently detailed in [10] and [19].

In [10] Laver established that if $\gamma < \kappa$ and there is a (κ, κ, γ)-saturated κ-ideal (which entails that κ must be a regular limit cardinal) and $\beta^\gamma < \kappa$ for every $\beta < \kappa$, then $\kappa \longrightarrow (\kappa, \alpha)^2$ holds for every $\alpha < \gamma^+$. He then showed, starting with a measurable cardinal κ, how to cleverly augment the forcing for adding many Cohen subsets of a $\gamma < \kappa$ to retain such κ-ideals with κ newly accessible, a paradigmatic instance being a $(2^{\aleph_1}, 2^{\aleph_1}, \aleph_1)$-saturated $2^{\aleph_1}$-ideal with $\beta < 2^{\aleph_1}$ implying $\beta^{\aleph_0} < 2^{\aleph_1}$. From this one has the consistency of $2^{\aleph_1} \longrightarrow (2^{\aleph_1}, \alpha)^2$ for every $\alpha < \omega_2$, and this is sharp in two senses, indicative of what Laver was getting at: A classical Sierpiński observation is that $2^{\aleph_1} \longrightarrow (\omega_2)^2$ fails, and the well-known Erdős-Rado Theorem implies that $(2^{\aleph_1})^+ \longrightarrow ((2^{\aleph_1})^+, \omega_2)^2$ holds. Years later, [Todorcevic, 1986] established the consistency, relative only to the existence of a weakly compact cardinal, of $2^{\aleph_0} \longrightarrow (2^{\aleph_0}, \alpha)^2$ for every $\alpha < \omega_1$, as well as of $2^{\aleph_1} \longrightarrow (2^{\aleph_1}, \alpha)^2$ for every $\alpha < \omega_2$.

In [19] Laver established the consistency of a substantial version of his saturation property holding for a κ-ideal with κ a *successor* cardinal, thereby establishing the consistency of a partition property for such κ. In the late 1960s, while having a κ^+-saturated κ-ideal for some κ had been seen to be equi-consistent to having a measurable cardinal, Kunen had shown that the consistency strength, were κ posited to be a *successor* cardinal, was far stronger. In a *tour de force*, Kunen [1978] in 1972 established: *If κ is a huge cardinal, then in a forcing extension $\kappa = \omega_1$ and there is an $\aleph_2$-saturated ω_1-ideal.* In the large cardinal hierarchy huge cardinals are consistency-wise much stronger than the better known supercompact cardinals, and Kunen had unabashedly appealed to the strongest embedding hypothesis to date for carrying out a forcing construction. From the latter 1970s on, Kunen's argument, as variously elaborated and amended, would become and remain a prominent tool for producing strong phenomena at successor cardinals, though dramatic developments in the 1980s would show how weaker large cardinal hypotheses suffice to get $\aleph_2$-saturated ω_1-ideals themselves. Laver in 1976 was to first amend Kunen's argument, getting [19]: *If κ is a huge cardinal, then in a forcing extension $\kappa = \omega_1$ and there is an $(\aleph_2, \aleph_2, \aleph_0)$-saturated ω_1-ideal.* Not only had Laver mastered Kunen's sophisticated argument with elementary embedding, but he had managed to augment it, introducing "Easton supports".

Laver [19] (see also [Kanamori, 1986b]) established that the newly parametrized saturation property has a partition consequence: *If $\kappa^{<\kappa} = \kappa$ and there is a $(\kappa^+, \kappa^+, {<}\kappa)$-saturated (with the expected meaning) κ-ideal, then $\kappa^+ \longrightarrow (\kappa + \kappa + 1, \alpha)$ for every $\alpha < \kappa^+$.* This partition relation is thus satisfied at measurable cardinals κ, and with CH holding in Laver's [19] model it satisfies

$$\omega_2 \longrightarrow (\omega_1 + \omega_1 + 1, \alpha) \text{ for every } \alpha < \omega_2\,.$$

This result stood for decades as best possible for successor cardinals larger than ω_1. Then Matthew Foreman and Hajnal in [2003] extended the ideas to a stronger conclusion, albeit from a stronger ideal hypothesis. A κ-ideal I is *λ-dense iff* there is a family D of λ subsets of κ not in I such that for any subset X of κ not in I, there is a $Y \in D$ almost contained in X, i.e. $Y - X$ is in I. This is a natural notion of density for the Boolean algebra $\mathcal{P}(\kappa)/I$, and evidently a κ-dense κ-ideal is $(\kappa^+, \kappa^+, {<}\kappa)$-saturated. Foreman and Hajnal managed to prove that if $\kappa^{<\kappa} = \kappa$ and there is a κ-dense κ-ideal, then $\kappa^+ \longrightarrow (\kappa^2 + 1, \alpha)$ for every $\alpha < \kappa^+$. Central work by Woodin in the late 1980s had established the existence of an ω_1-dense ω_1-ideal relative to large cardinals, and so one had the corresponding improvement, $\omega_2 \longrightarrow (\omega_1^2 + 1, \alpha)$ for every $\alpha < \omega_2$, of the Laver [19] result and the best possible to date for ω_2.

4 Consolidations

In the later 1970s and early 1980s Laver, by then established at the University of Colorado at Boulder, went from strength to strength in exhibiting capability and willingness to engage with *au courant* concepts and questions over a broad range. In addition to the saturated ideals work, Laver established pivotal, consolidating results, each in a single incisive paper, and in what follows we deal with these and frame their significance.

4.1 Indestructibility

In a move that exhibited an exceptional insight into what might be proved about supercompact cardinals, Laver in 1976 established their possible "indestructibility" under certain forcings. This seminal result, presented in a short 4-page paper, would not only become part and parcel of method about supercompact cardinals but would become a concept to be investigated in its own right for large cardinals in general.

In 1968, Robert Solovay and William Reinhardt (cf. [Solovay *et al.*, 1978]) formulated the large cardinal concept of supercompactness as a generalization of the classical concept of measurability once its elementary embedding characterization was attained. A cardinal κ is *supercompact iff* for every $\lambda \geq \kappa$, κ is λ-supercompact, where in turn κ is λ*-supercompact iff* there is an elementary embedding $j\colon V \to M$ such that the least ordinal moved by j is κ and moreover M is closed under arbitrary sequences of length λ. That there is such a j is equivalent to having a *normal ultrafilter* over $\mathcal{P}_\kappa\lambda = \{x \subseteq \lambda \mid |x| < \kappa\}$; from such a j such a normal ultrafilter can be defined, and conversely, from such a normal ultrafilter U a corresponding elementary embedding j_U can be defined having the requsite properties. κ is κ-supercompact exactly when κ is measurable, as quickly seen from the embedding formulation of the latter.

In 1971, Silver established the relative consistency of having a measurable cardinal κ satisfying $\kappa^+ < 2^\kappa$. That this would require strong hypotheses had been known, and for Silver's argument having an elementary embedding j as given by the κ^{++}-supercompactness of κ suffices. Silver introduced two motifs that would become central to establishing consistency results from strong hypotheses. First, he forced the necessary structure of the model below κ, but *iteratively*, proceeding upward to κ. Second, in considering the j-image of the process he developed a *master condition* so that forcing through it would lead to an extension of j in the forcing extension, thereby preserving the measurability (in fact the κ^{++}-supercompactness) of κ.

Upon seeing Silver's argument as given e.g. in [Menas, 1976] and implementing the partial order approach from the Borel Conjecture work, Laver saw through to a generalizing synthesis, first establishing a means of universal anticipation below a supercompact cardinal and then applying it to render the supercompactness robust under further forcing. The first result exemplifies what reflection is possible at a supercompact cardinal: *Suppose that κ is supercompact. Then there is one function $f\colon \kappa \to V_\kappa$ such that for all $\lambda \geq \kappa$ and all sets x hereditarily of cardinality at most λ, there is a normal ultrafilter U over $\mathcal{P}_\kappa\lambda$ such that $j_U(f)(\kappa) = x$.* Such a function has been called a "Laver function" or "Laver diamond"; indeed, the proof is an elegant variant of the proof of the diamond principle $\diamondsuit$ in L which exploits elementary embeddings and definability of least counterexamples.

With this, Laver established his "indestructibility" result. A notion of forcing P is *κ-directed closed iff* whenever $D \subseteq P$ has size less that κ and is directed (i.e. any two members of D have a lower bound in D), D has a lower bound. Then: *Suppose that κ is supercompact. Then in a forcing extension κ is supercompact and remains so in any further extension via a κ-directed closed notion of forcing.* The forcing done is an iteration of forcings along a Laver function. To show that any further κ-directed closed forcing preserves supercompactness, master conditions are exploited to extend elementary embeddings.

For relative consistency results involving supercompact cardinals, Laver indestructibility leads to technical strengthenings as well as simplifications of proofs, increasing their perspicuity. At the outset as pointed out by Laver himself, while [Menas, 1976] had shown that for κ supercompact and $\lambda \geq \kappa$ there is a forcing extension in which κ remains supercompact and $2^\kappa \geq \lambda$, once a supercompact cardinal is "Laverized", from that single model 2^κ can be made arbitrarily large while preserving supercompactness. Much more substantially and particularly in arguments involving several large cardinals, Laver indestructibility was seen to set the stage after which one can proceed with iterations that preserve supercompactness without bothering with specific preparatory forcings. Laver indestructibility was thus applied in the immediately subsequent, central papers for large cardinal theory, [Magidor, 1977], [Foreman *et al.*, 1988a], and [Foreman *et al.*, 1988b].

The Laver function itself soon played a crucial role in a central relative consistency result. Taking on Shelah's proper forcing, the Proper Forcing Axiom (already mentioned at the end of §1) asserts that for any proper notion of forcing P and sequence $\langle D_\alpha \mid \alpha < \omega_1 \rangle$ of dense subsets of P, there is a filter F over P that meets every D_α. Early in 1979 Baumgartner (cf. [Devlin, 1983]) established: *Suppose that κ is supercompact. Then in a forcing extension $\kappa = \omega_2 = 2^{\aleph_0}$ and* PFA *holds.* Unlike for Martin's Axiom, to establish the consistency of PFA requires handling a proper class of forcings, and it sufficed to iterate proper forcings given along a Laver function, these anticipating all proper forcings through elementary embeddings. PFA is known to have strong consistency strength, and to this day Baumgartner's result, with its crucial use of a Laver function, stands as the bulwark for consistency.

Laver functions have continued to be specifically used in consistency proofs (e.g. [Cummings and Foreman, 1998, 2.6]) and have themselves become the subject of investigation for a range of large cardinal hypotheses ([Corazza, 2000]). As for the indestructibility of large cardinals, the concept has become part of the mainstream of large cardinals not only through application but through concerted investigation. [Gitik and Shelah, 1989] established a form of indestructibility for strong cardinals to answer a question about cardinal powers. [Apter and Hamkins, 1999] showed how to achieve universal indestructibility, indestructibility simultaneously for the broad range of large cardinals from weakly compact to supercompact cardinals. [Hamkins, 2000] developed a general kind of Laver function for any large cardinal and, with it, a general kind of Laver preparation forcing to achieve a broad range of new indestructibilities. Starting with [1998], Arthur Apter has pursued the indestructibility particularly of partially supercompact and strongly compact cardinals through over 20 articles. Recently, [Bagaria *et al.*, 2013] showed that very large cardinals, superstrong and above, are never Laver indestructible, so that there is a ceiling to indestructibility.

In retrospect, it is quite striking that Laver's modest 4-page paper should have had such an impact.

4.2 Eventual Dominance

Hugh Woodin in 1976, while still an undergraduate, made a remarkable reduction of a proposition ("Kaplansky's Conjecture") of functional analysis about the continuity of homomorphisms of Banach algebras to a set-theoretic asserton about embeddability into $\langle {}^{\omega}\omega, <^* \rangle$, the family of functions: $\omega \to \omega$ ordered by eventual dominance (i.e. $f <^* g$ *iff* $f(n) < g(n)$ for sufficiently large n). Solovay, the seasoned veteran, soon established the consistency of the set-theoretic assertion, and thereby, the relative consistency of the proposition.[9] In the process, Solovay raised a question, which Laver [15] by 1978 answered affirmatively by establishing the relative consistency of: $\aleph_1 < 2^{\aleph_0}$ and every linear ordering of size $\leq 2^{\aleph_0}$ is embeddable into $\langle {}^{\omega}\omega, <^* \rangle$.

Linear orderings of size $\leq \aleph_1$ are in any case embeddable into $\langle {}^{\omega}\omega, <^* \rangle$, yet if to a model of CH one e.g. adjoins many Cohen reals then ω_2 is still not embeddable into $\langle {}^{\omega}\omega, <^* \rangle$. Martin's Axiom (MA) implies that every well-ordering of size $< 2^{\aleph_0}$ is embeddable into $\langle {}^{\omega}\omega, <^* \rangle$, yet Kunen in incisive 1975 work had shown that MA is consistent with the existence of a linear ordering of size $2^{\aleph_0}$ *not* being embeddable into $\langle {}^{\omega}\omega, <^* \rangle$. Schematically proceeding as for the consistency of MA itself, Laver [15] in fact operatively showed: *For any cardinal* κ *satisfying* $\kappa^{<\kappa} = \kappa$, *there is a c.c.c. forcing extension in which* $2^{\aleph_0} = \kappa$ *and the saturated linear order of size* $2^{\aleph_0}$ *(i.e. the extant size* $2^{\aleph_0}$ *linear order into which every other linear order of size* $\leq 2^{\aleph_0}$ *embeds) is embeddable into* $\langle {}^{\omega}\omega, <^* \rangle$.

Laver's construction would have to do with the classical work of Hausdorff on order types and gaps at the beginnings of set theory. For a linear order $\langle L, < \rangle$ and $A, B \subseteq L$, $\langle A, B \rangle$ is a *gap iff* every element of A is $<$ any element of B yet there is no member of L $<$-between A and B. Such a gap is a (κ, λ^*)-*gap iff* A has $<$-increasing order type κ and B has $<$-decreasing order type λ. Hausdorff famously constructed what is now well-known as a "Hausdorff gap", an (ω_1, ω_1^*)-gap in $\langle {}^{\omega}\omega, <^* \rangle$ which is not fillable in any forcing extension preserving $\aleph_1$.

---

[9] See [Dales and Woodin, 1987] for an account of Kaplansky's Conjecture, Woodin's reduction, and Woodin's own version of the relative consistency incorporating Martin's Axiom.

To establish his theorem Laver proceeded, in an iterative way with finite support, to adjoin $f_\alpha \in {}^\omega\omega$ so that $\langle\{f_\alpha \mid \alpha < \kappa\}, <^*\rangle$ will be the requisite saturated linear order. At stage β, if there is a gap $\langle A, B\rangle$ with $A \cup B = \{f_\alpha \mid \alpha < \beta\}$, Laver adjoined a generic f_β to fill the gap. As Laver astutely pointed out, his construction would have to avoid prematurely creating a Hausdorff gap, and it does so by iteratively creating a saturated linear order *generically* with finite support. Although Laver does not mention it, his construction affirmatively answered, consistency-wise, the first question of [Hausdorff, 1907, §6]: Is there a pantachie with no (ω_1, ω_1^*) gaps? (For Hausdorff a *panachie* is a maximal linear sub-ordering of $\langle {}^\omega\mathbb{R}, <^*\rangle$, i.e. with the functions being real-valued, but Laver's construction can be adapted.) Historically, Hausdorff's question was the first in ongoing mathematics whose positive answer entailed $2^{\aleph_0} = 2^{\aleph_1}$ and hence the failure of the Continuum Hypothesis.

On topic, Woodin soon augmented Laver's construction to incorporate MA as well. This sharpened the situation, since as mentioned earlier Kunen had shown the consistency of MA and the proposition that there is a linear ordering of size $2^{\aleph_0}$ not embeddable into $\langle {}^\omega\omega, <^*\rangle$. Baumgartner [1984, 4.5] later pointed out that the Proper Forcing Axiom directly implies this proposition.

A decade later Laver [33] pursued the study the space of functions : $\omega_1 \to \omega_1$ under eventual dominance modulo finite sets.

4.3 κ-Suslin Trees

Laver and Shelah [18] showed: *If κ is weakly compact, then in a forcing extension $\kappa = \omega_2$,* CH, *and the $\aleph_2$-Suslin Hypothesis holds.* The proof establishes an analogous result for the successor of any regular cardinal less than κ. Laver had first established the result with "weakly compact" replaced by "measurable", and then Shelah refined the argument. This was the first result appropriately affirming a higher Suslin hypothesis, and as such would play an important, demarcating role in the investigation of generalized Martin's axioms.

A *κ-Aronszajn tree* is a tree with height κ all of whose levels have size less that κ yet there no chain (linearly ordered subset) of size κ; a *κ-Suslin tree* is an κ-Aronszajn tree with no antichain (subset of pairwise incomparable elements) of size κ as well; and the *κ-Suslin Hypothesis* asserts that there are *no* κ-Suslin trees. Without the "κ-" it is to be understood that $\kappa = \aleph_1$.

A classical 1920 question of Mikhail Suslin was shown to be equivalent to the Suslin Hypothesis (SH), and Nathan Aronszajn observed in the early 1930s that in any case there are Aronszajn trees. In the post-Cohen era the investigation of SH led to formative developments in set theory: Stanley Tennenbaum showed how to force $\neg$SH, i.e. to adjoin a Suslin tree; he and Solovay showed how to force $\neg$CH + SH with an inaugural multiple forcing argument, one that straightforwardly modified gives the stronger $\neg$CH + MA; Jensen showed that $V = L$ implies that there is a Suslin tree, the argument leading to the isolation of the diamond principle $\diamondsuit$; and Jensen established the consistency of CH + SH, the argument motivating Shelah's eventual formulation of proper forcing.

With this esteemed, central work at $\kappa = \aleph_1$, Laver one level up faced the $\aleph_2$-Suslin Hypothesis. A contextualizing counterpoint was Silver's deduction through forcing developed by Mitchell (cf. [1972, p.41]) that if κ is weakly compact, then in a forcing extension $\kappa = \omega_2$ and there are no $\aleph_2$-Aronszajn trees at all. But here CH fails, and indeed CH implies that there is an $\aleph_2$-Aronszajn tree. So, the indicated approach would be to start with CH, do forcing that adjoins no new reals and yet destroys all $\aleph_2$-Suslin trees, perhaps using a large cardinal.

In the Solovay-Tennenbaum approach, Suslin trees were destroyed one at a time by forcing through long chains; the conditions for a forcing were just the members of a Suslin tree under the tree ordering, and so one has the c.c.c., which can be iterated with finite support. One level up, one would have to have countably closed forcing (for preseving CH) that, iterated with countable support, would maintain the $\aleph_2$-c.c. (for preserving e.g. the necessary cardinal structure). However, Laver [18, p.412] saw that there could be countably closed $\aleph_2$-Suslin trees whose product may not have the $\aleph_2$-c.c.

Laver then turned to the clever idea of destroying an $\aleph_2$-Suslin tree not by injecting a long chain but a large *antichain*, simply forcing with antichains under inclusion. But for this approach too, Laver astutely saw a problem. For a tree T, with its αth level denoted T_α, a *κ-ascent path* is a sequence of functions $\langle f_\alpha \mid \alpha \in A\rangle$ where A is an unbounded subset of $\{\alpha \mid T_\alpha \neq \emptyset\}$, each $f_\alpha \colon \kappa \to T_\alpha$, and: if $\alpha < \beta$ are both in A, then for sufficiently large $\xi < \kappa$, $f_\alpha(\xi)$ precedes $f_\beta(\xi)$ on the tree. Laver [18, p.412] noted that if an $\aleph_2$-Suslin tree has an ω-ascent path, then the forcing for adjoining a large antichain does not satisfy the $\aleph_2$-c.c., and showed that it is relatively consistent to have an $\aleph_2$-Suslin tree with an ω-ascent path. In the subsequent elaboration of higher Suslin trees, the properties and constructions of trees with ascent paths became a significant topic in itself; cf. [Cummings, 1997] from which the terminology is drawn.

Laver saw how, then, to proceed. With conceptually resonating precedents like [Mitchell, 1972], Laver first (Levy) collapsed a large cardinal κ to render it ω_2 and then carried out the iterative injection of large antichains to destroy $\aleph_2$-Suslin trees. The whole procedure is countably closed so that $2^{\aleph_0} = \aleph_1$ is preserved, and the initial collapse incorporates the κ-c.c. throughout to preserve κ as a cardinal.

Especially with this result in hand, the question arises, analogous to MA implying SH, whether there is a version of MA adapted to $\aleph_2$ that implies the $\aleph_2$-Suslin Hypothesis. Laver in 1973 was actually the first to propose a generalized Martin's axiom; Baumgartner in 1975 proposed another; and then Shelah [1978] did also (cf. [Tall, 1994, p.216]). These various axioms are consistent (relative to ZFC) and can be incorporated into the Laver-Shelah construction. However, none of them can *imply* the $\aleph_2$-Suslin Hypothesis, since [Shelah and Stanley, 1982] soon showed, as part of extensive work on forcing principles and morasses, that CH + $\aleph_2$-Suslin Hypothesis implies that the (real) $\aleph_2$ is inaccessible in L. In particular, *some* large cardinal hypothesis is necessary to implement Laver-Shelah.

The Laver idea of injecting large antichains rather than long chains stands resilient; while generalized Martin's axioms do not apply to such forcings, the $\aleph_2$-Suslin Hypothesis can be secured. It is still open whether, analogous to Jensen's consistency of CH + SH, it is consistent to have CH + $2^{\aleph_1} = \aleph_2$ + $\aleph_2$-Suslin Hypothesis.

With respect to ($\aleph_1$-)Suslin trees, Shelah [1984] in the early 1980s showed that forcing to add a single Cohen real actually adjoins a Suslin tree. This was a surprising result about the fragility of SH that naturally raised the question about other generic reals. After working off and on for several years, Laver finally clarified the situation with respect to Sacks and random reals.

As set out in Carlson-Laver [27], Laver showed that if CH, then adding a Sacks real forces $\diamondsuit$, and hence that a Suslin tree exists, i.e. $\neg$SH. Tim Carlson specified a strengthening of MA, which can be shown consistent, and then showed that if it holds, then adding a Sacks real forces $\text{MA}_{\aleph_1}$, Martin's Axiom for meeting $\aleph_1$ dense sets, and hence SH. Finally, Laver [24] showed that if $\text{MA}_{\aleph_1}$ holds, then adding any number of random reals does *not* adjoin a Suslin tree, i.e. SH is maintained.

4.4 Nonregular Ultrafilters

With his experience with saturated ideals and continuing interest in strong properties holding low in the cumulative hierarchy, Laver [20] in 1982 established substantial results about the existence of nonregular ultrafilters over ω_1. This work became a pivot point for possibility, as we emphasize by first describing the wake of emerging results, including Laver [9] on constructibility, and then the related subsequent work, tucking in a reference to the joint Foreman-Laver [26] on downwards transfer.

For present purposes, an ultrafilter U over κ which is uniform (i.e. every element of U has size κ) is *regular iff* there are κ sets in U any infinitely many of which have empty intersection. Regular ultrafilters were considered at the beginnings of the study of ultraproduct models in the early 1960s, in substantial part as they ensure large ultrapowers, e.g. if U over κ is regular, then its ultrapower of ω must have size 2^κ. With the expansion of set theory through the 1960s, the regularity of ultrafilters became topical, and [Prikry, 1970] astutely established by isolating a combinatorial principle that holds in L that if $V = L$, then every ultrafilter over ω_1 is regular.

Can there be, consistently, a uniform nonregular ultrafilter over ω_1? Given the experience of saturated ideals and large cardinals, perhaps one can similarly collapse a large cardinal e.g. to ω_1 while retaining the ultrafilter property and the weak completeness property of nonregularity. This was initially stimulated by a result of [Kanamori, 1976], that if there were such a nonregular ultrafilter over ω_1, then there would be one with the large cardinal-like property of being *weakly normal*: If $\{\alpha < \omega_1 \mid f(\alpha) < \alpha\} \in U$, then there is a $\beta < \omega_1$ such that $\{\alpha < \omega_1 \mid f(\alpha) < \beta\} \in U$. Using this, [Ketonen, 1976] showed in fact that if there were such an ultrafilter, then $0^\#$ exists. [Magidor, 1979] was first to establish the existence of a nonregular ultrafilter, showing that if there is a huge cardinal, then e.g. in a forcing extension there is a uniform ultrafilter U over ω_2 such that its ultrapower of ω has size only $\aleph_2$ and hence is nonregular.

Entering the fray, Laver first provided incisive commentary in a two-page paper [9] on Prikry's result [1970] about regular ultrafilters in L. Jensen's principle $\diamondsuit^*$ is a strengthening of $\diamondsuit$ that he showed holds in L. Laver established: *Assume $\diamondsuit^*$. Then for every $\alpha < \omega_1$, there is a partition $\{\xi \mid \alpha < \xi < \omega_1\} = X_{\alpha 0} \cup X_{\alpha 1}$ such that for any function $h\colon \omega_1 \to 2$ there is an $\aleph_1$ size subset of $\{X_{\alpha h(\alpha)} \mid \alpha < \omega_1\}$ such that any countably many of these has empty intersection.* Thus, while Prikry had come up with a new combinatorial principle holding in L and used it to establish that every uniform ultrafilter over ω_1 is regular there, Laver showed that the combinatorial means had already been isolated in L, one that led to a short, elegant proof! Laver's proof, as does Prikry's, generalizes to get analogous results at all successor cardinals in L.

Laver [20] subsequently precluded the possibility that saturated ideals themselves could account for nonregular ultrafilters. First he characterized those κ-c.c. forcings that preserve κ^+-saturated κ-ideals, a result rediscovered and exploited by [Baumgartner and Taylor, 1982]. Laver then applied this to show that if there is an $\aleph_2$-saturated ω_1-ideal, then in a forcing extension there is such an ideal and moreover every uniform ultrafilter over ω_1 is regular.

Laver [20] then answered the pivotal question by showing that, consistently, there can be a uniform nonregular ultrafilter over ω_1. Woodin had recently shown that starting from strong determinacy hypotheses a ZFC model can be constructed which satisfies: $\diamondsuit$ + "There is an ω_1-dense ω_1-ideal".[10] From this, Laver [20] established that it follows that there is a uniform nonregular ultrafilter over ω_1. In fact, he applied $\diamondsuit$ to show that the filter dual to such an ideal can be extended to an ultrafilter by just adding $\aleph_1$ sets and closing off. Such an ultrafilter must be nonregular, and in fact the size of its ultrapower of ω is only $\aleph_1$.

With this achievement establishing nonregular ultrafilters on the landscape, they later figured in central work that reduced the strong hypotheses needed to get strong properties to hold low in the cumulative hierarchy. Reorienting

[10] In the 1990s Woodin would reduce the hypothesis to (just) the Axiom of Determinacy.

large cardinal theory, [Foreman *et al.*, 1988a] reduced the sufficient hypothesis for getting the consistency of an $\aleph_2$-saturated ω_1-ideal from Kunen's initial huge cardinal to just having a supercompact cardinal. Moreover, it was established in [Foreman *et al.*, 1988b] that if there is a supercompact cardinal, then in a forcing extension there is a nonregular ultrafilter over ω_1, and that analogous results hold for successors of regular cardinals. It was noted in [Kanamori, 1986a] that both this and the Laver result could be refined to get ultrafilters with "finest partitions", which made evident that the size of their ultrapowers is small.

In extending work done by the summer of 1992, Foreman [1998] showed that if there is a huge cardinal, then in a forcing extension there is an $\aleph_1$-dense ideal over ω_2 in a strong sense, from which it follows that there is a uniform ultrafilter over ω_2 such that its ultrapower of ω has size only $\aleph_1$.

Earlier, Foreman and Laver [26] by 1988 had incisively refined Kunen's original argument for getting an $\aleph_2$-saturated ω_1-ideal from a huge cardinal by incorporating Foreman's thematic κ-centeredness into the forcing to further get strong downwards transfer properties. A prominent such property was that every graph of size and chromatic number $\aleph_2$ has a subgraph of size and chromatic number $\aleph_1$. Foreman [1998] showed that having a nonregular ultrafilter over ω_2 directly implies this graph downwards transfer property. This work still stands in terms of consistency strength in need not just of supercompactness but hugeness to get strong propositions low in the cumulative hierarchy.

4.5 Products of Infinitely Many Trees

Laver [22] by 1983 established a striking partition theorem for infinite products of trees which, separate of being of considerable combinatorial interest, answered a specific question about possibilities for product forcing. The theorem is the infinite generalization of the Halpern-Laüchli Theorem [1966], a result to which Laver in 1969 had arrived at independently in a reformulation, in presumably his first substantive result in set theory. He worked off and on for many years on the infinite possibility, and so finally establishing it must have been a particularly satisfying achievement.

For present purposes, a *perfect tree* is a tree of height ω such that every element has incomparable successors, and $T(n)$ denotes the n-level of a tree T. For $A \subseteq \omega$ and a sequence of trees $\langle T_i \mid i < d\rangle$, let $\bigotimes^A \langle T_i \mid i < d\rangle = \bigcup_{n\in A} \Pi_{i<d} T_i(n)$, the set of d-tuples across the trees at the levels indexed by A. Finally, for $d \leq \omega$ let HL_d be the proposition:

> If $\langle T_i \mid i < d\rangle$ is a sequence of perfect trees and $\bigotimes^\omega \langle T_i \mid i < d\rangle = G_0 \cup G_1$, then there are $j < 2$, infinite $A \subseteq \omega$, and downwards closed perfect subtrees T_i' of T_i for $i < d$ such that $\bigotimes^A \langle T_i' \mid i < d\rangle \subseteq G_j$.

That HL_d holds for $d < \omega$ is essentially the Halpern-Läuchli Theorem [1966], which was established and applied to get a model for the Boolean Prime Ideal Theorem together with the failure of the Axiom of Choice.[11] Laver in 1969 from different motivations (see below) and also David Pincus by 1974[12] arrived at a incisive "dense set" formulation from which HD_d readily follows. For a sequence of trees $\langle T_i \mid i < d\rangle$, $\langle X_i \mid i < d\rangle$ is *n-dense iff* for some $m \geq n$, $X_i \subseteq T_i(m)$ for $i < d$, and moreover, for $i < d$ every member of $T_i(n)$ is below some member of X_i. For $\vec{x} = \langle x_i \mid i < d\rangle \in \bigotimes^\omega \langle T_i \mid i < d\rangle$, $\langle X_i \mid i < d\rangle$ is $\vec{x}$*-n-dense iff* it is n-dense in $\langle (T_i)_{x_i} \mid i < d\rangle$, where $(T_i)_{x_i}$ is the subtree of T_i consisting of the elements comparable with x_i. Let LP_d be the proposition:

> If $\langle T_i \mid i < d\rangle$ is a sequence of perfect trees and $\bigotimes^\omega \langle T_i \mid i < d\rangle = G_0 \cup G_1$, then either (a) for all $n < \omega$ there is an n-dense $\langle X_i \mid i < d\rangle$ with $\bigotimes^\omega \langle X_i \mid i < d\rangle \subseteq G_0$, or (b) for some $\vec{x} = \langle x_i \mid i < d\rangle$ and all $n < \omega$ there is an $\vec{x}$-n-dense $\langle X_i \mid i < d\rangle$ with $\bigotimes^\omega \langle X_i \mid i < d\rangle \subseteq G_1$.

LP_d and HD_d for finite d are seen to be mutually derivable, but unaware of HD_d Laver in 1969 had astutely formulated and proved LP_d for finite d in order to establish a conjecture of Galvin. In the late 1960s at Berkeley, Galvin had proved that if the rationals are partitioned into finitely many cells, then there is a subset of the same order type η whose members are in at most two of the cells. Galvin then conjectured that if the r-element sequences of rationals are partitioned into finitely many cells, then there are sets of rationals $X_0, X_1, \ldots, X_{r-1}$ each of order type η such that the members of $\Pi_{i<r} X_i$ are in at most $r!$ of the cells. Laver while a graduate student affirmed this, soon after he had established Fraïssé's Conjecture.[13] Notably, Keith Milliken [1979], in his UCLA thesis with Laver on the committee, applied LP_d to derive a "pigeonhole principle", actually a partition theorem in terms of "strongly embedded trees" rather than perfect subtrees.

Finally to Laver's [22] result after all this set up, he after years of returning to it finally established HL_ω, the infinite generalization of Halpern-Läuchli. With its topicality, Tim Carlson (cf. [Carlson and Simpson, 1984]) also established HL_ω in a large context of "dual Ramsey theorems". HL_ω is seen as an infinitary Ramsey theorem, but in any case, Laver had an explicit motivation from forcing, for Baumgartner had raised the issue of HL_ω in the late 1970s. Extending the combinatorics for Sacks reals and HL_d for finite d, Baumgartner had observed that HL_ω implies that when adding κ Sacks reals "side-by-side", i.e. with product forcing, any subset of ω contains or is disjoint from an infinite subset of ω in the ground model. This now became an impressive fact about the stability of product Sacks forcing.

---

[11] cf. [Halpern and Levy, 1971].

[12] cf. [Pincus and Halpern, 1981].

[13] All this is noted in [Erdős and Hajnal, 1974, p.275]. Laver latterly published his proof of Galvin's conjecture from LP_d as Theorem 2 in [22].

In retrospect, what slowed Laver's progress to HL_ω was his inability to establish the ostensibly stronger LP_ω despite numerous attempts. He finally saw that by patching together a technical weakening of LP_ω, he could get to HL_ω. As he moved on to further triumphs, the one problem he would bequeath to set theory is the infinitary generalization of his earliest result in the combinatorics of the infinite: Does LP_ω hold?

5 Embeddings of Rank into Rank

Some time in the mid-1980s, Laver [23] began tinkering with elementary embeddings $j\colon V_\delta \to V_\delta$, combining them and looking at how they move the ordinals. On the one hand, that there are such embeddings at all amounts to asserting a consistency-wise very strong hypothesis, and on the other hand, there was an algebraic simplicity in the play of endomorphisms and ordinals. Laver persisted through a proliferation of embeddings and ordinals moved to get at patterns and issues about algebras of embeddings. He [29] then made enormous strides in discerning a normal form, and, with it, getting at the freeness of the algebras, as well as the solvability of their word problems. Subsequently, Laver [31] was able to elaborate the structure of iterated embeddings and formulate new finite algebras of intrinsic interest. Laver not only brought in distinctively algebraic incentives into the study of strong hypotheses in set theory, but opened up separate algebraic vistas that stimulated a new cottage industry at this intersection of higher set theory and basic algebra. Moving on however, Laver [34][36] considerably clarified the situation with respect to even stronger embedding hypotheses, and eventually he [39] returned, remarkably, to something basic about forcing, that the ground model is definable in any generic extension.

In what follows, we delve forthwith into strong elementary embeddings and successively describe Laver's work. There is less in the way of historical background and less that can be said about the algebraic details, so the comparative brevity of this section belies to some extent the significance and depth of this work. It is assumed and to be implicit in the notation that elementary embeddings j are not the identity, and so, as their domains satisfy enough set-theoretic axioms, they have a *critical point* $\mathrm{cr}(j)$, a least ordinal α such that $\alpha < j(\alpha)$.

5.1 Algebra of Embeddings

Kunen in 1970 had famously delimited the large cardinal hypotheses by establishing an outright inconsistency in ZFC, that there is no elementary embedding $j\colon V \to V$ of the universe into itself. The existence of large cardinals as strong axioms of infinity had turned on their being critical points of elementary embeddings $j\colon V \to M$ with M being larger and larger inner models, and Kunen showed that there is a limit to such formulations with M being V itself. Of course, it is all in the proof, and with $\kappa = \mathrm{cr}(j)$ and λ the supremum of $\kappa < j(\kappa) < j^2(\kappa) < \ldots$, Kunen had actually showed that having a certain combinatorial object in $V_{\lambda+2}$ leads to a contradiction. Several hypotheses just skirting Kunen's inconsistency were considered, the simplest being that $\mathcal{E}_\lambda \neq \emptyset$ for some limit λ, where $\mathcal{E}_\lambda = \{j \mid j\colon V_\lambda \to V_\lambda \text{ is elementary}\}$. The λ here is taken anew, but from Kunen's argument it is understood that if $j \in \mathcal{E}_\lambda$ and $\kappa = \mathrm{cr}(j)$, the supremum of $\kappa < j(\kappa) < j^2(\kappa) < \ldots$ would have to be λ.

Laver [23] in 1985 explored $\mathcal{E}_\lambda$, initially addressing definability issues, under two binary operations. Significantly, in this he worked a conceptual shift from critical points as large cardinals to the embeddings themselves and their interactions. One operation was *composition*: if $j, k \in \mathcal{E}_\lambda$, then $j \circ k \in \mathcal{E}_\lambda$. The other, possible as embeddings are sets of ordered pairs, was *application*: if $j, k \in \mathcal{E}_\lambda$, then $j \cdot k = \bigcup_{\alpha<\lambda} j(k \cap V_\alpha) \in \mathcal{E}_\lambda$ with $\mathrm{cr}(j \cdot k) = j(\mathrm{cr}(k))$. Application was first exploited by [Martin, 1980], with these laws easily checked:

$$(\Sigma) \qquad i \circ (j \circ k) = (i \circ j) \circ k; \quad (i \circ j) \cdot k = i \cdot (j \cdot k);$$
$$i \cdot (j \circ k) = (i \cdot j) \circ (i \cdot k); \text{ and } \ i \circ j = (i \cdot j) \circ i.$$

From these follows the *left distributive law* for application: $i \cdot (j \cdot k) = (i \cdot j) \cdot (i \cdot k)$. A basic question soon emerged as to whether these are the only laws, and Laver a few years later in 1989 showed this. For $j \in \mathcal{E}_\lambda$, Let $\mathcal{A}_j$ be the closure of $\{j\}$ in $\langle \mathcal{E}_\lambda, \cdot \rangle$, and let $\mathcal{P}_j$ be the closure of $\{j\}$ in $\langle \mathcal{E}_\lambda, \cdot, \circ \rangle$. Laver [29] established: *$\mathcal{A}_j$ is the free algebra $\mathcal{A}$ with one generator satisfying the left distributive law, and $\mathcal{P}_j$ is the free algebra $\mathcal{P}$ with one generator satisfying Σ.*

Freeness here has the standard meaning. For $\mathcal{P}_j$ and $\mathcal{P}$, let W be the set of terms in one constant a in the language of $\cdot$ and $\circ$. Define an equivalence relation $\equiv$ on W by stipulating that $u \equiv v$ *iff* there is a sequence $u = u_0, u_1, \ldots, u_n = v$ with each u_{i+1} obtained from u_i by replacing a subterm of u_i by a term equivalent to it according to one of the laws of Σ. Then $\cdot$ and $\circ$ are well-defined for equivalence classes, and Laver's result asserts that the resulting structure on $W/{\equiv}$ and $\mathcal{P}_j$ are isomorphic via the map induced by sending the equivalence class of a to j.

For $u, v \in W$, define $u <_L v$ *iff* u is an iterated left divisor of v, in the sense that for some $w_1, \ldots, w_{n+1} \in W$,

$$v \equiv ((\ldots(u \cdot w_1) \cdot w_2) \ldots \cdot w_n) \cdot w_{n+1}\ , \text{ or}$$
$$v \equiv ((\ldots(u \cdot w_1) \cdot w_2) \ldots \cdot w_n) \circ w_{n+1}\ .$$

When in the mid-1980s Laver was trying to understand the proliferation of critical points of members of $\mathcal{A}_j$, he had worked with equivalence relations on embeddings based on partial agreement and had shown that *if $\mathcal{E}_\lambda \neq \emptyset$ for some λ, then $<_L$ is irreflexive*, i.e. $u <_L u$ always fails. Assuming, then, the irreflexivity of $<_L$, Laver [29] showed that every equivalence class in $W/\equiv$ has a unique member in a certain normal form; that the lexicographic ordering of these normal forms is a linear ordering; that this lexicographic ordering then agrees with $<_L$; and hence that $<_L$ on W is a linear ordering. This structuring of the freeness leads to *the solvability of the word problem for* $W/\equiv$, i.e. there is an effective procedure for deciding whether or not $u \equiv v$ for arbitrary $u, v \in W$. For $\mathcal{A}$, with just one operation, there is no normal form, but Laver showed that $\mathcal{P}$ is conservative over $\mathcal{A}$ in that two terms in the language of $\cdot$ are equivalent as per the laws Σ exactly when they are equivalent as per just the left distributive law. Considerable interest was generated by a hypothesis bordering on the limits of consistency entailing solvability in finitary mathematics, particularly because of the peculiar and enticing possibility that some strong hypothesis may be necessary. By 1990 Laver [30] had extended his normal form result systematically to get, for any $p <_L q \in \mathcal{P}$, a unique, recursively defined "p-division form" equivalent to q, so that there is a $<_L$-largest p_0 with $p \cdot p_0 \leq_L q$, and one can conceptualize the process as a division algorithm.

Patrick Dehornoy, having been pursuing similar initiatives, made important contributions.[14] He first provided in some 1989 work an alternative proof [1992a] of Laver's [29] freeness and solvability results, one that does with less than the irreflexivity of $<_L$ at the cost of foregoing normal forms. Then in 1991, he [1992b, 1994] established that irreflexivity outright in ZFC by following algebraic incentives and bringing out a realization of $\mathcal{A}$ within the Artin braid group B_∞ with infinitely many strands. Consequently, the various structure results obtained about $W/\equiv$ by Laver [29] became theorems of ZFC.

The braid group connection was soon seen explicitly. The braid group B_n, with $2 \leq n \leq \infty$, is generated by elements $\{\sigma_i \mid 0 < i < n\}$ satisfying $\sigma_i\sigma_j = \sigma_j\sigma_i$ for $|i - j| > 1$ and $\sigma_i\sigma_{i+1}\sigma_i = \sigma_{i+1}\sigma_i\sigma_{i+1}$. Define the "Dehornoy bracket" on B_∞ by: $g[h] = g\,\mathrm{sh}(h)\,\sigma_1\,(\mathrm{sh}(g))^{-1}$, where sh is the shift homomorphism given by $\mathrm{sh}(\sigma_i) = \sigma_{i+1}$. The Dehornoy bracket is left distributive, and one can assign to each $u \in A$ a $\overline{u} \in B_\infty$ by assigning the generator of A to σ_1, and recursively, $\overline{uv} = \overline{u}[\overline{v}]$. For the irreflexivity of $<_L$, assume that $u <_L u$. Then the corresponding assertion about $\overline{u}$ leads to a "σ_1-positive" element, an element with an occurrence of σ_1 but none of σ_1^{-1}, which represents the identity of B_∞. But one argues that this cannot happen, that B_∞ is torsion-free. Laver's student David Larue [1994] provided a straightforward argument of this last, and so a shorter, more direct proof of the irreflexivity of $<_L$.

The investigation of elementary embeddings continued full throttle, and this led strikingly to new connections and problems in finitary mathematics. Randall Dougherty [1993], coming forthwith to the scene, persisted with a detailed investigation of the proliferation of critical points. Fixing a $j \in \mathcal{E}_\lambda$

[14]See [Dehornoy, 2010] for an expository account of the eventually developed theory from his perspective, one from which material in what follows is drawn.

with $\kappa = \mathrm{cr}(j)$, define a corresponding function f on ω by:

$$f(n) = |\{\mathrm{cr}(k) \mid k \in \mathcal{A}_j \ \wedge\ j^n(\kappa) < \mathrm{cr}(k) < j^{n+1}(\kappa)\}| \ .$$

Laver had seen that $f(0) = 0$, $f(1) = 0$, and $f(2) = 1$, but that $f(3)$ is suddenly large because of application. Dougherty [93] established a large lower bound for $f(3)$ and showed moreover that f eventually dominates the Ackermann function, and hence cannot be primitive recursive.

In tandem, with the $f(n)$'s not even evidently finite, Laver [31] duly established that $f(n)$ is finite for all n. Let $j_{[1]} = j$, and $j_{[n+1]} = j_{[n]} \cdot j$. Then Laver showed that the sequence $\langle \mathrm{cr}(j_{[n]}) \mid n \in \omega \rangle$ enumerates, increasingly with repetitions, the first ω critical points of embeddings in $\mathcal{A}_j$. A general result of John Steel implies that the supremum of this sequence is $\lambda = \sup_n j^n(\kappa)$, so that the sequence must have all the critical points. Hence the Laver-Steel conclusion is that indeed the $f(n)$'s are finite.

Toward his result about the critical points of the $j_{[n]}$'s, Laver [31] worked with finite left distributive algebras, algebras which can be presented without reference to elementary embeddings. For $k \in \omega$, let $A_k = \{1, 2, \ldots, 2^k\}$ with the operation $*_k$ given by the cycling $a *_k 1 = a + 1$ for $1 \leq a < 2^k$ and $2^k *_k 1 = 1$, and the left distributive $a *_k (b *_k 1) = (a *_k b) *_k (a *_k 1)$. Then the $\langle A_k, *_k \rangle$ turn out to be *the* finite algebras satisfying the left distributive law as well as the cycling laws $a *_k 1 = a + 1$ for $1 \leq a < 2^k$ and $2^k *_k 1 = 1$. The multiplication table or "Laver table" for $k = 3$ is as follows:

A_3	1	2	3	4	5	6	7	8
1	2	4	6	8	2	4	6	8
2	3	4	7	8	3	4	7	8
3	4	8	4	8	4	8	4	8
4	5	6	7	8	5	6	7	8
5	6	8	6	8	6	8	6	8
6	7	8	7	8	7	8	7	8
7	8	8	8	8	8	8	8	8
8	1	2	3	4	5	6	7	8

Each element in A_k is periodic with period a power of 2. Laver in his asymptotic analysis of $\mathcal{A}_j$ showed that *for each $a \in \omega$, the period of a in A_k tends to infinity with k*, and e.g. *for every k the period of 2 in A_k is at least that of 1*.

Remarkably, such observations about these finite algebas are not known to hold just in ZFC. Appealing to the former's earlier work, Dougherty and Thomas Jech [1997] did show that the number of computations needed to guarantee that the period of 1 in A_k grows faster in k than the Ackermann function.

While Dehornoy's work had established the irreflexivity of $<_L$ and so the solvability the word problem in ZFC, it is a striking circumstance that Laver's arguments using a strong large cardinal hypothesis still stand for establishing some seemingly basic properties of the finite algebras A_k.

In [32] Laver impressively followed the trail into braid group actions. Dehornoy [1994] had shown that the ordering $<_L$ naturally induces linear orderings of the braid groups. With the positive braids being those not having any inverses of the strands appearing, Laver applied his structural analysis of $<_L$ to show that for the braid groups B_n with n finite the Dehornoy ordering actually well-orders the positive braids. This would stand as a remarkable fact that would frame the emerging order theory of braid groups, and with a context set, e.g. [Carlucci *et al.*, 2011] investigated unprovability results according to the order type of long descending sequences in the Dehornoy order.

Laver's last papers [38][40][41] were to be on left-distributivity. With John Moody, Laver [38] stated conjectures about the free left distributive algebras $\mathcal{A}^{(k)}$ extending $\mathcal{A}$ by having k generators. These conjectures, about how a comparison process must terminate, would establish the following, still open for $\mathcal{A} = \mathcal{A}^{(1)}$: If $w \in \mathcal{A}^{(k)}$, the set $\{u \in A^{(k)} \mid \exists v(u \cdot v = w)\}$ of (direct) left divisors of w is well-ordered by $<_L$. With his student Sheila Miller, Laver [40] applied his division algorithm [30] to get at the possibility of comparisons and well-orderings, establishing that in $\mathcal{A}$, if $ab = cd$ and a and b have no common left divisors and c and d have no common left divisors, the $a = c$ and $b = d$. Laver and Miller [41] further simplify the division algorithm and provide a mature account of the theory of left distributive algebras.

5.2 Implications Between Very Large Cardinals

In the later 1990s Laver [34][36][39], moving on to higher pastures, developed the definability theory of elementary embedding hypotheses even stronger than $\mathcal{E}_\lambda \neq \emptyset$, getting into the upper reaches near Kunen's inconsistency, reaches first substantially broached by Woodin for consistency strength in the 1980s.

Kunen's inconsistency argument showed in a sharp form that in ZFC there is no elementary embedding $j\colon V_{\lambda+2} \to V_{\lambda+2}$. Early on, the following "strongest hypotheses" approaching the known inconsistency were considered, the last setting the stage for Laver's investigation of the corresponding algebra of embeddings.

$E_\omega(\lambda)$: There is an elementary $j\colon V_{\lambda+1} \to V_{\lambda+1}$.

$E_1(\lambda)$: There is an elementary $j\colon V \to M$ with $\mathrm{cr}(j) < \lambda = j(\lambda)$ and $V_\lambda \subseteq M$.

$E_0(\lambda)$: There is an elementary $j\colon V_\lambda \to V_\lambda$. ($\mathcal{E}_\lambda \neq \emptyset$.)

In all of these it is understood that, with $\mathrm{cr}(j) = \kappa$ the corresponding large cardinal, λ must be the supremum of $\kappa < j(\kappa) < j^2(\kappa) < \ldots$ by Kunen's argument. Thrown up *ad hoc* for stepping back from inconsistency, these strong hypotheses were not much investigated except in connection with a substantial application by [Martin, 1980] to determinacy.

In work dating back to his first abstract [23] on embeddings, Laver [34] established a hierarchy up through $E_\omega(\lambda)$ with definability. In the language for second-order logic with $\in$, a formula is Σ^1_0 if it contains no second-order quantifiers and is Σ^1_n if it is of the form $\exists X_1 \forall X_2 \cdots Q_n X_n \Phi$ with second-order variables X_i and Φ being Σ^1_0. A $j\colon V_\lambda \to V_\lambda$ is Σ^1_n *elementary* if for any Σ^1_n Φ in one free second-order variable and $A \subseteq V_\lambda$, $V_\lambda \models \Phi(A) \leftrightarrow \Phi(j(A))$.

It turns out that $E_1(\lambda)$ above is equivalent to having an elementary $j\colon V_\lambda \to V_\lambda$ which is Σ^1_1. Also, if there is an elementary $j\colon V_\lambda \to V_\lambda$ which is Σ^1_n for every n, then j witnesses $E_\omega(\lambda)$. Incorporating these notational anticipations, define:

$E_n(\lambda)$: There is a Σ^1_n elementary $j\colon V_\lambda \to V_\lambda$.

Next, say that for parametrized large cardinal hypotheses, $\Psi_1(\lambda)$ *strongly implies* $\Psi_2(\lambda)$ if for every λ, $\Psi_1(\lambda)$ implies $\Psi_2(\lambda)$ and moreover there is a $\lambda' < \lambda$ such that $\Psi_2(\lambda')$.

Taking compositions and inverse limits of embeddings, Laver [23][34] established that in the sequence,

$$E_0(\lambda),\ E_3(\lambda),\ E_5(\lambda),\ \ldots,\ E_\omega(\lambda)$$

each hypothesis strongly implies the previous ones, each $E_{n+2}(\lambda)$ in fact providing for many $\lambda' < \lambda$ such that $E_n(\lambda')$ as happens in the hierarchy of large cardinals. Martin had essentially shown that any Σ^1_{2n+1} elementary $j\colon V_\lambda \to V_\lambda$ is Σ^1_{2n+2}, so Laver's results complete the hierarchical analysis for second-order definability.

In [34] Laver reached a bit higher in his analysis, and in [36] he went up to a very strong hypothesis formulated by Woodin:

$W(\lambda)$: There is an elementary $j\colon L(V_{\lambda+1}) \to L(V_{\lambda+1})$
with $\mathrm{cr}(j) < \lambda$.

That $W(\lambda)$ holds for some λ, just at the edge of the Kunen inconsistency, was formulated by Woodin in 1984, and, in the first result securing a mooring for the Axiom of Determinacy (AD) in the large cardinals, shown by him to imply that the axiom holds in the inner model $L(\mathbb{R})$, $\mathbb{R}$ the reals. Just to detail, $L(\mathbb{R})$ and $L(V_{\lambda+1})$ are constructible closures, where the *constructible closure* of a set A is the class $L(A)$ given by $L_0(A) = A$; $L_{\alpha+1}(A) = \mathrm{def}(L_\alpha(A))$, the first-order definable subsets of $L_\alpha(A)$; and $L(A) = \bigcup_\alpha L_\alpha(A)$. Woodin, in original work, developed and pursued an analogy between $L(\mathbb{R})$ and $L(V_{\lambda+1})$ taking V_λ to be the analogue of ω and $V_{\lambda+1}$ to be analogue of $\mathbb{R}$ and established AD-like consequences for $L(V_{\lambda+1})$ from $W(\lambda)$.[15]

Consider an elaboration of $W(\lambda)$ according to the constructible hierarchy $L(V_{\lambda+1}) = \bigcup L_\alpha(V_{\lambda+1})$:

$W_\alpha(\lambda)$: There is an elementary $j\colon L_\alpha(V_{\lambda+1}) \to L_\alpha(V_{\lambda+1})$ with $\mathrm{cr}(j) < \lambda$.

Laver in his [34] topped his hierarchical analysis there by showing that $W_1(\lambda)$ strongly implies $E_\omega(\lambda)$, again in a strong sense. In [36] Laver impressively engaged with some of Woodin's work with $W(\lambda)$ to extend hierarchical analysis into the transfinite, showing that $W_{\lambda^+ + \omega + 1}(\lambda)$ strongly implies $W_{\lambda^+}(\lambda)$ and analogous results with the "λ^+" replaced e.g. by the supremum of all second-order definable prewellorderings of $V_{\lambda+1}$.

In the summarizing [37], Laver from his perspective set out the landmarks of the work on elementary embeddings as well as provided an outline of Woodin's work on the AD-like consequences of $W(\lambda)$, and stated open problems for channeling the further work. Notably, Laver's speculations reached quite high, beyond $W(\lambda)$; [Woodin, 2011b, p.117] mentioned "Laver's Axiom", an axiom providing for an elementary embedding to provide an analogy with the strong determinacy axiom $\mathrm{AD}_{\mathbb{R}}$.

In what turned out to be his last paper in these directions, Laver [39] in 2004 established results about the large cardinal propositions $E_n(\lambda)$ for $n \leq \omega$ and forcing. For each $n \leq \omega$, let $E_n(\kappa, \lambda)$ be $E_n(\lambda)$ further parametrized by specifying the large cardinal $\kappa = \mathrm{cr}(j)$. Laver established: *If $V[G]$ is a forcing extension of V via a forcing poset of size less than κ and $n \leq \omega$, then $V[G] \models \exists\lambda E_n(\kappa, \lambda)$ implies $V \models \exists\lambda E_n(\kappa, \lambda)$.* The converse direction follows by well-known arguments about small forcings preserving large cardinals. The Laver direction is not surprising, on general grounds that consistency strength should not be created by forcing. However, as Laver notes by counterexamples, a λ satisfying $E_n(\kappa, \lambda)$ in $V[G]$ need not satisfy $E_n(\kappa, \lambda)$ in V, and a j witnessing $E_n(\kappa, \lambda)$ in $V[G]$ need not satisfy $j \restriction V_\lambda \in V$.

Laver established his result by induction on n deploying work from [34], and what he needed at the basis and first proved is the lemma: *If $V[G]$ is a forcing extension of V via a forcing poset of size less than κ and j witnesses $E_0(\kappa, \lambda)$, then $j \restriction V_\alpha \in V$ for every $\alpha < \lambda$.* Laver came up with a proof of this using a result $(**)$ he proved about models of ZFC that does not involve large cardinals, and this led to a singular development.

[15] This work would remain unpublished by Woodin. On the other hand, in his latest work [2011b] on suitable extenders Woodin considerably developed and expanded the $L(V_{\lambda+1})$ theory with $W(\lambda)$ in his quest for an ultimate inner model.

As Laver [39] described it, Joel Hamkins pointed out how the methods of his [2003], also on extensions not creating large cardinals, can establish (∗∗) in a generalized form, and Laver wrote this out as a preferred approach. Motivated by [Hamkins, 2003], Laver [39], and Woodin independently and in his scheme of things, established *ground model definability: Suppose that V is a model of* ZFC, *$P \in V$, and $V[G]$ is a generic extension of V via P. Then in $V[G]$, V is definable from a parameter.* (With care, the parameter could be made $\mathcal{P}^{V[G]}(P)$ through Hamkins' work.)

The ground model is definable in any generic extension! Although a parameter is necessary, this is an illuminating result about forcing as method. Was this issue raised decades earlier at the inception of forcing? In truth, for a particular forcing a term $\dot{V}$ can be introduced into the forcing language for assertions about the ground model, so there may not have been an earlier incentive. The argumentation for ground model definability provided *one* formula that defines the ground model in any generic extension in terms of a corresponding parameter. Like Laver indestructibility, this available uniformity stimulated renewed investigations and conceptualizations involving forcing.

Motivated by ground model definability, Hamkins and Jonas Reitz formulated the Ground Axiom: The universe of sets is not the forcing extension of any inner model W by a (nontrivial) forcing $P \in W$. [Reitz, 2007] investigated this axiom, and [Hamkins *et al.*, 2008] established its consistency with $V \neq$ HOD. [Fuchs *et al.*, 2011] then extended the investigations into "set-theoretic geology", digging into the remains of a model of set theory once the layers created by forcing are removed. On his side, Woodin [2011a, §8] used ground model definability to formalize a conception of the "generic-multiverse"; the analysis here dates back to 2004. The definability is a basic ingredient in his latest work [Woodin, 2010] toward an ultimate inner model.

Ground model definability serves as an apt and worthy capstone to a remarkable career. It encapsulates the several features of Laver's major results that made them particularly compelling and potent: it has a succinct basic-sounding statement, it nonetheless requires a proof of substance, and it gets to a new plateau of possibilities. With it, Laver circled back to his salad days.

6 Envoi

Let me indulge in a few personal reminiscences, especially to bring out more about Rich Laver.

A long, long time ago, I was an aspiring teenage chess master in the local San Francisco chess scene. It was a time fraught with excitement and inventiveness, as well as encounters with eclectic, quirky personalities. In one tournament, I had a gangly opponent who came to the table with shirt untucked and opened 1.g4, yet I still managed to lose. He let on that he was a graduate student in mathematics, which mystified me at the time (what's new in subtraction?).

During my Caltech years, I got wind that Rich Laver was on the UC Berkeley team that won the national collegiate chess championship that year. I eventually saw a 1968 game he lost to grandmaster Pal Benko when the latter was trotting out his gambit, a game later anthologized in [Benko, 1974]. A mutual chess buddy mentioned that Laver had told him that his thesis result could be explained to a horse—only years later did I take in that he had solved Fraïssé's Conjecture.[16]

In 1971 when Rich was a post-doc at Bristol and heard that I was up at King's College, Cambridge, he started sending me postcards. In one he suggested meeting up at the big Islington chess tourney (too complicated) and in others he mentioned his results and problems about partition relation consequences of MA. I has just getting up to speed, and still could not take it in.

A few years later, I was finally up and running, and when I sent him my least function result for nonregular ultrafilters, he was very complementary and I understood then that we were on par. When soon later I was writing up the Solovay-Reinhardt work on large cardinals, Laver pointedly counseled me against the use of the awkward "n-hypercompact" for "n-huge", and I forthwith used the Kunen term.

By then at Boulder, Rich would gently suggest going mountain climbing, but I would hint at a constitutional reluctance. He did mention how he was a member of a party that took Paul Erdős up a Flatiron (mountain) near Boulder and how Erdős came in his usual light beige clothes and sandals. In truth, our paths rarely crossed as I remained on the East Coast. Through the 1980s Rich would occasionally send me preprints, sometimes with pencil scribblings. One time, he sent me his early thinking about embeddings of rank into rank. Regrettably, I did not follow up.

The decades went by with our correspondence turning more and more to chess, especially fanciful problems and extraordinary grandmaster games. In a final email to me, which I can now time as well after the onset of Parkinson's, Rich posed the following chess problem: Start with the initial position and play a sequence of legal moves until Black plays 5...NxR mate. I eventually figured out that the White king would have to be at f2, and so sent him: 1. f3, Nf6 2. Kf2, Nh5 3. d3, Ng3 4. Be3, a6 5. Qe1, NxR mate. But then, Rich wrote back, now do it with an intervening check! This new problem kicked around in my mental attic for over a year, and one bright day I saw: 1. f3, Nf6 2. e4, Nxe4 3. Qe2, Ng3 4. Qxe7ch, QxQch 5. Kf2, NxR mate. But by then it was too late to write him.

[16]According to artsandsciences.colorado.edu/magazine/2012/12/by-several-calculations-a-life-well-lived/ the crucial point came to Laver in an epiphanous moment while he, mountain climbing, was stranded for a night "on a ledge in darkness" at Yosemite.

Doctoral Students of Richard Laver

Stephen Grantham, *An analysis of Galvin's tree game*, 1982.

Carl Darby, *Countable Ramsey games and partition relations*, 1990.

Janet Barnett, *Cohen reals, random reals and variants of Martin's Axiom*, 1990.

Emanuel Knill, *Generalized degrees and densities for families of sets*, 1991.

David Larue, *Left-distributive algebras and left-distributive idempotent algebras*, 1994.

Rene Schipperus, *Countable partition ordinals*, 1999.

Sheila Miller, *Free left distributive algebras*, 2007.

In addition to having these doctoral students at Boulder, Laver was on the thesis committees of, among many: Keith Devlin (Bristol), Maurice Pouzet (Lyon), Keith Milliken (UCLA), Joseph Rebholz (UCLA), Carl Morgenstern (Boulder), Stewart Baldwin (Boulder), Steven Leth (Boulder), Kai Hauser (Caltech), Mohammed Bekkali (Boulder), and Serge Burckel (Caen).

Publications of Richard Laver

[1] Well-quasi-ordering scattered order types. In Richard Guy *et al.*, editors, *Combinatorial Structures and their Applications. Proceedings of the Calgary International Conference*, page 231. Gordon and Breach, New York, 1970.

[2] On Fraïssé's order type conjecture. *Annals of Mathematics*, 93:89-111, 1971.

[3] An order type decomposition theorem. *Annals of Mathematics*, 98:96-119, 1973.

[4] (with James Baumgartner, Fred Galvin, and Ralph McKenzie) Game theoretic versions of partition relations. In András Hajnal *et al.*, editors, *Infinite and Finite Sets. Keszthely (Hungary), 1973*, volume I, Colloquia Mathematica Societatis János Bolyai 10, pages 131-135. North-Holland, Amsterdam, 1975.

[5] On strong measure zero sets. In András Hajnal *et al.*, editors, *Infinite and Finite Sets. Keszthely (Hungary), 1973*, volume II, Colloquia Mathematica Societatis János Bolyai 10, pages 1025-1027. North-Holland, Amsterdam, 1975.

[6] Partition relations for uncountable cardinals $\leq 2^{\aleph_0}$. In András Hajnal *et al.*, editors, *Infinite and Finite Sets. Keszthely (Hungary), 1973*, volume II, Colloquia Mathematica Societatis János Bolyai 10, pages 1029-1042. North-Holland, Amsterdam, 1975.

[7] Well-quasi-orderings and sets of finite sequences. *Mathematical Proceedings of the Cambridge Philosophical Society*, 79:1-10, 1976.

[8] On the consistency of Borel's conjecture. *Acta Mathematica*, 137:151-169, 1976.

[9] A set in L containing regularizing families for ultrafilters. *Mathematika*, 24:50-51, 1977.

[10] A saturation property on ideals. *Compositio Mathematica*. 36:233-242, 1978.

[11] Making the supercompactness of κ indestructible under κ-directed closed forcing. *Israel Journal of Mathematics*, 29:385-388, 1978.

[12] Better-quasi-orderings and a class of trees. In *Studies in Foundations and Combinatorics*, Advances in Mathematics Supplementary Studies 1, pages 31-48. Academic Press, New York, 1978.

[13] (with Vance Faber and Ralph McKenzie) Coverings of groups by abelian subgroups. *Canadian Journal of Mathematics*, 30:933-945, 1978.

[14] Strong saturation properties of ideals (abstract). *The Journal of Symbolic Logic* 43:371, 1978.

[15] Linear orders in $(\omega)^\omega$ under eventual dominance. In Maurice Boffa *et al.*, editors, *Logic Colloquium '78*, Studies in Logic and the Foundations of Mathematics 97, pages 299-302. North-Holland, Amsterdam, 1979.

[16] (with James Baumgartner) Iterated perfect-set forcing. *Annals of Mathematical Logic*, 17:271-288, 1979.

[17] (with Jean Larson and George McNulty) Square-free and cube-free colorings of the ordinals. *Pacific Journal of Mathematics*, 89:137-141, 1980.

[18] (with Saharon Shelah) The $\aleph_2$-Souslin Hypothesis. *Transactions of the American Mathematical Society*, 264:411-417, 1981.

[19] An $(\aleph_2, \aleph_2, \aleph_0)$-saturated ideal on ω_1. In Dirk van Dalen *et al.*, editors, *Logic Colloquium '80*, Studies in Logic and the Foundations of Mathematics 108, 1982, pages 173-180. North-Holland, Amsterdam, 1982.

[20] Saturated ideals and nonregular ultrafilters. In George Metakides, editor, *Patras Logic Symposion*, Studies in Logic and the Foundations of Mathematics 109, pages 297-305. North-Holland, Amsterdam, 1982.

[21] Products of infinitely many perfect trees. *Journal of the London Mathematical Society*, 29:385-396, 1984.

[22] Precipitousness in forcing extensions. *Israel Journal of Mathematics*, 48:97-108, 1984.

[23] Embeddings of a rank into itself. *Abstracts of Papers presented to the American Mathematical Society* 7:6, 1986.

[24] Random reals and Souslin trees. *Proceedings of the American Mathematical Society*, 100:531-534, 1987.

[25] (with Marcia Groszek) Finite groups of OD-conjugates. *Periodica Mathematica Hungarica*, 18:87-97, 1987.

[26] (with Matthew Foreman) Some downwards transfer properties for $\aleph_2$. *Advances in Mathematics*, 67:230-238, 1988.

[27] (with Tim Carlson) Sacks reals and Martin's Axiom. *Fundamenta Mathematicae*, 133:161-168, 1989.

[28] (with Krzysztof Ciesielski) A game of D. Gale in which the players have limited memory. *Periodica Mathematica Hungarica*, 22:153-158, 1990.

[29] The left distributive law and the freeness of an algebra of elementary embeddings. *Advances in Mathematics*, 91:209-231, 1992.

[30] A division algorithm for the free left distributive algebra. In Juha Oikkonen and Jouko Väänänen, editors, *Logic Colloquium '90*, Lecture Notes in Logic 2, pages 155-162. Springer, Berlin, 1993.

[31] On the algebra of elementary embeddings of a rank into itself. In *Advances in Mathematics*, 110:334-346, 1995.

[32] Braid group actions on left distributive structures, and well orderings in the braid groups. *Journal of Pure and Applied Algebra*, 108:81-98, 1996.

[33] Adding dominating functions mod finite. *Periodica Mathematica Hungarica*, 35:35-41, 1997.

[34] Implications between strong large cardinal axioms. *Annals of Pure and Applied Logic*, 90:79-90, 1997.

[35] (with Carl Darby) Countable length Ramsey games. In Carlos Di Prisco *et al.*, editors, *Set Theory: Techniques and Applications*, pages 41-46. Kluwer, Dordrecht, 1998.

[36] Reflection of elementary embedding axioms on the $L[V_{\lambda+1}]$ hierarchy. *Annals of Pure and Applied Logic*, 107:227-238, 2001.

[37] On very large cardinals. In Gábor Hálasz *et al.*, editors, *Paul Erdős and his Mathematics*, volume II, Bolyai Society Mathematical Studies 11, pages 453-469. Springer, Berlin, 2002.

[38] (with John Moody) Well-foundedness conditions connected with left distributivity. *Algebra Universalis*, 47:65-68, 2002.

[39] Certain very large cardinals are not created in small forcing extensions. *Annals of Pure and Applied Logic*, 149:1-6, 2007.

[40] (with Sheila Miller) Left division in the free left distributive algebra on one generator. *Journal of Pure and Applied Algebra*, 215(3):276-282, 2011.

[41] (with Sheila Miller) The free one-generated left distributive algebra: basics and a simplified proof of the division algorithm. *Central European Journal of Mathematics*, 11(12):2150-2175, 2013.

References

[Apter and Hamkins, 1999] Arthur W. Apter and Joel D. Hamkins. Universal indestructibility. *Kobe Journal of Mathematics*, 16:119–130, 1999.

[Apter, 1998] Arthur W. Apter. Laver indestructibility and the class of compact cardinals. *The Journal of Symbolic Logic*, 63:149–157, 1998.

[Bagaria *et al.*, 2013] Joan Bagaria, Joel D. Hamkins, Konstantinos Tsaprounis, and Toshimichi Usuba. Superstrong and other large cardinals are never Laver indestructible, 2013. http://arxiv.org/abs/1307.3486.

[Bartoszyński and Judah, 1995] Tomek Bartoszyński and Haim Judah. *Set Theory. On the Structure of the Real Line.* A K Peters, Wellesley, 1995.

[Baumgartner and Hajnal, 1973] James Baumgartner and András Hajnal. A proof (involving Martin's Axiom) of a partition relation. *Fundamenta Mathematicae*, 78:193–203, 1973.

[Baumgartner and Taylor, 1982] James E. Baumgartner and Alan D. Taylor. Saturation properties of ideals in generic extensions. II. *Transactions of the American Mathematical Society*, 271:587–609, 1982.

[Baumgartner, 1983] James E. Baumgartner. Iterated forcing. In Adrian Mathias, editor, *Surveys in Set Theory*, volume 83 of *London Mathematical Society Lecture Note Series*, pages 1–59. Cambridge University Press, Cambridge, 1983.

[Baumgartner, 1984] James E. Baumgartner. Applications of the Proper Forcing Axiom. In *Handbook of Set-Theoretic Topology*, pages 913–959. North-Holland, Amsterdam, 1984.

[Benko, 1974] Pal Benko. *The Benko Gambit.* Batsford, London, 1974.

[Borel, 1919] Émile Borel. Sur la classification des ensembles de mesure nulle. *Bulletin de la Société Mathématique de France*, 47:97–125, 1919.

[Cantor, 1895] Georg Cantor. Beiträge zur Begründrung der transfiniten Mengenlehre. I. *Mathematische Annalen*, 46:481–512, 1895. Translated in *Contributions to the Founding of the Theory of Transfinite Numbers*, Open Court, Chicago 1915; reprinted Dover, New York, 1965.

[Carlson and Simpson, 1984] Timothy J. Carlson and Stephen G. Simpson. A dual form of Ramsey's Theorem. *Advances in Mathematics*, 53:265–290, 1984.

[Carlucci *et al.*, 2011] Lorenzo Carlucci, Patrick Dehornoy, and Andreas Weiermann. Unprovability results involving braids. *Proceedings of the London Mathematical Society*, 102:159–192, 2011.

[Corazza, 2000] Paul Corazza. The Wholeness Axiom and Laver sequences. *Annals of Pure and Applied Logic*, 105:157–260, 2000.

[Cummings and Foreman, 1998] James Cummings and Matthew Foreman. The tree property. *Advances in Mathematics*, 133:1–32, 1998.

[Cummings, 1997] James Cummings. Souslin trees which are hard to specialize. *Proceedings of the American Mathematical Society*, 125:2435–2441, 1997.

[Dales and Woodin, 1987] H. Garth Dales and W. Hugh Woodin. *An Introduction to Independence for Analysts*, volume 115 of *London Mathematical Society Lecture Note Series.* Cambridge University Press, Cambridge, 1987.

[Dehornoy, 1992a] Patrick Dehornoy. An alternative proof of Laver's results on the algebra generated by an elementary embedding. In Haim Judah, Winfried Just, and W. Hugh Woodin, editors, *Set Theory of the Continuum*, volume 26 of *Mathematical Sciences Research Institute Publications*, pages 27–33. Springer, Berlin, 1992.

[Dehornoy, 1992b] Patrick Dehornoy. Preuve de la conjecture d'irreflexivité pour les structures distributives libres. *Comptes Rendues de l'AcadÕmie des Sciences de Paris*, 314:333–336, 1992.

[Dehornoy, 1994] Patrick Dehornoy. Braid groups and left distributive operations. *Transactions of the American Mathematical Society*, 345:115–150, 1994.

[Dehornoy, 2010] Patrick Dehornoy. Elementary embeddings and algebra. In Matthew Foreman and Akihiro Kanamori, editors, *Handbook of Set Theory*, volume 2, pages 737–774. Springer, Dordrecht, 2010.

[Devlin, 1983] Keith J. Devlin. The Yorkshireman's guide to proper forcing. In Adrian Mathias, editor, *Surveys in Set Theory*, volume 83 of *London Mathematical Society Lecture Note Series*, pages 60–115. Cambridge University Press, Cambridge, 1983.

[Dougherty and Jech, 1997] Randall Dougherty and Thomas J. Jech. Finite left-distributive algebras and embedding algebras. *Advances in Mathematics*, 130:213–241, 1997.

[Dougherty, 1993] Randall Dougherty. Critical points in an algebra of elementary embeddings. *Annals of Pure and Applied Logic*, 65:211–241, 1993.

[Dushnik and Miller, 1940] Ben Dushnik and Edwin Miller. Concerning similarity transformations of linearly ordered sets. *Bulletin of the American Mathematical Society*, 46:322–326, 1940.

[Erdős and Hajnal, 1974] Paul Erdős and András Hajnal. Solved and unsolved problems in set theory. In Leon Henkin, editor, *Proceedings of the Tarski Symposium*, volume 25 of *Proceeding of Symposia in Pure Mathematics*, pages 269–287. American Mathematical Society, Providence, 1974.

[Foreman and Hajnal, 2003] Matthew Foreman and András Hajnal. A partition relation for successors of large cardinals. *Mathematische Annalen*, 325:583–623, 2003.

[Foreman *et al.*, 1988a] Matthew Foreman, Menachem Magidor, and Saharon Shelah. Martin's Maximum, saturated ideals and nonregular ultrafilters. Part I. *Annals of Mathematics*, 127:1–47, 1988.

[Foreman *et al.*, 1988b] Matthew Foreman, Menachem Magidor, and Saharon Shelah. Martin's Maximum, saturated ideals and nonregular ultrafilters. Part II. *Annals of Mathematics*, 127:521–545, 1988.

[Foreman, 1998] Matthew Foreman. An $\aleph_1$-dense ideal on $\aleph_2$. *Israel Journal of Mathematics*, 108:253–290, 1998.

[Fraïssé, 1948] Roland Fraïssé. Sur la comparaison des types d'ordres. *Comptes rendus hebdomadaires des séances de l'Académie des Sciences, Paris*, 226:1330–1331, 1948.

[Fuchs *et al.*, 2011] Gunter Fuchs, Joel D. Hamkins, and Jonas Reitz. Set-theoretic geology. http://arxiv.org/abs/1107.4776, 2011.

[Gitik and Shelah, 1989] Moti Gitik and Saharon Shelah. On certain indestructibility of strong cardinals and a question of Hajnal. *Archive for Mathematical Logic*, 28:35–42, 1989.

[Halpern and Läuchli, 1966] James D. Halpern and Hans Läuchli. A partition theorem. *Transactions of the American Mathematical Society*, 124:360–367, 1966.

[Halpern and Levy, 1971] James D. Halpern and Azriel Levy. The Boolean Prime Ideal does not imply the Axiom of Choice. In Dana S. Scott, editor, *Axiomatic Set Theory*, volume 13(1) of *Proceedings of Symposia in Pure Mathematics*, pages 83–134. Americal Mathematical Society, Providence, 1971.

[Hamkins *et al.*, 2008] Joel D. Hamkins, Jonas Reitz, and W. Hugh Woodin. The Ground Axiom is consistent with $V \neq \mathrm{HOD}$. *Proceedings of the American Mathematical Society*, 136:2943–2949, 2008.

[Hamkins, 2000] Joel D. Hamkins. The lottery preparation. *Annals of Pure and Applied Logic*, 101:103–146, 2000.

[Hamkins, 2003] Joel D. Hamkins. Extensions with the approximation and cover properties have no new large cardinals. *Fundamenta Mathematicae*, 180:237–277, 2003.

[Hausdorff, 1907] Felix Hausdorff. Untersuchungen über Ordnungstypen IV, V. *Berichte über die Verhandlungen der Königlich Sächsischen Gesellschaft der Wissenschaften zu Leipzig, Mathematisch-Physische Klasse*, 59:84–159, 1907. Translated with an introduction in Jacob Plotkin, editor, *Hausdorff on Ordered Sets*, American Mathematical Society, 2005, pp. 97-171.

[Hausdorff, 1908] Felix Hausdorff. Grundzüge einer Theorie der geordneten Mengen. *Mathematische Annalen*, 65:435–505, 1908. Translated with an introduction in Jacob Plotkin, editor, *Hausdorff on Ordered Sets*, American Mathematical Society, 2005, pp. 181-258.

[Higman, 1952] Graham Higman. Ordering by divisibility in abstract algebras. *Proceedings of the London Mathematical Society*, 2:326–336, 1952.

[Ishiu and Moore, 2009] Tetsuya Ishiu and Justin T. Moore. Minimality of non-σ-scattered orders. *Fundamenta Mathematicae*, 205:29–44, 2009.

[Judah and Shelah, 1990] Haim Judah and Saharon Shelah. The Kunen-Miller chart (Lebesgue measure, the Baire property, Laver reals and preservation theorems for forcing). *The Journal of Symbolic Logic*, 55:909–927, 1990.

[Judah *et al.*, 1990] Haim Judah, Saharon Shelah, and W. Hugh Woodin. The Borel conjecture. *Annals of Pure and Applied Logic*, 50:255–269, 1990.

[Kanamori *et al.*, 2012] Akihiro Kanamori, John Woods, and Dov Gabbay, editors. *Sets and Extensions in the Twentieth Century*, volume 6 of *Handbook of the History of Logic*. Elsevier, Amsterdam, 2012.

[Kanamori, 1976] Akihiro Kanamori. Weakly normal ultrafilters and irregular ultrafilters. *Transactions of the American Mathematical Society*, 220:393–399, 1976.

[Kanamori, 1986a] Akihiro Kanamori. Finest partitions for ultrafilters. *The Journal of Symbolic Logic*, 51:327–332, 1986.

[Kanamori, 1986b] Akihiro Kanamori. Partition relations for successor cardinals. *Advances in Mathematics*, 59:152–169, 1986.

[Ketonen, 1976] Jussi Ketonen. Nonregular ultrafilters and large cardinals. *Transactions of the American Mathematical Society*, 224:61–73, 1976.

[Kozen, 1988] Dexter Kozen. A finite model theorem for the propositional μ-calculus. *Studia Logica*, 47:233–241, 1988.

[Kruskal, 1960] Joseph Kruskal. Well-quasi-ordering, the tree theorem, and Vázsonyi's conjecture. *Transactions of the American Mathematical Society*, 95:210–225, 1960.

[Kunen, 1978] Kenneth Kunen. Saturated ideals. *The Journal of Symbolic Logic*, pages 65–76, 1978.

[Larue, 1994] David M. Larue. On braid words and irreflexivity. *Algebra Universalis*, 31:104–112, 1994.

[Magidor, 1977] Menachem Magidor. On the singular cardinals problem II. *Annals of Mathematics*, 106:517–547, 1977.

[Magidor, 1979] Menachem Magidor. On the existence of nonregular ultrafilters and cardinality of ultrapowers. *Transactions of the American Mathematical Society*, 249:97–111, 1979.

[Marcone and Montalbán, 2009] Alberto Marcone and Antonio Montalbán. On Fraïssé's conjecture for linear orders of finite Hausdorff rank. *Annals of Pure and Applied Logic*, 160:355–367, 2009.

[Marcone, 2005] Alberto Marcone. WQO and BQO theory in subsystems of second order arithmetic. In Stephen G. Simpson, editor, *Reverse Mathematics 2001*, volume 21 of *Lectures Notes in Logic*, pages 303–330. A K Peters, Wellesley, 2005.

[Martin, 1980] D. Anthony Martin. Infinite games. In Olli Lehto, editor, *Proceedings of the International Congress of Mathematicians*, volume 1, pages 269–273. Academia Scientiarum Fennica, 1980.

[Martinez-Ranero, 2011] Carlos Martinez-Ranero. Well-quasi-ordering Aronszajn lines. *Fundamenta Mathematicae*, 213:197–211, 2011.

[Menas, 1976] Telis Menas. Consistency results concerning supercompactness. *Transactions of the American Mathematical Society*, 223:61–91, 1976.

[Miller, 1980] Arnold W. Miller. There are no Q-points in Laver's model for the Borel conjecture. *Proceedings of the American Mathematical Society*, 78:103–106, 1980.

[Milliken, 1979] Keith R. Milliken. A Ramsey theorem for trees. *Journal of Combinatorial Theory, Series A*, 26:215–237, 1979.

[Mitchell, 1972] William J. Mitchell. Aronszajn trees and the independence of the transfer property. *Annals of Pure and Applied Logic*, 5:21–46, 1972.

[Montalbán, 2005] Antonio Montalbán. Up to equimorphism, hyperarithmetic is recursive. *The Journal of Symbolic Logic*, 70:360–378, 2005.

[Montalbán, 2006] Antonio Montalbán. Equivalence between Fraïssé's conjecture and Jullien's theorem. *Annals of Pure and Applied Logic*, 139:1–42, 2006.

[Moore, 2009] Justin T. Moore. A universal Aronszajn line. *Mathematical Research Letters*, 16:121–131, 2009.

[Nash-Williams, 1965] Crispin St. J. A. Nash-Williams. On well-quasi-ordering infinite trees. *Proceedings of the Cambridge Philosophical Society*, 61:697–720, 1965.

[Pincus and Halpern, 1981] David Pincus and James D. Halpern. Partitions of products. *Transactions of the American Mathematical Society*, pages 549–568, 1981.

[Prikry, 1970] Karel Prikry. On a problem of Gillman and Keisler. *Annals of Mathematical Logic*, 2:179–187, 1970.

[Prikry, 1972] Karel Prikry. On a problem of Erdös, Hajnal, and Rado. *Discrete Mathematics*, 2:51–59, 1972.

[Reitz, 2007] Jonas Reitz. The Ground Axiom. *The Journal of Symbolic Logic*, 72:1299–1317, 2007.

[Rothberger, 1952] Fritz Rothberger. On the property C and a problem of Hausdorff. *Canadian Journal of Mathematics*, 4:111–116, 1952.

[Shelah and Stanley, 1982] Saharon Shelah and Lee J. Stanley. S-forcing, I. A "black-box" theorem for morasses, with applications to super-Souslin trees. *Israel Journal of Mathematics*, 43:185–224, 1982.

[Shelah, 1978] Saharon Shelah. A weak generalization of MA to higher cardinals. *Israel Journal of Mathematics*, 30:297–306, 1978.

[Shelah, 1982] Saharon Shelah. Better quasi-orders for uncountable cardinals. *Israel Journal of Mathematics*, 42:177–226, 1982.

[Shelah, 1984] Saharon Shelah. Can you take Solovay's inaccessible away? *Israel Journal of Mathematics*, 48:1–47, 1984.

[Shore, 1993] Richard Shore. On the strength of Fraïssé's conjecture. In John Crossley, Jeffrey Remmel, Richard Shore, and Moss Sweedler, editors, *Logic Methods. In Honor of Anil Nerode's Sixtieth Birthday*, volume 12 of *Progress in Computer Science and Applied Logic*, pages 782–813. Birkhauser, Boston, 1993.

[Sierpiński, 1946] Wacław Sierpiński. Sur les types d'ordre de puissance du continu. *Revista de Ciencias, Lima*, 48:305–307, 1946.

[Sierpiński, 1950] Wacław Sierpiński. Sur les types d'ordre des ensembles linéaires. *Fundamenta Mathematicae*, 37:253–264, 1950.

[Solovay *et al.*, 1978] Robert M. Solovay, William N. Reinhardt, and Akihiro Kanamori. Strong axioms of infinity and elementary embeddings. *Annals of Mathematical Logic*, 13:73–116, 1978.

[Steprāns, 2012] Juris Steprāns. History of the continuum in the 20th Century. In Akihiro Kanamori, John Woods, and Dov Gabbay, editors, *Sets and Extensions in the Twentieth Century*, volume 6 of *Handbook of the History of Logic*, pages 73–144. Elsevier, Amsterdam, 2012.

[Tall, 1994] Franklin D. Tall. Some applications of a generalized Martin's axiom. *Topology and its Applications*, 57:215–248, 1994.

[Todorcevic, 1983] Stevo Todorcevic. Forcing positive partition relations. *Transactions of the American Mathematical Society*, 280:703–720, 1983.

[Todorcevic, 1986] Stevo Todorcevic. Reals and positive partition relations. In Ruth Barcan Marcus, Georg Dorn, and Paul Weingartner, editors, *Logic, Methodology and Philosophy of Science VII*, pages 159–169. North-Holland, Amsterdam, 1986.

[Todorcevic, 2007] Stevo Todorcevic. Lipschitz maps on trees. *Journal de l'Institut de Mathématiques de Jussieu*, 6:527–566, 2007.

[Woodin, 2010] W. Hugh Woodin. Suitable extender models I. *Journal of Mathematical Logic*, 10:101–339, 2010.

[Woodin, 2011a] W. Hugh Woodin. The Continuum Hypothesis, the generic-multiverse of sets, and the Ω Conjecture. In Juliette Kennedy and Roman Kossak, editors, *Set Theory, Arithmetic, and Foundations of Mathematics: Theorems, Philosophies*, volume 36 of *Lecture Notes in Logic*, pages 13–42. Cambridge University Press, Cambridge, 2011.

[Woodin, 2011b] W. Hugh Woodin. Suitable extender models II: Beyond ω-huge. *Journal of Mathematical Logic*, 11:115–436, 2011.

19 Mathias and Set Theory

1 Introduction

Adrian Richard David Mathias (born 12 February 1944) has cut quite a figure on the "surrealist landscape" of set theory ever since it became a modern and sophisticated field of mathematics, and his 70th birthday occasions a commemorative account of his mathematical *oeuvre*. It is of particular worth to provide such an account, since Mathias is a set theorist distinctive in having both established a range of important combinatorial and consistency results as well as in carrying out definitive analyses of the axioms of set theory.

Setting out, Mathias secured his set-theoretic legacy with the *Mathias real*, now squarely in the pantheon of generic reals, and the eventual rich theory of happy families developed in its surround. He then built on and extended this work in new directions including those resonant with the Axiom of Determinacy, and moreover began to seriously take up social and cultural issues in mathematics. He reached his next height when he scrutinized how Nicolas Bourbaki and particularly Saunders Mac Lane attended to set theory from their mathematical perspectives, and in dialectical engagement investigated how their systems related to mainstream axiomatic set theory. Then in new specific research, Mathias made an incisive set-theoretic incursion into dynamics. Latterly, Mathias refined his detailed analysis of the axiomatics of set theory to weaker set theories and minimal axiomatic sufficiency for constructibility and forcing.

We discuss Mathias' mathematical work and writings in roughly chronological order, bringing out their impact on set theory and its development. We describe below how, through his extensive travel and varied working contexts, Mathias has engaged with a range of stimulating issues. For accomplishing this, Mathias' webpage, https://www.dpmms.cam.ac.uk/~ardm/ , proved to be a valuable source for articles and details about their contents and appearance. Also, discussions and communications with Mathias provided detailed information about chronology and events. The initial biographical sketch which follows forthwith is buttressed by this information.

Mathias came up to the University of Cambridge and read mathematics at Trinity College, receiving his B.A. in 1965. This was a heady time for set theory, with Paul Cohen in 1963 having established the independence of the Axiom of Choice and of the Continuum Hypothesis with inaugural uses of his method of forcing. With the infusion of new model-theoretic and combinatorial methods,

---

Republished from *Mathematical Logic Quarterly* 20 (2016), pp.278-294, with permission from Wiley-VCH GmbH.

This text corresponds to an invited lecture given on Mathias Day, August 27, 2015, at the Fifth European Set Theory conference held at the Isaac Newton Institute, Cambridge, England. My thanks to the organizers of the conference for the invitation and for stimulating this text. My thanks also to Adrian Mathias for providing biographical details.

set theory was transmuting into a modern, sophisticated field of mathematics. With the prospect of new investigative possibilities opening up, Mathias with enterprise procured a studentship to study with Ronald Jensen at Bonn, at the time the leading set theorist in Europe.

With Jensen and through reading and study, Mathias proceeded to assimilate a great deal of set theory, and this would lay the groundwork not only for his lifelong research and writing but also for a timely survey of the entire subject. In the summer of 1967, he with Jensen ventured into the New World, participating in a well-remembered conference held at U.C.L.A., July 10 to August 5, which both summarized the progress and focused the energy of a new field opening up. There, he gave his first paper [46], a contribution to the emerging industry of independences among consequences of the Axiom of Choice.[1]

At the invitation of Dana Scott, Jensen and Mathias spent the autumn and winter of 1967-8 at Stanford University, and the year flowing into the summer proved to be arguably Mathias' most stimulative and productive in set theory. Mathias benefitted from the presence at Stanford of Harvey Friedman and Kenneth Kunen and from the presence at nearby Berkeley of Jack Silver and Karel Příkrý. With inspirations and initiatives from the U.C.L.A. conference, Mathias worked steadily to complete a survey of all of the quickly emerging set theory, "Surrealist landscape with figures". This was in typescript *samizdat* circulation by the summer, and proved to be of enormous help to aspiring set theorists. Almost as an aside, Mathias and Kopperman [32] established model-theoretic results about groups.

What came to occupy center stage were developments stimulated by a seminar at Stanford in the autumn of 1967 and cumulating in dramatic results established back in Bonn in the summer of 1968. As described in §19.2, Mathias established decisive results about the partition relation $\omega \longrightarrow (\omega)^{\omega}_{2}$, coming up with the key concept of *capturing* for the Mathias real and establishing that the relation holds in Solovay's model. Mathias quickly announced [41] his results and in four weeks penned *On a generalization of Ramsey's theorem* [42], submitted in August 1968 to Trinity College for a research fellowship.

Unsuccessful, Mathias forthwith spent the academic year 1968-9 at the University of Wisconsin, Madison, at the invitation of Kunen. There Mathias became familiar with the work of Kunen and David Booth, particularly on combinatorics of ultrafilters over ω. In late 1968 Mathias again applied for research fellowships at Cambridge with [42], and by April 1969 he was notified that he was successful at Peterhouse. June 1969 he spent at Monash University, Australia, and by then he had come to an elaboration of his work, also

[1] Mathias [46] established, with forcing, that Tarski's Order Extension Principle, that every partially ordered set can be extended to a total ordering, is independent of Mostowski's Ordering Principle, that every set has a total ordering.

described in §19.2, in terms of happy families, Ramsey ultrafilters generically included in them, and Mathias forcing adapted to the situation.

In October 1969 Mathias duly took up his fellowship at Peterhouse. In 1970 he with [42] was admitted to the Ph.D. by the University of Cambridge; from October 1970 for five years he held a university assistant lectureship; and from 1972 he held a Peterhouse teaching fellowship. Mathias would spend just over two decades as a fellow at Peterhouse, in a position of considerable amenity and stability. In 1971, with energy and enthusiasm he organized the Cambridge Summer School of Logic, August 1-21, a conference that arguably rivaled the U.C.L.A. conference, and brought out a proceedings [44]. In 1972-1974, Mathias edited the *Mathematical Proceedings of the Cambridge Philosophical Society.* This journal had been published as the *Proceedings of the Cambridge Philosophical Society* since 1843 and had published mathematical papers; Mathias during his tenure managed to attach the "mathematical" to its title.[2] In 1973, Mathias worked at the Banach Center at Warsaw to fully update his 1968 survey. The revision process however got out of hand as set theory was expanding too rapidly, and so he would just publish his survey, with comments added about recent work, as *Surrealist landscape with figures* [51]. In late 1974, Mathias was finally able to complete a mature, seasoned account of his work on $\omega \longrightarrow (\omega)^{\omega}_{2}$ and Mathias reals, and this appeared with a publication delay as *Happy families* [48] in 1977. In 1978, Mathias organized another summer school at Cambridge, August 7-25, and brought out a corresponding compendium [52]. The academic year 1979-1980 Mathias spent as *Hochschulassistent* to Jensen at the Mathematical Institute at Freiburg.

In ongoing research through this period and into the mid-1980s, Mathias obtained important results on filters and in connection with the Axiom of Determinacy, building on his work on $\omega \to (\omega)^{\omega}_{2}$. Moreover, Mathias began to take up topics and themes in rhetorical pieces about mathematics as set in society and culture. §3 describes this far-ranging work.

In the 1980s an unhappy climate developed in Peterhouse, and in 1990 Mathias did not have his fellowship renewed. This was a remarkable turn of events, not the least for setting in motion a memorable journey, a veritable wanderer fantasy that started in the last year of his fellowship: The academic year 1989-90 Mathias spent at the Mathematical Sciences Research Institute, Berkeley, and the Spring of 1991 he was Visiting professor at the University of California at Berkeley. For 1991-2 Mathias was Extraordinary Professor at the University of Warsaw. For much of 1992-3 he was *Dauergast* at the research institute at Oberwolfach in Germany. The years 1993-6 Mathias spent at Centre de Recerca Matemàtica of the Institute d'Estudis Catalans at the suggestion of Joan Bagaria. 1996-7 he was in Wales. For 1997-8 Mathias was at the Universidad de los Andes, at the suggestion of Carlos Montenegro.

[2] Mathias, personal communication.

At the beginning of these wanderings, Mathias engaged in a controversy with the distinguished mathematician Saunders Mac Lane. This stimulated Mathias to fully take up the investigation of the scope and limits of "Mac Lane set theory", something that he had started to do while still at Peterhouse. Over several years in the mid-1990s, the timing too diffuse to chronicle here, Mathias provided a rich, definitive analysis of weak set theories around Mac Lane and Kripke-Platek. §4 describes these developments, which established Mathias as a major figure in the fine axiomatization of set theory.

During his three-year period in Barcelona, Mathias was stimulated by colleagues and circumstances to pursue a set-theoretic approach to a basic iteration problem in dynamics. After establishing the subject in 1996, he would elaborate and refine it well into the next millennium. §5 describes this work, which decisively put the face of well-foundedness on dynamics.

In early 1999, Mathias landed on the island of La Réunion in the Indian Ocean, the most remote of the overseas départements of France, and obtained tenure at the university in 2000. At last there was stability again, though he was retired from the professoriate in 2012, with *France d'outre mer* also being subject to the French university rules for mandatory retirement.

Through these years, Mathias would continue to travel, and at Barcelona during a "set theory year" 2003-4, Mathias newly investigated set theories weaker than Kripke-Platek and intended to be axiomatic bases for developing constructibility, specifically in being able to carry the definition of the truth predicate for bounded formulas. Thus in continuing engagement with axiomatics, Mathias soon addressed the problem of finding the weakest system that would support a smooth, recognizable theory of forcing. He met with remarkable success in the theory of rudimentary recursion and provident sets. §6 describes this work, his finest and at the same time deepest work on the axiomatics of set theory.

In the fullness of time, Mathias returned to his first tussle with ill-suited set theories, the "ignorance of Bourbaki". After pointing out mathematical pathologies, Mathias gave full vent to his disapproval of Bourbaki's logic and influence in a lengthy piece. Though there is polemic, one also sees both Mathias' passionate advocacy of set theory in the face of detractors, and his hope for a kind of regeneration of mathematics through competition rather than centralization. The last section, §7, gets at these matters, very much to be considered part of his mathematical *oeuvre*.

On July 18, 2015, Adrian Mathias was admitted to the degree of Doctor of Science by the University of Cambridge.[3] Mathias was honored at the Fifth

[3]The Doctor of Science is a higher doctorate of the university. According to the *Statutes and Ordinances of the University of Cambridge* (p.519): "In order to qualify for the degree of Doctor of Science or Doctor of Letters a candidate shall be required to give proof of distinction by some original contribution to the advancement of science and of learning."

European Set Theory Conference held at the Isaac Newton Institute at Cambridge by having August 27, 2015 declared as "Mathias Day", a day given over to talks on his work and its influence.

2 $\omega \longrightarrow (\omega)^\omega_2$ and Mathias Reals

Recall that in the Erdős-Rado partition calculus from the 1950s, $[X]^\gamma = \{y \subseteq X \mid y \text{ has order type } \gamma\}$ for X a set of ordinals, and that the partition relation for ordinals

$$\beta \longrightarrow (\alpha)^\gamma_\delta$$

asserts that for any partition $f\colon [\beta]^\gamma \to \delta$, there is an $H \in [\beta]^\alpha$ *homogeneous* for f, i.e. $|f``[H]^\gamma| \leq 1$.

Frank Ramsey [80] established Ramsey's Theorem, that for $0 < r, k < \omega$,

$$\omega \longrightarrow (\omega)^r_k\,,$$

and the Finite Ramsey Theorem, that for any $0 < r, k, m < \omega$, there is an $n < \omega$ such that

$$n \longrightarrow (m)^r_k\,.$$

In 1934, the youthful Paul Erdős, still at university in Hungary, and György Szekeres popularized the Finite Ramsey Theorem with a seminal application to a combinatorial problem in geometry.[4] Erdős forthwith, in a letter to Richard Rado, asked whether the "far-reaching" generalization $\omega \longrightarrow (\omega)^\omega_2$ of Ramsey's Theorem can be established, and by return post Rado provided the now well-known counterexample deploying a well-ordering of the reals—the first result of Ramsey theory after Ramsey's.[5] With this, Ramsey theory and the partition calculus as developed by Erdős and his collaborators would focus on finite exponents, i.e. partitions of $[\beta]^\gamma$ for finite γ.

In the post-Cohen climate of Axiom of Choice independence results, the young Harvey Friedman, in a seminar conducted by Dana Scott at Stanford on partition relations in the autumn of 1967, newly raised the possibility of establishing the *consistency* of $\omega \longrightarrow (\omega)^\omega_2$. Alerted to a related possibility raised by Scott, that *definable* partitions of $[\omega]^\omega$ might have infinite homogeneous sets, Fred Galvin and Karel Příkrý at Berkeley established, by the winter of 1967, the now well-known and widely applied Galvin-Příkrý Theorem ([22]). $Y \subseteq [\omega]^\omega$ is *Ramsey iff* $\exists x \in [\omega]^\omega([x]^\omega \subseteq Y \text{ or } [x]^\omega \subseteq [\omega]^\omega - Y)$. Nash-Williams, Cohen, and Ehrenfeucht in early contexts had established that open sets are Ramsey. Reconstruing the classical notion of a set of reals being Borel, Galvin and Příkrý established that Borel partitions have large homogeneous sets: *If*

[4] Cf. their [21].

[5] Cf. Erdős [20].

Y is Borel, then Y is Ramsey. Analytic ($\mathbf{\Sigma}^1_1$) sets are classically the projection of Borel subsets of the plane, and Silver [82] at Berkeley forthwith improved the result to: *If Y is analytic, then Y is Ramsey.*

For Mathias, this past would be a prologue. In June of 1968, back in Bonn, he returned to the possibility of $\omega \longrightarrow (\omega)^\omega_2$, i.e. that *every* $Y \subseteq [\omega]^\omega$ is Ramsey. With ideas, results, and concepts in the air, he would in a few weeks put together the workings of a known pivotal *model* with a newly tailored *genericity* concept to achieve a decisive result.[6]

Mathias had learned from Silver of Solovay's celebrated 1964 model ([83]) in which every set of reals is Lebesgue measurable, and discussions with Jensen led to its further understanding. Solovay's model was remarkable for its early sophistication and revealed what standard of argument was possible with forcing. Starting with an inaccessible cardinal, Solovay first passed to a generic extension given by the Lévy collapse of the cardinal to render it ω_1 and then to a desired inner model, which can most simply be taken to be the constructible closure $L(\mathbb{R})$ of the reals. The salient point here is that in the generic extension $V[G]$,

$$(*) \quad \text{if } Y \text{ is a set of reals ordinal definable from a real } r\text{, then there is a formula } \varphi(\,\cdot\,,\,\cdot\,) \text{ such that: } x \in Y \textit{ iff } V[r][x] \models \varphi[r,x].$$

Solovay used this to get the Lebesgue measurability of Y with an infusion of random reals.

After Příkrý came up with Příkrý forcing for measurable cardinals in the summer of 1967, Mathias started exploring a version of this forcing for ω around the end of 1967. *Mathias forcing* has as conditions $\langle s, A\rangle$, where $s \subseteq \omega$ is finite, $A \subseteq \omega$ is infinite, and $\max(s) < \min(A)$, ordered by:

$$\langle t, B\rangle \leq \langle s, A\rangle \quad \textit{iff} \quad s \text{ is an initial segment of } t \text{ and } B \cup (t - s) \subseteq A.$$

A condition $\langle s, A\rangle$ is to determine a new, generic subset of ω through initial segments s, the further members to be restricted to A. A generic subset of ω thus generated is a *Mathias real.*

Mathias opined that $\omega \longrightarrow (\omega)^\omega_2$ should hold in Solovay's inner model if every infinite subset of a Mathias generic real is also Mathias generic, and that this indeed should be the case from his previous work. In the small hours of July 7, 1968, everything fell into place when Mathias came up with the property of *capturing*, a sort of well-foundedness notion. A Mathias condition $\langle s, A\rangle$ *captures* a dense set Δ of conditions *iff* every infinite subset B of A has a finite initial segment t such that $\langle s \cup t, A - (\max t + 1)\rangle \in \Delta$. Mathias saw that

[6] Mathias [42, §0] describes the progression, from which much of what follows, as well as the previous paragraph, are drawn.

for any $\langle s, A\rangle$ and dense set Δ, there is an infinite subset A' of A such that $\langle s, A'\rangle$ captures Δ. With this, Mathias proved (a) *For any condition* $\langle s, A\rangle$ *and formula* ψ *of the forcing language, there is an infinite* $B \subseteq A$ *such that* $\langle s, B\rangle$ *decides* ψ, i.e. s need not be extended to decide formulas, as well as (b) *If* x *is a Mathias real over a model* M *and* $y \subseteq x$ *is infinite, then* y *is a Mathias real over* M. Thus, $\omega \longrightarrow (\omega)^\omega_2$ was confirmed in Solovay's inner model: For a set Y of reals definable from a real r as in (*), by (a) there is in $V[r]$ a Mathias condition $\langle \emptyset, A\rangle$ that decides $\varphi(r, c)$, where c is the canonical name for a Mathias real. There is surely in $V[G]$ a real $x \subseteq A$ Mathias generic over $V[r]$, and by (b) x confirms that Y is Ramsey.

With energy and initiative Mathias, in four weeks starting July 8, penned *On a generalization of a theorem of Ramsey* [42], providing a comprehensive account of Solovay's model, Mathias reals, and the consistency of $\omega \longrightarrow (\omega)^\omega_2$, as well a forcing proof of Silver's analytic-implies-Ramsey result and a range of results about Mathias reals that exhibited their efficacy and centrality.

By June of 1969, Mathias had uncovered a rich elaboration of his work. As formulated in Booth [10], an ultrafilter U over ω is *Ramsey iff* for any $f : [\omega]^2 \longrightarrow 2$ there is an $H \in U$ homogeneous for f. For a filter F, *F-Mathias forcing* is Mathias forcing with the additional proviso that conditions $\langle s, A\rangle$ are to satisfy $A \in F$. Mathias realized that a real x Mathias over a ground model V generates a Ramsey ultrafilter F on $\mathcal{P}(\omega) \cap V$ given by $F = \{X \in \mathcal{P}(\omega) \cap V \mid x - X \text{ is finite}\}$ and that generically adjoining x to V is *equivalent* to first generically adjoining the corresponding F, without adjoining any reals, and then doing F-Mathias forcing over $V[F]$ to adjoin x. A *happy family* is, in one formulation, a set A of infinite subsets of ω such that $\mathcal{P}(\omega) - A$ is an ideal, and: whenever $X_i \in A$ with $X_{i+1} \subseteq X_i$ for $i \in \omega$ there is a $Y \in A$ which diagonalizes the X_i's, i.e. its increasing enumeration f satisfies $f(i+1) \in X_i$ for every $i \in \omega$. The set of infinite subsets of ω is a happy family, as is a Ramsey ultrafilter. A happy family is just the sort through which one can force a Ramsey ultrafilter without adjoining any reals. Mathias thus had a tripartite elaboration: One entertains happy families by overlaying Ramsey ultrafilters, which themselves can be reduced to Mathias reals. With this, Mathias could give systematic generalizations not only of his earlier results but those of Galvin-Příkrý and Silver.

Coordinating this work with Příkrý forcing, Mathias [45] established a corresponding characterization, that if U is a normal ultrafilter over a measurable cardinal κ, then a countable $x \subseteq \kappa$ is Příkrý generic *iff* for any $X \in U$, $x - X$ is finite. This "Mathias property" would become a pivotal feature of all generalized Příkrý forcings (cf. [23]).

Mathias eventually laid out his theory with all its trimmings and trappings in *Happy families* [48], proving the Ramseyness results as well as further, 1969 results in the elaborated context. For H a happy family, say that a set $Y \subseteq [\omega]^\omega$

is *H-Ramsey iff* $\exists x \in H([x]^\omega \subseteq Y$ or $[x]^\omega \subseteq [\omega]^\omega - Y)$. Mathias established that every analytic set is H-Ramsey for every happy family H. At the other end, Mathias established results about Solovay's inner model if one started with a Mahlo cardinal: In the Lévy collapse of a Mahlo cardinal to ω_1, every set of reals in $L(\mathbb{R})$ is in fact H-Ramsey for every happy family H, and also, in $L(\mathbb{R})$ there are no maximal almost disjoint families of subsets of ω.

To frame [48] at one end, it is worth mentioning that Mathias in [48], for the happy-families improvement of Silver's analytic-implies-Ramsey, returns to a classical characterization of analytic sets. The Luzin-Sierpiński investigation of analytic sets in the 1910s was the first occasion where well-foundedness was explicit and instrumental, and today one can vouchsafe the sense of analytic sets as given by well-founded relations on finite sequences of natural numbers. Mathias in his researches would continue to engage with the theory of analytic sets, and his work would itself draw out his later contention that set theory itself is ultimately the study of well-foundedness.

To elaborate on the Mahlo cardinal results at the other end of [48], Mathias made a distinctive advance by incorporating elementary substructures into the mix to establish results about the $L(\mathbb{R})$ of the Lévy collapse of a Mahlo cardinal instead of just an inaccessible cardinal. Mathias' result [48, §5] that, in this Lévy collapse, every set of reals in $L(\mathbb{R})$ is H-Ramsey for every happy family H would eventually be complemented 30 years later in terms of consistency by Todd Eisworth [18], who showed, applying a later Henle-Mathias-Woodin [53] result, that if the Continuum Hypothesis holds and every set of reals in $L(\mathbb{R})$ is U-Ramsey for every Ramsey ultrafilter U, then ω_1 is Mahlo in L. A prominent open problem addresses Mathias' first, 1968 result that in the Solovay Lévy collapse of an inaccessible to ω_1, every set of reals in $L(\mathbb{R})$ is Ramsey. Is the consistency strength of having an inaccessible cardinal necessary? Halbeisen-Judah [24], among others, considered this question and established related results.

For Mathias' other Mahlo cardinal result, a family of infinite subsets of ω is *almost disjoint* if distinct members have finite intersection, and is *maximal almost disjoint* (MAD) if moreover no proper extension is almost disjoint. Mathias observed that MAD families generate happy families, and showed that MAD families cannot be analytic. He moreover built on his previous work to show that in the Lévy collapse of a Mahlo cardinal to ω_1, there are no MAD families in $L(\mathbb{R})$. Almost a half a century later, Asger Törnquist [86] showed that in Solovay's original Lévy collapse of just an inaccessible cardinal to ω_1, there are no MAD families in $L(\mathbb{R})$. The question remains whether it is consistent, relative to ZF, whether there are no MAD families.

As to the core of Mathias' [48], the extent to which a piece of mathematics is pursed and extended by others is a measure of both its mathematical significance and depth. Early on, topological proofs of Silver's analytic-implies-

Ramsey result were found by Erik Ellentuck [19] and of Mathias' analytic-implies-U Ramsey for Ramsey ultrafilters U were found by Alain Louveau [36] and Keith Milliken [78]. The Ellentuck Theorem is "an infinite dimensional Ramsey theorem" in what is now considered optimal form, deploying what is now widely known as the *Ellentuck topology*, with open sets of form $O_{\langle a,S\rangle} = \{x \in {}^{\omega}\omega \mid a \subseteq x \subseteq S\}$ for a Mathias condition $\langle a, S\rangle$. Andreas Blass [7] and Claude Laflamme [34] took the Mathias [48] theory to a next level of generalization, extending it to non-Ramsey ultrafilters and corresponding Mathias-type generic reals.

In their [14], Timothy Carlson and Stephen Simpson established "dual" Ramsey theorems. Their Dual Ramsey Theorem asserts that, with $(\omega)^k$ the set of partitions of ω into k cells, if $k < \omega$ and $(\omega)^k = C_1 \cup C_2 \cup \ldots \cup C_n$ with the C_i's Borel, then there is an i and some k-cell partition H such that all its k-cell coarsenings lie in C_i (partitions are taken to be equivalence relations on ω, and the topology here is the product topology on $2^{\omega\times\omega}$). They established a dual Ellentuck theorem and introduced a dual Mathias forcing. Carlson [13] subsequently generalized a large part of the Ramsey theory at the time to Ramsey spaces, structures that satisfy a corresponding Ellentuck theorem. A *Ramsey space* is a space of infinite sequences with a topology such that every set with the property of Baire is Ramsey and no open set is meager. Stevo Todorčević [85] is a magisterial axiomatic account of abstract Ramsey spaces, with corresponding combinatorial forcing and Ellentuck theorems.

Returning to the original Mathias real, it has taken a fitting place in the pantheon of generic reals. Mathias forcing, in a filter form, was deployed in the classic Martin-Solovay [40, p.153], in a proof that in modern terms can be construed as showing that Martin's Axiom implies $\mathfrak{p} = \mathfrak{c}$, i.e. that the pseudo-intersection number is the cardinality of the continuum. Mathias forcing has since become common fare in the study of cardinal invariants of the continuum. After Richard Laver [35] famously established the consistency of Borel's Conjecture with a paradigmatic countable support iteration featuring Laver reals, he noted (p.168) that, after all, Mathias reals could have been deployed instead. James Baumgartner worked this out in an incisive account [5] of iterated forcing for the 1978 Cambridge conference organized by Mathias. Mathias reals in various forms would occur in a range of work on ultrafilters, e.g. by Blass and Saharon Shelah [9] on ultrafilters with small generating sets.

3 Varia

In his two decades at Peterhouse, Mathias pursued research that built on and resonated with his Ramseyness work as well as forged new directions. Moreover, he began to articulate ways of thinking and points of view about mathematics as set in society and culture.

Following on his incisive incorporation of Ramsey ultrafilters, Mathias made contributions to an emerging theory of filters and ultrafilters. A filter F over ω is a *p-point iff* whenever $X_i \in F$ for $i \in \omega$, there is a $Y \in F$ such that $Y - X_i$ is finite for every $i \in \omega$. The *Rudin-Keisler ordering* $\leq_{\mathrm{RK}}$ is defined generally by: For filters $F, G \subseteq \mathcal{P}(I)$, $F \leq_{\mathrm{RK}} G$ *iff* there is an $f\colon I \to I$ such that $F = f_*(G)$, where $f_*(G) = \{X \subseteq I \mid f^{-1}(X) \in G\}$. These concepts emerged in the study of the Stone-Čech compactification $\beta\mathbb{N}$, identifiable with the ultrafilters over ω. Walter Rudin showed that the Continuum Hypothesis (CH) implies that there is a p-point in $\beta\mathbb{N} - \mathbb{N}$, and since p-points are topologically invariant and there are non-p-points, $\beta\mathbb{N} - \mathbb{N}$ is not homogeneous. Answering an explicitly-posed question, Mathias [43] showed that CH *implies that there is an ultrafilter U over ω with no p-point below it in the Rudin-Keisler ordering.* For this, Mathias used properties of analytic sets.

With a neat observation, Mathias [47] showed that $\omega \longrightarrow (\omega)^{\omega}_2$ implies that the filters over ω are as curtailed as they can be. Let $Fr = \{X \subseteq \omega \mid \omega - X \text{ is finite}\}$—the *Fréchet filter.* A filter over ω is *feeble*, being as far from being an ultrafilter as can be, *iff* there is a finite-to-one $f\colon \omega \to \omega$ such that $f_*(F) = Fr$. Mathias [47] pointed out that for filters F over ω extending Fr, F is feeble *iff* a corresponding $P^F \subseteq \mathcal{P}(\omega)$ is Ramsey. Hence, $\omega \to (\omega)^{\omega}_2$ *implies that every filter over ω extending Fr is feeble*! Being feeble would become a pivotal concept for filters,[7] especially as it was soon seen that a filter over ω extending Fr is feeble *iff* it is meager in the usual topology on $\mathcal{P}(\omega)$. In Solovay's model or under the assumption of the *Axiom of Determinacy*, every filter extending Fr is feeble forthwith because every set of reals has the Baire property.

Already in the early 1970's, whether ZFC (Zermelo-Fraenkel set theory with the Axiom of Choice) implies that there are p-point ultrafilters over ω became a focal question. As an approach to the question, the author [28] asked whether, if not such an ultrafilter, at least coherent filters exist, where in the above terms, a filter over ω is *coherent iff* it extends Fr, is a p-point, and is not feeble. Remarkably, Mathias [50] quickly established that *if $0^\sharp$ does not exist or $2^{\aleph_0} \leq \aleph_{\omega+1}$, then there are coherent filters.* The elegant proof depended on a covering property for families of sets, with Jensen's recent Covering Theorem for L providing the ballast with $0^\sharp$. Around 1978, Shelah famously established[8] that it is consistent relative to ZFC that there are no p-point ultrafilters. Still, the question of whether ZFC implies that there are coherent filters has remained. Adding to the grist, the 1990 [27], with Mathias a co-author, contains a range of results about and applications of coherent filters. This paper came about as a result of another co-author, Winfried Just, having rediscovered the main results of Mathias' [50] around 1986.

Continuing his engagement with analytic sets, Mathias with Andrzej Os-

[7] Cf. Blass [8, §6].

[8] Cf. Wimmers [87] or Shelah [81, VI§§3,4].

tazewski and Michel Talagrand in [77] addressed the following proposition queried by Rogers and Jayne: Given a non-Borel analytic set A, there is a compact set K such that $K \cap A$ is not Borel. They showed that Martin's Axiom + $\neg$CH implies this proposition, and that $V = L$ implies its negation.

With his work on $\omega \longrightarrow (\omega)^\omega_2$ in hand, Mathias pursued the study of analogous partition properties for uncountable cardinals as they became topical. The situating result would be the Kechris-Woodin [30] characterization of the Axiom of Determinacy (AD). With Θ the supremum of ordinals ξ such that there is a surjection: $P(\omega) \to \xi$ and the demarcating limit of the effect of AD, they showed that assuming $V = L(\mathbb{R})$, AD is equivalent to Θ being the limit of cardinals κ having the *strong partition property*, i.e. $\kappa \longrightarrow (\kappa)^\kappa_\alpha$ for every $\alpha < \kappa$.

With James Henle, Mathias [25] lifted results about "smooth" functions from *Happy Families* [48, §6] to the considerably more complex situation of uncountable κ satisfying the strong partition property. They achieved a remarkably strong continuity for functions $[\kappa]^\kappa \to [\kappa]^\kappa$ and applied it to get results about the Rudin-Keisler ordering in this context.

With Henle and Hugh Woodin, Mathias [26] elaborated on the *Happy Families* [48, §4] forcing for the happy family of infinite subsets of ω, newly dubbed the "Hausdorff extension". They showed that assuming $\omega \to (\omega)^\omega_2$, the Hausdorff extension not only adjoins no new real but no new sets of ordinals at all. With this, they could show, applying [25] and pointing out that AD + $V = L(\mathbb{R})$ implies $\omega \to (\omega)^\omega_2$, the following: If AD, then in the Hausdorff extension of $L(\mathbb{R})$, Θ is still the limit of cardinals κ having the strong partition property *and* there is a Ramsey ultrafilter over ω. In particular, AD fails, and so $V = L(\mathbb{R})$ is necessary for the Kechris-Woodin characterization.

In a determinacy capstone of sorts, Mathias [53] squarely took up a ZF (Zermelo-Fraenkel set theory) issue about ordinals and their subsets that arose in the investigation of AD and provided a penetrating exegesis. An ordinal is *unsound iff* it has subsets A_n for $n \in \omega$ such that *un*countably many ordinals are realized as ordertypes of sets of form $\bigcup\{A_n \mid n \in a\}$ for some $a \subseteq \omega$. Woodin had asked whether there is an unsound ordinal, and eventually showed that AD implies that there is one less than ω_2. While the issue remains unsettled in ZF, Mathias [53] showed that: (a) If ω_1 is regular, then every ordinal less than $\omega_1^{\omega+2}$ is sound, and (b) $\aleph_1 \leq 2^{\aleph_0}$, i.e. there is a uncountable well-orderable set of reals, *iff* $\omega_1^{\omega+2}$ is exactly the least unsound ordinal. The proofs proceed through an intricate combinatorial analysis of indecomposable ordinals and a generalization of the well-known Milnor-Rado paradox. An interesting open question is the following: In Solovay's model in which every set of reals is Lebesgue measurable, starting specifically from L as the ground model, is every ordinal sound?

During this period, no doubt with the fellowship of Peterhouse an intellectual stimulus, Mathias became rhetorically engaged with various sociological and cultural aspects of mathematics. An opening shot was an address given at the Logic Colloquium '76, of which a whiff remains in the proceedings [49, p.543] in which Mathias likened postures in mathematics to stances in religion.[9] In *Logic and terror* [54], read to the Perne Club[10] on February 12, 1978, Mathias deftly potted some history in the service of drawing out first the surround of the Law of the Excluded Middle $A \vee \neg A$ and of the Law of Contradiction $A \wedge \neg A$, and then in connection with the latter, the tension between (formal) logic and (Hegelian) dialectic in the Soviet Union, eventually resolved by Stalin's back-door admittance of the former (published as [54] and [55]).

The Ignorance of Bourbaki [58], read to the Quintics Club[11] on October 29, 1986, would set in motion initiatives in Mathias' later work (published as [56, 57] and translated as [60] and [68]). He argued that Bourbaki, in their setting up of a foundations for mathematics, ignored Gödel's work on incompleteness until much later and that the set theory that they did fix on was inadequate. With more deftly potted history, Mathias suggested that the reason for these was Bourbaki's underlying accentuation of the geometrical over the arithmetical. Extending his reach, Mathias took on the distinguished mathematician Saunders Mac Lane in this view, newly stressing that while weak set theories are adequate for set formation, they are not for recursive definition—a theme to be subsequently much elaborated by Mathias.

Toward the end of what would be his time at Peterhouse, Mathias, exhibiting research breadth and prescience, made a first observation on a conjecture of Erdős, one which would become a focal problem more than two decades later. Erdős around 1932, while still an undergraduate, made one of his earliest conjectures, in number theory: For any sequence $x_1, x_2, x_3, \ldots$ with each x_n either -1 or $+1$ and any integer C, there exist positive m and d such that $C < |\Sigma_{k=1}^{m} x_{kd}|$. This is a remarkably simple assertion, and the problem of trying to affirm it came to be known as the Erdős Discrepancy Problem. Mathias around 1986 affirmed the conjecture for $C = 1$, and latterly published the proof [59] in the proceedings of a 1993 conference celebrating Erdős' 80th birthday. Much later in 2010, the Polymath Project took up solving the Erdős Discrepancy Problem as a project, Polymath5; Polymath Project was started by Timothy Gowers to carry out collaborative efforts online to solve problems.

[9]With his recent conversion to Catholicism, Mathias considered that [49, p.543] "parallels may be drawn between Platonism and Catholicism, which are both concerned with what is true; between intuitionism and Protestant presentations of Christianity, which are concerned with the behaviors of mathematicians and the morality of individuals; between formalism and atheism, which deny any need for postulating external entities; and between category theory and dialectical materialism."

[10]The Perne Club is a club of Peterhouse where papers are read to senior and junior members of the College on historical and philosophical matters.

[11]The Quintics Club is an undergraduate mathematics club for the junior members of five colleges of the University of Cambridge of which Peterhouse is one.

Polymath5 [79] describes aspects of the collaborative effort, collaborative effort on the Erdős conjecture are described, and under Annotated Bibliography, the Mathias paper [59] is annotated: "This one page paper established that the maximal length of [the] sequence for the case where $C = 1$ is 11, and is the starting point for our experimental studies." Extending the polymath project work, Boris Konev and Alexei Lisitsa [31] showed that every sequence of length at least 1161 satisfies the conjecture in the case $C = 2$. Finally and suddenly, Terence Tao [84] in September 2015 announced a proof of the Erdős conjecture.

4 Mac Lane Set Theory

At the beginning of his worldly wanderings, Mathias engaged in a notable controversy—in the classical sense of an exchange carried out in published articles—with Mac Lane. This stimulated Mathias to fully take up the investigation of the scope and limits of "Mac Lane set theory", something that he had started to do while still at Peterhouse.

The publication of Mac Lane's *Mathematics: Form and Function* [37] met with a noticeable lack of response or even acknowledgement. One exception was Mathias [58], which took issue with Mac Lane's advocacy of a weak set theory and his contention that set theory cannot serve a substantial foundational role because of independence results. Mac Lane's set theory, ZBQC, is ZFC without Replacement and with Separation restricted to the Δ_0 (bounded quantifier) formulas. Mathias pointed out that this is inadequate for iterative constructions. The spatial (geometric) and temporal (arithmetical, iterative) are two modes that posit "an essential *bimodality* of mathematical thought", and set theory buttresses the latter as a study of well-foundedness. In reply, Mac Lane [38] maintained that his ZBQC does better fit what most mathematicians do; that for other work "there are other foundations"; that there is "no need for a single foundation". To Mathias' veering toward set theory as a foundation for mathematics "in ontological terms", Mac Lane opined: "Each mathematical notion is protean, thus deals with different realities, so does not have an ontology".

A decade later, Mathias [62] in a last rhetorical article divided "[o]pponents of a full-blooded set-theoretic account of the foundations of mathematics" among three categories: Those who "may hold, with Mac Lane, that . . . [ZBQC] suffices for 'all important mathematics' "; those who "may hold, with the early Bourbakistes, that ZC [Zermelo set theory, including Foundation and Choice] suffices"; and those who "may accept the axioms of ZFC, but deny the relevance of large cardinals to ordinary mathematics". Focusing on the last, Mathias formulated in palatable terms and described the essential involvement of large cardinal properties in several "strong statements of analysis": $\mathbf{\Sigma}^1_2$ sets have the perfect set property; $\mathbf{\Pi}^1_1$ sets are determined; $\mathbf{\Sigma}^1_2$ sets are universally Baire; and $\mathbf{\Sigma}^1_2$ sets are determined. Moreover, Mathias issued challenges to find various direct proofs, e.g. of $\mathbf{\Sigma}^1_2$ sets being universally Baire implying that $\mathbf{\Pi}^1_1$ sets are determined, that does not proceed through large cardinals—a potent kind of argument insisting on purity of proofs.[12] Mathias' [62] was followed by Mac Lane's [39] in the manner of a reply. He simply wrote contrarily that "Mathias has not produced any counter examples of actual [sic] mathematics which requires the use of a stronger [than Δ_0] separation", and argued for ZBQC being bolstered by an equiconsistency with a suitable categorical foundation recently established with the Mitchell-Benabou language.

Having taken on the question of the adequacy of weak set theories for mathematics in his exchange with Mac Lane, Mathias during a period of relative stability in the mid-1990s at the Centre de Recerca Matemàtica at Barcelona, would become deeply engaged with the analysis of the strength of various set theories. *Slim models of Zermelo set theory* [64] resonates with his initial insistence on adequate set theories supporting recursive definitions. Let Z be Zermelo set theory (without Choice) and KP, Kripke-Platek set theory.[13]

To specify for here and later, as Mathias has it,

Z has Extensionality, Empty Set, Pairing, Union,
Power Set, Foundation, Infinity, and Separation; and
KP has Extensionality, Empty Set, Pairing, Union, Π_1 Foundation,
Δ_0 Separation, and Δ_0 Collection.

With Foundation for A being the assertion $A \neq \emptyset \longrightarrow \exists x \in A(x \cap A = \emptyset)$, Foundation for sets A—the usual Foundation—and Foundation for classes A are easily seen to be equivalent in the presence of Separation, as first observed by Gödel. KP is usually formulated with Foundation for all classes A, but Mathias considers it befitting to have just Π_1 Foundation, which expectedly is Foundation for Π_1-defined classes. Δ_0 Separation and Δ_0 Collection are analogously the schemes restricted to Δ_0 formulas and so forth for other classes of formulas in the Lévy hierarchy, where Collection generally is the following the following scheme, a variant of Replacement:

$$\forall x \in a \exists y \varphi(x,y) \longrightarrow \exists b \forall x \in a \exists y \in b \varphi(x,y).$$

Affirming that Z is weak as a vehicle for recursive definitions while KP is "orthogonal" in providing a natural setting for such, Mathias [64] established that in Z + KP, one can recursively define a supertransitive (i.e. closed under subsets) inner (i.e. containing all ordinals) model that exhibits evident failures of Replacement, starting with the class $V_\omega = \mathrm{HF}$ of hereditarily finite sets not

[12] That every $\mathbf{\Sigma}^1_2$ set is universally Baire is equivalent to every set of ordinals having a sharp, and that every $\mathbf{\Pi}^1_1$ set is determined is equivalent to every real having a sharp. Hence, the first implies the second through large cardinals hypotheses.

[13] See [17] for the historical particulars on Zermelo's axiomatization and [4] for Kripke-Platek set theory.

being a set. Bringing together some classical ideas, he deployed a simple yet potent scheme for controlling entry into inner models according to growth rates of functions: $\omega \to \omega$.

In the magisterial *The strength of Mac Lane set theory* [65], Mathias provided a rich and definitive analysis of set theories around Mac Lane's and Kripke-Platek, bringing in the full weight of basic themes and delineating a range of theories according to relative consistency, motivating concepts, and basic set-existence principles and techniques. With remarkable energy and both syntactic and semantic finesse, Mathias set out sharp results in minimal settings, elaborating and refining a half a century of work.

First, Mathias carries out von Neumann's classical construction, with minimal hypotheses, of the inner model of well-founded sets, to effect the relative consistency of the Axiom of Foundation as well as Transitive Containment, viz. that every set is a member of a transitive set. Transitive Containment is a consequence of KP, a first step toward Replacement. Mathias takes MAC to be Mac Lane's ZBQC together with Transitive Containment; takes M to be MAC without Choice; and next takes up what is to become a focal Axiom H:

$$\forall u \exists t (\bigcup t \subseteq t \wedge \forall z (\bigcup z \subseteq z \wedge |z| \leq |u| \longrightarrow z \subseteq t))$$

That is, for any set u, there is a transitive set t of which every transitive set of size at most $|u|$ is a subset. Axiom H is a second step toward Replacement. Mathias approximates having Mostowski collapses, i.e. transitizations of well-founded, extensional relations, and makes isomorphic identifications to get the relative consistency of having Axiom H, e.g. Con(MAC) implies Con(MAC + H). Drawing out the centrality of Axiom H, Mathias shows that over a minimal set theory H is equivalent to actually having Mostowski collapses, and with a minimal Skolem hull argument, that over MAC, H subsumes KP, being equivalent, quite notably, to Σ_1 Separation together with Δ_0 Collection.

Mathias next makes the steep ascent to the first height, parsimoniously building the constructible hierarchy L and the relative consistency of AC without a direct recursive definition. Working in M, he simulates Gödel's recursive set-by-set generation along well-orderings, takes transitizations, and makes identifications as before for adjoining H, to define L. Working out condensation, Mathias then gets to his first significantly new result, that

$$\mathrm{Con(M)} \longrightarrow \mathrm{Con(M + KP} + V = L),$$

completing the circle to its first announcement in his [58]. With Z weak for recursive definitions, this approach notably provides the first explicit proof of Con(Z) $\longrightarrow$ Con(Z + AC), once announced by Gödel. This speaks to a historical point: Between 1930 and 1935, Gödel tried to generate L recursively along well-orderings, themselves provided along the way in "autonomous progression".

It was only after his embrace of Replacement in 1935, giving the von Neumann ordinals as canonical versions of all well-orderings, that he could rigorously get an inner model built on the spine of the ordinals.[14]

Setting out on a further climb, Mathias next layers, with the mediation of L, the region between M + KP and Z + KP in terms of Σ_n Separation. The thrust is that with $V = L$, Σ_n Separation implies the consistency of Σ_{n-1} Separation, and that Σ_n Separation implies the same in the sense of L. With H a steady pivot, Mathias proceeds in a minimal setting to construct Σ_n hulls based on L-least witnesses, and there is a notable refinement of analysis deploying "fine structure" lemmas, adapted from work of Sy Friedman, to control quantifier complexity.

Proceeding in a different direction that draws out a subtle interplay between power set and recursion, Mathias develops, partly with the purpose of getting sharp independence results, a theory subsuming M + KP first isolated for "power-admissible sets" by Harvey Friedman. The germ is to incorporate $\forall x \subseteq y$ and $\exists x \subseteq y$ as part of bounded quantification, setting up a new basis for a Lévy-type hierarchy, the Takahashi hierarchy of $\Sigma_n^{\mathcal{P}}$ formulas. Working through a subtle syntactical analysis, Mathias develops normal forms and situates $\Delta_0^{\mathcal{P}}$ Separation. The focus becomes $\mathrm{KP}^{\mathcal{P}}$, which is M + KP + $\Pi_1^{\mathcal{P}}$ Foundation + $\Delta_0^{\mathcal{P}}$ Collection. Bringing in the Gandy Basis Theorem, standard parts of admissible sets, and forcing over ill-founded models, Mathias is able to establish the surprising result that, unlike for KP,

$$\mathrm{KP}^{\mathcal{P}} + V = L \text{ proves the consistency of } \mathrm{KP}^{\mathcal{P}},$$

and delimitative results, e.g. even $\mathrm{KP}^{\mathcal{P}}$ + AC + "every cardinal has a successor" does not prove H (nor therefore Σ_1 Separation).

Mathias attends, lastly, to systems type-theoretic in spirit, working the theme that MAC with its Power Set and Δ_0 Separation, is latently in this direction. Mathias shows that e.g. in MAC "strong stratifiable Σ_1 Collection" is provable, a narrow bridge to Quine's NF. Also, Mathias provides the first explicit proof of a result implicit in John Kemeny's 1949 thesis, that MAC is equiconsistent with the simple theory of types (together with the Axiom of Infinity).

Although definitively developed with a wide range of themes and concepts and a great deal of detail, one can discern in Mathias' [65] a base line speaking to his earlier engagements about and larger conception of set theory. As he retrospectively wrote [65, 10.7], "The purpose of my paper ... is to study the relation of Mac Lane's system, which encapsulates in set-theoretic terms his mathematical world, to the Kripke-Platek system that gives a standard formalization of a certain kind of abstract recursion." As a reviewer noted,

[14]See Kanamori [29, §4].

"Monumental in conception and rich in results, this paper merits the attention not just of set-theorists, but of all mathematicians concerned with the broader foundations of mathematics."[15]

Mathias' later [72] can be seen as extending the sharp deductive analysis of [65] to encompass Replacement and thus full ZF. Consider the scheme

$$\text{(Repcoll)} \qquad \forall y \in u \exists ! z \varphi(y, z) \longrightarrow \exists w \forall y \in u \forall z (\varphi(y, z) \rightarrow z \in w)\,,$$

with passive parameters allowed in φ and y possibly a vector of variables. Loosely speaking, any class function restricted to a set has range included in a set. Instances of this scheme have hypotheses stronger than those of Collection and conclusions weaker than those of Replacement. Replacement implies Collection over ZF, since if $\forall u \exists z \varphi$ and a is a set, one can for each $x \in a$ functionally specify the least rank of a witnessing z. A. K. Simpson had asked about the strength of Repcoll, and Mathias [72] answered his question by showing that, unexpectedly,

The theory M + Repcoll implies ZF,

where, to recall, M is the relatively weak theory, MAC without Choice. Actually, Infinity does not play a role here, and can be effaced from both sides. Also, if Infinity is retained, then Transitive Containment can be effaced from the left side. The thrust of the proof is to show that in M, every set has a rank, and, using this together with Δ_0 Separation and Repcoll, to work inductively up the Lévy hierarchy of formulas to full Separation and that exact ranges of class functions of sets are sets as well.

We further tuck in here work connected by a thread, of even later vintage. Mac Lane had envisioned his foundational set theory as sufficient for mathematics, inclusive of category theory, and the Mac Lane–Mathias controversy had turned on the possible necessity of strong, even large cardinal, hypotheses. In recent affirmations, the very strong large cardinal principle, Vopěnka's Principle, has been pressed to display categorical consequences, e.g. that all reflective classes in locally presentable categories are small-orthogonality classes. In collaborative work with Joan Bagaria, Carles Casacuberta and Jiří Rosický, Mathias [2, 3] (a) sharpened this result by reducing the hypothesis to having a proper class of supercompact cardinals yet still drawing a substantial conclusion; (b) got categorical equivalents to Vopěnka's Principle; and (c) showed as a consequence that [3] "the existence of cohomological localizations of simplicial sets, a long-standing open problem in algebraic topology, is implied by the existence of arbitrarily large supercompact cardinals".

[15] Bell [6].

5 Dynamics

During his time at Barcelona, Mathias was also stimulated by colleagues and circumstances to pursue a set-theoretic approach to a basic iteration problem in dynamics, having once been alerted to such a possibility in the late 1970s. Then, chaotic dynamics, with the stuff of period orbits, strange attractors, and the like, was quite the rage, with the straightforward mathematical context of iterating functions providing simply posed problems. Mathias took up a basic issue cast in general terms; saw the applicability of descriptive set theory; and established substantial results that revealed a remarkable structure for recurrent points and "long delays" in dynamics. With [61] an initial article, Mathias, once established at La Réunion, put together the main account [63] as well as produced [70, 67, 69] containing refinements and further solutions. Taking an initial cue from dynamics, Mathias [63] started by setting up a transfinite context:

Let χ be a Polish space (complete, separable metric space) and $f\colon \chi \to \chi$ a continuous function. $\langle \chi, f\rangle$ is a dynamical system, a topological space with a continuous function acting on it. Define a relation $\curvearrowright_f$ on χ by:

$$x \curvearrowright_f y \textit{ iff } \text{there is an increasing } \alpha\colon \omega \to \omega \text{ with } \lim_{n\to\infty} f^{\alpha(n)}(x) = y,$$

i.e. y is a cluster point of the f-iterates of x. This is basic to topological dynamics which focuses on recurrent points, point b such that $b \curvearrowright b$. Let

$$\omega_f(x) = \{y \mid x \curvearrowright_f y\} \text{ and}$$
$$\Gamma_f(X) = \bigcup\{\omega_f(x) \mid x \in X\},$$

both being $\curvearrowright_f$-closed as $\curvearrowright_f$ is easily seen to be a transitive relation. For $a \in \chi$, recursively define

$$\begin{aligned} A^0(a,f) &= \omega_f(a)\,, \\ A^{\beta+1}(a,f) &= \Gamma_f(A^\beta(a,f))\,, \text{and} \\ A^\lambda(a,f) &= \textstyle\bigcap_{\nu<\lambda} A^\nu(a,f) \text{ for limit } \lambda. \end{aligned}$$

Then $A^0(a,f) \supseteq A^1(a,f) \supseteq A^2(a,f) \supseteq \ldots$ again by the transitivity of $\curvearrowright_f$. Let $\theta(a,f)$ by the least ordinal θ such that $A^{\theta+1}(a,f) = A^\theta(a,f)$, and let $A(a,f) = A^{\theta(a,f)}(a,f)$. The thrust of Mathias' work is to investigate the closure ordinal $\theta(a,f)$ as providing the dynamic sense of $\curvearrowright_f$.

Mathias first established that $\theta(a,f) \le \omega_1$, *with the inequality being strict when $A(a,f)$ is Borel.* With the result, and context, reflective of familiar paths for analytic sets, he associated to each $x \in \omega_f(a)$ a tree of $\curvearrowright_f$-descending finite sequences, so that $x \notin A(a,f)$ *iff* the tree is well-founded. Then he adapted to $\curvearrowright_f$ the Kunen proof of the Kunen-Martin Theorem on bounding ranks of well-founded trees. Notably, Mathias' argument works for any transitive relation in place of $\curvearrowright_f$, and so it can be seen as a nice incorporation of well-foundedness into the study of transitivity.

Particular to $\curvearrowright_f$ and dynamics, Mathias [63] established a striking result about recurrent points, points b such that $b \curvearrowright_f b$. With an intricate metric construction of a recurrent point, he showed that $y \in A(a,f)$ *iff for some* z, $a \curvearrowright_f z \curvearrowright_f z \curvearrowright_f y$, so that in particular there are recurrent points in $\omega_f(a)$ exactly when $A(a,f) \neq \emptyset$.

More particular still with χ being Baire space, ${}^\omega\omega$, Mathias showed that if $s\colon {}^\omega\omega \to {}^\omega\omega$ is the (backward) shift function given by $s(g)(n) = g(n+1)$, then for each $\zeta < \omega_1$ there is an $a \in {}^\omega\omega$ such that $\theta(a,s) = \zeta$, a "long delay". For this, he carefully embedded countable well-founded trees into the graph of $\curvearrowright_f$. Approaching the issue of whether there can be χ, f and a such that $\theta(a,f)$ is actually ω_1, Mathias adapted his embedding apparatus to ill-founded trees and carried out a Cantor-Bendixson analysis on the hyperarithmetic hierarchy to provide an effective answer: *There is a recursive* $a \in {}^\omega\omega$ *such that* $\theta(a,s) = \omega_1^{\mathrm{CK}}$, *the first non-recursive ordinal.* For this Mathias was inspired by Kreisel's construction of a recursively coded closed set whose Cantor-Bendixson sequences of derivations stabilizes at ω_1^{CK}.

Mathias' formulations and results, being of evident significance for dynamics, soon attracted those working in the area. In particular, Lluís Alsedà, Moira Chas, and Jaroslav Smítal [1] set off Mathias' results against a known backdrop and established a characterization for the closed unit interval of reals of positive topological entropy. This work led (p.1721) to new observations by Alexander Sharkovskii, well-known for his pioneering work on periodic points for dynamical systems.

In [70], Mathias answered questions left open in [63]. Using Baire space ${}^\omega\omega$ and the shift function s, he showed that *there is a recursive real* a *such that* $A^1(a,s)$ *is not even Borel, and* $A^2(a,s)$ *is empty.* In [63], Mathias had approached, but did not resolve, whether there could be a some χ, f, and a with the longest delay $\theta(a,f) = \omega_1$ and indeed Alsedà, Chas and Smítal [1, p.1720] conjectured no. In the best result in this subject, certainly the one with the most involved proof, Mathias established that *with Baire space and the shift function* s, *there is a recursive real* b *giving the longest delay,* $\theta(b,s) = \omega_1$. In subsequently written papers [67, 69], Mathias set out ways for extending the results of [70] to general χ and f.

Also stimulated by Mathias' work, his colleague Christian Delhommé at La Réunion has developed it in generalizing directions. For instance, he [15] extended Mathias' [63] embedding of countable well-founded trees into the graph of $\curvearrowright_s$ to countable binary relations, appropriately retracted, in a broadly general setting.

While these may be the outward landmarks, there is a great deal of details and elaboration of concepts in [63], as further pursued in [70, 67, 69]. The subject that Mathias uncovered from a simple dynamics issue has remarkable depth and richness, as his penetrating results and constructions have shown. As such, it is a testament both to Mathias' mathematical prowess as well as to his insistence on well-foundedness as the bedrock of set theory.

6 Weaker Set Theories

In continuing travels but also with the stability afforded by La Réunion, Mathias, with his definitive work on Mac Lane set theory as providing a broad context, pursued themes in the axiomatics of set theory with renewed energy. He newly illuminated the interstices of deductible possibilities and refined systematic interconnections and minimal axiomatic sufficiency in connection with constructibility and forcing.

Mathias' *Weak systems of Gandy, Jensen, and Devlin* [71], written during the "set theory year" 2003-4 at Barcelona, provides a definitive analysis of set theories weaker than Kripke-Platek (KP) for lack of full Δ_0 Collection. Mathias' earlier [65] was initially stimulated by the question of the adequacy of Mac Lane's set theory for ongoing mathematics, and here he was initially stimulated by the question of the adequacy of a basic set theory in Keith Devlin's *Constructibility* [16] for the investigation of constructibility in terms of Jensen's rudimentary set functions and the J_α hierarchy.

Mathias first variegates the landscape between ReS, which is KP without Δ_0 Collection, and KP with a range of systems and variants, focal ones being the following. DB "Devlin Basic" is ReS augmented with having Cartesian products. GJ "Gandy-Jensen" is DB augmented with Rudimentary Replacement: for Δ_0 formulas ϕ,

$$\forall x \exists w \forall v \in x \exists t \in w \forall u (u \in t \longleftrightarrow u \in x \wedge \phi(u, v)) .$$

Further augmentations with restricted versions of Δ_0 Collection are formulated, getting closer to full KP. Devlin's original system is DB augmented with Infinity and Foundation for all Classes. GJ axiomatizes the closure under the rudimentary functions for Jensen's fine structure investigations; a transitive set is closed under the rudimentary functions *iff* it models GJ – Π_1 Foundation.

Mathias next sets out the crucial set formations that can be effected in the various systems, and then establishes independences through over a dozen models. In particular, he applied his "slim model" [64] technique to get a supertransitive inner model of DB but not GJ, the model showing that DB cannot prove the existence of $[\omega]^3$.

Mathias' distinctive contribution is to incorporate into the various systems an axiom S asserting for all x the existence of $S(x) = \{y \mid y \subseteq x \text{ is finite}\}$. Proceeding analogously to [65, §6], Mathias carries out a subtle syntactical analysis of formulas having $\forall y \in S(x)$ and $\exists y \in S(x)$ as part of bounded quantification and sets up a corresponding hierarchy of Σ_n^S formulas. With that, he augments the various systems with Infinity and S and establishes corresponding Separation, Collection, etc. for formulas in the new hierarchy as well as semantic independences. With this preparation Mathias confirms in exacting detail the flaws in Devlin's book, *Constructibility*, especially the inadequacy of his basic set theory for formulating the satisfaction predicate for Δ_0 formulas. GJ + Infinity does work, as it axiomatizes the rudimentary functions. Answering a call for doing without the theory of rudimentary functions, Mathias shows that what also works for a parsimonious development of constructibility is Devlin's system (DB + Infinity + Foundation for all Classes) as augmented by S, as well as the particularly enticing subsystem MW "Middle Way": DB + Infinity + $\forall a \forall k \in \omega([a]^k \in V)$.

Already in the mid-1990s, deeply engaged in axiomatics, Mathias became interested in the problem of finding the weakest system that would support a smooth, recognizable theory of forcing. Through a period of germination in the new millennium proceeding through his contextualizing work on weak systems for constructibility, Mathias developed the concepts of rudimentary recursion and provident set, finally to meet with a remarkable success, in his maturity, which secures the axiomatic and methodological essence of a fundamental technique in set theory. With this work proceeding in forward and circling strides, [73] provides an overview, and then [12, 76] systematically set out the details in full.

The *Rudimentary recursion, gentle functions and provident sets* [12], with Nathan Bowler, worked towards an optimal theory for forcing, drawing on previous axiomatics and getting at the recursions just sufficient for formulating forcing. In KP, one has that if G is a total Σ_1^1 function, then so is F given by $F(x) = G(F \restriction x)$, and such recursions handily suffice for forcing. The first move toward purity of method is to work only with *rudimentarily recursive* functions, i.e. those F as given above but defined from rudimentary G.

As described in [12], Mathias systematically developed a theory of rudimentary recursion in weak set theories and toward their use in forcing. A significant complication was that the composition of two rudimentarily recursive functions is not necessarily rudimentarily recursive, and to finesse this, Bowler developed the *gentle functions*, functions $H \circ F$ where H is rudimentary and F is rudimentarily recursive. The composition of gentle functions *is* gentle, and this considerably simplified the formulation of forcing. That formulation also to require having the forcing partial order as a parameter, the general theory was extended to the p-rudimentarily recursive functions, functions F given by

$F(x) = G(p, F \restriction x)$ with p as a parameter in the rudimentary recursion.

Forcing is to be done over the provident sets. A set is *provident iff* it is non-empty, transitive, closed under pairing and for all $x, p \in A$ and p-rudimentarily recursive F, $F(x) \in A$. Mathias variously characterized the provident sets, and ramified them as cumulative unions of transitive sets in a canonical fashion. The Jensen rudimentary functions are nine, and can be but together into one set formation function T such that for transitive u, $\bigcup_{n\in\omega} T^n(u)$ is rudimentarily closed. With this, given a transitive set c, define c_ν and P^c_ν by simultaneous recursion:

$$\begin{aligned} &c_0 = \emptyset\,, &&c_{\nu+1} = c \cap \{x \mid x \subseteq c_\nu\}\,, &&c_\lambda = \textstyle\bigcup_{\nu<\lambda} c_\nu \\ &P^c_0 = \emptyset\,, &&P^c_{\nu+1} = T(P^c_\nu) \cup c_{\nu+1} \cup \{c_\nu\}\,, &&P^c_\lambda = \textstyle\bigcup_{\nu<\lambda} P^c_\nu\,. \end{aligned}$$

This master recursion gradually builds up rudimentary closed levels relative to c. Mathias showed that for transitive c and indecomposable ordinal θ, P^c_θ is provident. Moreover, one can define the *provident closure* of any non-empty M by: $\mathrm{Prov}(M) = \bigcup\{P^c_\theta \mid c$ is the transitive closure of some finite subset of $M\}$, where θ is the least indecomposable ordinal not less than the set-theoretic rank of M. In particular, if M is already provident, $\mathrm{Prov}(M) = M$ exhibits a canonical ramification. With this, one gets a finite set of axioms Prov warranting the recursion so that the transitive models of Prov are exactly the provident sets. J_ν is provident *iff* $\omega\nu$ is indecomposable, so that $J_\omega = V_\omega = \mathrm{HF}$ is provident, the only provident set not satisfying Infinity, and so are $J_{\omega^2}, J_{\omega^3} \ldots$.

With the above theory, *Provident sets and rudimentary set forcing* [76] duly carries out a parsimonious development of forcing. The Shoenfield-Kunen approach[16] is taken, carefully tailored to be effected in Prov + Infinity. Although the overall scheme to be followed is thus straightforward, the progress step by step reveals many eddies of deduction and astute choices of terms, drawing out the methodological necessity and sufficiency of provident sets. Mathias eventually shows, in a rich surround of textured results established in his context, that if M is provident, $P \in M$ is a forcing partial order, and G is P-generic over M, then $M[G]$ is provident, with $M[G] = \mathrm{Prov}(M \cup \{G\})$. Moreover, he shows of the various axioms of set theory like Power Set that they persist from M to $M[G]$. Mathias' [76] is a veritable paean to formalism and forcing, one that exhibits an intricate melding of axiomatics and technique in set theory. As such, it together with his [12] is Mathias' arguably most impressive accomplishment in axiomatics.

[16]Cf. Kunen [33, VII].

7 Bourbaki

In the fullness of time and with his remarkable work on axiomatics in hand, Mathias latterly took up the lance once again against the windmill of the "ignorance of Bourbaki" [57], circling back to his first tussle with ill-suited set theories. This time there would be telling mathematical pathologies exposed as well as a remarkable arching argument inset in French history and praxis.

Bourbaki, in their first book, the 1954-1957 *Théorie des Ensembles*, developed a theory of sets in a logical formalism as the axiomatic basis of their structural exposition of mathematics. They adapted for purposes of quantification the Hilbert ε-operator, which for each formula φ introduces a term $\varepsilon x \varphi$ replicating the entire formula. It will be remembered that Hilbert in the 1920's had introduced these terms, presumably motivated by the use of ideal points in mathematics; he had $\varphi(t) \rightarrow \varphi(\varepsilon x \varphi)$ for terms t, and defined the quantifiers by $\exists x \varphi \longleftrightarrow \varphi(\varepsilon x \varphi)$ and $\forall x \varphi \longleftrightarrow \varphi(\varepsilon x \neg \varphi)$. Mathias [66] underlined the cumulating complications in Bourbaki's rendition by showing that their definition of the cardinal number 1, itself awkward, when written out in their formalism would have length 4,523,659,424,929 !

Earlier in 1948, André Weil, on behalf of Bourbaki, gave an invited address to the Association of Symbolic Logic on set theory as a "foundations of mathematics for the working mathematician" (cf. [11]). Bourbaki's system has Extensionality, Separation, Power Set, ordered pair as primitive with axioms to match, and Cartesian products. From Separation and Power set, one gets singletons $\{x\}$. Mathias [74] showed, to blunt purpose yet with considerable dexterity, that there is a model of Bourbaki's system in which the Pairing Axiom, having $\{x, y\}$, fails!

Hilbert, Bourbaki and the scorning of logic [75] is Mathias' mature criticism of Bourbaki, one that works out rhetorically a line of argument through history, mathematics, and education. There is a grand sweep, but also a specificity of mathematical detail and a sensitivity to systemic influence, no doubt heightened by his teaching years at La Réunion.

Mathias' line of argument is as follows:

(a) Hilbert in 1922 proposed an alternative treatment of first-order logic using his ε-operator,

(b) which, despite its many unsatisfactory aspects, was adopted by Bourbaki for their exposition of mathematics

(c) and by Godement for his classic *Cours d'Algèbre*, though leading him to express distrust of logic.

(d) It is this distrust, intensified to a phobia by the vehemence of Dieudonné's writings,

(e) and fostered by, for example, the errors and obscurities of a well-known undergraduate text, Jacqueline Lelong-Ferrand and Jean-Marie Arnaudiès' *Cours de mathèmatiques*,

(f) that has, it is suggested, led to the exclusion of logic from the CAPES examination—"tout exposé de logique formelle est exclu".

(g) Centralist rigidity has preserved the underlying confusion and consequently flawed teaching;

(h) the recovery will start when mathematicians adopt a post-Gödelian treatment of logic.

For (a) and (b), Mathias recounts in detail Hilbert's engagement with logic and his program for establishing the consistency of mathematics and how Bourbaki adopted wholesale the awkward Hilbert ε-operator approach to logic. According to Mathias, with the appearance of Gödel's work on incompleteness "the hope of a single proof of the consistency and completeness of mathematics, in my view the only justification for basing an encyclopaedic account of mathematics on Hilbert's operator, had been dashed". For (c), (d), and (e) to be regarded as case studies, Mathias quotes at considerable length from the sources, revealing most particularly in the last case tissues of confusion about truth vs. provability and metalanguage vs. uninterpreted formalism. For (f) and (g), Mathias is unrestrained about the over-centralized French educational system and how in its rigidity it has perpetuated the Bourbaki hostility to logic and set theory.

In the articulation of (h), Mathias synthetically puts forth layer by layer a larger vision of mathematics and its regeneration, one that straddles his earlier work and writing. He first observes, through inner minutes of Bourbaki, that there was uncertainty and even dissension in the tribe about the adequacy of their adopted set theory. Mathias then recapitulates, with specific examples from his own writings, mathematics that the Bourbakistes would not be able to encompass. Calling Bourbaki as well as Mac Lane structuralists, Mathias regards structuralism and set theory as on opposite sides of the divide between taking equality up to isomorphism as good enough and not, and recalls his exchange with Mac Lane. Mathias then emphasizes how the divide brings out a dual nature of mathematics, and how mathematics is impoverished by the disregard of one side or the other. Mathias personally defines set theory as the study of well-foundedness; as such, it is highly successful in meeting the call for recursive constructions—and it is to be pointed out that his own work is a particular and abiding testament to this. In a concluding peroration, Mathias points to the stultifying effects of centralization and bureaucratization across centuries and cultures and calls for intellectual independence within community, as at Oxford and Cambridge, with all contributing through competition to the regeneration of mathematics and more broadly, culture.

Adrian Mathias is a estimable mathematician with a remarkable range of results addressing problems that emerged in the course of a wide-ranging engagement with mathematics. From his work with Mathias reals to the dynamics of iterated maps, he has uncovered a great deal of structure and brought forth new understanding. But also, Mathias can be seen, quite distinctively, as a fine analyst of social and cultural aspects of mathematics and the axiomatic basis of concepts and methods of set theory. His work on the last attains an exceptional virtuosity, and with that, one can count Mathias—-whose work is going from strength to strength—as one of the very few who through logical axiomatic analysis has contributed to meaning in mathematics.

References

[1] Luuís Alsedà, Moira Chas, and Jarslav Smítal. On the structure of the ω-limit sets for continuous maps of the interval. *International Journal of Bifurcation and Chaos*, 9:1719–1729, 1999.

[2] Joan Bagaria, Carles Casacuberta, and Adrian R.D. Mathias. Epireflections and supercompact cardinals. *Journal of Pure and Applied Algebra*, 213:1208–1215, 2009.

[3] Joan Bagaria, Carles Casacuberta, Adrian R.D. Mathias, and Jiří Rosický. Definable orthogonality classes in accessible categories are small. *Journal of the European Mathematical Society*, 17:549–589, 2015.

[4] Jon Barwise. *Admissible Sets and Structures.* Perspectives in Mathematical Logic. Springer-Verlag, 1975.

[5] James E. Baumgartner. Iterated forcing. In Adrian Mathias, editor, *Surveys in Set Theory*, volume 83 of *London Mathematical Society Lecture Note Series*, pages 1–59. Cambridge University Press, Cambridge, 1983.

[6] John L. Bell. Review of "The strength of Mac Lane set theory". *Mathematical Reviews*, 2002. MR1846761.

[7] Andreas Blass. Selective ultrafilters and homogeneity. *Annals of Pure and Applied Logic*, 38:215–255, 1988.

[8] Andreas Blass. Combinatorial cardinal characteristics of the continuum. In Matthew Foreman and Akihiro Kanamori, editors, *Handbook of Set Theory*, pages 395–489. Springer, Dordrecht, 2010.

[9] Andreas Blass and Saharon Shelah. Ultrafilters with small generating sets. *Israel Journal of Mathematics*, 65:259–271, 1989.

[10] David Booth. Ultrafilters on a countable set. *Annals of Mathematical Logic*, 2:1–24, 1970.

[11] Nicolas Bourbaki. Foundations of mathematics for the working mathematician. *The Journal of Symbolic Logic*, 14:1–8, 1949.

[12] Nathan Bowler and Adrian R.D. Mathias. Rudimentary recursion, gentle recursion, and provident sets, 2015.

[13] Timothy Carlson. Some unifying principles of Ramsey theory. *Discrete Mathematics*, 68:117–169, 1988.

[14] Timothy Carlson and Stephen Simpson. A dual form of Ramsey's theorem. *Advances in Mathematics*, 53:265–290, 1984.

[15] Christian Delhommé. Representation in the s's attacks of Baire space, 2009. Preprint.

[16] Keith J. Devlin. *Constructibility.* Perspectives in Mathematical Logic. Springer, Berlin, 1984.

[17] Hans-Dieter Ebbinghaus. *Ernst Zermelo, An Approach to his Life and Work.* Springer, 2015. Second revised edition.

[18] Todd Eisworth. Selective ultrafilters and $\omega \to (\omega)^\omega$. *Proceedings of the American Mathematical Society*, 127, 1999.

[19] Erik Ellentuck. A new proof that analytic sets are Ramsey. *The Journal of Symbolic Logic*, 39:163–165, 1974.

[20] Paul Erdős. My joint work with Richard Rado. In C. A. Whitehead, editor, *Surveys in Combinatorics 1987*, volume 123 of *London Mathematical Society Theory Lecture Note Series*, pages 53–80. Cambridge University Press, Cambridge, 1987.

[21] Paul Erdős and György Szekeres. A combinatorial problem in geometry. *Compositio Mathematica*, 2:463–470, 1935.

[22] Fred Galvin and Karel Prikry. Borel sets and Ramsey's theorem. *The Journal of Symbolic Logic*, 38:193–198, 1973.

[23] Moti Gitik. Prikry-Type Forcings. In Matthew Foreman and Akihiro Kanamori, editors, *Handbook of Set Theory*, pages 1351–1447. Springer, Dordrecht, 2010.

[24] Lorenz Halbeisen and Haim Judah. Mathias absoluteness and the Ramsey property. *The Journal of Symbolic Logic*, 61:177–194, 1996.

[25] James M. Henle and Adrian R.D. Mathias. Supercontinuity. *Mathematical Proceedings of the Cambridge Philosophical Society*, 92:1–15, 1982.

[26] James M. Henle, Adrian R.D. Mathias, and W. Hugh Woodin. A barren extension. In Carlos A. Di Prisco, editor, *Methods in Mathematical Logic. Proceedings, Caracas 1983*, volume 1130 of *Lecture Notes in Mathematics*, pages 195–207. Springer-Verlag, Berlin, 1985.

[27] Winfried Just, Adrian R.D. Mathias, Karel Prikry, and Petr Simon. On the existence of large p-ideals. *The Journal of Symbolic Logic*, 55:457–465, 1990.

[28] Akihiro Kanamori. Some combinatorics involving ultrafilters. *Fundamenta Mathematicae*, 100:145–155, 1978.

[29] Akihiro Kanamori. In praise of Replacement. *The Bulletin of Symbolic Logic*, 18, 2013.

[30] Alexander S. Kechris and W. Hugh Woodin. Equivalence of partition properties and determinacy. *Proceedings of the National Academy of Sciences U.S.A.*, 80:1783–1786, 1983.

[31] Boris Konev and Alexei Lisitsa. The SAT attack on the Erdos Discrepancy Conjecture, 2014. arXiv:1402.2184.

[32] Ralph D. Kopperman and Adrian R.D. Mathias. Some problems in group theory. In Jon Barwise, editor, *The Syntax and Semantics of Infinitary Languages*, volume 72 of *Lecture Notes in Mathematics*, pages 131–138. Springer-Verlag, Berlin, 1968.

[33] Kenneth Kunen. *Set Theory. An Introduction to Independence Proofs.* North-Holland, Amsterdam, 1980.

[34] Claude Laflamme. Forcing with filters and complete combinatorics. *Annals of Pure and Applied Logic*, 42:125–163, 1989.

[35] Richard Laver. On the consistency of Borel's conjecture. *Acta Mathematica*, 137:151–169, 1976.

[36] Alain Louveau. Démonstration topologique de théorèms de Silver et Mathias. *Bulletin des Sciences Mathématiques*, 98:97–102, 1974.

[37] Saunders MacLane. *Mathematics: Form and Function.* Springer-Verlag, 1986.

[38] Saunders MacLane. Is Mathias an ontologist? In Haim Judah, Winfried Just, and W. Hugh Woodin, editors, *Set Theory of the Continuum*, Mathematical Sciences Research Institute publication #26, pages 119–120. Springer-Verlag, New York, 1992.

[39] Saunders MacLane. Contrary statements about mathematics. *Bulletin of the London Mathematical Society*, 32:527, 2000.

[40] D. Anthony Martin and Robert M. Solovay. Internal Cohen extensions. *Annals of Mathematical Logic*, 2:143–178, 1970.

[41] Adrian R.D. Mathias. On a generalization of Ramsey's theorem. *Notices of the American Mathematical Society*, 15, 1968. Abstract 68T-E19, received 18 July 1968.

[42] Adrian R.D. Mathias. *On a Generalization of Ramsey's Theorem*. PhD thesis, University of Cambridge, 1970. Submitted in August 1968 to Trinity College for a fellowship.

[43] Adrian R.D. Mathias. Solution of problems of Choquet and Puritz. In Wilfrid Hodges, editor, *Conference in Mathematical Logic—London '70*, volume 255 of *Lecture Notes in Mathematics*, pages 204–210. Springer-Verlag, Berlin, 1972.

[44] Adrian R.D. Mathias, editor. *Cambridge Summer School in Mathematical Logic*, volume 337 of *Lecture Notes in Mathematics*. Springer-Verlag, Berlin, 1973. Edited with Hartley Rogers, Jr.

[45] Adrian R.D. Mathias. On sequences generic in the sense of Prikry. *Journal of the Australian Mathematical Society*, 15:409–414, 1973.

[46] Adrian R.D. Mathias. The order extension principle. In Thomas J. Jech, editor, *Axiomatic Set Theory*, volume 13(2) of *Proceedings of Symposia in Pure Mathematics*, pages 179–183. American Mathematical Society, Providence, 1974.

[47] Adrian R.D. Mathias. A remark on rare filters. In András Hajnal, Richard Rado, and Vera T. Sós, editors, *Infinite and Finite Sets, Keszthely (Hungary), 1973, volume III*, volume 10 of *Colloquia Mathematica Societatis János Bolyai*, pages 1095–1097. North-Holland, Amsterdam, 1975.

[48] Adrian R.D. Mathias. Happy families. *Annals of Mathematical Logic*, 12:59–111, 1977.

[49] Adrian R.D. Mathias. The real line and the universe. In Robin O. Gandy and J. Martin E. Hyland, editors, *Logic Colloquium '76*, volume 87 of *Studies in Logic and the Foundations of Mathematics*, pages 531–546. North-Holland, Amsterdam, 1977.

[50] Adrian R.D. Mathias. $0^\sharp$ and p-point problem. In Gert H. Müller and Dana S. Scott, editors, *Higher Set Theory. Proceedings, Oberwolfach, Germany 1977*, volume 669 of *Lecture Notes in Mathematics*, pages 375–384. Springer-Verlag, Berlin, 1978.

[51] Adrian R.D. Mathias. Surrealist landscape with figures (a survey of recent results in set theory). *Periodica Mathematica Hungarica*, 10:109–175, 1979.

[52] Adrian R.D. Mathias, editor. *Surveys in Set Theory*, volume 87 of *London Mathematical Society Lecture Note Series.* Cambridge University Press, Cambridge, 1983.

[53] Adrian R.D. Mathias. Unsound ordinals. *Mathematical Proceedings of the Cambridge Philosophical Society*, 96:391–411, 1984.

[54] Adrian R.D. Mathias. Logic and terror. *Jahrbuch. Kurt-Gödel Gesellschaft*, 1990:117–130, 1991.

[55] Adrian R.D. Mathias. Logic and terror. *Physis—Rivista Internazionale di Storia della Scienza*, 28:557–578, 1991.

[56] Adrian R.D. Mathias. The ignorance of Bourbaki. *Physis—Rivista Internazionale di Storia della Scienza*, 28:887–904, 1991.

[57] Adrian R.D. Mathias. The ignorance of Bourbaki. *The Mathematical Intelligencer*, 14:4–13, 1992.

[58] Adrian R.D. Mathias. What is Mac Lane missing? In Haim Judah, Winfried Just, and W. Hugh Woodin, editors, *Set Theory of the Continuum*, Mathematical Sciences Research Institute publication #26, pages 113–118. Springer-Verlag, New York, 1992.

[59] Adrian R.D. Mathias. On a conjecture of Erdős and Čudakov. In Béla Bollobás and Andrew Thomason, editors, *Combinatorics, Geometry and Probability. A Tribute to Paul Erdős*, pages 487–488. Cambridge University Press, Cambridge, 1997.

[60] Adrian R.D. Mathias. Bourbaki tévútjai. *Természet Világa*, pages 121–127, 1998.

[61] Adrian R.D. Mathias. Recurrent points and hyperarithmetic sets. In Carlos A. Di Prisco, editor, *Set Theory: Techniques and Applications. Curaçao 1995 and Barcelona 1996 Conferences*, pages 157–174. Kluwer Academic Publishers, Dordrecht, 1998.

[62] Adrian R.D. Mathias. Strong statements of analysis. *Bulletin of the London Mathematical Society*, 32:513–526, 2000.

[63] Adrian R.D. Mathias. Delays, recurrence and ordinals. *Proceedings of the London Mathematical Society*, 82:257–298, 2001.

[64] Adrian R.D. Mathias. Slim models of Zermelo set theory. *The Journal of Symbolic Logic*, 66:487–496, 2001.

[65] Adrian R.D. Mathias. The strength of Mac Lane set theory. *Annals of Pure and Applied Logic*, 110:107–234, 2001.

[66] Adrian R.D. Mathias. A term of length 4 523 659 424 929. *Synthèse*, 133:75–86, 2002.

[67] Adrian R.D. Mathias. Choosing an attacker by a local derivation. *Acta Universitatis Carolinae. Mathematica et Physica*, pages 59–65, 2004. 32nd Winter School of Abstract Analysis.

[68] Adrian R.D. Mathias. Ignorancia de Bourbaki. *La Gaceta de la Real Matemàtica Espanola*, 7:727–748, 2004.

[69] Adrian R.D. Mathias. A scenario for transferring high scores. *Acta Universitatis Carolinae. Mathematica et Physica*, pages 67–73, 2004. 32nd Winter School of Abstract Analysis.

[70] Adrian R.D. Mathias. Analytic sets under attack. *Mathematical Proceedings of the Cambridge Philosophical Society*, 138:465–485, 2005.

[71] Adrian R.D. Mathias. Weak systems of Gandy, Jensen and Devlin. In Joan Bagaria and Stevo Todorcevic, editors, *Set Theory*, Trends in Mathematics, pages 149–224. Birkhäuser, Basel, 2006.

[72] Adrian R.D. Mathias. A note on the schemes of replacement and collection. *Archives of Mathematical Logic*, 46:43–50, 2007.

[73] Adrian R.D. Mathias. Set forcing over models of Zermelo or Mac Lane. In *One Hundred Years of Axiomatic Set Theory*, volume 17 of *Cahiers du Centre de Logique*, pages 41–66. Academia Bruylandt, Louvain-la-Neuve, 2010.

[74] Adrian R.D. Mathias. Unordered pairs in the set theory of Bourbaki 1949. *Archiv der Mathematik*, 94:1–10, 2010.

[75] Adrian R.D. Mathias. Hilbert, Bourbaki and the scorning of logic. In Chi Tat Chong *et al.*, editor, *Infinity and Truth*, volume 25 of *Lecture Notes Series. Institute for Mathematical Sciences. National University of Singapore*, pages 47–156. World Scientific Publishing, Hackensack, 2014.

[76] Adrian R.D. Mathias. Provident sets and rudimentary set forcing, 2015.

[77] Adrian R.D. Mathias, Andrzej Ostawsezski, and Michel Talagrand. On the existence of an analytic set meeting each compact set in a Borel set. *Mathematical Proceedings of the Cambridge Philosophical Society*, 84:5–10, 1978.

[78] Keith R. Milliken. Completely separable families and Ramsey's theorem. *Journal of Combinatorial Theory, Series A*, 19:318–334, 1975.

[79] Polymath5. Erdős discrepancy problem. On Michael Nielsen's Polymath webpage.

[80] Frank P. Ramsey. On a problem of formal logic. *Proceedings of the London Mathematical Society*, 30:264–286, 1930.

[81] Saharon Shelah. *Proper Forcing.* Lecture Notes in Mathematics #940. Springer-Verlag, Berlin, 1982.

[82] Jack Silver. Every analytic set is Ramsey. *The Journal of Symbolic Logic*, 35:60–64, 1970.

[83] Robert M. Solovay. A model of set theory in which every set of reals is Lebesgue measurable. *Annals of Mathematics*, 92:1–56, 1970.

[84] Terence Tao. The Erdos discrepancy problem, 2015. arXiv:1509.05363.

[85] Stevo Todorcevic. *Introduction to Ramsey spaces.* Annals of Mathematics Studies. Princeton University Press, Princeton, 2010.

[86] Asger Törnquist. Definability and almost disjoint families, 2015. arXiv:1503.07577.

[87] Edward L. Wimmers. The Shelah p-point independence theorem. *Israel Journal of Mathematics*, 43:28–48, 1982.